U0150776

nature

The Living Record of Science

《自然》百年科学经典

英汉对照版（平装本）

第十卷（下）

总顾问：李政道（Tsung-Dao Lee）

英方主编：Sir John Maddox
Sir Philip Campbell
中方主编：路甬祥

2002-2007

外语教学与研究出版社 · 麦克米伦教育 · 自然科研

FOREIGN LANGUAGE TEACHING AND RESEARCH PRESS · MACMILLAN EDUCATION · NATURE RESEARCH

北京 BEIJING

Original English Text © Springer Nature Limited
Chinese Translation © Foreign Language Teaching and Research Press

This edition is published under arrangement with Macmillan Publishers (China) Limited. It is for sale in the People's Republic of China only, excluding Hong Kong SAR, Macao SAR and Taiwan Province, and may not be bought for export therefrom.

图书在版编目（CIP）数据

《自然》百年科学经典．第十卷．下，2002–2007 ：英汉对照 ／（英）约翰·马多克斯（John Maddox），（英）菲利普·坎贝尔（Philip Campbell），路甬祥主编．-- 北京 ：外语教学与研究出版社，2019.12（2023.2 重印）
ISBN 978-7-5213-1474-8

Ⅰ．①自… Ⅱ．①约… ②菲… ③路… Ⅲ．①自然科学－文集－英、汉 Ⅳ．①N53

中国版本图书馆 CIP 数据核字 (2020) 第 021651 号

地图审图号：GS（2019）3264 号

出 版 人　王　芳
项目统筹　章思英
项目负责　刘晓楠　黄小斌
责任编辑　王丽霞
责任校对　黄小斌
封面设计　孙莉明　曹志远
版式设计　孙莉明
出版发行　外语教学与研究出版社
社　　址　北京市西三环北路 19 号（100089）
网　　址　http://www.fltrp.com
印　　刷　北京华联印刷有限公司
开　　本　787×1092　1/16
印　　张　34
版　　次　2020 年 4 月第 1 版　2023 年 2 月第 2 次印刷
书　　号　ISBN 978-7-5213-1474-8
定　　价　168.00 元

购书咨询：（010）88819926　电子邮箱：club@fltrp.com
外研书店：https://waiyants.tmall.com
凡印刷、装订质量问题，请联系我社印制部
联系电话：（010）61207896　电子邮箱：zhijian@fltrp.com
凡侵权、盗版书籍线索，请联系我社法律事务部
举报电话：（010）88817519　电子邮箱：banquan@fltrp.com
物料号：314740001

《自然》百年科学经典（英汉对照版）

总 顾 问：李政道（Tsung-Dao Lee）

英方主编：Sir John Maddox　　中方主编：路甬祥

Sir Philip Campbell

编审委员会

英方编委	中方编委（以姓氏笔画为序）
Philip Ball	许智宏
Vikram Savkar	赵忠贤
David Swinbanks	滕吉文

本卷审稿专家（以姓氏笔画为序）

于　军	王　宇	王晓良	尹凤玲	冯珑珑	邢　松	刘冬生
许　冰	李　然	肖景发	张颖奇	陆培祥	陈含章	陈继征
季江徽	周济林	郑旭峰	胡永云	胡松年	夏俊卿	徐　栋
徐文堪	高树基	曹　俊	曹庆宏	常　江	曾长青	蔡荣根
裴端卿	黎　卓	潘　雷				

编译委员会

本卷翻译工作组稿人（以姓氏笔画为序）

王丽霞　王晓蕾　王耀杨　刘　明　刘晓楠　关秀清　李　琦
何　铭　周家斌　郭红锋　黄小斌　蔡则怡

本卷翻译人员（以姓氏笔画为序）

王耀杨　毛晨晖　田晓阳　吕　静　吕孟珍　任　奕　刘项琨
刘皓芳　齐红艳　安宇森　孙惠南　李　平　李　梅　李　辉
肖　莉　何　钧　汪　浩　张瑶楠　金世超　周　杰　周家斌
郭思彤　梁恩思　韩　然　谭秀慧

本卷校对人员（以姓氏笔画为序）

刘雨佳　张玉光　陈思原　周少贞　贺舒雅　夏洁媛　顾海成
郭思彤　黄小斌　蔡则怡　潘卫东　Eric Leher（澳）

Contents
目录

Volume X

(2002-2007)

Infrared Radiation from an Extrasolar Planet

D. Deming *et al.*

Editors's Note

Many of the planets that have been found outside our solar system are "hot Jupiters", with masses similar to that of Jupiter and orbiting very close to their parent stars. Some of these planets' orbits are aligned such that the planet periodically crosses the face of the star (from our perspective) and is then eclipsed by it. Here Drake Deming and colleagues report infrared observations of the planet HD 209458b while it was neither crossing nor eclipsed, so that light from the planet contributed to the total light from the system. By subtracting the light when the planet was in eclipse, they were able for the first time to detect light coming directly from the atmosphere of an extrasolar planet.

A class of extrasolar giant planets—the so-called "hot Jupiters" (ref. 1)—orbit within 0.05 AU of their primary stars (1 AU is the Sun–Earth distance). These planets should be hot and so emit detectable infrared radiation[2]. The planet HD 209458b (refs 3, 4) is an ideal candidate for the detection and characterization of this infrared light because it is eclipsed by the star. This planet has an anomalously large radius (1.35 times that of Jupiter[5]), which may be the result of ongoing tidal dissipation[6], but this explanation requires a non-zero orbital eccentricity (~0.03; refs 6, 7), maintained by interaction with a hypothetical second planet. Here we report detection of infrared (24 μm) radiation from HD 209458b, by observing the decrement in flux during secondary eclipse, when the planet passes behind the star. The planet's 24-μm flux is 55 ± 10 μJy (1σ), with a brightness temperature of $1{,}130 \pm 150$ K, confirming the predicted heating by stellar irradiation[2,8]. The secondary eclipse occurs at the midpoint between transits of the planet in front of the star (to within ± 7 min, 1σ), which means that a dynamically significant orbital eccentricity is unlikely.

OPERATING cryogenically in a thermally stable space environment, the Spitzer Space Telescope[9] has sufficient sensitivity to detect hot Jupiters at their predicted infrared flux levels[8]. We observed the secondary eclipse (hereafter referred to as "the eclipse") of HD 209458b with the 24-μm channel of the Multiband Imaging Photometer for Spitzer (MIPS)[10]. Our photometric time series observations began on 6 December 2004 at 21:29 UTC (Coordinated Universal Time), and ended at approximately 03:23 UTC on 7 December 2004 (5 h 54 min duration). We analyse 1,696 of the 1,728 10-s exposures so acquired, rejecting 32 images having obvious flaws. The Supplementary Information contains a sample image, together with information on the noise properties of the data.

来自一颗系外行星的红外辐射

戴明等

编者按

太阳系外已经发现的行星中，大部分都是“热木星”。它们的质量和木星相近，并以非常近的距离围绕它们的主星公转。从我们的视角来看，这些行星当中的一部分有着“对齐”的公转轨道，这种“对齐”使得它们会周期性地先从主星面前经过，接着被主星所遮掩。本文中，德雷克·戴明和他的同事们报告了他们对行星 HD 209458b 在既没有经过恒星表面，也没有被恒星所遮掩时进行的红外观测。这时，来自该系统的所有光线中有一部分是行星所贡献的。通过扣除行星在被主星遮掩时的光线，他们第一次得以探测到直接来自一颗系外行星大气层的光线。

所谓的“热木星”[1]，是一类太阳系外巨行星。它们在距离主星 0.05 个天文单位的范围内（1 个天文单位指的是太阳到地球的平均距离）围绕主星公转。由于靠近主星，这些行星的温度较高，所以可以发出可探测的红外辐射 [2]。行星 HD 209458b[3,4] 因为会被其主星所遮掩，所以是探测并刻画这种红外辐射的理想候选体。这颗行星的半径较大，略显反常，是木星的 1.35 倍 [5]。这可能是正在发生的潮汐耗散作用的结果 [6]，但是这种解释需要假设存在第二颗行星，以使 HD 209458b 始终保持非零的轨道偏心率（约 0.03[6,7]）。这里我们报告通过观测行星经过恒星背面（即次食）时流量降低的程度而探测到的来自 HD 209458b 的红外辐射（24 微米）。这颗行星的 24 微米流量为 55 ± 10 μJy (1σ)，表面亮温度为 1,130 ± 150 K。这确认了预测存在的由恒星辐射所产生的加热 [2,8]。行星在经过恒星表面的时候会产生凌星现象。由于次食恰好发生在两次凌星的中间时刻（正负偏差在 7 分钟以内，1σ），这就意味着该行星不太可能有一个在动力学上显著的轨道偏心率。

在热稳定的太空环境中低温运行的斯皮策空间望远镜 [9]，它的精度足够在理论预测的红外流量水平对热木星进行探测 [8]。我们利用斯皮策多波段成像光度计（MIPS）[10] 的 24 微米通道，对 HD 209458b 的次食进行了观测。我们的测光时序观测开始于 2004 年 12 月 6 日 UTC 时间（协调世界时）21:29，大致结束于 2004 年 12 月 7 日 UTC 时间 03:23（持续时间 5 小时 54 分钟）。在观测获得的 1,728 张 10 秒曝光图片中，我们舍弃 32 张有明显缺陷的图片，并对剩下的 1,696 张进行研究。补充信息中包含了一张示例图片，并附上了与数据噪声属性相关的信息。

We first verify that circumstellar dust does not contribute significantly to the stellar flux. Summing each stellar image over a 13 × 13 pixel synthetic aperture (33 × 33 arcsec), we multiply the average sum by 1.15 to account for the far wings of the point spread function (PSF)[11], deriving a flux of 21.17 ± 0.11 mJy. The temperature of the star is close to 6,000 K (ref. 12). At a distance of 47 pc (ref. 13), a model atmosphere[14] predicts a flux of 22 mJy, agreeing with our observed flux to within an estimated ~2 mJy error in absolute calibration. We conclude that the observed flux is dominated by photospheric emission, in agreement with a large Spitzer study of planet-bearing stars at this wavelength[11].

Our time series analysis is optimized for high relative precision. We extract the intensity of the star from each image using optimal photometry with a spatial weighting function[15]. Selecting the Tiny Tim[16] synthetic MIPS PSF for a 5,000-K source at 24 μm, we spline-interpolate it to 0.01 pixel spacing, rebin it to the data resolution, and centre it on the stellar image. The best centring is judged by a least-squares fit to the star, fitting to within the noise level. The best-centred PSF becomes the weighting function in deriving the stellar photometric intensity. We subtract the average background over each image before applying the weights. MIPS data includes per-pixel error estimates[17], which we use in the optimal photometry and to compute errors for each photometric point. The optimal algorithm[15] predicts the signal-to-noise ratio (SNR) for each photometric point, and these average to 119. Our data are divided into 14 blocks, defined by pre-determined raster positions of the star on the detector. To check our SNR, we compute the internal scatter within each block. This gives SNR in the range from 95 to 120 (averaging 111), in excellent agreement with the optimal algorithm. For each point we use the most conservative possible error: either the scatter within that block or the algorithm estimate, whichever is greater. We search for correlations between the photometric intensities and small fluctuations in stellar position, but find none. We also perform simple aperture photometry on the images, and this independent procedure confirms our results, but with 60% greater errors.

The performance of MIPS at 24 μm is known to be excellent[18]. Only one instrument quirk affects our photometry. The MIPS observing sequence obtains periodic bias images, which reset the detector. Images following resets have lower overall intensities (by ~0.1–1%), which recover in later images. The change is common to all pixels in the detector, and we remove it by dividing the stellar intensities by the average zodiacal background in each image. We thereby remove variations in instrument/detector response, both known and unknown. The best available zodiacal model[19] predicts a background increase of 0.18% during the ~6 h

我们首先确认了星周尘对恒星流量没有明显贡献。利用一个大小为 13 × 13 像素的合成孔径（33 × 33 角秒），我们在每一张恒星图像上进行求和。考虑到点扩散函数（PSF）线翼远端的贡献，我们又对和的平均值乘以系数 1.15[11]，最终得到其流量为 21.17 ± 0.11 mJy。这颗恒星的温度接近 6,000 K[12]，在 47 pc[13] 的距离上，利用大气模型 [14] 估计的流量为 22 mJy。这个值与我们观测到的流量相吻合，差别在绝对定标时对误差的估计（约为 2 mJy）以内。由此我们得出结论，观测到的流量中光球层发射占主导。在此波长上，该结论与一个利用斯皮策望远镜对带有行星的恒星所进行的大型研究相吻合 [11]。

我们的时序分析对高相对精度进行了优化。在从每幅图像中抽取恒星强度的过程中，我们利用了带有空间权重函数的优化测光算法 [15]。我们选择了利用 Tiny Tim 软件包 [16] 合成 MIPS 点扩散函数，并将源的温度设定为 5,000 K，波长设定为 24 微米。接着，我们在将该点扩散函数样条插值至 0.01 像素的间距，并在重新合并至数据分辨率精度后，将其中心与恒星图像对齐。中心对齐的最好结果是由对恒星的最小二乘拟合来判定的，拟合程度为噪声范围以内。中心对齐最好的点扩散函数成为在计算恒星测光强度过程中的权重函数。在利用权重以前，我们从每幅图像中都扣除了背景平均值。MIPS 数据中包含了每个像素的误差估计 [17]，我们在优化测光算法和计算每个测光数据点的误差中都使用了这些误差估计。我们使用的优化算法 [15] 可以估计每个测光数据点的信噪比（SNR），它们的平均值为 119。我们的数据被分为 14 个区块，它们是利用恒星在传感器上预定的光栅位置来确定的。为了检验我们的信噪比，我们在每个区块的内部计算了内部的弥散。计算结果给出的信噪比范围为 95 到 120（平均值为 111），与优化算法给出的结果很好地吻合。在每一个测光点的误差选择上，我们采用了最为保守的方法：选择该点所在的区块内的弥散值和算法给出的误差估计中偏大的结果。在测光强度和细微的恒星位置波动之间，我们也没有找到相关性。我们在恒星图像上还使用了简单的孔径测光法来提取恒星强度。这种方法与前述方法之间是相互独立的，它也确认了我们得出的结果，但是得出的误差却高出 60%。

众所周知，MIPS 在 24 微米处的性能非常优秀 [18]，只有一处与仪器相关的问题影响测光精度。MIPS 在观测过程中，会周期性地拍摄偏置图像。该过程会重置传感器。传感器在重置过程后所拍摄的图像在总体强度上会有所偏小（约 0.1 ~ 1%），但会随着拍摄过程而逐渐恢复。该变化过程在传感器所有像素上都有体现，因此我们对其改正的方法为在每一幅图像上用得到的恒星强度除以平均黄道背景。由此，我们移除了仪器/设备上已知和未知的变化。在可用的黄道模型中，最好的模型 [19] 在

of our photometry. Because the star will not share this increase, we remove a 0.18% linear baseline from the stellar photometry. Note that the eclipse involves both a decrease and increase in flux, and its detection is insensitive to monotonic linear baseline effects.

To detect weak signals reliably requires investigating the nature of the errors. We find that shot noise in the zodiacal background is the dominant source of error; systematic effects are undetectable after normalizing any individual pixel to the total zodiacal background. All of our results are based on analysis of the 1,696 individual photometric measurements versus heliocentric phase from a recent ephemeris[20] (Fig. 1a). We propagate the individual errors (not shown on Fig. 1a) through a transit curve fit to calculate the error on the eclipse depth. Because about half of the 1,696 points are out of eclipse, and half are in eclipse, and the SNR $\approx$ 111 per point, the error on the eclipse depth should be $\sim 0.009 \times 2^{0.5}/848^{0.5} = 0.044\%$ of the stellar continuum. Model atmospheres for hot Jupiters[2,8,21-24] predict eclipse depths in the range from 0.2–0.4% of the stellar continuum, so we anticipate a detection of 4–9σ significance. The eclipse is difficult to discern by eye on Fig. 1a, because the observed depth (0.26%) is a factor of 4 below the scatter of individual points. We use the known period (3.524 days) and radii[5] to fit an eclipse curve to the Fig. 1a data, varying only the eclipse depth, and constraining the central phase to 0.5. This fit detects the eclipse at a depth of $0.26\% \pm 0.046\%$, with a reduced χ^2 of 0.963, denoting a good fit. Note that the 5.6σ significance applies to the aggregate result, not to individual points. The eclipse is more readily seen by eye on Fig. 1b, which presents binned data and the best-fit eclipse curve. The data are divided into many bins, so the aggregate 5.6σ significance is much less for a single bin (SNR $\approx$ 1 per point). Nevertheless, the dip in flux due to the eclipse is apparent, and the observed duration is approximately as expected. As a check, we use a control photometric sequence (Fig. 1b) to eliminate false positive detection of the eclipse due to instrument effects. We also plot the distribution of points in intensity for both the in-eclipse and out-of-eclipse phase intervals (Fig. 1c). This shows that the entire distribution shifts as expected with the eclipse, providing additional discrimination against a false positive detection.

我们约 6 小时的测光观测中所预测的背景强度升高 0.18%。由于恒星本身并不受这种增加的影响，所以我们在恒星测光中移除了一条 0.18% 的线性基线。需要指出的是，次食过程既包含了流量下降的过程也包含了流量上升的过程，而探测次食现象并不受单调的线性基线影响。

探究误差的本质在可靠地探测较弱信号的过程中是必需的。我们发现黄道背景中的散粒噪声是误差的最主要来源；把所有像素以黄道背景总和为标准进行归一化后，来源于系统误差的影响均无法探测到。我们所有的结果都是从 1,696 张独立测光观测中获得的。这些测光观测对应于该系统最近的一个星历测定工作所得出的日心相位 [20]（图 1a）。为了计算次食时掩食深度的误差，我们通过拟合凌星曲线来传递每个测光数据点的误差（误差并没有在图 1a 中标注）。因为在 1,696 个数据点中，次食内和次食外的点各占一半左右，并且每个点的信噪比约为 111，所以次食时的掩食深度误差应为恒星连续谱的 $\sim 0.009 \times 2^{0.5} / 848^{0.5} = 0.044\%$。热木星的大气模型 [2,8,21-24] 对次食时的掩食深度的预计范围为恒星连续谱的 0.2 ~ 0.4%，所以我们估计本次探测的显著程度应在 $4 \sim 9\sigma$ 之间。由于观测到的次食深度（0.26%）仅为单独数据点的弥散程度的 1/4 左右，在图 1a 上很难用肉眼识别出次食的形状。我们使用已知的周期（3.524 天）和半径 [5] 在图 1a 的数据上进行次食曲线的拟合。拟合过程中，我们将相位中心固定至 0.5，只调整次食的掩食深度。该拟合结果较好，探测到的次食深度为 $0.26\% \pm 0.046\%$，约化 χ^2 为 0.963。需要指出的是，5.6σ 的显著程度适用于累计的结果，并不适用于单独的数据点。图 1b 展示了合并以后的数据和最佳拟合的次食曲线，在图 1b 上更容易用肉眼观测到次食。因为数据点被分隔至很多个区间中，所以累计的 5.6σ 显著程度对每一个单独的区间要小很多（每个点信噪比约为 1）。尽管如此，由次食所造成的流量下降是明显的，而且观测到的次食时长也与预期基本相符。作为检查手段，我们使用了一个对照测光序列（图 1b）来消除由于可能的仪器效应所造成的次食假阳性探测。我们还按照强度区间，画出了次食相位内和次食相位外的数据点的分布（图 1c）。该图展示了整个分布会根据次食相位的不同而移动，这是与预期相吻合的，也提供了探测到的次食并非假阳性结果的额外证据。

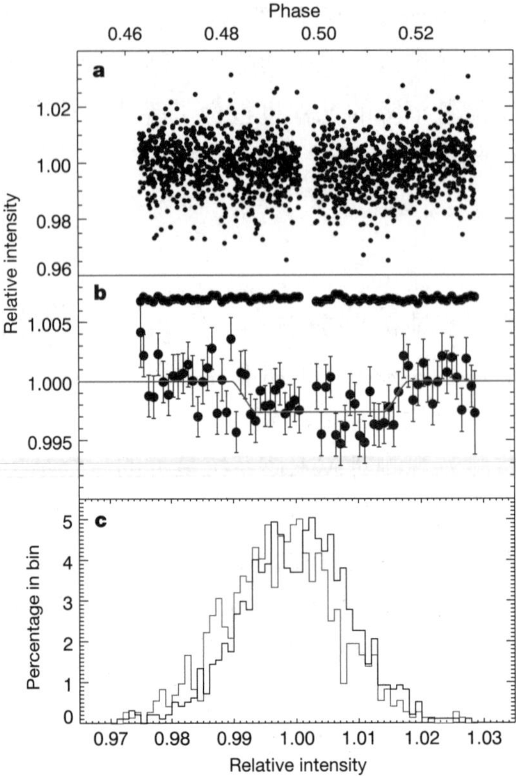

Fig. 1. Observations showing our detection of the secondary eclipse in HD 209458. **a**, Relative intensities versus heliocentric phase (scale at top) for all 1,696 data points. The phase is corrected for light travel time at the orbital position of the telescope. Error bars are suppressed for clarity. The gap in the data near phase 0.497 is due to a pause for telescope overhead activity. The secondary eclipse is present, but is a factor of ~4 below the ~1% noise level of a single measurement. **b**, Intensities from **a**, averaged in bins of phase width 0.001 (scale at top), with 1σ error bars computed by statistical combination from the errors of individual points. The red line is the best-fit secondary eclipse curve (depth = 0.26%), constrained to a central phase of 0.5. The points in blue are a control sequence, summing intensities over a 10×10-pixel region of the detector, to beat down the random errors and reveal any possible systematic effects. The control sequence uses the same detector pixels, on average, as those where the star resides, but is sampled out of phase with the variations in the star's raster motion during the MIPS photometry cycle. **c**, Histograms of intensity (lower abscissa scale) for the points in **a**, with bin width 0.1%, shown separately for the out-of-eclipse (black) and in-eclipse intervals (red).

We further illustrate the reality of the eclipse on Fig. 2. Now shifting the eclipse curve in phase, we find the best-fitting amplitude and χ^2 at each shift. This determines the best-fit central phase for the eclipse, and also further illustrates the statistical significance of the result. The thick line in Fig. 2a shows that the maximum amplitude (0.26%) is obtained at exactly phase 0.5 (which is also the minimum of χ^2). Further, we plot the eclipse "amplitude" versus central phase using 100 sets of synthetic data, consisting of gaussian noise with dispersion matching the real data, but without an eclipse. The amplitude (0.26%) of the eclipse in the real data stands well above the statistical fluctuations in the synthetic data.

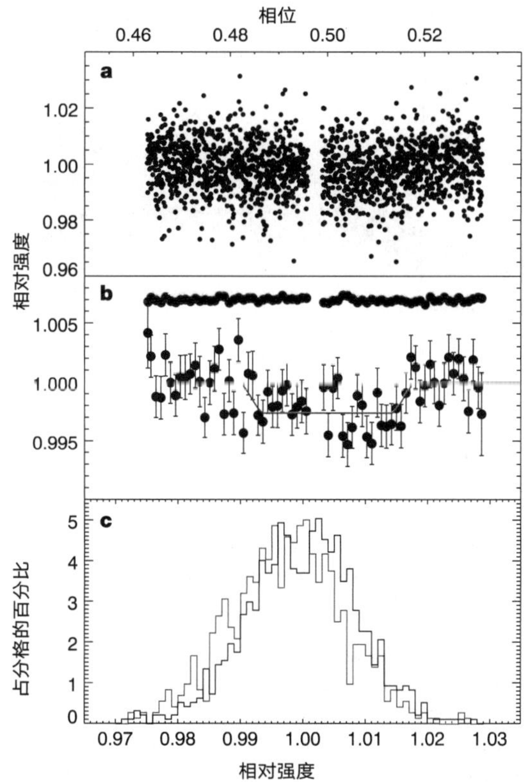

图 1. 观测展示了我们探测到的 HD 209458 系统中的次食。**a**，我们画出 1,696 个数据点的相对强度与该系统日心坐标系相位的关系图（横坐标在图的上方）。我们根据望远镜轨道位置的光传播时间对相位进行了修正。为将图像展示得更明了，我们没有画误差棒。相位 0.497 附近的数据间断是由于望远镜的非观测性活动造成的暂停产生的。系统的次食是存在的，但是仅为单独数据点噪声水平（约 1%）的 1/4 左右。**b**，将图 **a** 中的强度以 0.001 相位分格平均于每个小格中（横坐标在图的上方），同时也画出了 1σ 误差棒。误差棒是用统计的方法将单独点的误差合并计算得到的。红色线为最佳拟合的次食曲线（掩食深度为 0.26%），其中心相位被限制于 0.5。通过对传感器上 10 × 10 像素区域的强度求和，我们得到了一个对照序列，在图中是以蓝色显示的。我们使用该对照序列来减小随机误差的影响，同时揭示可能存在的系统误差影响。对照序列在传感器上使用的像素与恒星星像所处的像素基本相同，但在 MIPS 测光观测周期对它们采样时，恒星在传感器上的光栅位置是恰好错开的。**c**，用图 **a** 中的强度数据画出的直方图（横坐标位于图的下方），分格大小为 0.1%。次食区间内和次食区间外分别用红色和黑色标出。

我们利用图 2 来更深入地阐明次食的实际情况。我们移动次食曲线至不同的相位位置，并在每个移动的位置上找到拟合最好的幅度和 χ^2。该操作可以确定最佳拟合的次食中心相位，并可以进一步阐释结果的统计显著性。图 2a 中的粗线显示最大的幅度（0.26%）是准确地在 0.5 相位处获得的（该位置也是 χ^2 最小值获得处）。更进一步，我们用 100 组合成数据画出次食的“幅度”相对于中心相位的图像。这些数据包含了与真实数据相当的高斯噪声弥散，但并不含有次食事件。真实数据中次食的掩食深度（0.26%）显著高于合成数据的统计波动。

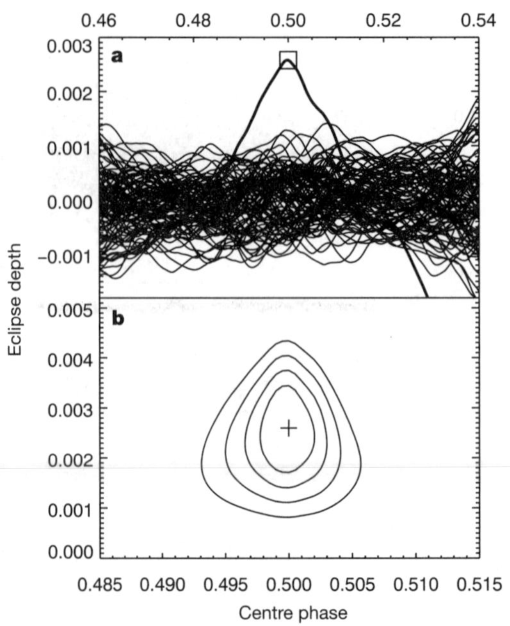

Fig. 2. Amplitude of the secondary eclipse versus assumed central phase, with confidence intervals for both. **a**, The darkest line shows the amplitude of the best-fit eclipse curve versus the assumed central phase (scale at top). The overplotted point marks the fit having smallest χ^2, which also has the greatest eclipse amplitude. The numerous thinner black lines show the effect of fitting to 100 synthetic data sets containing no eclipse, but having the same per-point errors as the real data. Their fluctuations in retrieved amplitude versus phase are indicative of the error in eclipse amplitude, and are consistent with $\sigma = 0.046\%$. Note that the eclipse amplitude found in the real data (0.26%) stands well above the error envelope at phase 0.5. **b**, Confidence intervals at the 1, 2, 3 and 4σ levels for the eclipse amplitude and central phase (note expansion of phase scale, at bottom). The plotted point marks the best fit (minimum χ^2) with eclipse depth of 0.26%, and central phase indistinguishable from 0.5. The centre of the eclipse occurs in our data at Julian day 2453346.5278.

Figure 2b shows confidence intervals on the amplitude and central phase, based on the χ^2 values. The phase shift of the eclipse is quite sensitive to eccentricity (e) and is given[25] as $\Delta t = 2Pe\cos(\omega)/\pi$, where P is the orbital period, and ω is the longitude of periastron. The Doppler data alone give $e = 0.027 \pm 0.015$ (Laughlin, G., personal communication), and allow a phase shift as large as ± 0.017 (87 min). We find the eclipse centred at phase 0.5, and we checked the precision using a bootstrap Monte Carlo procedure[26]. The 1σ phase error from this method is 0.0015 (~7 min), consistent with Fig. 2b. A dynamically significant eccentricity, $e \approx 0.03$ (refs 6, 7), constrained by our 3σ limit of $\Delta t < 21$ min, requires $|(\omega - \pi/2)| < 12$ degrees and is therefore only possible in the unlikely case that our viewing angle is closely parallel to the major axis of the orbit. A circular orbit rules out a promising explanation for the planet's anomalously large radius: tidal dissipation as an interior energy source to slow down planetary evolution and contraction[7]. Because the dynamical time for tidal decay to a circular orbit is short, this scenario posited the presence of a perturbing second planet in the system to continually force the eccentricity—a planet that is no longer necessary with a circular orbit for HD 209458b.

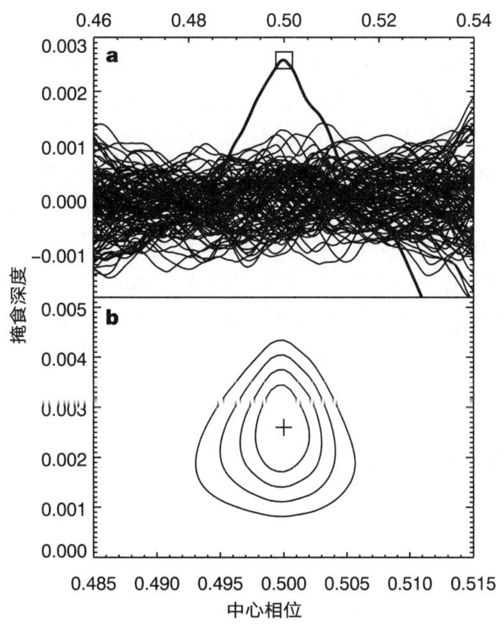

图 2. 次食深度与推定的中心相位关系图和二者各自的置信区间图。**a**，颜色最深的线展示了最佳拟合的次食曲线的幅度和推定的中心相位之间的关系（横坐标位于图的上方）。图中覆盖在粗线上方的点代表 χ^2 最小的拟合，该拟合结果的掩食深度也最大。旁边多条稍细的黑线展示了利用 100 组人造数据获得的结果。这些数据的单点误差和真实数据相同，但并不包含次食事件。人造数据拟合得到的掩食深度和相位之间的波动与 $\sigma = 0.046\%$ 相符，且可以大概标明从真实数据中得到的掩食深度的误差范围。可以看到，真实数据中得到的次食深度（0.26%）明显高于 0.5 相位处的误差范围。**b**，次食深度和中心相位分别在 1、2、3、4σ 处的置信区间（横坐标位于下方，请注意相位尺度的放大）。图中画出的点代表最佳拟合结果（最小的 χ^2），此时的次食深度为 0.26%，中心相位基本位于 0.5。在我们的数据中，次食中心发生在儒略日 2453346.5278。

图 2b 展示了幅度和中心相位基于 χ^2 的置信区间。次食的相移对于偏心率（e）较为敏感[25]，其表达式为 $\Delta t = 2Pe\cos(\omega)/\pi$，其中 P 为轨道周期，ω 为近星点经度。单独的多普勒数据给出的偏心率是 $e = 0.027 \pm 0.015$（劳克林，个人交流），该值所允许的最大相移为 ± 0.017（87 分钟）。我们发现掩食的中心在相位 0.5 处，然后利用自举蒙特卡罗方法来检测该结果的精度[26]。该方法所给出的 1σ 相位误差为 0.0015（约为 7 分钟），与图 2b 一致。利用前述结果可知 3σ 限制所对应的 $\Delta t < 21$ 分钟。在该约束条件下，若该系统有一个动力学显著的偏心率 $e \approx 0.03$[6,7]，则要求有 $|\omega - \pi/2| < 12°$。该条件成立的可能性不高，因为它只有在我们的观测方向和行星公转轨道的半长径方向较为平行的特殊情况下才可以满足。由于潮汐耗散作为内部能量源可以减慢行星演化和收缩，它是该行星拥有异常大半径的有效解释，但是该行星的圆轨道排除了这种可能性[7]。因为潮汐对轨道圆化的动力学时标较短，因此如果该行星有动力学显著的偏心率，则该偏心率需假定系统中还有第二个持续对 HD 209458b 摄动的行星来维持。但对于圆轨道的 HD 209458b 来说，第二颗行星便不再必要。

The infrared flux from the planet follows directly from our measured stellar flux (21.2 mJy) and the eclipse depth (0.26%), giving 55 ± 10 μJy. The error is dominated by uncertainty in the eclipse depth. Using the planet's known radius[5] and distance[13], we obtain a brightness temperature $T_{24} = 1{,}130 \pm 150$ K, confirming heating by stellar irradiation[2]. Nevertheless, T_{24} could differ significantly from the temperature of the equivalent blackbody (T_{eq}), that is, one whose bolometric flux is the same as the planet. Without measurements at shorter wavelengths, a model atmosphere must be used to estimate T_{eq} from the 24-μm flux. One such model is shown in Fig. 3, having $T_{eq} = 1{,}700$ K. This temperature is much higher than T_{24} (1,130 K) due to strong, continuous H_2O vapour absorption at 24 μm. The bulk of the planetary thermal emission derives ultimately from re-radiated stellar irradiation, and is emitted at 1–4 μm, between H_2O bands. However, our 24-μm flux error admits a range of models, including some with a significantly lower T_{eq} (for example, but not limited to, models with reflective clouds or less H_2O vapour).

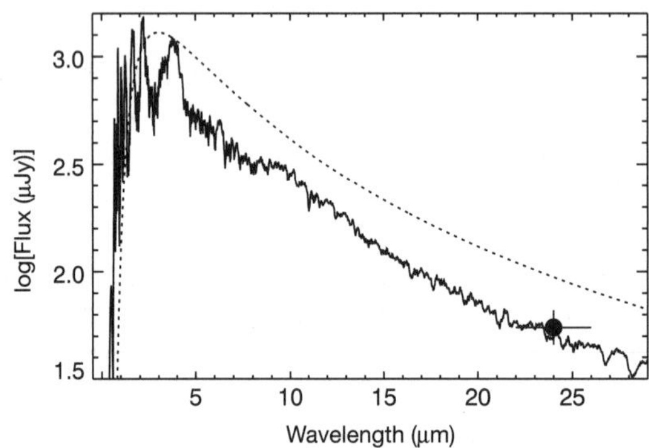

Fig. 3. Flux from a model atmosphere shown in comparison to our measured infrared flux at 24 μm. A theoretical spectrum (solid line) shows that planetary emission (dominated by absorbed and re-radiated stellar radiation) should be very different from a blackbody. Hence, models are required to interpret the 24-μm flux measurement in terms of the planetary temperature. The model shown has $T_{eq} = 1{,}700$ K and was computed from a one-dimensional plane-parallel radiative transfer model, considering a solar system abundance of gases, no clouds, and the absorbed stellar radiation re-emitted on the day side only. Note the marked difference from a 1,700-K blackbody (dashed line), although the total flux integrated over the blackbody spectrum is equal to the total flux integrated over the model spectrum. (The peaks at short wavelength dominate the flux integral in the atmosphere model, note log scale in the ordinate.) The suppressed flux at 24 μm is due to water vapour opacity. This model lies at the hot end of the range of plausible models consistent with our measurement, but the error bars admit models with cooler T_{eq}.

Shortly after submission of this Letter, we became aware of a similar detection for the TrES-1 transiting planet system[27] using Spitzer's Infrared Array Camera[28]. Together, these Spitzer results represent the first measurement of radiation from extrasolar planets. Additional Spitzer observations should rapidly narrow the range of acceptable models, and reveal the atmospheric structure, composition, and other characteristics of close-in extrasolar giant planets.

(**434**, 740-743; 2005)

在测量出该行星主星的红外流量（21.2 mJy）和掩食深度（0.26%）后，来自行星的红外流量便可以直接得出，为 55 ± 10 μJy。误差主要来源于掩食深度的不确定性。利用该行星已知的半径[5]和距离[13]，我们可以得出其亮温度为 T_{24} = 1,130 ± 150 K。这个结果证实了恒星辐射的加热作用[2]。尽管如此，T_{24} 也有可能与等效的黑体温度（T_{eq}），即用行星的热流量计算出的温度，有着显著的差别。在缺乏短波长区域测量的情况下，我们必须使用行星大气模型结合 24 微米处的流量来估计 T_{eq}。图 3 中展示了我们使用的一种模型，它得出的 T_{eq} = 1,700 K。该温度明显高出 T_{24}（1,130 K），主要原因是在 24 微米处有着水蒸气连续且较强的吸收。行星热辐射的主要部分最终的源头是对恒星辐射的再辐射，主要集中在 1～4 微米之间，介于水的波段之间。但是，我们所计算出的 24 微米处流量的误差使得很多种模型都有可能，其中就包括一些可以产生明显更低的 T_{eq} 的模型（比如有着反射性云层或者更少水蒸气的模型，但并不局限于此）。

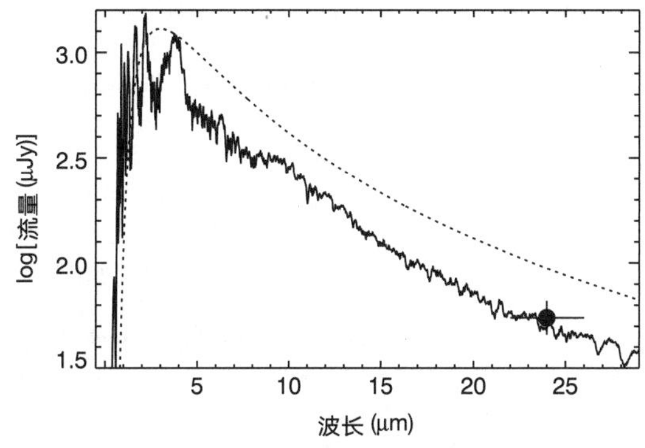

图 3. 利用大气模型得到的流量与我们在 24 微米处测量得到流量的比较。该理论光谱（实线）表明行星辐射（由吸收和再辐射恒星辐射占主导）应与黑体非常不同。因此，合理的模型必须能够用行星温度来解释 24 微米处测量到的流量。图中所展示的是一个一维平面平行辐射转移模型的结果。在计算中，我们设定的条件为已知太阳系的气体丰度，无云，且只有行星的向昼侧会吸收恒星辐射并再辐射。此时得到的 T_{eq} = 1,700 K。注意一个 1,700 K 的黑体所发出的辐射（虚线）与模型的不同，尽管二者积分得到的总流量是相等的。（短波长附近的尖峰在大气模型的积分总流量中占据主导地位，请注意纵坐标为对数坐标）。24 微米处流量偏低是由于水蒸气的不透明度造成的。该模型在一系列与我们的测量相符合的可能模型中是偏热的，但是测量得到的误差棒是允许有其他 T_{eq} 较低的模型存在的。

在我们刚提交这篇文章后不久，我们了解到其他研究组使用了斯皮策望远镜[28]的红外阵列相机对 TrES−1 凌星行星系统有过相似的观测[27]。与该工作一起，我们使用斯皮策望远镜得出的结果代表着首次对系外行星辐射的测量。更多使用斯皮策望远镜的观测应该会迅速地将可信模型的范围缩窄，同时也会揭示出密近太阳系外巨行星的大气结构、成分和其他的特性等信息。

（梁恩思 翻译；周济林 审稿）

Drake Deming[1], Sara Seager[3], L. Jeremy Richardson[2] & Joseph Harrington[4]

[1] Planetary Systems Laboratory and Goddard Center for Astrobiology, Code 693; [2] Exoplanet and Stellar Astrophysics Laboratory, Code 667, NASA's Goddard Space Flight Center, Greenbelt, Maryland 20771, USA

[3] Department of Terrestrial Magnetism, Carnegie Institution of Washington, 5241 Broad Branch Road NW, Washington DC 20015, USA

[4] Center for Radiophysics and Space Research, Cornell University, 326 Space Sciences Bldg, Ithaca, New York 14853-6801, USA

Received 3 February; accepted 28 February 2005; doi:10.1038/nature03507.
Published online 23 March 2005.

References:

1. Collier-Cameron, A. Extrasolar planets: what are hot Jupiters made of? *Astron. Geophys.* **43**, 421-425 (2002).
2. Seager, S. & Sasselov, D. D. Extrasolar giant planets under strong stellar irradiation. *Astrophys. J.* **502**, L157-L161 (1998).
3. Charbonneau, D., Brown, T. M., Latham, D. W. & Mayor, M. Detection of planetary transits across a sun-like star. *Astrophys. J.* **529**, L45-L48 (2000).
4. Henry, G. W., Marcy, G. W., Butler, R. P. & Vogt, S. S. A transiting "51 Peg-like" planet. *Astrophys. J.* **529**, L41-L44 (2000).
5. Brown, T. M., Charbonneau, D., Gilliland, R. L., Noyes, R. W. & Burrows, A. Hubble Space Telescope time-series photometry of the transiting planet of HD 209458. *Astrophys. J.* **552**, 699-709 (2001).
6. Bodenheimer, P., Lin, D. N. C. & Mardling, R. A. On the tidal inflation of short-period extrasolar planets. *Astrophys. J.* **548**, 466-472 (2001).
7. Laughlin, G. *et al.* A comparison of observationally determined radii with theoretical radius predictions for short-period transiting extrasolar planets. *Astrophys. J.* (in the press).
8. Burrows, A., Sudarsky, D. & Hubeny, I. in *The Search for Other Worlds: Proc. 14th Annu. Astrophys. Conf. in Maryland* (eds Holt, S. & Deming, D.) Vol. 713, 143-150 (American Institute of Physics, Melville, New York, 2003).
9. Werner, M. W. *et al.* The Spitzer Space Telescope mission. *Astrophys. J. Suppl.* **154**, 1-9 (2004).
10. Rieke, G. H. *et al.* The Multiband Imaging Photometer for Spitzer (MIPS). *Astrophys. J. Suppl.* **154**, 25-29 (2004).
11. Beichman, C. A. *et al.* Planets and IR excesses: preliminary results from a *Spitzer* MIPS survey of solar-type stars. *Astrophys. J.* (in the press).
12. Ribas, A. H., Solano, E., Masana, E. & Gimenez, A. Effective temperatures and radii of planet-hosting stars from IR photometry. *Astron. Astrophys.* **411**, L501-L504 (2003).
13. Perryman, M. A. C. (ed.) *The* Hipparcos *and Tycho Catalogues* (ESA SP-1200, European Space Agency, Noordwijk, 1997).
14. Kurucz, R. *Solar Abundance Model Atmospheres for 0, 1, 2, 4, and 8 km/s* CD-ROM 19 (Smithsonian Astrophysical Observatory, Cambridge, Massachusetts, 1994).
15. Horne, K. An optimal extraction algorithm for CCD spectrososcopy. *Publ. Astron. Soc. Pacif.* **98**, 609-617 (1986).
16. Krist, J. in *Astronomical Data Analysis Software and Systems IV* (eds Shaw, R. A., Payne, H. E. & Hayes, J. J. E.) Vol. 77, 349-352 (Astronomical Society of the Pacific, San Francisco, 1995).
17. Gordon, K. D. Reduction algorithms for the Multiband Imaging Photometer for SIRTF. *Publ. Astron. Soc. Pacif.* (in the press).
18. Rieke, G. H. *et al.* in *Proc. SPIE: Optical, Infrared, and Millimeter Space Telescopes* (ed. Mather, J. C.) Vol. 5487, 50-61 (SPIE, Bellingham, Washington, 2004).
19. Kelsall, T. *et al.* The COBE Diffuse Infrared Background Experiment (DIRBE) search for the cosmic infrared background. II. Model of the interplanetary dust cloud. *Astrophys. J.* **508**, 44-73 (1998).
20. Wittenmyer, R. A. *The Orbital Ephemeris of HD 209458b.* Master's thesis, San Diego State Univ. (2003).
21. Goukenleuque, C., Bezard, B., Joguet, B., Lellouch, E. & Freedman, R. A radiative equilibrium model of 51 Peg b. *Icarus* **143**, 308-323 (2000).
22. Seager, S., Whitney, B. A. & Sasselov, D. D. Photometric light curves and polarization of close-in extrasolar giant planets. *Astrophys. J.* **540**, 504-520 (2000).
23. Barman, T. S., Hauschildt, P. H. & Allard, F. Irradiated planets. *Astrophys. J.* **556**, 885-895 (2001).
24. Sudarsky, D., Burrows, A. & Hubeny, I. Theoretical spectra and atmospheres of extrasolar giant planets. *Astrophys. J.* **588**, 1121-1148 (2003).
25. Charbonneau, D. in *Scientific Frontiers in Research on Extrasolar Planets* (eds Deming, D. & Seager, S.) Vol. 294, 449-456 (Astronomical Society of the Pacific, San Francisco, 2003).
26. Press, W. H., Teukolsky, S. A., Vettering, W. T. & Flannery, B. P. *Numerical recipes in C* 2nd edn (Cambridge Univ. Press, Cambridge, 1992).
27. Alonso, R. *et al.* TrES-1, the transiting planet of a bright K0V star. *Astrophys. J.* **613**, L153-L156 (2004).
28. Charbonneau, D. *et al.* Detection of thermal emission from an extrasolar planet. *Astrophys. J.* (in the press).

Supplementary Information accompanies the paper on www.nature.com/nature.

Acknowledgements. We thank G. Laughlin for communicating the latest orbital eccentricity solutions from the Doppler data and for his evaluation of their status. We acknowledge informative conversations with D. Charbonneau, G. Marcy, B. Hansen, K. Menou and J. Cho. This work is based on observations made with the

Spitzer Space Telescope, which is operated by the Jet Propulsion Laboratory, California Institute of Technology, under contract to NASA. Support for this work was provided directly by NASA, and by its Origins of Solar Systems programme and Astrobiology Institute. We thank all the personnel of the Spitzer telescope and the MIPS instrument, who ultimately made these measurements possible. L.J.R. is a National Research Council Associate at NASA's Goddard Space Flight Center.

Competing interests statement. The authors declare that they have no competing financial interests.

Correspondence and requests for materials should be addressed to D.D. (Leo.D.Deming@nasa.gov).

Origin of the Cataclysmic Late Heavy Bombardment Period of the Terrestrial Planets

R. Gomes *et al.*

Editor's Note

The Moon seems to have experienced a large increase in strikes by comets and asteroids about 700 million years after the formation of the solar system. It is hard to understand why this is so; but here Alessandro Morbidelli and colleagues propose an explanation. They show that if the giant planets migrated slowly (Jupiter inward, Saturn, Uranus and Neptune outward), they could hit a "resonance" between the orbits of Jupiter and Saturn that would lead to very rapid migration. This gravitational churning of the solar system would cause objects from the disk of debris outside the orbit of Neptune to be pulled into the inner solar system, increasing the rate of impacts on planets and moons.

The petrology record on the Moon suggests that a cataclysmic spike in the cratering rate occurred ~700 million years after the planets formed[1]; this event is known as the Late Heavy Bombardment (LHB). Planetary formation theories cannot naturally account for an intense period of planetesimal bombardment so late in Solar System history[2]. Several models have been proposed to explain a late impact spike[3-6], but none of them has been set within a self-consistent framework of Solar System evolution. Here we propose that the LHB was triggered by the rapid migration of the giant planets, which occurred after a long quiescent period. During this burst of migration, the planetesimal disk outside the orbits of the planets was destabilized, causing a sudden massive delivery of planetesimals to the inner Solar System. The asteroid belt was also strongly perturbed, with these objects supplying a significant fraction of the LHB impactors in accordance with recent geochemical evidence[7,8]. Our model not only naturally explains the LHB, but also reproduces the observational constraints of the outer Solar System[9].

PREVIOUS work[9] explains the current orbital architecture of the planetary system by invoking an initially compact configuration in which Saturn's orbital period was less than twice that of Jupiter. After the dissipation of the gaseous circumsolar nebula, Jupiter's and Saturn's orbits diverged as a result of their interaction with a massive disk of planetesimals, and thus the ratio of their orbital periods, P_S/P_J, increased. When the two planets crossed their mutual 1:2 mean motion resonance (1:2 MMR, that is, $P_S/P_J = 2$) their orbits became eccentric. This abrupt transition temporarily destabilized the giant planets, leading to a short phase of close encounters among Saturn, Uranus and Neptune. As a result of these encounters, and of the interactions of the ice giants with the disk, Uranus

类地行星激变晚期重轰击期的起源

戈梅斯等

编者按

月球似乎在太阳系形成之后大约7亿年时，遭受到的彗星和小行星的撞击大大增加。这一现象的原因很难解释。但本文中亚历山德罗·莫尔比代利和他的同事就这一现象给出了一个解释。他们的研究显示，如果巨行星缓慢迁移(木星向内迁移，土星、天王星和海王星向外迁移)，它们将使得木星和土星的轨道之间形成一种“共鸣”，从而导致非常迅速的迁移。这种太阳系重力的扰动会造成海王星轨道外碎屑盘的天体被吸引到内太阳系，从而增加了其对行星和月球造成影响的概率。

针对月球岩石学记录的研究表明，在行星形成后的大约7亿年，月球陨击率激增[1]，这一事件被称为晚期重轰击(LHB)。行星形成理论尚不能合理解释为何在太阳系演化历史如此晚的阶段，会突然出现这样强的星子轰击期[2]。为此，研究人员提出了若干理论解释这一晚期的撞击突增[3-6]，然而目前还尚无一种理论建立在太阳系演化的自洽框架之内。本文中，我们提出LHB是由于巨行星的快速迁移触发的，该时期出现于长时间的宁静期之后。在这次迁移爆发的过程中，位于行星轨道外侧的星子盘变得不稳定，从而导致星子突然向太阳系内侧运动。同时，小行星带也受到强烈的扰动，其中的天体构成了LHB撞击体的绝大部分，这与最近发现的地球化学证据相符[7-8]。我们的模型不仅自然解释了LHB，而且重现了外太阳系的观测约束[9]。

已有的研究[9]通过引入一个最初致密的构型，在这一构型中土星的轨道周期小于木星轨道周期的两倍，从而可以解释当前行星系统的轨道体系。当气态环日星云耗散后，木星、土星与大质量星子盘发生相互作用，导致二者的轨道发生背离，所以二者的轨道周期之比P_S/P_J也随之增加。当木星和土星越过其1:2平运动共振(MMR)时(即$P_S/P_J = 2$)，这两颗行星轨道变得越来越椭。这一突然转变暂时使得巨行星变得不稳定，从而导致土星、天王星以及海王星在短时间轨道发生近距离交会。由于上述交会以及带有盘的冰巨行星的相互作用，天王星和海王星到达目前的日心

and Neptune reached their current heliocentric distances and Jupiter and Saturn evolved to their current orbital eccentricities[9]. The main idea of this Letter is that the same planetary evolution could explain the LHB, provided that Jupiter and Saturn crossed the 1:2 MMR roughly 700 Myr after they formed. Thus, our goal is to determine if there is a generic mechanism that could delay the migration process.

In previous studies[9-12], planet migration started immediately because planetesimals were placed close enough to the planets to be violently unstable. Although this type of initial condition was reasonable for the goals of those studies, it is unlikely. Planetesimal-driven migration is probably not important for planet dynamics as long as the gaseous massive solar nebula exists. The initial conditions for the migration simulations should represent the system that existed at the time the nebula dissipated. Thus, the planetesimal disk should contain only those particles that had dynamical lifetimes longer than the lifetime of the solar nebula. In planetary systems like those we adopt from ref. 9, we find that they had to be beyond ~15.3 AU (Fig. 1), leading to the initial conditions illustrated in Fig. 2a.

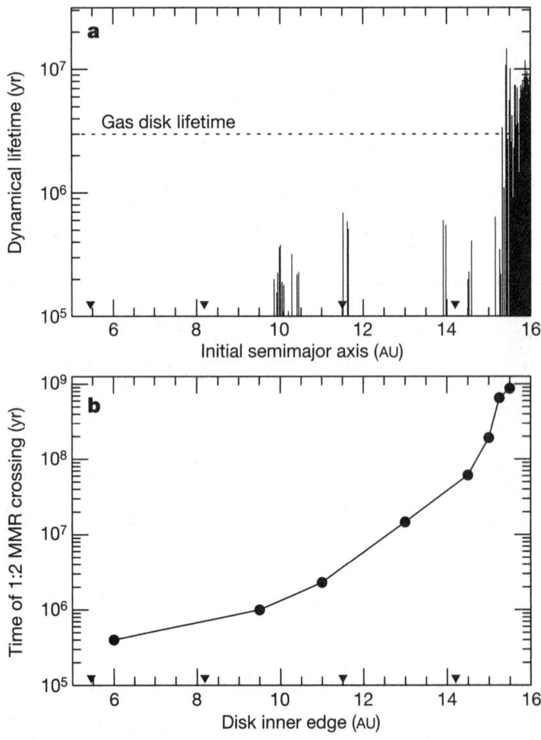

Fig. 1. Disk location and LHB timing. **a**, The histogram reports the average dynamical lifetime of massless test particles placed in a planetary system (shown as triangles) with Jupiter, Saturn and the ice giants on nearly circular, co-planar orbits at 5.45, 8.18, 11.5 and 14.2 AU, respectively. Initially, we placed 10 particles with $e = i = 0$ (where e is eccentricity and i is inclination) and random mean anomaly at each semimajor axis. Stable Trojans of the planets have been removed from this computation. Each vertical bar in the plot represents the average lifetime for those 10 particles. We define "dynamical lifetime" as the time required for a particle to encounter a planet within a Hill radius. A comparison between the histogram and the putative lifetime of the gaseous nebula[20] argues that, when the latter dissipated, the inner edge of

距，而木星和土星轨道则经过演化形成目前的偏心率[9]。本文的主要观点是，以上行星演化理论同样也可以用于解释 LHB，不过需要满足以下条件：木星和土星需要在形成后大约 7 亿年越过其 1:2 平运动共振。因此，我们的研究目的是确定是否存在可以推迟迁移进程的一般机制。

在以往的研究中[9-12]，星子与行星的距离被认为足够近，从而使行星变得非常不稳定，进而导致行星瞬间开始迁移。尽管这种初始条件对那些研究而言也是合理的，但是实际上却不可能发生。只要大量的气态大质量太阳星云还存在，那么星子驱动的迁移对于行星动力学而言可能就不会很重要。用于仿真模拟迁移过程的初始条件应该表征星云耗散时刻的系统状态。因此，星子盘应该只含有动力学寿命长于太阳星云的那些粒子。在类似被采纳的行星系统中[9]，我们发现这些星子的距离超过 ~15.3 AU(图 1)，图 2a 对相应的初始条件进行了说明。

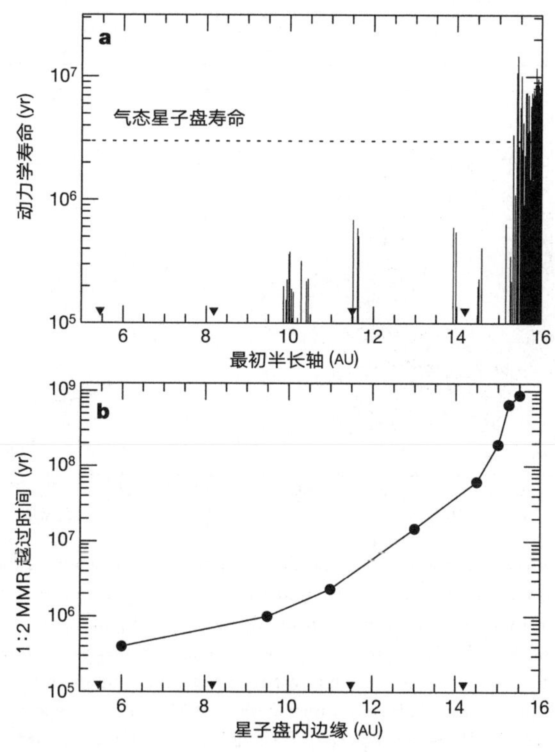

图 1. 星子盘的位置和 LHB 发生时间。**a**，直方图给出了无质量检验粒子的平均动力学寿命；这些粒子被置于与木星、土星和冷巨行星几乎共面的圆形轨道上，它们的轨道距离分别为 5.45 AU、8.18 AU、11.5 AU、14.2 AU，如图中三角所示。最初，我们放置了 10 颗粒子，其偏心率 e 和倾角 i 均为零，每个半长轴处的平近点角随机取值。在该计算中，我们没有考虑稳定的特洛伊型小行星。图中的每条垂直线代表着 10 颗粒子的平均寿命。我们将“动力学寿命”定义为粒子在希尔半径内交会行星所需要的时间。通过对比直方图和推定的气体星云的寿命可知[20]，当星云耗散后，星子盘内边缘将超过最外层冰

the planetesimal disk had to be about 1–1.5 AU beyond the outermost ice giant. **b**, Time at which Jupiter and Saturn crossed the 1:2 MMR, as a function of the location of the planetesimal disk's inner edge, as determined from our first set of migration simulations. In all cases, the disk had a surface density equivalent to 1.9 M_E per 1 AU annulus. The outer edge of the disk was varied so that the total mass of the disk was 35 M_E. The disk was initially very dynamically cold, with $e = 0$ and $i < 0.5°$. A comparison between **a** and **b** shows that a disk that naturally should exist when the nebula dissipated would produce a 1:2 MMR crossing at a time comparable to that of the LHB event.

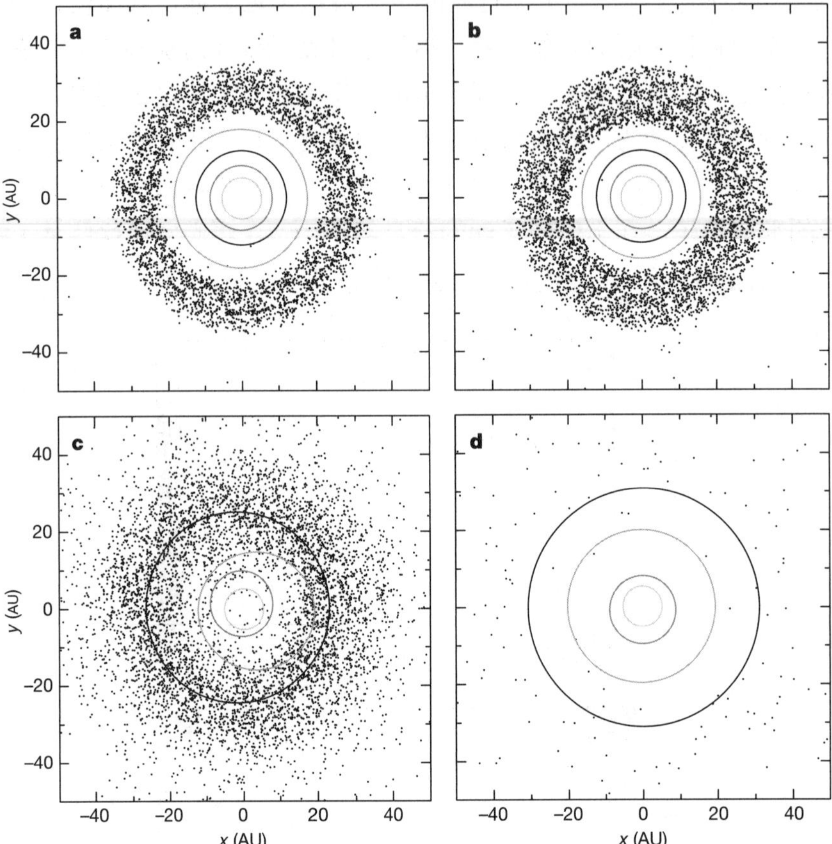

Fig. 2. The planetary orbits and the positions of the disk particles, projected on the initial mean orbital plane. The four panels correspond to four different snapshots taken from our reference simulation. In this run, the four giant planets were initially on nearly circular, co-planar orbits with semimajor axes of 5.45, 8.18, 11.5 and 14.2 AU. The dynamically cold planetesimal disk was 35 M_E, with an inner edge at 15.5 AU and an outer edge at 34 AU. Each panel represents the state of the planetary system at four different epochs: **a**, the beginning of planetary migration (100 Myr); **b**, just before the beginning of LHB (879 Myr); **c**, just after the LHB has started (882 Myr); and **d**, 200 Myr later, when only 3% of the initial mass of the disk is left and the planets have achieved their final orbits.

In this configuration, the initial speed of migration would be dependent on the rate at which disk particles evolve onto planet-crossing orbits. The time at which Jupiter and Saturn cross their 1:2 MMR depends on: (1) their initial distance from the location of the resonance, (2) the surface density of the disk near its inner edge, and (3) the relative location of the inner edge of the disk and the outer ice giant. On the basis of the above arguments, we initially

巨行星大约 1 ~ 1.5 AU。**b**，图中给出了木星和土星越过 1∶2 MMR 的时间与星子盘内边缘位置的函数关系，这一关系由我们第一组迁移的仿真模拟所确认。在所有的情况下，星子盘的密度等价于每 1 AU 环带 1.9 M_E。星子盘在最初动力学上是冷的，其偏心率 $e = 0$，倾角 $i < 0.5°$。比较图 **a** 和图 **b** 可知，当星云耗散后，自然存在的星子盘就会在与 LHB 事件相当的时间里越过 1∶2 MMR。

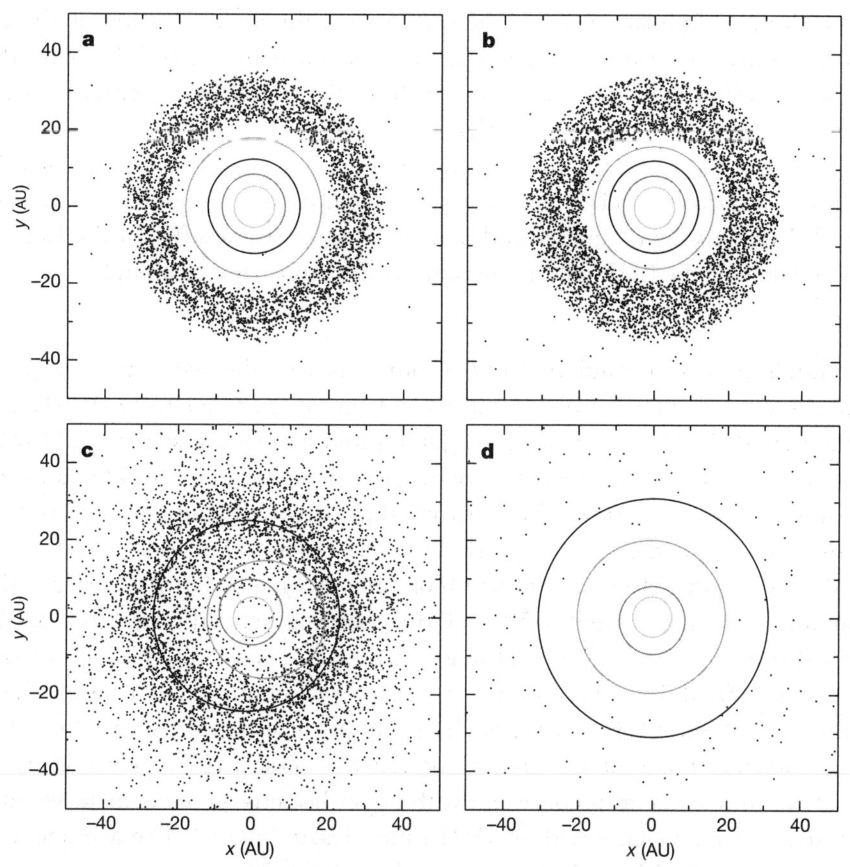

图 2. 在初始平均轨道平面上投影得到的行星轨道和星子盘粒子的位置。上面的四幅图片分别对应我们模拟过程中的四张快照。在我们的模拟中，四颗巨行星初始位置位于近圆形共面轨道上，其半长轴分别为 5.45 AU、8.18 AU、11.5 AU 和 14.2 AU。动力学上的冷星子盘的质量为 35 M_E，内边缘位于 15.5 AU，外边缘位于 34 AU。四幅图分别给出了行星系统在四个不同时期的状态：**a**，行星迁移的开始（1 亿年）；**b**，LHB 开始前不久（8.79 亿年）；**c**，LHB 开始后不久（8.82 亿年）；**d**，LHB 开始后的 2 亿年，这时星子盘只剩下 3% 的初始质量，各行星已经进入它们最终的轨道。

在这一构型中，行星迁移的初始速度将取决于星子盘粒子逸出至行星交会轨道的速率。木星和土星轨道越过 1∶2 MMR 的时间取决于：(1) 它们与共振位置的初始距离；(2) 靠近星子盘内边缘的表面密度；(3) 星子盘内边缘与外侧冰巨行星的相对位置。基于以上参数，我们进行了八个仿真模拟，其中星子盘内边缘的位置设为唯

performed a series of eight simulations where the location of the inner edge of the disk was set as the unique free parameter (Fig. 1). As expected, we found a strong correlation between the location of the inner edge and the time of the 1:2 MMR crossing. For disks with inner edges near 15.3 AU (see above), we find crossing times between 192 Myr and 880 Myr (since the beginning of the simulation).

We also performed eight simulations where we varied the initial location of the ice giants by ~1 AU, Saturn's location by ~0.1 AU, the total mass of the disk by 5 Earth masses (5 M_E), and its initial dynamical state by pushing the particles' eccentricities up to 0.1 and inclinations up to 3.5°. We found that we can delay the resonant crossing to 1.1 Gyr since the beginning of the simulation, although longer times are clearly possible for more extreme initial conditions. Therefore, we can conclude that the global instability caused by the 1:2 MMR crossing of Jupiter and Saturn could be responsible for the LHB, because the estimated date of the LHB falls in the range of the times that we found.

Figures 2 and 3 show the evolution of one of our runs from the first series of eight. Initially, the giant planets migrated slowly owing to leakage of particles from the disk (Fig. 3a). This phase lasted 880 Myr, at which point Jupiter and Saturn crossed the 1:2 MMR. After the resonance crossing event, the orbits of the ice giants became unstable and they were scattered into the disk by Saturn. They disrupted the disk and scattered objects all over the Solar System, including the inner regions. The solid curve in Fig. 3b shows the amount of material that struck the Moon as a function of time. A total of 9×10^{21} g struck the Moon after resonance crossing—roughly 50% of this material arrived in the first 3.7 Myr and 90% arrived before 29 Myr. The total mass is consistent with the estimate[4] of 6×10^{21} g, which was determined from the number and size distribution of lunar basins that formed around the time of the LHB epoch[1]. Such an influx spike happened in all our runs. The amount of cometary material delivered to the Earth is $\sim 1.8 \times 10^{23}$ g, which is about 6% of the current ocean mass. This is consistent with upper bounds on the cometary contribution to the Earth's water budget, based on D/H ratio measurement[13]. The average amount of material accreted by the Moon during this spike was $(8.4 \pm 0.3) \times 10^{21}$ g.

The above mass delivery estimate corresponds only to the cometary contribution to the LHB, as the projectiles originated from the external massive, presumably icy, disk. However, our scheme probably also produced an in flux of material from the asteroid belt. As Jupiter and Saturn moved from 1:2 MMR towards their current positions, secular resonances (which occur when the orbit of an asteroid processes at the same rate as a planet) swept across the entire belt[14]. These resonances can drive asteroids onto orbit with eccentricities and inclinations large enough to allow them to evolve into the inner Solar System and hit the Moon[4].

一的自由参量(图 1)。正如预期的那样，我们发现越过 1:2 MMR 的时间与星子盘内边缘的位置密切相关。当内边缘为 15.3 AU 时(见前述)，相应的越过时间介于 1.92 到 8.80 亿年之间(从模拟启动开始计)。

同时，我们也进行了另外八个仿真模拟。在这些模拟中，我们设定冰巨行星的起始位置变化约 1 AU，土星的位置变化约 0.1 AU，星子盘的总质量变化 5 倍地球质量(5 M_E)，并通过推动粒子的偏心率至 0.1 且推动倾角至 3.5° 来改变星子盘的起始动力学状态。通过这些模拟，我们发现可以将共振穿越的时间推迟到 11 亿年(从启动模拟开始计算)。当然，如果采用更为极端的初始条件，也可能得到更长的时间。因此，我们得出以下结论：越过木星和土星的 1∶2 MMR 引起的全局不稳定性将会导致 LHB，因为推算出的 LHB 发生时间恰好落在我们模拟的时间范围内。

图 2 和图 3 展示了第一批八个模拟中的一次演化过程。最初，由于星子盘中粒子的渗漏，巨行星迁移得很缓慢，参见图 3a。这个阶段持续了 8.8 亿年，同时木星和土星在该时间越过了 1∶2 MMR。共振穿越事件发生后，冰巨行星的轨道变得不稳定，并且通过土星散射入星子盘；这导致了星子盘的瓦解，并将其中的天体散射到太阳系各处，包括靠近太阳的区域。图 3b 中的实线给出了撞击月球的物质数量随时间的变化。共振穿越之后，总共有质量为 9×10^{21} g 的物质撞击了月球——其中，在前 370 万年有大约 50% 的物质达到月球，而在前 2,900 万年有 90% 的物质达到月球。通过分析 LHB 时期形成的月球盆地的数量和尺寸分布 [1]，计算出的撞击物质总质量与预测结果一致，为 6×10^{21} g[4]。在我们所有的模拟中，均出现了这样的内向通量的峰值。该过程大约向地球释放了质量为 1.8×10^{23} g 的彗星物质，这大体相当于目前地球上海水总质量的 6%。该结果也与地球水量平衡中彗星贡献的上界相符，后者是基于 D/H 比值测量得到的 [13]。在整个轰击峰值时期，月球增加的物质质量平均为 $(8.4 \pm 0.3) \times 10^{21}$ g。

以上输运质量的估计仅仅是对应在 LHB 时期的彗星物质贡献，即这些彗星抛射物质源自外部大质量的，推测可能是冰质的盘。然而，我们的体系中也存在源自小行星带的物质内向通量。随着木星和土星从 1∶2 MMR 移动到目前的位置，当小行星带中的一颗小行星的轨道进动速率(这里指近星点进动速率)与一颗行星的进动速率相同时，就会发生扫过整个小行星带的长期共振 [14]。这些长期共振驱使部分小行星进入偏心率和倾角较大的轨道，从而使得这些小行星足以进入太阳系内部区域并撞击月球 [4]。

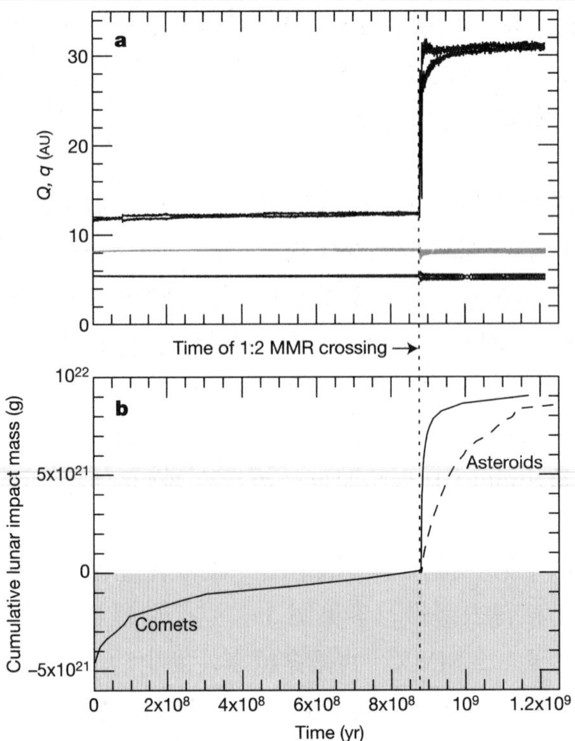

Fig. 3. Planetary migration and the associated mass flux towards the inner Solar System from a representative simulation. **a**, The evolution of the four giant planets. Each planet is represented by a pair of curves—the top and bottom curves are the aphelion and perihelion distances, Q and q, respectively. Jupiter and Saturn cross the 1:2 MMR at 880 Myr. The subsequent interaction between the planets and the disk led to the current planetary configuration as shown in ref. 9. **b**, The cumulative mass of comets (solid curve) and asteroids (dashed curve) accreted by the Moon. We have offset the comet curve so that the value is zero at the time of 1:2 MMR crossing. Thus, $\sim 5 \times 10^{21}$ g of comets was accreted before resonant crossing and 9×10^{21} g of cometary material would have struck the Moon during the LHB. Although the terrestrial planets were not included in our cometary simulations, we estimated the amount of material accreted by the Moon directly from the mass of the planetesimal disk by combining the particles' dynamical evolution with the analytic expressions in ref. 21. The impact velocity of these objects ranged from 10 to 36 km s^{-1} with an average of 21 km s^{-1}. Estimating the asteroidal flux first requires a determination of the mass of the asteroid belt before resonant crossing. This value was determined by first combining the percentage of asteroids remaining in the belt at the end of a simulation ($\sim$10%, very sensitive to planet migration rate and initial asteroid distribution) with estimates of the current mass of the belt to determine the initial asteroid belt mass ($\sim 5 \times 10^{-3}$ M_E). The flux was then again determined by combining the particles' dynamical evolution with the analytic expressions in ref. 21. The dashed curve shows a simulation where class 2 particles dominate. The average asteroidal impact velocity is 25 km s^{-1}.

We investigated the role of asteroid impactors in our LHB model by the following numerical integrations. The orbits of an asteroid belt, composed of 1,000 massless particles with semimajor axes between 2.0 and 3.5 AU, were integrated under the gravitational influence of the Sun, Venus, Earth, Mars, Jupiter and Saturn. Because formation models[15,16] predict that the asteroid belt was partially depleted and dynamically excited well before the LHB, we set the particles' eccentricities between 0 and 0.3 and inclinations between 0° and 30°, but kept the perihelion distances, q, > 1.8 AU and aphelion distances, Q, < 4 AU. Jupiter

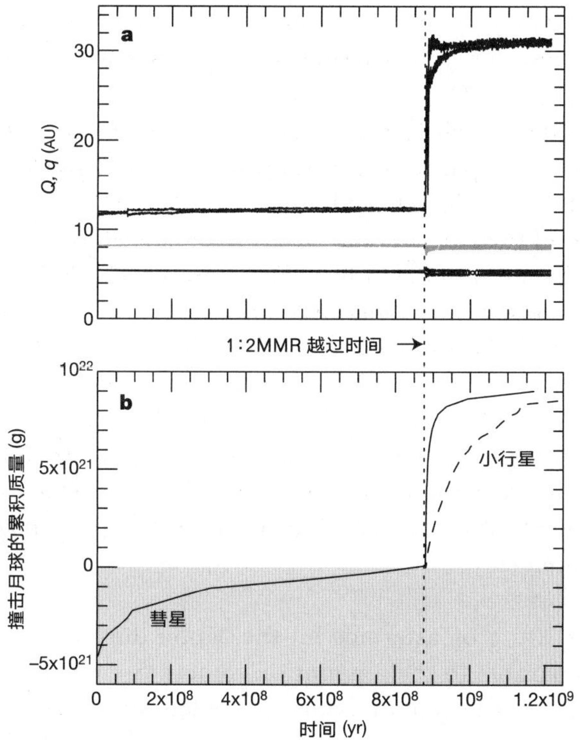

图 3. 一次典型模拟中的行星迁移以及进入内太阳系相关的质量通量。**a**，图中给出了四颗巨行星的演化。图中每颗行星由两条曲线表示，上下两条曲线分别代表远日点(Q)和近日点(q)距离。木星和土星在 8.80 亿年越过 1:2 MMR。随后，行星和星子盘之间的相互作用形成了现在的行星系统构型[9]。**b**，图中给出了彗星和小行星撞击月球的累积质量，分别用实线和虚线表示。我们对彗星曲线进行了偏移，使得 1:2 MMR 穿越发生时刻的累积质量为零。因此，在共振穿越发生前大约质量为 5×10^{21} g 的彗星物质落在月球上，而在 LHB 期间总共有质量为 9×10^{21} g 的彗星物质撞击了月球。尽管我们在彗星模拟过程中并没有将类地行星也包含进来，但我们可以将粒子的动力学演化和参考文献 21 中的解析表达式结合起来，进而估算出星子盘物质撞击月球的质量。这些小型天体的撞击速度介于 10 至 36 km/s，平均速度为 21 km/s。为了估计小行星流量，需要首先确定共振穿越前的小行星带的质量。我们可以将模拟结束后的小行星在带中的剩余比例(约 10%，对行星迁移率和小行星初始分布十分敏感)，与小行星带当前的估算质量综合起来，确定小行星带的初始质量(约 $5\times10^{-3}M_E$)。为了确定流量，我们可以将粒子动力学演化和参考文献 21 的解析表达式结合起来。图中虚线表明第二类粒子起主导作用的模拟结果。小行星的平均撞击速度为 25 km/s。

下面，我们借助数值积分的方法研究了 LHB 模型中小行星撞击体的作用。在太阳、金星、地球、火星、木星和土星的引力场的影响下，对以下小行星带进行积分：这是一条由 1,000 颗无质量粒子组成的小行星带，其中粒子的轨道半长轴介于 2.0 与 3.5 AU 之间。由于形成模型预测在 LHB 发生之前，这条小行星带就已经部分被清空并处于动力学上充分激发态[15-16]，因此我们将粒子的偏心率设在 0 到 0.3 之间，倾角

and Saturn were forced to migrate at rates that varied from run to run (adopted from ref. 9) by adding a suitably chosen drag-force term to their equations of motion.

We find that objects that reach Earth-crossing orbits follow one of two general paths. Some, referred to as class 1 particles, get trapped in the periapse secular resonance with Saturn (which affects eccentricities) and are driven directly onto Earth-crossing orbits. Other particles, referred to as class 2, stay in the asteroid belt, but are dynamically excited by resonant sweeping onto unstable orbits. These objects slowly leak out of the asteroid belt and can evolve into the inner Solar System. The two classes produce impact spikes with different temporal behaviours. Roughly 50% of class 1 particles arrive in the first 10 Myr, while 90% arrive within ~30 Myr. Conversely, the median arrival time for class 2 particles is ~50 Myr and 90% arrive within ~150 Myr. Class 2 particles dominated in our runs (Fig. 3). However, a preliminary investigation into this issue shows that this result is probably sensitive to the exact evolution of the giant planets and the dynamical state of the asteroid belt. Thus, the best we can conclude is that the impact spike due to asteroids is between these two extremes.

We find that (3–8) $\times 10^{21}$ g of asteroids hit the Moon during our simulations (Fig. 3). This amount is comparable to the amount of comets. So, our model predicts that the LHB impactors should have been a mixture of comets and asteroids. Unfortunately, we cannot say with any certainty the exact ratio of comets to asteroids in our model because, although the amount of cometary material is fairly well constrained (probably better than a factor of 2), the amount of asteroidal material is not well known (and could be outside the range reported above), because we do not have good estimates of the mass distribution in the asteroid belt before the LHB. It should also be noted that this ratio is probably a function of impactor size, because comets and asteroids probably have different size distributions. This ratio probably also varied with time. Within the first ~30 Myr comets dominated according to these simulations, but the last impactors were asteroidal. This is consistent with recent cosmochemical findings suggesting that some of the Moon's basins were formed by asteroids[7,8].

Our results support a cataclysmic model for the lunar LHB. Although many aspects of the LHB are not well known[1], our simulations reproduce two of the main characteristics attributed to this episode: (1) the 700 Myr delay between the LHB and terrestrial planet formation, and (2) the overall intensity of lunar impacts. Our model predicts a sharp increase in the impact rate at the beginning of the LHB. Unfortunately, the available lunar data are not yet capable of addressing this prediction.

Our model also has the advantage of supplying impactors that are a mixture of comets and asteroids. Our model predicts that the asteroid belt was depleted by a factor of ~10 during the LHB. This depletion does not contradict collisional evolution models[17,18]. On the contrary, the late secular resonance sweeping could explain why we do not see a large number of asteroid families that were produced during the LHB[18]. Our model predicts that the LHB lasted from between ~10 Myr and ~150 Myr. Correspondingly, the drop-off in

则介于 0 到 30°，近日点距离 q 满足 $q > 1.8$ AU，远日点距离 Q 满足 $Q < 4$ AU。通过在运动方程中加入合适的阻力项，木星和土星就会被迫以不断变化的速率迁移[9]。

我们发现，进入越地轨道的天体大致上遵循两条路径中的一条。一些天体被称为第一类粒子，它们会由于土星对偏心率的影响而被近星点长期共振俘获，而被直接拽入越地轨道。另一些粒子被称为第二类粒子，这类粒子位于小行星带中，但是会由共振清除作用被动力地激发到非稳定轨道上。这些小天体会逐渐脱离小行星带，然后进入内太阳系。以上两类粒子产生的撞击在时间分布上也不同。在第一类粒子中会有大约 50% 在最初的 1 千万年内发生撞击，大约 90% 会在约 3 千万年内发生撞击。对于第二类粒子，半数到达时间约为 5 千万年，90% 会在约 1.5 亿年内到达。如图 3 所示，在我们的计算机模拟中，第二类粒子起着主导作用。然而，我们的初步研究表明，以上问题的结果可能对巨行星的具体演化和小行星带的动力学状态十分敏感。因此，我们得出的最可信的结论是，小行星的撞击高峰期位于以上两个极值之间。

在模拟中，我们发现总质量为 $(3 \sim 8) \times 10^{21}$ g 的小行星撞击了月球，如图 3 所示。这一质量可以与彗星的质量相比较。因此，我们的模型预测 LHB 撞击体应该包含彗星和小行星。但是，我们并不能确定模型中彗星和小行星的确切比例。尽管可以较为准确地限定彗星物质的质量（可能优于因子 2），但是尚不清楚小行星物质的质量（可能在上述范围之外），这是因为我们尚不能很好地估计出 LHB 之前小行星带的质量分布。需要指出的是，彗星和小行星二者质量的比值可能也是撞击体尺寸的函数，因为彗星和小行星可能具有不同的尺寸分布。同时，这一比值也可能随时间变化。根据我们的模拟可知，在最初的约 3 千万年，彗星起主导作用；但是最后一批撞击体主要由小行星构成。这与最近的宇宙化学发现相符，其认为部分月球盆地是由于小行星撞击形成的[7,8]。

我们的结果支持月球 LHB 的灾变模型。尽管对于 LHB，有许多问题亟待解决，但是我们的模拟重现了这一事件的两个主要特征：(1) LHB 要比地球形成晚 7 亿年；(2) 月球撞击的总强度。我们的模型也预测在 LHB 刚开始的时候，撞击率迅速增加。然而，目前的月球数据尚不能证实这一预测。

我们的模型另一个优点在于指出撞击体是彗星和小行星共同构成的。该模型预测，在 LHB 时期，整个小行星带经过耗损只剩下原先的约 1/10。这种耗损与碰撞演化模型并不矛盾[17,18]。相对地，晚期的长期共振清除机制可以解释为什么我们现在没有观察到大量产生于 LHB 时期的小行星族[18]。我们的模型预测，LHB 持续的时间介于 1 千万 ~ 1.5 亿年之间。相应的，这段时间内撞击率的衰减可以相当迅速

impact rates could be quite fast (with 50% of the impacts occurring in the first 3.7 Myr and 90% in 29 Myr) or moderately slow (with 50% of the impacts occurring in the first 50 Myr and 90% in 150 Myr). We are unable to pinpoint more exact values because the duration and the drop-off of the LHB depends on the relative contributions of class 1 asteroids, class 2 asteroids, and comets, which in turn are very sensitive to the pre-LHB orbital structure of the asteroid belt.

Most importantly, our scheme for the LHB is the result of a generic migration-delaying mechanism, followed by an instability, which is itself induced by a deterministic mechanism of orbital excitation of the planets[9]. This revised planetary migration scheme naturally accounts for the currently observed planetary orbits[9], the LHB, the present orbital distribution of the main-belt asteroids and the origin of Jupiter's Trojans[19].

(**435**, 466-469; 2005)

R. Gomes[1,2], H. F. Levison[2,3], K. Tsiganis[2] & A. Morbidelli[2]

[1] ON/MCT and GEA/OV/UFRJ, Ladeira do Pedro Antonio, 43 Centro 20.080-090, Rio de Janeiro, RJ, Brazil

[2] Observatoire de la Côte d' Azur, CNRS, BP 4229, 06304 Nice Cedex 4, France

[3] Department of Space Studies, Southwest Research Institute, 1050 Walnut Street, Suite 400, Boulder, Colorado, USA

Received 6 December 2004; accepted 18 April 2005.

References:

1. Hartmann, W. K., Ryder, G., Dones, L. & Grinspoon, D. in *Origin of the Earth and Moon* (eds Canup, R. & Righter, K.) 493-512 (Univ. Arizona Press, Tucson, 2000).
2. Morbidelli, A., Petit, J.-M., Gladman, B. & Chambers, J. A plausible cause of the Late Heavy Bombardment. *Meteorit. Planet. Sci.* **36**, 371-380 (2001).
3. Zappala, V., Cellino, A., Gladman, B. J., Manley, S. & Migliorini, F. Asteroid showers on Earth after family break-up events. *Icarus* **134**, 176-179 (1998).
4. Levison, H. F. *et al.* Could the lunar "Late Heavy Bombardment" have been triggered by the formation of Uranus and Neptune? *Icarus* **151**, 286-306 (2001).
5. Chambers, J. E. & Lissauer, J. J. A new dynamical model for the lunar Late Heavy Bombardment. *Lunar Planet. Sci. Conf.* **XXXIII**, abstr. 1093 (2002).
6. Levison, H. F., Thommes, E. W., Duncan, M. J., Dones, L. A. in *Debris Disks and the Formation of Planets: A Symposium in Memory of Fred Gillett (11-13 April 2002, Tucson, Arizona)* (eds Caroff, L., Moon, L. J., Backman, D. & Praton, E.) 152-167 (ASP Conf. Ser. 324, Astronomical Society of the Pacific, San Francisco, 2005).
7. Kring, D. A. & Cohen, B. A. Cataclysmic bombardment throughout the inner Solar System 3.9-4.0 Ga. *J. Geophys. Res. Planets* **107**(E2), 4-10 (2002).
8. Tagle, R. LL-ordinary chondrite impact on the Moon: Results from the 3.9 Ga impact melt at the landing site of Apollo 17. *Lunar Planet. Sci. Conf.* **XXXVI**, abstr. 2008 (2005).
9. Tsiganis, K., Gomes, R., Morbidelli, A. & Levison, H. F. Origin of the orbital architecture of the giant planets of the Solar System. *Nature* doi:10.1038/nature03539 (this issue).
10. Fernandez, J. A. & Ip, W.-H. Some dynamical aspects of the accretion of Uranus and Neptune—The exchange of orbital angular momentum with planetesimals. *Icarus* **58**, 109-120 (1984).
11. Hahn, J. M. & Malhotra, R. Orbital evolution of planets embedded in a planetesimal disk. *Astron. J.* **117**, 3041-3053 (1999).
12. Gomes, R. S., Morbidelli, A. & Levison, H. F. Planetary migration in a planetesimal disk: Why did Neptune stop at 30 AU? *Icarus* **170**, 492-507 (2004).
13. Morbidelli, A. *et al.* Source regions and timescales for the delivery of water to Earth. *Meteorit. Planet. Sci.* **35**, 1309-1320 (2000).
14. Gomes, R. S. Dynamical effects of planetary migration on the primordial asteroid belt. *Astron. J.* **114**, 396-401 (1997).
15. Wetherill, G. W. An alternative model for the formation of the asteroids. *Icarus* **100**, 307-325 (1992).
16. Petit, J., Morbidelli, A. & Chambers, J. The primordial excitation and clearing of the asteroid belt. *Icarus* **153**, 338-347 (2001).
17. Davis, D. R., Ryan, E. V. & Farinella, P. Asteroid collisional evolution: results from current scaling algorithms. *Planet. Space Sci.* **42**, 599-610 (1994).
18. Bottke, W. *et al.* The fossilized size distribution of the main asteroid belt. *Icarus* **175**(1), 111-140 (2005).
19. Morbidelli, A., Levison, H. F., Tsiganis, K. & Gomes, R. Chaotic capture of Jupiter's Trojan asteroids in the early Solar System. *Nature* doi:10.1038/nature03540 (this issue).

(50% 的撞击发生于开始的 370 万年，90% 的撞击发生于 2,900 万年内)，或者适当减缓(50% 的撞击发生于开始的 5 千万年，90% 的撞击发生于 1.5 亿年内)。我们并不能更为精确的确定以上时间，因为 LHB 的持续和减弱依赖于第一类小行星、第二类小行星以及彗星的相对贡献，而这三者对 LHB 发生前的小行星带轨道结构颇为敏感。

最为重要的是，我们的 LHB 体系是由一般的迁移延迟机制造成的，随后出现的不稳定性本身又是由行星轨道的激发的决定性机制所诱发的 [9]。这一修正的行星迁移体系可以自然地将目前我们观测到的行星轨道 [9]、LHB、目前的主带小行星轨道分布以及木星–特洛伊族小行星的起源 [19] 纳入其中。

(金世超 翻译；季江徽 审稿)

20. Haisch, K. E., Lada, E. A. & Lada, C. J. Disk frequencies and lifetimes in young clusters. *Astrophys. J.* **553**, L153-L156 (2001).

21. Wetherill, G. W. Collisions in the asteroid belt. *J. Geophys. Res.* **72**, 2429-2444 (1967).

Acknowledgements. R.G. thanks Conselho Nacional de Desenvolvimento Científico e Tecnológico for support for his sabbatical year in the OCA observatory in Nice. K.T. was supported by an EC Marie Curie Individual Fellowship. A.M. and H.F.L. thank the CNRS and the NSF for funding collaboration between the OCA and the SWRI groups. H.F.L. was supported by NASA's Origins and PG&G programmes.

Author Information. Reprints and permissions information is available at npg.nature.com/reprintsandpermissions. The authors declare no competing financial interests. Correspondence and requests for materials should be addressed to A.M. (morby@obs-nice.fr).

Experimental Investigation of Geologically Produced Antineutrinos with KamLAND

T. Araki *et al.*

Editor's Note

The KamLAND anti-neutrino detector at the Kamioka mine in Japan was the first detector sensitive enough to detect neutrino oscillations—the spontaneous transformation of a neutrino from one type to another—by looking at anti-neutrinos from nuclear reactors. As Takeo Araki and colleagues note here, the detector was also sensitive enough to detect neutrinos produced in the Earth's interior from radioactive decay of uranium or thorium. Previous estimates of the energy released by such decays suggested that it accounted for fully half of the total energy dissipated inside the planet, making radioactivity a significant contributor to the heat of the deep earth. The results here broadly agree with the earlier figures, and put an improved upper limit on its value.

The detection of electron antineutrinos produced by natural radioactivity in the Earth could yield important geophysical information. The Kamioka liquid scintillator antineutrino detector (KamLAND) has the sensitivity to detect electron antineutrinos produced by the decay of ^{238}U and ^{232}Th within the Earth. Earth composition models suggest that the radiogenic power from these isotope decays is 16 TW, approximately half of the total measured heat dissipation rate from the Earth. Here we present results from a search for geoneutrinos with KamLAND. Assuming a Th/U mass concentration ratio of 3.9, the 90 per cent confidence interval for the total number of geoneutrinos detected is 4.5 to 54.2. This result is consistent with the central value of 19 predicted by geophysical models. Although our present data have limited statistical power, they nevertheless provide by direct means an upper limit (60 TW) for the radiogenic power of U and Th in the Earth, a quantity that is currently poorly constrained.

THE Kamioka liquid scintillator antineutrino detector (KamLAND) has demonstrated neutrino oscillation using electron antineutrinos ($\bar{\nu}_e$s) with energies of a few MeV from nuclear reactors[1,2]. Additionally, KamLAND is the first detector sensitive enough to measure $\bar{\nu}_e$s produced in the Earth from the ^{238}U and ^{232}Th decay chains. Using $\bar{\nu}_e$s to study processes inside the Earth was first suggested by Eder[3] and Marx[4], and has been reviewed a number of times[5-10]. As $\bar{\nu}_e$s produced from the ^{238}U and ^{232}Th decay chains have exceedingly small interaction cross-sections, they propagate undisturbed in the Earth's interior, and their measurement near the Earth's surface can be used to gain information on their sources. Although the detection of $\bar{\nu}_e$s from ^{40}K decay would also be of great interest in geophysics, with possible applications in the interpretation of geo-magnetism, their energies are too

KamLAND 对地质来源的反中微子的实验研究

荒木岳夫等

编者按

KamLAND 反中微子探测器位于日本的神冈矿井下。它是第一个足够灵敏到通过监测来自反应堆的反中微子而探测到中微子振荡(即一种类型的中微子自发转变为另一种类型)的探测器。如荒木岳夫及其同事所指出的，该探测器也足够灵敏到可探测地球内部铀和钍放射性衰变产生的中微子。之前的估算表明，这些衰变产生的能量约占地球释放的总能量的一半，这使得放射性成为地球深部热能的一个主要来源。本文的结果与之前的数据大致吻合，并给出了更好的上限值。

探测地球中天然放射性产生的电子反中微子可以提供重要的地球物理信息。神冈液体闪烁体反中微子探测器(简称 KamLAND)可以探测到地球中放射元素 ^{238}U 和 ^{232}Th 衰变产生的电子反中微子(地球中微子)。地球组成模型表明，这些同位素衰变的放射性生热为 16 TW，大约占地球总散失热率测量值的一半。本文介绍了 KamLAND 探测地球中微子的结果。假设 Th/U 的质量丰度比为 3.9，探测到 90% 置信区间内地球中微子总数为 4.5 到 54.2 个。该结果与地球物理模型预测的中心值(19 个)一致。尽管现有数据统计量有限，这些数据仍然以直接的方式给出了此前知之甚少的地球中 U 和 Th 放射性生热的上限(60 TW)。

神冈液体闪烁体反中微子探测器(KamLAND)通过对来自核反应堆的几兆电子伏的电子反中微子($\bar{\nu}_e$)的探测，证明了反应堆中微子振荡现象[1,2]。另外，KamLAND 也是第一个足够灵敏并探测到地球中 ^{238}U 和 ^{232}Th 衰变链产生的 $\bar{\nu}_e$ 的探测器。埃德[3]和马克思[4]第一次提出可利用 $\bar{\nu}_e$ 来测量地球内部性质，之后该观点被引述了数次[5-10]。^{238}U 和 ^{232}Th 衰变链产生的 $\bar{\nu}_e$ 具有极其小的相互作用截面，因此它们不受干扰地在地球内部传播，在地球表面附近对 $\bar{\nu}_e$ 进行测量可获得其产生来源的信息。尽管探测 ^{40}K 衰变产生的 $\bar{\nu}_e$ 对地球物理学也具有重要意义，可能用以解释地磁

low to be detected with KamLAND. The antineutrino flux above our detection threshold from other long-lived isotopes is expected to be negligible.

The Radiogenic Earth

The total power dissipated from the Earth (heat flow) has been measured with thermal techniques[11] to be 44.2 ± 1.0 TW. Despite this small quoted error, a more recent evaluation[12] of the same data (assuming much lower hydrothermal heat flow near mid-ocean ridges) has led to a lower figure of 31 ± 1 TW. On the basis of studies of chondritic meteorites[13] the calculated radiogenic power is thought to be 19 TW, 84% of which is produced by ^{238}U and ^{232}Th decay. Some models of mantle convection suggest that radiogenic power is a larger fraction of the total power[14,15].

^{238}U and ^{232}Th decay via a series of well-established α and β^- processes[16] terminating in the stable isotopes ^{206}Pb and ^{208}Pb, respectively. Each β^- decay produces a daughter nucleus, an electron and a $\bar{\nu}_e$. The $\bar{\nu}_e$ energy distribution is well established[17], and includes a correction for the electromagnetic interaction between the electron and the charge distribution of the daughter nucleus. Figure 1 shows the expected $\bar{\nu}_e$ distribution, $\mathrm{d}n(E_\nu)/\mathrm{d}E_\nu$, as a function of $\bar{\nu}_e$ energy, E_ν, for the ^{238}U and ^{232}Th decay chains.

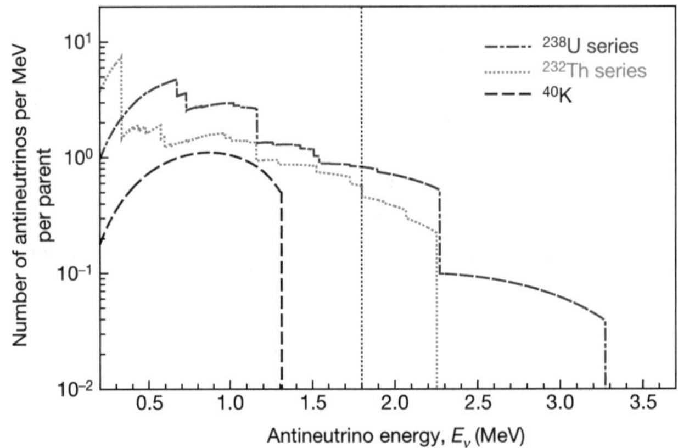

Fig. 1. The expected ^{238}U, ^{232}Th and ^{40}K decay chain electron antineutrino energy distributions. KamLAND can only detect electron antineutrinos to the right of the vertical dotted black line; hence it is insensitive to ^{40}K electron antineutrinos.

Ignoring the negligible neutrino absorption, the expected $\bar{\nu}_e$ flux at a position $\mathbf{r}$ for each isotope is given by:

$$\frac{\mathrm{d}\phi(E_\nu, \mathbf{r})}{\mathrm{d}E_\nu} = A\frac{\mathrm{d}n(E_\nu)}{\mathrm{d}E_\nu}\int_{V_\oplus} \mathrm{d}^3\mathbf{r}' \frac{a(\mathbf{r}')\rho(\mathbf{r}')P(E_\nu, |\mathbf{r}-\mathbf{r}'|)}{4\pi|\mathbf{r}-\mathbf{r}'|^2} \tag{1}$$

where A is the decay rate per unit mass, the integral is over the volume of the Earth, $a(\mathbf{r}')$ is the isotope mass per unit rock mass, $\rho(\mathbf{r}')$ is the rock density, and $P(E_\nu, |\mathbf{r}-\mathbf{r}'|)$ is the $\bar{\nu}_e$

现象，但是它们的能量太低，不能被 KamLAND 探测到。来自其他长寿命同位素且能量高于我们探测阈值的反中微子被认为可忽略不计。

地球的放射能

地球释放的总能量（热流）已通过热学方法[11]测得，为 44.2 ± 1.0 TW。尽管以上结果给出的误差很小，最近利用相同数据的推算[12]（假定洋中脊附近热液的热流低得多）给出一个更低的值 31 ± 1 TW。基于对球粒陨石的研究[13]，计算出的放射性生热约为 19 TW，其中 84% 来自 ^{238}U 和 ^{232}Th 衰变。一些地幔对流模型给出的放射性生热占总能量的比例更高[14,15]。

^{238}U 和 ^{232}Th 经过一系列已知的 α 衰变和 β^- 衰变后[16]，最终分别变成稳定同位素 ^{206}Pb 和 ^{208}Pb。每次 β^- 衰变产生一个子核、一个电子和一个 $\bar{\nu}_e$。这个 $\bar{\nu}_e$ 的能谱已经比较精确[17]，包括对电子和子核电荷分布之间电磁相互作用的修正。图 1 是来自 ^{238}U 和 ^{232}Th 衰变链的预期 $\bar{\nu}_e$ 分布，即 $\mathrm{d}n(E_\nu)/\mathrm{d}E_\nu$，该结果是 $\bar{\nu}_e$ 能量 (E_ν) 的函数。

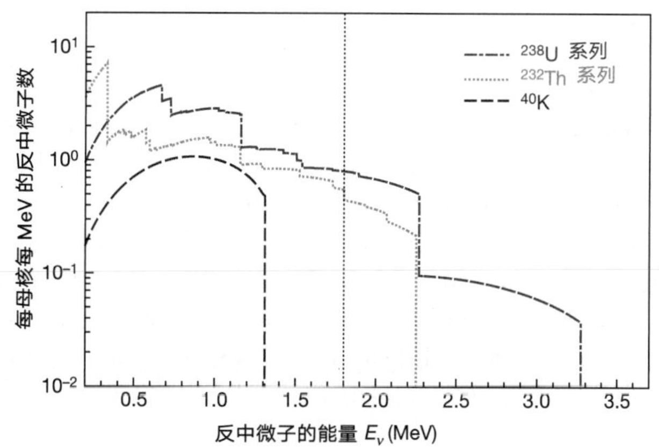

图 1. 从 ^{238}U、^{232}Th 和 ^{40}K 衰变链产生的电子反中微子的预期能量分布。KamLAND 只能探测到黑竖点虚线右侧部分的电子反中微子，因此对 ^{40}K 电子反中微子不敏感。

忽略微不足道的中微子吸收，每种同位素在位置 **r** 处的 $\bar{\nu}_e$ 预期通量如下：

$$\frac{\mathrm{d}\phi(E_\nu,\mathbf{r})}{\mathrm{d}E_\nu}=A\frac{\mathrm{d}n(E_\nu)}{\mathrm{d}E_\nu}\int_{V_\oplus}\mathrm{d}^3\mathbf{r}'\frac{a(\mathbf{r}')\rho(\mathbf{r}')P(E_\nu,|\mathbf{r}-\mathbf{r}'|)}{4\pi|\mathbf{r}-\mathbf{r}'|^2} \tag{1}$$

其中 A 是单位质量的衰变率，积分针对整个地球的体积进行，$a(\mathbf{r}')$ 是每单位岩体质量的同位素质量，$\rho(\mathbf{r}')$ 是岩石密度，$P(E_\nu,|\mathbf{r}-\mathbf{r}'|)$ 是 $\bar{\nu}_e$ 穿过 $|\mathbf{r}-\mathbf{r}'|$ 距离后的“存活”

"survival" probability after travelling a distance $|\mathbf{r}-\mathbf{r}'|$. This probability derives from the now accepted phenomenon of neutrino oscillation, and can be written, for two neutrino flavours as[18]

$$P(E_\nu, L) \cong 1-\sin^2 2\theta_{12}\sin^2\left(\frac{1.27\Delta m_{12}^2[\mathrm{eV}^2]L[\mathrm{m}]}{E_\nu[\mathrm{MeV}]}\right) \tag{2}$$

where $L=|\mathbf{r}-\mathbf{r}'|$. The neutrino oscillation parameters $\Delta m_{12}^2 = 7.9^{+0.6}_{-0.5}\times 10^{-5}$ eV^2 and $\sin^2 2\theta_{12} = 0.82 \pm 0.07$ are also determined with KamLAND[2] using reactor $\bar{\nu}_e$s with energies above those of geoneutrinos, combined with solar neutrino experiments[19]. Corrections from three flavour neutrino oscillation ($<5\%$) and "matter effects"[20] ($\sim 1\%$) are ignored. For typical geoneutrino energies, the approximation $P(E_\nu, |L|) = 1-0.5\sin^2 2\theta_{12}$ only affects the accuracy of the integral in equation (1) at 1% owing to the distributed $\bar{\nu}_e$ production points. This approximation, used in this paper, neglects energy spectrum distortions.

Geoneutrino Detection

KamLAND is located in the Kamioka mine, 1,000 m below the summit of Mt Ikenoyama, Gifu prefecture, Japan (36° 25′ 36″ N, 137° 18′ 43″ E). It detects electron antineutrinos in $\sim$1 kton of liquid scintillator via neutron inverse β-decay,

$$\bar{\nu}_e + p \rightarrow e^+ + n \tag{3}$$

which has a well-established cross-section[21] as a function of E_ν. Scintillation light from the e^+, "prompt event", gives an estimate of the incident $\bar{\nu}_e$ energy, $E_\nu \approx E_{e+} + 0.8$ MeV (neglecting the small neutron recoil), where E_{e+} is the kinetic energy of the positron plus the electron–positron annihilation energy. With a mean time of $\sim$200 μs, the neutron is captured by a proton, producing a deuteron and a 2.2 MeV γ-ray. The detection of scintillation light from this 2.2 MeV γ-ray is referred to as the "delayed event". The spatial and temporal coincidences between the prompt and delayed events provide a powerful tool for reducing backgrounds, which generally limit the sensitivity in low energy neutrino studies.

A reference model[22] is constructed using seismic data to divide the Earth into continental crust, oceanic crust, mantle, core and sediment. Some of these regions are further sub-divided, with each sub-region having different U and Th concentrations. This model assumes that U and Th are absent from the core. The expected geoneutrino flux at KamLAND, including a suppression factor of 0.59 due to neutrino oscillations, is 2.34×10^6 cm^{-2} s^{-1} and 1.98×10^6 cm^{-2} s^{-1} from the ^{238}U and ^{232}Th decay chains, respectively. Including the detection cross-section, the number of geoneutrinos expected at KamLAND from ^{238}U and ^{232}Th decay is 3.85×10^{-31} $\bar{\nu}_e$ per target proton per year, 79% of which is due to ^{238}U. Figure 2 shows that a large fraction of the expected geoneutrino flux originates in the area surrounding KamLAND. The effect of local geology was studied extensively in the context of the reference model[22] and was found to produce less than a 10% error on the total expected flux.

概率。该概率从已被证实的中微子振荡现象中得来。对两种中微子味道，该概率可写为[18]

$$P(E_\nu, L) \simeq 1 - \sin^2 2\theta_{12} \sin^2\left(\frac{1.27\Delta m_{12}^2[\mathrm{eV}^2]L[\mathrm{m}]}{E_\nu[\mathrm{MeV}]}\right) \tag{2}$$

其中 $L = |\mathbf{r} - \mathbf{r}'|$。利用能量高于地球中微子的反应堆 $\bar{\nu}_e$ 的 KamLAND 实验[2]，并结合太阳中微子实验[19]，确定了中微子振荡参数的大小：$\Delta m_{12}^2 = 7.9^{+0.6}_{-0.5} \times 10^{-5}\ \mathrm{eV}^2$，$\sin^2 2\theta_{12} = 0.82 \pm 0.07$。三味中微子振荡效应（$< 5\%$）和“物质效应”[20]（$\sim 1\%$）的修正可忽略。对典型的地球中微子能量，取 $P(E_\nu, |L|) = 1 - 0.5\sin^2 2\theta_{12}$ 的近似对公式（1）积分精度的影响只有 1%，主要来自分散的 $\bar{\nu}_e$ 产生地点。本文中使用这种近似，忽略了能谱的变形。

地球中微子探测

KamLAND 实验位于日本岐阜县池野山地下 1,000 米处的神冈矿井中（北纬 36°25′36″，东经 137°18′43″）。它通过中子反 β 衰变过程，在约 1 千吨的液体闪烁体中探测电子反中微子，

$$\bar{\nu}_e + p \rightarrow e^+ + n \tag{3}$$

该过程的反应截面非常清楚[21]，是 E_ν 的函数。来自 e^+ 的闪烁光被称为“快信号”，它给出入射 $\bar{\nu}_e$ 的能量的估计值，$E_\nu \approx E_{e^+} + 0.8\ \mathrm{MeV}$（忽略了小的中子反冲）。此处 E_{e^+} 是正电子动能加上电子-正电子的湮灭能。经过平均约 200 μs 的时间，中子被质子俘获，产生一个氘核和一个 2.2 MeV 的 γ 光子。探测到的 2.2 MeV γ 光子的闪烁光被称为“慢信号”。快慢信号在时间和位置上的符合提供了一个减少本底的强大工具。在低能中微子研究中，本底通常会制约实验灵敏度。

利用地震数据构建的一个参考模型[22]将地球划分为陆壳、洋壳、地幔、地核和沉积层。部分上述区域可进一步细分，每个子区域有不同的 U 和 Th 丰度。这个模型假定地核里不含 U 和 Th。预期 KamLAND 实验站点来自 ^{238}U 和 ^{232}Th 衰变链的地球中微子通量分别为 $2.34 \times 10^6\ \mathrm{cm}^{-2} \cdot \mathrm{s}^{-1}$ 和 $1.98 \times 10^6\ \mathrm{cm}^{-2} \cdot \mathrm{s}^{-1}$，其中考虑了来自中微子振荡的平均压低因子 0.59。考虑探测截面后，KamLAND 预期能捕获的来自 ^{238}U 和 ^{232}Th 衰变链的地球中微子事例率为 3.85×10^{-31} 个 $\bar{\nu}_e$ 每靶质子每年，其中 79% 来自 ^{238}U。从图 2 中可以看出，预期的地球中微子通量中，很大一部分在 KamLAND 附近区域产生。在参考模型中细致地研究了当地的地质效应[22]，发现导致的误差小于总预期通量的 10%。

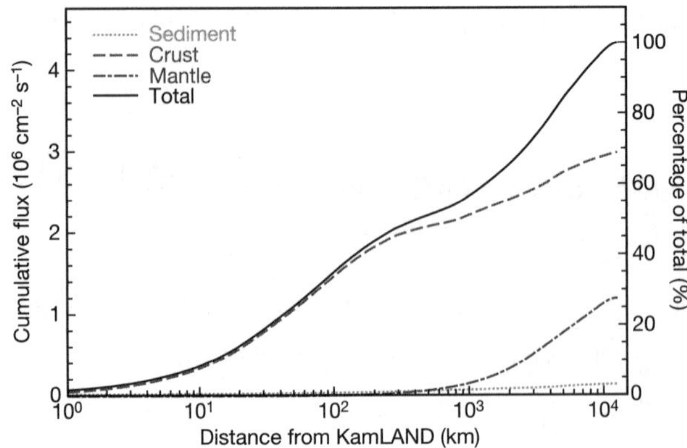

Fig. 2. The expected total ^{238}U and ^{232}Th geoneutrino flux within a given distance from KamLAND[22]. Approximately 25% and 50% of the total flux originates within 50 km and 500 km of KamLAND, respectively. The line representing the crust includes both the continental and the almost negligible oceanic contribution.

The data presented here are based on a total detector live-time of 749.1 ± 0.5 d after basic cuts to ensure the reliability of the data. The number of target protons is estimated at $(3.46 \pm 0.17) \times 10^{31}$ on the basis of target proton density and a spherical fiducial scintillator volume with 5 m radius, resulting in a total exposure of $(7.09 \pm 0.35) \times 10^{31}$ target proton years. The overall efficiency for detecting geoneutrino candidates with energies between 1.7 and 3.4 MeV in the fiducial volume is estimated to be 0.687 ± 0.007. The energy range reaches below the inverse β-decay threshold owing to the detector energy resolution.

Backgrounds for geoneutrino candidates are dominated by $\bar{\nu}_e$s from nuclear reactors in the vicinity of the detector, and by α-particle induced neutron backgrounds due to radioactive contamination within the detector. Reactor $\bar{\nu}_e$s reach substantially higher energies, as shown in Fig. 3. Therefore, the oscillation parameters in ref. 2 were determined by analysing $\bar{\nu}_e$s with energies greater than 3.4 MeV, where there is no signal from the geoneutrinos. Using these parameters, the number of nuclear reactor $\bar{\nu}_e$ background events used by the "rate only" analysis discussed below is determined to be 80.4 ± 7.2.

The α-particle-induced neutron background is due to the $^{13}C(\alpha,n)^{16}O$ reaction where the α-particle is produced in ^{210}Po decay with a kinetic energy of 5.3 MeV. The ^{210}Po is produced by the decay of ^{210}Pb, which has a half-life of 22 yr. The ^{210}Pb resulted from the decay of ^{222}Rn contamination, and is distributed throughout the detector. The neutrons in the $^{13}C(\alpha,n)^{16}O$ reaction are produced with kinetic energy up to 7.3 MeV. Owing to scintillation light quenching for high ionization density, only about one-third of this energy is converted into "visible" energy as the neutrons thermalize. The thermal neutrons are captured by protons with a mean capture time of $\sim$200 μs, producing a delayed signal identical to that from neutron inverse β-decay. The number of ^{13}C nuclei in the fiducial volume is determined from the measured $^{13}C/^{12}C$ ratio in the KamLAND scintillator. On the basis of the $^{13}C(\alpha,n)^{16}O$ reaction cross-section[23], the α-particle energy loss in the scintillator[24], and the number of ^{210}Po decays, the total number of neutrons produced is

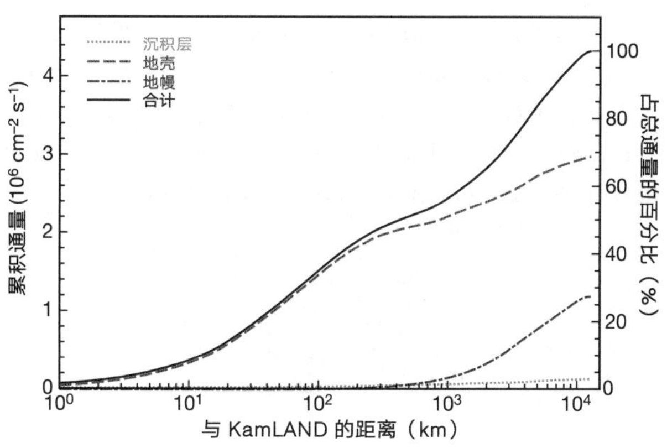

图 2. 在 KamLAND 不同距离范围内产生的 ^{238}U 和 ^{232}Th 地球中微子通量 [22]。大约 25% 和 50% 的总通量来自 KamLAND 附近 50 km 和 500 km 范围内。图中地壳的贡献包括陆壳和几乎可忽略的洋壳的贡献。

本文中的数据经过了一些基本挑选，以确保数据的可靠性，总的探测器有效时间为 749.1 ± 0.5 天。基于平均靶质子密度和半径为 5 m 的闪烁体球形有效体积，总的靶质子数估计为 $(3.46 \pm 0.17) \times 10^{31}$ 个，因此总的曝光量为 $(7.09 \pm 0.35) \times 10^{31}$ 靶质子 · 年。选取有效体积内能量在 1.7 MeV 到 3.4 MeV 之间的地球中微子候选者，总探测效率为 0.687 ± 0.007。能量范围取到反 β 衰变的阈值以下是因为探测器存在能量分辨精度。

地球中微子候选者中的本底主要来自探测器附近核反应堆产生的电子反中微子，以及探测器中放射性杂质产生的 α 粒子引发的中子本底。如图 3 所示，反应堆中微子可达到的能量比地球中微子要高得多。因此参考文献 2 里的振荡参数通过分析能量大于 3.4 MeV 的 $\bar{\nu}_e$ 得到，此能量区间内无地球中微子信号。利用这些参数，得到反应堆中微子本底数为 80.4 ± 7.2，下面将要介绍的“事例率”分析法中用到此值。

α 粒子引发的中子本底来自 $^{13}C(\alpha,n)^{16}O$ 反应，其中 α 粒子来自 ^{210}Po 衰变，动能为 5.3 MeV。^{210}Po 来自半衰期为 22 年的 ^{210}Pb 衰变。^{210}Pb 来自 ^{222}Rn 衰变，而 ^{222}Rn 污染分布在整个探测器内。$^{13}C(\alpha,n)^{16}O$ 反应产生的中子最大动能为 7.3 MeV。由于闪烁光在高电离密度下的淬灭效应，当中子慢化时，只有约三分之一的能量转化为“可见”能量。热中子被质子所俘获，平均俘获时间约为 200 μs，产生一个与中子反 β 衰变相同的慢信号。^{13}C 原子核在探测器有效体积内的数量通过测量 KamLAND 闪烁体的 $^{13}C/^{12}C$ 比值得到。基于 $^{13}C(\alpha,n)^{16}O$ 的反应截面 [23]、α 粒子在闪烁体里的能量损失 [24] 以及 ^{210}Po 衰变数，预期产生的总中子数为 93 ± 22，误差主要由 $^{13}C(\alpha,n)^{16}O$

expected to be 93 ± 22. This error is dominated by estimated 20% and 14% uncertainties in the total $^{13}C(\alpha,n)^{16}O$ reaction cross-section and the number of ^{210}Po decays, respectively. The neutron energy distribution is calculated using the measured neutron angular distributions in the centre of mass frame[25,26]. Including the efficiency for passing the $\bar{\nu}_e$ candidate cuts, the number of (α,n) background events is estimated to be 42 ± 11.

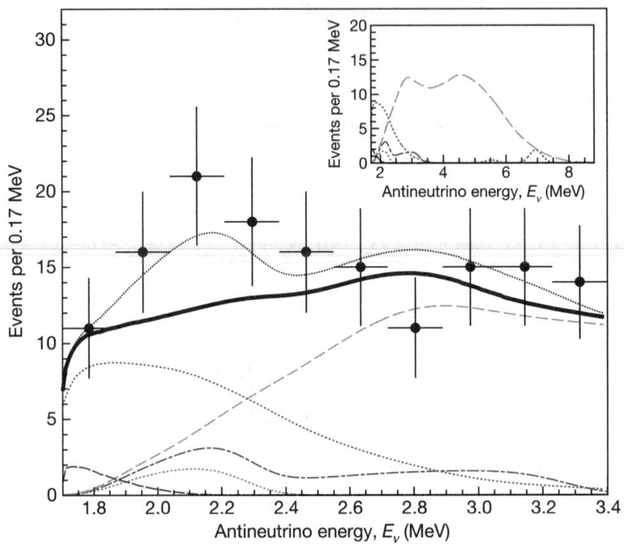

Fig. 3. $\bar{\nu}_e$ energy spectra in KamLAND. Main panel, experimental points together with the total expectation (thin dotted black line). Also shown are the total expected spectrum excluding the geoneutrino signal (thick solid black line), the expected signals from ^{238}U (dot-dashed red line) and ^{232}Th (dotted green line) geoneutrinos, and the backgrounds due to reactor $\bar{\nu}_e$ (dashed light blue line), $^{13}C(\alpha,n)^{16}O$ reactions (dotted brown line), and random coincidences (dashed purple line). Inset, expected spectra extended to higher energy. The geoneutrino spectra are calculated from our reference model, which assumes 16 TW radiogenic power from ^{238}U and ^{232}Th. The error bars represent ± 1 standard deviation intervals.

There is a small contribution to the background from random coincidences, $\bar{\nu}_e$s from the β^- decay of long lived nuclear reactor fission products, and radioactive isotopes produced by cosmic rays. Using an out-of-time coincidence cut from 10 ms to 20 s, the random coincidence background is estimated to be 2.38 ± 0.01 events. Using the expected $\bar{\nu}_e$ energy spectrum[27] for long lived nuclear reactor fission products, the corresponding background is estimated to be 1.9 ± 0.2 events. The most significant background due to radioactive isotopes produced by cosmic rays is from the β^- decay $^9Li \rightarrow 2\alpha + n + e^- + \bar{\nu}_e$, which has a neutron in the final state. On the basis of events correlated with cosmic rays, the estimated number of background events caused by radioactive ^9Li is 0.30 ± 0.05. Other backgrounds considered and found to be negligible include spontaneous fission, neutron emitters and correlated decays in the radioactive background decay chains, fast neutrons from cosmic ray interactions, (γ,n) reactions and solar ν_e induced break-up of ^2H. The total background is estimated to be 127 ± 13 events (1σ error).

The total number of observed $\bar{\nu}_e$ candidates is 152, with their energy distribution shown in

反应截面的 20% 不确定度和 ^{210}Po 衰变数的 14% 不确定度决定。通过在质心系下测量的中子角分布计算得到中子的能量分布 [25,26]，再考虑 $\bar{\nu}_e$ 事例挑选的效率，估算出 (α,n) 本底大约有 42 ± 11 个。

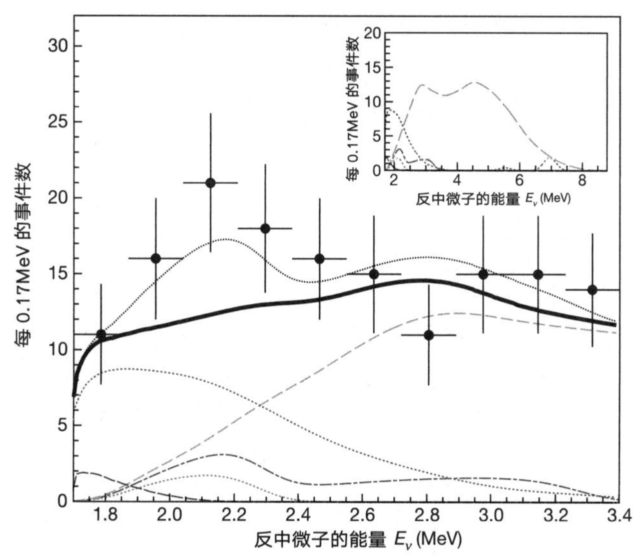

图 3. KamLAND 测得的电子反中微子能谱。主图为实验数据点和总的预期能谱（细黑点线）。同时显示了除地球中微子信号外所有预期能谱的总和（粗黑实线）、^{238}U 地球中微子预期信号（红点划线）、^{232}Th 地球中微子预期信号（绿点线）、反应堆中微子本底（浅蓝虚线）、$^{13}C(\alpha,n)^{16}O$ 反应本底（棕点线）以及随机符合的本底（紫虚线）。小插图显示了延伸到更高能量下的预期能谱。地球中微子能谱从我们的参考模型计算而来，其中假定了 ^{238}U 和 ^{232}Th 共产生 16 TW 的放射性热能。误差棒代表 ±1 倍标准偏差。

其他较小的本底还包括随机符合本底、长寿命的反应堆裂变产物发生 β^- 衰变产生的 $\bar{\nu}_e$ 以及宇宙线产生的放射性同位素。根据信号时间窗以外的 10 ms 到 20 s 之间的符合情况来估算，约有 2.38 ± 0.01 个随机符合事例。根据长寿命的反应堆裂变产物的预期 $\bar{\nu}_e$ 能谱 [27]，相应的本底估计为 1.9 ± 0.2 个事例。宇宙线产生的最显著的放射性同位素本底来自末态有一个中子的 β^- 衰变 $^9\mathrm{Li}\rightarrow2\alpha+n+e^-+\bar{\nu}_e$。基于与宇宙线关联的事例，由放射性的 ^{9}Li 产生的预期本底事例为 0.30 ± 0.05 个。其他考虑过但发现可忽略不计的本底还包括自发裂变、放射性本底衰变链中的中子发射和关联衰变、宇宙线产生的快中子、(γ,n) 反应以及太阳中微子引发的 ^{2}H 分解。总本底估计为 127 ± 13 个事例（1σ 误差）。

观测到的 $\bar{\nu}_e$ 候选者总数为 152 个，其能量分布显示在图 3 中。考虑到地球中微

Fig. 3. Including the geoneutrino detection systematic errors, parts of which are correlated with the background estimation errors, a "rate only" analysis gives 25^{+19}_{-18} geoneutrino candidates from the ^{238}U and ^{232}Th decay chains. Dividing by the detection efficiency, live-time, and number of target protons, the total geoneutrino detected rate obtained is $5.1^{+3.9}_{-3.6} \times 10^{-31}$ $\bar{\nu}_e$ per target proton per year.

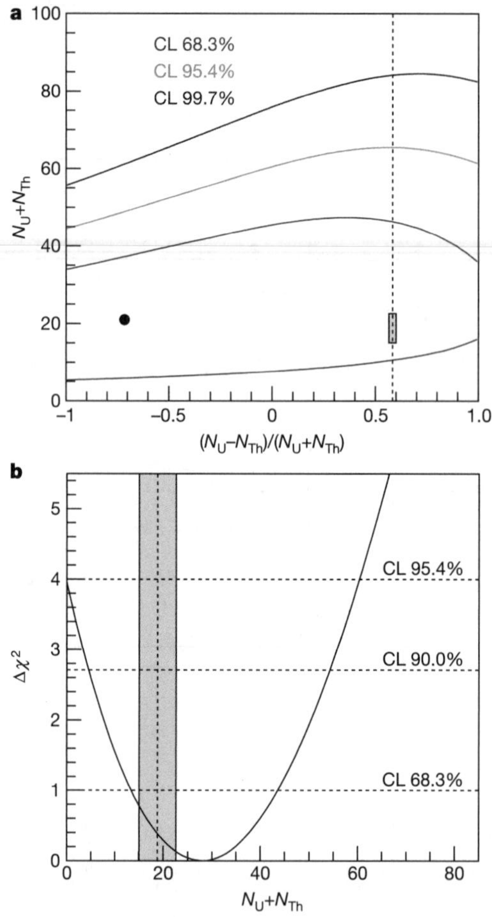

Fig. 4. Confidence intervals for the number of geoneutrinos detected. Panel **a** shows the 68.3% confidence level (CL; red), 95.4% CL (green) and 99.7% CL (blue) contours for detected ^{238}U and ^{232}Th geoneutrinos. The small shaded area represents the prediction from the geophysical model. The vertical dashed line represents the value of $(N_U-N_{Th})/(N_U+N_{Th})$ assuming the mass ratio, Th/U = 3.9, derived from chondritic meteorites, and accounting for the ^{238}U and ^{232}Th decay rates and the $\bar{\nu}_e$ detection efficiencies in KamLAND. The dot represents our best fit point, favouring 3 ^{238}U geoneutrinos and 18 ^{232}Th geoneutrinos. Panel **b** shows $\Delta\chi^2$ as a function of the total number of ^{238}U and ^{232}Th geoneutrino candidates, fixing the normalized difference to the chondritic meteorites constraint. The grey band gives the value of N_U+N_{Th} predicted by the geophysical model.

We also perform an un-binned maximum likelihood analysis of the $\bar{\nu}_e$ energy spectrum between 1.7 and 3.4 MeV, using the known shape of the signal and background spectra. As the neutrino oscillation parameters do not significantly affect the expected shape of the

子探测的系统误差(其中一部分与本底估算误差相关联)，采用事例率分析法，^{238}U 和 ^{232}Th 衰变链产生的地球中微子数为 25^{+19}_{-18}。该事例数除以探测器效率、取数时间以及靶质子数，探测到的地球中微子总事例率为每年每靶质子 $5.1^{+3.9}_{-3.6}\times10^{-31}$ 个 $\bar{\nu}_e$。

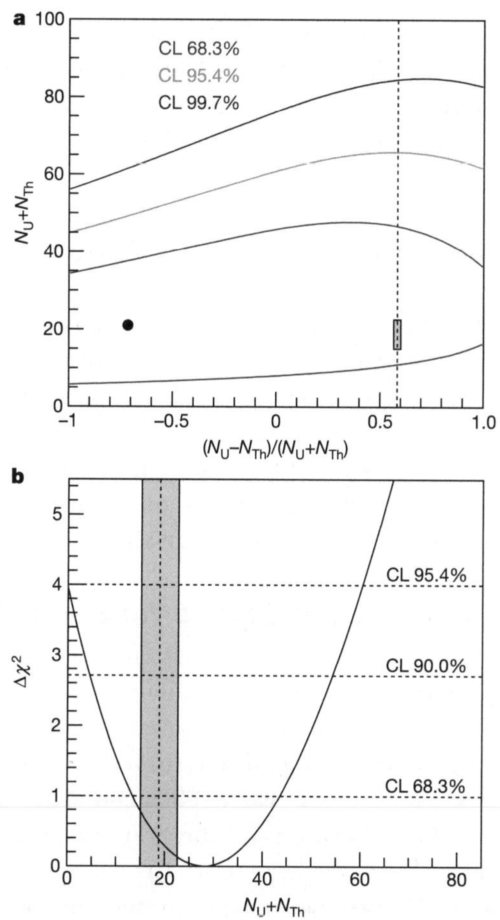

图 4. 探测到的地球中微子事例数的置信区间。图 **a** 是探测到的 ^{238}U 和 ^{232}Th 地球中微子数分别在 68.3% 置信度(红色)、95.4% 置信度(绿色)和 99.7% 置信度(蓝色)下的等值线。小的阴影区是地球物理模型的预期，竖虚线表示基于从球粒陨石推出的 Th/U 质量比 3.9，并考虑了 ^{238}U 和 ^{232}Th 衰变率以及 KamLAND 对 $\bar{\nu}_e$ 的探测效率等情况下的 (N_U-N_{Th}) 与 (N_U+N_{Th}) 的比值。黑点为最佳拟合点，倾向于 ^{238}U 地球中微子数为 3 个，^{232}Th 地球中微子数为 18 个。图 **b** 是 $\Delta\chi^2$ 随 ^{238}U 和 ^{232}Th 地球中微子总数的函数分布，其中归一化后的差别固定到球粒陨石的限制值。灰色区域带是地球物理模型预期的 N_U+N_{Th} 值。

我们也利用已知的信号能谱和本底能谱的形状，对 1.7 ~ 3.4 MeV 区间的 $\bar{\nu}_e$ 能谱进行了一项不分能量区间的最大似然分析。中微子振荡参数对预期的地球中微子能谱形状影响很小，因此采用了无振荡的谱形。而反应堆中微子本底的谱形则包括

geoneutrino signal, the un-oscillated shape is assumed. However, the oscillation parameters are included in the reactor background shape. Figure 4a shows the confidence intervals for the number of observed ^{238}U and ^{232}Th geoneutrinos. Based on a study of chondritic meteorites[28], the Th/U mass ratio in the Earth is believed to be between 3.7 and 4.1, and is known better than either absolute concentration. Assuming a Th/U mass ratio of 3.9, we estimate the 90% confidence interval for the total number of ^{238}U and ^{232}Th geoneutrino candidates to be 4.5 to 54.2, as shown in Fig. 4b. The central value of 28.0 is consistent with the "rate only" analysis. At this point, the value of the fit parameters are $\Delta m_{12}^2 = 7.8 \times 10^{-5}$ eV^2, $\sin^2 2\theta_{12} = 0.82$, $p_\alpha = 1.0$, and $q_\alpha = 1.0$, where these last two parameters are defined in the Methods section. The 99% confidence upper limit obtained on the total detected ^{238}U and ^{232}Th geoneutrino rate is 1.45×10^{-30} $\bar{\nu}_e$ per target proton per year, corresponding to a flux at KamLAND of 1.62×10^7 cm^{-2} s^{-1}. On the basis of our reference model, this corresponds to an upper limit on the radiogenic power from ^{238}U and ^{232}Th decay of 60 TW.

As a cross-check, an independent analysis[29] has been performed using a partial data set, including detection efficiency, of 2.6×10^{31} target proton years. In this analysis, the ^{13}C(α,n)^{16}O background was verified using the minute differences in the time structures of scintillation light from different particle species. Scintillation light in the prompt part of $\bar{\nu}_e$ events is caused by positrons, whereas scintillation light in the prompt part of ^{13}C(α,n)^{16}O background events is caused by neutron thermalization. This alternative analysis produced a slightly larger geoneutrino signal, which is consistent with the results presented here.

Discussion and Future Prospects

In conclusion, we have performed the first experimental study of antineutrinos from the Earth's interior using KamLAND. The present measurement is consistent with current geophysical models, and constrains the $\bar{\nu}_e$ emission from U and Th in the planet to be less than 1.45×10^{-30} $\bar{\nu}_e$ per target proton per year at 99% confidence limits, corresponding to a flux of 1.62×10^7 cm^{-2} s^{-1}. There is currently a programme underway to reduce the ^{210}Pb content of the detector. This should help to reduce the substantial systematic error due to the ^{13}C(α,n)^{16}O background. Further background reduction will require a new detector location, far away from nuclear reactors. The reported investigation of geoneutrinos should pave the way to future and more accurate measurements, which may provide a new window for the exploration of the Earth.

Methods

As shown in Fig. 5, KamLAND[1] consists of 1 kton of ultrapure liquid scintillator contained in a transparent nylon/EVOH (ethylene vinyl alcohol copolymer) composite film balloon suspended in non-scintillating oil. Charged particles deposit their kinetic energy in the scintillator; some of this energy is converted into scintillation light. The scintillation light is then detected by an array of 1,325 17-inch-diameter photomultiplier tubes (PMTs) and 554 20-inch-diameter PMTs mounted on the inner surface of an 18-m-diameter spherical stainless-steel containment vessel. A 3.2-

了振荡参数。图 4a 显示了观测到的 ^{238}U 和 ^{232}Th 地球中微子数的置信区间。基于对球粒陨石的研究 [28]，地球上 Th/U 的质量比被认为在 3.7 ~ 4.1 之间，相比于 Th 和 U 各自的绝对丰度，人们了解的 Th/U 质量比是更为准确的。假设 Th/U 的质量比为 3.9，我们估计 90% 置信区间内 ^{238}U 和 ^{232}Th 地球中微子总数为 4.5 到 54.2，如图 4b 所示。中心值 28.0 与前面事例率分析方法的结果一致。此处拟合参数的值分别为 $\Delta m_{12}^2 = 7.8 \times 10^{-5}$ eV^2、$\sin^2 2\theta_{12} = 0.82$、$p_\alpha = 1.0$、$q_\alpha = 1.0$，其中最后两个参数的定义在下面的方法章节中介绍。探测到的 ^{238}U 和 ^{232}Th 地球中微子总事例率的 99% 置信度上限为 1.45×10^{-30} 个 $\bar{\nu}_e$ 每靶质子每年，对应到达 KamLAND 的地球中微子通量为 1.62×10^7 cm$^{-2} \cdot$ s^{-1}。基于我们的参考模型，这相应于 ^{238}U 和 ^{232}Th 衰变的放射性功率上限为 60 TW。

作为验证，使用总量为 2.6×10^{31} 个靶质子 · 年的部分数据（包含探测器效率）进行了独立分析 [29]。在分析中，^{13}C$(\alpha,n)^{16}$O 本底通过不同粒子种类产生闪烁光的微小时间分布差异得到了验证。$\bar{\nu}_e$ 事例的快信号是由正电子产生的闪烁光，而 ^{13}C$(\alpha,n)^{16}$O 本底事例的快信号是中子慢化产生的闪烁光。此分析得到稍微大一点的地球中微子信号，与本文给出的结果一致。

讨论和前景

综上所述，我们利用 KamLAND 探测器，对来自地球内部的反中微子进行了首次实验研究。测量结果与现有的地球物理模型一致：在 99% 的置信限，限制了地球中 U 和 Th 发射的 $\bar{\nu}_e$ 数小于每靶质子每年 1.45×10^{-30} 个，对应于通量 1.62×10^7 cm$^{-2} \cdot$ s^{-1}。目前正在执行计划以减少探测器中 ^{210}Pb 的含量，这将帮助降低由 ^{13}C$(\alpha,n)^{16}$O 本底带来的较大的系统误差。进一步降低本底需要一个新的远离核反应堆的实验点。本文报告的对地球中微子的研究应该为未来更准确的测量铺平了道路，这可能为探索地球提供了一个新的窗口。

方　法

如图 5 所示，KamLAND[1] 由装在透明的尼龙/EVOH（乙烯–乙烯醇共聚物）复合膜气球里的 1 千吨超纯净的液体闪烁体组成，气球悬浮在不产生闪烁光的矿物油中。带电粒子在液体闪烁体里沉积能量，这些能量的一部分转化为闪烁光，被分布在四周的 1,325 个直径为 17 英寸的光电倍增管（PMT）和 554 个直径为 20 英寸的 PMT 探测到。PMT 安装在直径为 18 米的球形不锈钢容器的内表面。包围着球形容器的是装有 225 个直径为 20 英寸的 PMT 的 3.2

kton water-Cherenkov detector with 225 20-inch-diameter PMTs surrounds the containment sphere. This outer detector tags cosmic-ray muons and absorbs γ-rays and neutrons from the surrounding rock.

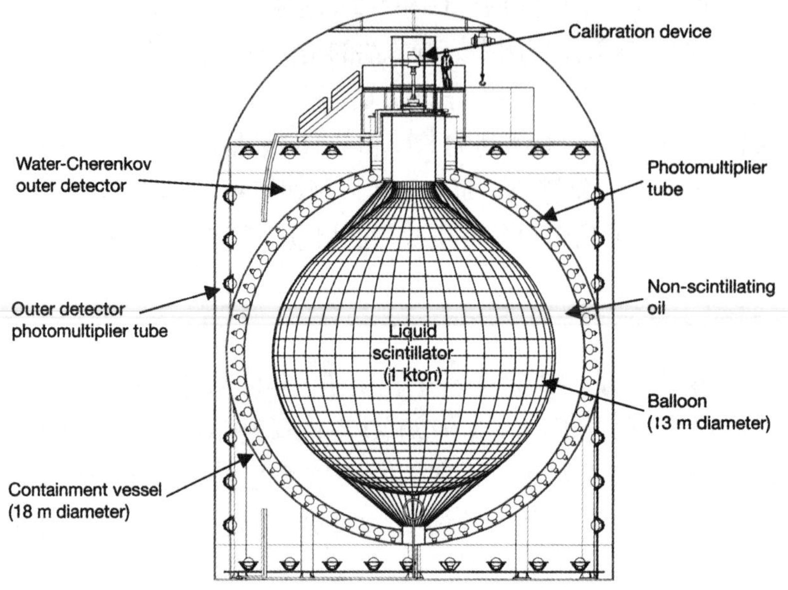

Fig. 5. Schematic diagram of the KamLAND detector.

The arrival times of photons at the PMTs allow us to determine the location of particle interactions inside the detector, and the amount of detected light after correcting for spatial variation of the detector response allows us to determine the particle's energy. The event location and energy determination is calibrated with γ-ray sources deployed vertically down the centre of the detector. To be classified as a $\bar{\nu}_e$ candidate, the time coincidence between the prompt and delayed events (ΔT) is required to satisfy $0.5\ \mu s < \Delta T < 500\ \mu s$. The position of the prompt ($\mathbf{r}_p$) and delayed ($\mathbf{r}_d$) events with respect to the centre of the detector are required to satisfy $|\mathbf{r}_p| < 5$ m; $|\mathbf{r}_d| < 5$ m and $|\mathbf{r}_p - \mathbf{r}_d| < 1.0$ m: The energy of the electron antineutrino is required to satisfy $1.7\ \text{MeV} < E_\nu < 3.4\ \text{MeV}$ and the energy of the delayed event (E_d) is required to satisfy $1.8\ \text{MeV} < E_d < 2.6\ \text{MeV}$.

Given N_U and N_{Th} geoneutrinos detected from the ^{238}U and ^{232}Th decay chains, the expected energy distribution of the candidates is

$$\begin{aligned}\frac{\mathrm{d}\overline{N}(E_\nu)}{\mathrm{d}E_\nu} &= N_U \frac{\mathrm{d}P_U(E_\nu)}{\mathrm{d}E_\nu} + N_{Th} \frac{\mathrm{d}P_{Th}(E_\nu)}{\mathrm{d}E_\nu} + \frac{\mathrm{d}N_r(E_\nu;\, \Delta m_{12}^2,\, \sin^2 2\theta_{12})}{\mathrm{d}E_\nu} \\ &\quad + P_\alpha \frac{\mathrm{d}N_\alpha(E_\nu/q_\alpha)}{\mathrm{d}E_\nu} + \sum_k \frac{\mathrm{d}N_k(E_\nu)}{\mathrm{d}E_\nu}\end{aligned} \tag{4}$$

where $\mathrm{d}P_U(E_\nu)/\mathrm{d}E_\nu$ and $\mathrm{d}P_{Th}(E_\nu)/\mathrm{d}E_\nu$ are the normalized expected geoneutrino spectra from ^{238}U and ^{232}Th decay chains. The third term on the right hand side of equation (4) is the energy spectrum of the expected $\bar{\nu}_e$ reactor background, which is a function of the neutrino oscillation parameters Δm_{12}^2

千吨水切伦科夫探测器，这个外部探测器用来标记宇宙线 μ 子并吸收来自周围岩石的 γ 射线和中子。

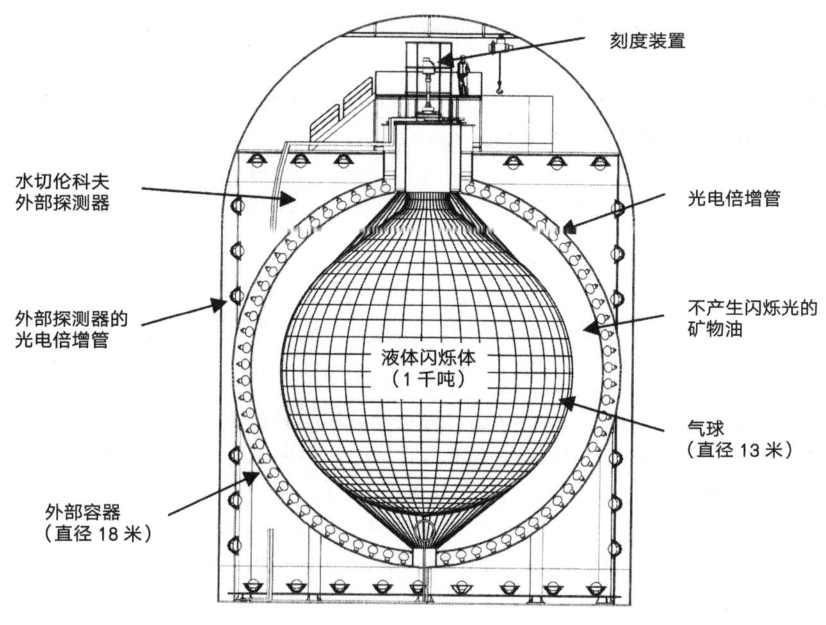

图 5. KamLAND 探测器的示意图。

根据光子到达 PMT 的时间分布可确定探测器内粒子发生相互作用的位置。修正探测器响应随位置的变化后，可通过探测到的总光子数确定粒子的能量。事例的位置和能量通过垂伸至探测器中心的 γ 源进行刻度。对 $\bar{\nu}_e$ 事例的挑选，要求满足以下条件：快慢信号之间的时间符合 (ΔT) 满足 $0.5\ \mu s < \Delta T < 500\ \mu s$；相对于探测器中心的快信号 ($\mathbf{r}_p$) 和慢信号 ($\mathbf{r}_d$) 的位置分别满足 $|\mathbf{r}_p| < 5$ m，$|\mathbf{r}_d| < 5$ m 以及 $|\mathbf{r}_p - \mathbf{r}_d| < 1.0$ m；电子反中微子的能量满足 $1.7\ \text{MeV} < E_\nu < 3.4\ \text{MeV}$；慢信号能量 E_d 满足 $1.8\ \text{MeV} < E_d < 2.6\ \text{MeV}$。

假定有 N_U 和 N_{Th} 个探测到的地球中微子分别来自 ^{238}U 和 ^{232}Th 衰变链，预期探测到的全部 $\bar{\nu}_e$ 候选者的能量分布为：

$$\frac{d\bar{N}(E_\nu)}{dE_\nu} = N_U \frac{dP_U(E_\nu)}{dE_\nu} + N_{Th} \frac{dP_{Th}(E_\nu)}{dE_\nu} + \frac{dN_r(E_\nu; \Delta m_{12}^2, \sin^2 2\theta_{12})}{dE_\nu} + P_\alpha \frac{dN_\alpha(E_\nu/q_\alpha)}{dE_\nu} + \sum_k \frac{dN_k(E_\nu)}{dE_\nu} \tag{4}$$

其中，$dP_U(E_\nu)/dE_\nu$ 和 $dP_{Th}(E_\nu)/dE_\nu$ 分别是 ^{238}U 和 ^{232}Th 衰变链产生的归一化后的预期地球中微子能谱，等式右边第三项是反应堆中微子本底的能谱，它是中微子振荡参数 Δm_{12}^2 和 $\sin^2 2\theta_{12}$ 的函数。$dN_\alpha(E_\nu/q_\alpha)/dE_\nu$ 是 $^{13}C(\alpha,n)^{16}O$ 本底经过能量和事例率的缩放因子 q_α 和 p_α 修正后的

and $\sin^2 2\theta_{12}$. $dN_\alpha(E_\nu/q_\alpha)/dE_\nu$ is the energy spectrum of the expected $^{13}C(\alpha,n)^{16}O$ background with energy and rate scaling factors q_α and p_α, respectively. The sum is over the other known backgrounds where $dN_k(E_\nu)/dE_\nu$ is the expected energy spectrum of the background. All expected spectra include energy smearing due to the detector energy resolution. Integrating equation (4) between 1.7 and 3.4 MeV gives the total number of expected candidates, $\overline{N}$.

The number of geoneutrinos from the ^{238}U and ^{232}Th decay chains is determined from an unbinned maximum likelihood fit. The log likelihood is defined by

$$\log L = -\frac{(N-\overline{N})^2}{2\left(\overline{N}+\sigma_{\overline{N}}^2\right)} + \sum_{i=1}^{N} \log \frac{1}{\overline{N}} \frac{d\overline{N}(E_i)}{dE_\nu} - \frac{(p_\alpha - 1)^2}{2\sigma_p^2} - \frac{(q_\alpha - 1)^2}{2\sigma_q^2} - \frac{\chi^2(\Delta m_{12}^2, \sin^2 2\theta_{12})}{2} \quad (5)$$

where N is the total number of observed candidates and $\sigma_{\overline{N}}$ is the error on $\overline{N}$. E_i is the energy of the ith $\overline{\nu}_e$ candidate. $\sigma_p = 0.24$ and $\sigma_q = 0.1$ are the fractional errors on q_α and p_α, respectively. The term $\chi^2(\Delta m^2, \sin^2 2\theta)$ provides a constraint on the neutrino oscillation parameters from the KamLAND reactor measurements and the solar neutrino results[30]. $\log L$ is maximized at different values of N_U and N_{Th} by varying Δm_{12}^2, $\sin^2 2\theta_{12}$, p_α and q_α. The best fit point for N_U and N_{Th} corresponds to the maximum $\log L$. A $\Delta\chi^2$ is defined by

$$\Delta\chi^2 = 2(\log L_{max} - \log L) \quad (6)$$

where $\log L_{max}$ is the $\log L$ at the best fit point. The confidence intervals are calculated from this $\Delta\chi^2$.

(**436**, 499-503; 2005)

T. Araki[1], S. Enomoto[1], K. Furuno[1], Y. Gando[1], K. Ichimura[1], H. Ikeda[1], K. Inoue[1], Y. Kishimoto[1], M. Koga[1], Y. Koseki[1], T. Maeda[1], T. Mitsui[1], M. Motoki[1], K. Nakajima[1], H. Ogawa[1], M. Ogawa[1], K. Owada[1], J.-S. Ricol[1], I. Shimizu[1], J. Shirai[1], F. Suekane[1], A. Suzuki[1], K. Tada[1], S. Takeuchi[1], K. Tamae[1], Y. Tsuda[1], H. Watanabe[1], J. Busenitz[2], T. Classen[2], Z. Djurcic[2], G. Keefer[2], D. Leonard[2], A. Piepke[2], E. Yakushev[2], B. E. Berger[3], Y. D. Chan[3], M. P. Decowski[3], D. A. Dwyer[3], S. J. Freedman[3], B. K. Fujikawa[3], J. Goldman[3], F. Gray[3], K. M. Heeger[3], L. Hsu[3], K. T. Lesko[3], K.-B. Luk[3], H. Murayama[3], T. O'Donnell[3], A. W. P. Poon[3], H. M. Steiner[3], L. A. Winslow[3], C. Mauger[4], R. D. McKeown[4], P. Vogel[4], C. E. Lane[5], T. Miletic[5], G. Guillian[6], J. G. Learned[6], J. Maricic[6], S. Matsuno[6], S. Pakvasa[6], G. A. Horton-Smith[7], S. Dazeley[8], S. Hatakeyama[8], A. Rojas[8], R. Svoboda[8], B. D. Dieterle[9], J. Detwiler[10], G. Gratta[10], K. Ishii[10], N. Tolich[10], Y. Uchida[10], M. Batygov[11], W. Bugg[11], Y. Efremenko[11], Y. Kamyshkov[11], A. Kozlov[11], Y. Nakamura[11], H. J. Karwowski[12], D. M. Markoff[12], K. Nakamura[12], R. M. Rohm[12], W. Tornow[12], R. Wendell[12], M.-J. Chen[13], Y.-F. Wang[13] & F. Piquemal[14]

[1] Research Center for Neutrino Science, Tohoku University, Sendai 980-8578, Japan

[2] Department of Physics and Astronomy, University of Alabama, Tuscaloosa, Alabama 35487, USA

[3] Physics Department, University of California at Berkeley and Lawrence Berkeley National Laboratory, Berkeley, California 94720, USA

[4] W. K. Kellogg Radiation Laboratory, California Institute of Technology, Pasadena, California 91125, USA

[5] Physics Department, Drexel University, Philadelphia, Pennsylvania 19104, USA

[6] Department of Physics and Astronomy, University of Hawaii at Manoa, Honolulu, Hawaii 96822, USA

[7] Department of Physics, Kansas State University, Manhattan, Kansas 66506, USA

[8] Department of Physics and Astronomy, Louisiana State University, Baton Rouge, Louisiana 70803, USA

[9] Physics Department, University of New Mexico, Albuquerque, New Mexico 87131, USA

预期能谱；最后一项是对已知其他所有本底能谱 $dN_k(E_\nu)/dE_\nu$ 的求和。所有的能谱都包含了由探测器能量分辨引起的能量弥散。对公式(4)在 1.7 ~ 3.4 MeV 能量区间内进行积分给出总的预期 $\overline{\nu}_e$ 事例数 $\overline{N}$。

^{238}U 和 ^{232}Th 衰变链产生的地球中微子数目由不分能量区间的最大似然法拟合得到。对数似然函数的定义如下，

$$\log L = -\frac{(N-\overline{N})^2}{2\left(\overline{N}+\sigma_{\overline{N}}^2\right)} + \sum_{i=1}^{N}\log\frac{1}{\overline{N}}\frac{d\overline{N}(E_i)}{dE_i} - \frac{(p_\alpha-1)^2}{2\sigma_p^2} - \frac{(q_\alpha-1)^2}{2\sigma_q^2} - \frac{\chi^2(\Delta m_{12}^2, \sin^2 2\theta_{12})}{2} \tag{5}$$

其中，N 是总的观测事例数，$\sigma_{\overline{N}}$ 是 $\overline{N}$ 的误差，E_i 是第 i 个 $\overline{\nu}_e$ 事例的能量，$\sigma_p = 0.24$ 和 $\sigma_q = 0.1$ 分别是 p_α 和 q_α 的相对误差。$\chi^2(\Delta m^2, \sin^2 2\theta)$ 项对 KamLAND 反应堆中微子测量和太阳中微子测量 [30] 得到的振荡参数提供了一个约束。通过改变 Δm_{12}^2、$\sin^2 2\theta_{12}$、p_α 和 q_α 这四个参数，将 $\log L$ 在不同 N_U 和 N_{Th} 值处最大化，最大的 $\log L$ 即是 N_U 和 N_{Th} 的最佳拟合点。$\Delta\chi^2$ 的定义如下

$$\Delta\chi^2 = 2(\log L_{max} - \log L) \tag{6}$$

其中 $\log L_{max}$ 是 $\log L$ 在最佳拟合点的值，而置信区间由 $\Delta\chi^2$ 计算得到。

（韩然 翻译；曹俊 审稿）

[10] Physics Department, Stanford University, Stanford, California 94305, USA
[11] Department of Physics and Astronomy, University of Tennessee, Knoxville, Tennessee 37996, USA
[12] Physics Department, Duke University, Durham, North Carolina 27008, USA, and Physics Department, North Carolina State, Raleigh, North Carolina 27695, USA, and Physics Department, University of North Carolina, Chapel Hill, North Carolina 27599, USA
[13] Institute of High Energy Physics, Beijing 100039, China
[14] CEN Bordeaux-Gradignan, IN2P3-CNRS and University Bordeaux I, F-33175 Gradignan Cedex, France

Received 25 May; accepted 4 July 2005.

References:

1. Eguchi, K. *et al.* First results from KamLAND: Evidence for reactor antineutrino disappearance. *Phys. Rev. Lett.* **90**, 021802 (2003).
2. Araki, T. *et al.* Measurement of neutrino oscillation with KamLAND: Evidence of spectral distortion. *Phys. Rev. Lett.* **94**, 081801 (2005).
3. Eder, G. Terrestrial neutrinos. *Nucl. Phys.* **78**, 657-662 (1966).
4. Marx, G. Geophysics by neutrinos. *Czech. J. Phys. B* **19**, 1471-1479 (1969).
5. Avilez, C., Marx, G. & Fuentes, B. Earth as a source of antineutrinos. *Phys. Rev. D* **23**, 1116-1117 (1981).
6. Krauss, L. M., Glashow, S. L. & Schramm, D. N. Antineutrino astronomy and geophysics. *Nature* **310**, 191-198 (1984).
7. Kobayashi, M. & Fukao, Y. The Earth as an antineutrino star. *Geophys. Res. Lett.* **18**, 633-636 (1991).
8. Raghavan, R. S. *et al.* Measuring the global radioactivity in the Earth by multidetector antineutrino spectroscopy. *Phys. Rev. Lett.* **80**, 635-638 (1998).
9. Rothschild, C. G., Chen, M. C. & Calaprice, F. P. Antineutrino geophysics with liquid scintillator detectors. *Geophys. Res. Lett.* **25**, 1083-1086 (1998).
10. Mantovani, F., Carmignani, L., Fiorentini, G. & Lissia, M. Antineutrinos from Earth: A reference model and its uncertainties. *Phys. Rev. D* **69**, 013001 (2004).
11. Pollack, H. N., Hurter, S. J. & Johnson, J. R. Heat flow from the Earth's interior: analysis of the global data set. *Rev. Geophys.* **31**, 267-280 (1993).
12. Hofmeister, A. M. & Criss, R. E. Earth's heat flux revised and linked to chemistry. *Tectonophysics* **395**, 159-177 (2005).
13. McDonough, W. F. & Sun, S.-s. The composition of the Earth. *Chem. Geol.* **120**, 223-253 (1995).
14. Jackson, M. J. & Pollack, H. N. On the sensitivity of parameterized convection to the rate of decay of internal heat sources. *J. Geophys. Res.* **89**, 10103-10108 (1984).
15. Richter, F. M. Regionalized models for the thermal evolution of the Earth. *Earth Planet. Sci. Lett.* **68**, 471-484 (1984).
16. Firestone, R. B. *Table of Isotopes* 8th edn (John Wiley, New York, 1996).
17. Behrens, H. & Jänecke, J. *Landolt-Börnstein - Group I, Elementary Particles, Nuclei and Atoms* Vol. 4 (Springer, Berlin, 1969).
18. McKeown, R. D. & Vogel, P. Neutrino masses and oscillations: triumphs and challenges. *Phys. Rep.* **394**, 315-356 (2004).
19. Ahmed, S. N. *et al.* Measurement of the total active ^8B solar neutrino flux at the Sudbury Neutrino Observatory with enhanced neutral current sensitivity. *Phys. Rev. Lett.* **92**, 181301 (2004).
20. Wolfenstein, L. Neutrino oscillations in matter. *Phys. Rev. D* **17**, 2369-2374 (1978).
21. Vogel, P. & Beacom, J. F. Angular distribution of neutron inverse beta decay, $\bar{\nu}_e + p \rightarrow e^+ + n$. *Phys. Rev. D* **60**, 053003 (1999).
22. Enomoto, S. *Neutrino Geophysics and Observation of Geo-neutrinos at KamLAND.* Thesis, Tohoku Univ. (2005); available at ⟨http://www.awa.tohoku.ac.jp/KamLAND/publications/Sanshiro_thesis.pdf⟩.
23. JENDL Japanese Evaluated Nuclear Data Library. ⟨http://wwwndc.tokai.jaeri.go.jp/jendl/jendl.html⟩ (2004).
24. Apostolakis, J. Geant—Detector description and simulation tool. ⟨http://wwwasd.web.cern.ch/wwwasd/geant/index.html⟩ (2003).
25. Walton, R. B., Clement, J. D. & Borlei, F. Interaction of neutrons with oxygen and a study of the $C^{13}(\alpha,n)O^{16}$ reaction. *Phys. Rev.* **107**, 1065-1075 (1957).
26. Kerr, G. W., Morris, J. M. & Risser, J. R. Energy levels of ^{17}O from $^{13}C(\alpha,\alpha_0)^{13}C$ and $^{13}C(\alpha,n)^{16}O$. *Nucl. Phys. A* **110**, 637-656 (1968).
27. Kopeikin, V. I. *et al.* Inverse beta decay in a nonequilibrium antineutrino flux from a nuclear reactor. *Phys. Atom. Nuclei* **64**, 849-854 (2001).
28. Rocholl, A. & Jochum, K. P. Th, U and other trace elements in carbonaceous chondrites: Implications for the terrestrial and solar-system Th/U ratios. *Earth Planet. Sci. Lett.* **117**, 265-278 (1993).
29. Tolich, N. *Experimental Study of Terrestrial Electron Anti-neutrinos with KamLAND* Thesis, Stanford Univ. (2005); available at ⟨http://www.awa.tohoku.ac.jp/KamLAND/publications/Nikolai_thesis.pdf⟩.
30. KamLAND collaboration. Data release accompanying the 2nd KamLAND reactor result. ⟨http://www.awa.tohoku.ac.jp/KamLAND/datarelease/2ndresult.html⟩ (2005).

Acknowledgements. We thank E. Ohtani and N. Sleep for advice and guidance. The KamLAND experiment is supported by the COE program of the Japanese Ministry of Education, Culture, Sports, Science, and Technology, and by the United States Department of Energy. The reactor data were provided courtesy of the following associations in Japan: Hokkaido, Tohoku, Tokyo, Hokuriku, Chubu, Kansai, Chugoku, Shikoku and Kyushu Electric Power Companies, Japan Atomic Power Co. and Japan Nuclear Cycle Development Institute. Kamioka Mining and Smelting Company provided services for activity at the experimental site.

Author Information. Reprints and permissions information is available at npg.nature.com/reprintsandpermissions. The authors declare no competing financial interests. Correspondence and requests for materials should be addressed to S.E. (sanshiro@awa.tohoku.ac.jp) or N.T. (nrtolich@lbl.gov).

First Fossil Chimpanzee

S. McBrearty and N. G. Jablonski

Editor's Note

Fossil hominins are rare, but fossils of chimpanzees—our closest relatives—were entirely absent until this report of the remains of chimpanzee teeth from Kenya. The teeth are around half a million years old, and were buried alongside fossils attributable to our own genus, *Homo*. Most hominin fossils in Africa are found in the Rift Valley, whereas chimps are today confined to forested west and central Africa. But this discovery suggested that hominins and chimps once shared their environment. It still leaves open the question why chimpanzee fossils are so scarce, and why the fossil record of our other relation, the gorilla, is completely blank.

There are thousands of fossils of hominins, but no fossil chimpanzee has yet been reported. The chimpanzee (*Pan*) is the closest living relative to humans[1]. Chimpanzee populations today are confined to wooded West and central Africa, whereas most hominin fossil sites occur in the semi-arid East African Rift Valley. This situation has fuelled speculation regarding causes for the divergence of the human and chimpanzee lineages five to eight million years ago. Some investigators have invoked a shift from wooded to savannah vegetation in East Africa, driven by climate change, to explain the apparent separation between chimpanzee and human ancestral populations and the origin of the unique hominin locomotor adaptation, bipedalism[2-5]. The Rift Valley itself functions as an obstacle to chimpanzee occupation in some scenarios[6]. Here we report the first fossil chimpanzee. These fossils, from the Kapthurin Formation, Kenya, show that representatives of *Pan* were present in the East African Rift Valley during the Middle Pleistocene, where they were contemporary with an extinct species of *Homo*. Habitats suitable for both hominins and chimpanzees were clearly present there during this period, and the Rift Valley did not present an impenetrable barrier to chimpanzee occupation.

THE Kapthurin Formation forms the Middle Pleistocene portion of the Tugen Hills sequence west of Lake Baringo (Figs 1 and 2). It consists of a package of fluvial, lacustrine and volcanic sediments ~125 m thick, exposed over ~150 km^2 (refs 7–9) that contains numerous palaeontological and archaeological sites[9-11]. It is divided into five members informally designated K1–K5 (ref. 7), and the sequence is well calibrated by $^{40}Ar/^{39}Ar$ dating[12].

第一个黑猩猩化石

麦克布里雅蒂，查邦斯基

编者按

古人类化石是很罕见的，但是黑猩猩的化石——我们的近亲——在这篇来自肯尼亚的黑猩猩牙齿化石的报道之前是完全没有的。这些牙齿大约有 50 万年的历史，和我们人属的化石一起被埋葬。非洲的大部分古人类化石都是在东非大裂谷发现的，而今天的黑猩猩只能生活在非洲西部和中部的森林里。但本文发现表明，古人类和黑猩猩曾经共享它们的生存环境。为什么黑猩猩化石如此稀少？为什么我们的另一个亲戚大猩猩的化石记录是完全空白的？这些问题仍然悬而未决。

我们迄今已发现数以千计的古人类化石，但还没有黑猩猩化石被发现的报道。黑猩猩（黑猩猩属）是现存的与人类亲缘关系最近的动物[1]。今天的黑猩猩种群局限于树木繁茂的非洲西部和中部，而大多数古人类化石遗址都出现在半干旱的东非大裂谷。这种情形更加刺激了人们关于 500 万到 800 万年前人类和黑猩猩世系分叉脱离的有关原因的推测。一些研究者提出，在东非，气候变化引起的从森林到热带稀树草原植被的变化，导致黑猩猩与人类祖先种群之间的明显分离，以及古人类独特的运动方式适应——直立行走[2-5]。在某些情况下，东非大裂谷本身对黑猩猩的生存居住起到了障碍的作用[6]。在此我们报道第一个黑猩猩化石。来自肯尼亚卡普图林组的这些化石，显示黑猩猩属在中更新世期间就在东非大裂谷出现了，此时当地还存在着一种灭绝了的人属成员。很显然，这期间的生活环境适于黑猩猩与古人类生存，而且对于黑猩猩的生存居住，东非大裂谷并非不可逾越的障碍。

卡普图林组构成巴林戈湖西部图根丘陵层序的中更新世部分（图 1、图 2）。它由一套 ~ 125 米厚的河流、湖泊和火山沉积构成，出露约 150 平方千米（参考文献 7 ~ 9），其中有许多古生物和考古遗址[9-11]。它被非正式地划分为 K1 ~ K5（参考文献 7）5 层，且层序由 $^{40}Ar/^{39}Ar$ 测年标度校准[12]。

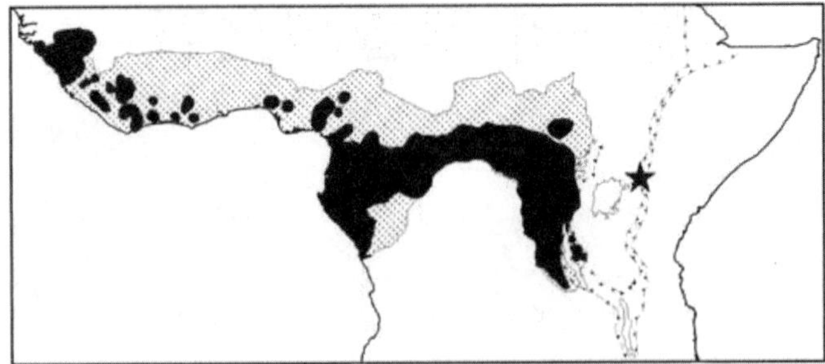

Fig. 1. Map showing current (solid black) and historical (stippled) ranges of *Pan* in equatorial Africa relative to major features of the eastern and western Rift Valleys. The Kapthurin Formation, Kenya, in the Eastern Rift Valley is marked by a star.

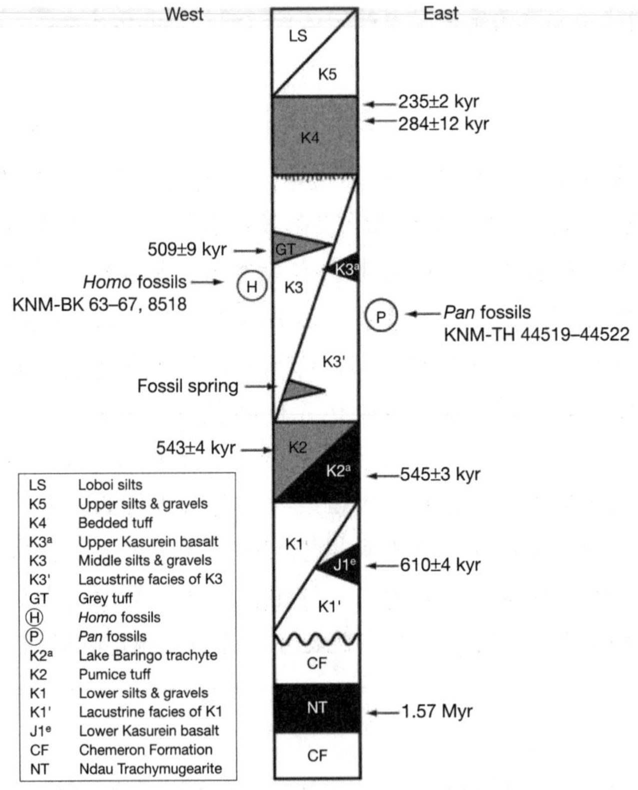

Fig. 2. Idealized stratigraphic column of the Kapthurin Formation, Kenya.

Hominin fossils attributed to *Homo erectus* or *Homo rhodesiensis* have been found in the fluvial sediments of K3 (refs 11, 13, 14). The new chimpanzee fossils were discovered at Locality (Loc.) 99 in K3′, the lacustrine facies of the same geological member. Loc. 99 consists of ~80 m^2 of exposures at an outcrop ~1 km northeast of site GnJh-19 where hominin mandible KNM-BK (Kenya National Museum-Baringo Kapthurin) 8518 was found[14]. Two chimpanzee fossils, KNM-TH (Kenya National Museum-Tugen Hills) 45519 and KNM-TH 45520, were found in surface context within an area of ~12 m^2 within Loc.

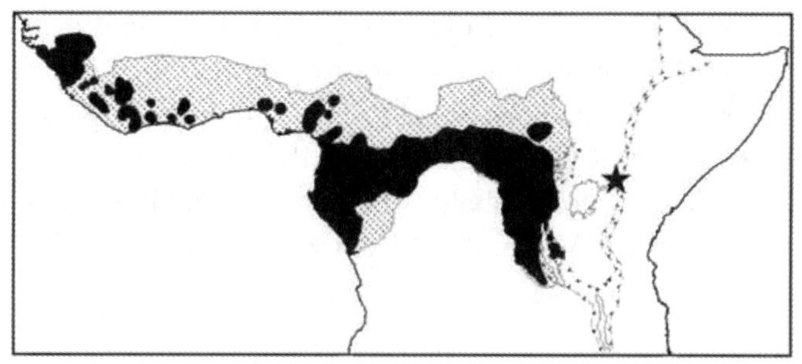

图 1. 与东非大裂谷西部与东部的主要特征相比，黑猩猩属在非洲赤道地区当前范围（实心黑色区）和历史范围（画点区），东非大裂谷肯尼亚卡普图林组用星形标出。

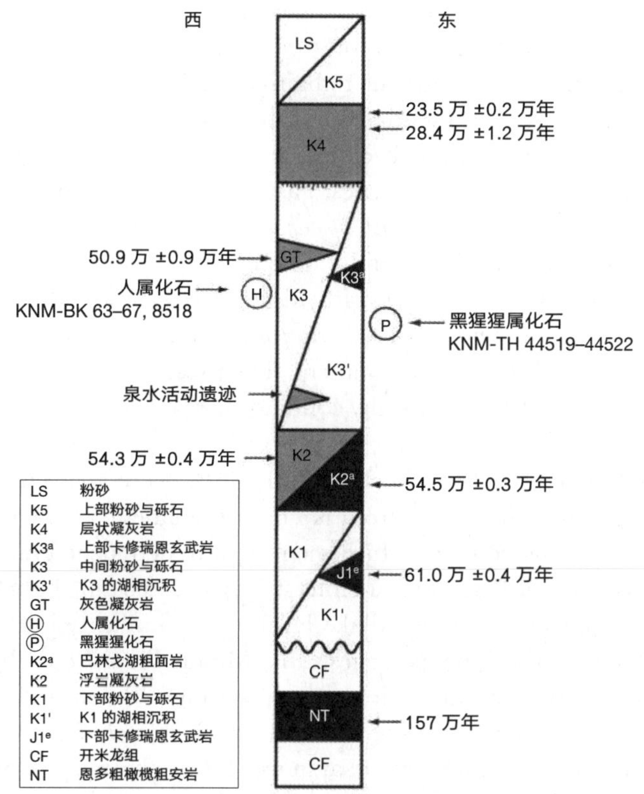

图 2. 肯尼亚卡普图林组的理想地层柱状图

属于直立人或罗得西亚人的化石已经在 K3 的河流沉积中发现（参考文献 11、13、14）。新的黑猩猩化石是在 99 号地点 K3 层中的湖相沉积（编为 K3′）中发现的。在 GnJh-19 遗址东北约 1 千米的断面上，99 号点有约 80 平方米的出露，在那里古人类下颌骨 KNM-BK（肯尼亚国家博物馆–巴林戈卡普图林）8518 被发现[14]。两个黑猩猩化石，KNM-TH（肯尼亚国家博物馆–图根丘陵）45519 和 KNM-TH 45520，

99; additional specimens (KNM-TH 45521 and KNM-TH 45522) were recovered from sieved superficial sediments within the same restricted area. The age of the chimpanzee fossils is constrained by $^{40}Ar/^{39}Ar$ dates of 545 ± 3 kyr (thousand years) on underlying K2 and 284 ± 12 kyr on overlying K4 (ref. 12). Because they are derived from a position low in this stratigraphic interval, they are probably closer to the maximum age of 545 kyr. *Homo* fossils KNM-BK 63-67 and KNM-BK 8518 from K3 are bracketed by $^{40}Ar/^{39}Ar$ dates of 543 ± 4 kyr and 509 ± 9 kyr[12] (Fig. 2).

K3′ sediments are exposed in an outcrop of ~1 km^2 in the eastern portion of the Kapthurin Formation. They consist of black and red zeolitized clays interbedded with sands and heavily altered volcanics. Sedimentary and geochemical features of the clays indicate that they were laid down in a shallow body of water that alternated between fresh and intensely saline-alkaline, probably as a response to changes in outflow geometry controlled by local volcanism[15]. Additional intermittent sources of fresh water are suggested by localized ephemeral stream channel features and the remains of an extensive fossil spring. Loc. 99 has produced fragmentary fossils representing suids, bovids, rodents, cercopithecoid primates and catfish. Eight additional faunal collecting areas in K3′ have also produced elephants, hippopotami, carnivores, crocodiles, turtles, gastropods and additional micromammals. Many K3′ taxa, notably hippopotami (*Hippopotamus*), crocodiles, catfish (*Clarias*), gastropods and turtles, reflect local aquatic conditions. The bulk of K3′ non-aquatic fauna, including a colobine monkey, the elephant, the bovids *Kobus*, *Tragelaphus* and specimens probably belonging to *Syncerus*, and the suids *Potamochoerus porcus* (bushpig) and the extinct *Kolpochoerus majus*[16], are consistent with a closed environment. The presence of the cane rat (*Thryonomys*) indicates localized patches of moist, marshy conditions.

Remains of *Homo* (KNM-BK 63-67 and KNM-BK 8518) were recovered at sites GnJh-01 and GnJh-19 by previous workers[11,13,14] from K3 fluvial sediments to the west that represent a system of braided streams, some of which seem to have debouched into the lake. Fluvial K3 deposits and lacustrine K3′ deposits are interstratified, indicating a shoreline that shifted in position in response to alterations in lake levels. The similarity in the array of fossils encountered in K3 and K3′ sediments suggests that Middle Pleistocene *Pan* and *Homo* lived, or at least died, in broadly similar environmental settings. Taken together, the evidence suggests a locally wooded habitat on the shore of an alternately fresh and saline-alkaline lake, fluctuating lake levels, ephemeral nearshore fluvial channels, a nearby freshwater spring, and a semi-arid climatic regime. These conditions are not unlike those found near the shore of Lake Baringo today, although dense human populations have eliminated much of the woodland that formerly supported chimpanzees and the faunal community of which they were a part.

The chimpanzee specimens comprise a minimum of three teeth, probably from the same individual. Two of these are right and left upper central permanent incisors (I^1; KNM-TH 45519 and KNM-TH 45521, respectively). They exhibit broad, spatulate and moderately worn crowns, with thin dental enamel (Fig. 3). The lingual tubercle is large and flanked at the base by deep mesial and distal foveae, characteristic of *Pan*. This feature imparts great

在 99 号点的地表约 12 平方米范围内发现。另外的标本（KNM-TH 45521 和 KNM-TH 45522）在相同的区域内，通过筛选表层堆积物获得。黑猩猩化石的年龄由 $^{40}Ar/^{39}Ar$ 测年限定在下部 K2 和上部 K4 层位之间，对应年代（54.5 万 ± 0.3 万年～28.4 万 ± 1.2 万年）（参考文献 12）。因为它们源于该地层间隔的低位，所以其年龄比较接近于最大年龄值 54.5 万年。人属化石 KNM-BK 63-67 和 KNM-BK 8518 由 $^{40}Ar/^{39}Ar$ 测年框定在 54.3 万 ± 4.3 万年和 50.9 万 ± 0.9 万年之间 [12]（图 2）。

在卡普图林组东部约 1 平方千米的露头里出露 K3′ 沉积。它们由黑色与红色沸石化黏土构成，并与砂及强变质火山岩互层。黏土的沉积学与地球化学特征显示它们是在淡水与强盐碱水之间交替变化的浅水体中沉积下来的。这种交替可能是作为被当地火山作用控制的流出物质几何学的一种响应 [15]。另外，局部的短暂溪流通道痕迹和泉水活动的相关遗迹提示有额外的间歇性淡水来源。99 号点已经出土的化石碎片分别代表猪科、牛科、啮齿动物、猕猴科灵长类和鲶鱼。在 K3′ 另外 8 个动物化石采集区也出土了象、河马、食肉动物、鳄鱼、海龟、腹足动物及另外的小型哺乳动物。K3′ 的许多动物类，特别是河马、鳄鱼、鲶鱼、腹足动物及海龟，反映了当地的水生环境。K3′ 的大部分非水生动物，包括非洲产疣猴、大象、牛羚、薮羚及可能属于非洲野牛属的标本 [16]，以及猪科的河猪（丛林猪）和灭绝了的阴野猪，是与封闭环境一致的。藤鼠的出现显示出局部有潮湿沼泽的小环境。

人属化石（KNM-BK 63-67 与 KNM-BK 8518）是前人 [11,13,14] 在 GnJh-01 与 GnJh-19 遗址，从 K3 河流沉积到代表辫流系统（其中某些辫流河似乎已经流到了湖里）的西部获得。河流相 K3 沉积与湖相 K3′ 沉积是互层的，这显示水滨线随着湖面变化而产生推移。K3 与 K3′ 沉积中，化石的排列类似，提示出在中更新世的黑猩猩属与人属，是生活在或者至少是死在大致相似的环境里。综上所述，证据提示出这样一个局部树木繁茂的湖岸生活环境：湖泊淡水与盐碱交替变化，湖面上下涨落，季节性近岸河道，靠近淡水泉，气候具有半干旱特征。这些环境条件与今天在靠近巴林戈湖的水岸发现的情况没有什么不同，尽管浓密的人口毁灭了以前黑猩猩赖以为生的很多林地以及它们所属的动物群落。

黑猩猩标本至少有 3 颗牙齿，可能都来自同一个个体的。其中两颗分别是左、右上颌中门齿（I^1，编号 KNM-TH 45519 和 KNM-TH 45521）。其齿冠宽阔，竹片状，磨损中等，釉质薄（图 3）。舌侧结节大，两侧基底连有近远中凹，这是黑猩猩属的

thickness to the labiolingual profiles of the teeth, and clearly distinguishes them from known hominins. The mesial and distal marginal ridges are well formed. The distal corners of the incisal edges are slightly chipped and the labial enamel surfaces exhibit pre-mortem wear as well as slight post-mortem surface weathering. The roots have closed apices and are straight, conical and relatively short. The incisal edges and lingual tubercles exhibit dentinal exposure resulting from wear. Measurements of the specimens, with comparisons to those of extant species of *Pan*, are provided in Table 1. The upper incisors are nearly identical to those of modern *Pan* in all aspects of morphology except their shorter root length. The sub-parallel mesial and distal margins of the incisors bestow a quadrate, rather than triangular, outline to the crowns, a feature that among living chimpanzees is considered to be more common among living *P. troglodytes* than *P. paniscus*[17]. The enamel and cementum coverings are in good condition and the perikymata on the labial surfaces of the crowns and the periradicular striae on the lingual surfaces of the roots can be easily seen. Several of the perikymata near the cervices of the teeth are faintly incised, indicating mild enamel hypoplasia having occurred at about the age of 5 years[18]. The well-matched mesial interproximal wear facets of the Kapthurin Formation *Pan* incisors (KNM-TH 45519 and KNM-TH 45521), the comparable degree of wear on their incisal edges, and the continuity of the enamel hypoplasia on their crowns and the incremental markings on their roots suggest that the two teeth are antimeres.

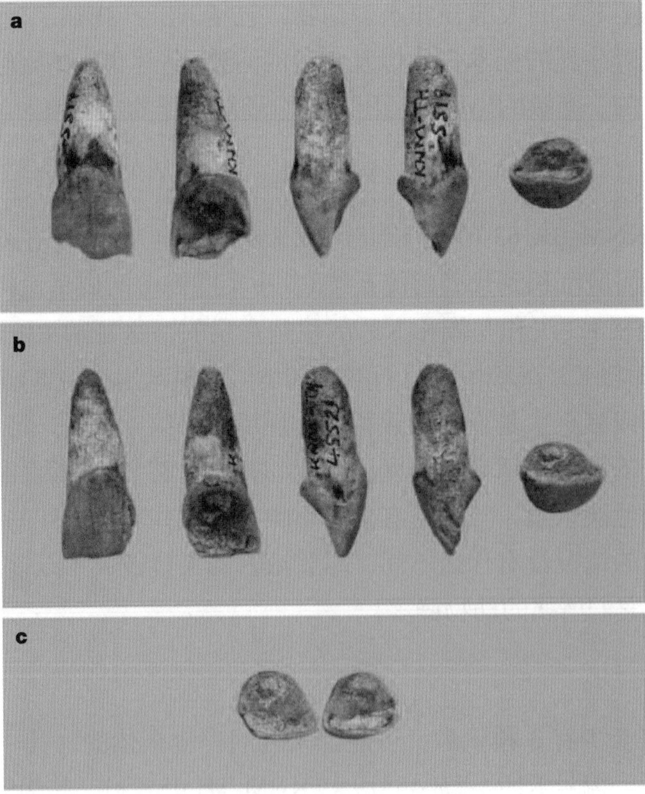

特征。较大的舌侧结节增加了齿冠的唇舌径，并使它们明显有别于已知的古人类。近远中边缘脊发育明显。门齿切缘的近中角有轻微缺口，齿冠唇侧表面有生前磨损和死后轻微风化痕迹。牙根直，尖端封闭，呈圆锥形，较短。门齿切缘和舌侧结节因为磨损而有齿质暴露。化石黑猩猩与其他现生黑猩猩属的测量数据对比列在了表1中。除了其牙根较短外，化石黑猩猩上中门齿的各方面形态特征与现生黑猩猩属几乎是一样的。门齿的近似平行的近远中边缘使牙冠排成的轮廓呈方形而不是三角形，这个特征在现生的黑猩猩属黑猩猩种中要比黑猩猩属倭黑猩猩种中更普遍[17]。釉质和牙骨质状态良好，牙冠唇侧外面釉面横纹和齿根舌侧表面的根周带清晰可见。部分靠近齿颈线的釉面横纹有轻微蚀刻，显示在大约5岁时牙釉质出现轻微发育不全[18]。黑猩猩属的门齿(KNM-TH 45519和KNM-TH 45521)相匹配的近中邻接面、程度相近的切缘磨耗、齿冠上对应的釉质发育不良、齿根上对应的生长线都提示这两颗牙齿源自同一个体的两侧。

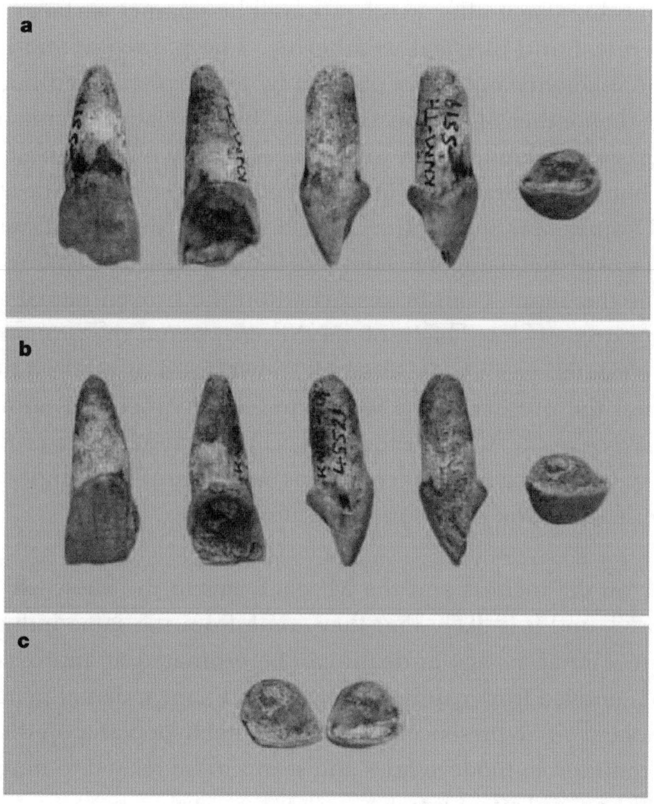

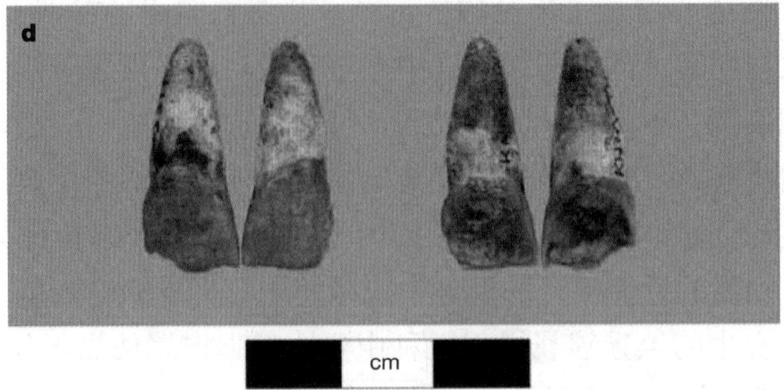

Fig. 3. Central upper incisors of *Pan* from the Kapthurin Formation, Kenya. **a**, KNM-TH 45519. From left to right: labial, lingual, mesial, distal and incisal views. **b**, KNM-TH 45521. Images are in the same sequence as for the previous specimen. **c**, Enlargement of the incisal edge of KNM-TH 45519 (left) and KNM-TH 45521 (right), showing the extreme thinness of the enamel characteristic of modern chimpanzees. **d**, Labial and lingual views of KNM-TH 45519 and KNM-TH 45521.

The third tooth is a lightly worn crown of a left upper permanent molar (KNM-TH 45520) (Fig. 4). It can be problematic to distinguish first from second upper molars in *Pan*, but we identify KNM-TH 45520 as an M^1, judging from the relatively large size of its hypocone, as this cusp is known to decrease in size from M^1 to M^3 (ref. 19). The Kapthurin Formation M^1 is an extremely low molar crown that has lost most of the enamel on its mesial and lingual faces due to breakage after fossilization. The enamel surfaces are pockmarked as a result of chemical and physical weathering. The paracone and metacone are of approximately equal heights and are separated by a sharply incised buccal groove. The hypocone is lower than either of the buccal cusps, but is relatively large and well defined. A shallow trigon basin is delimited by a weak and obliquely oriented postprotocrista (crista obliqua). A deep but short distal fovea lies between the postprotocrista and the low distal marginal ridge. Despite marring of the enamel surface, perikymata are visible on the buccal and distal faces of the paracone, but there is no evidence of enamel hypoplasia. The relative thinness of the enamel can be discerned on the broken mesial and lingual faces of the tooth. The extremely low height of the M^1 crown and the pronounced thinness of the enamel distinguish the tooth from those of known fossil or modern hominins. Among living chimpanzees, the presence of a well-expressed hypocone is more common in *P. troglodytes* than in *P. paniscus*[20]. A fourth tooth (KNM-TH 45522), the crown and proximal roots of a tooth that may be plausibly identified as an aberrant right upper third molar (M^3), will be described elsewhere and is not further discussed here.

The state of wear on the incisors and the M^1 conforms to the known sequence of dental emergence in *Pan*[19,21], and it is likely that they come from the same individual. If they do represent the same animal, its age at death can be estimated at approximately 7–8 years based on standards derived from captive animals[22] and known dental maturation schedules for mandibular molars[23]. The presence of linear enamel hypoplasia on the incisors, but not on the molars, is common in modern apes and seems to be related to nutritional stress that is experienced by the animal after weaning[24].

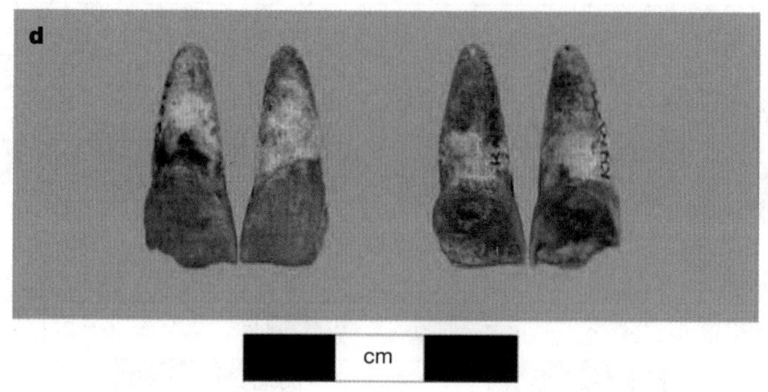

图 3. 来自肯尼亚卡普图林组黑猩猩属的上中门齿。**a**，KNM-TH 45519。从左至右：唇侧，舌侧，近中，远中，切面(咬合面)。**b**，KNM-TH 45521。该牙齿的图片排列顺序与 KNM-TH 45519 一致。**c**，KNM-TH 45519(左)与 KNM-TH 45521(右)放大的切缘，显示出现代黑猩猩的牙釉质非常薄的特征。**d**，KNM-TH 45519 与 KNM-TH 45521 的唇侧和舌侧。

第三颗牙齿是左侧上颌臼齿(KNM-TH 45520)(图 4)轻度磨损的齿冠。要把黑猩猩属上颌第一臼齿和第二臼齿分开是成问题的，但是我们把 KNM-TH 45520 视为上颌第一臼齿，这主要是因为这颗牙齿的次尖较大，而次尖从上颌第一臼齿到上颌第三臼齿是不断减小的(参考 19)。卡普图林组上颌第一臼齿具有较低的齿冠，其近中和舌侧的大部分釉质因石化后遭受的破坏而丧失。釉质表面因为受到物理与化学风化而有麻子坑。前尖和后尖高度接近，被一条较深的颊侧沟分开。次尖比前尖和后尖都低，但尺寸仍然较大、轮廓清晰。发育较弱的斜脊在近中侧定义了一个较浅的三角座凹。斜脊和较低的远中边缘脊之间有一个深但短的远中凹。尽管釉质表面被损伤，但前尖颊侧和远中侧仍保留较明显的釉面横纹，无釉质发育不良。从破损的牙齿近中面和舌侧面可以看出其釉质较薄。较低的齿冠和较薄的釉质使得化石黑猩猩这颗上颌第一臼齿与已知化石及现代人类区分开。在现生黑猩猩中，发育较好的次尖在黑猩猩属黑猩猩种上比在黑猩猩属倭黑猩猩种上更普遍 [20]。第四颗牙齿(KNM-TH 45522)，其牙冠和近齿冠的齿根部分似乎可以被认定为是一颗奇怪的上颌第三臼齿。这颗牙齿会在其他地方描述，在此不做更进一步的讨论。

门齿与上颌第一臼齿的磨耗级别符合黑猩猩属牙齿萌出顺序 [19,21]，且它们很可能来自同一个个体。如果它们确实是代表同一个个体，那么根据从圈养黑猩猩 [22] 身上得来的标准和已知的下颌骨臼齿的牙齿成熟时间表，可以估计这个化石黑猩猩的死亡年龄约为 7～8 岁 [23]。门齿上出现线状牙釉质发育不良，但臼齿上没有出现，这在现代猿类上很常见，而且似乎是与动物断奶后所经历的营养压力有关 [24]。

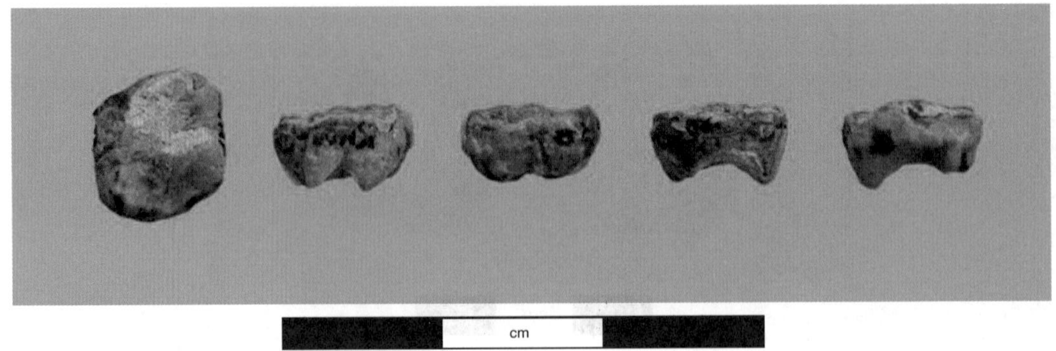

Fig. 4. Upper left first molar (KNM-TH 45520). From left to right: occlusal, labial, lingual, mesial and distal views. Note the thinness of the enamel on the broken mesial face of the paracone in the mesial view.

The morphology of the Kapthurin Formation teeth, especially the pronounced lingual tubercle on the incisors, the thickness of the bases of the incisors, the lowness of the molar crown, and the thinness of the enamel on all the teeth clearly supports their attribution to *Pan* rather than *Homo*. Specific diagnosis of isolated teeth within *Pan*, however, must be approached with caution, and for this reason we assign the Kapthurin Formation specimens to *Pan* sp. indet. Nonmetric characters that have been suggested as diagnostic criteria for *P. troglodytes*, such as a more quadrilateral outline shape to the upper central incisor crowns[17] and a better expressed hypocone on the maxillary molars[19,20], seem to suggest more similarity for the Kapthurin Formation fossils to *P. troglodytes* than to *P. paniscus*, but these features are variably expressed among the living species and subspecies of *Pan*[19,25]. Although mean tooth size is known to be significantly smaller in *P. paniscus* than in *P. troglodytes*[17,25,26], size ranges overlap (Table 1). Furthermore, apart from the present specimens, we lack a fossil record for the Pliocene and Pleistocene from which to assess past variability within the genus, and it is feasible that the Kapthurin Formation fossils represent members of an extinct lineage within the genus *Pan*.

Table 1. Dimensions of the Kapthurin Formation fossil chimpanzee teeth

Sample	Tooth	Mesiodistal dimension (mm)	Mesiodistal range (mm)	Buccolingual dimension (mm)	Buccolingual range (mm)
KNM-TH 45519	Right I^1	10.46	—	9.12	—
KNM-TH 45521	Left I^1	10.50	—	9.33	—
P. troglodytes (male)	I^1	12.6 (n = 14)	10.5–13.5	10.1 (n = 15)	9.0–11.3
P. troglodytes (female)	I^1	11.9 (n = 51)	10.0–13.4	9.6 (n = 50)	8.3–11.7
P. paniscus (male)	I^1	10.3 (n = 15)	8.9–11.9	7.9 (n = 15)	7.2–9.2
P. paniscus (female)	I^1	10.4 (n = 20)	9.0–11.5	7.6 (n = 21)	6.8–8.5
KNM-TH 45520	Left M^1	9.7 (estimate)	—	Damage prevents measurement	—
P. troglodytes (male)	M^1	10.3 (n = 19)	9.3–11.2	11.7 (n = 19)	10.7–13.2
P. troglodytes (female)	M^1	10.1 (n = 51)	9.0–11.9	10.9 (n = 50)	7.0–12.8

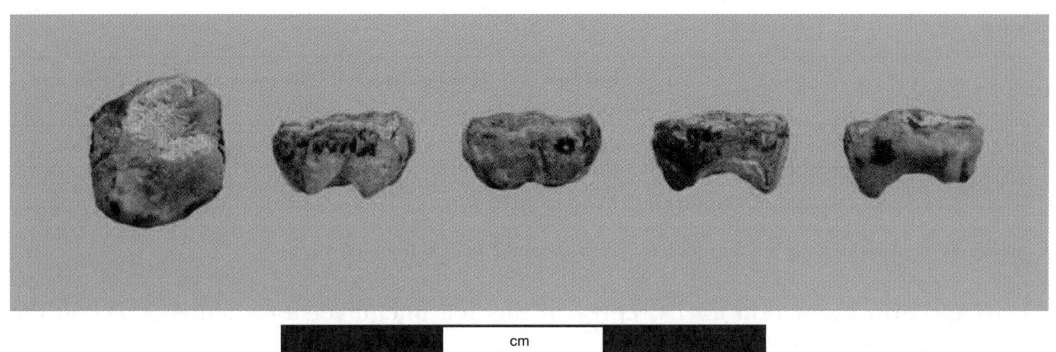

图 4. 左上第一臼齿(KNM-TH 45520)。从左至右：咬合面、颊侧、舌侧、近中侧、远中侧。注意在近中侧视图中前尖破裂的近中面上较薄的釉质。

卡普图林组牙齿的形态，尤其是门齿明显的舌侧结节，较厚的齿冠底部，低的臼齿齿冠，所有牙齿较薄的釉质，都明显地支持它们是属于黑猩猩属而不是人属成员。但是对黑猩猩属单颗牙齿的种一水平上的判断，必须要谨慎，因此我们将卡普图林组标本定为黑猩猩未定种。已经被提出来的非测量特征，比如上中门齿齿冠[17]的轮廓形状要更像四边形，上颌臼齿的次尖发育明显[19,20]，这似乎都提示卡普图林组化石更类似于黑猩猩属黑猩猩种，而与黑猩猩属倭黑猩猩种不同[19,25]。然而，这些特征在黑猩猩属的现生种和亚种上变异较大。虽然已经知道黑猩猩属倭黑猩猩种牙齿的平均尺寸要明显小于黑猩猩属黑猩猩种的[17,25,26]，但仍有重叠(表 1)。此外，除了本文报道的标本外，我们还缺乏用来评估该属过去变异程度的上新世和更新世的化石材料。卡普图林组化石可能代表了黑猩猩属一个已灭绝的谱系成员。

表 1. 卡普图林组黑猩猩牙齿化石的尺寸

标　本	牙齿	近远中径（毫米）	近远中径范围（毫米）	颊舌径（毫米）	颊舌径范围（毫米）
KNM-TH 45519	右 I^1	10.46	–	9.12	–
KNM-TH 45521	左 I^1	10.50	–	9.33	–
黑猩猩属黑猩猩种(雄性)	I^1	12.6 (n = 14)	10.5 ~ 13.5	10.1 (n = 15)	9.0 ~ 11.3
黑猩猩属黑猩猩种(雌性)	I^1	11.9 (n = 51)	10.0 ~ 13.4	9.6 (n = 50)	8.3 ~ 11.7
黑猩猩属倭黑猩猩种(雄性)	I^1	10.3 (n = 15)	8.9 ~ 11.9	7.9 (n = 15)	7.2 ~ 9.2
黑猩猩属倭黑猩猩种(雌性)	I^1	10.4 (n = 20)	9.0 ~ 11. 5	7.6 (n = 21)	6.8 ~ 8.5
KNM-TH 45520	左 M^1	9.7 (估计)	–	因损坏而无法进行测量	–
黑猩猩属黑猩猩种(雄性)	M^1	10.3 (n = 19)	9.3 ~ 11.2	11.7 (n = 19)	10.7 ~ 13.2
黑猩猩属黑猩猩种(雌性)	M^1	10.1 (n = 51)	9.0 ~ 11.9	10.9 (n = 50)	7.0 ~ 12.8

Continued

Sample	Tooth	Mesiodistal dimension (mm)	Mesiodistal range (mm)	Buccolingual dimension (mm)	Buccolingual range (mm)
P. paniscus (male)	M^1	8.5 (n = 7)	7.9–9.4	9.5 (n = 6)	9.2–10.4
P. paniscus (female)	M^1	8.3 (n = 6)	7.6–8.8	9.7 (n = 6)	9.3–10.4

Comparative dimensions are given for modern *P. troglodytes* and *P. paniscus* from ref. 19.

The Kapthurin Formation fossils represent the first unequivocal evidence of *Pan* in the fossil record, and they demonstrate the presence of chimpanzees in the eastern Rift Valley of Kenya, ~600 km east of the limit of their current range (Fig. 1). The Rift Valley clearly did not pose a physiographical or ecological barrier to chimpanzee occupation. Chimpanzee habitat is now highly fragmented, in part by human activities, but in historic times chimpanzees ranged over a wide belt of equatorial Africa from southern Senegal to western Uganda and Tanzania (Fig. 1). Although much of this region is rainforest, chimpanzees currently also occupy dry forest, woodland and dry savannah, particularly near the eastern edge of their range[27-29]. The modern Baringo region ecosystem is a mosaic of semi-arid *Acacia* bushland and riverine woodland, with a significant substratum of perennial and annual grasses[30]. The Tugen Hills palaeosol carbon isotope record indicates that the woodland and grassland components of the vegetation have been present there from 16 Myr[30]. Representatives of both *Homo* and *Pan* are present in the same stratigraphic interval of the Kapthurin Formation at sites only ~1 km apart, and faunal data suggest that they occupied broadly similar environments in the Middle Pleistocene. This evidence shows that in the past chimpanzees occupied regions in which the only hominoid inhabitants were thought to have been members of the human lineage. Now that chimpanzees are known to form a component of the Middle Pleistocene fauna in the Rift Valley, it is quite possible that they remain to be recognized in other portions of the fossil record there, and that chimpanzees and hominins have been sympatric since the time of their divergence.

(**437**, 105-108; 2005)

Sally McBrearty[1] & Nina G. Jablonski[2]

[1] Department of Anthropology, University of Connecticut, Box U-2176, Storrs, Connecticut 06269, USA

[2] Department of Anthropology, California Academy of Sciences, 875 Howard Street, San Francisco, California 94103, USA

Received 31 January; accepted 4 July 2005.

References:

1. Ruvolo, M. E. Molecular phylogeny of the hominoids: inferences from multiple independent DNA sequence data sets. *Mol. Biol. Evol.* **14**, 248-265 (1997).
2. Darwin, C. *The Descent of Man and Selection in Relation to Sex* (John Murray, London, 1871).
3. Washburn, S. L. in *Changing Perspectives on Man* (ed. Rothblatt, B.) 193-201 (Univ. Chicago Press, Chicago, 1968).
4. Kortlandt, A. *New Perspectives on Ape and Human Evolution* (Univ. Amsterdam, Amsterdam, 1972).
5. Pilbeam, D. & Young, N. Hominoid evolution: synthesizing disparate data. *C. R. Palevol.* **3**, 305-321 (2004).
6. Coppens, Y. East side story: the origin of mankind. *Sci. Am.* **270**, 88-95 (1994).
7. Martyn, J. *The Geologic History of the Country Between Lake Baringo and the Kerio River, Baringo District, Kenya* (PhD dissertation, Univ. London, 1969).
8. Tallon, P. in *Geological Background to Fossil Man* (ed. Bishop, W. W.) 361-373 (Scottish Academic Press, Edinburgh, 1978).

续表

标 本	牙齿	近远中径（毫米）	近远中径范围（毫米）	颊舌径（毫米）	颊舌径范围（毫米）
黑猩猩属倭黑猩猩种（雄性）	M^1	8.5 (n = 7)	7.9 ~ 9.4	9.5 (n = 6)	9.2 ~ 10.4
黑猩猩属倭黑猩猩种（雌性）	M^1	8.3 (n = 6)	7.6 ~ 8.8	9.7 (n = 6)	9.3 ~ 10.4

参考文献 19 给出了现生黑猩猩属黑猩猩种和黑猩猩属倭黑猩猩种的比较尺寸规格。

卡普图林组标本是第一个明确的黑猩猩属的化石证据，它们证明了在肯尼亚东非大裂谷东部，也就是在它们目前范围界限以东约 600 千米的地方，曾经生活着黑猩猩（图 1）。东非大裂谷在地形上、生态上对黑猩猩的生活栖息无疑都不是一个障碍。黑猩猩的栖息地由于人类活动而变得很零散，但在历史上，黑猩猩分布在非洲从塞内加尔南部到乌干达西部及坦桑尼亚这一广阔赤道地带（图 1）。虽然这个区域大部分为雨林，但黑猩猩现也生活在干燥森林、林地及干燥热带稀树草原，特别是靠近它们分布范围的东部边缘[27-29]。现代巴林戈区的生态系统，由半干旱金合欢矮灌丛地与有着多年生和一年生草的显著林下层的河边林地嵌合而成[30]。图根丘陵古土壤碳同位素记录显示，该地区植被的林地与草地成分已经存在了 1,600 万年[30]。典型的人属和黑猩猩属成员出现在卡普图林组同一地层范围，遗址仅相距大约1千米，动物群种类提示它们在中更新世所处环境相似。该证据显示在过去黑猩猩占据的区域中人科居民仅有人类世系成员。现在认为黑猩猩构成了东非大裂谷中更新世动物群的一个组成部分，很有可能在化石记录的其他部分里，它们可能会被发现；而且黑猩猩和古人类自从分离后，分布区一直是重叠的。

（田晓阳 翻译；邢松 审稿）

9. McBrearty, S., Bishop, L. C. & Kingston, J. Variability in traces of Middle Pleistocene hominid behaviour in the Kapthurin Formation, Baringo, Kenya. *J. Hum. Evol.* **30**, 563-580 (1996).

10. McBrearty, S. in *Late Cenozoic Environments and Hominid Evolution: a Tribute to Bill Bishop* (eds Andrews, P. & Banham, P.) 143-156 (Geological Society, London, 1999).

11. McBrearty, S. & Brooks, A. The revolution that wasn't: a new interpretation of the origin of modern human behaviour. *J. Hum. Evol.* **39**, 453-563 (2000).

12. Deino, A. & McBrearty, S. $^{40}Ar/^{39}Ar$ chronology for the Kapthurin Formation, Baringo, Kenya. *J. Hum. Evol.* **42**, 185-210 (2002).

13. Leakey, M., Tobias, P. V., Martyn, J. E. & Leakey, R. E. F. An Acheulian industry with prepared core technique and the discovery of a contemporary hominid at Lake Baringo, Kenya. *Proc. Prehist. Soc.* **35**, 48-76 (1969).

14. Wood, B. A. & Van Noten, F. L. Preliminary observations on the BK 8518 mandible from Baringo, Kenya. *Am. J. Phys. Anthropol.* **69**, 117-127 (1986).

15. Renaut, R. W., Tiercelin, J.-J. & Owen, B. in *Lake Basins Through Space and Time* (eds Gierlowski-Kordesch, E. H. & Kelts, K. R.) 561-568 (Am. Assoc. Petrol. Geol., Tulsa, Oklahoma, 2000).

16. Bishop, L. C., Hill, A. P. & Kingston, J. in *Late Cenozoic Environments and Hominid Evolution: a Tribute to Bill Bishop* (eds Andrews, P. & Banham, P.) 99-112 (Geological Society, London, 1999).

17. Johanson, D. C. Some metric aspects of the permanent and deciduous dentition of the pygmy chimpanzee (*Pan paniscus*). *Am. J. Phys. Anthropol.* **41**, 39-48 (1974).

18. Dean, M. C. & Reid, D. J. Perikymata spacing and distribution on hominid anterior teeth. *Am. J. Phys. Anthropol.* **116**, 209-215 (2001).

19. Swindler, D. R. *Primate Dentition: An Introduction to the Teeth of Non-Human Primates* (CUP, Cambridge, 2002).

20. Kinzey, W. G. in *The Pygmy Chimpanzee* (ed. Susman, R. L.) 65-88 (Plenum, New York, 1984).

21. Smith, B. H., Crummett, T. L. & Brandt, K. L. Ages of eruption of primate teeth: a compendium for aging individuals and comparing life histories. *Yearb. Phys. Anthropol.* **37**, 177-232 (1994).

22. Kuykendall, K. L., Mahoney, C. J. & Conroy, G. C. Probit and survival analysis of tooth emergence ages in a mixed-longitudinal sample of chimpanzees (*Pan troglodytes*). *Am. J. Phys. Anthropol.* **89**, 379-399 (1992).

23. Anemone, R. L., Watts, E. S. & Swindler, D. R. Dental development of known-age chimpanzees, *Pan troglodytes* (Primates, Pongidae). *Am. J. Phys. Anthropol.* **86**, 229-241 (1991).

24. Skinner, M. F. & Hopwood, D. Hypothesis for the causes and periodicity of repetitive linear enamel hypoplasia in large, wild African (*Pan troglodytes* and *Gorilla gorilla*) and Asian (*Pongo pygmaeus*) apes. *Am. J. Phys. Anthropol.* **123**, 216-235 (2004).

25. Uchida, A. *Craniodental Variation Among the Great Apes* (Harvard Univ. Peabody Mus., Cambridge, Massachusetts, 1996).

26. Johanson, D. C. *An Odontological Study of the Chimpanzee with Some Implications for Hominoid Evolution* (PhD dissertation, Univ. Chicago, 1974).

27. Kormos, R., Boesch, C., Bakarr, M. I. & Butynski, T. M. *West African Chimpanzees: Status Survey and Conservation Action Plan* (IUCN Publication Unit, Cambridge, 2003).

28. McGrew, W. C., Baldwin, P. J. & Tutin, C. E. G. Chimpanzees in a hot, dry and open habitat: Mt. Assirik, Senegal. *J. Hum. Evol.* **10**, 227-244 (1981).

29. McGrew, W. C., Marchant, L. F. & Nishida, T. *Great Ape Societies* (CUP, Cambridge, 1996).

30. Kingston, J. D., Marino, B. & Hill, A. P. Isotopic evidence for Neogene hominid palaeoenvironments in the Kenya Rift Valley. *Science* **264**, 955-959 (1994).

Acknowledgements. We wish to thank B. Kimeu, N. Kanyenze and M. Macharwas, who found the chimpanzee fossils reported here. Research in the Kapthurin Formation is carried out with the support of an NSF grant to S.M., and under a research permit from the Government of the Republic of Kenya and a permit to excavate from the Minister for Home Affairs and National Heritage of the Republic of Kenya. Both of these are issued to A. Hill and the Baringo Paleontological Research Project, an expedition conducted jointly with the National Museums of Kenya. We also thank personnel of the Departments of Palaeontology, Ornithology and Mammalogy of the National Museums of Kenya, Nairobi; A. Zihlman; and Y. Hailie-Selassie, L. Jellema and M. Ryan for curation and access to specimens. We express gratitude to A. Hill for his comments on the manuscript. We also thank G. Chaplin for drafting Fig. 1, B. Warren for preparing Figs 3 and 4, and A. Bothell for help with submission of the figures. We are grateful to J. Kelley, J. Kingston, M. Leakey, R. Leakey, C. Tryon, A. Walker and S. Ward for discussions. We thank G. Suwa for his remarks.

Author Information. Reprints and permissions information is available at npg.nature.com/reprintsandpermissions. The authors declare no competing financial interests.

Correspondence and requests for materials should be addressed to S.M. (mcbrearty@uconn.edu).

Molecular Insights into Human Brain Evolution

R. S. Hill and C. A. Walsh

Editor's Note

The knowledge of the human genome is expected to have important consequences throughout biology—not least, to offer insights into how humans diverged from our ancestral species at the genetic level. This paper by two neuroscientists at Harvard Medical School uses information from the human genome sequence to make inferences about the evolution of the human brain. It argues that there has been significant genetic change relatively recently—that is, since the emergence of *Homo sapiens*—particularly in a gene called *FOXP2* associated with articulation and speech in humans. *FOXP2* is often now popularly called a "language gene", although that designation is far too simplistic.

Rapidly advancing knowledge of genome structure and sequence enables new means for the analysis of specific DNA changes associated with the differences between the human brain and that of other mammals. Recent studies implicate evolutionary changes in messenger RNA and protein expression levels, as well as DNA changes that alter amino acid sequences. We can anticipate having a systematic catalogue of DNA changes in the lineage leading to humans, but an ongoing challenge will be relating these changes to the anatomical and functional differences between our brain and that of our ancient and more recent ancestors.

SANTIAGO Ramon y Cajal, widely regarded as the founder of modern neuroscience, recognized as early as the turn of the twentieth century that the human brain was not just larger than that of our ancestors, but it differed in its circuitry as well. Over the course of the last century these differences have been extensively studied at a histological level, although specifying the exact changes that distinguish the human brain has been elusive.

> "The opinion generally accepted at that time that the differences between the brain of [non-human] mammals (cat, dog, monkey, etc) and that of man are only quantitative, seemed to me unlikely and even a little offensive to the human dignity... My investigations showed that the functional superiority of the human brain is intimately bound up with the prodigious abundance and unusual wealth of forms of the so-called neurons with short axon." (Ref. 1, translated by J. DeFelipe).

Comparative Differences in Brain Structure

Understanding the genetic changes that distinguish our brain from that of our ancestors starts with defining the key structural and functional differences between the human brain and that of other primates. Our brain is roughly three times the size of the chimpanzee

从分子角度洞察人类大脑的演化

希尔，沃尔什

编者按

人类基因组的知识预计将在整个生物学领域产生重要影响，尤其是，它将为我们了解在基因层面上人类如何与我们的祖先物种分化提供见解。哈佛医学院的两位神经科学家利用人类基因组序列的信息，对人类大脑的演化做出了推断。该研究认为，自智人出现以来，人类最近发生了重大的基因变化，特别是在一种被称为 *FOXP2* 的基因中，这种基因与人类的发音和说话有关。*FOXP2* 现在通常被称为“语言基因”，尽管这个称谓过于简单了。

基因组结构和序列相关知识的迅速发展使得我们能够用新的手段分析与人类大脑和其他哺乳动物大脑的差异相关的特定 DNA 改变。最近的研究将演化改变归因于信使 RNA 和蛋白质表达水平，以及会改变氨基酸序列的 DNA 变化。我们可以预期将来在导致现代人产生的谱系中会有 DNA 变化的系统编目，而我们的大脑和我们远古的以及更近的祖先的大脑在解剖和功能上存在差异，我们正在面临的一项挑战就是如何将这些 DNA 的变化与这些差异联系起来。

圣地亚哥·拉蒙·卡哈尔，被认为是现代神经科学的创始人，早在十九和二十世纪之交就认识到人类的大脑不仅仅是比我们祖先的大，而且脑回路也不同。虽然在上个世纪这些差异已经在组织学层面上得到了广泛的研究，但是确定能够区别人脑的确切的改变仍然是难乎其难。

“在那个年代人们普遍接受的观点是，[非人类] 哺乳动物（猫、狗、猴子等）与人类大脑之间的不同仅仅是数量上的差异，在我看来这不太可能，甚至感觉是对人类尊严的冒犯……我的研究表明人类大脑的功能优越性，是与数量惊人、种类非同寻常的具有短轴突的神经元紧密相关的。”（参考文献 1，德费利佩译）

大脑结构的比较解剖差异

要理解将我们的大脑同祖先的大脑区分开来的遗传改变，需要首先确定人类的大脑与其他灵长类动物的大脑之间关键结构和功能上的不同。在七百万年至八百万

brain, our nearest living relative, from which we diverged 7–8 million years ago, and about twice the size of pre-human hominids living as recently as 2.5 million years ago[2]. The increased size particularly affects the cerebral cortex, the largest brain structure and seat of most higher cognitive functions. The cortex is a multi-layered sheet that is smooth in rodents, but folded in mammals with larger cortices (Fig. 1), allowing more cortex to squeeze into the limited volume of the head.

The enlarged cortex of great apes reflects a longer period of neuronal formation during pre-natal development, so that each dividing progenitor cell undergoes more cell cycles before stopping cell division[3]. Cortical progenitors undergo 11 rounds of cell division in mice[4], at least 28 in the macaque[3], and probably far more in human. In addition to making a larger cortex, the longer period of neurogenesis adds novel neurons to the cortex, so that the cortical circuit diagram differs between primates and other mammals (Fig. 1). Upper cortical layers, generated late in neurogenesis, are over-represented in the primate cerebral cortex, especially in humans[5]. Additionally, special cell types, such as spindle cells (specialized, deep-layer neurons[6]), are unique to primates. The upper-layer neurons that are so unusually common in great apes represent either locally projecting neurons—the "neurons with short axon"of Cajal—or neurons that connect the cortex to itself, but do not project out of the cortex (Fig. 1).

The cerebral cortex shows remarkable local specialization, reflected as functionally distinct cortical "areas" that are essentially a map of the behaviours and capabilities most essential to each species. For example, whereas rodents show relatively larger areas that respond to odours and sensation from the whiskers, they have small areas subserving their limited vision. In contrast, primates are highly visual, with more than a dozen distinct functional areas analysing various features of a visual scene. Recent work has compared functionally homologous visual regions between humans and macaques, suggesting that some areas are quite similar, whereas other visual areas have been either added or greatly modified during the course of evolution[7]. Primates also have particularly large areas of the frontal lobes anterior to the motor cortex (prefrontal cortex), whereas prefrontal cortex is tiny in non-primates. Prefrontal areas regulate many social behaviours and are preferentially enlarged in great apes. Although it has long been thought that prefrontal cortex is especially enlarged in humans, recent work suggests that other great apes may have equivalent proportions of prefrontal cortex[8].

The human cerebral cortex also shows functional asymmetries, with most of us being right handed and having language function preferentially localized in the left hemisphere. Chimpanzees do not show such strong asymmetry in handedness[9], although their brains show some asymmetries in frontal and temporal lobes (which correspond to language areas in humans)[10]. Recent evidence suggests that the left–right asymmetries of the human cerebral cortex are accompanied by asymmetric gene expression during early fetal development[11], although it is not known whether asymmetries of gene expression are seen in non-human primates. There is some evidence from fossil skulls for cortical asymmetry in human predecessors as well[12].

年前我们与关系最近的现生亲戚黑猩猩发生分化，我们大脑的大小是其3倍左右，而相比于近至250万年前的现代人之前的古人类，我们的大脑是其大脑的2倍[2]。大脑尺寸的增加尤其会影响大脑皮层，后者是大脑中最大的结构，也是大部分高级认知功能的所在地。大脑皮层在啮齿动物中是光滑的多层薄片，但在具有较大面积大脑皮层的哺乳动物中是折叠的（图1），这样可以将更多的皮层塞入有限的头部空间。

大猿类大脑皮层的增大反映了出生前发育过程中一个较长的神经元形成时期，因此每个将要分裂的祖细胞在细胞分裂停止之前都要经历更多的细胞周期[3]。大脑皮层祖细胞在小鼠中要经历11轮细胞分裂[4]，在猕猴中至少28轮[3]，而在人类中可能更多。除了形成一个更大的大脑皮层之外，更长的神经发生期还会在大脑皮层上增加新的神经元，因此灵长类动物和其他哺乳动物的大脑皮层回路图是不同的（图1）。在神经发生晚期生成的上部皮层层次，在灵长类尤其是人类的大脑皮层中占有优势的比例[5]。除此之外，特殊的细胞类型，如梭形细胞（一类特化的深层神经元[6]），是灵长类独有的。在大猿中极其常见的上层神经元，要么是局部延伸的神经元，即卡哈尔的“短轴突神经元”，要么是将自己连接到皮层但并不延伸到皮层之外的神经元（图1）。

大脑皮层有显著的区域特化，反映为功能截然不同的皮层“区域”，本质上就是对于每个物种最为根本的行为和能力的映射。例如，尽管啮齿类对气味和来自胡须的感觉做出反应的区域相对较大，但服务于其有限视力的区域却很小。相比之下，灵长类更加依赖视觉，有十几处不同的功能区域分析视觉场景中各种不同的特征。最近的研究比较了人和猕猴功能上同源的视觉区域，结果表明，尽管有些区域十分相似，但其他的视觉区域在演化过程中要么被添加要么被大幅度修改[7]。在运动皮层（前额叶皮层）前方灵长类也具有特别大的额叶区域，而在非灵长类中前额叶皮层很小。前额叶区调控着许多社会行为，在大猿当中该区域得到了优先增大。尽管长期以来人们一直认为，人的前额叶皮层尤其增大，但是最近的研究表明，其他的大猿可能具有相同比例的前额叶皮层[8]。

人类大脑皮层也表现出功能的不对称性，就像我们大部分人都是右利手，而语言功能被优先置于左半球。在使用右手或者左手的习惯上，黑猩猩在利手方面并没有表现出强烈的不对称性[9]，尽管它们的大脑在额叶和颞叶（与人类的语言区域相对应）上也表现出一些不对称性[10]。新的证据表明，人类大脑皮层的左右不对称性与胎儿早期发育过程中基因的不对称表达有关[11]，尽管在非人灵长类中还不知道是否也存在基因的不对称表达。也有来自化石头骨的证据表明在人类先祖中也存在皮层不对称性[12]。

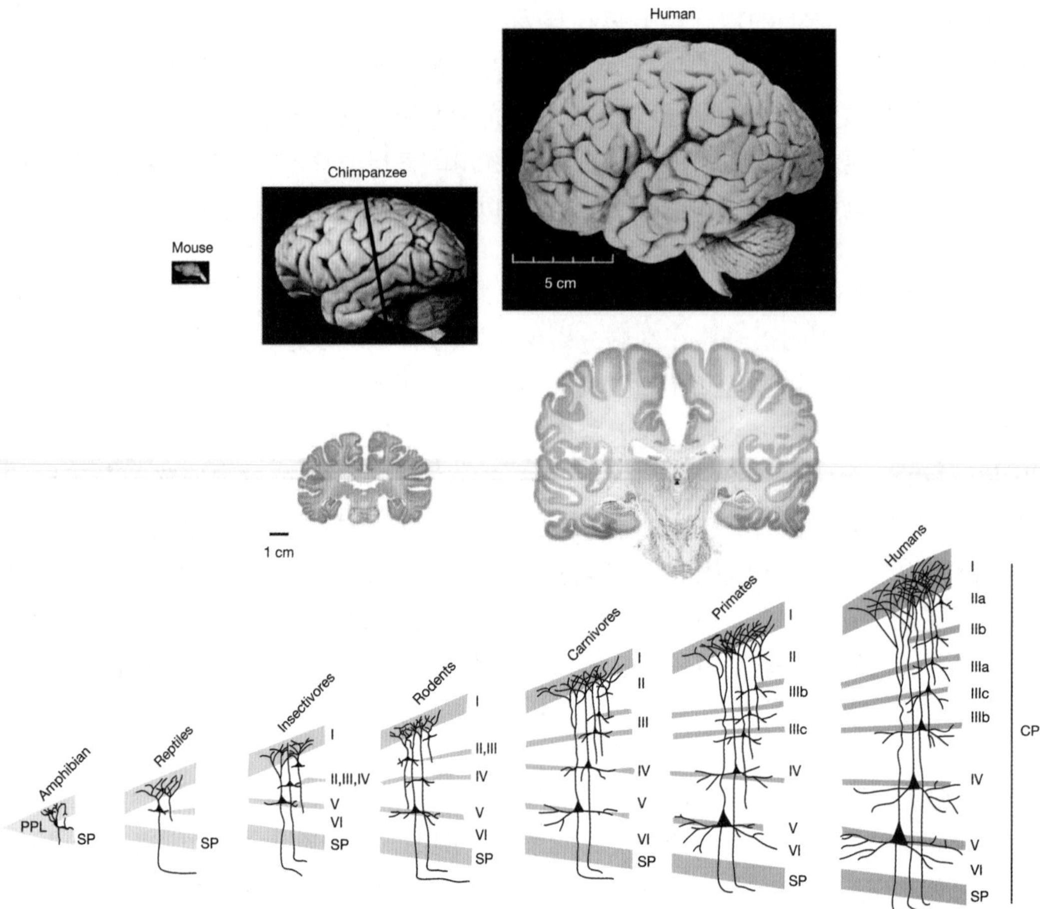

Fig. 1. Differences in cerebral cortical size are associated with differences in the cerebral cortex circuit diagram. The top panel shows side views of the brain of a rodent (mouse), a chimpanzee and a human to show relative sizes. The middle panel shows a cross-section of a human and chimpanzee brain, with the cellular composition of the cortex illustrated in the bottom panel (adapted from ref. 5). The cerebral cortex derives from two developmental cell populations: the primordial plexiform layer (PPL) and the cortical plate (CP). The primordial plexiform layer seems to be homologous to simple cortical structures in Amphibia and Reptilia, and appears first temporally during mammalian brain development. The cortical plate develops as a second population that splits the primordial plexiform layer into two layers (layer I at the top and the subplate (SP) at the bottom; numbering follows the scheme of ref. 31). Cortical-plate-derived cortical layers are added developmentally from deeper first (VI, V) to more superficial (III, II) last. Cortical-plate-derived cortical layers are progressively elaborated in mammals with larger brains (for example, insectivores have a single layer II/III/IV that is progressively subdivided into II, III, IV, then IIa, IIb, and so on), so that humans have a larger proportion of these late-derived neurons, which project locally or elsewhere within the cortex. Images from the top and middle panels are from the Comparative Brain Atlas (http://www.brainmuseum.org).

Evolutionary Mechanisms

What sorts of genetic changes underlie diverse brain shape and size? Approaches to this question have come increasingly into focus, although the answers themselves await further

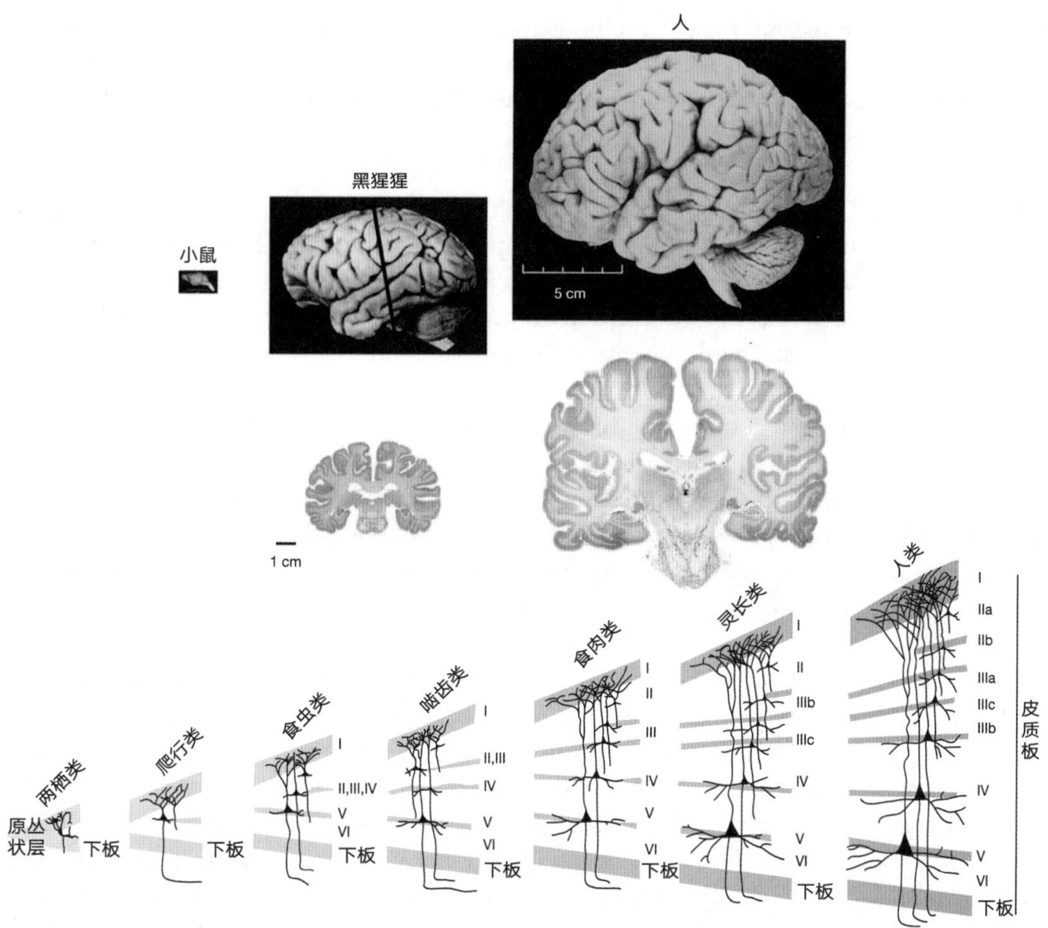

图 1. 大脑皮层大小的不同与大脑皮层回路图的不同相关。上面一排为啮齿动物（小鼠）、黑猩猩以及人类大脑的侧视图，显示相对大小。中间一排为人和黑猩猩大脑的截面，其皮层细胞组成在下面一排图中示意说明（修改自参考文献 5）。大脑皮层源自两种发育细胞群体：原丛状层（PPL）和皮质板（CP）。原丛状层似乎与两栖纲和爬行纲中简单的皮层结构同源，并且在哺乳动物大脑发育过程中首先出现。皮质板作为第二个细胞群体发育，将原丛状层分成两层（在顶部的 I 层和在底部的下板（SP）；编号遵循参考文献 31 的方案）。源自皮质板的皮层层次随着发育从较深层的最初（VI、V）到较浅层的最后（III、II）被逐渐添加。源自皮质板的皮层层次在具有较大大脑的哺乳动物中被逐步精心打造（例如，食虫类有一个单层的 II/III/IV，随着演化过程会被逐步细分成 II、III、IV，然后再分成 IIa、IIb 等），因此人类拥有很大比例的这些晚期衍生的神经元，它们在皮层中局部延伸或向别处延伸。上面和中间一排的图片引自大脑对比图集（http://www.brainmuseum.org）。

演化机制

什么类型的基因改变决定了大脑形态和大小的不同呢？尽管答案本身仍然需要进一步的研究，但是解决问题的途径已经引起越来越多的关注。演化蜕变的三

work. Three major mechanisms of evolutionary changes include: (1) addition or subtraction of entire genes to or from the genome; (2) alterations in levels or patterns of gene expression; and (3) alterations in the coding sequence of genes. Recent evidence suggests roles for all of these mechanisms.

The recent completion of sequencing the chimpanzee genome emphasizes the highly similar composition of the human and chimpanzee genomes[13]. There is evidence for inactivation of genes, especially many olfactory receptor genes, by their conversion into pseudogenes[14]. However, there is currently little evidence to suggest that the addition of novel genes is a major mechanism in human brain evolution[13].

Recent studies suggest that human brain evolution is associated with changes in gene expression specifically within the brain as opposed to other tissues such as liver. A few studies suggest more-accelerated gene expression changes in the brain along the human lineage compared with the chimpanzee lineage[15]. Although the studies differ in design and principal conclusions, they share support for an increase in expression level in a subset of brain-expressed genes in the lineage leading to humans[16,17].

There is also accumulating evidence that some neural genes underwent important changes in their coding sequence over the course of recent brain evolution, although the proportion of neural genes that were targets of positive selection is still in debate. Genes strongly influenced by natural selection can be identified by comparing DNA changes that occur in different, closely related species, for example in different primate species. Synonymous DNA substitutions do not alter the amino acid sequence because they occur at degenerate sites in the codon (such as a CGT to CGG change, as both codons encode arginine). Because synonymous changes do not alter the biochemical properties of the encoded protein, they are usually evolutionarily neutral. In contrast, non-synonymous DNA changes alter the amino acid sequence. The vast majority of non-synonymous DNA changes represent disabling mutations that cause disease, hence decreasing the fitness of the organism, and so most non-synonymous DNA changes are subject to negative, or purifying, selection. In contrast, on rare occasions non-synonymous DNA changes might make the protein work slightly better, hence increasing the fitness of the organism and becoming subject to positive selection (that is, advantageous changes propagated to future generations). A ratio of non-synonymous (K_A) to synonymous (K_S) changes $\ll 1$ is typical of most proteins where change is detrimental[18]; rare proteins show $K_A/K_S > 1$, which can indicate positive selection.

In order to test whether genes expressed in the brain were frequent targets of positive selection in primates, one study[19] analysed 200 brain-expressed genes, comparing them to 200 widely expressed genes. They compared K_A/K_S ratios between rats and mice and between humans and macaque monkeys. They concluded that genes involved in brain development or function had a higher tendency to be under positive selection between macaques and humans than between mice and rats. In contrast, systematic surveys of K_A/K_S ratios across much larger numbers of genes between chimpanzees and humans failed to show that neural genes, as a group, have higher K_A/K_S ratios than genes expressed

个主要机制如下：(1)从基因组中增加或者减除一个完整的基因；(2)基因表达水平或者模式上的改变；(3)基因编码序列的改变。新的证据表明以上所有这些机制都在起作用。

最近黑猩猩的基因组测序工作完成进一步表明人的基因组与黑猩猩基因组组成高度相似[13]。其中也存在基因失效转变成假基因的证据，尤其是许多嗅觉受体基因[14]。然而，目前还没有证据表明新基因的增加是人类大脑演化的主要机制[13]。

最近的研究表明人类大脑演化尤其与大脑内的基因表达变化有关，而与肝脏这样的其他的组织截然相反。一些研究发现与黑猩猩演化谱系相比，在人的演化谱系中大脑中的基因表达变化更加快速[15]。尽管这些研究在目的和主要结论上有所不同，但是他们都认为，在通往现代人的谱系中，大脑表达基因的一个子集的表达水平有所增加[16,17]。

越来越多的证据表明，一些神经基因在大脑最近的演化过程中编码序列发生了重大的变化，尽管作为正选择目标的神经基因所占的比例仍有争议。通过比较不同的或者紧密相关的物种中的 DNA 变化，就可以识别受自然选择强烈影响的基因，例如在不同灵长类物种之间。同义 DNA 替换并不会改变氨基酸序列，因为它们出现在密码子的简并位点(例如，CGT 变成 CGG，但两个密码子都编码精氨酸)。因为同义改变并不改变被编码蛋白质的生物化学性质，所以它们在演化上通常也是中性的。与之相反，非同义 DNA 变化会使氨基酸序列发生变化。绝大多数的非同义 DNA 变化都是失效突变，会引发疾病，从而降低生物的适应性，因此大部分非同义 DNA 变化往往会遭到负面选择或肃清。相比之下，在极少数情况下，非同义 DNA 变化可能会使蛋白质起到稍好的作用，从而提高生物的适应性，受到正选择(也就是说，这种有利的改变会传播给将来的世代)。对于大多数蛋白质来讲，非同义变化(K_A)与同义变化(K_S)的比值通常远小于 1，这些变化是有害的[18]；很少蛋白质显示 K_A/K_S 大于 1，这表明是正选择。

为了检验灵长类大脑中表达的基因是否是正选择的目标，一项研究[19]分析了 200 个大脑表达基因，并将它们同 200 个广泛表达的基因进行了比对。他们比较了大鼠和小鼠之间以及人和猕猴之间的 K_A/K_S 比率。他们得出结论，大脑发育或功能所涉及的基因，在猕猴和人之间比在小鼠和大鼠之间有更高的倾向受到正选择。相比之下，黑猩猩和人之间的更大数量基因的 K_A/K_S 比值的系统考察，并没有显示在这两个物种之间，作为一组的神经基因比大脑之外表达的基因有更高的 K_A/K_S 比值[20,21]。对 50 个 K_A/K_S 比值最高的基因进行分析后发现，它们中只有出奇的极少数

outside of the brain between these two species[20,21]. Analysis of the top 50 genes with the highest K_A/K_S ratios showed surprisingly few with known essential roles in the brain[20]. Analysis of the chimpanzee genome confirms that neural genes, as a group, have much lower average K_A/K_S ratios than genes expressed outside of the brain[13]. However, the more recent study suggested that a substantial fraction of the genes with the highest K_A/K_S ratios had roles in brain development or function[13]. These studies are most easily reconciled by suggesting that a small subset of neural genes may be targets for positive selection (see below), whereas neural genes as a whole are subject to intense negative selection due to the severe disadvantages conferred by mutations that disrupt brain function.

Correlation of Genetic Evolution with Human Brain Function

Whereas genome-wide analyses systematically highlight targets of positive genetic selection in the human lineage, there has been great interest in a subset of human genes that show positive evolutionary selection, and for which correlations between evolutionary patterns and gene function in humans are possible. For example, mutant alleles of *FOXP2* cause a severe disorder of articulation and speech in humans, yet subtle differences in *FOXP2* sequence between humans and non-humans show evidence of positive evolutionary selection by K_A/K_S ratio. Its involvement in speech production suggests that changes in *FOXP2* may have been important in the evolution of language[22,23]. Furthermore, analysis of *FOXP2*'s DNA sequence in diverse human populations suggests that the gene shows unusually low sequence diversity—that is, many human populations share a common ancestral sequence at the *FOXP2* locus. This evidence for a "selective sweep" (explained in detail in several recent reviews[2,24]) within humans suggests that evolutionary selection on this gene may have occurred very recently in human evolution; that is, after the appearance of *Homo sapiens*.

Two genes that cause microcephaly (small cerebral cortex) also show strong evidence for positive evolutionary selection. Microcephaly reduces the human brain to 50% or less of its normal mass; that is, to about the size of the brain of chimpanzees or our pre-human ancestors. Whereas marked mutations in abnormal spindle microcephaly (encoded by the *ASPM* locus) and microcephalin (encoded by the *MCPH1* locus) cause microcephaly, both genes show strong evidence that subtler sequence changes were subject to positive selection in the lineage leading to humans (manifested by a high K_A/K_S ratio)[25-29]. Although the precise functions of the two genes are unknown, both are highly expressed in dividing neural precursor cells in the cerebral cortex, and available evidence suggests roles in cell proliferation. Notably, just as neurons in the upper layers of the cerebral cortex (Fig. 1) are added last during development, and are most highly elaborated in humans and great apes, these upper-layer neurons are preferentially lost in many cases of microcephaly, supporting a requirement for microcephaly genes in the formation of the upper cortical layers.

AHI1, which is essential for axon pathfinding from the cortex to the spinal cord (and hence for normal coordination and gait), is another gene that causes a neurological disease

在大脑中有已知的重要功能[20]。对黑猩猩基因组的分析也确认了一点，神经基因，作为一组，相比在大脑外表达的基因有更低的 K_A/K_S 比值[13]。然而，更新的研究工作表明，有相当一部分具有最高 K_A/K_S 比值的基因在大脑发育和功能中发挥了作用[13]。这些研究最容易这样折中：神经基因的一个非常小的子集或许会是正选择的目标（见下文），而由于这些破坏大脑功能的基因突变所带来的严重的劣势，神经基因，作为一个整体，会遭受很强的负选择。

基因演化与人类大脑功能的相关性

由于基因组范围的分析系统地标明了人类谱系中基因正选择的目标，人类基因中显示正演化选择的一个子集引起了极大的关注，并且对于这些基因来讲，探讨人类中演化模式和基因功能的相关性是可能的。例如，在人类中 *FOXP2* 的突变等位基因会导致严重的语言和表达障碍，但是人类和非人类 *FOXP2* 基因序列的细微差异通过 K_A/K_S 比值显示该基因受到正演化选择。其参与语言产生的这个现象表明，*FOXP2* 中的变化在语言演化中起着很重要的作用[22,23]。况且，不同现代人群体的 *FOXP2* 的 DNA 序列表明该基因显示出非常低的序列多样性，也就是说，许多现代人群体在 *FOXP2* 位点上共有相同的祖先序列。这一人类内部"选择性清除"（在最近的几篇综述[2,24]中对此进行了详细解释）的证据表明，针对这一基因的演化选择在人类进化过程中可能出现得非常晚，也就是说，在智人出现之后。

导致小头畸形（大脑皮层小）的两个基因也显示出正演化选择强有力的证据。小头畸形会使人的大脑减小到正常体积的 50% 或更小，也就是说减小到大约相当于黑猩猩或现代人之前的祖先的脑的大小。尽管异常纺锤体小头畸形蛋白（由 *ASPM* 位点编码）和小头畸形素（由 *MCPH1* 位点编码）的显著突变会导致小头畸形，但是两个基因都显示出强有力的证据：较细微的序列变化在通往现代人的演化谱系中会受到正选择（表现为 K_A/K_S 比值很高）[25-29]。尽管两个基因的确切功能尚不清楚，但二者在要分裂的大脑皮层的神经先驱细胞中被高度表达，现有证据表明它们在细胞增殖中发挥了作用。值得注意的是，正如大脑皮层上层的神经元（图 1）是在发育的过程的最后被添加，并且在现代人和大猿中被最为精心打造一样，在很多小头畸形病例中，这些上层神经元会被专门丢弃，这证明了导致小头畸形的基因参与了皮层上部层次的形成。

AHI1 对于轴突从皮层到脊髓寻路是至关重要的基因（因此对正常的协调和步态也至关重要），该基因的突变也会导致神经系统疾病，但该基因在灵长类之间较细微

when mutated, but for which subtler changes between primate species suggest positive evolutionary selection in the lineage leading to humans[30]. Patients with *AHI1* mutations not only show mental retardation, but can also show symptoms characteristic of autism, such as antisocial behaviour. This raises the intriguing possibility that evolutionary differences in *AHI1* may relate not only to human patterns of gait, but potentially species-specific social behaviour.

The linkage of studies of gene function in humans with evolutionary analysis is just beginning, and is limited mainly by the rate at which the essential functional roles of genes in the human brain are elucidated. As a population, humans show many mutant alleles for every gene that has been extensively studied, so that the human population is likely to represent, to a first approximation, saturation mutagenesis, such that for each gene in the genome there is a human carrying a mutated allele for that gene. Many neurological diseases affect the very processes that define us evolutionarily as human: intelligence (mental retardation), social organization (autism and attention deficit disorder) and higher-order language (dyslexia). As the genes for these uniquely human disorders are characterized, they may give us new insight into our recent evolutionary history.

(**437**, 64-67; 2005)

Robert Sean Hill[1] & Christopher A. Walsh[1]

[1] Division of Neurogenetics and Howard Hughes Medical Institute, Beth Israel Deaconess Medical Center, and Department of Neurology, Harvard Medical School, Room 266, New Research Building, 77 Avenue Louis Pasteur, Boston, Massachusetts 02115, USA

References:

1. Ramon y Cajal, S. *Recuerdos de mi Vida* Vol. 2 *Historia de mi Labour Científica* 345-346 (Moya, Madrid, 1917).
2. Carroll, S. B. Genetics and the making of *Homo sapiens*. *Nature* **422**, 849-857 (2003).
3. Kornack, D. R. & Rakic, P. Changes in cell-cycle kinetics during the development and evolution of primate neocortex. *Proc. Natl Acad. Sci. USA* **95**, 1242-1246 (1998).
4. Takahashi, T., Nowakowski, R. S. & Caviness, V. S. Jr. The cell cycle of the pseudostratified ventricular epithelium of the embryonic murine cerebral wall. *J. Neurosci.* **15**, 6046-6057 (1995).
5. Marin-Padilla, M. Ontogenesis of the pyramidal cell of the mammalian neocortex and developmental cytoarchitectonics: a unifying theory. *J. Comp. Neurol.* **321**, 223-240 (1992).
6. Allman, J., Hakeem, A. & Watson, K. Two phylogenetic specializations in the human brain. *Neuroscientist* **8**, 335-346 (2002).
7. Orban, G. A., Van Essen, D. & Vanduffel, W. Comparative mapping of higher visual areas in monkeys and humans. *Trends Cogn. Sci.* **8**, 315-324 (2004).
8. Semendeferi, K., Lu, A., Schenker, N. & Damasio, H. Humans and great apes share a large frontal cortex. *Nature Neurosci.* **5**, 272-276 (2002).
9. Hopkins, W. D. & Cantalupo, C. Handedness in chimpanzees (*Pan troglodytes*) is associated with asymmetries of the primary motor cortex but not with homologous language areas. *Behav. Neurosci.* **118**, 1176-1183 (2004).
10. Cantalupo, C. & Hopkins, W. D. Asymmetric Broca's area in great apes. *Nature* **414**, 505 (2001).
11. Sun, T. *et al.* Early asymmetry of gene transcription in embryonic human left and right cerebral cortex. *Science* **308**, 1794-1798 (2005).
12. Broadfield, D. C. *et al.* Endocast of Sambungmacan 3 (Sm 3): a new *Homo erectus* from Indonesia. *Anat. Rec.* **262**, 369-379 (2001).
13. The Chimpanzee Sequencing and Analysis Consortium. Initial sequence of the chimpanzee genome and comparison with the human genome. *Nature* doi:10.1038/nature04072 (this issue).
14. Gilad, Y., Man, O. & Glusman, G. A comparison of the human and chimpanzee olfactory receptor gene repertoires. *Genome Res.* **15**, 224-230 (2005).
15. Enard, W. *et al.* Intra- and interspecific variation in primate gene expression patterns. *Science* **296**, 340-343 (2002).
16. Caceres, M. *et al.* Elevated gene expression levels distinguish human from non-human primate brains. *Proc. Natl Acad. Sci. USA* **100**, 13030-13035 (2003).
17. Uddin, M. *et al.* Sister grouping of chimpanzees and humans as revealed by genome-wide phylogenetic analysis of brain gene expression profiles. *Proc. Natl Acad. Sci. USA* **101**, 2957-2962 (2004).

的变化表明，在通向现代人的人类演化谱系中它是受到正选择的[30]。*AHI1* 基因突变的病人不仅表现为智力低下，而且还可能表现出自闭症的典型症状，比如反社会行为。这就产生了一种令人不解的可能性，即在 *AHI1* 基因上的演化差异可能不仅与人类的步态模式有关，而且还可能与物种特定的社会行为有关。

人类基因功能的研究与演化分析之间的结合只是刚刚开始，并且主要受限于阐明这些基因在人类大脑中的根本功能的研究速度。作为一个种群，人类被广泛研究的每一个基因都显示出许多突变等位基因，因此人类种群很可能大致上代表了饱和突变的发生，使得对于基因组中的每个基因，都有一个人携带着这个基因的突变等位基因。许多神经系统疾病会影响这个在演化意义上定义我们为人类的过程：智慧（智力低下），社会组织（自闭症和注意力缺陷症），高级语言（阅读障碍）。随着这些人类特有的疾病的基因的特征都得到详细的描绘，它们可能会让我们对人类最近的演化历史有一个全新的了解。

（吕静 翻译；张颖奇 审稿）

18. Goldman, N. & Yang, Z. A codon-based model of nucleotide substitution for protein-coding DNA sequences. *Mol. Biol. Evol.* **11**, 725-736 (1994).

19. Dorus, S. *et al.* Accelerated evolution of nervous system genes in the origin of *Homo sapiens*. *Cell* **119**, 1027-1040 (2004).

20. Nielsen, R. *et al.* A scan for positively selected genes in the genomes of humans and chimpanzees. *PLoS Biol.* **3**, e170 (2005).

21. Clark, A. G. *et al.* Inferring nonneutral evolution from human-chimp-mouse orthologous gene trios. *Science* **302**, 1960-1963 (2003).

22. Lai, C. S., Fisher, S. E., Hurst, J. A., Vargha-Khadem, F. & Monaco, A. P. A forkhead-domain gene is mutated in a severe speech and language disorder. *Nature* **413**, 519-523 (2001).

23. Enard, W. *et al.* Molecular evolution of *FOXP2*, a gene involved in speech and language. *Nature* **418**, 869-872 (2002).

24. Gilbert, S. L., Dobyns, W. B. & Lahn, B. T. Genetic links between brain development and brain evolution. *Nature Rev. Genet.* **6**, 581-590 (2005).

25. Bond, J. *et al.* *ASPM* is a major determinant of cerebral cortical size. *Nature Genet.* **32**, 316-320 (2002).

26. Evans, P. D., Anderson, J. R., Vallender, E. J., Choi, S. S. & Lahn, B. T. Reconstructing the evolutionary history of microcephalin, a gene controlling human brain size. *Hum. Mol. Genet.* **13**, 1139-1145 (2004).

27. Evans, P. D. *et al.* Adaptive evolution of *ASPM*, a major determinant of cerebral cortical size in humans. *Hum. Mol. Genet.* **13**, 489-494 (2004).

28. Kouprina, N. *et al.* Accelerated evolution of the *ASPM* gene controlling brain size begins prior to human brain expansion. *PLoS Biol.* **2**, E126 (2004).

29. Zhang, J. Evolution of the human *ASPM* gene, a major determinant of brain size. *Genetics* **165**, 2063-2070 (2003).

30. Ferland, R. J. *et al.* Abnormal cerebellar development and axonal decussation due to mutations in *AHI1* in Joubert syndrome. *Nature Genet.* **36**, 1008-1013 (2004).

31. Marin-Padilla, M. Dual origin of the mammalian neocortex and evolution of the cortical plate. *Anat. Embryol.* **152**, 109-126 (1978).

Acknowledgements. This work was supported by grants from the NINDS and Cure Autism Now. We thank M. Ruvolo and D. Reich for comments on an earlier version of this manuscript, and J. DeFilipe for the translation of the Cajal quotation. Owing to space limitations we were unable to cite directly some of the relevant work in this field. C.A.W. is an Investigator of the Howard Hughes Medical Institute.

Author Information. Reprints and permissions information is available at npg.nature.com/reprintsandpermissions. The authors declare no competing financial interests. Correspondence and requests for materials should be addressed to C.A.W. (cwalsh@bidmc.harvard.edu).

Anthropogenic Ocean Acidification over the Twenty-first Century and Its Impact on Calcifying Organisms

J. C. Orr *et al.*

Editor's Note

Acidification of the oceans is the other great environmental change besides raised temperatures that increased atmospheric carbon dioxide will cause. It could have profound effects on animals such as molluscs and coral, as a shift in the sea's carbon chemistry will make it harder to make calcium carbonate shells. This paper by James Orr and colleagues was one of those that raised the alarm, showing that the shells of planktonic molluscs begin to dissolve in water mimicking the ocean predicted for 2100 by a "business as usual" emission scenario. The effect is greater in cold water—placing corals in a climate-change pincer, as those reefs least at risk from bleaching in warmer water would be most at risk from acidification.

Today's surface ocean is saturated with respect to calcium carbonate, but increasing atmospheric carbon dioxide concentrations are reducing ocean pH and carbonate ion concentrations, and thus the level of calcium carbonate saturation. Experimental evidence suggests that if these trends continue, key marine organisms—such as corals and some plankton—will have difficulty maintaining their external calcium carbonate skeletons. Here we use 13 models of the ocean–carbon cycle to assess calcium carbonate saturation under the IS92a "business-as-usual" scenario for future emissions of anthropogenic carbon dioxide. In our projections, Southern Ocean surface waters will begin to become undersaturated with respect to aragonite, a metastable form of calcium carbonate, by the year 2050. By 2100, this undersaturation could extend throughout the entire Southern Ocean and into the subarctic Pacific Ocean. When live pteropods were exposed to our predicted level of undersaturation during a two-day shipboard experiment, their aragonite shells showed notable dissolution. Our findings indicate that conditions detrimental to high-latitude ecosystems could develop within decades, not centuries as suggested previously.

OCEAN uptake of CO_2 will help moderate future climate change, but the associated chemistry, namely hydrolysis of CO_2 in seawater, increases the hydrogen ion concentration $[H^+]$. Surface ocean pH is already 0.1 unit lower than preindustrial values. By the end of the century, it will become another 0.3–0.4 units lower[1,2] under the IS92a scenario, which translates to a 100–150% increase in $[H^+]$. Simultaneously, aqueous CO_2 concentrations $[CO_2(aq)]$ will increase and carbonate ion concentrations $[CO_3^{2-}]$ will decrease, making it more difficult for marine calcifying organisms to form biogenic calcium

21 世纪人类活动引起的海洋酸化效应及其对钙质生物的影响

奥尔等

编者按

大气中二氧化碳的持续增加，除了使得全球气温升高，也造成了另一个巨大的环境变化——海洋酸化。海洋酸化改变了海洋碳化学，使得贝类和珊瑚等动物产生碳酸钙壳体变得更加困难，因而产生深远的影响。詹姆斯·奥尔及其同事的这篇论文敲响了警钟，他们的研究表明，在模拟"照常排放"情景的 2100 年海洋的水体中，浮游软体动物的外壳会开始溶解。这样的效应在冷水中效果更大——就像是把珊瑚放在气候变化的困境里，因为那些冷水种珊瑚礁虽然躲开了在温暖的海水中白化的风险，却躲不了酸化的危机。

当今的海洋表层水中，碳酸钙处于饱和状态，然而随着大气二氧化碳浓度的升高，海水 pH 值会持续下降，碳酸根离子浓度也将不断降低，结果将导致海水中碳酸钙的饱和度下降。我们的实验证明，倘若这种趋势持续下去，海洋中的关键生物，如珊瑚和一些钙质浮游生物将很难保住他们的钙质外壳。我们利用 13 种海洋–碳循环模型，模拟在 IS92a"照常排放"情景下，人为排放二氧化碳对于海洋碳酸钙的饱和度的影响。结果显示，到 2050 年，南大洋表层水将处于文石不饱和状态（文石是碳酸钙的一种亚稳定形式），而到了 2100 年，这种不饱和状态将扩展至整个南大洋并进入太平洋亚北极区海域。根据为期两天的船载实验，将翼足类暴露到模拟的不饱和度下，其文石外壳将遭受显著溶解。这说明，海洋酸化危害高纬度地区生态系统的情况也许几十年内就会出现，而不需要早先提出的几个世纪之久。

海洋对 CO_2 的吸收有助于减轻未来气候的变化，然而，与此相关的化学过程，即海水中 CO_2 的水解，却会使氢离子浓度（$[H^+]$）升高。目前表层海水的 pH 值已经比工业革命前降低了 0.1 个单位。根据 IS92a 排放情景，到 21 世纪末，pH 值可能还要降低 0.3 ~ 0.4 个单位，也就是说 $[H^+]$ 将升高 100% ~ 150%。同时，海水溶解的 CO_2 浓度（$[CO_2(aq)]$）也将升高，而碳酸根离子浓度（$[CO_3^{2-}]$）则相应降低，由此导致海洋钙质生物难以沉淀生物源的碳酸钙（$CaCO_3$）。大量实验证明，在低纬度海

carbonate ($CaCO_3$). Substantial experimental evidence indicates that calcification rates will decrease in low-latitude corals[3-5], which form reefs out of aragonite, and in phytoplankton that form their tests (shells) out of calcite[6,7], the stable form of $CaCO_3$. Calcification rates will decline along with $[CO_3^{2-}]$ owing to its reaction with increasing concentrations of anthropogenic CO_2 according to the following reaction:

$$CO_2 + CO_3^{2-} + H_2O \rightarrow 2HCO_3^- \quad (1)$$

These rates decline even when surface waters remain supersaturated with respect to $CaCO_3$, a condition that previous studies have predicted will persist for hundreds of years[4,8,9].

Recent predictions of future changes in surface ocean pH and carbonate chemistry have primarily focused on global average conditions[1,2,10] or on low latitude regions[4], where reef-building corals are abundant. Here we focus on future surface and subsurface changes in high latitude regions where planktonic shelled pteropods are prominent components of the upper-ocean biota in the Southern Ocean, Arctic Ocean and subarctic Pacific Ocean[11-15]. Recently, it has been suggested that the cold surface waters in such regions will begin to become undersaturated with respect to aragonite only when atmospheric CO_2 reaches 1,200 p.p.m.v., more than four times the preindustrial level ($4 \times CO_2$) of 280 p.p.m.v. (ref. 9). In contrast, our results suggest that some polar and subpolar surface waters will become undersaturated at $\sim 2 \times CO_2$, probably within the next 50 years.

Changes in Carbonate

We have computed modern-day ocean carbonate chemistry from observed alkalinity and dissolved inorganic carbon (DIC), relying on data collected during the CO_2 Survey of the World Ocean Circulation Experiment (WOCE) and the Joint Global Ocean Flux Study (JGOFS). These observations are centred around the year 1994, and have recently been provided as a global-scale, gridded data product GLODAP (ref. 16; see Supplementary Information). Modern-day surface $[CO_3^{2-}]$ varies meridionally by more than a factor of two, from average concentrations in the Southern Ocean of 105 $\mu mol\ kg^{-1}$ to average concentrations in tropical waters of 240 $\mu mol\ kg^{-1}$ (Fig. 1). Low $[CO_3^{2-}]$ in the Southern Ocean is due to (1) low surface temperatures and CO_2-system thermodynamics, and (2) large amounts of upwelled deep water, which contain high $[CO_2(aq)]$ from organic matter remineralization. These two effects reinforce one another, yielding a high positive correlation of present-day $[CO_3^{2-}]$ with temperature (for example, $R^2 = 0.92$ for annual mean surface maps). Changes in $[CO_3^{2-}]$ and $[CO_2(aq)]$ are also inextricably linked to changes in other carbonate chemistry variables (Supplementary Fig. S1).

区，利用文石来形成珊瑚礁的珊瑚[3-5]以及利用方解石（$CaCO_3$ 的稳固晶型）来形成外壳的浮游植物[6,7]的钙化速率都将降低。随着人类活动形成的 CO_2 浓度的不断升高，[CO_3^{2-}] 因与 CO_2 发生反应而减少，导致钙化速率下降。CO_3^{2-} 与 CO_2 的反应方程式如下：

$$CO_2+CO_3^{2-}+H_2O \rightarrow 2HCO_3^- \tag{1}$$

钙化速率降低的趋势将会持续几百年[4,8,9]，即使表层水体中 $CaCO_3$ 如早先预测的仍为过饱和。

近期对未来大洋表层 pH 值及碳酸盐化学变化的趋势预测研究主要集中于全球平均环境条件[1,2,10]或低纬度地区[4]，在这些地区造礁珊瑚非常丰富。本文则聚焦未来高纬度地区表层和次表层海水中将发生的变化，浮游带壳翼足类是这些区域（南大洋、北冰洋以及太平洋亚北极区）上层海洋主要的生物群组成[11-15]。近来有人提出，由于上述地区表层海水较冷，只有当大气 CO_2 浓度达到 1,200 ppmv 时，水环境才可能变为文石不饱和状态，而这一浓度是工业革命前 CO_2 浓度 280 ppmv 的 4 倍多（参考文献 9）。相反，我们的研究结果则表明，未来的 50 年内，当 CO_2 浓度约为现今的两倍时，极地、副极地的表层海水就会达到文石不饱和状态。

碳酸盐的变化

我们根据观测到的碱度及溶解无机碳（DIC）数据计算了现今海洋中的碳酸盐化学系统，其中所用数据来自世界大洋环流实验（WOCE）以及全球海洋通量联合研究（JGOFS）的 CO_2 研究结果。上述两项调查是在 1994 年前后进行的，并于近期形成了一个全球网格化产品 GLODAP（参考文献 16，见补充信息）。当代表层海水 [CO_3^{2-}] 在南北方向上变化，最大差异可达 2 倍以上，从南大洋平均浓度为 105 μmol · kg^{-1} 到热带海域的 240 μmol · kg^{-1}（图 1）。南大洋表层水 [CO_3^{2-}] 较低的原因有：(1) 表层温度以及 CO_2 体系的热力学活性较低；(2) 南大洋有大量向上涌升的深层水，此深层水积累了大量有机质再矿化而来的高 [CO_2(aq)]。两种效应相互促进，使得 [CO_3^{2-}] 与温度之间表现为非常好的正相关关系（比如，表层海洋年均分布图中 R^2 可以达到 0.92）。[CO_3^{2-}] 和 [CO_2(aq)] 的变化也与其他碳酸盐化学参数的变化有着密不可分的关系（附图 S1）。

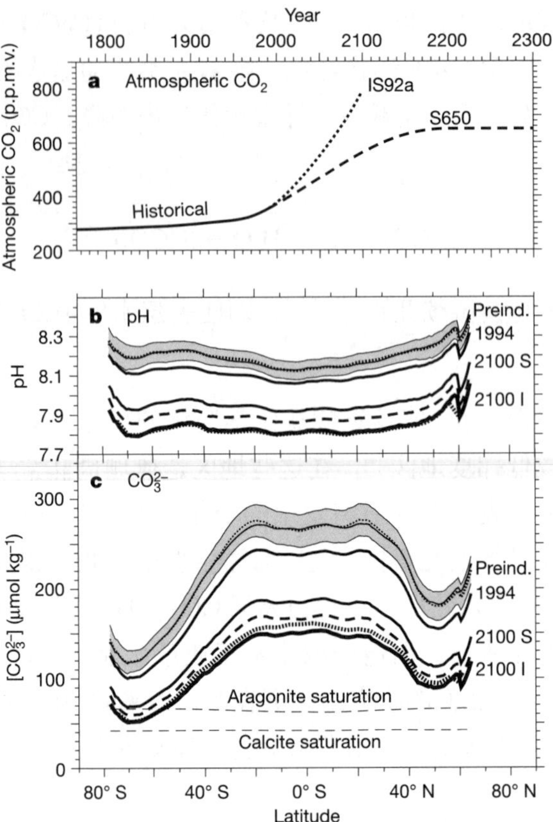

Fig. 1. Increasing atmospheric CO_2 and decreasing surface ocean pH and [CO_3^{2-}]. **a**, Atmospheric CO_2 used to force 13 OCMIP models over the industrial period ("Historical") and for two future scenarios: IS92a ("I" in **b** and **c**) and S650 ("S" in **b** and **c**). **b**, **c**, Increases in atmospheric CO_2 lead to reductions in surface ocean pH (**b**) and surface ocean [CO_3^{2-}] (**c**). Results are given as global zonal averages for the 1994 data and the preindustrial ("Preind.") ocean. The latter were obtained by subtracting data-based anthropogenic DIC (ref. 17) (solid line in grey-shaded area), as well as by subtracting model-based anthropogenic DIC (OCMIP median, dotted line in grey-shaded area; OCMIP range, grey shading). Future results for the year 2100 come from the 1994 data plus the simulated DIC perturbations for the two scenarios; results are also shown for the year 2300 with S650 (thick dashed line). The small effect of future climate change simulated by the IPSL climate–carbon model is added as a perturbation to IS92a in the year 2100 (thick dotted line); two other climate–carbon models, PIUB-Bern and Commonwealth Scientific and Industrial Research Organisation (CSIRO), show similar results (Fig. 3a). The thin dashed lines indicating the [CO_3^{2-}] for sea water in equilibrium with aragonite and calcite are nearly flat, revealing weak temperature sensitivity.

We also estimated preindustrial [CO_3^{2-}] from the same data, after subtracting data-based estimates of anthropogenic DIC (ref. 17) from the modern DIC observations and assuming that preindustrial and modern alkalinity fields were identical (see Supplementary Information). Relative to preindustrial conditions, invasion of anthropogenic CO_2 has already reduced modern surface [CO_3^{2-}] by more than 10%, that is, a reduction of 29 μmol kg^{-1} in the tropics and 18 μmol kg^{-1} in the Southern Ocean. Nearly identical results were found when, instead of the data-based anthropogenic CO_2 estimates, we used simulated anthropogenic CO_2, namely the median from 13 models that participated in the second

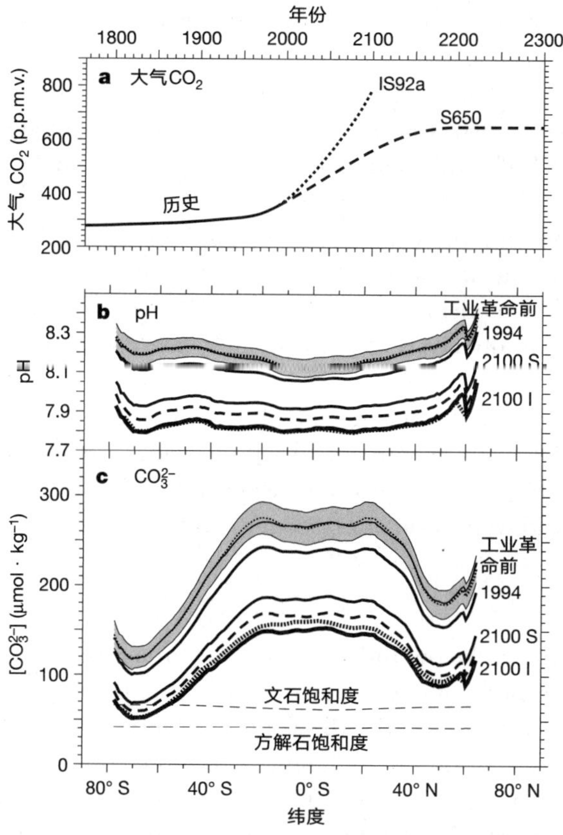

图 1. 大气 CO_2 含量的不断升高与大洋表层 pH 和 $[CO_3^{2-}]$ 的持续降低。图 **a** 为用来驱动 13 个 OCMIP 模型在工业革命时代（"历史"）以及两种未来情景——IS92a（图 **b** 和 **c** 中的"I"）和 S650（图 **b** 和 **c** 中的"S"）排放情景下的大气 CO_2 浓度逐时变化。图 **b** 和 **c** 为随着大气 CO_2 浓度的上升，表层大洋 pH（图 **b**）和表层大洋 $[CO_3^{2-}]$（图 **c**）降低的南北分布图。图中给出了 1994 年及工业革命前全球海洋中的纬向平均值，后者是通过分别减去由数据获得的人为排放的 DIC（参考文献 17）（灰色阴影内的实线）和模拟的人为排放的 DIC 而得到的（灰色阴影内的点线表示 OCMIP 中值，阴影表示 OCMIP 的变化范围）。2100 年的结果是根据 1994 年的数据加上两种排放情景下 DIC 变化量的模拟值得到的；图中还给出了 S650 排放情景下 2300 年的情形（粗虚线）。IPSL 气候–碳元素模型得到的未来气候变化产生的微弱干扰也被加到了 IS92a 排放情景下 2100 年的值上（粗点线），由另外两个模型 PIUB–伯尔尼和 CSIRO（澳大利亚联邦科学与工业研究组织）模型得到的结果与此相似（图 3a）。细虚线表示海水中文石与方解石平衡时的 $[CO_3^{2-}]$，该线近于水平，说明该参数对温度不敏感。

我们还利用上述数据，假定工业革命以前海洋的碱度与现今的量级相同（见补充信息），根据现代海洋 DIC 的观测数据再减去用观测数据估算出的人为排放的 DIC（参考文献 17）之后，得到了工业革命以前的 $[CO_3^{2-}]$ 的估计值。相对于工业革命以前的水平，人为排放 CO_2 的剧增已使现今海表 $[CO_3^{2-}]$ 降低了 10% 以上，具体为热带地区降低了 29 μmol · kg^{-1}，南大洋降低了 18 μmol · kg^{-1}。当我们用模型估计的人为排放 CO_2（利用参与二期大洋碳循环模型国际对比项目（即 OCMIP-2）的 13 个模型模拟得到

phase of the Ocean Carbon-Cycle Model Intercomparison Project, or OCMIP-2 (Fig. 1c).

To quantify future changes in carbonate chemistry, we used simulated DIC from ocean models that were forced by two atmospheric CO_2 scenarios: the Intergovernmental Panel on Climate Change (IPCC) IS92a "continually increasing" scenario (788 p.p.m.v. in the year 2100) and the IPCC S650 "stabilization" scenario (563 p.p.m.v. in the year 2100) (Fig. 1). Simulated perturbations in DIC relative to 1994 (the GLODAP reference year) were added to the modern DIC data; again, alkalinity was assumed to be constant. To provide a measure of uncertainty, we report model results as the OCMIP median $\pm 2\sigma$. The median generally outperformed individual models in OCMIP model–data comparison (Supplementary Fig. S2). By the year 2100, as atmospheric CO_2 reaches 788 p.p.m.v. under the IS92a scenario, average tropical surface $[CO_3^{2-}]$ declines to 149 ± 14 μmol kg^{-1}. This is a 45% reduction relative to preindustrial levels, in agreement with previous predictions[4,8]. In the Southern Ocean (all waters south of 60° S), surface concentrations dip to 55 ± 5 μmol kg^{-1}, which is 18% below the threshold where aragonite becomes undersaturated (66 μmol kg^{-1}).

These changes extend well below the sea surface. Throughout the Southern Ocean, the entire water column becomes undersaturated with respect to aragonite. During the twenty-first century, under the IS92a scenario, the Southern Ocean's aragonite saturation horizon (the limit between undersaturation and supersaturation) shoals from its present average depth of 730 m (Supplementary Fig. S3) all the way to the surface (Fig. 2). Simultaneously, in a portion of the subarctic Pacific, the aragonite saturation horizon shoals from depths of about 120 m to the surface. In the North Atlantic, surface waters remain saturated with respect to aragonite, but the aragonite saturation horizon shoals dramatically; for example, north of 50° N it shoals from 2,600 m to 115 m. The greater erosion in the North Atlantic is due to deeper penetration and higher concentrations of anthropogenic CO_2, a tendency that is already evident in present-day data-based estimates[17,18] and in models[19,20] (Supplementary Figs S4 and S5). Less pronounced changes were found for the calcite saturation horizon. For example, in the year 2100 the average calcite saturation horizon in the Southern Ocean stays below 2,200 m. Nonetheless, in 2100 surface waters of the Weddell Sea become slightly undersaturated with respect to calcite.

In the more conservative S650 scenario, the atmosphere reaches $2 \times CO_2$ in the year 2100, 50 years later than with the IS92a scenario. In 2100, Southern Ocean surface waters generally remain slightly supersaturated with respect to aragonite. However, the models also simulate that the Southern Ocean's average aragonite saturation horizon will have shoaled from 730 m to 60 m, and that the entire water column in the Weddell Sea will have become undersaturated (Fig. 2). In the north, all surface waters remain saturated under the S650 scenario. North of 50° N, the annual average aragonite saturation horizon shoals from 140 m to 70 m in the Pacific, whereas it shoals by 2,000 m to 610 m in the North Atlantic. Therefore, under either scenario the OCMIP models simulated large changes in surface and

的中值）代替用实测数据估算的数值进行扣除时，得到了基本相同的结果（图 1c）。

为了量化海水碳酸盐体系未来的变化趋势，我们以两种大气 CO_2 排放情景——联合国政府间气候变化专门委员会（IPCC）的 IS92a“持续增加型”排放情景（到 2100 年为 788 ppmv）和 IPCC S650“稳定型”排放情景（到 2100 年为 563 ppmv）——为准，用模型得出了 DIC 在两种排放情景下各自的模拟结果（图 1）。同样仍假定碱度不随时间变化的条件下，我们将模拟出的 DIC 相对于 1994 年（GLODAP 的参考年份）的扰动值加到现代观测到的 DIC 数值上。为给出一个不确定度的衡量标准，我们将模拟结果用 OCMIP 中值 ± 2σ 来表示。该中值通常要优于 OCMIP 模拟–数据对照中所用的任一单个模型的分析结果（附图 S2）。根据 IS92a 排放情景，到 2100 年，大气 CO_2 浓度将达到 788 ppmv，热带海洋表层的 [CO_3^{2-}] 的平均值将降低为 149 ± 14 μmol · kg^{-1}。该值比工业革命以前的浓度低 45%，这与之前的预测结果是一致的 [4,8]。在南大洋（60° S 以南的所有水体）中，表层浓度降至 55 ± 5 μmol · kg^{-1}，这一浓度比文石在海洋中能保持不溶解的极限 [CO_3^{2-}]（66 μmol · kg^{-1}）还低 18%。

这些变化还将波及海洋表层以下的水体，到时整个南大洋的全水柱将都处于文石不饱和状态。根据 IS92a 排放情景，在 21 世纪，南大洋文石的饱和深度（指文石在水体中处于不饱和状态和过饱和状态之间交界的深度）将从现在的平均 730 m 深（附图 S3）一直升到海洋表层（意指全水柱文石将都处于不饱和状态）（图 2）。同时，亚北极太平洋的部分海域的文石饱和深度也将由 120 m 深处逐渐升至表层。而在北大西洋，表层水体仍处于文石过饱和状态，只是其文石饱和深度也明显变浅；比如，50° N 以北海域将由 2,600 m 深处升至 115 m 处。北大西洋海水的侵蚀作用愈加强烈，是因为人为排放 CO_2 的浓度越来越高，侵蚀的范围也越来越深，无论是根据实测数据得出的估算值 [17,18] 还是从模拟结果 [19,20] 来看，这种趋势都十分明显（附图 S4 和 S5）。而方解石饱和深度的变化则相对而言不那么明显。例如，到 2100 年南大洋方解石平均饱和深度仍将位于 2,200 m 以下。尽管如此，2100 年时威德尔海表层水体中的方解石也将呈轻微不饱和状态。

而在更为保守的 S650 排放情景下，到 2100 年大气中 CO_2 浓度将是现在的两倍，这比 IS92a 排放情景要晚 50 年的时间。到 2100 年时，南大洋表层水体中文石仍略处于过饱和状态。然而，模型结果同样显示，南大洋的平均文石饱和深度将由 730 m 处上升到 60 m 处，同时整个威德尔海的全水柱都变得文石不饱和（图 2）。在 S650 排放情景下，北半球的表层水体仍全部处于过饱和状态下。在 50° N 以北海域，文石的年均饱和深度在太平洋会从 140 m 处上升到 70 m 处，而在北大西洋则由 2,000 m 处上升到 610 m 处。因此，不管在何种排放情景下，OCMIP 模型的模拟结果都显示，表

subsurface $[CO_3^{2-}]$. Yet these models account for only the direct geochemical effect of increasing atmospheric CO_2 because they were all forced with prescribed modern-day climate conditions.

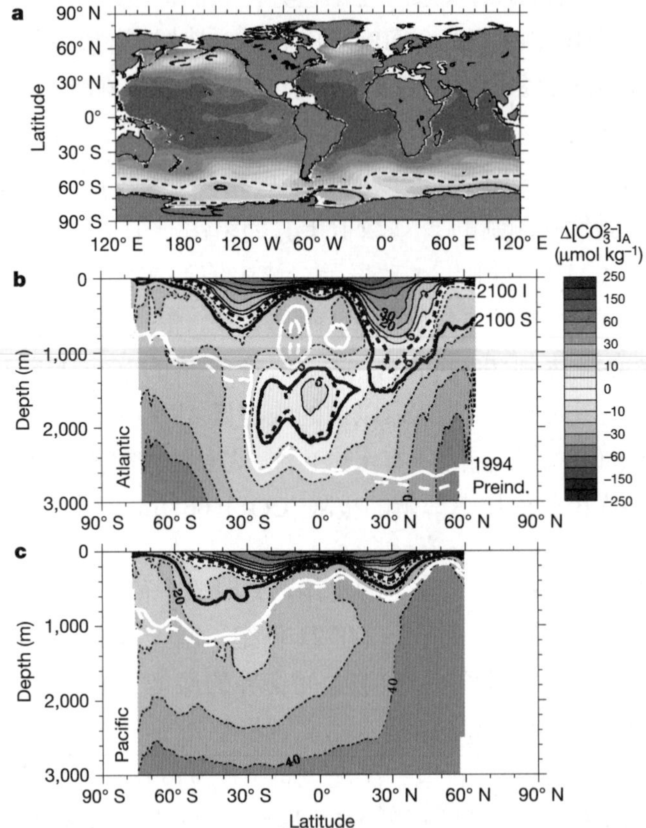

Fig. 2. The aragonite saturation state in the year 2100 as indicated by $\Delta[CO_3^{2-}]_A$. The $\Delta[CO_3^{2-}]_A$ is the *in situ* $[CO_3^{2-}]$ minus that for aragonite-equilibrated sea water at the same salinity, temperature and pressure. Shown are the OCMIP-2 median concentrations in the year 2100 under scenario IS92a: **a**, surface map; **b**, Atlantic; and **c**, Pacific zonal averages. Thick lines indicate the aragonite saturation horizon in 1765 (Preind.; white dashed line), 1994 (white solid line) and 2100 (black solid line for S650; black dashed line for IS92a). Positive $\Delta[CO_3^{2-}]_A$ indicates supersaturation; negative $\Delta[CO_3^{2-}]_A$ indicates undersaturation.

In addition to this direct geochemical effect, ocean $[CO_3^{2-}]$ is also altered by climate variability and climate change. To quantify the added effect of future climate change, we analysed results from three atmosphere–ocean climate models that each included an ocean carbon-cycle component (see Supplementary Information). These three models agree that twenty-first century climate change will cause a general increase in surface ocean $[CO_3^{2-}]$ (Fig. 3), mainly because most surface waters will be warmer. However, the models also agree that the magnitude of this increase in $[CO_3^{2-}]$ is small, typically counteracting less than 10% of the decrease due to the geochemical effect. High-latitude surface waters show the smallest increases in $[CO_3^{2-}]$, and even small reductions in some cases. Therefore, our analysis suggests that physical climate change alone will not substantially alter high-latitude surface $[CO_3^{2-}]$ during the twenty-first century.

层、次表层水体的 [CO_3^{2-}] 将发生明显变化。不过，这些模型仅考虑了大气 CO_2 升高后地球化学方面的直接变化，因为所有模型都是以现在的气候条件为基础的。

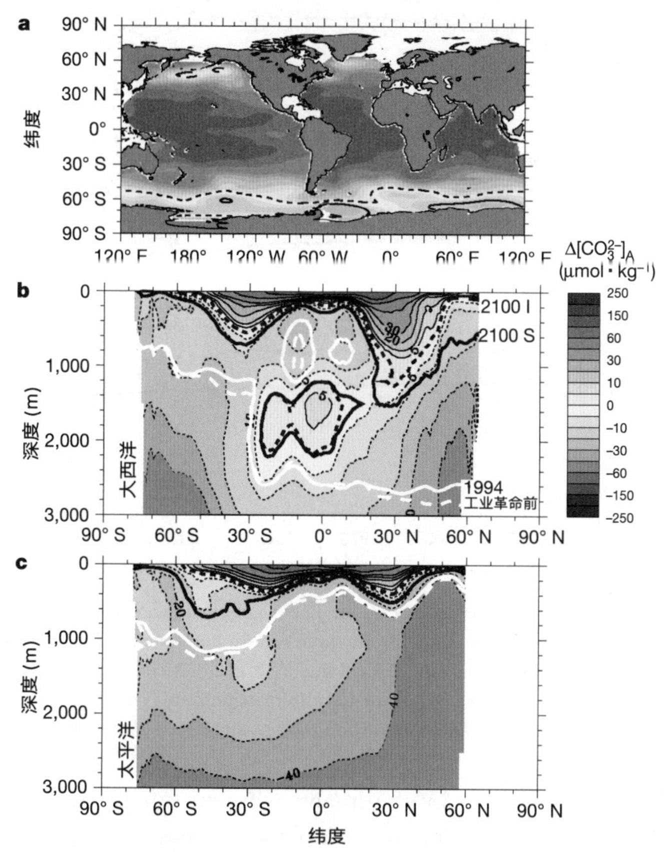

图 2. 以 Δ[CO_3^{2-}]$_A$ 估算的 2100 年时文石的饱和状态。Δ[CO_3^{2-}]$_A$ 指在盐度、温度和压力均相同的情况下原位 [CO_3^{2-}] 减去海水中文石饱和所需的 [CO_3^{2-}]。图中展示的是在 IS92a 排放情景下，2100 年时 OCMIP-2 估算浓度的中值：**a** 图为表层图；**b** 图为大西洋的纬向平均值；**c** 图为太平洋的纬向平均值。其中粗线分别表示 1765 年（工业革命前，白色虚线）、1994 年（白色实线）和 2100 年（黑色实线为 S650 排放情景下的情形，黑色虚线为 IS92a 排放情景下的情形）的文石饱和深度。Δ[CO_3^{2-}]$_A$ 为正，表示过饱和；Δ[CO_3^{2-}]$_A$ 为负，表示不饱和。

除地球化学方面的直接变化以外，海洋 [CO_3^{2-}] 还受气候变异性和气候变化的影响。为了量化未来气候变化对海洋 [CO_3^{2-}] 的影响，我们对三个海气气候模型的模拟结果作了分析，这三个模型中每一个都包含海洋碳循环的一个组分（见补充信息）。上述三个模型的结果一致显示，21 世纪气候变化将使大洋表层 [CO_3^{2-}] 普遍升高（图 3），这主要是由大部分表层水体的温度将升高所致。不过，模拟结果还显示，[CO_3^{2-}] 升高的幅度很小，一般仅仅可抵消地球化学变化引起的 [CO_3^{2-}] 降低值的 10%。高纬地区的表层水体中 [CO_3^{2-}] 升高最少，部分条件下甚至还会降低。因此，我们分析认为，在 21 世纪，仅仅物理层面上的气候变化不会使高纬地区表层海水的 [CO_3^{2-}] 发生很大变化。

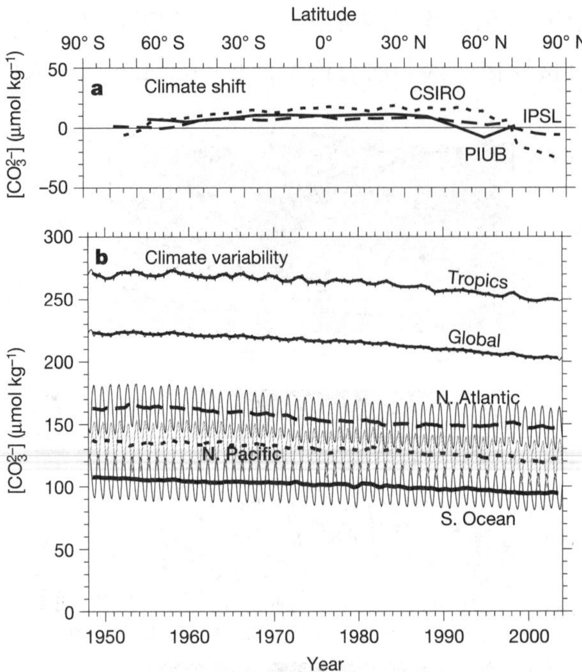

Fig. 3. Climate-induced changes in surface [CO_3^{2-}]. **a**, The twenty-first century shift in zonal mean surface ocean [CO_3^{2-}] due to climate change alone, from three atmosphere–ocean climate models—CSIRO-Hobart (short dashed line), IPSL-Paris (long dashed line) and PIUB-Bern (solid line)—that each include an ocean carbon-cycle component (see Supplementary Information). **b**, The regional-scale seasonal and interannual variability is simulated by an ocean carbon-cycle model forced with reanalysed climate forcing.

Climate also varies seasonally and interannually, whereas our previous focus has been on annual changes. To illustrate how climate variability affects surface [CO_3^{2-}], we used results from an ocean carbon-cycle model forced with the daily National Centers for Environmental Prediction (NCEP) reanalysis fields[21] over 1948–2003 (see Supplementary Information). These fields are observationally based and vary on seasonal and interannual timescales. Simulated interannual variability in surface ocean [CO_3^{2-}] is negligible when compared with the magnitude of the anthropogenic decline (Fig. 3b). Seasonal variability is also negligible except in the high latitudes, where surface [CO_3^{2-}] varies by about ± 15 μmol kg^{-1} when averaged over large regions. This is smaller than the twenty-first-century's transient change (for example, ~50 μmol kg^{-1} in the Southern Ocean). However, high-latitude surface waters do become substantially less saturated during winter, because of cooling (resulting in higher [$CO_2(aq)$]) and greater upwelling of DIC-enriched deep water, in agreement with previous observations in the North Pacific[22]. Thus, high-latitude undersaturation will be first reached during winter.

Our predicted changes may be compared to those found in earlier studies, which focused on surface waters in the tropics[8] and in the subarctic Pacific[22,23]. These studies assumed thermodynamic equilibrium between CO_2 in the atmosphere and the surface

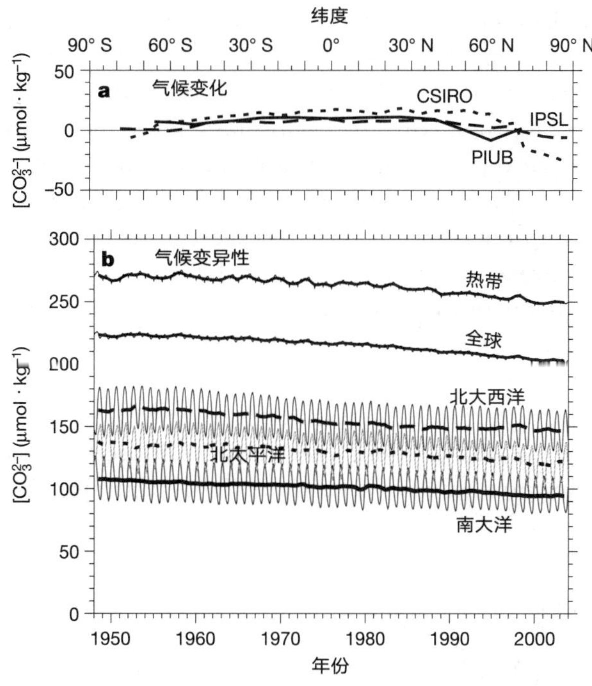

图 3. 气候变化引起的表层 [CO$_3^{2-}$] 改变。**a** 图为根据三个海气气候模型——CSIRO-霍巴特（短虚线），IPSL-巴黎（长虚线）和 PIUB-伯尔尼（实线）——得到的 21 世纪仅由气候变化引起的大洋表层 [CO$_3^{2-}$] 纬向平均值的变化，这三个模型中每一个都包含一个碳循环组分（见补充信息）。**b** 图为气候变异性影响下的海洋碳循环模型对区域变化尺度与季节变化、年际变化尺度上的 [CO$_3^{2-}$] 波动的模拟结果。

气候还存在季节变化以及年际变化，我们之前仅分析了一年内的变化。为了进一步阐释长期气候波动对表层水体中 [CO$_3^{2-}$] 的影响，我们采用代入基于美国国家环境预报中心（NCEP）处理过的 1948 ~ 2003 年间每天天气场 [21]，运行海洋碳循环模型（见补充信息）。这些工作是根据观测数据得到的，其结果随季节和年际时间尺度上的变化而变化，与人为排放造成的 [CO$_3^{2-}$] 降低的量级相比，模型得到的大洋表层 [CO$_3^{2-}$] 的年际变化几乎可以忽略不计（图 3b）。除高纬度地区以外，季节变化也可忽略不计。如果考虑大范围内的平均值的话，高纬海域的表层 [CO$_3^{2-}$] 的季节变化量可以达到 ±15 μmol · kg^{-1}。该值要小于 21 世纪的短期变化（例如，南大洋为 50 μmol · kg^{-1} 左右）。不过，冬季时，由于水体温度降低（导致了 [CO_2(aq)] 升高）与更多富含 DIC 的深水的上涌，高纬度地区表层水体的文石饱和程度确实要低得多，这与之前在北太平洋观测到的结果 [22] 是一致的。因此，高纬度海域的海水文石不饱和状态将最先在冬季出现。

我们所预测出的变化可与早期主要集中于热带 [8] 和亚北极区太平洋海域 [22,23] 的表层水体的研究结果相对照。这些研究均假定，在原位碱度、温度和盐度下，大气 CO_2 和表层水体中溶解的 CO_2 之间处于热力学平衡状态。如果假设两者真的达到

waters at their *in situ* alkalinity, temperature and salinity. If, in the equilibrium approach, the p_{CO_2} is taken only to represent seawater p_{CO_2}, then the results agree with our non-equilibrium approach when the sets of carbonate chemistry constants are identical (Fig. 4). However, assuming equilibrium with the atmosphere leads to the prediction that future undersaturation will occur too soon (at lower atmospheric CO_2 levels), mainly because the anthropogenic transient in the ocean actually lags that in the atmosphere. For example, with the equilibrium approach, we predict that average surface waters in the Southern Ocean become undersaturated when atmospheric CO_2 is 550 p.p.m.v. (in the year 2050 under IS92a), whereas our non-equilibrium approach, which uses models and data, indicates that undersaturation will occur at 635 p.p.m.v. (in the year 2070). Despite these differences, both approaches indicate that the Southern Ocean surface waters will probably become undersaturated with respect to aragonite during this century. Conversely, both of these approaches disagree with a recent assessment[9] that used a variant of the standard thermodynamic equilibrium approach, where an incorrect input temperature was used inadvertently.

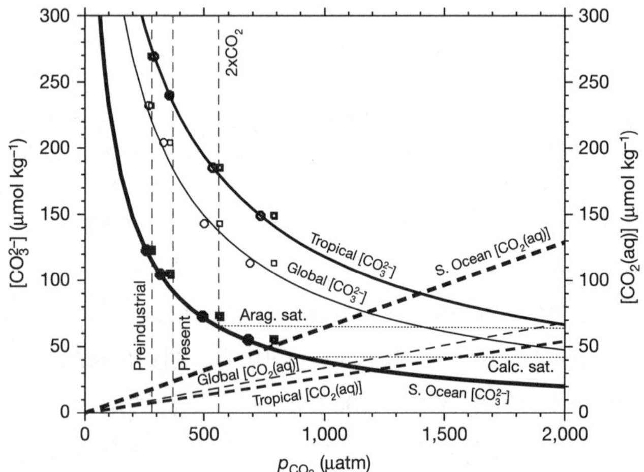

Fig. 4. Key surface carbonate chemistry variables as a function of p_{CO_2}. Shown are both [CO_3^{2-}] (solid lines) and [CO_2(aq)] (dashed lines) for average surface waters in the tropical ocean (thick lines), the Southern Ocean (thickest lines) and the global ocean (thin lines). Solid and dashed lines are calculated from the thermodynamic equilibrium approach. For comparison, open symbols are for [CO_3^{2-}] from our non-equilibrium, model-data approach versus seawater p_{CO_2} (open circles) and atmospheric p_{CO_2} (open squares); symbol thickness corresponds with line thickness, which indicates the regions for area-weighted averages. The nearly flat, thin dotted lines indicate the [CO_3^{2-}] for seawater in equilibrium with aragonite ("Arag. sat.") and calcite ("Calc. sat.").

Uncertainties

The three coupled climate–carbon models show little effect of climate change on surface [CO_3^{2-}] (compare Fig. 3a to Fig. 1) partly because air–sea CO_2 exchange mostly compensates for the changes in surface DIC caused by changes in marine productivity and circulation.

了热力学平衡，那么模型计算所用的二氧化碳分压(p_{CO_2})只能取海水中的p_{CO_2}的数值。在这种情况下，其他碳酸盐体系组分同时保持不变的条件下，平衡假设计算出的结果与我们用非平衡假设得到的结果才是一致的(图 4)。然而，采用海气热力学平衡假设(同时考虑大气p_{CO_2}的真实增加值)会导致海水预计的不饱和状态在未来会更早地出现(即大气 CO_2 浓度还较低时就会出现)，这主要是因为实际上海洋对人为排放 CO_2 的响应要滞后于大气。例如，在平衡假设下，我们预计当大气 CO_2 浓度为 550 ppmv(根据 IS92a 排放情景应为 2050 年)时，南大洋的表层水体总体上将转变为不饱和状态。而根据我们采用的非平衡假设下的模型和数据来做的研究则显示，在大气 CO_2 浓度达到 635 ppmv(2070 年)时，海水的不饱和状态才会出现。尽管存在上述差异，两种方法的结果却都显示，本世纪南大洋的表层水体极有可能成为文石不饱和水体。反之，这两种方法都挑战了最近利用标准热动力平衡法得到的结论[9]，在该方法中，研究者因疏忽而输入了错误的温度值。

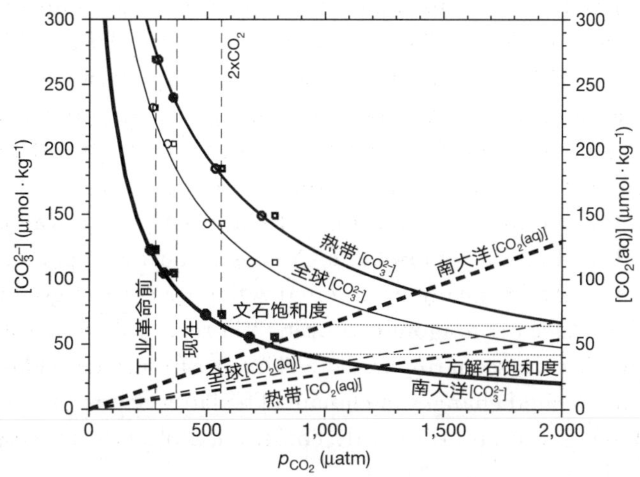

图 4. 主要的表层碳酸盐化学变量，以p_{CO_2}的函数表示。图中所示为 $[CO_3^{2-}]$(实线)和 $[CO_2(aq)]$(虚线)在热带海域(粗线)、南大洋(最粗线)和全球海洋(细线)表层水体中的平均值。实线和虚线是热动力平衡法计算而来的。为了展开比较，空心符号表示由我们的非平衡数据模拟法得到的 $[CO_3^{2-}]$ 与海水的p_{CO_2}(空心圆圈)和大气的p_{CO_2}(空心方框)的比值；符号的粗细与线的粗细相对应，表示该区域面积所占的权重。图中近乎水平的细点线表示文石和方解石处于平衡状态时海水中的 $[CO_3^{2-}]$。

不确定性

三个气候–碳耦合模型结果显示，气候变化对表层 $[CO_3^{2-}]$ 的影响很小(比较图 3a 与图 1)，部分原因是大气与海洋之间的 CO_2 交换在很大程度上补偿了由海洋生产力和大洋循环的变化引起的表层 DIC 的变化。在缺乏类似补偿机制的次表层水体中，

In subsurface waters where such compensation is lacking, these models could under- or over-predict how much [CO_3^{2-}] will change as a result of changes in overlying marine productivity. However, the models project a consistent trend, which only worsens the decline in subsurface [CO_3^{2-}]; that is, all coupled climate models predict increased evaporation in the tropics and increased precipitation in the high latitudes[24]. This leads to greater upper ocean stratification in the high latitudes, which in turn decreases nutrients (but not to zero) and increases light availability (owing to more shallow mixed layers). Thus, at $2 \times CO_2$ there is a 10% local increase in surface-to-deep export of particulate organic carbon (POC) in the Southern Ocean using the Institut Pierre Simon Laplace (IPSL)-Paris model[25]. Subsequent remineralization of this exported POC within the thermocline would increase DIC, which would only exacerbate the decrease in high-latitude subsurface [CO_3^{2-}]. For the twenty-first century, these uncertainties appear small next to the anthropogenic DIC invasion (see Supplementary Information).

The largest uncertainty by far, and the only means to limit the future decline in ocean [CO_3^{2-}], is the atmospheric CO_2 trajectory. To better characterize uncertainty due to CO_2 emissions, we compared the six illustrative IPCC Special Reports on Emission Scenarios (SRES) in the reduced complexity, Physics Institute University of Bern (PIUB)-Bern model. Under the moderate SRES B2 scenario, average Southern Ocean surface waters in that model become undersaturated with respect to aragonite when atmospheric CO_2 reaches 600 p.p.m.v. in the year 2100 (Fig. 5). For the three higher-emission SRES scenarios (A1FI, A2 and A1B), these waters become undersaturated sooner (between the years 2058 and 2073); for the two lower-emission scenarios (A1T and B1), these waters remain slightly supersaturated in 2100. Thus, if atmospheric CO_2 rises above 600 p.p.m.v., most Southern Ocean surface waters will become undersaturated with respect to aragonite. Yet, even below this level, the Southern Ocean's aragonite saturation horizon will shoal substantially (Fig. 2). For a given atmospheric CO_2 scenario, predicted changes in surface ocean [CO_3^{2-}] are much more certain than the related changes in climate. The latter depend not only on the model response to CO_2 forcing, but also on poorly constrained physical processes, such as those associated with clouds.

上述模型可能会低估或高估上层海洋生产力的变化所引起的 [CO_3^{2-}] 的变化量。尽管如此，上述模型为我们指出了表层 [CO_3^{2-}] 连续的变化趋势，而该降低趋势只会在次表层水体中更加严重。也就是说，所有耦合气候模型的预测结果都是：热带地区的蒸发量将不断增大，而在高纬地区降雨量则不断增多 [24]。这将导致高纬度海区上层水体层化现象加重，进而使其营养盐含量降低(但不会降到零)，同时对光的利用率增加(因为上混合层变浅)。因此，根据法国国家科研中心的皮埃尔 · 西蒙 · 拉普拉斯(IPSL)–巴黎模型 [25]，当 CO_2 浓度达到现今的两倍时，南大洋中由表层向深处输出的颗粒态有机碳(POC)将升高 10%。这些输出的 POC 在温跃层中的再矿化将使 DIC 增加，进而加剧高纬度地区次表层 [CO_3^{2-}] 的降低。不过，在 21 世纪，这种不确定性的影响与人为排放 DIC 的侵蚀相比还是较小(见补充信息)。

到目前为止，最大的不确定性就是大气中 CO_2 的变化趋势，了解该趋势也是限制未来海洋 [CO_3^{2-}] 下降的唯一手段。为了更好地表征 CO_2 排放带来的不确定性，我们利用简化的伯尔尼大学物理研究所(PIUB)–伯尔尼模型，对 IPCC 的六份解释性排放情景特别报告(SRES)作了比较。在适中的 SRES B2 排放情景下，到 2100 年，大气 CO_2 浓度达到 600 ppmv 时，该模型得到的南大洋表层水体将处于文石不饱和状态(图 5)。对于三个排放量较高的 SRES 排放情景(A1FI、A2 和 A1B)，水体达到不饱和的时间则早得多(在 2058 ~ 2073 年间)；而在其他两个排放量较低的排放情景(A1T 和 B1)下，上述水体到 2100 年还将处于轻微的过饱和状态。所以，倘若大气 CO_2 浓度升至 600 ppmv 以上，南大洋的大部分表层水体都将处于文石不饱和状态。然而，即便 CO_2 低于这一水平，南大洋中文石饱和深度也将明显变浅(图 2)。在某个给定大气 CO_2 排放情景下，所预测的大洋表层 [CO_3^{2-}] 变化相对于气候变化可能带来的影响则要确定得多，后者并不仅仅取决于模型对 CO_2 强迫的响应，还受物理过程的影响，如云中发生的相关过程。

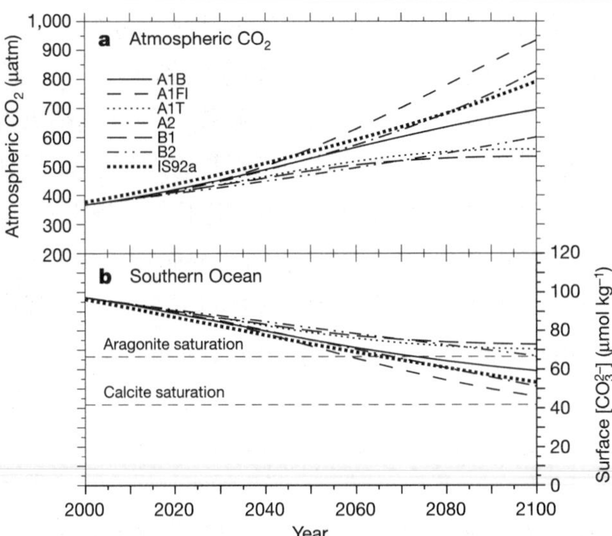

Fig. 5. Average surface $[CO_3^{2-}]$ in the Southern Ocean under various scenarios. Time series of average surface $[CO_3^{2-}]$ in the Southern Ocean for the PIUB-Bern reduced complexity model (see Fig. 3 and Supplementary Information) under the six illustrative IPCC SRES scenarios. The results for the SRES scenarios A1T and A2 are similar to those for the non-SRES scenarios S650 and IS92a, respectively.

Ocean CO_2 Uptake

With higher levels of anthropogenic CO_2 and lower surface $[CO_3^{2-}]$, the change in surface ocean DIC per unit change in atmospheric CO_2 (μmol kg^{-1} per p.p.m.v.) will be about 60% lower in the year 2100 (under IS92a) than it is today. Simultaneously, the $CO_3^{2-}/CO_2(aq)$ ratio will decrease from 4:1 to 1:1 in the Southern Ocean (Fig. 4). These decreases are due to the well-understood anthropogenic reduction in buffer capacity[26], already accounted for in ocean carbon-cycle models.

On the other hand, reduced export of $CaCO_3$ from the high latitudes would increase surface $[CO_3^{2-}]$, thereby increasing ocean CO_2 uptake and decreasing atmospheric CO_2. Owing to this effect, ocean CO_2 uptake could increase by 6–13 petagrams (Pg) C over the twenty-first century, based on one recent model study[27] that incorporated an empirical, CO_2-dependant relationship for calcification[7]. Rates of calcification could decline even further, to zero, if waters actually became undersaturated with respect to both aragonite and calcite. We estimate that the total shutdown of high-latitude aragonite production would lead to, at most, a 0.25 Pg C yr^{-1} increase in ocean CO_2 uptake, assuming that 1 Pg C yr^{-1} of $CaCO_3$ is exported globally[28], that up to half of that is aragonite[9,29], and that perhaps half of all aragonite is exported from the high latitudes. The actual increase in ocean CO_2 uptake could be much lower because the aragonite fraction of the $CaCO_3$ may be only 0.1 based on low-latitude sediment traps[30], and the latitudinal distribution of aragonite export is uncertain. Thus, increased CO_2 uptake from reduced export of aragonite will provide little compensation for decreases in ocean CO_2 uptake due to reductions in buffer capacity. Of

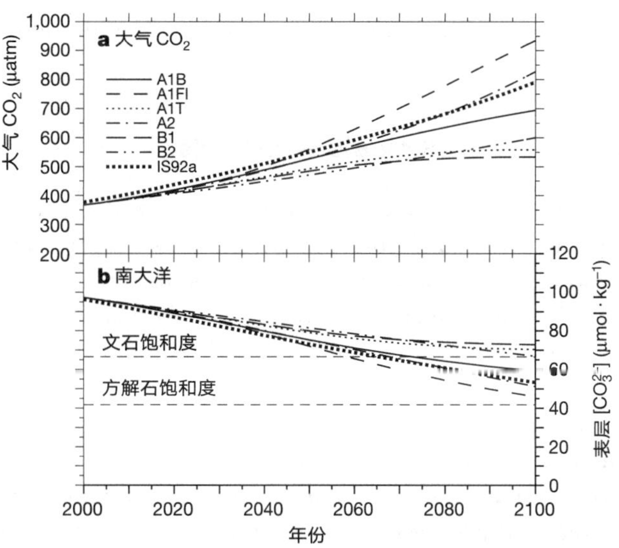

图 5. 各种排放情景下，南大洋表层水体 $[CO_3^{2-}]$ 的平均值。利用 PIUB–伯尔尼简化模型（见图 3 及补充信息）得到的，在六种 IPCC SRES 排放情景下，南大洋表层平均 $[CO_3^{2-}]$ 的时间序列值。SRES A1T 和 A2 排放情景下的结果分别与非 SRES 的 S650 排放情景和 IS92a 排放情景相似。

海洋对 CO_2 的吸收

到 2100 年（根据 IS92a 排放情景），由于人为排放的 CO_2 含量较高而表层海水 $[CO_3^{2-}]$ 较低，大气 CO_2 浓度每改变一个单位（μmol · kg^{-1} 每 ppmv），大洋表层 DIC 发生的变化量将比现今的低 60%。与此同时，南大洋中 $CO_3^{2-}/CO_2(aq)$ 的比值由 4∶1 降至 1∶1（图 4）。这种下降是由人类排放引起的海水的缓冲能力降低所致 [26]，海洋碳循环模型中已经考虑到了这一点。

另一方面，高纬度地区 $CaCO_3$ 输出量的减少会使表层海水中 $[CO_3^{2-}]$ 增加，从而促使海洋对 CO_2 的吸收能力增强，进而导致大气 CO_2 降低。根据最近的一项模型研究 [27]，基于该效应，21 世纪海洋对 CO_2 的吸收能力可提高 6 ~ 13 Pg C（皮克碳），该模型包含经验的 CO_2 与钙化过程的相互关系 [7]。倘若水体中文石和方解石均变为不饱和状态，钙化速率可能下降得更多，甚至变为 0。据我们估计，假设全球碳酸钙的输出通量为 1 Pg C · yr^{-1}[28]（其中一半的输出通量假设由文石贡献 [9,29]，而其中文石一半又假设是由高纬度海域输出），那么高纬海域文石生产的消失，至多可使海洋对 CO_2 的吸收能力提高 0.25 Pg C · yr^{-1}。然而海洋对 CO_2 吸收能力的实际提高可能小得多，因为根据低纬海域沉积物捕获器的数据研究 [30]，文石在输出 $CaCO_3$ 中所占的比例大约仅为 0.1，并且文石输出量的纬度分布也不确定。因此，由文石输出的减少所带来的海洋对 CO_2 吸收能力的增强，对因缓冲能力下降而导致的吸收能力的减弱所起到

greater concern are potential biological impacts due to future undersaturation.

Biological Impacts

The changes in seawater chemistry that we project to occur during this century could have severe consequences for calcifying organisms, particularly shelled pteropods: the major planktonic producers of aragonite. Pteropod population densities are high in polar and subpolar waters. Yet only five species typically occur in such cold water regions and, of these, only one or two species are common at the highest latitudes[31]. High-latitude pteropods have one or two generations per year[12,15,32], form integral components of food webs, and are typically found in the upper 300 m where they may reach densities of hundreds to thousands of individuals per m^3 (refs 11, 13–15). In the Ross Sea, for example, the prominent subpolar–polar pteropod *Limacina helicina* sometimes replaces krill as the dominant zooplankton, and is considered an overall indicator of ecosystem health[33]. In the strongly seasonal high latitudes, sedimentation pulses of pteropods frequently occur just after summer[15,34]. In the Ross Sea, pteropods account for the majority of the annual export flux of both carbonate and organic carbon[34,35]. South of the Antarctic Polar Front, pteropods also dominate the export flux of $CaCO_3$ (ref. 36).

Pteropods may be unable to maintain shells in waters that are undersaturated with respect to aragonite. Data from sediment traps indicate that empty pteropod shells exhibit pitting and partial dissolution as soon as they fall below the aragonite saturation horizon[22,36,37]. *In vitro* measurements confirm such rapid pteropod shell dissolution rates[38]. New experimental evidence suggests that even the shells of live pteropods dissolve rapidly once surface waters become undersaturated with respect to aragonite[9]. Here we show that when the live subarctic pteropod *Clio pyramidata* is subjected to a level of undersaturation similar to what we predict for Southern Ocean surface waters in the year 2100 under IS92a, a marked dissolution occurs at the growing edge of the shell aperture within 48 h (Fig. 6). Etch pits formed on the shell surface at the apertural margin (which is typically ~7-μm-thick) as the < 1-μm exterior (prismatic layer) peeled back (Fig. 6c), exposing the underlying aragonitic rods to dissolution. Fourteen individuals were tested. All of them showed similar dissolution along their growing edge, even though they all remained alive. If *C. pyramidata* cannot grow its protective shell, we would not expect it to survive in waters that become undersaturated with respect to aragonite.

的补偿作用很小。人们更需要关注的是未来的不饱和状态将引起的潜在的生物效应。

生物学影响

我们所预计的本世纪海水化学性质的变化将对钙质生物，特别是带壳的翼足类（文石的主要浮游生产者）有重大影响。在极地和副极地海域翼足类的种群密度相当高。不过在这些冷水区翼足类一般也只有 5 个种，其中仅有一到两个种在最高纬度的海域较常见[31]。高纬地区的翼足类每年繁殖一到两代[12,15,32]，形成了完整的食物网，而且它们一般生活在表层 300 m 处，在这里每立方米海水中可分布成千上万只翼足类（参考文献 11，13 ~ 15）。例如，在罗斯海中，主要的极地–副极地翼足类蟠虎螺（*Limacina helicina*）有时会取代磷虾在浮游动物中占据主导地位，这被认为是衡量生态系统健康状况的最全面的指标[33]。在受季节变化影响很大的高纬地区，翼足类沉降输出的高峰一般出现在夏季刚刚结束[15,34]时。在罗斯海中，翼足类占碳酸盐和有机碳年输出通量的绝大部分[34,35]。在南极锋面以南，翼足类也是 $CaCO_3$ 输出通量的主要来源（参考文献 36）。

在文石不饱和水体中，翼足类可能无法维持其壳体。来自沉积物捕获器的数据显示，一旦降落至文石饱和面以下，空的翼足类壳体上就会存在蚀损斑，而且壳体也将被部分溶解[22,36,37]。体外实验也证实翼足类壳体的溶解速率非常快[38]。新的实验证据表明，在文石不饱和的表层水体中，即使是活体，翼足类的壳体溶解速度也非常快[9]。如图所示，当活着的亚北极区翼足类动物——尖菱蝶螺（*Clio pyramidata*）暴露于与 IS92a 排放情景下预测的 2100 年南大洋表层水体的不饱和度相当的环境中时，48 小时内其壳口生长边缘就会发生明显的溶解（图 6）。蚀坑形成于壳口边缘（通常厚度 7 μm 左右）的壳体表面，有小于 1 μm 的外壳（柱状层）向后剥蚀（图 6c），使下面的文石柱体暴露出来而受到溶解。在本实验中，我们共研究了 14 个个体。尽管所有个体都是活着的，但是沿其生长边缘仍都出现了类似的溶解现象。我们认为，倘若尖菱蝶螺不能让其保护壳继续生长，它将无法在文石不饱和的水体中生存。

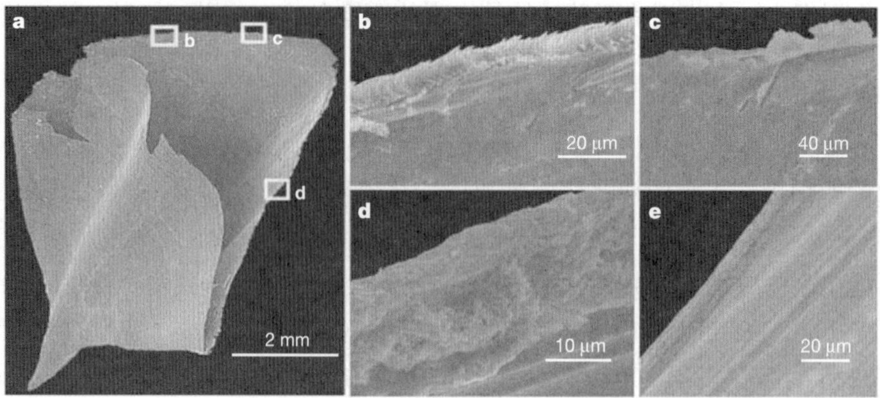

Fig. 6. Shell dissolution in a live pteropod. **a–d**, Shell from a live pteropod, *Clio pyramidata*, collected from the subarctic Pacific and kept in water undersaturated with respect to aragonite for 48 h. The whole shell (**a**) has superimposed white rectangles that indicate three magnified areas: the shell surface (**b**), which reveals etch pits from dissolution and resulting exposure of aragonitic rods; the prismatic layer (**c**), which has begun to peel back, increasing the surface area over which dissolution occurs; and the aperture region (**d**), which reveals advanced shell dissolution when compared to a typical *C. pyramidata* shell not exposed to undersaturated conditions (**e**).

If the response of other high-latitude pteropod species to aragonite undersaturation is similar to that of *C. pyramidata*, we hypothesize that these pteropods will not be able to adapt quickly enough to live in the undersaturated conditions that will occur over much of the high-latitude surface ocean during the twenty-first century. Their distributional ranges would then be reduced both within the water column, disrupting vertical migration patterns, and latitudinally, imposing a shift towards lower-latitude surface waters that remain supersaturated with respect to aragonite. At present, we do not know if pteropod species endemic to polar regions could disappear altogether, or if they can make the transition to live in warmer, carbonate-rich waters at lower latitudes under a different ecosystem. If pteropods are excluded from polar and subpolar regions, their predators will be affected immediately. For instance, gymnosomes are zooplankton that feed exclusively on shelled pteropods[33,39]. Pteropods also contribute to the diet of diverse carnivorous zooplankton, myctophid and nototheniid fishes[40-42], North Pacific salmon[43,44], mackerel, herring, cod and baleen whales[45].

Surface dwelling calcitic plankton, such as foraminifera and coccolithophorids, may fare better in the short term. However, the beginnings of high-latitude calcite undersaturation will only lag that for aragonite by 50–100 years. The diverse benthic calcareous organisms in high-latitude regions may also be threatened, including cold-water corals which provide essential fish habitat[46]. Cold-water corals seem much less abundant in the North Pacific than in the North Atlantic[46], where the aragonite saturation horizon is much deeper (Fig. 2). Moreover, some important taxa in Arctic and Antarctic benthic communities secrete magnesian calcite, which can be more soluble than aragonite. These include gorgonians[46], coralline red algae and echinoderms (sea urchins)[47]. At $2 \times CO_2$, juvenile echinoderms stopped growing and produced more brittle and fragile exoskeletons in a subtropical six-

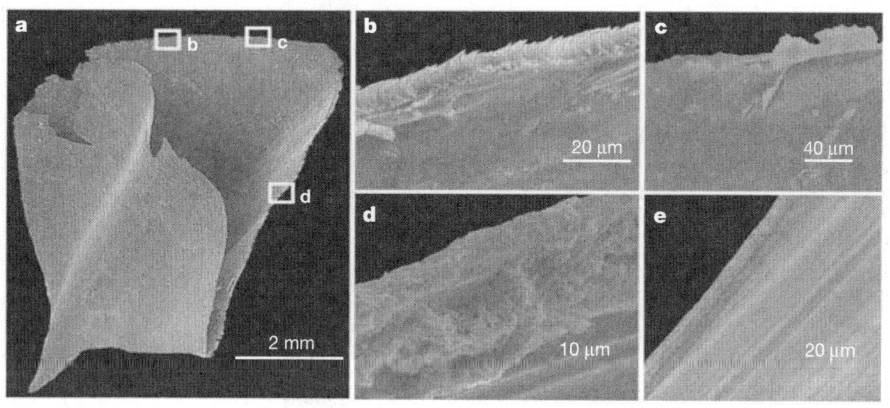

图 6. 一个活体翼足类壳体的溶解。图 a ~ d 为活体翼足类尖菱蝶螺的壳体，采自亚北极区太平洋海域，将其置于文石不饱和水中放置 48 小时。整个壳体（图 **a**）出现的白色矩形，对应着三处放大区域：壳体表面（图 **b**），出现深蚀坑并导致文石柱体暴露出来；柱状层（图 **c**），开始向后剥蚀，导致易被溶解的表面积增大；口缘区域（图 **d**），与未暴露到不饱和环境下的典型尖菱蝶螺（图 **e**）相比，最外缘的壳已被溶解。

倘若高纬翼足类的其他种对文石不饱和的响应也与尖菱蝶螺一致，我们就可以认为，这些翼足类将无法足够快地适应这种变化，以在不饱和环境中继续生存，而在 21 世纪大部分高纬海域的表层水体将呈现为不饱和状态。因此，到那时翼足类的分布范围将大大减小，不管是在垂向上（扰乱其垂向迁移的固有模式）还是纬向上（导致它们向仍处于文石饱和状态的低纬表层水域迁移）。目前我们并不清楚，极地海域的翼足类到底会全部消失，还是可以迁移至较温暖的水域生存，要知道低纬海域的富碳酸盐水体处于完全不同的生态系统之下。倘若极地和副极地海域的翼足类都不存在了，那么以它们为食的那些动物很快也将受到影响。例如，海天使就是一种专以带壳翼足类为食的浮游动物 [33,39]。翼足类还是其他许多食肉浮游动物，如蛇鼻鱼与南极鱼 [40–42]、北太平洋鲑鱼 [43,44]、鲭鱼、鲱鱼、鳕鱼以及须鲸等 [45] 的食物来源。

生活于表层的方解石质浮游生物，如有孔虫和颗石藻在短期内还不会受到影响。然而，高纬海域的方解石不饱和也仅会比文石不饱和滞后 50 ~ 100 年。除此之外，高纬地区的各种底栖钙质生物也将受到威胁，其中就包括冷水种珊瑚，而它又是许多鱼类基本生存居所的提供者 [46]。北太平洋海域的冷水种珊瑚远没有北大西洋丰富 [46]，因为北大西洋中文石的饱和深度要深得多（图 2）。不仅如此，南北两极底栖生物群中的一些重要种属分泌的是镁方解石，它比文石更易溶解。这些种属包括柳珊瑚 [46]、红珊瑚藻以及棘皮动物（海胆）[47] 等。根据在副热带地区进行的为期 6 个月的养殖实验发现，当大气 CO_2 浓度达到现今的两倍时，棘皮动物的幼体就会停

month manipulative experiment[48]. However, the responses of high-latitude calcifiers to reduced [CO_3^{2-}] have generally not been studied. Yet experimental evidence from many lower-latitude, shallow-dwelling calcifiers reveals a reduced ability to calcify with a decreasing carbonate saturation state[9]. Given that at $2 \times CO_2$, calcification rates in some shallow-dwelling calcareous organisms may decline by up to 50% (ref. 9), some calcifiers could have difficulty surviving long enough even to experience undersaturation. Certainly, they have not experienced undersaturation for at least the last 400,000 years[49], and probably much longer[50].

Changes in high-latitude seawater chemistry that will occur by the end of the century could well alter the structure and biodiversity of polar ecosystems, with impacts on multiple trophic levels. Assessing these impacts is impeded by the scarcity of relevant data.

(**437**, 681-686; 2005)

James C. Orr[1], Victoria J. Fabry[2], Olivier Aumont[3], Laurent Bopp[1], Scott C. Doney[4], Richard A. Feely[5], Anand Gnanadesikan[6], Nicolas Gruber[7], Akio Ishida[8], Fortunat Joos[9], Robert M. Key[10], Keith Lindsay[11], Ernst Maier-Reimer[12], Richard Matear[13], Patrick Monfray[1]†, Anne Mouchet[14], Raymond G. Najjar[15], Gian-Kasper Plattner[7,9], Keith B. Rodgers[1,16]†, Christopher L. Sabine[5], Jorge L. Sarmiento[10], Reiner Schlitzer[17], Richard D. Slater[10], Ian J. Totterdell[18]†, Marie-France Weirig[17], Yasuhiro Yamanaka[8] & Andrew Yool[18]

[1] Laboratoire des Sciences du Climat et de l'Environnement, UMR CEA-CNRS, CEA Saclay, F-91191 Gif-sur-Yvette, France
[2] Department of Biological Sciences, California State University San Marcos, San Marcos, California 92096-0001, USA
[3] Laboratoire d'Océanographie et du Climat: Expérimentations et Approches Numériques (LOCEAN), Centre IRD de Bretagne, F-29280 Plouzané, France
[4] Woods Hole Oceanographic Institution, Woods Hole, Massachusetts 02543-1543, USA
[5] National Oceanic and Atmospheric Administration (NOAA)/Pacific Marine Environmental Laboratory, Seattle, Washington 98115-6349, USA
[6] NOAA/Geophysical Fluid Dynamics Laboratory, Princeton, New Jersey 08542, USA
[7] Institute of Geophysics and Planetary Physics, UCLA, Los Angeles, California 90095-4996, USA
[8] Frontier Research Center for Global Change, Yokohama 236-0001, Japan
[9] Climate and Environmental Physics, Physics Institute, University of Bern, CH-3012 Bern, Switzerland
[10] Atmospheric and Oceanic Sciences (AOS) Program, Princeton University, Princeton, New Jersey 08544-0710, USA
[11] National Center for Atmospheric Research, Boulder, Colorado 80307-3000, USA
[12] Max Planck Institut für Meteorologie, D-20146 Hamburg, Germany
[13] CSIRO Marine Research and Antarctic Climate and Ecosystems CRC, Hobart, Tasmania 7001, Australia
[14] Astrophysics and Geophysics Institute, University of Liege, B-4000 Liege, Belgium
[15] Department of Meteorology, Pennsylvania State University, University Park, Pennsylvania 16802-5013, USA
[16] LOCEAN, Université Pierre et Marie Curie, F-75252 Paris, France
[17] Alfred Wegener Institute for Polar and Marine Research, D-27515 Bremerhaven, Germany
[18] National Oceanography Centre Southampton, Southampton SO14 3ZH, UK
† Present addresses: Laboratoire d'Etudes en Géophysique et Océanographie Spatiales, UMR 5566 CNES-CNRS-IRD-UPS, F-31401 Toulouse, France (P.M.); AOS Program, Princeton University, Princeton, New Jersey 08544-0710, USA (K.B.R.); The Met Office, Hadley Centre, FitzRoy Road, Exeter EX1 3PB, UK (I.J.T.)

Received 15 June; accepted 29 July 2005.

止生长，同时其所形成的外骨骼更加脆弱和易碎[48]。不过高纬海域方解石钙质生物对 [CO_3^{2-}] 降低的响应目前还未得到广泛研究。尽管如此，来自低纬海域浅水钙质生物的大量实验证据已表明，随着碳酸盐饱和度的降低，方解石钙质生物的钙化能力会不断下降[9]。假设 CO_2 升高一倍，某些浅水钙质生物的钙化速率就可能降低 50%（参考文献 9），因此，在这种条件下某些种群可能都无法存活至水体方解石真正达到不饱和状态的那一天。当然，在过去 40 万年[49]甚至更长的时间[50]里，它们还从未经历过方解石不饱和环境。

本世纪末将要发生的高纬海域海水化学性质的变化可能会极大地改变极地生态系统的结构及其生物多样性，同时会对多个营养级都产生影响。由于缺乏相关数据，对上述影响的量化评估还无法实现。

（齐红艳 翻译；高树基 审稿）

References:

1. Haugan, P. M. & Drange, H. Effects of CO_2 on the ocean environment. *Energy Convers. Mgmt* **37**, 1019-1022 (1996).
2. Brewer, P. G. Ocean chemistry of the fossil fuel CO_2 signal: the haline signal of "business as usual". *Geophys. Res. Lett.* **24**, 1367-1369 (1997).
3. Gattuso, J.-P., Frankignoulle, M., Bourge, I., Romaine, S. & Buddemeier, R. W. Effect of calcium carbonate saturation of seawater on coral calcification. *Glob. Planet. Change* **18**, 37-46 (1998).
4. Kleypas, J. A. *et al.* Geochemical consequences of increased atmospheric carbon dioxide on coral reefs. *Science* **284**, 118-120 (1999).
5. Langdon, C. *et al.* Effect of elevated CO_2 on the community metabolism of an experimental coral reef. *Glob. Biogeochem. Cycles* **17**, 1011, doi:10.1029/2002GB001941 (2003).
6. Riebesell, U. *et al.* Reduced calcification of marine plankton in response to increased atmospheric CO_2. *Nature* **407**, 364-367 (2000).
7. Zondervan, I., Zeebe, R., Rost, B. & Riebesell, U. Decreasing marine biogenic calcification: A negative feedback on rising atmospheric p_{CO_2}. *Glob. Biogeochem. Cycles* **15**, 507-516 (2001).
8. Broecker, W. S. & Peng, T.-H. Fate of fossil fuel carbon dioxide and the global carbon budget. *Science* **206**, 409-418 (1979).
9. Feely, R. A. *et al.* The impact of anthropogenic CO_2 on the $CaCO_3$ system in the oceans. *Science* **305**, 362-366 (2004).
10. Caldeira, K. & Wickett, M. E. Anthropogenic carbon and ocean pH. *Nature* **425**, 365 (2003).
11. Urban-Rich, J., Dagg, M. & Peterson, J. Copepod grazing on phytoplankton in the Pacific sector of the Antarctic Polar Front. *Deep-Sea Res. II* **48**, 4223-4246 (2001).
12. Kobayashi, H. A. Growth cycle and related vertical distribution of the thecosomatous pteropod *Spiratella "Limacina" helicina* in the central Arctic Ocean. *Mar. Biol.* **26**, 295-301 (1974).
13. Pakhomov, E. A., Verheye, H. M., Atkinson, A., Laubscher, R. K. & Taunton-Clark, J. Structure and grazing impact of the mesozooplankton community during late summer 1994 near South Georgia, Antarctica. *Polar Biol.* **18**, 180-192 (1997).
14. Fabry, V. J. Aragonite production by pteropod molluscs in the subarctic Pacific. *Deep-Sea Res. I* **36**, 1735-1751 (1989).
15. Bathmann, U., Noji, T. T. & von Bodungen, B. Sedimentation of pteropods in the Norwegian Sea in autumn. *Deep-Sea Res.* **38**, 1341-1360 (1991).
16. Key, R. M. *et al.* A global ocean carbon climatology: Results from Global Data Analysis Project (GLODAP). *Glob. Biogeochem. Cycles* **18**, 4031, doi:10.1029/2004GB002247 (2004).
17. Sabine, C. L. *et al.* The ocean sink for anthropogenic CO_2. *Science* **305**, 367-370 (2004).
18. Gruber, N. Anthropogenic CO_2 in the Atlantic Ocean. *Glob. Biogeochem. Cycles* **12**, 165-191 (1998).
19. Sarmiento, J. L., Orr, J. C. & Siegenthaler, U. A perturbation simulation of CO_2 uptake in an ocean general circulation model. *J. Geophys. Res.* **97**, 3621-3645 (1992).
20. Orr, J. C. *et al.* Estimates of anthropogenic carbon uptake from four three-dimensional global ocean models. *Glob. Biogeochem. Cycles* **15**, 43-60 (2001).
21. Kalnay, E. *et al.* The NCEP/NCAR 40-year reanalysis project. *Bull. Am. Meteorol. Soc.* **77**, 437-471 (1996).
22. Feely, R. A. *et al.* Winter-summer variations of calcite and aragonite saturation in the northeast Pacific. *Mar. Chem.* **25**, 227-241 (1988).
23. Feely, R. A., Byrne, R. H., Betzer, P. R., Gendron, J. F. & Acker, J. G. Factors influencing the degree of saturation of the surface and intermediate waters of the North Pacific Ocean with respect to aragonite. *J. Geophys. Res.* **89**, 10631-10640 (1984).
24. Sarmiento, J. L. *et al.* Response of ocean ecosystems to climate warming. *Glob. Biogeochem. Cycles* **18**, 3003, doi:10.1029/2003GB002134 (2004).
25. Bopp, L. *et al.* Potential impact of climate change of marine export production. *Glob. Biogeochem. Cycles* **15**, 81-99 (2001).
26. Sarmiento, J. L., Le Quéré, C. & Pacala, S. Limiting future atmospheric carbon dioxide. *Glob. Biogeochem. Cycles* **9**, 121-137 (1995).
27. Heinze, C. Simulating oceanic $CaCO_3$ export production in the greenhouse. *Geophys. Res. Lett.* **31**, L16308, doi:10.1029/2004GL020613 (2004).
28. Iglesias-Rodriguez, M. D. *et al.* Representing key phytoplankton functional groups in ocean carbon cycle models: Coccolithophorids. *Glob. Biogeochem. Cycles* **16**, 1100, doi:10.1029/2001GB001454 (2002).
29. Berner, R. A. in *The Fate of Fossil Fuel CO_2 in the Oceans* (eds Andersen, N. R. & Malahoff, A.) 243-260 (Plenum, New York, 1977).
30. Fabry, V. J. Shell growth rates of pteropod and heteropod molluscs and aragonite production in the open ocean: Implications for the marine carbonate system. *J. Mar. Res.* **48**, 209-222 (1990).
31. Bé, A. W. H. & Gilmer, R. W. *Oceanic Micropaleontology* Vol. 1 (ed. Ramsey, A.) 733-808 (Academic, London, 1977).
32. Dadon, J. R. & de Cidre, L. L. The reproductive cycle of the Thecosomatous pteropod *Limacina retroversa* in the western South Atlantic. *Mar. Biol.* **114**, 439-442 (1992).
33. Seibel, B. A. & Dierssen, H. M. Cascading trophic impacts of reduced biomass in the Ross Sea, Antarctica: Just the tip of the iceberg? *Biol. Bull.* **205**, 93-97 (2003).
34. Accornero, A., Manno, C., Esposito, F. & Gambi, M. C. The vertical flux of particulate matter in the polyna of Terra Nova Bay. Part II. Biological components. *Antarct. Sci.* **15**, 175-188 (2003).
35. Collier, R., Dymond, J., Susumu Honjo, S. M., Francois, R. & Dunbar, R. The vertical flux of biogenic and lithogenic material in the Ross Sea: moored sediment trap observations 1996-1998. *Deep-Sea Res. II* **47**, 3491-3520 (2000).
36. Honjo, S., Francois, R., Manganini, S., Dymond, J. & Collier, R. Particle fluxes to the interior of the Southern Ocean in the western Pacific sector along 170°W. *Deep-Sea Res. II* **47**, 3521-3548 (2000).
37. Betzer, P. R., Byrne, R., Acker, J., Lewis, C. S. & Jolley, R. R. The oceanic carbonate system: a reassessment of biogenic controls. *Science* **226**, 1074-1077 (1984).
38. Byrne, R. H., Acker, J. G., Betzer, P. R., Feely, R. A. & Cates, M. H. Water column dissolution of aragonite in the Pacific Ocean. *Nature* **312**, 321-326 (1984).

39. Lalli, C. M. Structure and function of the buccal apparatus of *Clione limacina* (Phipps) with a review of feeding in gymnosomatous pteropods. *J. Exp. Mar. Biol. Ecol.* **4**, 101-118 (1970).

40. Foster, B. A. & Montgomery, J. C. Planktivory in benthic nototheniid fish in McMurdo Sound, Antarctica. *Environ. Biol. Fish.* **36**, 313-318 (1993).

41. Pakhomov, E., Perissinotto, A. & McQuaid, C. D. Prey composition and daily rations of myctophid fishes in the Southern Ocean. *Mar. Ecol. Prog. Ser.* **134**, 1-14 (1996).

42. La Mesa, M., Vacchi, M. & Sertorio, T. Z. Feeding plasticity of *Trematomus newnesi* (Pisces, Nototheniidae) in Terra Nova Bay, Ross Sea, in relation to environmental conditions. *Polar Biol.* **23**, 38-45 (2000).

43. Willette, T. M. *et al.* Ecological processes influencing mortality of juvenile pink salmon (*Oncorhynchus gorbuscha*) in Prince William Sound, Alaska. *Fish. Oceanogr.* **10**, 14-41 (2001).

44. Boldt, J. L. & Haldorson, L. J. Seasonal and geographical variation in juvenile pink salmon diets in the Northern Gulf of Alaska and Prince William Sound. *Trans. Am. Fisheries Soc.* **132**, 1035-1052 (2003).

45. Lalli, C. M. & Gilmer, R. *Pelagic Snails* (Stanford Univ. Press, Stanford, 1989).

46. Freiwald, A., Fosså, J. H., Grehan, A., Koslow, T. & Roberts, J. M. *Cold-water Coral Reefs: Out of Sight—No Longer Out of Mind* (No. 22 in Biodiversity Series, UNEP-WCMC, Cambridge, UK, 2004).

47. Dayton, P. K. in *Polar Oceanography, Part B: Chemistry, Biology and Geology* (ed. Smith, W. O.) 631-685 (Academic, San Diego, 1990).

48. Shirayama, Y. & Thornton, H. Effect of increased atmospheric CO_2 on shallow-water marine benthos. *J. Geophys. Res.* **110**, C09S09, doi:10.1029/2004JC002561 (2005).

49. Petit, J. R. *et al.* Climate and atmospheric history of the past 420,000 years from the Vostok ice core, Antarctica. *Nature* **399**, 429-436 (1999).

50. Pearson, P. N. & Palmer, M. R. Middle Eocene seawater pH and atmospheric carbon dioxide concentrations. *Science* **284**, 1824-1826 (1999).

Supplementary Information is linked to the online version of the paper at www.nature.com/nature.

Acknowledgements. We thank M. Gehlen for discussions, and J.-M. Epitalon, P. Brockmann and the Ferret developers for help with analysis. All but the climate simulations were made as part of the OCMIP project, which was launched in 1995 by the Global Analysis, Integration and Modelling (GAIM) Task Force of the International Geosphere–Biosphere Programme (IGBP) with funding from NASA (National Aeronautics and Space Administration). OCMIP-2 was supported by the European Union Global Ocean Storage of Anthropogenic Carbon (EU GOSAC) project and the United States JGOFS Synthesis and Modeling Project funded through NASA. The interannual simulation was supported by the EU Northern Ocean Carbon Exchange Study (NOCES) project, which is part of OCMIP-3.

Author Information. Reprints and permissions information is available at npg.nature.com/reprintsandpermissions. The authors declare no competing financial interests. Correspondence and requests for materials should be addressed to J.C.O. (orr@cea.fr).

Characterization of the 1918 Influenza Virus Polymerase Genes

J. K. Taubenberger *et al.*

Editor's Note

The 1918 "Spanish" influenza pandemic killed more people—about 50 million—than the First World War. There are fears that some recent flu strains could be similarly lethal. Here US virologist Jeffrey Taubenberger and colleagues analyse virus samples extracted from the lung tissue of one victim. They find that the lethal virus was not the result of an animal and human strain merging their DNA, but instead a bird strain that evolved to infect humans. The 1918 virus has some similarities to H5N1 avian flu, which can kill humans but cannot spread between them, and another avian strain, H7N7, which has also killed people, helping researchers identify what separates the relatively mild flu strains from the killers.

The influenza A viral heterotrimeric polymerase complex (PA, PB1, PB2) is known to be involved in many aspects of viral replication and to interact with host factors[1], thereby having a role in host specificity[2,3]. The polymerase protein sequences from the 1918 human influenza virus differ from avian consensus sequences at only a small number of amino acids, consistent with the hypothesis that they were derived from an avian source shortly before the pandemic. However, when compared to avian sequences, the nucleotide sequences of the 1918 polymerase genes have more synonymous differences than expected, suggesting evolutionary distance from known avian strains. Here we present sequence and phylogenetic analyses of the complete genome of the 1918 influenza virus[4-8], and propose that the 1918 virus was not a reassortant virus (like those of the 1957 and 1968 pandemics[9,10]), but more likely an entirely avian-like virus that adapted to humans. These data support prior phylogenetic studies suggesting that the 1918 virus was derived from an avian source[11]. A total of ten amino acid changes in the polymerase proteins consistently differentiate the 1918 and subsequent human influenza virus sequences from avian virus sequences. Notably, a number of the same changes have been found in recently circulating, highly pathogenic H5N1 viruses that have caused illness and death in humans and are feared to be the precursors of a new influenza pandemic. The sequence changes identified here may be important in the adaptation of influenza viruses to humans.

INFLUENZA A viruses cause annual outbreaks in humans and domestic animals. Periodically, new strains emerge in humans that cause global pandemics. The severe "Spanish" influenza pandemic of 1918–1919 infected hundreds of millions, and resulted in the death of approximately 50 million people[12]. We have previously used phylogenetic

1918 年流感病毒聚合酶基因的特点

陶本伯格等

编者按

1918 年的“西班牙”流感大流行造成的死亡人数超过第一次世界大战，约为 5,000 万。人们担心最近的一些流感病毒株可能具有同样的致命性。在这里，美国病毒学家杰弗里·陶本伯格及其同事分析了从一名受害者的肺组织中提取的病毒样本。他们发现这种致命的病毒不是动物和人类病毒株整合 DNA 的结果，而是由禽流感病毒株进化而来感染人类的。1918 年的流感病毒与 H5N1 禽流感病毒（它们可以感染并致人死亡但不能在人与人之间传播）以及另一种禽类病毒 H7N7（同样可以致人死亡）有一些相似之处，这有助于研究人员确定相对温和的流感病毒株与高致病性病毒株的区别。

已知甲型流感病毒异三聚体聚合酶复合体（PA、PB1、PB2）在病毒复制的多个方面发挥作用，并和宿主因子发生作用[1]，因此在宿主特异性中发挥作用[2,3]。1918 年人流感病毒的聚合酶蛋白序列只有少数氨基酸与禽流感病毒共有序列不同，从而支持了它们在大流行前不久由禽流感病毒产生的假说。但是，与禽流感序列相比较，1918 年病毒的聚合酶基因的核苷酸序列同义性差异大于预期，提示其与已知的禽流感毒株存在进化距离。这里我们给出了 1918 年流感病毒整个基因组序列并进行了系统发育分析[4-8]，提出 1918 年病毒不是一种基因重组病毒（如 1957 年和 1968 年大流行的病毒[9,10]），而更可能是能够感染人类的完全类似禽流感（病毒）的病毒。这些数据支持了先期系统发育研究结果，并表明 1918 年病毒来源于禽流感[11]。聚合酶蛋白中总共 10 个氨基酸的突变将 1918 年病毒以及随后的人类流感病毒序列同禽流感病毒序列区别开来。显而易见地，许多相同的突变在最近流行的高致病性 H5N1 病毒中被找到，该病毒已经造成了人类的疾病和死亡，并让人担心这是新的流感大流行的先兆。这里鉴定的序列突变可能对于流感病毒适应人类的过程非常重要。

甲型流感病毒每年都在人类和家畜中引起疫情暴发。新的毒株周期性地在人类中出现并引起全球大流行。1918 ~ 1919 年严重的“西班牙”流感大流行波及了数以亿计的人口，并导致了大约五千万人的死亡[12]。我们之前已经用系统发育分析来

analyses to help understand the origin of the pandemic virus[8,11]; functional studies to understand the pathogenicity of the 1918 virus are underway[6,13-17]. Recent data have shown that viral constructs bearing the 1918 haemagglutinin gene are pathogenic in a mouse model, but the genetic basis of this observation has not yet been mapped[6,13-17]. The overall goals of this project have been to understand the origin and unusual virulence of the 1918 influenza virus.

The influenza virus A polymerase functions as a heterotrimer formed by the PB2, PB1 and PA proteins (see ref. 1 for a review). An additional small open reading frame has recently been identified, coding for a peptide (PB1-F2) that is thought to play a role in virus-induced cell death[18]. It is not yet clear how the polymerase complex must change to adapt to a new host[3]. A single amino acid change in PB2, E627K, was shown (1) to be important for mammalian adaptation[2,3], (2) to distinguish highly pathogenic avian influenza (HPAI) H5N1 viruses in mice[19], and (3) to be present in the single fatal human infection during the HPAI H7N7 outbreak in the Netherlands in 2003 (ref. 20), and in some recent H5N1 isolates from humans in Vietnam and Thailand and wild birds in China[21-23].

The open reading frame sequences of segment 1 (PB2), segment 2 (PB1) and segment 3 (PA) of A/Brevig Mission/1/1918, and theoretical translations of the four identified reading frames, are shown in Supplementary Fig. 1a–c. The 1918 PB2 protein contained five changes from the avian consensus sequence (Table 1). Of these, A199S is in the area mapped as the PB1 binding site, and the L475M change is in a nuclear localization signal[24-26]. Three other changes at residues 567, 627 and 702 occur at sites that are not in known functional domains.

Table 1. Amino acid residues distinguishing human and avian influenza polymerases

Gene	Residue no.	Avian	1918	Human H1N1	Human H2N2	Human H3N2	Classical swine	Equine
PB2	199	A	S	S	S	S	S	A
PB2	475	L	M	M	M	M	M	L
PB2	567	D	N	N	N	N	D	D
PB2	627	E	K	K	K	K	K	E
PB2	702	K	R	R*	R	R	R	K
PB1	375	N/S/T†	S	S	S	S	S	S
PA	55	D	N	N	N	N	N	N
PA	100	V	A	A	A	A	V	A
PA	382	E	D	D	D	D	D	E
PA	552	T	S	S	S	S	S	T

* All human H1N1 PB2 sequences have an Arg residue at position 702, except that two out of three A/PR/8/34 sequences have a Lys residue.

† The majority of avian sequences have an Asn residue at position 375 of PB1, 18% have a Ser residue, 13% a Thr residue.

帮助了解这次大流行病毒的来源[8,11]；了解 1918 年流感病毒致病性的功能研究正在进行中[6,13-17]。最近的数据显示带有 1918 年血凝素基因的重组病毒在小鼠模型中具有致病性，但是这个现象的基因基础尚未明确[6,13-17]。这个项目的总体目标就是了解 1918 年流感病毒的起源和超常毒力。

甲型流感病毒聚合酶以 PB2、PB1 和 PA 蛋白异三聚体的形式发挥功能（见文献 1 综述）。最近发现了一个小的开放阅读框，它编码一种被认为在病毒诱导的细胞死亡中发挥作用的肽（PB1-F2）[18]。目前还不清楚聚合酶复合体必须如何改变以适应新的宿主[3]。PB2 中的单个氨基酸突变——E627K 被发现：(1) 对哺乳动物的适应性非常重要[2,3]；(2) 可用于鉴别小鼠中的高致病性禽流感病毒 H5N1[19]；(3) 出现在 2003 年荷兰高致病性禽流感 H7N7 暴发期间的单例人致死感染（文献 20），以及最近从越南和泰国的人类及中国的野生鸟类中分离出的一些 H5N1 中[21-23]。

A/布雷维格教区/1/1918，PB2、PB1 和 PA 的开放阅读框序列以及四个已识别的阅读框的理论蛋白翻译详见补充材料图 1a ~ 1c。1918 年 PB2 蛋白含有五个相对于禽流感病毒共有序列的突变（表 1）。其中，A199S 位于 PB1 结合位点区域，L475M 突变位于核定位信号中[24-26]。567、627 和 702 三个突变都位于目前已知的功能区域之外。

表 1. 区分人类和禽类流感聚合酶的氨基酸残基

基因	残基号	禽类	1918	人 H1N1	人 H2N2	人 H3N2	猪	马
PB2	199	A	S	S	S	S	S	A
PB2	475	L	M	M	M	M	M	L
PB2	567	D	N	N	N	N	D	D
PB2	627	E	K	K	K	K	K	E
PB2	702	K	R	R*	R	R	R	K
PB1	375	N/S/T†	S	S	S	S	S	S
PA	55	D	N	N	N	N	N	N
PA	100	V	A	A	A	A	V	A
PA	382	E	D	D	D	D	D	E
PA	552	T	S	S	S	S	S	T

* 除了三个 A/PR/8/34 序列中的两个含有赖氨酸残基，所有的人类 H1N1 PB2 序列在位点 702 含有精氨酸残基。
† 大部分禽类序列在 PB1 的位点 375 含有天冬酰胺残基，18% 含有丝氨酸残基，13% 含有苏氨酸残基。

The 1918 PB1 protein differed from the avian consensus by seven residues (one of which is shown in Table 1; see also Supplementary Fig. 2). Of these, K54R is in the overlapping binding domains for complementary (c)RNA and viral (v)RNA. Changes at residues 375, 383 and 473 all occur in between the four conserved polymerase motifs in the cRNA binding domain[27], and changes at residues 576, 645 and 654 occur in the vRNA binding domain[28].

Seven changes were noted in the 1918 PA protein compared with the avian consensus (four of which are shown in Table 1, the other three being C241Y, K312R and I322V). The C241Y change occurs in a nuclear localization signal, but the other six changes (at residues 55, 100, 312, 322, 382 and 552) occur at sites outside of known functional domains[24-26].

Representative phylogenetic analyses of the three polymerase genes are shown in Figs 1–3. The 1918 human pandemic viral polymerase genes were compared to representative avian influenza genes with regards to transition/transversion (T_i/T_v) ratio, synonymous/non-synonymous (S/N) ratio, and the numbers of differences at fourfold degenerate sites (defined in ref. 11). T_i/T_v ratios for most comparisons using the 1918 viral genes and representative sequences of either North American or Eurasian avian genes yielded values between 2 and 4. This range was similar to that observed for comparisons of various avian genes with one another, except for the PB1 gene. For PB1, comparisons of the 1918 viral gene with avian virus PB1 genes was always close to 2, whereas comparisons of various avian genes with one another were in the range of 6–10. There were fewer transversions in comparisons between avian PB1 genes than in comparisons between avian PB1 and 1918 human virus PB1, probably reflecting that transversions more often lead to non-synonymous changes.

S/N ratios for most comparisons using the 1918 viral genes and representative sequences of either North American or Eurasian avian genes usually yielded values in the range of 7–16 for both the PA and PB2 genes, as is the case for most avian versus avian PA and PB2 gene comparisons. Like the T_i/T_v ratios, the S/N ratios were somewhat higher with the PB1 gene (most of the comparisons yielded ratios in the range of 16–25), owing to a smaller number of non-synonymous changes in comparisons of avian PB1 genes with one another. These findings may reflect a more conservative evolution of PB1 in birds.

A subset of synonymous differences occurs at sites that are fourfold degenerate (that is, where a substitution with any base does not result in an amino acid replacement). As these sites are not subject to selective pressure at the protein level, base substitutions at many fourfold degenerate sites may accumulate rapidly. If influenza virus genes have been evolving in birds for long enough to reach evolutionary stasis, as is suggested by the high S/N ratios described above, one would predict that at many of the sites where fourfold degeneracy is possible, all four bases would be present in the avian clade unless the constraints of RNA secondary structure limit the accumulation of synonymous changes. In fact, when avian sequences from geographically distinct lineages (North American versus European) were compared, the per cent difference at fourfold degenerate sites yielded values in the 27–38% range. In contrast, calculating the per cent difference at fourfold degenerate

在蛋白序列上，1918 病毒株的 PB1 蛋白与禽流感株有七个氨基酸残基的差异(其中一个在表 1 中显示，也可见于补充材料图 2)。其中，K54R 位于互补(c)RNA 和病毒(v)RNA 的重叠结合域。375、383 和 473 处的突变都位于 cRNA 结合域的四个保守的聚合酶基序之间 [27]，而 576、645 和 654 处的突变位于 vRNA 结合域 [28]。

1918 病毒株 PA 蛋白与禽流感株在蛋白序列上的差别一共有七处(其中四个在表 1 中显示，其他三个是 C241Y、K312R 和 I322V)。C241Y 突变发生在核定位信号中，但是其他六个突变(55、100、312、322、382 和 552)均在已知的功能区域之外 [24-26]。

三个聚合酶基因的进化分析呈现在图 1 ~ 图 3 中。1918 年人类大流行病毒聚合酶基因与代表性的禽流感病毒基因进行了转换/颠换(T_i/T_v)比、同义/非同义(S/N)比和四倍简并位点(在文献 11 中定义)差别数目的对比。大部分 1918 年病毒基因和北美或者欧亚禽流感病毒基因的代表性序列进行对比获得的 T_i/T_v 数值都在 2 和 4 之间。该范围与不同禽类携带禽流感病毒基因之间对比所得到的结果类似，除了 PB1 基因。对 PB1 基因来说，1918 年病毒基因与禽流感基因的对比始终接近 2，而不同禽类携带禽流感基因之间的对比都在 6 ~ 10 的范围内。禽流感 PB1 基因之间对比得到的颠换数少于禽流感和 1918 年人流感病毒 PB1 基因之间对比得出的颠换数，很可能反映了颠换更常导致非同义突变。

对 PA 和 PB2 基因来说，大部分用 1918 年病毒基因和北美或者欧亚任一禽类流感病毒基因的代表性序列进行对比获得的 S/N 比都在 7 ~ 16 的范围内，大部分禽流感病毒之间对比的结果范围也是一样。就像 T_i/T_v 比一样，PB1 基因的 S/N 比稍微高一些(大部分对比得出的数值在 16 ~ 25 范围内)，因为相对于其他来说，禽流感病毒 PB1 基因的非同义突变数量少一些。这些发现可能反映了禽流感病毒 PB1 进化过程中的保守性更高。

一部分同义突变发生在四倍简并位点(即任何碱基的置换都不会造成氨基酸的替换)。由于这些位点不会在蛋白质水平面临选择压力，许多四倍简并位点的碱基置换可能会迅速地累积。如果流感病毒基因如上面所描述的高 S/N 比所揭示的那样，已经在鸟类中存在足够长的时间并达到进化停滞，我们能够预计在可能出现四倍简并的许多位点，所有四种碱基都可能出现在禽类分支上，除非 RNA 二级结构的限制约束了这种同义突变的积累。事实上，当对比地理上不同谱系(北美和欧亚)的禽类来源流感病毒序列时，四倍简并位点的百分比差别都在 27% ~ 38% 之间。相反的，对比 1918 年病毒的 PA、PB1、PB2 基因序列和禽流感病毒序列时，计算出四倍简

sites in comparisons of the 1918 viral PA, PB1 and PB2 gene sequences with avian sequences yielded consistently higher values (range 41–51%) for all three genes. As with the other 1918 genes[11], this suggests that the donor source of the 1918 virus was in evolutionary isolation from those avian influenza viruses currently represented in the databases.

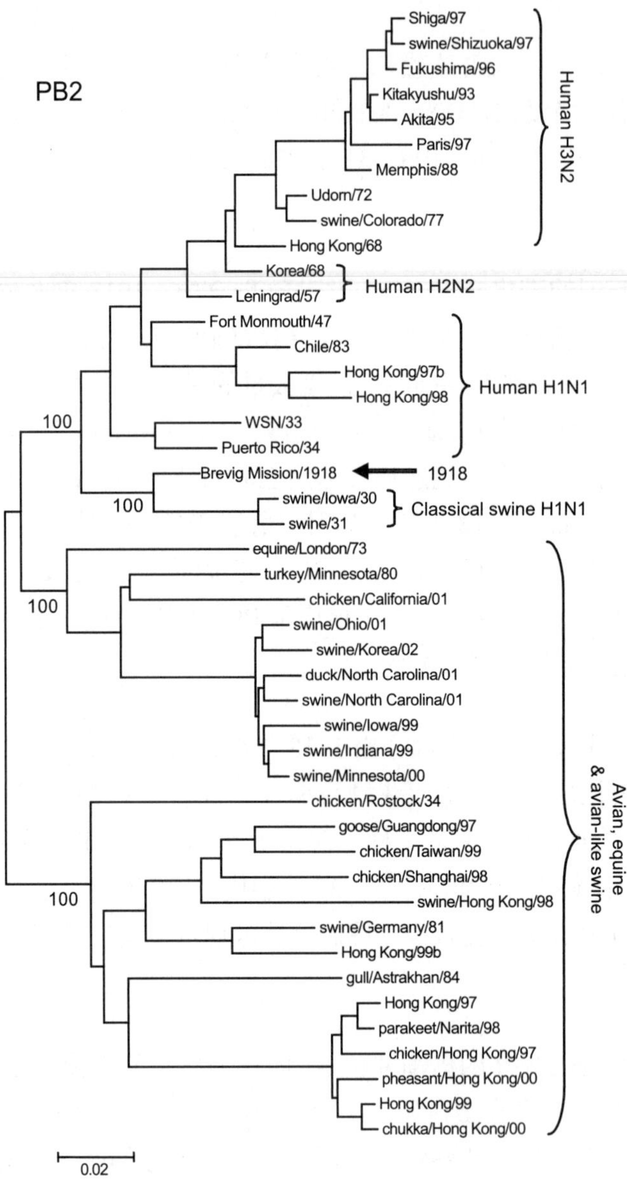

Fig. 1. Phylogenetic tree of the PB2 gene. Sequences were aligned and analysed for phylogenetic relationships using the NJ algorithm, with the proportion of sequence differences as the distance measure. Bootstrap values (100 replications) for key nodes are shown (for clarity, identical and nearly identical sequences have been removed from the trees). Major clades are identified with large brackets. The arrow identifies the position of the 1918 PB2 gene sequence. A distance bar is shown below the tree. Influenza strain abbreviations used in the analyses are listed in Supplementary Table 1.

并位点的百分比差别在三种基因中都处于高水平(41%～51%)。和其他的 1918 年病毒基因一样 [11]，这提示 1918 年病毒的供体来源与目前数据库中显示的禽流感病毒在进化上是隔离的。

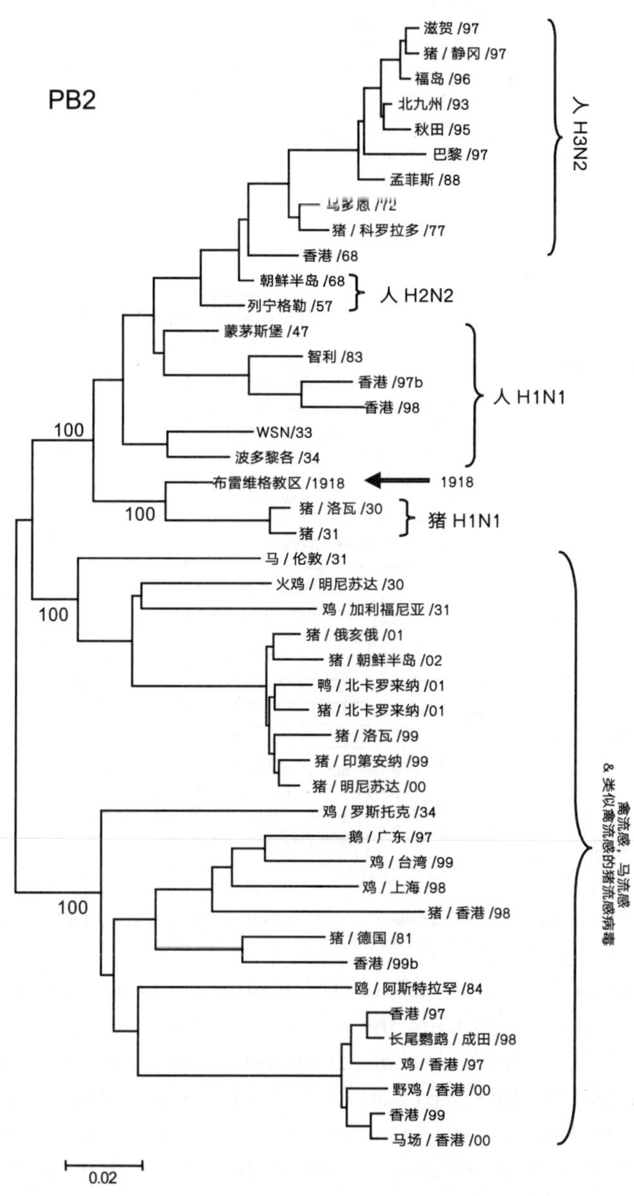

图 1. PB2 基因的系统发育树。用 NJ 算法排比和分析序列的分子进化关系，其距离的度量是序列差别的比例。图上标注了关键节点的自展值(100 个重复)(为了清晰起见，相同的和几乎相同的序列从树中移除)。主要分支用大括号标记出来。箭头指出了 1918 年病毒 PB2 基因序列的位置。距离的比例尺在树下面标出。分析中使用的流感病毒株系的缩写见补充信息表 1。

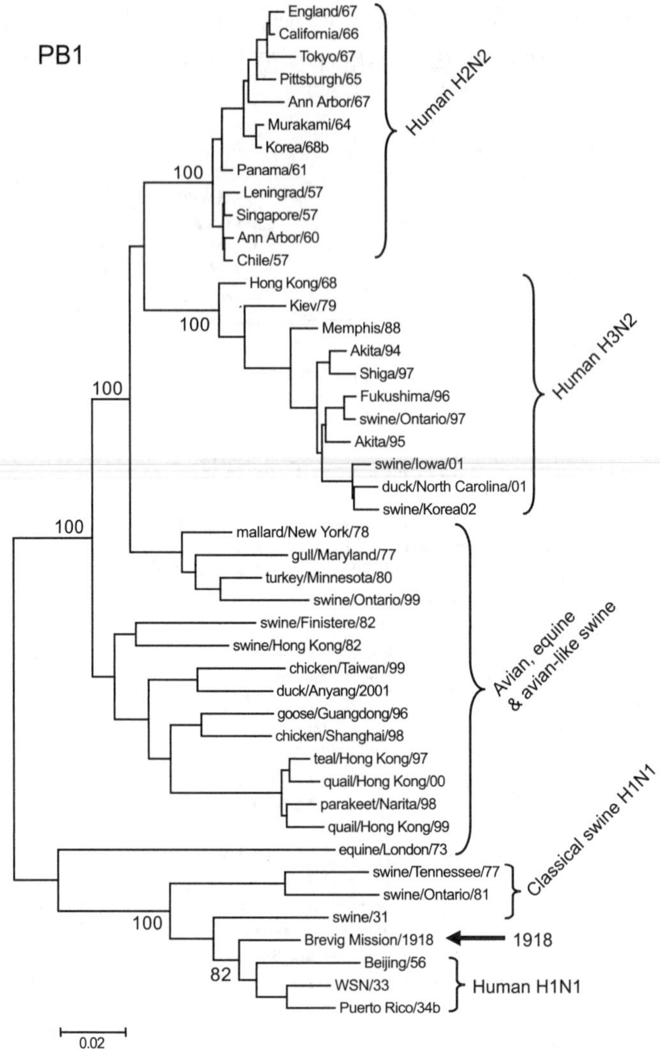

Fig. 2. Phylogenetic tree of the PB1 gene. Sequences were aligned and analysed as detailed in the legend to Fig. 1.

Emphasizing the avian-like nature of the 1918 influenza virus polymerase proteins, out of 19 total amino acid changes from the avian consensus, there are only 10 amino acid positions (out of 2,232 total codons) that consistently distinguish the 1918 and subsequent human polymerase proteins PB2, PB1 and PA from their avian influenza counterparts (these are defined as changes from avian sequences in the 1918 virus that are maintained without change in subsequent human viruses) (Table 1). It is likely that these changes have an important role in human adaptation. Seven of these ten changes were previously noted in an alignment between avian and human influenza polymerases[3]. What follows is a comparison between the 1918 virus changes and recent H5N1 isolates, in order to evaluate possible examples of parallel evolution in the adaptation of avian influenza viruses to humans.

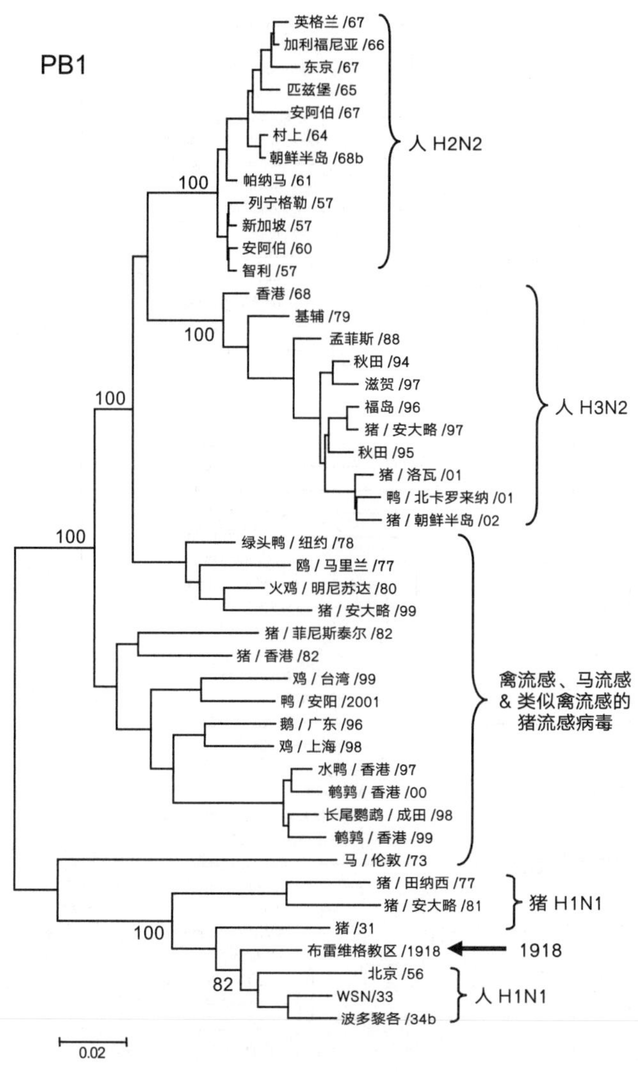

图 2. PB1 基因的系统发育树。如图 1 图注中详细所述进行序列的排比和分析。

在和禽流感病毒共有序列不同的 19 个氨基酸中，只有 10 个氨基酸位置（总共 2,232 个密码子）始终如一地在 1918 年及之后的人类流感病毒聚合酶蛋白 PB2、PB1 和 PA 中与它们的禽流感病毒对应部分保持差别（这些被定义为 1918 年病毒中禽类序列发生的突变，其在后续人类病毒中一直保持稳定）（表 1），这也突出了 1918 年流感病毒聚合酶蛋白的类似禽流感病毒的性质。很有可能这些改变在适应人类的过程中具有重要作用。在之前比对禽流感和人流感聚合酶时就发现了这 10 个突变中的 7 个 [3]。随后就是比较 1918 年病毒变化和最近的 H5N1 病毒分离株，目的是评估禽流感病毒在适应人类的过程中发生平行进化的可能实例。

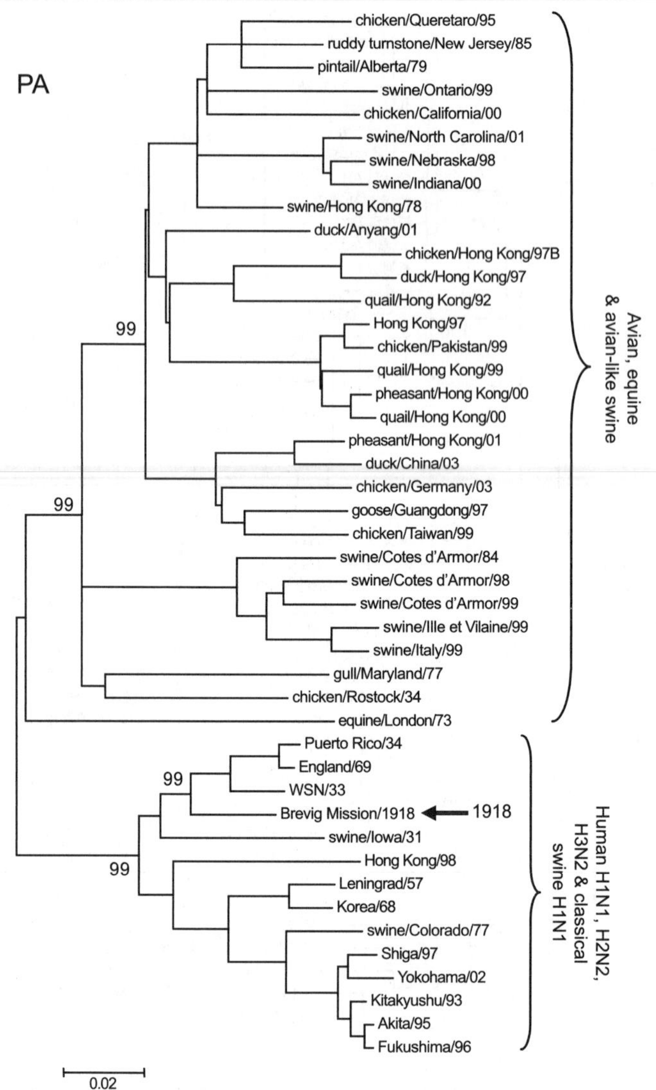

Fig. 3. Phylogenetic tree of the PA gene. Sequences were aligned and analysed as detailed in the legend to Fig. 1.

In the PB2 protein, five changes distinguish the human isolates from avian sequences (Table 1). Out of 253 available PB2 sequences from human H1N1, H2N2 and H3N2 isolates, these five changes are almost completely preserved, with the exception that two recent H3N2 isolates have the avian Lys residue at position 702. Only a small number of avian influenza isolates show any of these five changes, and it is intriguing that almost all of these isolates are from HPAI H5N1 or H7N7 viruses, or from the H9N2 lineage that infected a small number of humans in China in the late 1990s (ref. 29). Only 5 out of 282 available avian PB2 sequences have a Ser residue at position 199, four of these being 1997 H5N1 isolates from Hong Kong. The A199S change was also found in 5 out of 18 H5N1 strains isolated from humans (all five were from the 1997 Hong Kong outbreak). Of the avian viruses, 36 out of 336 have an Arg residue at position 702, 30 of which are H9N2 isolates from China

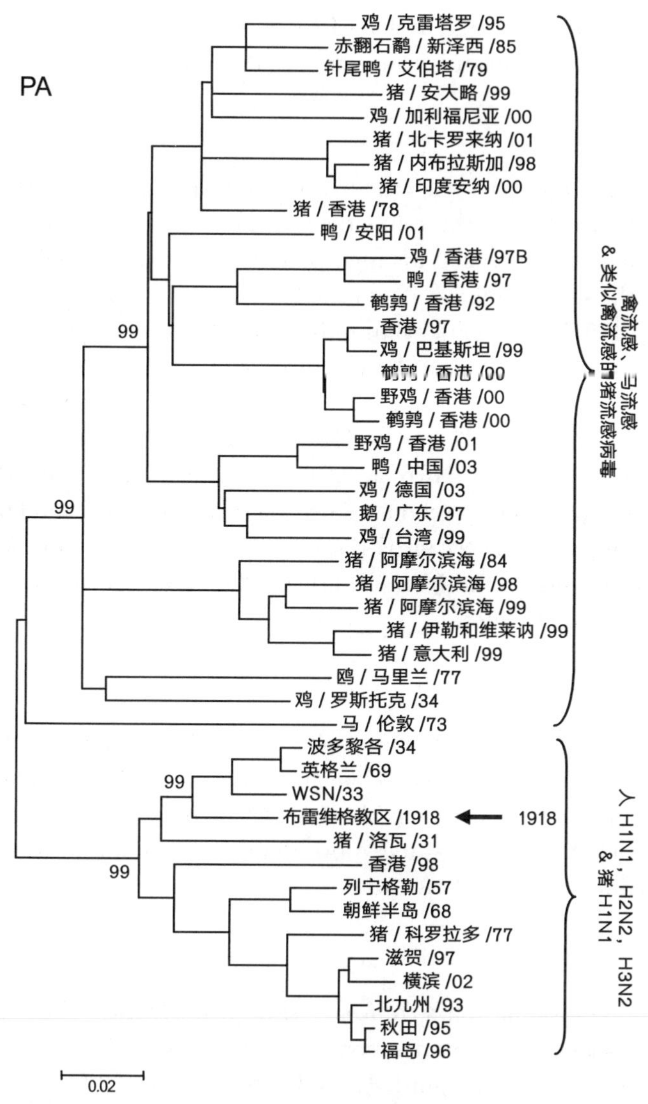

图 3. PA 基因的系统发育树。如图 1 图注中详细所述，进行序列的排比和分析。

在 PB2 蛋白中，人流感病毒分离株和禽病毒序列有 5 处差别(表 1)。在从人类 H1N1、H2N2 和 H3N2 病毒分离株获得的 253 个 PB2 序列中，这 5 处突变几乎完全保留，除了最近分离的两株 H3N2 在位点 702 具有禽流感的赖氨酸残基。只有少量的禽流感病毒分离株含有 5 种突变中的任意一种，而且有趣的是几乎所有含有这些突变的毒株都是从高致病性禽流感 H5N1 或者 H7N7 病毒中分离出来的，或者是从 20 世纪 90 年代晚期感染中国少部分人的 H9N2 病毒株中分离的(文献 29)。获得的 282 个禽流感 PB2 序列中只有 5 个在位点 199 是丝氨酸残基，其中 4 个来自香港的 1997 年 H5N1 病毒株。从人类中分离的 18 个 H5N1 株系中的 5 个发现了 A199S 突变(所有 5 个都来自 1997 年香港流感疫情)。336 个禽流感病毒中，只有 36 个在

around 1996–2000, and 5 are H5N1 isolates from Hong Kong in 1997 and 2001. Out of 18 available 1997 H5N1 strains isolated from humans, three have the K702R change.

Perhaps most interestingly, the 1918 virus and subsequent human isolates have a Lys residue at position 627. This residue has been implicated in host adaptation[2,3], and has previously been shown to be crucial for high pathogenicity in mice infected with the 1997 H5N1 virus[19]. Of the avian isolates, 19 out of 345 have a Lys residue at position 627, 18 of which are HPAI H5N1 or H7N7 avian influenza viruses. Sixteen of these were recently characterized H5N1 isolates from a die-off of wild waterfowl around Qinghai Lake in western China in 2005 (ref. 21). In human H5N1 isolates, 11 out of 37 have the E627K change: A/Hong Kong/483/1997 and A/Hong Kong/485/1997, four out of six isolates from Vietnam in 2004 (ref. 22), and two out of three isolates from Thailand in 2004 (ref. 23). The E627K mutation was seen in six out of seven H5N1 isolates from Thai tigers in 2004, and was also present in the H7N7 virus responsible for the single human fatality during the HPAI H7N7 outbreak in the Netherlands in 2003 (ref. 20). It was not noted in the contemporaneous chicken isolates.

At position 475, only one out of 355 avian isolates has a Met residue (an H5N1 HPAI virus from 2004). Similarly, only one out of 345 avian viruses has an Asn residue at position 567. None of the human H5N1 isolates has the L475M or the D567N changes. None of the available H5N1 or H7N7 sequences has more than one of the proposed human-adaptive PB2 changes determined for the 1918 virus.

The PA protein shows a similar pattern: four residues consistently differ between 1918 and subsequent human isolates and the avian consensus sequence (Table 1). Three other changes (C241Y, K312R and I322V) distinguish 1918, H1N1 and H2N2 human isolates, but most H3N2 isolates have the avian amino acid at these positions. Of 295 available sequences from human H1N1, H2N2 and H3N2 isolates, all have Asn at position 55 (except A/WSN/33), Ala at position 100 and Ser at position 552. Only 5 out of 295 human isolates have the avian Glu residue at position 382. Notably, these five isolates make up a minor clade of recent H3N2 isolates that have a number of unusual changes from typical human H3N2 viruses[30]. When avian influenza sequences are analysed, none (out of 209 sequences) has Asn at position 55 or Ser at position 552. Only 8 out of 209 avian PA protein sequences show the V100A change: six recent H6N2 isolates from chickens in California, and two HPAI 2002 H5N1 duck isolates from China. Of the 209 avian sequences, five have an Asp residue at position 382, including two HPAI H5N2 isolates from chickens in Mexico in 1994.

位点 702 是精氨酸残基，其中的 30 个是 1996 年 ~ 2000 年左右来自中国的 H9N2 分离株，5 个则是来自 1997 年和 2000 年香港的 H5N1 分离株。从人类中获得的 18 个 1997 年 H5N1 病毒株中，三个含有 K702R 突变。

可能最有趣的是，1918 年病毒和之后的人类分离株都在位点 627 含有一个赖氨酸残基。该残基与宿主的适应性相关 [2,3]，并且在之前的研究中已经被证实该位点对于 1997 年 H5N1 病毒感染小鼠时的高致病性非常关键 [19]。在 345 个禽流感分离株中，19 个在位点 627 含有赖氨酸残基，其中 18 个都是高致病性禽流感——H5N1 或者 H7N7 禽流感病毒。其中的 16 个是最近从 2005 年中国西部青海湖周围死去的野生水禽中分离出来的 H5N1 病毒株（文献 21）。在 37 个人类 H5N1 分离株中，11 个具有 E627K 突变：A/香港/483/1997 和 A/香港/485/1997，2004 年从越南分离出的 6 个病毒株中的 4 个（文献 22），以及 2004 年从泰国分离出的 3 个病毒株中的 2 个（文献 23）。该 E627K 突变在 2004 年从泰国老虎身上分离的 7 个 H5N1 毒株的 6 个中发现，并且存在于引起 2003 年荷兰高致病性禽流感 H7N7 大暴发时唯一人类死亡病例的 H7N7 病毒中（文献 20）。在同时期的鸡分离株中未发现该突变。

在位点 475，355 个禽类分离株中只有 1 个含有甲硫氨酸残基（2004 年分离的一个 H5N1 高致病性禽流感病毒）。类似的，345 个禽流感病毒株中只有 1 个在位点 567 含有天冬酰胺残基。人类 H5N1 分离株无 L475M 或者 D567N 突变。在获得的 H5N1 或者 H7N7 序列中，没有一个含有超过一种像 1918 年病毒那样的人类适应性 PB2 突变。

PA 蛋白表现出类似的模式：1918 年大流行及之后的人类分离株和禽类分离株相应序列之间有固定的 4 个残基的差别（表 1）。三个其他的改变（C241Y、K312R、I322V）将 1918 年、H1N1 和 H2N2 人类分离株区别开来，但是大部分 H3N2 分离株在这些位点都有禽流感的氨基酸。在 295 个从人类 H1N1、H2N2 和 H3N2 分离株中获得的序列中，全都在位点 55 含有天冬酰胺残基（除了 A/WSN/33），在位点 100 含有丙氨酸残基，位点 552 含有丝氨酸残基。295 个人类分离株中只有 5 个在位点 382 含有禽流感的谷氨酸残基。值得注意的是，这 5 个分离株形成了近期 H3N2 株的一个小的分支，与经典的人类 H3N2 毒株相比有一些不寻常的改变 [30]。当分析禽流感病毒序列时，没有一个序列（总共 209 个序列）在位点 55 含有天冬酰胺残基或者在位点 552 含有丝氨酸残基。209 个禽类 PA 蛋白序列中只有 8 个含有 V100A 突变：6 个来自近期加利福尼亚的鸡 H6N2 分离株，2 个来自中国 2002 年高致病性禽流感 H5N1 鸭分离株。在这 209 个禽类序列中，5 个在位点 382 含有天冬氨酸残基，包括 2 个从 1994 年墨西哥鸡中分离出的高致病性禽流感 H5N2 病毒。

The PB1 gene segment was replaced by reassortment in both the 1957 and 1968 pandemics[9]. We compared the PB1 protein from the 1918 human virus with those of the avian-derived PB1 segments from the 1957 and 1968 pandemics. Human H1N1, H2N2 and H3N2 viruses derived from the 1918, 1957 and 1968 pandemics, respectively, each possessed a uniquely derived avian-like PB1 gene segment, and so we sought to identify any parallel changes that might shed light on human adaptation. The three human pandemic PB1 proteins differ from the avian consensus by only 4–7 residues each (Supplementary Fig. 2). Only one of these changes is shared among the pandemic isolates: an N375S change. This change to a serine residue is also found in swine and equine influenza A isolates. With few exceptions, all human influenza PB1 proteins have Ser at this site. Of 230 human influenza sequences, only two H1N1 isolates (A/FM/47 and A/Beijing/1956) and the "minor clade" H3N2 isolates described above have the avian Asn residue[30]. In contrast, although this residue is maintained in almost all mammalian isolates, it is variable among avian PB1 proteins. Of 293 avian isolates, 66% have the consensus Asn residue at position 375, 18% have a Ser residue and 12% have a Thr residue.

The data presented here highlight the marked conservation of the PB1 protein in avian influenza viruses. PB1 functions as an RNA-dependent RNA polymerase, and so it is reasonable to hypothesize that its enzymatic function is optimal in this conserved form. In humans, the PB1 proteins experience linear change over time. Indeed, PB1 in humans acquires ~0.4 amino acid changes per year. As there is such strong antigenic selection on human viruses, it is possible that although the observed changes in PB1 are selectively beneficial with respect to antigenicity, they are mildly deleterious to enzyme function. Such complex fitness trade-offs are thought to be commonplace in RNA virus evolution. Supporting this hypothesis, a recent study examining combinations of avian and human influenza polymerases showed that the most efficient influenza transcriptional activity *in vitro* was seen with an avian-derived PB1, even if the PB2, PA and NP proteins were from a human virus[3]. Acquiring an avian PB1 by reassortment might provide a replicative advantage to the new virus, possibly explaining why both of the last two pandemics and the 1918 influenza virus all had very avian-like PB1 proteins.

Both the 1957 and 1968 pandemic influenza viruses were avian/human reassortants in which 2–3 avian gene segments were reassorted with the then-circulating, human-adapted virus[9,10]. Unlike the 1957 and 1968 pandemics, however, the 1918 virus was most likely not a human/avian reassortant virus, but rather an avian-like virus that adapted to humans *in toto*[8,11]. On the basis of amino acid replacement rates in human influenza virus polymerase genes, it is possible that these segments were circulating in human influenza viruses as early as 1900. However, proof that the 1918 virus did not retain gene segments from the previously circulating human influenza A strain would require discovery of a sample of the pre-1918 virus from archival material. The donor source, although avian-like at the protein level, may have come from a subset of avian influenza viruses not currently represented in the sequence databases and may have been in evolutionary isolation.

在 1957 年和 1968 年的大流行中，PB1 基因片段都发生了重组[9]。我们比较了 1918 年人类病毒的 PB1 蛋白与 1957 年和 1968 年大流行时禽类来源的 PB1 片段。分别来自 1918 年、1957 年和 1968 年大流行的人类 H1N1、H2N2 和 H3N2 病毒，每个都含有独特起源的类似禽流感的 PB1 基因片段，因此我们试图找到任何可能有助于人类适应性的平行突变。这三个人类大流行的 PB1 蛋白与禽类共有序列都仅有 4 ~ 7 个残基的差别(补充信息图 2)。只有一个突变在三个大流行分离株中都存在：N375S 突变。在猪和马的甲型流感分离株中也发现了这种到丝氨酸残基的突变。除了少数几个例外，所有的人类流感 PB1 蛋白在这个位点都是丝氨酸，在 230 个人类流感病毒序列中，只有两个 H1N1 分离株(A/FM/47 和 A/北京/1956)以及上面所述的 H3N2 分离株微小分支含有禽类天冬酰胺残基[30]。相反，尽管该残基在几乎所有的哺乳动物分离株中都存在，其在禽类 PB1 蛋白中是可变的。在 293 个禽类分离株中，66% 在位点 375 含有相同的天冬酰胺残基，18% 含有丝氨酸残基，12% 含有苏氨酸残基。

这里给出的数据凸显出了禽流感病毒 PB1 蛋白的显著保守性。PB1 作为一种 RNA 依赖的 RNA 聚合酶发挥功能，因此有理由假设在这种保守的形式下其酶功能是最佳的。在感染的人类中，PB1 蛋白随时间经历了线性的改变。实际上，PB1 在人群中每年获得大约 0.4 个氨基酸的改变。由于人类病毒存在很强的抗原性选择，尽管 PB1 的改变在抗原性方面具有选择优势，但它们在一定程度上降低了酶功能。这种复杂的适应性权衡被认为在 RNA 病毒的进化中非常普遍。最近一个研究支持了这个假设，它研究了禽流感和人流感聚合酶的组合，结果显示即使 PB2、PA 和 NP 蛋白都来源于人流感，含有禽类来源的 PB1 时，病毒的体外转录活性最有效率[3]。通过重组获得禽流感 PB1 蛋白可能给新病毒提供复制优势，这可能解释了为什么后两次大流行和 1918 年的流感病毒全都含有类似禽流感的 PB1 蛋白。

1957 年和 1968 年大流行的流感病毒都是禽类/人类的重组病毒，其中 2 ~ 3 个禽流感基因片段都与那时流行的适应人类的病毒发生了重组[9,10]。但是，不像 1957 年和 1968 年大流行，1918 年的病毒很可能不是禽类/人类的重组病毒，而是完全适应人类的类似禽流感的病毒[8,11]。根据人类流感病毒聚合酶基因氨基酸替换的速度，有可能这些片段早在 1900 年就在人类流感病毒中传播。但是，要证明 1918 年病毒是否保留先前传播的人类甲型流感病毒的基因片段，需要在档案材料中发现 1918 年之前的病毒样本。其来源尽管在蛋白质水平类似于禽流感，但可能是目前序列数据库中未显示的禽流感病毒亚类，并且已经出现了进化隔离。

The fact that amino acid changes identified in the 1918 analysis are also seen in HPAI strains of H5N1 and H7N7 avian viruses that have caused fatalities in humans is intriguing, and suggests that these changes may facilitate virus replication in human cells and increase pathogenicity. It is possible that the high pathogencity of the 1918 virus was related to its emergence as a human-adapted avian influenza virus. These changes may reflect a process of parallel evolution as avian influenza A viruses mutate in response to adaptational pressures, and suggest that the genetic basis of avian influenza virus adaptation to humans can be mapped.

Methods

RNA isolation, amplification and sequencing. RNA was isolated from frozen 1918 human lung tissue using Trizol (Invitrogen) according to the manufacturer's instructions. Each fragment was reverse transcribed, amplified, and sequenced at least twice. Reverse transcription polymerase chain reaction (RT–PCR), isolation of products and sequencing have been previously described[4]. Lists of primers and primer sequences are available upon request. Replicate RT–PCR reactions from independently produced RNA preparations gave identical sequence results. The 2,280-nucleotide complete coding sequence of PB2 was amplified in 33 overlapping fragments. The 2,274-nucleotide coding sequence of PB1 was amplified in 33 overlapping fragments. The 2,151-nucleotide coding sequence of PA was amplified in 32 overlapping fragments. The PCR products ranged in size from 77–138 bp.

Phylogenetic analyses. Phylogenetic analyses of the three polymerase genes were done using standard methods. We generated trees using the neighbour-joining (NJ) algorithm, with proportion of differences as the distance measure using MEGA version 2.1. Character evolution was analysed with the MacClade program after a parsimony analysis using PAUP version 4.0 beta, using ACTRAN as the optimization method. Trees were also generated using maximum-likelihood with midpoint rooting. All algorithms generated comparable trees, with major clades representing human, classical swine and avian-like viruses (NJ trees shown in Figs 1–3; complete data set available upon request). Polymerase segment sequences used in this analysis were obtained from GenBank and the Influenza Sequence Databank (ISD). (See Supplementary Table 1 for a list of sequences used.) For the PB2 gene, 83 sequences were used, all of which were full length. For the PB1 gene, 91 sequences were used, three of which were not full length. For the PA gene, 105 sequences were used, six of which were not full length.

(**437**, 889-893; 2005)

Jeffery K. Taubenberger[1], Ann H. Reid[1]†, Raina M. Lourens[1]†, Ruixue Wang[1], Guozhong Jin[1] & Thomas G. Fanning[1]

[1] Department of Molecular Pathology, Armed Forces Institute of Pathology, Rockville, Maryland 20850, USA

† Present addresses: Board on Life Sciences, The National Academies, 6th Floor, 500 Fifth Street N.W., Washington DC 20001, USA (A.H.R.); University of Iowa, Roy J. and Lucille A. Carver College of Medicine, 200 CMAB, Iowa City, Iowa 52242, USA (R.M.L.)

Received 30 June; accepted 19 September 2005.

1918 年病毒分析中得到的氨基酸改变也在导致人类死亡的 H5N1 和 H7N7 高致病性禽流感株中发现，这非常有意思；同时这也表明上述改变可能促进病毒在人细胞中复制并增加致病性。有可能 1918 年病毒的高致病性与其适应人类的禽流感病毒本质有关。这些改变可能反映了平行进化的过程，即甲型流感病毒（禽流感）应对适应压力做出的突变。它也表明我们可以绘制出禽流感病毒适应人类的遗传基础。

方　法

RNA 分离、扩增和测序　根据生产商的说明，用 Trizol 试剂（Invitrogen）从冰冻的 1918 年人肺组织中提取病毒 RNA。每个片段都进行至少两次逆转录、扩增和测序。逆转录聚合酶链式反应（RT–PCR）、产物的分离和测序之前已经描述过 [4]。引物和引物序列的列表可以向我们索取。从单独制备的 RNA 进行重复的 RT–PCR 反应得到了相同的序列产物。PB2 的 2,280 个核苷酸的完整编码序列在 33 个重叠的片段中扩增出来。PB1 的 2,274 个核苷酸的编码序列在 33 个重叠的片段中扩增出来。PA 的 2,151 个核苷酸的编码序列在 32 个重叠的片段中扩增出来。PCR 产物的大小在 77 ~ 138 bp 之间。

系统发育分析　三个聚合酶基因的系统发育分析用标准的方法进行。我们用邻接（NJ）算法绘制进化树，使用 MEGA 2.1 版将差异的比例作为距离度量。使用 ACTRAN 作为优化方法，在使用 PAUP 4.0 beta 版进行简易分析之后用 MacClade 程序分析特征进化。进化树也可用带有中点根的最大似然产生。所有的算法产生了可以比较的树，其主要的分支代表人类病毒、猪病毒和类似禽类的病毒（NJ 树在图 1 ~ 图 3 中显示；详细的数据可以向我们索取）。本分析中使用的聚合酶基因片段序列从 GenBank 和流感序列数据库获得。（使用的序列列表详见补充信息表 1）对于 PB2 基因，使用了 83 个序列，所有都是全长的。对于 PB1 基因，使用了 91 个序列，其中三个不是全长的。对于 PA 基因，使用了 105 个序列，其中六个不是全长的。

（毛晨晖 翻译；陈继征 审稿）

References:

1. Fodor, E. & Brownlee, G. G. in *Influenza* (ed. Potter, C. W.) 1-29 (Elsevier, Amsterdam, 2002).

2. Subbarao, E. K., London, W. & Murphy, B. R. A single amino acid in the PB2 gene of influenza A virus is a determinant of host range. *J. Virol.* **67**, 1761-1764 (1993).

3. Naffakh, N., Massin, P., Escriou, N., Crescenzo-Chaigne, B. & van der Werf, S. Genetic analysis of the compatibility between polymerase proteins from human and avian strains of influenza A viruses. *J. Gen. Virol.* **81**, 1283-1291 (2000).

4. Reid, A. H., Fanning, T. G., Hultin, J. V. & Taubenberger, J. K. Origin and evolution of the 1918 "Spanish" influenza virus hemagglutinin gene. *Proc. Natl Acad. Sci. USA* **96**, 1651-1656 (1999).

5. Reid, A. H., Fanning, T. G., Janczewski, T. A. & Taubenberger, J. K. Characterization of the 1918 "Spanish" influenza virus neuraminidase gene. *Proc. Natl Acad. Sci. USA* **97**, 6785-6790 (2000).

6. Basler, C. F. *et al.* Sequence of the 1918 pandemic influenza virus nonstructural gene (NS) segment and characterization of recombinant viruses bearing the 1918 NS genes. *Proc. Natl Acad. Sci. USA* **98**, 2746-2751 (2001).

7. Reid, A. H., Fanning, T. G., Janczewski, T. A., McCall, S. & Taubenberger, J. K. Characterization of the 1918 "Spanish" influenza virus matrix gene segment. *J. Virol.* **76**, 10717-10723 (2002).

8. Reid, A. H., Fanning, T. G., Janczewski, T. A., Lourens, R. & Taubenberger, J. K. Novel origin of the 1918 pandemic influenza virus nucleoprotein gene segment. *J. Virol.* **78**, 12462-12470 (2004).

9. Kawaoka, Y., Krauss, S. & Webster, R. G. Avian-to-human transmission of the PB1 gene of influenza A viruses in the 1957 and 1968 pandemics. *J. Virol.* **63**, 4603-4608 (1989).

10. Scholtissek, C., Rohde, W., Von Hoyningen, V. & Rott, R. On the origin of the human influenza virus subtypes H2N2 and H3N2. *Virology* **87**, 13-20 (1978).

11. Reid, A. H., Taubenberger, J. K. & Fanning, T. G. Evidence of an absence: the genetic origins of the 1918 pandemic influenza virus. *Nature Rev. Microbiol.* **2**, 909-914 (2004).

12. Johnson, N. P. & Mueller, J. Updating the accounts: global mortality of the 1918–1920 "Spanish" influenza pandemic. *Bull. Hist. Med.* **76**, 105-115 (2002).

13. Geiss, G. K. *et al.* Cellular transcriptional profiling in influenza A virus-infected lung epithelial cells: the role of the nonstructural NS1 protein in the evasion of the host innate defense and its potential contribution to pandemic influenza. *Proc. Natl Acad. Sci. USA* **99**, 10736-10741 (2002).

14. Tumpey, T. M. *et al.* Existing antivirals are effective against influenza viruses with genes from the 1918 pandemic virus. *Proc. Natl Acad. Sci. USA* **99**, 13849-13854 (2002).

15. Tumpey, T. M. *et al.* Pathogenicity and immunogenicity of influenza viruses with genes from the 1918 pandemic virus. *Proc. Natl Acad. Sci. USA* **101**, 3166-3171 (2004).

16. Kash, J. C. *et al.* The global host immune response: contribution of HA and NA genes from the 1918 Spanish influenza to viral pathogenesis. *J. Virol.* **78**, 9499-9511 (2004).

17. Kobasa, D. *et al.* Enhanced virulence of influenza A viruses with the haemagglutinin of the 1918 pandemic virus. *Nature* **431**, 703-707 (2004).

18. Chen, W. *et al.* A novel influenza A virus mitochondrial protein that induces cell death. *Nature Med.* **7**, 1306-1312 (2001).

19. Shinya, K. *et al.* PB2 amino acid at position 627 affects replicative efficiency, but not cell tropism, of Hong Kong H5N1 influenza A viruses in mice. *Virology* **320**, 258-266 (2004).

20. Fouchier, R. A. *et al.* Avian influenza A virus (H7N7) associated with human conjunctivitis and a fatal case of acute respiratory distress syndrome. *Proc. Natl Acad. Sci. USA* **101**, 1356-1361 (2004).

21. Chen, H. *et al.* Avian flu: H5N1 virus outbreak in migratory waterfowl. *Nature* **436**, 191-192 (2005).

22. Li, K. S. *et al.* Genesis of a highly pathogenic and potentially pandemic H5N1 influenza virus in eastern Asia. *Nature* **430**, 209-213 (2004).

23. Puthavathana, P. *et al.* Molecular characterization of the complete genome of human influenza H5N1 virus isolates from Thailand. *J. Gen. Virol.* **86**, 423-433 (2005).

24. Toyoda, T., Adyshev, D. M., Kobayashi, M., Iwata, A. & Ishihama, A. Molecular assembly of the influenza virus RNA polymerase: determination of the subunit-subunit contact sites. *J. Gen. Virol.* **77**, 2149-2157 (1996).

25. Masunaga, K., Mizumoto, K., Kato, H., Ishihama, A. & Toyoda, T. Molecular mapping of influenza virus RNA polymerase by site-specific antibodies. *Virology* **256**, 130-141 (1999).

26. Ohtsu, Y., Honda, Y., Sakata, Y., Kato, H. & Toyoda, T. Fine mapping of the subunit binding sites of influenza virus RNA polymerase. *Microbiol. Immunol.* **46**, 167-175 (2002).

27. Biswas, S. K. & Nayak, D. P. Mutational analysis of the conserved motifs of influenza A virus polymerase basic protein 1. *J. Virol.* **68**, 1819-1826 (1994).

28. Gonzalez, S. & Ortin, J. Distinct regions of influenza virus PB1 polymerase subunit recognize vRNA and cRNA templates. *EMBO J.* **18**, 3767-3775 (1999).

29. Guo, Y. J. *et al.* Characterization of the pathogenicity of members of the newly established H9N2 influenza virus lineages in Asia. *Virology* **267**, 279-288 (2000).

30. Holmes, E. C. *et al.* Whole-genome analysis of human influenza A virus reveals multiple persistent lineages and reassortment among recent H3N2 viruses. *PLoS Biol.* **3**, e300 (2005).

Supplementary Information is linked to the online version of the paper at www.nature.com/nature.

Acknowledgements. The research described in this report was done using stringent safety precautions to protect the laboratory workers, the environment and the public from this virus. The intention of this research is to provide the basis for understanding how influenza pandemic strains form and to help ascertain the risk of future influenza pandemics. This study was partially supported by a grant to J.K.T. from the National Institutes of Health, and by intramural funds from the Armed Forces Institute of Pathology. The opinions contained herein are the private views of the authors and are not to be construed as official or as reflecting the views of the US Department of the Army or the US Department of Defense.

Author Contributions. J.K.T. planned the project, and A.H.R., R.M.L., R.W. and G.J. generated the sequence data. J.K.T., A.H.R. and T.G.F. performed data analysis. J.K.T. wrote the manuscript.

Author Information. Coding sequences of the PB2, PB1 and PA genes have been deposited in GenBank under accession numbers DQ208309, DQ208310 and DQ208311, respectively. Reprints and permissions information is available at npg.nature.com/reprintsandpermissions. The authors declare no competing financial interests. Correspondence and requests for materials should be addressed to J.K.T. (taubenberger@afip.osd.mil).

Implications for Prediction and Hazard Assessment from the 2004 Parkfield Earthquake

W. H. Bakun *et al.*

Editor's Note

Whether earthquakes can be reliably predicted is still a contentious matter among geoscientists. This paper from a team mostly at the US Geological Survey argues that prediction is still very remote, if it is possible at all. It describes an earthquake that occurred on the Californian San Andreas fault in 2004, centred on the small community of Parkfield which has experienced rather regular quakes for the past 150 years. For this reason Parkfield is extremely well monitored for seismic activity, and the previous quasi-regular seismicity led to predictions of an earthquake before 1993 that was long "overdue" by 2004. Despite this intensive study, the timing of the 2004 quake was unanticipated, and it was accompanied by no obvious precursory activity.

Obtaining high-quality measurements close to a large earthquake is not easy: one has to be in the right place at the right time with the right instruments. Such a convergence happened, for the first time, when the 28 September 2004 Parkfield, California, earthquake occurred on the San Andreas fault in the middle of a dense network of instruments designed to record it. The resulting data reveal aspects of the earthquake process never before seen. Here we show what these data, when combined with data from earlier Parkfield earthquakes, tell us about earthquake physics and earthquake prediction. The 2004 Parkfield earthquake, with its lack of obvious precursors, demonstrates that reliable short-term earthquake prediction still is not achievable. To reduce the societal impact of earthquakes now, we should focus on developing the next generation of models that can provide better predictions of the strength and location of damaging ground shaking.

EARTHQUAKE prediction is the Holy Grail of seismology. Although the ability to predict the time and location of earthquakes remains elusive, predicting their effects, such as the strength and geographical distribution of shaking, is routine practice. The extent to which earthquake phenomena can accurately be predicted will ultimately depend on how well the underlying physical conditions and processes are understood. To understand earthquakes requires observing them up close and in detail—a difficult task because they are at present largely unpredictable, and so knowing where to put the instrumentation needed to make such observations is a challenge. The 40-km-long Parkfield section of the San Andreas fault was recognized two decades ago as a promising earthquake physics laboratory and an intensive experiment was established to record the next segment-rupturing earthquake there and provide the much-needed detailed

从 2004 年帕克菲尔德地震看地震预测和灾害评估的启示

巴昆等

编者按

地震能否被可靠地预测，仍然是地球科学家之间争论的事情。这篇论文来自一个研究团队，团队中多数研究人员在美国地质调查局工作。论文认为，假使确实有可能预测地震，但现在看来预测仍然非常遥不可及。这篇论文描述了 2004 年在加利福尼亚州圣安德烈亚斯断层上发生的地震，震中位于帕克菲尔德这个小社区。过去 150 年来，帕克菲尔德发生过具有一定规律的地震。因此，帕克菲尔德的地震活动得到了非常好的监测，根据先前准规律的地震活动性预测，1993 年前将发生一次地震，但到 2004 年地震才姗姗来迟。尽管研究人员进行了深入研究，但 2004 年地震发生的时间却在意料之外，而且此次地震没有伴随明显的前兆活动。

获得接近大地震的高质量测量并不容易：研究人员必须在正确的地点、正确的时间，使用正确的仪器。2004 年 9 月 28 日，当位于圣安德烈亚斯断层上的美国加利福尼亚州帕克菲尔德地震发生时，这种汇合发生了，这一地震正是发生在为监测帕克菲尔德地区地震而设计的密集地震台网的中心区域。得到的数据揭示了地震过程中一些前所未见的方面。在此，我们将展示这些数据，并与以前的帕克菲尔德地震数据结合，这为我们提供了有关地震物理和地震预测的信息。2004 年帕克菲尔德地震，由于缺乏明显的前兆，我们认为可靠的短期地震预测仍然无法实现。目前，为了降低地震的社会影响，我们应该集中精力开发下一代模型，以便更好地预测破坏性地面震动的强度和位置。

地震预测是地震学的“圣杯”。尽管预测地震的时间和位置仍然难以实现，但是预测地震造成的影响，例如震动的强度和地理位置分布，已是常规做法。地震现象能够精确预测的程度最终取决于对内在的物理条件及过程的理解程度。要了解地震，就需要对它们进行近距离细致的观测，这是一项艰巨的任务。因为目前地震在很大程度上是不可预测的，所以要知道在何处放置进行这种观测所需的仪器是一个挑战。圣安德烈亚斯断层的 40 km 长的帕克菲尔德段在 20 年前被公认是一个有希望的地震物理实验室，在此进行的密集实验用于记录下一个断层段破裂引起的地震，并提供迫切需要的详细观测结果。2004 年 9 月 28 日发生的预期矩震级 $M_W = 6.0$ 的地震（发震

observations. The occurrence of the anticipated moment magnitude $M_w = 6.0$ earthquake on 28 September 2004 (origin time 17:15:24 Coordinated Universal Time, UTC; epicentre location 35.815° N, 120.374° W; depth 7.9 km) fulfilled that promise.

The Parkfield section of the San Andreas fault is bounded on the northwest by a 150-km-long creeping section, where numerous small earthquakes occur, and on the southeast by hundreds of kilometres of locked fault where few earthquakes have been detected in the twentieth century (Fig. 1). The 1857 $M_w = 7.9$ Fort Tejon earthquake ruptured the locked fault southeast of Parkfield and is thought to have initiated near Parkfield[1]. On the Parkfield section, the motion of the Pacific plate relative to the North America plate is partly accommodated by repeating $M_w = 6.0$ earthquakes. The historical record of earthquakes at Parkfield includes at least six such events since 1857 (ref. 2; Supplementary Text 1, Supplementary Fig. 1 and Supplementary Table 1).

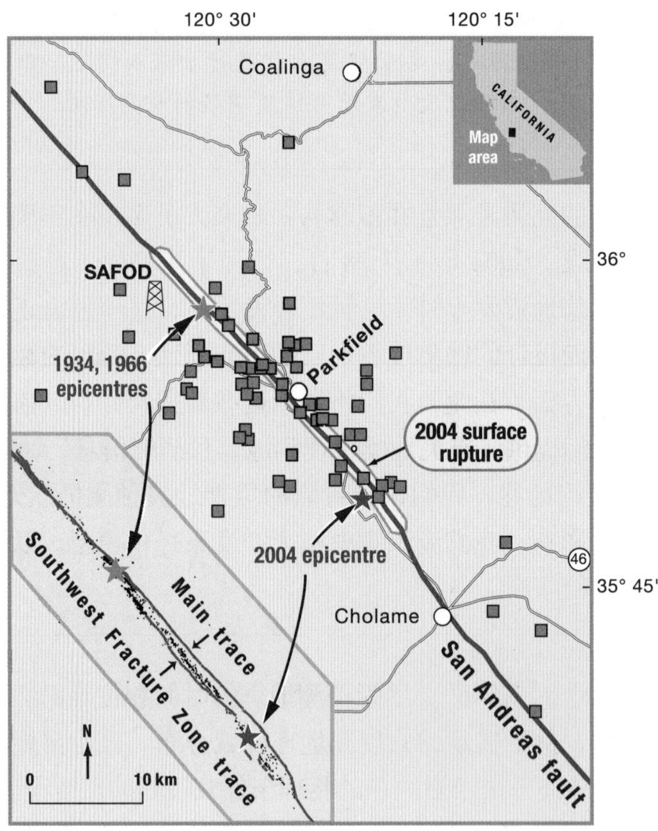

Fig. 1. Location of the 2004 Parkfield earthquake. The zone of surface rupture (yellow) is shown along the San Andreas fault (red lines). Locations of seismographs, strainmeters, creepmeters, magnetometers, and continuous GPS stations shown as squares. The strong-motion sensors (not shown here) are located on Fig. 4. Lower inset (same scale, in pale green) shows epicentres of 2004 aftershocks (black dots) plotted relative to fault traces[30]. Upper inset, map location.

The simple setting and apparent regularity of Parkfield earthquakes[3] offered the rationale

时刻 UTC 时间 17:15:24；震中位置 35.815° N，120.374° W；深度 7.9 km）实现了这一目标。

圣安德烈亚斯断层上，与帕克菲尔德段相邻的西北段是长为 150 km 的以小震密集活动为特征的蠕滑段，而东南则是长达数百千米的强闭锁段，20 世纪在这里几乎没有检测到地震（图 1）。在此强闭锁段上，上一次大地震是 1857 年发生的 $M_W = 7.9$ 蒂洪堡地震，该地震从帕克菲尔德附近开始破裂[1]。在帕克菲尔德断层段上，太平洋板块相对北美板块的运动，由多次 $M_W = 6.0$ 地震来部分调节，自 1857 年以来，帕克菲尔德地震的历史记录中至少包括 6 次这样的地震（参考文献 2，补充文本 1、补充图 1 和补充表 1）。

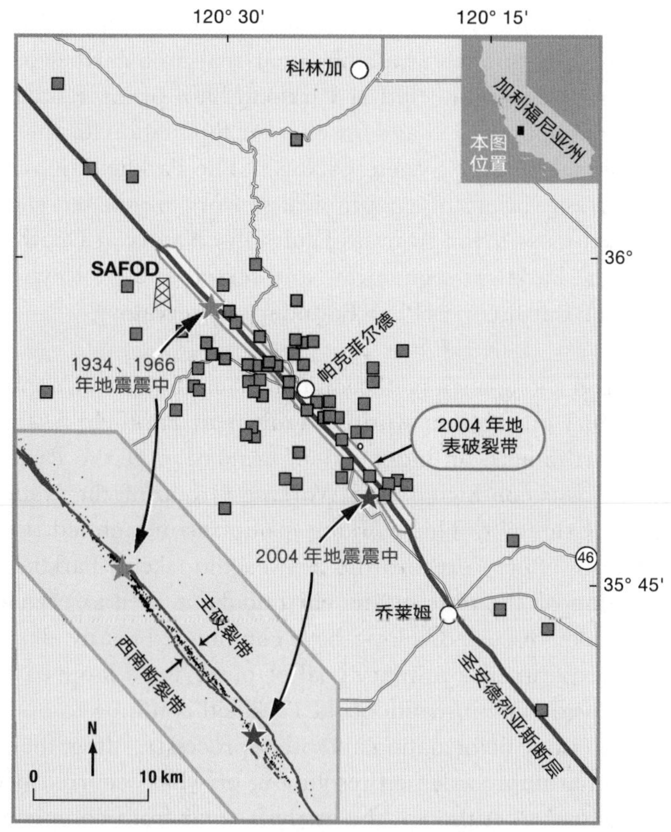

图 1. 2004 年帕克菲尔德地震的位置。地表破裂带（黄色）沿圣安德烈亚斯断层（红线）展布。地震仪、应变仪、蠕变仪、地磁仪和连续 GPS 站的位置以方块表示，强运动传感器的位置（未在此图标出）在图 4 中标出。图中下方插图（等比例尺，浅绿色）显示 2004 年余震震中（一系列黑点）相对于断层线的位置[30]，图中上方的插图显示了本图的位置。

帕克菲尔德地震具有简单的环境和明显的规律性[3]，这为科学的地震预测提供

for the only scientific earthquake prediction officially recognized by the United States government[4] and an opportunity to place instruments in the region before the anticipated earthquake. The primary goal of the Parkfield Earthquake Prediction Experiment[5] was to obtain a detailed understanding of the processes leading up to the anticipated earthquake; a secondary goal was to issue a public warning shortly before the earthquake. A variety of sensors were deployed in a dense network designed specifically to record the build-up of strain in the surrounding crust, monitor earthquakes and slip on the fault, and detect any precursors that might foreshadow a large earthquake. Complementary arrays of strong-ground-motion sensors were deployed to record shaking near the earthquake rupture zone[6].

A significant development of the Parkfield experiment has been the collaboration of federal, state and local officials to develop a protocol for issuing short-term earthquake alerts[7]; the protocol provided a template for communication between scientists and emergency responders and subsequently served as a prototype for volcanic hazard warning protocols[8]. Innovations in the collection, transmission and storage of Parkfield data included a pioneering effort to provide publicly available, near-real-time earth science data streams over the internet. The systems pioneered at Parkfield have become standard elements of seismic monitoring throughout the US and have set the foundation for the installation of the USGS Advanced National Seismic System (ANSS)[9]. The Parkfield dense instrumentation network, which includes a variety of geophysical sensors, motivated the placement of a scientific borehole, Earthscope's San Andreas Fault Observatory at Depth (SAFOD), at the northwestern end of the Parkfield segment[10,11] and served as a prototype for the geodetic networks that are part of Earthscope's Plate Boundary Observatory[11].

In 1985, the USGS issued a long-term prediction that an earthquake of approximately $M_w = 6$ would occur before 1993 on the San Andreas fault near Parkfield[4]. After the prediction window closed, *sans* earthquake, an independent evaluation of the Parkfield Earthquake Prediction Experiment was conducted by a Working Group of the National Earthquake Prediction Evaluation Council[12]. This working group recommended that monitoring be continued as a long-term effort to record the next earthquake at Parkfield. The failure of the long-term prediction of the time of the earthquake as well as certain aspects of the 2004 Parkfield earthquake (discussed below), have confounded nearly all simple earthquake models. However, the experiment's primary goal of observing a large earthquake near the rupture has been achieved. The instruments at Parkfield continue to operate and provide important data on postseismic deformation and other processes. Altogether, these recordings are providing a picture, at unprecedented resolution, of what occurred before, during and after the 2004 earthquake[13,14]. Although the analysis is far from complete, it is clear even now that the observations have important implications for nearly all areas of seismic hazard analysis and loss reduction.

了依据，该次地震预测成为唯一一次美国政府官方认可的预测[4]，从而有机会于预期地震发生前在该地区放置相关检测仪器。帕克菲尔德地震预测实验场[5]的首要目标是详细了解导致预期地震发生的过程，第二个目标是在地震前及时发布公共警报。各种传感器的安装形成了密集地震台网，专门用来记录周围地壳的应变积累，监测地震和断层滑动，探测可能预示大地震的任何前兆。在地震破裂带附近，部署强地面运动传感器作为补充阵列[6]，用来记录震动。

帕克菲尔德地震预测实验场有意义的工作进展之一是在美国联邦政府、州政府和地方政府的通力合作[7]下，建立了发布地震短期警报的预案。该预案提供了科学家和应急反应部门之间的实时通信联络模板，并以此作为火山灾害预警[8]的范例。在帕克菲尔德实验场，数据的收集、传输、存储等方面取得了开拓性进展，包括公共信息服务、准实时地球科学资料的网内数据流程等。实验场开发的数据系统已成为全美国地震监测的标准，并为美国地质调查局建立“国家级地震监测台网系统”(ANSS)奠定了基础[9]。帕克菲尔德密集仪器网络，包括各种地球物理传感器，催生了美国一系列科学项目的建立和实施，如在帕克菲尔德段的西北部安排了地震科学深钻，实施了地球透镜计划中的圣安德烈亚斯断层深部探测计划(SAFOD)[10,11]，实验场的应变观测台网作为板块边界观测计划的组成部分和地形变观测台网的向导[11]。

1985 年，美国地质调查局发布了一个长期地震预测，认为在 1993 年前帕克菲尔德附近的圣安德烈亚斯断层上将发生 $M_W = 6$ 左右的地震[4]。然而，当预测时间窗关闭后，地震并未发生。美国国家地震预测评审委员会组织专门工作组对帕克菲尔德地震预测实验场进行了独立评估[12]。该工作组提出尽管在预报期内未发生地震，但实验场的各种观测应继续下去，作为对帕克菲尔德未来地震的长期观测。地震长期预测在时间上的失败以及 2004 年帕克菲尔德地震的某些方面(将在后面章节讨论)使几乎所有的简单的地震模型失灵。然而，实验的首要目标，即在破裂带附近观测大地震已经实现。帕克菲尔德的观测仪器一直在持续运行，可以为震后形变及其他过程提供重要数据资料。总之，这些记录以前所未有的分辨率提供了 2004 年地震发生前、发生时和发生后的图像[13,14]。尽管分析远未结束，但目前很清楚的是观测几乎对地震灾害分析和减少损失等所有方面都有重要意义。

Precursors and Earthquake Prediction

The idea that detectable precursory processes precede earthquakes dates back to at least the seventeenth century[15]. However, with the exception of foreshocks, unambiguous and repeatable instrumental observations of such phenomena remain elusive. As noted, identical foreshocks preceded both the 1934 and 1966 Parkfield earthquakes by 17 min (ref. 16). However, such foreshocks did not precede the 1901 or 1922 events and so the Parkfield prediction experiment, which was designed to record potential foreshocks as well as other precursory signals, treated all precursors in a probabilistic manner[7]. At present, with the exception of an ambiguous low-level strain of $< 10^{-8}$ that occurred during the 24 h before the main shock, there is no evidence of any short-term precursory signal, either seismic or aseismic[13]. Even microseismicity, detectable at the $M = 0$ threshold in the epicentral region, was absent during the six days before the main shock[13]. Notable precursory signals are not evident in the magnetic field, telluric electric field, apparent resistivity, or creep observations[13]. This lack of short-term precursors emphasizes the difficulty of reliable short-term earthquake prediction (up to a few weeks before).

Subtle strain changes of a few nanostrain were recorded on several instruments in the 24 h before the earthquake. Such changes place important constraints on earthquake nucleation physics but are too small to provide a reliable basis for issuing public warnings that a damaging earthquake is imminent. The dense instrumentation arrays continue to uncover new processes, such as deep tremor under the locked section of the fault southeast of the Parkfield segment[17]. Future hopes for prediction will rest on whether such processes are precursory or simply commonplace.

Fault Structure and Segment Boundaries

The similar magnitude and rupture extent of the last six Parkfield earthquakes supports the concept of fault segmentation and the role of segment boundaries in influencing the rupture extent and magnitude of earthquakes. The nature of segment boundaries, however, is controversial. Fault geometry, rheological and frictional properties of materials, pore fluids and stress conditions have all been proposed to explain segment boundaries[18].

Lindh and Boore[19] suggested a fault-geometry-based explanation for the location of the boundaries of the Parkfield segment: a 5° bend in the fault trace to the northwest and a right-stepover to the southeast appeared to offer geometric obstacles that could limit earthquake rupture. The 2004 aftershocks relocated with a three-dimensional velocity model do not appear to show these features extending to depth (Fig. 2). Aftershocks and earlier seismicity[20] at depths below 6–7 km seem to be confined to a narrow band (see Supplementary Fig. 4). Because the fault seems straighter at depth, where large Parkfield earthquakes nucleate, than it does at the surface, it is possible that fault geometry is not the controlling factor in the location of the Parkfield segment boundaries[20]. However, the

前兆和地震预测

地震发生前具有可检测出的前兆过程，这一观点至少可追溯到 17 世纪[15]。然而，除了前震以外，明确的、可重复观测到的前兆现象仍然难以捉摸。根据记录，1934 年和 1966 年两次帕克菲尔德地震发生前 17 分钟，有完全相同的前震活动出现（参考文献 16），而 1901 年和 1922 年两次地震前却没有出现这样的前震活动。因此，专门设计用来记录潜在的前震活动及其他前兆信号的帕克菲尔德地震预测实验场，所有前兆现象都被视为是以一定的概率出现[7]。当今，除了在主震前 24 小时有十分模糊的低水平应变异常（$<10^{-8}$）活动外，未发现地震学或其他学科的任何短期前兆信号[13]。主震前 6 天，甚至阈值为 $M=0$ 的微震，在震中区也未观测到[13]。在地磁场、地电场、视电阻率和蠕变观测也都未见到值得注意的明显前兆信号[13]。这种短期前兆的缺乏突显了可靠的短期地震预测的困难（地震发生前几周内）。

在地震发生前 24 小时内，多台仪器记录到了非常微弱的仅为几个纳米的应变变化。这些变化在地震成核物理过程中是重要的地震约束条件。但由于这些变化太小，无法为发布即将来临的破坏性地震的公众警报提供可靠的依据。密集的仪器台阵还继续揭示新的事件过程，如在帕克菲尔德东南断层闭锁段下的深部颤动[17]。未来预测的成功与否将取决于这些事件是前兆还是正常的扰动。

断层结构和分段边界

最近 6 次的帕克菲尔德地震的震级和破裂长度都相似，这一点支撑了断层分段性的概念，以及分段边界在影响地震破裂长度和震级大小时起的作用。然而，断层分段边界性是有争议的。断层几何学、介质流变性与断层摩擦特性、孔隙流体与应力条件都被提出用以解释分段边界[18]。

林德和布勒[19]建议基于断层几何学解释帕克菲尔德断层段边界的位置：断层西北端的 5° 弯曲和东南端的右移，似乎提供了限制地震破裂的几何障碍体。然而，用三维速度模型对 2004 年余震重定位的结果似乎没有显示这些表层弯曲特征延伸到深度（图 2）。余震和早期地震活动[20]在 6～7 km 深度处似乎局限在狭窄的条带内（见补充材料图 4）。由于断层在深处看起来比在地表更直，帕克菲尔德大型地震在深处成核，因此断层几何结构可能不是帕克菲尔德断层分段边界位置的控制因素[20]。然而，复杂的地表几何结构可能是由孕震层深度的分段边界变形引起的。

complex surface trace geometry may result from deformation associated with the segment boundaries at seismogenic depth. That is, irregularities in the surface trace may reflect the presence of the boundaries at depth rather than being the primary cause of these boundaries.

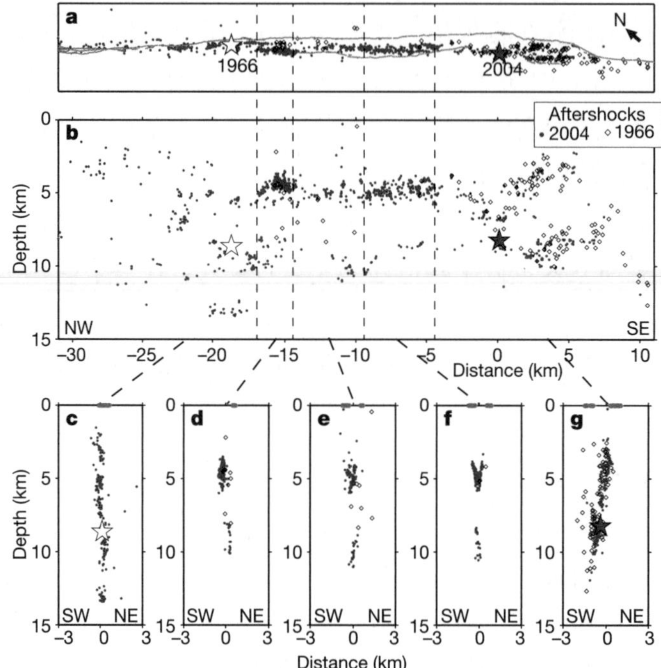

Fig. 2. Spatial distribution of Parkfield aftershocks. Locations are listed in Supplementary Table S2. Aftershocks in 2004 are shown as red dots, those in 1966 as black diamonds. Aftershocks were relocated (Supplementary Text 3) using a three-dimensional velocity model[48] and the double difference relocation technique[44]. San Andreas fault traces (purple lines) and the 1966 (open star) and the 2004 (red star) main shock hypocentres are also shown. **a**, Map view. **b**, Along-fault section. **c–g**, Cross sections for the fault sections shown in **a** and **b**. The purple dots at zero depth indicate the traces shown in **a**. The aftershocks in **f** reveal multiple strands[49] activated by the main shock.

An alternative explanation for the boundaries of the Parkfield segment is based on fault zone rheology. This segment is adjoined on the northwest by a creeping section where, perhaps, stable sliding precludes large earthquakes and on the southeast by a locked section, which may fail only in infrequent great earthquakes[19,21]. The reasons for creep to the northwest and locking to the southeast are not clear. Properties of materials and fluid overpressure adjacent to the fault have been proposed to explain creeping and locked fault segments[20,22]. A better knowledge of the materials and conditions within the fault zone obtained from SAFOD[10] should help to discriminate between these possibilities. Ultimately, a combination of factors, including deep fault geometry, fault rheology, and stress level, may be necessary to explain why a fault creeps or is locked and what constitutes a segment boundary.

也就是说，断层在地表的不规则性可能反映出深度边界的存在，而不是产生这些边界的主要原因。

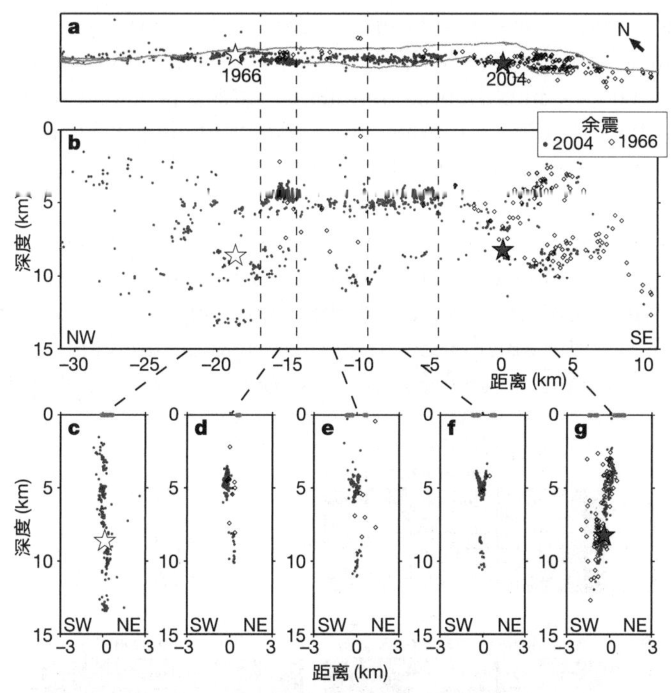

图 2. 帕克菲尔德余震的空间分布。位置在补充表 S2 中列出。2004 年的余震用红点表示，1966 年的余震用黑色菱形表示。利用三维速度模型 [48] 和双差重定位技术 [44] 重新定位余震（补充文本 3）。图中还显示了圣安德烈亚斯断层地表轨迹（紫色线），以及 1966 年（空心五角星）和 2004 年（红色五角星）的主要震源位置。图 **a** 为平面示意图。图 **b** 为沿断层的剖面。图 **c**～**g** 为图 **a** 和 **b** 中所示断层正交剖面，在 0 深度处的紫色点表示在图 **a** 中的地表断层。在图 **f** 中的余震表明主震破裂有多个分支 [49]。

对帕克菲尔德分段边界的另一种解释是基于断层带的流变特性。帕克菲尔德分段在西北部与蠕滑段毗连，该蠕滑段稳定的滑动可能会阻止大地震的发生；而在东南部与闭锁段相连，该闭锁段可能比较少发生大地震 [19,21]。西北部蠕滑和东南部闭锁的原因尚不清楚。介质特性和流体高压注入等被用于解释蠕滑段和闭锁段 [20,22]。由 SAFOD 计划所获得的对断层带内介质性质和环境条件等的进一步认识 [10] 将有助于区别这些可能性。最终，包括深层断层几何学、断层流变性和应力水平等的多因素综合研究对于解释为什么断层是蠕滑或是闭锁，以及哪些因素构成了断层分段边界的条件等是很有必要的。

Seismic and Aseismic Slip

Over the long term, both seismic and aseismic slip along plate boundaries like the San Andreas fault accommodate the relative motion between the plates. To apply even the simplest mechanical model for the build-up of strain and its sudden release in earthquakes, one must account for slip on the fault that occurs aseismically. Aseismic slip has been recognized as a potentially important component of the slip budget on strike-slip faults in the San Francisco Bay area[23] and on megathrust faults along subduction plate boundaries in the Pacific Northwest region of the United States[24] and in Japan[25]. How seismic and aseismic slip are distributed over a fault and how much and where aseismic slip occurs during the times between large earthquakes are, however, not well resolved. Measurements of slip beginning shortly after the 1966 Parkfield earthquake[13,26] appear to provide some insight into these questions. The region of maximum slip in the 2004 event appears to partially fill a deficit in the distribution of slip that had accumulated beginning with the 1966 earthquake[13], but some slip-deficient regions apparently remain (Fig. 3).

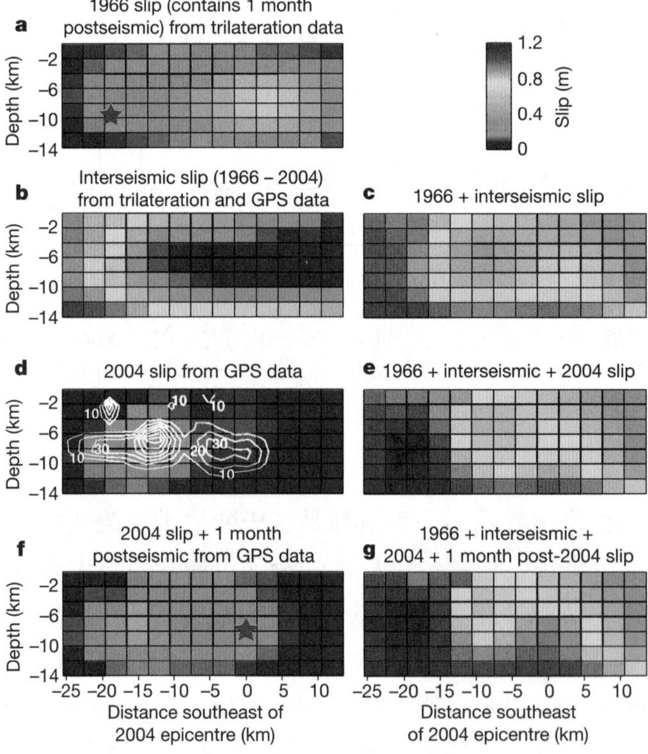

Fig. 3. Distribution of slip on the San Andreas fault since 1966 estimated from geodetic data. **a**, **b**, **d**, **f**, Slip; **c**, **e**, **g**, accumulated slip. Slip in the 2004 earthquake (**d**) concentrated near an area with an apparent slip deficit (compare **c** and **d**). In **d** we overlay contours of slip, estimated using both geodetic and seismologic data, giving a higher-resolution image of the slip distribution with a peak slip of 77 cm (ref. 50). These slip models illustrate how slip in earthquakes (coseismic and postseismic) combines with aseismic slip between earthquakes to generate the cumulative offset across the fault. Slip values are listed in Supplementary Table S3.

地震滑动和无震滑动

从长期来看，沿着像圣安德烈亚斯断层这样的板块边界带，地震滑动和无震滑动都调节着板块之间的相对运动。因此，即使用最简单的力学模型来建立地震应变积累和瞬间释放，人们也必须考虑断层上发生的无震滑动。无论是在旧金山湾地区[23]走滑断层的滑动预估方面，还是沿着美国西北太平洋地区[24]和日本[25]俯冲板块边界的巨型逆冲断层的滑动预估方面，无震滑动都被认为是潜在的重要组成部分。然而，地震滑动和无震滑动在断层上是如何分布的，在大地震间隔期间有多少无震滑动，以及在何处发生无震滑动，都没有得到很好的解决。1966 年帕克菲尔德地震后不久很快开始的滑动测量[13,26]似乎可以对这些问题提供一些见解。2004 年地震的最大滑动地区似乎部分地填补了自 1966 年地震开始累积的滑移分布亏损[13]，但一些滑动亏损地区显然仍然存在(图 3)。

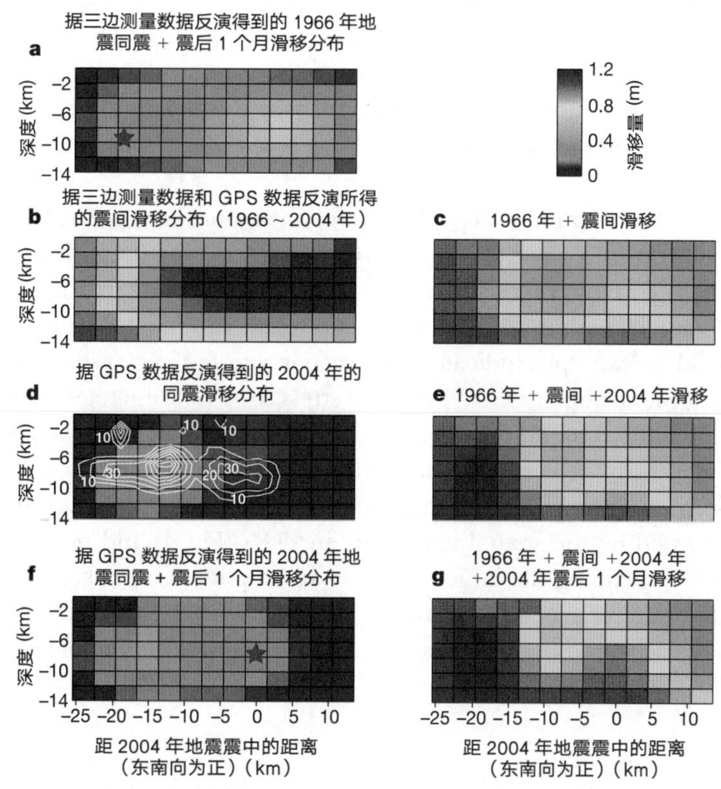

图 3. 由大地测量数据估计的自 1966 年以来圣安德烈亚斯断层的滑移量分布。图 **a**、**b**、**d**、**f** 表示滑移分布，图 **c**、**e**、**g** 表示累积滑移。2004 年地震(图 **d**)中的滑移集中分布在滑移明显亏损的地区附近(比较图 **c** 和图 **d**)。图 **d** 中的位移等值线是利用大地测量和地震学数据联合反演所得的高精度位错图像分布绘制而成，最大位错为 77 cm(参考文献 50)。这些位错模型说明了地震引起的位错(同震和震后)如何与震间的无震滑动共同作用构成了断层两侧的累积位错分布。滑移量值列在补充表 S3 中。

Postseismic surface slip of 35–45 cm was observed using alinement arrays following the 1966 Parkfield earthquake[27]. The 2004 Parkfield event, however, is the first at this location for which the geodetic data were recorded during and after the earthquake with sufficient temporal and spatial resolution to enable separation of the coseismic and postseismic signals. Postseismic slip equal to about 60% of the coseismic slip occurred in the first month after the 2004 event (Fig. 3f). Alinement array data from the 2004 earthquake suggest that near-surface slip will reach 20–50 cm over the next 2–5 yr (ref. 28), comparable to what was seen after the 1966 event[27]. Additionally, the Global Positioning System (GPS) data suggest that postseismic effects may persist for a decade, and that ultimately, the slip associated with this earthquake (coseismic plus postseismic) will balance the estimated slip deficit that existed on the fault at the time of the earthquake.

Prediction of Damaging Ground Motion

Most of the catastrophic damage in earthquakes occurs close to the earthquake source, but relatively few recordings of strong shaking close to an earthquake have been made. The ground motion near the 2004 earthquake[29] was recorded at eight sites within 1 km of the rupture and at 40 sites between 1 and 10 km from the rupture (Fig. 4), nearly doubling the global data set of strong-motion records within those distances. These records show the wavefield in unprecedented detail[14] and reveal large spatial variations in shaking amplitude.

The peak horizontal acceleration (PGA) for two of the records are greater than 1.0g (where g is the acceleration due to gravity) with one of these exceeding the instrument's recording capacity of 2.5g. Recordings of accelerations greater than 1g are rare, but they may not be anomalous at locations within a few kilometres of fault ruptures. Preliminary seismic slip models (see Fig. 3d for example) indicate slip was concentrated in two small regions, near the stations recording the strongest PGA. The local stress drop in these regions of concentrated slip appears to be more than an order of magnitude larger than the average stress drop of 0.2 MPa associated with the very smooth geodetic slip model (Fig. 3d). Temporal variations in rupture propagation, however, probably also influenced the radiation of the strongest shaking, as illustrated by the spatial variability in PGA (Fig. 4) and peak ground velocity close to the fault. Explaining the large variations in amplitude over distances of just a few kilometres continues to challenge our understanding of earthquake rupture dynamics and our ability to predict ground motions near the rupture (Fig. 4b and Supplementary Fig. S3).

1966 年帕克菲尔德地震后，研究人员利用线性台阵观测到 35 ~ 45 cm 的震后地表滑动[27]。然而，2004 年帕克菲尔德地震是第一次在该地点对地震和震后进行大地测量观测记录，数据的时间和空间分辨率足以将同震和震后信号分离开来。2004 年震后一个月的滑移量约为同震滑移的 60%(图 3f)。2004 年地震的线性台阵数据表明，在未来 2 ~ 5 年内，近地表滑移量将达到 20 ~ 50 cm(参考文献 28)，与 1966 年地震震后的情况相当[27]。此外，全球定位系统(GPS)数据表明，震后效应可能会持续 10 年。最终，与这次地震相关的滑动(同震 + 震后)将平衡地震时断层上存在的预估滑移亏损。

破坏性的地面运动的预测

地震灾害大多发生在震源附近，但关于震源附近发生强烈震动的记录相对较少。2004 年地震附近的地面运动[29]，在断裂带 1 km 内有 8 个台站记录到，在距离断裂带 1 ~ 10 km 范围内有 40 个台站记录到(图 4)，几乎是全球关于这些距离内发生强地面运动记录数据集的两倍。这些记录比以往都详细地显示了波场[14]，揭示了震动振幅在空间上的大幅度变化。

两个台站记录的峰值水平加速度(PGA)大于 1.0g(g 为重力加速度)，其中一个超过仪器的记录范围 2.5g。加速度大于 1g 的记录很少，但它们在断层破裂数千米范围内可能并没有异常。初步的地震滑移模型(如图 3d 所示)表明，滑动集中在两个小区域内，靠近记录最强 PGA 的台站。与非常平滑的大地测量数据测得的位错模型所得的平均应力降 0.2 Mpa 相比，这些滑动集中区域的局部应力降大了至少一个量级(图 3d)。然而，如 PGA 的空间变化(图 4)和靠近断层的强地面运动峰值速度所示，破裂传播的时间变化也可能影响最强震动的辐射。对仅几千米距离内振幅的巨大变化的解释，继续挑战着我们对地震破裂动力学的理解以及预测破裂附近地面运动的能力(图 4b 和补充图 S3)。

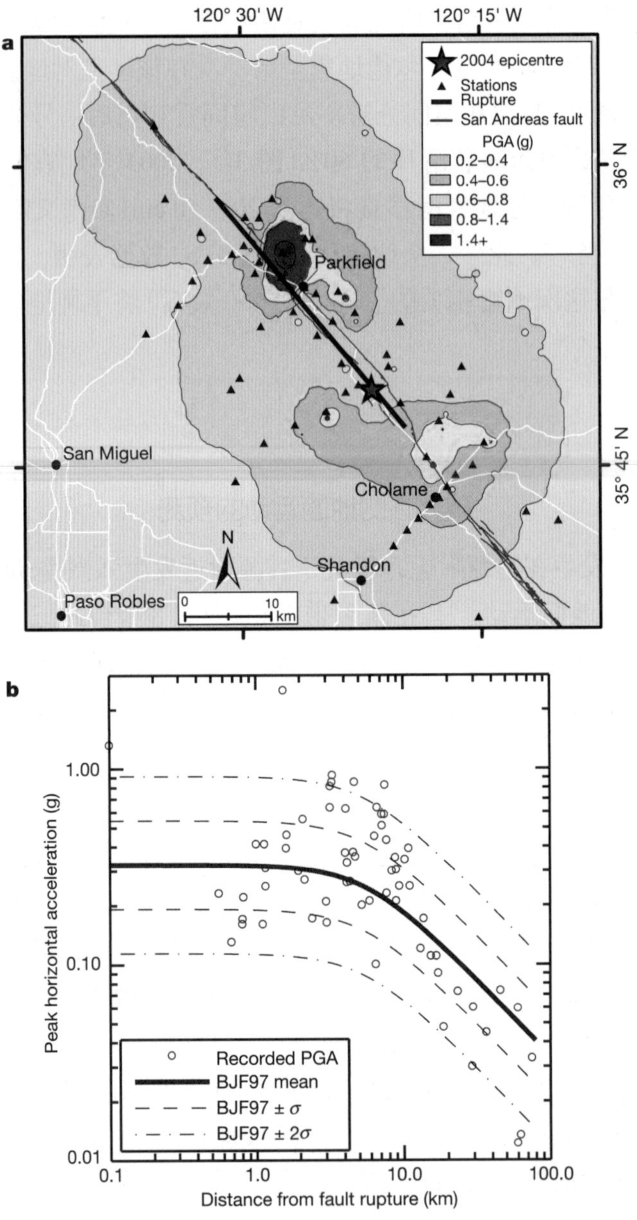

Fig. 4. Horizontal PGA from ShakeMap[51]. **a**, Map view. Station locations shown by triangles. The thin red line delineates the Alquist-Priolo fault traces[30] and the thick black line is the fault trace based on aftershock locations. **b**, Horizontal PGA as a function of distance from the fault rupture. The distances are based on the approximate projection of the fault to the ground surface. The mean (solid line) and $\pm 1\sigma$ (dashed lines) and $\pm 2\sigma$ (dash-dotted lines) for the Boore–Joyner–Fumal 1997 attenuation relation[52] are shown. The PGA which exceeded the 2.5g limit of the instrument is plotted at 2.5g.

Spatial variations in the intensity of shaking, such as peak ground acceleration and peak ground velocity, are often attributed to four factors: differences in soil conditions among sites, differences in the wave propagation to the sites, complexities of the rupture geometry, and heterogeneity of slip on the fault. Analysis of the strong-motion records from the 2004

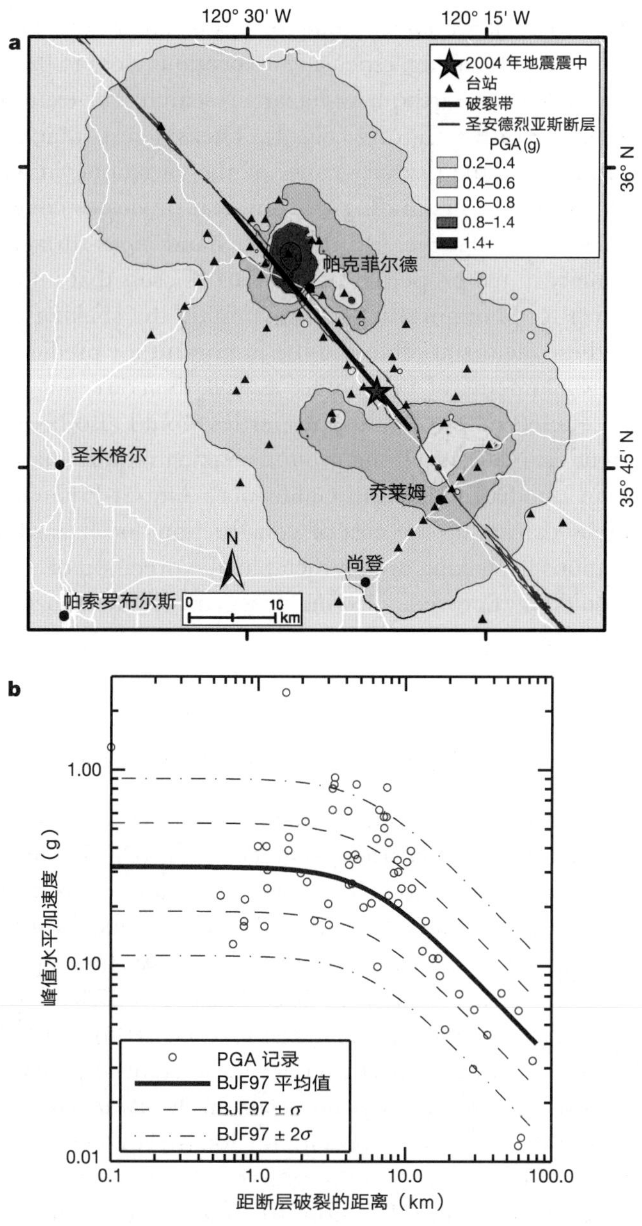

图 4. 震动图中的水平 PGA[51]。图 **a** 为地图，三角形表示观测台站位置，细红线表示阿尔奎斯特–普里奥洛断层 [30]，粗黑线表示基于余震位置勾勒出的断层。图 **b** 为水平 PGA，它是离断层破裂距离的函数。距离的计算基于断层在地表的近似投影。图中显示了布勒–乔伊纳–菲马尔 1997 衰减关系 [52]：平均值(实线)、±1σ(虚线) 和 ±2σ(点划线)。在图中超过仪器 2.5g 限值的 PGA 用 2.5g 表示。

震动强度的空间变化，如强地面运动峰值加速度和强地面运动峰值速度，通常归因于四个因素：观测台站之间土壤条件的差异、波向各个台站传播的差异、破裂几何结构的复杂性和断层滑动的不均匀性。对 2004 年地震的强地面运动记录的分

earthquake should lead to a fuller understanding of how each of these factors contributes to the spatial variability in strong shaking, especially at locations close to the fault rupture. This has significant ramifications for earthquake hazard research. For example, the variability in PGA was greatest close to the rupture (Fig. 4). This suggests that complexities in the seismic source may have been the primary cause of the variations, in which case research with a greater emphasis on understanding the physical processes controlling complexity of the source would be most effective. On the other hand, near-surface soil conditions at the site and heterogeneity in the properties of the Earth's crust that influence seismic-wave propagation are known to be important for determining the shaking at locations farther from the rupture. If these factors are found to be important for predicting the distribution of shaking within a few kilometres of the rupture as well, then directing additional resources towards developing detailed maps of these properties would also be effective. The large uncertainties in current estimates of strong ground shaking require that societal guidelines, including the Uniform Building Code and California's Alquist-Priolo Fault Zoning Act[30], be conservative, thereby driving up the cost of construction and hazard mitigation. To the extent that such variations in shaking are predictable, the precision of seismic hazard maps and building codes could be improved, allowing necessarily limited hazard mitigation funds to be used more effectively.

Long-term Non-randomness of Earthquakes

The notion that large earthquakes tend to occur as similar-size "characteristic" events on fixed segments of a fault and that these segments are identifiable from geologic and geophysical data arose in the 1980s (refs 3, 19 and 31) and remains central to fault-based Probabilistic Seismic Hazard Analysis (PSHA). PSHA also includes other approaches such as smoothed seismicity models (see ref. 32), random events and multiple segment ruptures (see ref. 23). The sequence of earthquakes at Parkfield since 1857 has long been considered a prime example of the recurrence of a characteristic earthquake[3,5,23]. Two classes of characteristic earthquakes have been considered for these events. In the first, events have the same faulting mechanism and magnitude, and occur on the same fault segment[31]; this class of characteristic behaviour is most appropriate for long-term forecasting of earthquakes and is often inferred from paleoseismic investigations. In the second class, the events also have the same epicentre and rupture direction[3]. If events in the second class were further constrained to have the same rupture time history and distribution of slip, then this class of recurrent behaviour would imply low variability in the distribution of strong ground shaking among the recurrences of characteristic events.

The 1934, 1966 and 2004 Parkfield earthquakes are remarkably similar in size (see Fig. 5 for example) and location of rupture, albeit not in epicentre or rupture propagation direction. The aftershocks of the 1966 and 2004 earthquakes delineate many of the same fault structures (Fig. 2). Furthermore, most of the observations available for the Parkfield earthquakes in 1881, 1901 and 1922 are consistent with the hypothesis that these earlier earthquakes were similar in size and general location to the later events[3,5]. Owing to limited

析应能使人们更全面地了解这些因素如何分别对强震动的空间变化产生影响，尤其是在断层破裂附近的地方。这对地震灾害研究具有重要的影响。例如，在靠近破裂的地方，PGA 的变化最大(图 4)。这表明，震源的复杂性可能是造成这些变化的主要原因，在这种情况下，着重理解控制震源复杂性的物理过程的相关研究将是最有效的。另一方面，众所周知，台站近地表土壤条件和影响地震波传播的地壳介质的非均匀性，对于确定距离破裂较远位置的震动是非常重要的。如果这些因素被发现对于预测破裂带几千米范围内的震动分布也很重要，那么把更多的资源用于绘制具有这些介质属性的详细分布图，也将是有效的。目前，强震动的估计存在很大的不确定性，这需要保守的社会规范，包括统一建筑法规和加利福尼亚州的阿尔奎斯特–普里奥洛断层分区法[30]，从而提高了建设成本和减灾的成本。在这种程度下，震动变化是可预测的，地震灾害图的精度和建筑法规的精度可以得到改善，有限的减灾资金可以得到更有效的使用。

地震的长期非随机性

大地震往往在断层固定段落上以相似规模的“尺度特征”发生，关于这些断层分段的段落可以从地质和地球物理数据中识别的观点是从 20 世纪 80 年代形成的(参考文献 3、19 和 31)，并一直在基于断层的地震危险性概率分析(PSHA)中居于核心地位。PSHA 还包括其他方法，如平滑地震活动模型(见参考文献 32)、随机地震事件和多段破裂(见参考文献 23)。自 1857 年以来，帕克菲尔德一连串的地震长期被认为是特征地震重复发生的主要例子[3,5,23]。其中，两类特征地震被考虑进来。在第一类中，地震具有相同的震源机制和震级，并且发生在同一断层段上[31]；这类特征地震行为最适合长期地震预测，通常从古地震调查中推断而来。在第二类中，地震也具有相同的震中和破裂方向[3]。如果第二类地震又具有相同的破裂时间历史和滑动分布，那么这类复发行为将意味着一组重复特征地震引起的强地面震动具有很小的分布变化。

1934 年、1966 年和 2004 年帕克菲尔德地震在大小(以图 5 为例)和破裂位置上极为相似，尽管在震中或破裂传播方向上不同。1966 年和 2004 年的余震勾勒出许多相同的断层结构(图 2)。此外，1881 年、1901 年和 1922 年帕克菲尔德地震的大多数观测结果都符合这样一个假设，即这些早期地震的大小及大体位置与后来的地震相似[3,5]。由于对这些早期地震的观测有限，对帕克菲尔德主震的严格定义必须限

observations of these earlier events, a rigorous definition of the Parkfield main shocks must be limited to their overall size and their location based on rupture along the Parkfield segment[33]. Michael and Jones' definition[33] was designed to encompass the Parkfield main shocks through 1966; the 2004 main shock also satisfies their definition. Thus, the Parkfield earthquakes are consistent with the first class of characteristic earthquake behaviour. However, the variability in the spatial distribution of slip for the last three events[34,35] and the different direction of rupture propagation in the 2004 event invalidates the application of the second class of characteristic behaviour to the Parkfield earthquakes.

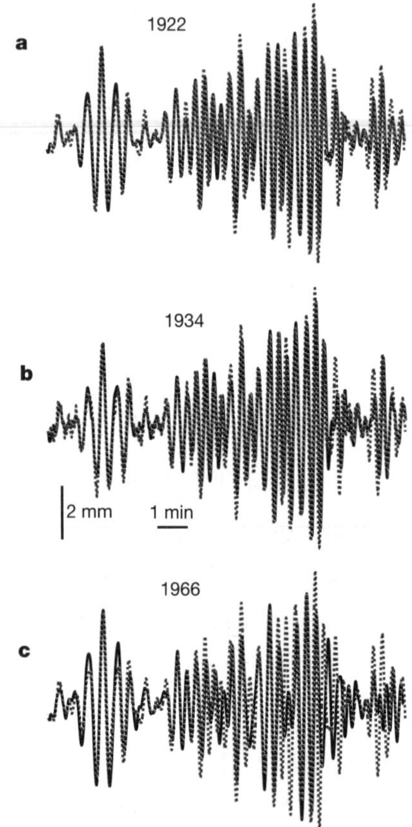

Fig. 5. Seismograms for Parkfield earthquakes at De Bilt, the Netherlands. North–south seismogram for the 2004 Parkfield earthquake (red dashed line) is plotted relative to the 1922, 1934, and 1966 Parkfield earthquakes in **a**, **b**, and **c** respectively. The 1922, 1934, and 1966 events were recorded by horizontal Galitzin seismographs. The 2004 signal, recorded by a three-component, broadband digital station located at the same site as the Galitzin seismograph, was digitally filtered to simulate a Galitzin seismograph. The similar amplitudes and waveforms imply the same seismic moment, focal mechanism, and teleseismic wave path.

We note that the six Parkfield earthquakes since 1857 have occurred with statistically significant (albeit imperfect) regularity in time—more regular than random but not sufficiently periodic to be predictable in any useful way beyond long-term statistical forecasts (Supplementary Text 2). This limited regularity underlies most of the long-term

制在沿着帕克菲尔德段破裂的总体尺寸和位置上[33]。迈克尔和琼斯的定义[33]旨在涵盖 1966 年的帕克菲尔德主震，2004 年的主震也满足他们的定义。因此，帕克菲尔德地震符合第一类特征地震行为。然而，最近的三次地震[34,35]的滑动空间分布的变化，以及 2004 年地震破裂传播方向的不同，使得第二类特征行为不再适用于帕克菲尔德地震。

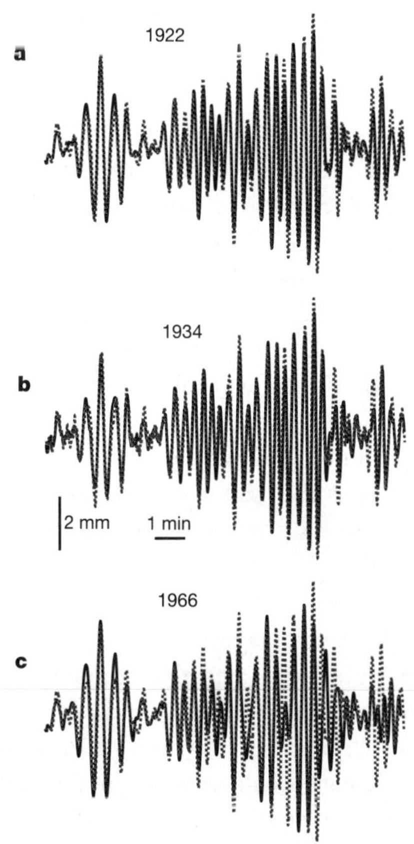

图 5. 帕克菲尔德地震在荷兰德比尔特测得的震波图。图 **a**、**b**、**c** 分别展示了 2004 年帕克菲尔德地震（红色虚线）相对于 1922 年、1934 年和 1966 年帕克菲尔德地震的北–南震波图。1922 年、1934 年和 1966 年地震都是由水平加利津地震仪记录的。2004 年的信号是由一个三分量的宽带数字观测站通过数字滤波模拟加利津地震仪记录的，该站与加利津地震仪位于同一站点。相似的振幅和波形意味着相同的地震矩、震源机制和远震地震波路径。

我们注意到，自 1857 年以来，6 次帕克菲尔德地震的发生时间在统计上具有显著（尽管不完美）的高于随机分布的规律性，但非完备的周期性，不可以用除长期统计预测外的任何有用方式进行预测（补充文本 2）。这种有限的规律性是多数长期地

prediction models proposed for the earthquakes[3,5]. The departures from perfectly regular occurrence of these earthquakes have been interpreted using physics-based variations upon the characteristic earthquake model. For example, the Parkfield Recurrence Model, used at the outset of the experiment in 1985, assumed a constant fault loading rate and failure threshold and allowed that main shocks could be triggered early by foreshocks, but did not allow for late events[5]. Consequently, it has been invalidated by the long time interval between the 1966 and 2004 main shocks. In a time-predictable model, the time between successive events is proportional to the slip of the prior event; in a slip-predictable model, the size of an earthquake is proportional to the time since the prior event[36]. Neither of these models is compatible with the sequence of Parkfield earthquake ruptures[34]. Various forms of fault interaction have been proposed to explain the variation in recurrence intervals of the earthquakes[37].

Statistical models of earthquake recurrence have also been applied to these events. The variability in the time between earthquakes implies a coefficient of variability (COV) of about 0.45 which is similar to the COV used in recent forecasts for the San Francisco Bay Area[23] but greater than that proposed in earlier models for the Parkfield sequence[38]. Earthquake activity over a wide range of smaller magnitudes ($M_w = 0$ to 5) also occurs at Parkfield. Clusters of microearthquakes that produce nearly identical waveforms repeatedly rupture small, fixed patches of the fault—some with remarkable regularity. Many of these clusters have characteristic recurrence times of months to years that scale with the magnitude of the repeating events. Changes in this recurrence time have been used to infer that slip rates over portions of the fault vary with time[39]. Similar to the Parkfield main shocks, models of these events suggest that they may balance their local slip budget with a mix of seismic and aseismic slip[40].

The earthquakes at Parkfield, both large and small, provide a fertile laboratory for testing and refining the characteristic earthquake concept by offering information on slip distribution, rupture dynamics and afterslip, and for testing models of earthquake recurrence and interaction, which are central to contemporary earthquake hazard assessment[23]. (The characteristic earthquake model can also be tested using global data sets. Kagan and Jackson[41] concluded that too few of Nishenko's[42] predicted gap-filling circum-Pacific earthquakes occurred in the first 5 yr.) Although the Parkfield earthquake history supports the use of characteristic events for earthquake forecasting, this topic remains controversial[42,43] and we must consider whether conclusions drawn from observations at Parkfield will transfer to other seismogenic regions. For instance, does the presence of aseismic slip to the north of and within the Parkfield segment yield unusual earthquake behaviour? Observations of the large amount of postseismic slip following the 2004 earthquake suggest that it may help to balance the slip budget. Faults that do not slip aseismically may exhibit more irregular behaviour because other large events are needed to balance the slip budget. There are other faults, however, with aseismic slip that produce small repeating events and may thus produce characteristic events similar to those observed at Parkfield. These include the Hayward and Calaveras faults in California and partially coupled subduction zones[44-46].

震预测模型的基础[3,5]。与地震完备的规律的偏离，可以基于特征地震模型用物理基础的变化来解释。例如，1985 年的实验场开始时使用的帕克菲尔德重复地震模型，假设了恒定的断层加载速率和固定的失稳阈值，并认为主震可能由前震触发，然而后来的地震并非如此[5]。因此，基于该重复地震模型给出的长期地震预测，被 1966 年和 2004 年两次主震间的长时间间隔所推翻。在时间可预测模型中，连续地震之间的时间与上一次地震的滑动量成比例；在震级可预测模型中，地震的大小与上一次地震以来的时间成比例[36]。这两种模型都不适合帕克菲尔德地震序列[34]。于是就提出了各种断层相互作用的理论，用于解释地震复发间隔的变化[37]。

应用地震复发的多种统计模型来分析帕克菲尔德地震序列时发现，地震之间的时间变化表明可变化性系数（COV）为 0.45 左右，与最近用于预测旧金山湾区的 COV[23] 相当，但高于早期模型对帕克菲尔德序列所评估出的 COV 值[38]。在帕克菲尔德也有较小震级（$M_W = 0 \sim 5$）的地震活动。微小地震群产生几乎相同的波形，并反复发生使小而固定的断层带破裂，其中一些具有显著的规律性。这些地震群中的许多地震具有几个月到几年的特征复发时间，特征复发时间与重复地震的震级成比例关系。复发时间的变化被用来推断断层段落的滑动速率随时间的变化[39]。与帕克菲尔德主震相似，这些地震的模型表明，它们以地震滑动和无震蠕滑的形式来调节局部滑动预估[40]。

帕克菲尔德地震，无论大小，都搭建了一个丰富的实验平台，通过提供有关滑动分布、破裂动力学和震后余滑的信息来检验和完善特征地震概念，并检验地震复发和相互作用的模型，这些模型在现代地震灾害评估中居于核心地位[23]。（特征地震模型也可以使用全球数据集进行测试。卡根和杰克逊[41] 得出的结论是，尼森科[42] 预测的填充空区的环太平洋地震前 5 年发生得太少。）尽管帕克菲尔德地震历史支持使用特征地震进行地震预测，但这一问题仍存在争议[42,43]，我们必须考虑从帕克菲尔德的观测中得出的结论是否可以用到其他孕震区。例如，在帕克菲尔德段的北部及其内部存在的无震滑动，会产生异常的地震行为吗？ 2004 年地震后，震后较大的滑动量观测结果表明，震后滑动可能有助于平衡滑动预估。不发生无震滑动的断层可能表现出更不规律的行为，因为需要其他大地震来平衡滑动预估。然而，还有其他一些进行无震滑动的断层，会产生小的重复地震，因此可能产生类似于在帕克菲尔德观测到的特征地震。这些断层包括加利福尼亚的海沃德断层和卡拉韦拉斯断层，以及部分耦合俯冲带[44-46]。

Implications for Future Research

The magnitude and rupture extent of the 2004 Parkfield earthquake were correctly anticipated, but its time of occurrence clearly was not. This suggests that long-term earthquake forecasts require models that include higher degrees of variability (for example, see ref. 23). Although the 2004 Parkfield earthquake was ideally located within a dense monitoring network specifically designed to detect foreshocks and other possible short-term precursors, no significant signals were detected. This documented absence of clear precursory activity sets stringent bounds on the processes that preceded this earthquake. Attempts to detect short-term precursory strain changes near several other recent $M_w = 5.3–7.3$ earthquakes in California and Japan have also failed[47]. These experiences demonstrate that reliable short-term earthquake prediction (up to a few weeks in advance) will be very difficult at best. Although the search for precursors should not be abandoned, we should thoroughly explore other ways to mitigate losses in earthquakes.

Earthquake loss mitigation begins with hazard assessment. Improved hazard assessment will require incorporating the observed variability in both earthquake sources and the resulting ground motions into probabilistic seismic hazard analysis. The history of events at Parkfield and the detailed observations of the 2004 event have revealed variability in intervals between earthquakes, variations in slip distributions of large events, and spatial variations in strong ground motion. Incorporating realistic variability into hazard assessments will entail sophisticated three-dimensional numerical models that can accurately explore many seismic cycles and include the build-up of strain via plate motions, dynamic stress changes during rupture, and postseismic deformation. Such models must be able to explain the interaction of aseismic and seismic slip, the segmentation of faults, and the strong spatial variations in the intensity of strong shaking. Complementary data from *in situ* studies of the Earth's crust, such as SAFOD[10], laboratory experiments that recreate the conditions of faults in the Earth, and continued seismic monitoring will be needed to constrain the numerical models.

The value of the unique long-term record of crustal deformation being collected at Parkfield suggests that the monitoring there should continue. Parkfield-like experiments embedded within broader monitoring networks in other locations can provide similarly valuable data for faults in other tectonic contexts. Additional selected faults in California, which are already contained within sparse monitoring networks, should be densely instrumented. Large earthquakes are also anticipated on known fault segments in China, Japan, Turkey and elsewhere, and international cooperation should be sought to develop comprehensive monitoring in these regions. The Parkfield experiment showed that diligence is required to maintain these specialized networks until a large earthquake occurs, and the detailed observations made at Parkfield demonstrate how valuable such perseverance can be for advancing our understanding of earthquakes.

(**437**, 969-974; 2005)

对未来研究的启示

2004 年帕克菲尔德地震的震级和破裂程度得到了正确的预测，但对发震时间的预测显然不准。这意味着地震长期预测需要具有更高变化维度的模型（比如参考文献 23）。虽然 2004 年帕克菲尔德地震十分理想地发生在专门设计用于监测前震和其他可能的短期前兆的密集监测台网内，但监测台网并未观测到有意义的信号。明确的前兆活动的缺失对这次地震发生前的过程研究提出了非常严峻的挑战。在加利福尼亚州和日本的几次最近发生的 $M_W = 5.3 \sim 7.3$ 的地震附近，检测短期前兆应变变化的尝试也失败了 [47]。这些经验表明，可靠的短期地震预测（震前几周）无论如何都是非常困难的。尽管我们不应放弃对地震前兆的探索，但我们更应该深入探索减轻地震损失的其他多种途径。

减少地震损失从灾害评估开始。改进后的灾害评估需要将观测到的震源变化和由此产生的地面运动都纳入概率地震灾害分析中。帕克菲尔德地震的历史和 2004 年地震的详细观测已经揭示了地震时间间隔的可变动性、大地震滑动分布的变化以及强地面震动的空间变化。在灾害评估中考虑这些现实的可变动性需要复杂的三维数值模型，以准确地模拟多个地震周期并考虑板块运动中的应变累积、破裂过程中的动态应力变化和震后变形。这些模型必须能够解释无震滑动和地震滑动的相互作用、断层的分段性以及强震动的剧烈空间变化，从已有地壳研究的补充数据，如 SAFOD[10]，重建地球中断层活动的条件和实验环境，且以连续的地震监测来约束数值模型。

在帕克菲尔德收集到的地壳形变的唯一长期记录表明，该地区的监测应该继续下去。在其他地方配备更为广泛监测台网的类似帕克菲尔德的地震预测实验场，可以为当地的构造环境提供同样有价值的资料。加利福尼亚州地区，对已处于稀疏监测台网中的选定的断层，应该加密布置仪器观测。中国、日本、土耳其和其他地区，大地震也被预测在某些已知断层的段落上，这些地区应寻求国际合作开展综合监测。帕克菲尔德地震实验场表明，在发生大地震之前，这些专门的网络需要付出艰辛来继续坚持。在帕克菲尔德地区的详细观测表明，这种长期观测的坚持不懈对于增进我们对地震的理解是多么有价值。

（孙惠南 郭思彤 翻译；尹凤玲 审稿）

W. H. Bakun[1], B. Aagaard[1], B. Dost[2], W. L. Ellsworth[1], J. L. Hardebeck[1], R. A. Harris[1], C. Ji[3], M. J. S. Johnston[1], J. Langbein[1], J. J. Lienkaemper[1], A. J. Michael[1], J. R. Murray[1], R. M. Nadeau[4], P. A. Reasenberg[1], M. S. Reichle[5], E. A. Roeloffs[6], A. Shakal[5], R. W. Simpson[1] & F. Waldhauser[7]

[1] US Geological Survey, Menlo Park, California 94025, USA

[2] Royal Netherlands Meteorological Institute (KNMI), Seismology Division, PO Box 201, 3730AE De Bilt, The Netherlands

[3] Division of Geological and Planetary Sciences, Caltech, Pasadena, California 91125, USA

[4] University of California, Berkeley Seismological Laboratory, Berkeley, California 94720, USA

[5] California Geological Survey, Sacramento, California 95814, USA

[6] US Geological Survey, Vancouver, Washington 98683, USA

[7] Lamont-Doherty Earth Observatory, Columbia University, Palisades, New York 10964, USA

Received 28 January; accepted 10 July 2005.

References:

1. Sieh, K. E. Slip along the San Andreas Fault associated with the great 1857 earthquake. *Bull. Seismol. Soc. Am.* **68**, 1421-1447 (1978).
2. Toppozada, T. R., Branum, D. M., Reichle, M. S. & Hallstrom, C. L. San Andreas fault zone, California; M ⩾ 5.5 earthquake history. *Bull. Seismol. Soc. Am.* **92**, 2555-2601 (2002).
3. Bakun, W. H. & McEvilly, T. V. Recurrence models and Parkfield, California, earthquakes. *J. Geophys. Res.* **89**, 3051-3058 (1984).
4. Shearer, C. F. Southern San Andreas fault geometry and fault zone deformation: implications for earthquake prediction (National Earthquake Prediction Council Meeting, March, 1985). *US Geol. Surv. Open-file Rep.* **85-507**, 173-174 (USGS, Reston, Virginia, 1985).
5. Bakun, W. H. & Lindh, A. G. The Parkfield, California, earthquake prediction experiment. *Science* **229**, 619-624 (1985).
6. Sherburne, R. W. Ground shaking and engineering studies near the San Andreas fault zone. *Calif. Geol.* **41**, 27-32 (1988).
7. Bakun, W. H. *et al.* Parkfield, California, earthquake prediction scenarios and response plans. *US Geol. Surv. Open-file Rep.* **87-192**, 1-45 (USGS, Reston, Virginia, 1987).
8. Hill, D. P. *et al.* Response plans for volcanic hazards in the Long Valley caldera and Mono Craters area, California. *US Geol. Surv. Open-file Rep.* **91-270**, 1-64 (USGS, Reston, Virginia, 1991).
9. USGS *Advanced National Seismic System* ⟨http://earthquake.usgs.gov/anss/⟩ (USGS Earthquake Hazards Program, 2000).
10. Hickman, S., Zoback, M. & Ellsworth, W. Introduction to the special section: preparing for the San Andreas Fault Observatory at depth. *Geophys. Res. Lett.* **31**, L12S01, doi:10.1029/2004GL020688 (2004).
11. Earthscope Project *Exploring the Structure and Evolution of the North American Continent* ⟨http://www.earthscope.org/⟩ (2003).
12. National Earthquake Prediction Evaluation Council Working Group. Earthquake research at Parkfield, California, 1993 and beyond—report of the NEPEC working group to evaluate the Parkfield earthquake prediction experiment. *US Geol. Surv. Circ.* **1116**, 1-14 (USGS, Reston, Virginia, 1994).
13. Langbein, J. *et al.* Preliminary report on the 28 September 2004, M6.0 Parkfield, California, earthquake. *Seismol. Res. Lett.* **76**, 10-26 (2005).
14. Shakal, A. *et al.* Preliminary analysis of strong-motion recordings from the 28 September 2004 Parkfield, California earthquake. *Seismol. Res. Lett.* **76**, 27-39 (2005).
15. Rikitake, T. *Earthquake Prediction* 7-26 (Elsevier, Amsterdam, Netherlands, 1976).
16. Bakun, W. H. & McEvilly, T. V. Earthquakes near Parkfield, California; comparing the 1934 and 1966 sequences. *Science* **205**, 1375-1377 (1979).
17. Nadeau, R. M. & Dolenc, D. Nonvolvanic tremors deep beneath the San Andreas fault. *Science* **307**, 389, doi:10.1126/science.1107142 (2004).
18. Harris, R. A. Numerical simulations of large earthquakes: Dynamic rupture propagation on heterogeneous faults. *Pure Appl. Geophys.* **161**, 2171-2181, doi:10.1007/s00024-004-2556-8(2004).
19. Lindh, A. G. & Boore, D. M. Control of rupture by fault geometry during the 1966 Parkfield earthquake. *Bull. Seismol. Soc. Am.* **71**, 95-116 (1981).
20. Eberhart-Phillips, D. & Michael, A. J. Three-dimensional velocity structure, seismicity, and fault structure in the Parkfield region, Central California. *J. Geophys. Res.* **98**, 15737-15758 (1993).
21. Liu, J., Klinger, Y., Sieh, K. & Rubin, C. Six similar sequential ruptures of the San Andreas Fault, Carrizo Plain, California. *Geology* **32**, 649-652 (2004).
22. Irwin, W. P. & Barnes, I. Effect of geologic structure and metamorphic fluids on seismic behaviour of the San Andreas fault system in central and northern California. *Geology* **3**, 713-716 (1975).
23. Working Group on California Earthquake Probabilities. Earthquake probabilities in the San Francisco Bay region: 2002-2031. *US Geol. Surv. Open-file Rep.* **03-214**, 1-340 (USGS, Reston, Virginia, 2003).
24. Dragert, H., Wang, K. L. & James, T. S. A silent slip event on the deeper Cascadia subduction interface. *Science* **292**, 1525-1528 (2001).
25. Kawasaki, I. *et al.* The 1992 Sanriku-Oki, Japan, ultra-slow earthquakes. *J. Phys. Earth* **43**, 105-116 (1995).
26. Harris, R. A. & Segall, P. Detection of a locked zone at depth on the Parkfield, California segment of the San Andreas fault. *J. Geophys. Res.* **92**, 7945-7962 (1987).
27. Lienkaemper, J. J. & Prescott, W. H. Historic surface slip along the San Andreas fault near Parkfield, California. *J. Geophys. Res.* **94**, 17647-17670 (1989).
28. Lienkaemper, J. J., Baker, B. & McFarland, F. S. Slip in the 2004 Parkfield, California earthquake measured on alinement arrays. *Eos* **85**(47), Abstr. S54B- 01 (2004).
29. CISN *Strong Motion Engineering Data Center* ⟨http://www.quake.ca.gov/cisn-edc/⟩ (2000).

30. California Geological Survey *Alquist-Priolo Earthquake Fault Zones* 〈http://www.consrv.ca.gov/CGS/rghm/ap/〉 (1998).

31. Schwartz, D. P. & Coppersmith, K. J. Fault behaviour and characteristic earthquakes; examples from the Wasatch and San Andreas fault zones. *J. Geophys. Res.* **89**, 5681-5698 (1984).

32. Frankel, A. D. *et al.* USGS national seismic hazard maps. *Earthq. Spect.* **16**, 1-19 (2000).

33. Michael, A. J. & Jones, L. M. Seismicity alert probabilities at Parkfield, California, revisited. *Bull. Seismol. Soc. Am.* **88**, 117-130 (1998).

34. Murray, J. R. & Segall, P. Testing time-predictable earthquake recurrence by direct measure of strain accumulation and release. *Nature* **419**, 287-291 (2002).

35. Segall, P. & Du, Y. How similar were the 1934 and 1966 Parkfield earthquakes? *J. Geophys. Res.* **98**, 4527-4538 (1993).

36. Shimazaki, K. & Nakata, T. Time-predictable recurrence model for large earthquakes. *Geophys. Res. Lett.* **7**, 279-282 (1980).

37. Ben-Zion, Y., Rice, J. R. & Dmowska, R. Interaction of the San Andreas fault creeping segment with adjacent great rupture zones, and earthquake recurrence at Parkfield. *J. Geophys. Res.* **98**, 2135-2144 (1993).

38. Savage, J. C. The Parkfield Prediction Fallacy. *Bull. Seismol. Soc. Am.* **83**, 1-6 (1993).

39. Nadeau, R. M. & McEvilly, T. V. Fault slip rates at depth from recurrence intervals of repeating earthquakes. *Science* **285**, 718-721 (1999).

40. Beeler, N. M., Lockner, D. A. & Hickman, S. H. A simple creep-slip and stick-slip model for repeating earthquakes and its application to micro-earthquakes at Parkfield. *Bull. Seismol. Soc. Am.* **91**, 1797-1804 (2001).

41. Kagan, Y. Y. & Jackson, D. D. New seismic gap hypothesis: Five years after. *J. Geophys. Res.* **100**, 3943-3959 (1995).

42. Nishenko, S. P. Circum-Pacific seismic potential—1989-1999. *Pure Appl. Geophys.* **135**, 169-259 (1991).

43. Main, I. *Is reliable earthquake prediction of individual earthquakes a realistic scientific goal?* 〈http://www.nature.com/nature/debates/earthquake/equake_contents.html〉 (Nature Debate 25 February to 8 April 1999).

44. Waldhauser, F. & Ellsworth, W. L. A double-difference earthquake location algorithm; method and application to the northern Hayward Fault, California. *Bull. Seismol. Soc. Am.* **90**, 1353-1368 (2000).

45. Schaff, D. P., Beroza, G. C. & Shaw, B. E. Post-seismic response of repeating earthquakes. *Geophys. Res. Lett.* **107**, B9, doi:10.1029/2001JB000633 (1998).

46. Uchida, N., Matsuzawa, T., Igarashi, T. & Hasegawa, A. Interplate quasi-static slip off Sanriku, NE Japan, estimated from repeating earthquakes. *Geophys. Res. Lett.* **30**, doi:10.1029/2003GL017452 (2003).

47. Johnston, M. J. S. & Linde, A. T. *Implications* of crustal strain during conventional, slow, and silent earthquakes. *Int. Handbk Earthq. Eng. Seismol.* **81A**, 589-605 (2002).

48. Michael, A. J. & Eberhart-Phillips, D. M. Relations among fault behaviour, subsurface geology, and three-dimensional velocity models. *Science* **253**, 651-654 (1991).

49. Waldhauser, F., Ellsworth, W. L., Schaff, D. P. & Cole, A. Streaks, multiplets, and holes: high-resolution spatio-temporal behaviour of Parkfield seismicity. *Geophys. Res. Lett.* **31**, L18608, doi:10.1029/2004GL020649 (2004).

50. Ji, C., Choi, K. K., King, N., Larson, K. M. & Hudnut, K. W. Co-seismic slip history and early afterslip of the Parkfield earthquake. *Eos* **85**(47), Abstr. S53D-04 (2004).

51. CISN ShakeMap. *Estimated Instrumental Intensity* 〈http://www.quake.ca.gov/shake/index.html〉 (1997).

52. Boore, D. M., Joyner, W. B. & Fumal, T. E. Equations for estimating horizontal response spectra and peak acceleration from Western North American earthquakes: A summary of recent work. *Seismol. Res Lett.* **68**, 128-153 (1997).

Supplementary Information is linked to the online version of the paper at www.nature.com/nature.

Acknowledgements. The Parkfield experiment has served as a model for the collaboration of federal and state agencies with researchers in academia and industry, and many, far too numerous to list here, have contributed to its successes. In particular, J. Davis, J. Filson, A. Lindh and T. McEvilly made the experiment happen. We thank T. Hanks, S. Hough, D. Jackson, Y. Kagan, A. Lindh, M. Rymer, W. Thatcher, D. Wald and M. L. Zoback for their comments and suggestions and L. Blair, J. Boatwright, M. Huang, and D. Wald for technical assistance.

Author Information. Reprints and permissions information is available at npg.nature.com/reprintsandpermissions. The authors declare no competing financial interests. Correspondence and requests for materials should be addressed to W.H.B. (bakun@usgs.gov).

A Haplotype Map of the Human Genome

The International HapMap Consortium*

Editor's Note

Three years after its official launch, the International HapMap Consortium revealed its completed haplotype map of the human genome. The map reveals the most common genetic differences found across the entire genome for 269 humans from four different populations. The research, led by geneticists David Altshuler and Peter Donnelly, groups single nucleotide polymorphisms (SNPs)—the single-letter differences in the DNA between individuals—into haplotypes, combinations of SNPs that have travelled together over evolutionary time. The goal is to understand the complex genetic changes underlying common diseases such as cardiovascular disease and cancer. Two years later the map was "upgraded" to a second-generation version of over 3 million SNPs.

Inherited genetic variation has a critical but as yet largely uncharacterized role in human disease. Here we report a public database of common variation in the human genome: more than one million single nucleotide polymorphisms (SNPs) for which accurate and complete genotypes have been obtained in 269 DNA samples from four populations, including ten 500-kilobase regions in which essentially all information about common DNA variation has been extracted. These data document the generality of recombination hotspots, a block-like structure of linkage disequilibrium and low haplotype diversity, leading to substantial correlations of SNPs with many of their neighbours. We show how the HapMap resource can guide the design and analysis of genetic association studies, shed light on structural variation and recombination, and identify loci that may have been subject to natural selection during human evolution.

DESPITE the ever-accelerating pace of biomedical research, the root causes of common human diseases remain largely unknown, preventative measures are generally inadequate, and available treatments are seldom curative. Family history is one of the strongest risk factors for nearly all diseases—including cardiovascular disease, cancer, diabetes, autoimmunity, psychiatric illnesses and many others—providing the tantalizing but elusive clue that inherited genetic variation has an important role in the pathogenesis of disease. Identifying the causal genes and variants would represent an important step in the path towards improved prevention, diagnosis and treatment of disease.

More than a thousand genes for rare, highly heritable "mendelian" disorders have been

* The full list of authors and affiliations has been removed. The original text is available in the *Nature* online archive.

人类基因组的单体型图谱

国际单体型图谱协作组 *

编者按

计划正式启动三年后，国际 HapMap 协作组发布了完整的人类基因组单体型图谱。这份图谱揭示了来自四个不同种群的 269 个人类个体在整个基因组范围内的最常见的遗传差异。由遗传学家戴维·阿特舒勒和彼得·唐纳利领导的这项研究，将单核苷酸多态性（SNP）——个体间 DNA 上单个字母的差异——集合成单体型，即随着时间共同演化的 SNP 组合。项目的目标是理解常见疾病如心血管疾病和癌症等背后的复杂的遗传变化。两年后这份图谱“升级”为包含超过 300 万个 SNP 的第二代版本。

遗传变异在人类疾病中有着重要作用，但这些作用大部分还未被阐明。在本文中我们报道了一个公开的人类基因组中常见变异的数据库，它包含超过 100 万个单核苷酸多态性（SNP）。这些位点的准确而全面的基因型是通过来自四个种群的 269 份 DNA 样本获得的，我们也在 10 个 500 kb 的区域中提取了基本上所有的常见 DNA 变异的信息。这些数据证明了重组热点、连锁不平衡的区块结构和低单体型密度的普遍性，进而证实了 SNP 与很多邻近 SNP 的大量关联。我们展示了 HapMap 资源如何指导遗传学关联研究的设计和分析，为结构变异和重组提供线索，以及鉴定在人类演化过程中可能经历过自然选择的基因座。

尽管生物医学研究的步伐不断加速，但常见人类疾病的根本原因仍然大部分未知，预防措施普遍不足，现有的治疗手段鲜有疗效。对于几乎所有的疾病——包括心血管疾病、癌症、糖尿病、自身免疫病、精神疾病和很多其他的疾病，家族史是最强的风险因素之一，提供了诱人又难以捉摸的线索，即遗传变异在疾病的发病机理中有着重要的作用。在通往更好地预防、诊断和治疗疾病的道路上，鉴定导致疾病的基因和变异将代表着重要的一步。

对于罕见的、高度遗传的“孟德尔”疾病，已有超过 1,000 个基因被鉴定出来。

* 本书略去了作者和单位名单。原文可从《自然》在线数据库中获得。

identified, in which variation in a single gene is both necessary and sufficient to cause disease. Common disorders, in contrast, have proven much more challenging to study, as they are thought to be due to the combined effect of many different susceptibility DNA variants interacting with environmental factors.

Studies of common diseases have fallen into two broad categories: family-based linkage studies across the entire genome, and population-based association studies of individual candidate genes. Although there have been notable successes, progress has been slow due to the inherent limitations of the methods; linkage analysis has low power except when a single locus explains a substantial fraction of disease, and association studies of one or a few candidate genes examine only a small fraction of the "universe" of sequence variation in each patient.

A comprehensive search for genetic influences on disease would involve examining all genetic differences in a large number of affected individuals and controls. It may eventually become possible to accomplish this by complete genome resequencing. In the meantime, it is increasingly practical to systematically test common genetic variants for their role in disease; such variants explain much of the genetic diversity in our species, a consequence of the historically small size and shared ancestry of the human population.

Recent experience bears out the hypothesis that common variants have an important role in disease, with a partial list of validated examples including *HLA* (autoimmunity and infection)[1], *APOE4* (Alzheimer's disease, lipids)[2], Factor V^{Leiden} (deep vein thrombosis)[3], *PPARG* (encoding PPARγ; type 2 diabetes)[4,5], *KCNJ11* (type 2 diabetes)[6], *PTPN22* (rheumatoid arthritis and type 1 diabetes)[7,8], insulin (type 1 diabetes)[9], *CTLA4* (autoimmune thyroid disease, type 1 diabetes)[10], *NOD2* (inflammatory bowel disease)[11,12], complement factor H (age-related macular degeneration)[13-15] and *RET* (Hirschsprung disease)[16,17], among many others.

Systematic studies of common genetic variants are facilitated by the fact that individuals who carry a particular SNP allele at one site often predictably carry specific alleles at other nearby variant sites. This correlation is known as linkage disequilibrium (LD); a particular combination of alleles along a chromosome is termed a haplotype.

LD exists because of the shared ancestry of contemporary chromosomes. When a new causal variant arises through mutation—whether a single nucleotide change, insertion/deletion, or structural alteration—it is initially tethered to a unique chromosome on which it occurred, marked by a distinct combination of genetic variants. Recombination and mutation subsequently act to erode this association, but do so slowly (each occurring at an average rate of about 10^{-8} per base pair (bp) per generation) as compared to the number of generations (typically 10^4 to 10^5) since the mutational event.

The correlations between causal mutations and the haplotypes on which they arose have long served as a tool for human genetic research: first finding association to a haplotype, and then subsequently identifying the causal mutation(s) that it carries. This was pioneered in studies of the *HLA* region, extended to identify causal genes for mendelian diseases

在这些疾病中，单个基因的变异对于导致疾病是必要且充分的。相比而言，常见疾病的研究已被证明更具挑战性，因为这些疾病被认为是源于很多不同的易感 DNA 变异与环境因素相互作用的组合效应。

常见疾病的研究分为两大类：基于家系的全基因组连锁研究和基于自然人群的对单个候选基因的关联研究。尽管这些研究已经有令人瞩目的成功，但是由于方法的内在限制，进展一直比较缓慢；连锁分析的效力通常很低，除非单个基因座能解释疾病的很大一部分，而一个或几个候选基因的关联研究则仅仅是检查了一个病人体内序列变异“宇宙”的很小一部分。

全面寻找遗传对疾病的影响将涉及在大量的病例和对照中研究所有的遗传差异。通过完整的基因组重测序，有可能最终实现这一点。同时，系统检测常见的遗传变异在疾病中的作用变得越来越可行；人类种群在历史上有着较小的规模和共同的祖先，因此这样的变异可以解释我们自身物种中很大一部分的遗传多样性。

最新的实践证明了常见变异在疾病中有着重要的作用这个假说，部分得到验证的例子包括 *HLA*（自身免疫病和感染）[1]、*APOE4*（阿尔茨海默病，脂质）[2]，V^{Leiden} 因子（深部静脉血栓形成）[3]、*PPARG*（编码 PPARγ；2 型糖尿病）[4,5]、*KCNJ11*（2 型糖尿病）[6]、*PTPN22*（类风湿性关节炎和 1 型糖尿病）[7,8]、胰岛素（1 型糖尿病）[9]、*CTLA4*（自身免疫性甲状腺病，1 型糖尿病）[10]、*NOD2*（炎性肠病）[11,12]、补体因子 H（老年性黄斑变性）[13-15] 和 *RET*（先天性巨结肠）[16,17]，此外还有很多。

某些个体如果在一个位点携带一个特定的 SNP 等位基因，则常常可预见地在邻近的其他变异位点也携带特定的等位基因，这个事实促进了对常见遗传变异的系统研究。这种关联被称为连锁不平衡（LD）；等位基因在一条染色体上特定的组合被称为单体型。

LD 存在是由于现在的染色体有着共同的祖先。当一种新的导致疾病的变异通过突变——不管是单核苷酸改变、插入/缺失或者结构变异——而产生时，它最初会局限于它所发生的染色体，以独特的遗传变异组合为特征。重组和突变随后削弱这种关联，但是相对于变异发生后的代数（通常 10^4 到 10^5），这种削弱是很缓慢的（分别以每代每碱基对（bp）大约 10^{-8} 的平均速率发生）。

导致疾病的变异与其所在的单体型的相关性早已被用作人类遗传研究的工具：首先发现与一个单体型的关联，然后鉴定该单体型含有的导致疾病的突变。这种方法最早应用在 *HLA* 区域的研究，后来拓展到孟德尔疾病（如囊性纤维化 [18] 和

(for example, cystic fibrosis[18] and diastrophic dysplasia[19]), and most recently for complex disorders such as age-related macular degeneration[13-15].

Early information documented the existence of LD in the human genome[20,21]; however, these studies were limited (for technical reasons) to a small number of regions with incomplete data, and general patterns were challenging to discern. With the sequencing of the human genome and development of high-throughput genomic methods, it became clear that the human genome generally displays more LD[22] than under simple population genetic models[23], and that LD is more varied across regions, and more segmentally structured[24-30], than had previously been supposed. These observations indicated that LD-based methods would generally have great value (because nearby SNPs were typically correlated with many of their neighbours), and also that LD relationships would need to be empirically determined across the genome by studying polymorphisms at high density in population samples.

The International HapMap Project was launched in October 2002 to create a public, genome-wide database of common human sequence variation, providing information needed as a guide to genetic studies of clinical phenotypes[31]. The project had become practical by the confluence of the following: (1) the availability of the human genome sequence; (2) databases of common SNPs (subsequently enriched by this project) from which genotyping assays could be designed; (3) insights into human LD; (4) development of inexpensive, accurate technologies for high-throughput SNP genotyping; (5) web-based tools for storing and sharing data; and (6) frameworks to address associated ethical and cultural issues[32]. The project follows the data release principles of an international community resource project (http://www.wellcome.ac.uk/doc_WTD003208.html), sharing information rapidly and without restriction on its use.

The HapMap data were generated with the primary aim of guiding the design and analysis of medical genetic studies. In addition, the advent of genome-wide variation resources such as the HapMap opens a new era in population genetics, offering an unprecedented opportunity to investigate the evolutionary forces that have shaped variation in natural populations.

The Phase I HapMap

Phase I of the HapMap Project set as a goal genotyping at least one common SNP every 5 kilobases (kb) across the genome in each of 269 DNA samples. For the sake of practicality, and motivated by the allele frequency distribution of variants in the human genome, a minor allele frequency (MAF) of 0.05 or greater was targeted for study. (For simplicity, in this paper we will use the term "common" to mean a SNP with MAF $\geqslant$ 0.05.) The project has a Phase II, which is attempting genotyping of an additional 4.6 million SNPs in each of the HapMap samples.

To compare the genome-wide resource to a more complete database of common variation—one in which all common SNPs and many rarer ones have been discovered and tested—a representative collection of ten regions, each 500 kb in length, was selected

畸形性骨发育不良[19])的致病基因的鉴定，最近则应用于复杂疾病，如老年性黄斑变性[13-15]。

早期的信息证实了人类基因组中LD的存在[20,21]；但是这些研究(由于技术原因)局限于数据不完整的少量区域，难以了解其一般的模式。随着人类基因组的测序和高通量基因组方法的发展，我们已经清楚，人类基因组一般比简单种群遗传模型[23]有着更多的LD[22]，此外，与之前的预估相比，LD在不同区域间的变化更大，也呈现更加分节段的结构[24-30]。这些现象说明，基于LD的方法总体来说将会有巨大的价值(因为邻近的SNP通常与很多附近的SNP关联)，也说明需要通过实验研究种群样本中高密度的多态性来确定LD关系。

国际HapMap计划在2002年10月启动，目的是创建一个公开的、全基因组范围的人类常见序列变异数据库，为指导临床表型的遗传学研究提供所需要的信息[31]。这个计划由于以下内容的汇集而已经变得可行：(1)人类基因组序列的获得；(2)设计基因型分型所需的常见SNP数据库的完善(该数据库被此计划进一步丰富)；(3)对人类LD的了解；(4)廉价而准确的高通量SNP基因型分型技术的发展；(5)基于网络的存储和分享数据的工具；以及(6)解决相关的伦理和文化问题的框架[32]。本计划遵循国际共享资源项目的数据释放原则(http://www.wellcome.ac.uk/doc_WTD003208.html)，迅速分享信息，不限制使用。

HapMap数据的主要目的是指导医学遗传学研究的设计和分析。另外，像HapMap这样的全基因组的变异资源将开启种群遗传学的新纪元，为研究在自然种群中塑造了变异的进化动力提供一个前所未有的机会。

第I阶段的HapMap

HapMap计划的第I阶段目标是，对于269个DNA样本的每一个，均能在全基因组范围内的每5,000个碱基中至少对一个常见SNP进行基因分型。为了实用，同时根据人类基因组中变异的等位基因频率分布，我们选择研究最小等位基因频率(MAF)大于或等于0.05的SNP。(为了简便，本文中我们使用“常见的”表示MAF ≥ 0.05的SNP)本计划的第II阶段将在各个HapMap样本中对另外460万个SNP进行基因分型。

为了比较全基因组范围的资源与更完整的常见变异数据库——该数据库中所有的常见SNP和很多相对罕见的SNP已经被发掘和检验，我们从ENCODE(DNA元

from the ENCODE (Encyclopedia of DNA Elements) Project[33]. Each 500-kb region was sequenced in 48 individuals, and all SNPs in these regions (discovered or in dbSNP) were genotyped in the complete set of 269 DNA samples.

The specific samples examined are: (1) 90 individuals (30 parent–offspring trios) from the Yoruba in Ibadan, Nigeria (abbreviation YRI); (2) 90 individuals (30 trios) in Utah, USA, from the Centre d'Etude du Polymorphisme Humain collection (abbreviation CEU); (3) 45 Han Chinese in Beijing, China (abbreviation CHB); (4) 44 Japanese in Tokyo, Japan (abbreviation JPT).

Because none of the samples was collected to be representative of a larger population such as "Yoruba", "Northern and Western European", "Han Chinese", or "Japanese" (let alone of all populations from "Africa", "Europe", or "Asia"), we recommend using a specific local identifier (for example, "Yoruba in Ibadan, Nigeria") to describe the samples initially. Because the CHB and JPT allele frequencies are generally very similar, some analyses below combine these data sets. When doing so, we refer to three "analysis panels" (YRI, CEU, CHB+JPT) to avoid confusing this analytical approach with the concept of a "population".

Important details about the design of the HapMap Project are presented in the Methods, including: (1) organization of the project; (2) selection of DNA samples for study; (3) increasing the number and annotation of SNPs in the public SNP map (dbSNP) from 2.6 million to 9.2 million (Fig. 1); (4) targeted sequencing of the ten ENCODE regions, including evaluations of false-positive and false-negative rates; (5) genotyping for the genome-wide map; (6) intense efforts that monitored and established the high quality of the data; and (7) data coordination and distribution through the project Data Coordination Center (DCC) (http://www.hapmap.org).

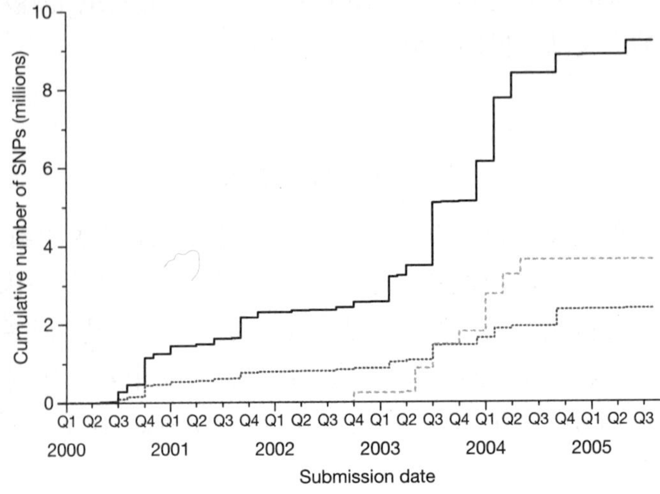

Fig. 1. Number of SNPs in dbSNP over time. The cumulative number of non-redundant SNPs (each mapped to a single location in the genome) is shown as a solid line, as well as the number of SNPs validated by genotyping (dotted line) and double-hit status (dashed line). Years are divided into quarters (Q1–Q4).

件百科全书）计划[33]中选择了10个具有代表性的区域，每个区域的长度是500 kb。先在48个个体中对这10个500 kb的区域进行测序，然后在全部269个DNA样本中对这些区域中所有的SNP（新发现的或已收录在dbSNP数据库的）进行基因分型。

检测的样本包括：(1) 尼日利亚伊巴丹市的90个（30个亲本–子孙三联家系）约鲁巴人（简写为YRI）；(2) 人类多态性研究中心收集的美国犹他州的90个（30个三联家系）个体（简写为CEU）；(3) 中国北京市的45个汉族中国人（简写为CHB）；(4) 日本东京市的44个日本人（简写为JPT）。

由于收集的这些样本并不是用来代表更大的种群如“约鲁巴人”“北欧和西欧人”“汉族中国人”或者“日本人”（更不代表“非洲”“欧洲”或者“亚洲”的所有种群），一开始我们建议使用特定的局部地区标识（例如，“尼日利亚伊巴丹市的约鲁巴人”）来描述样本。由于CHB和JPT等位基因频率大体上非常相似，下面的一些分析将其合并到一起。这样操作时，我们使用三个“分析小组”（YRI、CEU、CHB+JPT）的说法来避免将这种分析方法与“种群”的概念混淆。

关于HapMap计划设计的重要细节在方法部分中进行了描述，包括：(1) 计划的组织；(2) 用于研究的DNA样本的选择；(3) 公共SNP图谱（dbSNP数据库）中SNP的数量和注释的增长（从260万到920万）（图1）；(4) 对10个ENCODE区域的靶向测序，包括假阳性率和假阴性率的评估；(5) 全基因组图谱的基因分型；(6) 为建立和监控数据的高质量所付出的巨大努力；(7) 通过本计划的数据协调中心进行的数据协调和分配（http://www.hapmap.org）。

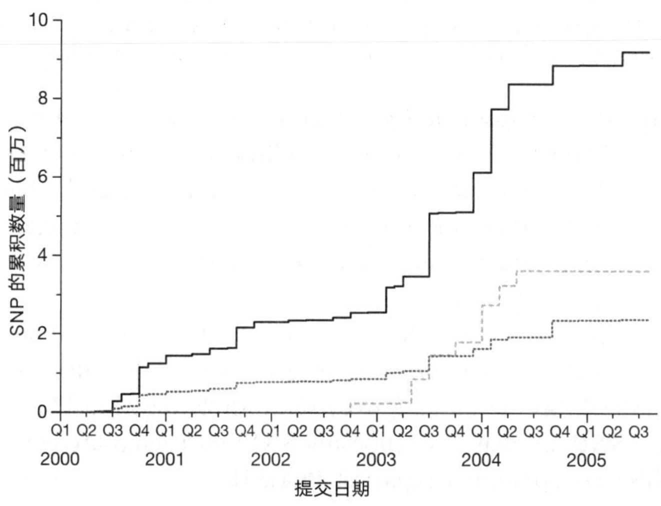

图1. dbSNP数据库中SNP数量随时间的变化。实线表示非冗余SNP（各个SNP比对到基因组的单一位置）的累积数量，虚线表示被基因分型验证的SNP数量，短划线表示“双打击”状态。每年分为四个季度（Q1～Q4）。

Description of the data. The Phase I HapMap contains 1,007,329 SNPs that passed a set of quality control (QC) filters (see Methods) in each of the three analysis panels, and are polymorphic across the 269 samples. SNP genotyping was distributed across centres by chromosomal region, with several technologies employed (Table 1). Each centre followed the same standard rules for SNP selection, quality control and data release; all SNPs were genotyped in the full set of 269 samples. Some centres genotyped more SNPs than required by the rules.

Table 1. Genotyping centres

Centre	Chromosomes	Technology
RIKEN	5, 11, 14, 15, 16, 17, 19	Third Wave Invader
Wellcome Trust Sanger Institute	1, 6, 10, 13, 20	Illumina BeadArray
McGill University and Génome Québec Innovation Centre	2, 4p	Illumina BeadArray
Chinese HapMap Consortium*	3, 8p, 21	Sequenom MassExtend, Illumina BeadArray
Illumina	8q, 9, 18q, 22, X	Illumina BeadArray
Broad Institute of Harvard and MIT	4q, 7q, 18p, Y, mtDNA	Sequenom MassExtend, Illumina BeadArray
Baylor College of Medicine with ParAllele BioScience	12	ParAllele MIP
University of California, San Francisco, with Washington University in St Louis	7p	PerkinElmer AcycloPrime-FP
Perlegen Sciences	5 Mb (ENCODE) on 2, 4, 7, 8, 9, 12, 18 in CEU	High-density oligonucleotide array

* The Chinese HapMap Consortium consists of the Beijing Genomics Institute, the Chinese National Human Genome Center at Beijing, the University of Hong Kong, the Hong Kong University of Science and Technology, the Chinese University of Hong Kong, and the Chinese National Human Genome Center at Shanghai.

Extensive, blinded quality assessment (QA) exercises documented that these data are highly accurate (99.7%) and complete (99.3%, see also Supplementary Table 1). All genotyping centres produced high-quality data (accuracy more than 99% in the blind QA exercises, Supplementary Tables 2 and 3), and missing data were not biased against heterozygotes. The Supplementary Information contains the full details of these efforts.

Although SNP selection was generally agnostic to functional annotation, 11,500 non-synonymous cSNPs (SNPs in coding regions of genes where the different SNP alleles code for different amino acids in the protein) were successfully typed in Phase I. (An effort was made to prioritize cSNPs in Phase I in choosing SNPs for each 5-kb region; all known non-synonymous cSNPs were attempted as part of Phase II.)

数据的描述　HapMap 第 I 阶段包含 1,007,329 个在三个分析小组中均通过了一套质控(QC)过滤(详见方法)，并且在 269 个样本中具有多态性的 SNP。SNP 基因分型任务按照染色体区域分配到各个中心，几种技术也部署到各个中心(表 1)。各个中心遵循相同的标准规则进行 SNP 挑选、质控和数据释放；所有的 SNP 都在整套 269 个样本中进行了基因分型。一些中心进行基因分型的 SNP 数量超出了规则所要求的。

表 1. 基因分型中心

中心	染色体	技术
日本理化研究所	5、11、14、15、16、17、19	Third Wave Invader
维康信托基金会桑格研究院	1、6、10、13、20	Illumina BeadArray
麦吉尔大学和基因组魁北克创新中心	2、4p	Illumina BeadArray
中国 HapMap 协作组 *	3、8p、21	Sequenom MassExtend, Illumina BeadArray
Illumina 公司	8q、9、18q、22、X	Illumina BeadArray
哈佛大学和麻省理工学院布罗德研究院	4q、7q、18p、Y、线粒体 DNA	Sequenom MassExtend, Illumina BeadArray
贝勒医学院和 ParAllele BioScience 公司	12	ParAllele MIP
加州大学旧金山分校和圣路易斯华盛顿大学	7p	PerkinElmer AcycloPrime-FP
Perlegen Sciences 公司	CEU 中 2、4、7、8、9、12、18 号染色体的 5 Mb 区域(ENCODE)	高密度寡核苷酸芯片

* 中国 HapMap 协作组包括北京基因组研究所、中国国家人类基因组北方研究中心、香港大学、香港科技大学、香港中文大学和中国国家人类基因组南方研究中心。

广泛的盲法质量评价(QA)发现这些数据是高度准确(99.7%)和完整的(99.3%，还可详见补充信息表 1)。所有的基因分型中心均产生高质量的数据(在盲法 QA 中准确度超过 99%，补充信息表 2 和 3)，并且缺失的数据没有相对杂合子的偏倚。补充信息包含这些工作的所有细节。

尽管选择 SNP 时通常并不知晓功能注释，但是在第 I 阶段，有 11,500 个非同义 cSNP(即位于基因编码区，且不同的等位基因编码蛋白质的不同氨基酸的 SNP)被成功分型。(对于各个 5 kb 区域，选择 SNP 时优先选择第 I 阶段中的 cSNP；所有已知的非同义 cSNP 作为第 II 阶段的一部分。)

Table 2. ENCODE project regions and genotyping

Region name	Chromosome band	Genomic interval (NCBI) (base numbers)†	Gene density (%)‡	Conservation score (%)§	Pedigree-based recombination rate (cM Mb^{-1})‖	Population-based recombination rate (cM Mb^{-1})¶	G+C content#	Available SNPs			Successfully genotyped SNPs††	Sequencing centre/genotyping centre(s)‡‡
								dbSNP☆	Sequence**	Total		
ENr112	2p16.3	51,633,239–52,133,238	0	3.8	0.8	0.9	0.35	1,570	1,762	3,332	2,275	Broad/McGill-GQIC
ENr131	2q37.1	234,778,639–235,278,638	4.6	1.3	2.2	2.5	0.43	1,736	1,259	2,995	1,910	Broad/McGill-GQIC
ENr113	4q26	118,705,475–119,205,474	0	3.9	0.6	0.9	0.35	1,444	2,053	3,497	2,201	Broad/Broad
ENm010	7p15.2	26,699,793–27,199,792	5.0	22.0	0.9	0.9	0.44	1,220	1,795	3,015	1,271	Baylor/UCSF-WU, Broad
ENm013*	7q21.13	89,395,718–89,895,717	5.5	4.4	0.4	0.5	0.38	1,394	1,917	3,311	1,807	Broad/Broad
ENm014*	7q31.33	126,135,436–126,632,577	2.9	11.2	0.4	0.9	0.39	1,320	1,664	2,984	1,966	Broad/Broad
ENr321	8q24.11	118,769,628–119,269,627	3.2	11.4	0.6	1.1	0.41	1,430	1,508	2,938	1,758	Baylor/Illumina
ENr232	9q34.11	127,061,347–127,561,346	5.9	8.3	2.7	2.6	0.52	1,444	1,523	2,967	1,324	Baylor/Illumina
ENr123	12q12	38,626,477–39,126,476	3.1	1.7	0.3	0.8	0.36	1,877	1,379	3,256	1,792	Baylor/Baylor
ENr213	18q12.1	23,717,221–24,217,220	0.9	7.4	1.2	0.9	0.37	1,330	1,459	2,789	1,640	Baylor/Illumina
Total	–	–	–	–	–	–	–	14,765	16,319	31,084	17,944	–

McGill-GQIC, McGill University and Génome Québec Innovation Centre.

* These regions were truncated to 500 kb for resequencing.

† Sequence build 34 coordinates.

‡ Gene density is defined as the percentage of bases covered either by Ensembl genes or human mRNA best BLAT alignments in the UCSC Genome Browser database.

§ Non-exonic conservation with mouse sequence was measured by taking 125 base non-overlapping sub-windows inside the 500,000 base windows. Sub-windows with less than 75% of their bases in a mouse alignment were discarded. Of the remaining sub-windows, those with at least 80% base identity were used to calculate the conservation

表 2. ENCODE 计划区域和基因分型

区域名称	染色体区带	基因组间距(NCBI)(碱基数量)†	基因密度(%)‡	保守性数值(%)§	基于家系的重组率($cM \cdot Mb^{-1}$)‖	基于种群的重组率($cM \cdot Mb^{-1}$)¶	G+C 含量#	已有的 SNP			成功基因分型的 SNP††	测序中心/基因分型中心 ‡‡
								dbSNP☆	测序 **	总计		
ENr112	2p16.3	51,633,239 ~ 52,133,238	0	3.8	0.8	0.9	0.35	1,570	1,762	3,332	2,275	布罗德/McGill-GQIC
ENr131	2q37.1	234,778,639 ~ 235,278,638	4.6	1.3	2.2	2.5	0.43	1,736	1,259	2,995	1,910	布罗德/McGill-GQIC
ENr113	4q26	118,705,475 ~ 119,205,474	0	3.9	0.6	0.9	0.35	1,444	2,053	3,497	2,201	布罗德/布罗德
ENm010	7p15.2	26,699,793 ~ 27,199,792	5	22.0	0.9	0.9	0.44	1,220	1,795	3,015	1,271	贝勒/UCSC-WU，布罗德
ENm013*	7q21.13	89,395,718 ~ 89,895,717	5.5	4.4	0.4	0.5	0.38	1,394	1,917	3,311	1,807	布罗德/布罗德
ENm014*	7q31.33	126,135,436 ~ 126,632,577	2.9	11.2	0.4	0.9	0.39	1,320	1,664	2,984	1,966	布罗德/布罗德
ENr321	8q24.11	118,769,628 ~ 119,269,627	3.2	11.4	0.6	1.1	0.41	1,430	1,508	2,938	1,758	贝勒/Illumina
ENr232	9q34.11	127,061,347 ~ 127,561,346	5.9	8.3	2.7	2.6	0.52	1,444	1,523	2,967	1,324	贝勒/Illumina
ENr123	12q12	38,626,477 ~ 39,126,476	3.1	1.7	0.3	0.8	0.36	1,877	1,379	3,256	1,792	贝勒/贝勒
ENr213	18q12.1	23,717,221 ~ 24,217,220	0.9	7.4	1.2	0.9	0.37	1,330	1,459	2,789	1,640	贝勒/Illumina
合计	–	–	–	–	–	–	–	14,765	16,319	31,084	17,944	–

McGill-GQIC，麦吉尔大学和基因组魁北克创新中心。

* 这些区域被削减到 500 kb 用于重测序。

† 34 版本序列坐标。

‡ 基因密度定义为被 Ensembl 基因或 UCSC 基因组浏览器数据库中人类 mRNA 最佳 BLAT 比对所覆盖的碱基的百分比。

§ 与小鼠序列的非外显子保守性是通过在 500,000 碱基窗口内取 125 碱基非重叠亚窗口来测定的。若一个亚窗口中小于 75% 的碱基比对到小鼠中，则该亚窗口被弃掉。剩下的亚窗口中，含有至少 80% 相同碱基的亚窗口用于计算保守性数值。小鼠比对中下列区域被弃掉：Ensembl 基因、所有的 GenBank mRNA Blastz 比对、FGenesh++ 基

score. The mouse alignments in regions corresponding to the following were discarded: Ensembl genes, all GenBank mRNA Blastz alignments, FGenesh++ gene predictions, Twinscan gene predictions, spliced EST alignments, and repeats.

‖ The pedigree-based sex-averaged recombination map is from deCODE Genetics[48].

¶ Recombination rate based on estimates from LDhat[46].

G+C content calculated from the sequence of the stated coordinates from sequence build 34.

☆ SNPs in dbSNP build 121 at the time the ENCODE resequencing began and SNPs added to dbSNP in builds 122–125 independent of the resequencing.

** New SNPs discovered through the resequencing reported here (not found by other means in builds 122–125).

†† SNPs successfully genotyped in all analysis panels (YRI, CEU, CHB+JPT).

‡‡ Perlegen genotyped a subset of SNPs in the CEU samples.

因预测、Twinscan 基因预测、剪接的 EST 比对以及重复。

‖ 基于家系的性别平均的重组图谱来自 deCODE Genetics[48]。

¶ 基于 LDhat[46] 估计的重组率。

G+C 含量根据来自序列版本 34 的相应序列计算。

☆ 在 ENCODE 重测序开始时 dbSNP 版本 121 中的 SNP，以及加入 dbSNP 的版本 122 ~ 125 中不依赖重测序结果的 SNP。

** 通过本文报道的重测序而新发现的 SNP(在版本 122 ~ 125 中通过其他方法未发现)。

†† 在所有分析小组(YRI、CEU、CHB+JPT)中被成功基因分型的 SNP。

‡‡ Perlegen 公司对 CEU 样本中一部分 SNP 进行了基因分型。

Across the ten ENCODE regions (Table 2), the density of SNPs was approximately tenfold higher as compared to the genome-wide map: 17,944 SNPs across the 5 megabases (Mb) (one per 279 bp).

More than 1.3 million SNP genotyping assays were attempted (Table 3) to generate the Phase I data on more than 1 million SNPs. The 0.3 million SNPs not part of the Phase I data set include 73,652 that passed QC filters but were monomorphic in all 269 samples. The remaining SNPs failed the QC filters in one or more analysis panels mostly because of inadequate completeness, non-mendelian inheritance, deviations from Hardy–Weinberg equilibrium, discrepant genotypes among duplicates, and data transmission discrepancies.

Table 3. HapMap Phase I genotyping success measures

SNP categories	Analysis panel		
	YRI	CEU	CHB+JPT
Assays submitted	1,273,716	1,302,849	1,273,703
Passed QC filters	1,123,296 (88%)	1,157,650 (89%)	1,134,726 (89%)
Did not pass QC filters*	150,420 (12%)	145,199 (11%)	138,977 (11%)
> 20% missing data	98,116 (65%)	107,626 (74%)	93,710 (67%)
> 1 duplicate inconsistent	7,575 (5%)	6,254 (4%)	10,725 (8%)
> 1 mendelian error	22,815 (15%)	13,600 (9%)	0 (0%)
< 0.001 Hardy–Weinberg *P*-value	12,052 (8%)	9,721 (7%)	16,176 (12%)
Other failures†	23,478 (16%)	17,692 (12%)	23,722 (17%)
Non-redundant (unique) SNPs	1,076,392	1,104,980	1,087,305
Monomorphic	156,290 (15%)	234,482 (21%)	268,325 (25%)
Polymorphic	920,102 (85%)	870,498 (79%)	818,980 (75%)
	All analysis panels		
Unique QC-passed SNPs	1,156,772		
Passed in one analysis panel	52,204 (5%)		
Passed in two analysis panels	97,231 (8%)		
Passed in three analysis panels	1,007,337 (87%)		
Monomorphic across three analysis panels	75,997		
Polymorphic in all three analysis panels	682,397		
MAF ⩾ 0.05 in at least one of three analysis panels	877,351		

* Out of 95 samples in CEU, YRI; 94 samples in CHB+JPT.

† "Other failures" includes SNPs with discrepancies during the data transmission process. Some SNPs failed in more than one way, so these percentages add up to more than 100%.

SNPs on the Phase I map are evenly spaced, except on Y and mtDNA. The Phase I data

在 10 个 ENCODE 区域中（表 2），SNP 的密度大约是全基因组图谱的十倍：5 Mb 区域中有 17,944 个 SNP(每 279 bp 一个）。

我们对超过 130 万个 SNP 进行了基因分型（表 3），得到了第 I 阶段的超过 100 万个 SNP 的数据。30 万个不在第 I 阶段数据集的 SNP 包括 73,652 个通过了 QC 过滤但是在所有 269 个样本中都是单态性的 SNP。剩下的 SNP 则是在一个或更多分析小组中没有通过 QC 过滤的，这大多是因为完整性不足、非孟德尔遗传、偏离哈迪–温伯格平衡、重复间基因型不一致以及数据传送不一致。

表 3. 第 I 阶段 HapMap 成功的基因分型

SNP 类别	分析小组		
	YRI	CEU	CHB+JPT
提交的试验	1,273,716	1,302,849	1,273,703
通过 QC 过滤	1,123,296 (88%)	1,157,650 (89%)	1,134,726 (89%)
没有通过 QC 过滤 *	150,420 (12%)	145,199 (11%)	138,977 (11%)
缺失的数据 >20%	98,116 (65%)	107,626 (74%)	93,710 (67%)
>1 个重复不一致	7,575 (5%)	6,254 (4%)	10,725 (8%)
>1 个孟德尔错误	22,815 (15%)	13,600 (9%)	0 (0%)
哈迪–温伯格平衡 P 值 <0.001	12,052 (8%)	9,721 (7%)	16,176 (12%)
其他失败 †	23,478 (16%)	17,692 (12%)	23,722 (17%)
非冗余（唯一）SNP	1,076,392	1,104,980	1,087,305
单态性的	156,290 (15%)	234,482 (21%)	268,325 (25%)
多态性的	920,102 (85%)	870,498 (79%)	818,980 (75%)
	所有分析小组		
特定的通过 QC 的 SNP	1,156,772		
在一个分析小组中通过	52,204 (5%)		
在两个分析小组中通过	97,231 (8%)		
在三个分析小组中通过	1,007,337 (87%)		
在三个分析小组中的单态性	75,997		
在所有三个分析小组中的多态性	682,397		
在三个分析小组的至少一个中 MAF ≥ 0.05	877,351		

* 得自 95 个 CEU 和 YRI 样本；94 个 CHB+JPT 样本。
† “其他失败”包括在数据传输过程中不一致的 SNP。一些 SNP 存在不止一种失败，所以这些比例总计超过 100%。

除 Y 染色体和线粒体 DNA 上，第 I 阶段图谱中的 SNP 是均匀分布的 第 I 阶

include a successful, common SNP every 5 kb across most of the genome in each analysis panel (Supplementary Fig. 1): only 3.3% of inter-SNP distances are longer than 10 kb, spanning 11.9% of the genome (Fig. 2; see also Supplementary Fig. 2). One exception is the X chromosome (Supplementary Fig. 1), where a much higher proportion of attempted SNPs were rare or monomorphic, and thus the density of common SNPs is lower.

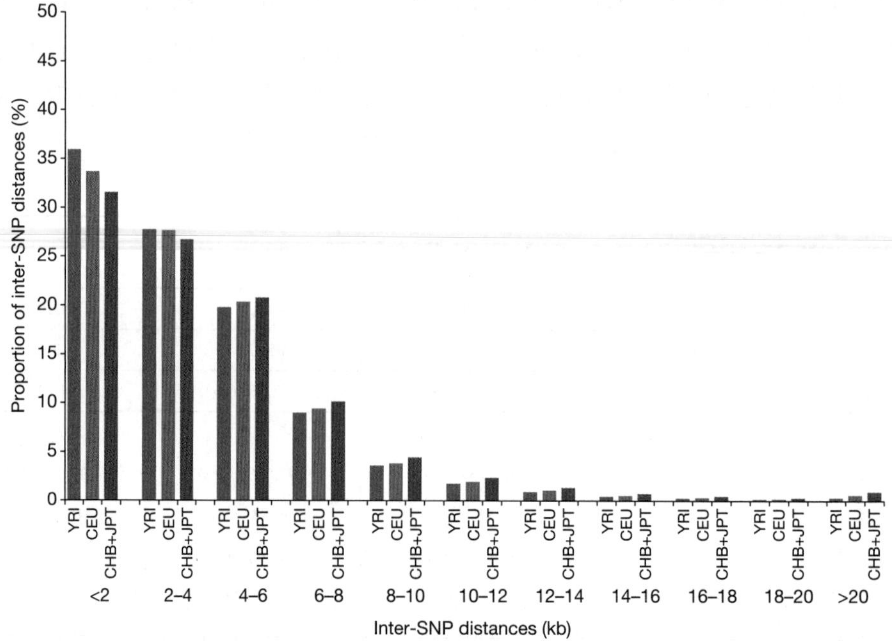

Fig. 2. Distribution of inter-SNP distances. The distributions are shown for each analysis panel for the HapMappable genome (defined in the Methods), for all common SNPs (with MAF ⩾ 0.05).

Two intentional exceptions to the regular spacing of SNPs on the physical map were the mitochondrial chromosome (mtDNA), which does not undergo recombination, and the non-recombining portion of chromosome Y. On the basis of the 168 successful, polymorphic SNPs, each HapMap sample fell into one of 15 (of the 18 known) mtDNA haplogroups[34] (Table 4). A total of 84 SNPs that characterize the unique branches of the reference Y genealogical tree[35-37] were genotyped on the HapMap samples. These SNPs assigned each Y chromosome to 8 (of the 18 major) Y haplogroups previously described (Table 4).

Table 4. mtDNA and Y chromosome haplogroups

MtDNA haplogroup	DNA sample*			
	YRI (60)	CEU (60)	CHB (45)	JPT (44)
L1	0.22	–	–	–
L2	0.35	–	–	–
L3	0.43	–	–	–
A	–	–	0.13	0.04

段的数据在各个分析小组中大多数的基因组范围内每 5 kb 有一个成功分型的常见 SNP(补充信息图 1)：只有 3.3% 的 SNP 间距超过 10 kb，占基因组的 11.9%(图 2；也可详见补充信息图 2)。X 染色体是一个例外(补充信息图 1)，该染色体上罕见的或者单多态性的 SNP 的比例高得多，因而常见 SNP 的密度较低。

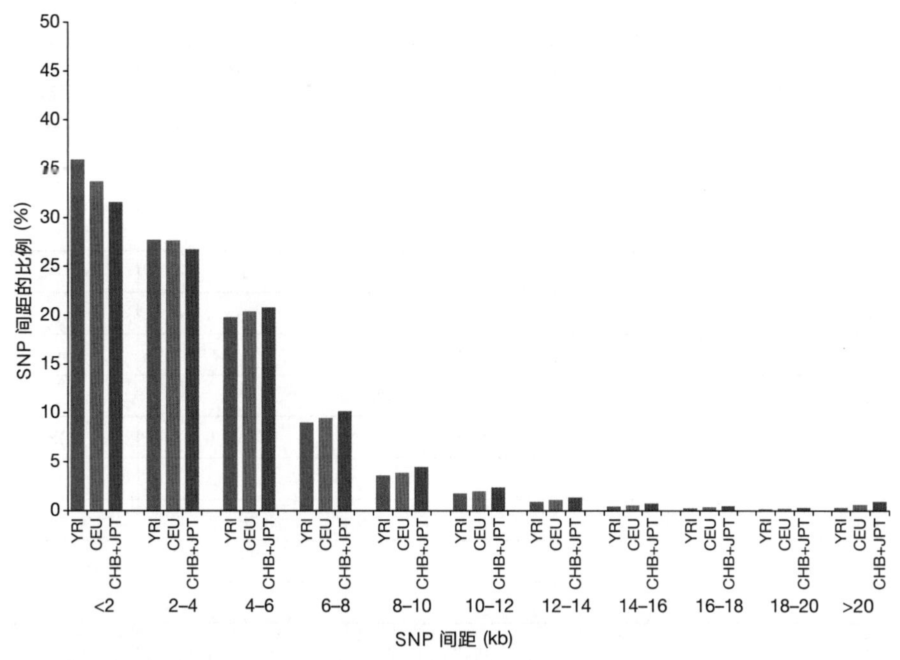

图 2. SNP 间距的分布。展示了各个分析小组中所有常见 SNP(MAF ≥ 0.05) 在可被 HapMap 比对的基因组(在方法中有定义) 中的分布。

物理图谱上 SNP 均匀间距的两个例外是不进行重组的线粒体染色体(mtDNA) 和 Y 染色体的非重组部分。基于 168 个成功分型的多态性的 SNP，各个 HapMap 样本分别归属于 18 个已知 mtDNA 单倍体型中的 15 个 [34](表 4)。总共 84 个表征参考 Y 染色体系统树独特分支的 SNP[35-37] 在 HapMap 样本中进行了基因分型。这些 SNP 将各个样本的 Y 染色体分配到 18 个主要的 Y 单体型中之前描述过的 8 个中(表 4)。

表 4. mtDNA 和 Y 染色体单体型

MtDNA 单体型	DNA 样本 *			
	YRI (60)	CEU (60)	CHB (45)	JPT (44)
L1	0.22	–	–	–
L2	0.35	–	–	–
L3	0.43	–	–	–
A	–	–	0.13	0.04

Continued

MtDNA haplogroup	DNA sample*			
	YRI (60)	CEU (60)	CHB (45)	JPT (44)
B	–	–	0.33	0.30
C	–	–	0.09	0.07
D	–	–	0.22	0.34
M/E	–	–	0.22	0.25
H	–	0.45	–	–
V	–	0.07	–	–
J	–	0.08	–	–
T	–	0.12	–	–
K	–	0.03	–	–
U	–	0.23	–	–
W	–	0.02	–	–
Y chromosome haplogroup	DNA sample*			
	YRI (30)	CEU (30)	CHB (22)	JPT (22)
E1	0.07	–	–	–
E3a	0.93	–	–	–
F, H, K	–	0.03	0.23	0.14
I	–	0.27	–	–
R1	–	0.70	–	–
C	–	–	0.09	0.09
D	–	–	–	0.45
NO	–	–	0.68	0.32

* Number of chromosomes sampled is given in parentheses.

Highly accurate phasing of long-range chromosomal haplotypes. Despite having collected data in diploid individuals, the inclusion of parent–offspring trios and the use of computational methods made it possible to determine long-range phased haplotypes of extremely high quality for each individual. These computational algorithms take advantage of the observation that because of LD, relatively few of the large number of possible haplotypes consistent with the genotype data actually occur in population samples.

The project compared a variety of algorithms for phasing haplotypes from unrelated individuals and trios[38], and applied the algorithm that proved most accurate (an updated version of PHASE[39]) separately to each analysis panel. (Phased haplotypes are available for download at the Project website.) We estimate that "switch" errors—where a segment of the maternal haplotype is incorrectly joined to the paternal—occur extraordinarily rarely in the trio samples (every 8 Mb in CEU; 3.6 Mb in YRI). The switch rate is higher in the CHB+JPT samples (one per 0.34 Mb) due to the lack of information from parent–

续表

MtDNA 单体型	DNA 样本 *			
	YRI (60)	CEU (60)	CHB (45)	JPT (44)
B	–	–	0.33	0.30
C	–	–	0.09	0.07
D	–	–	0.22	0.34
M/E	–	–	0.22	0.25
H	–	0.45	–	–
V	–	0.07	–	–
J	–	0.08	–	–
T	–	0.12	–	–
K	–	0.03	–	–
U	–	0.23	–	–
W	–	0.02	–	–
Y 染色体单体型	DNA 样本 *			
	YRI (30)	CEU (30)	CHB (22)	JPT (22)
E1	0.07	–	–	–
E3a	0.93	–	–	–
F, H, K	–	0.03	0.23	0.14
I	–	0.27	–	–
R1	–	0.70	–	–
C	–	–	0.09	0.09
D	–	–	–	0.45
NO	–	–	0.68	0.32

* 染色体样本数在括号中给出。

远距离染色体单体型的精准确定　尽管收集的是二倍体个体的数据，亲本–子代三联家系的纳入和计算方法的使用使我们能够非常高质量地确定各个个体远距离的单体型。这些计算算法利用了这样一个现象：由于 LD 现象的存在，可能与基因型数据一致的大量单体型中，只有很少会实际存在于种群样本中。

本计划比较了各种来自不相关个体和三联家系 [38] 的单体型确定算法，并且分别对各个分析小组应用了被证明最准确的算法（PHASE 的一个最新版本 [39]）。（确定的单体型数据可在本计划的网站上下载。）我们估计“转换”错误——母本单体型的一个片段错误地连到父本上——在三联家系样本中特别罕见（在 CEU 样本中每 8 Mb 一个；在 YRI 样本中每 3.6 Mb 一个）。由于缺少亲本–子代三联家系的信息，转换错

offspring trios, but even for the unrelated individuals, statistical reconstruction of haplotypes is remarkably accurate.

Estimating properties of SNP discovery and dbSNP. Extensive sequencing and genotyping in the ENCODE regions characterized the false-positive and false-negative rates for dbSNP, as well as polymerase chain reaction (PCR)-based resequencing (see Methods). These data reveal two important conclusions: first, that PCR-based sequencing of diploid samples may be biased against very rare variants (that is, those seen only as a single heterozygote), and second, that the vast majority of common variants are either represented in dbSNP, or show tight correlation to other SNPs that are in dbSNP (Fig. 3).

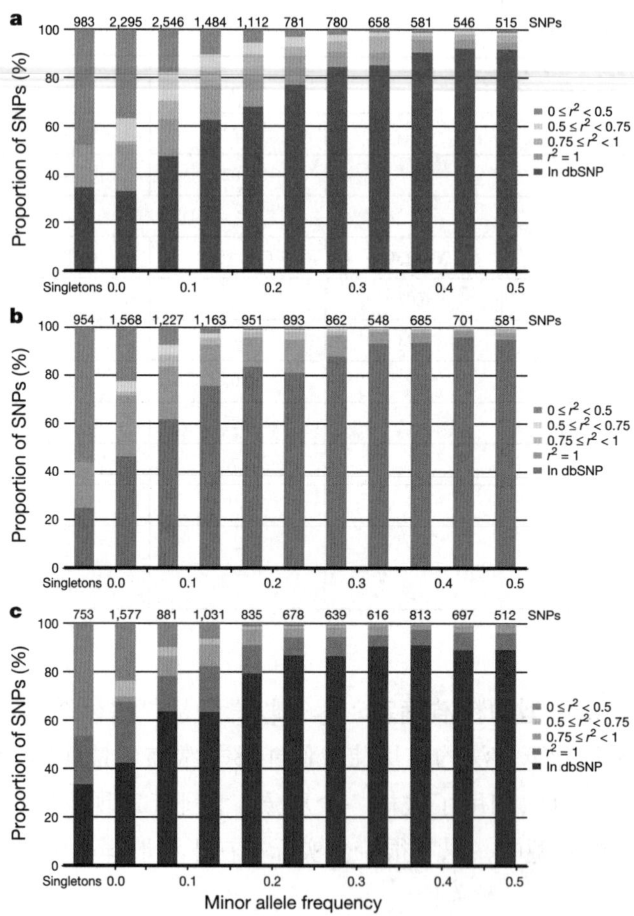

Fig. 3. Allele frequency and completeness of dbSNP for the ENCODE regions. **a**–**c**, The fraction of SNPs in dbSNP, or with a proxy in dbSNP, are shown as a function of minor allele frequency for each analysis panel (**a**, YRI; **b**, CEU; **c**, CHB+JPT). Singletons refer to heterozygotes observed in a single individual, and are broken out from other SNPs with MAF < 0.05. Because all ENCODE SNPs have been deposited in dbSNP, for this figure we define a SNP as "in dbSNP" if it would be in dbSNP build 125 independent of the HapMap ENCODE resequencing project. All remaining SNPs (not in dbSNP) were discovered only by ENCODE resequencing; they are categorized by their correlation (r^2) to those in dbSNP. Note that the number of SNPs in each frequency bin differs among analysis panels, because not all SNPs are polymorphic in all analysis panels.

误率在 CHB+JPT 样本中较高（每 0.34 Mb 一个），不过即使对于不相关的个体，单体型的统计学重建也是非常准确的。

SNP 发现和 dbSNP 数据库的特征估计 对 ENCODE 区域广泛的测序和基因分型，以及基于聚合酶链式反应（PCR）的重测序（详见方法），确定了 dbSNP 数据库的假阳性率和假阴性率。这些数据反映了两个重要的结论：第一，对二倍体样本进行基于 PCR 的测序，可能会在检测非常罕见的变异（即那些仅在单个杂合子中发现的变异）时产生偏倚；第二，大多数的常见变异要么收录在 dbSNP 中，要么与 dbSNP 中的其他 SNP 有紧密关联（图 3）。

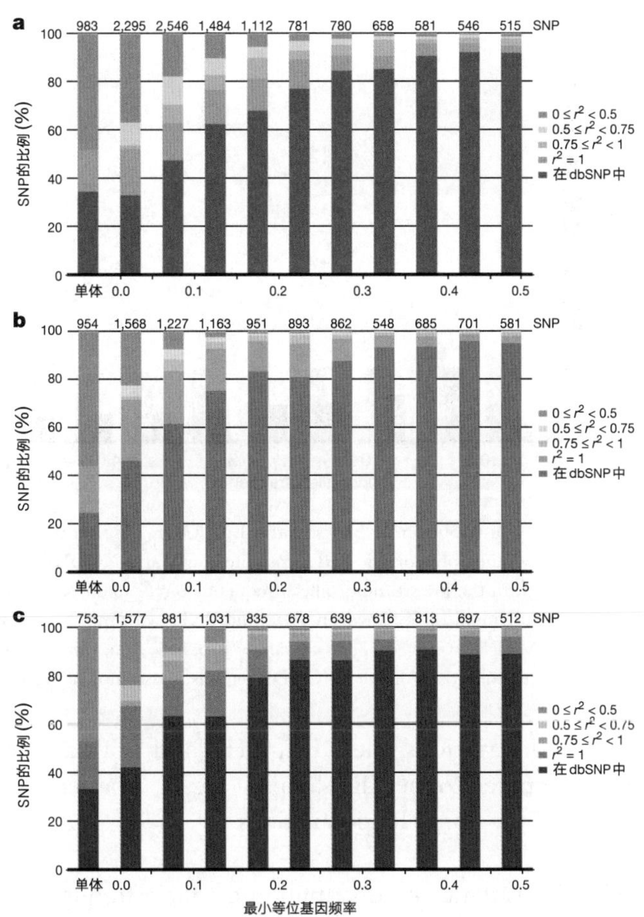

图 3. ENCODE 区域的 dbSNP 等位基因频率和完整性。**a**～**c**，根据各个分析小组中最小等位基因频率表示 dbSNP 中或者在 dbSNP 中有代表物的 SNP 所占的比例（**a**，YRI；**b**，CEU；**c**，CHB+JPT）。单体指的是仅在单个个体中观察到的杂合子，是从 MAF < 0.05 的 SNP 中单独分出去的。因为所有的 ENCODE SNP 已经被收录到 dbSNP 中，所以对于这张图，如果一个 SNP 不是基于 HapMap ENCODE 重测序计划而被收录在 dbSNP 版本 125 中，我们就定义它为“在 dbSNP 中”。所有剩下的 SNP（不在 dbSNP 中）只通过 ENCODE 重测序而被发现，并根据与 dbSNP 中的 SNP 的关联（r^2）而进一步分类。注意，各个频率块中 SNP 的数量在不同分析小组中是不同的，因为不是所有 SNP 在所有分析小组中都是多态性的。

Allele frequency distributions within population samples. The underlying allele frequency distributions for these samples are best estimated from the ENCODE data, where deep sequencing reduces bias due to SNP ascertainment. Consistent with previous studies, most SNPs observed in the ENCODE regions are rare: 46% had MAF < 0.05, and 9% were seen in only a single individual (Fig. 4). Although most varying sites in the population are rare, most heterozygous sites within any individual are due to common SNPs. Specifically, in the ENCODE data, 90% of heterozygous sites in each individual were due to common variants (Fig. 4). With ever-deeper sequencing of DNA samples the number of rare variants will rise linearly, but the vast majority of heterozygous sites in each person will be explained by a limited set of common SNPs now contained (or captured through LD) in existing databases (Fig. 3).

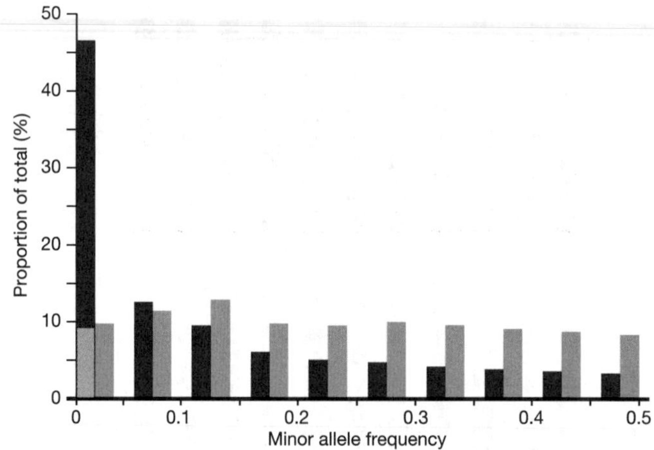

Fig. 4. Minor allele frequency distribution of SNPs in the ENCODE data, and their contribution to heterozygosity. This figure shows the polymorphic SNPs from the HapMap ENCODE regions according to minor allele frequency (blue), with the lowest minor allele frequency bin (< 0.05) separated into singletons (SNPs heterozygous in one individual only, shown in grey) and SNPs with more than one heterozygous individual. For this analysis, MAF is averaged across the analysis panels. The sum of the contribution of each MAF bin to the overall heterozygosity of the ENCODE regions is also shown (orange).

Consistent with previous descriptions, the CEU, CHB and JPT samples show fewer low frequency alleles when compared to the YRI samples (Fig. 5), a pattern thought to be due to bottlenecks in the history of the non-YRI populations.

In contrast to the ENCODE data, the distribution of allele frequencies for the genome-wide data is flat (Fig. 5), with much more similarity in the distributions observed in the three analysis panels. These patterns are well explained by the inherent and intentional bias in the rules used for SNP selection: we prioritized using validated SNPs in order to focus resources on common (rather than rare or false positive) candidate SNPs from the public databases. For a fuller discussion of ascertainment issues, including a shift in frequencies over time and an excess of high-frequency derived alleles due to inclusion of chimpanzee data in determination of double-hit status, see the Supplementary Information (Supplementary Fig. 3).

种群样本中的等位基因频率分布 对这些样本的潜在等位基因频率分布的最佳估计来自 ENCODE 数据，因为深度测序减少了确定 SNP 过程的偏倚。与之前的研究一致，ENCODE 区域内观察到的 SNP 大多数是罕见的：46% 的 MAF < 0.05，9% 仅在单个个体中存在（图 4）。尽管在种群中大多数变异位点是罕见的，但是在任何一个个体中，大多数杂合位点是常见 SNP。特别地，在 ENCODE 数据中，各个个体中 90% 的杂合位点是常见变异（图 4）。随着对 DNA 样本更加深度的测序，罕见变异的数量将会线性增长，但是每个个体中绝大多数的杂合位点仍可以用有限的一套已包含（或者通过 LD 而捕获）在已有数据库中的常见变异来解释（图 3）。

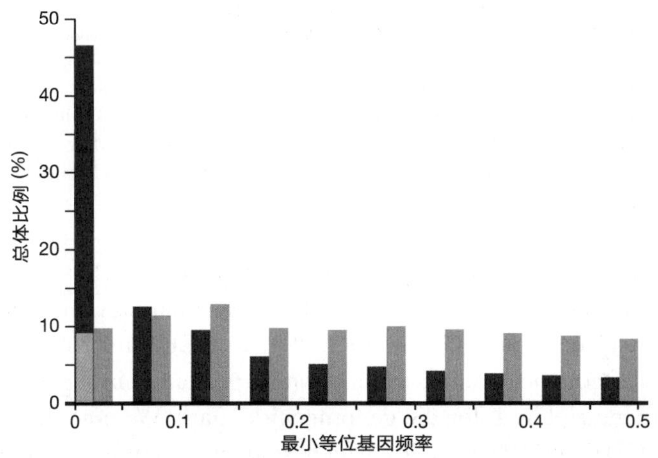

图 4. ENCODE 数据中 SNP 的最小等位基因频率分布和它们对杂合性的贡献。根据最小等位基因频率（蓝色）展示了来自 HapMap ENCODE 区域的多态性 SNP，最低的最小等位基因频率块（< 0.05）被分为单体（仅在一个个体中杂合的 SNP，用灰色表示）和在多于一个个体中杂合的 SNP。本分析中，MAF 是各分析小组的平均值。另外还展示了各个 MAF 块对 ENCODE 区域总体杂合性的贡献总和（橘色）。

与之前的描述一致，跟 YRI 样本相比，CEU、CHB 和 JPT 样本的低频等位基因更少（图 5），这种现象被认为是由于非 YRI 种群历史中的瓶颈效应。

与 ENCODE 数据相比，全基因组范围的等位基因频率分布是“平”的（图 5），在三个分析小组中的分布的相似性要高得多。这种模式可以用 SNP 筛选规则中固有的和人为的偏倚很好地进行解释：我们优先使用验证的 SNP，以便将资源集中于来自公共数据的常见的（而不是罕见或假阳性的）候选 SNP。对于确定 SNP 的问题的更加完整的讨论，包括频率随着时间的改变，以及在确定“双打击”状态时纳入黑猩猩数据而导致的高频率等位基因过量，详见补充信息（补充信息图 3）。

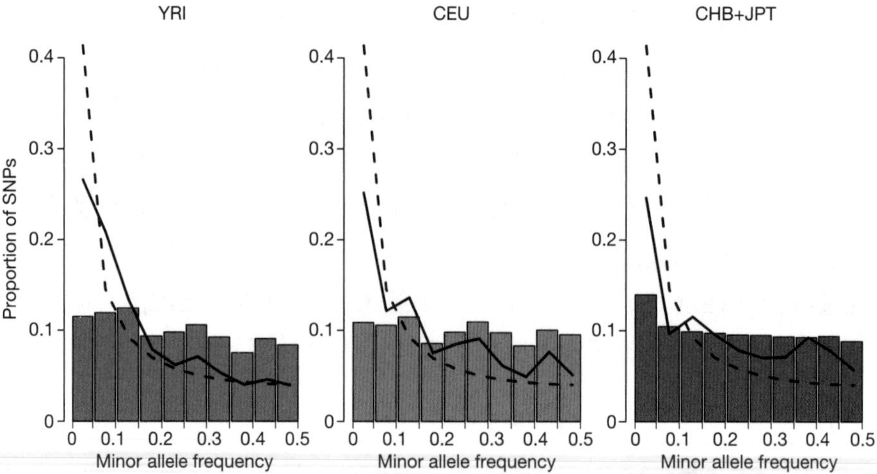

Fig. 5. Allele frequency distributions for autosomal SNPs. For each analysis panel we plotted (bars) the MAF distribution of all the Phase I SNPs with a frequency greater than zero. The solid line shows the MAF distribution for the ENCODE SNPs, and the dashed line shows the MAF distribution expected for the standard neutral population model with constant population size and random mating without ascertainment bias.

SNP allele frequencies across population samples. Of the 1.007 million SNPs successfully genotyped and polymorphic across the three analysis panels, only a subset were polymorphic in any given panel: 85% in YRI, 79% in CEU, and 75% in CHB+JPT. The joint distribution of frequencies across populations is presented in Fig. 6 (for the ENCODE data) and Supplementary Fig. 4 (for the genome-wide map). We note the similarity of allele frequencies in the CHB and JPT samples, which motivates analysing them jointly as a single analysis panel in the remainder of this report.

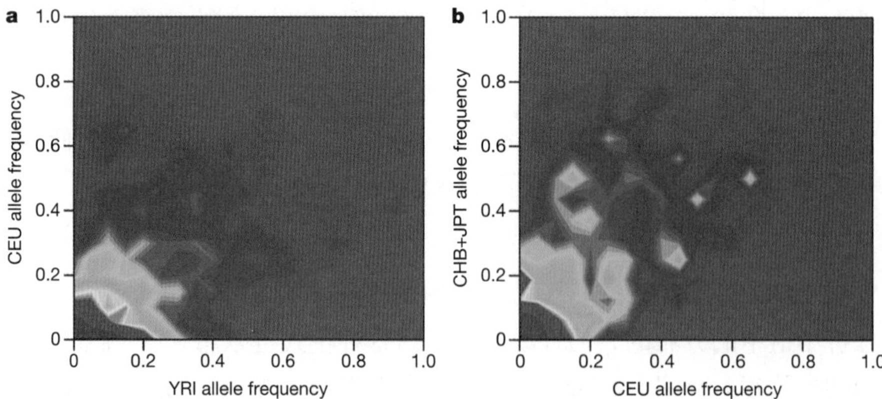

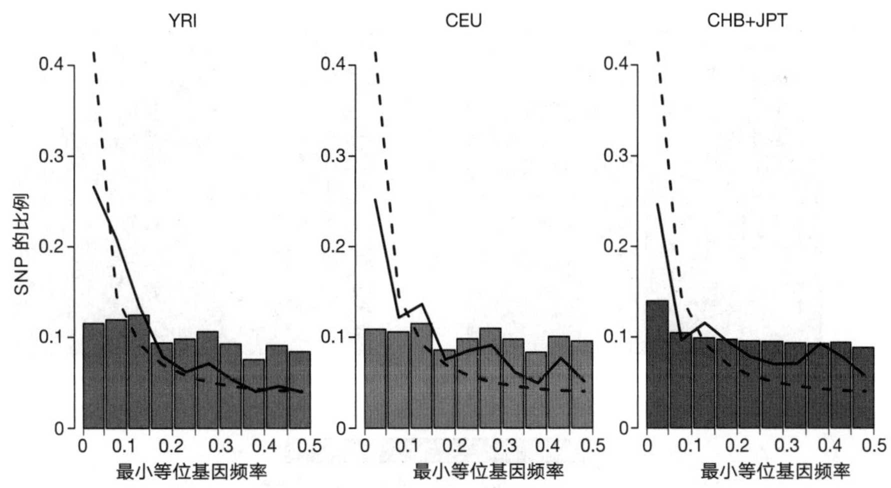

图 5. 常染色体 SNP 的等位基因频率分布。对于各个分析小组中所有第 I 阶段 HapMap 得到的频率大于零的 SNP，我们对其 MAF 分布作图（条形图）。实线表示 ENCODE SNP 的 MAF 分布，虚线表示基于恒定种群大小、随机交配以及无检测偏倚的标准中性种群模型得到的预期 MAF 分布。

种群样本间的 SNP 等位基因频率 三个分析小组共有 100.7 万成功基因分型并呈多态性的 SNP，其中只有一部分在三个分析小组中都是多态性的：YRI 中有 85%，CEU 中有 79%，CHB+JPT 中有 75%。种群间等位基因频率的联合分布见图 6（ENCODE 数据）和补充信息图 4（全基因组范围图谱）。我们注意到 CHB 和 JPT 样本等位基因频率相似，所以在这份报告的剩余部分将它们作为一个分析小组联合分析。

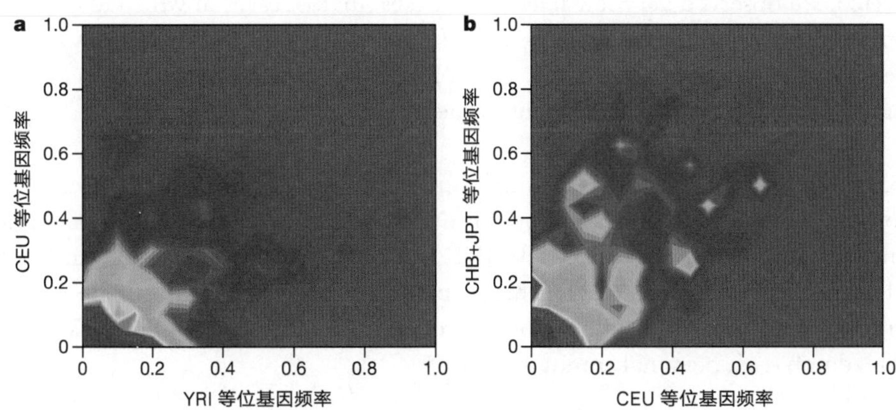

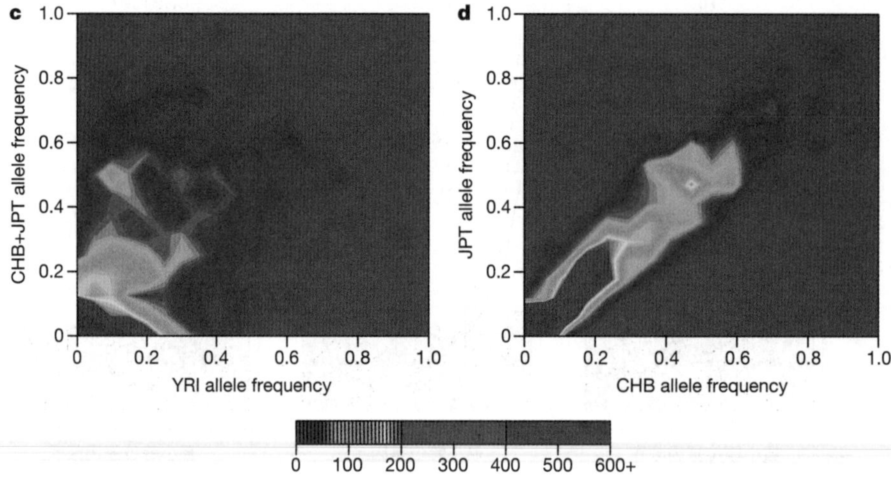

Fig. 6. Comparison of allele frequencies in the ENCODE data for all pairs of analysis panels and between the CHB and JPT sample sets. For each polymorphic SNP we identified the minor allele across all panels (**a–d**) and then calculated the frequency of this allele in each analysis panel/sample set. The colour in each bin represents the number of SNPs that display each given set of allele frequencies. The purple regions show that very few SNPs are common in one panel but rare in another. The red regions show that there are many SNPs that have similar low frequencies in each pair of analysis panels/sample sets.

A simple measure of population differentiation is Wright's F_{ST}, which measures the fraction of total genetic variation due to between-population differences[40]. Across the autosomes, F_{ST} estimated from the full set of Phase I data is 0.12, with CEU and CHB+JPT showing the lowest level of differentiation ($F_{ST} = 0.07$), and YRI and CHB+JPT the highest ($F_{ST} = 0.12$). These values are slightly higher than previous reports[41], but differences in the types of variants (SNPs versus microsatellites) and the samples studied make comparisons difficult.

As expected, we observed very few fixed differences (that is, cases in which alternate alleles are seen exclusively in different analysis panels). Across the 1 million SNPs genotyped, only 11 have fixed differences between CEU and YRI, 21 between CEU and CHB+JPT, and 5 between YRI and CHB+JPT, for the autosomes.

The extent of differentiation is similar across the autosomes, but higher on the X chromosome ($F_{ST} = 0.21$). Interestingly, 123 SNPs on the X chromosome were completely differentiated between YRI and CHB+JPT, but only two between CEU and YRI and one between CEU and CHB+JPT. This seems to be largely due to a single region near the centromere, possibly indicating a history of natural selection at this locus (see below; M. L. Freedman *et al.*, personal communication).

Haplotype sharing across populations. We next examined the extent to which haplotypes are shared across populations. We used a hidden Markov model in which each haplotype is modelled in turn as an imperfect mosaic of other haplotypes (see Supplementary Information)[42]. In essence, the method infers probabilistically which other haplotype in the sample is the closest relative (nearest neighbour) at each position along the chromosome.

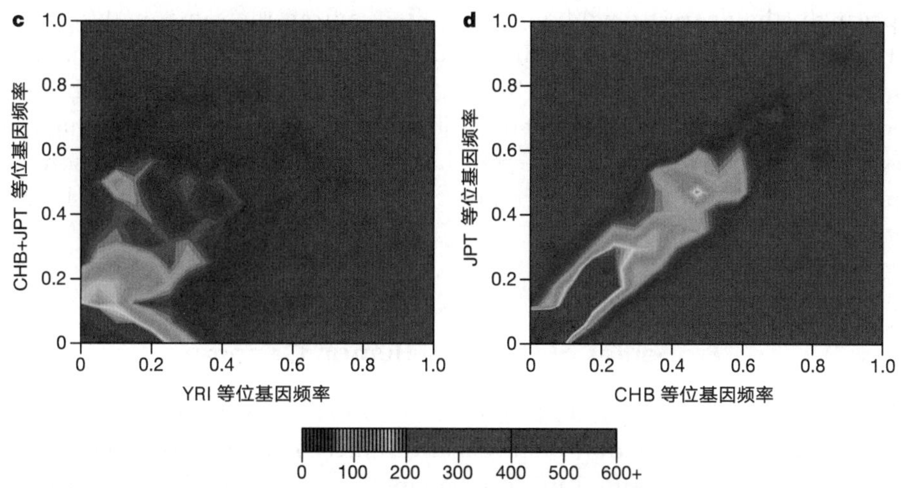

图 6. ENCODE 数据中的等位基因频率在所有分析小组间以及 CHB 和 JPT 样本集间的比较。对于各个多态性 SNP，我们鉴定其在所有分析小组的最小等位基因（**a** ~ **d**），然后计算这个等位基因在各个分析小组/样本集的频率。各个块中的颜色代表显示特定等位基因频率集的 SNP 的数量。紫色区域表明非常少的 SNP 是在一个分析小组中常见而在另一个分析小组中罕见的。红色区域表明很多 SNP 在各对分析小组/样品集中有类似的低频率。

种群差异的一个简单的衡量方法是赖特的 F_{ST}，它测量总的遗传差异中由种群间差异导致的比例 [40]。对于常染色体，从全套第 I 阶段数据估计的 F_{ST} 是 0.12，其中 CEU 和 CHB+JPT 表现出最低水平的差异（$F_{ST}=0.07$），而 YRI 和 CHB+JPT 则表现出最高的差异（$F_{ST}=0.12$）。这些数值略微高于之前的报道 [41]，但是变异类型的不同（SNP 相对于微卫星）和研究样本的差异使得比较变得困难。

正如预期的，我们观察到非常少的固定的差异（即不同的等位基因特定地出现在不同的分析小组中的情况）。在进行了基因分型的 100 万 SNP 中，对于常染色体，只有 11 个 SNP 在 CEU 和 YRI 间有固定的差异，CEU 和 CHB+JPT 间有 21 个，YRI 和 CHB+JPT 间有 5 个。

种群差异的程度在常染色体间是类似的，但是在 X 染色体上更高（$F_{ST}=0.21$）。有趣的是，X 染色体上有 123 个 SNP 在 YRI 和 CHB+JPT 间完全不同，而 CEU 和 YRI 间仅有两个，CEU 和 CHB+JPT 间仅有一个。这或许在很大程度上是由于着丝粒附近的一个区域，可能暗示着该基因座被自然选择的历史（见下文；弗里德曼等，个人交流）。

种群间共同的单体型 我们接下来检查了单体型在种群间一致的程度。我们使用的是隐马尔可夫模型，各个单体型依次被模型化为其他单体型的一个不完美镶嵌（详见补充信息）[42]。本质上，这个方法概率地推断在染色体的各个位置上，样本中的哪一个其他单体型是最近的近亲（近邻）。

Unsurprisingly, the nearest neighbour most often is from the same analysis panel, but about 10% of haplotypes were found most closely to match a haplotype in another panel (Supplementary Fig. 5). All individuals have at least some segments over which the nearest neighbour is in a different analysis panel. These results indicate that although analysis panels are characterized both by different haplotype frequencies and, to some extent, different combinations of alleles, both common and rare haplotypes are often shared across populations.

Properties of LD in the Human Genome

Traditionally, descriptions of LD have focused on measures calculated between pairs of SNPs, averaged as a function of physical distance. Examples of such analyses for the HapMap data are presented in Supplementary Fig. 6. After adjusting for known confounders such as sample size, allele frequency distribution, marker density, and length of sampled regions, these data are highly similar to previously published surveys[43].

Because LD varies markedly on scales of 1–100 kb, and is often discontinuous rather than declining smoothly with distance, averages obscure important aspects of LD structure. A fuller exploration of the fine-scale structure of LD offers both insight into the causes of LD and understanding of its application to disease research.

LD patterns are simple in the absence of recombination. The most natural path to understanding LD structure is first to consider the simplest case in which there is no recombination (or gene conversion), and then to add recombination to the model. (For simplicity we ignore genotyping error and recurrent mutation in this discussion, both of which seem to be rare in these data.)

In the absence of recombination, diversity arises solely through mutation. Because each SNP arose on a particular branch of the genealogical tree relating the chromosomes in the current populations, multiple haplotypes are observed. SNPs that arose on the same branch of the genealogy are perfectly correlated in the sample, whereas SNPs that occurred on different branches have imperfect correlations, or no correlation at all.

We illustrate these concepts using empirical genotype data from 36 adjacent SNPs in an ENCODE region (ENr131.2q37), selected because no obligate recombination events were detectable among them in CEU (Fig. 7). (We note that the lack of obligate recombination events in a small sample does not guarantee that no recombinants have occurred, but it provides a good approximation for illustration.)

毫不奇怪，最靠近的单体型大多来自同一分析小组，不过大约有 10% 的单体型与其他分析小组的单体型最匹配（补充信息图 5）。所有个体都有一些片段，其最靠近的单体型存在于其他分析小组中。这些结果说明，尽管分析小组具有不同的单体型频率，在某种程度上还具有不同的等位基因组合，但是常见的以及罕见的单体型在不同种群间经常是一致的。

人类基因组中 LD 的特征

传统上，LD 的描述集中于 SNP 对之间计算得到的测度，结果作为物理距离的函数取平均值。对 HapMap 数据的这种分析的例子在补充信息图 6 中有展示。对已知的混杂因素如样本量大小、等位基因频率分布、标志物密度和检测的区域长度进行校正后，这些数据跟之前发表的研究高度相似[43]。

因为 LD 在 1 ~ 100 kb 尺度上变化较大，而且经常是不连续的而非随距离平滑减少，因此取平均值会模糊 LD 结构的重要方面。对 LD 精细尺度结构更全面的探索，将不仅为 LD 的产生原因提供线索，也为它在疾病研究中的应用提供知识。

没有重组发生时 LD 模式是简单的 了解 LD 结构最自然的途径是首先考虑最简单的情况，即没有重组（或者基因转变）的情况，然后向模型中加入重组。（为了简便，在这部分分析中我们忽略了基因分型错误和频发突变，在这些数据中这两种情况看起来是罕见的。）

没有重组时，多样性只通过突变产生。因为各个 SNP 产生于系统树特定的分支，而该系统树与目前种群中的染色体相关，所以可观察到多个单体型。系统树相同分支上产生的 SNP 在样本中完美相关，而在不同分支上产生的 SNP 则不完美相关或者完全不相关。

我们利用实验观察到的基因型数据对这些概念进行了说明，这些数据来自一个 ENCODE 区域（ENr131.2q37）的 36 个相邻 SNP，选择它们是因为在 CEU 中没有检测到它们有专性重组事件（图 7）。（我们注意到，在小样本中没有专性重组事件并不能保证没有重组发生，但是它提供了一个很好的近似的例子。）

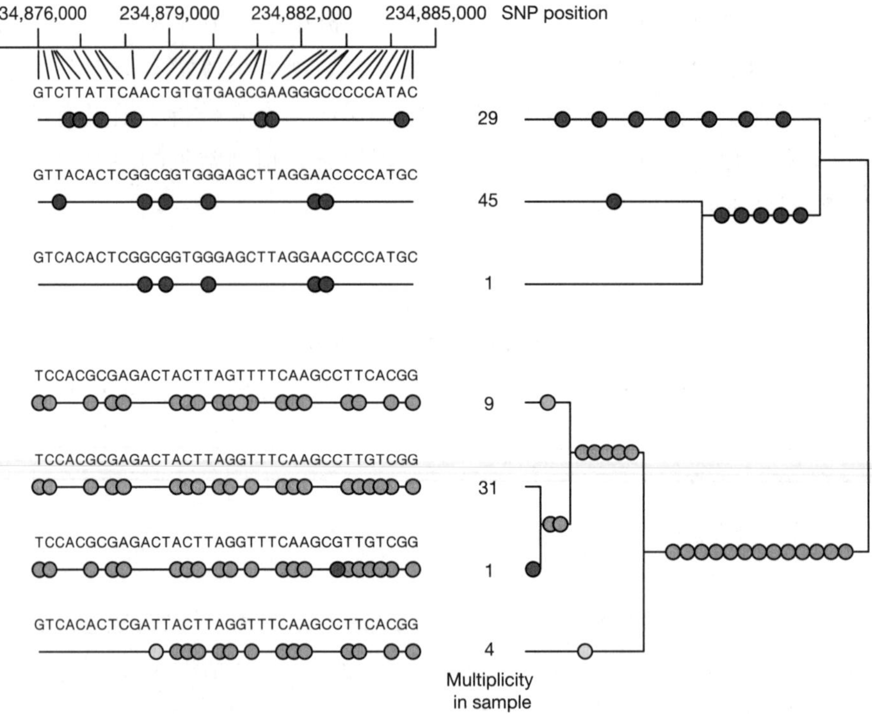

Fig. 7. Genealogical relationships among haplotypes and r^2 values in a region without obligate recombination events. The region of chromosome 2 (234,876,004–234,884,481 bp; NCBI build 34) within ENr131.2q37 contains 36 SNPs, with zero obligate recombination events in the CEU samples. The left part of the plot shows the seven different haplotypes observed over this region (alleles are indicated only at SNPs), with their respective counts in the data. Underneath each of these haplotypes is a binary representation of the same data, with coloured circles at SNP positions where a haplotype has the less common allele at that site. Groups of SNPs all captured by a single tag SNP (with $r^2 \geqslant 0.8$) using a pairwise tagging algorithm[53,54] have the same colour. Seven tag SNPs corresponding to the seven different colours capture all the SNPs in this region. On the right these SNPs are mapped to the genealogical tree relating the seven haplotypes for the data in this region.

In principle, 36 such SNPs could give rise to 2^{36} different haplotypes. Even with no recombination, gene conversion or recurrent mutation, up to 37 different haplotypes could be formed. Despite this great potential diversity, only seven haplotypes are observed (five seen more than once) among the 120 parental CEU chromosomes studied, reflecting shared ancestry since their most recent common ancestor among apparently unrelated individuals.

In such a setting, it is easy to interpret the two most common pairwise measures of LD: D' and r^2. (See the Supplementary Information for fuller definitions of these measures.) D' is defined to be 1 in the absence of obligate recombination, declining only due to recombination or recurrent mutation[27]. In contrast, r^2 is simply the squared correlation coefficient between the two SNPs. Thus, r^2 is 1 when two SNPs arose on the same branch of the genealogy and remain undisrupted by recombination, but has a value less than 1 when SNPs arose on different branches, or if an initially strong correlation has been disrupted by crossing over.

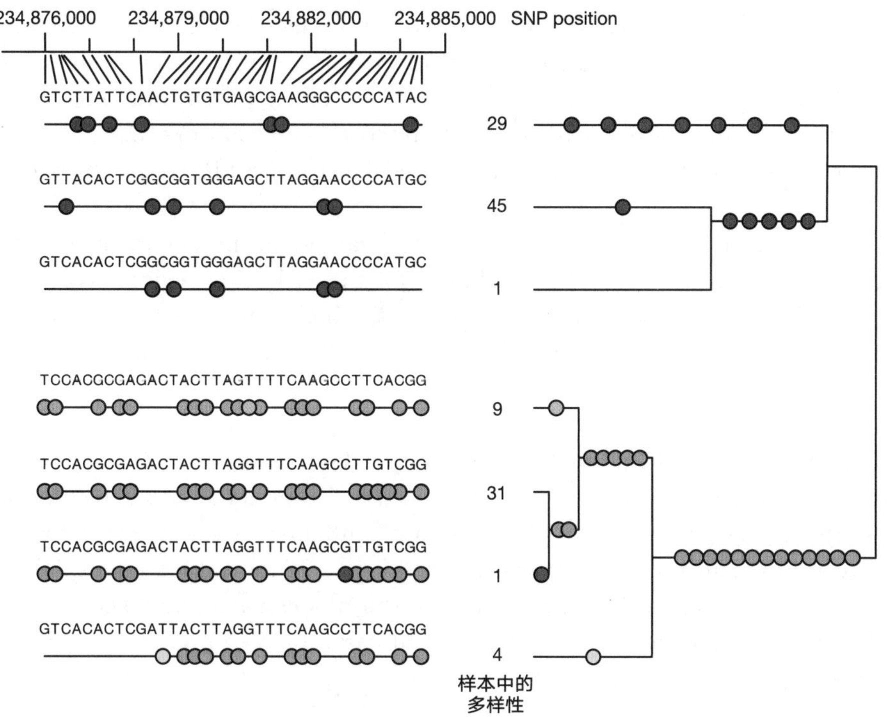

图 7. 没有专性重组事件的区域内单体型间的系谱关系和 r^2 值。2 号染色体 ENr131.2q37 区域（234,876,004 ~ 234,884,481 bp；NCBI 版本 34）包含 36 个 SNP，在 CEU 样本中没有专性重组事件。图的左边部分表示在这个区域中观察到的七个不同的单体型（仅在 SNP 处标示等位基因），以及它们在数据中各自的数量。在这些单体型各自的下面是相同数据的二进制表示，有颜色的圆圈表示单体型在该位点含有相对不常见等位基因的 SNP 位置。相同的颜色表示的是利用成对标签算法 [53,54] 通过单个标签 SNP 捕获（$r^2 \geqslant 0.8$）的各组 SNP。对应于七种不同颜色的七个标签 SNP 捕获了该区域所有的 SNP。在图的右边部分，这些 SNP 比对到系统树上，该系统树将该区域数据得到的七个单体型关联起来。

原则上，36 个这样的 SNP 能够产生 2^{36} 个不同的单体型。即使没有重组、基因转变或者频发突变，也能够形成多达 37 个不同的单体型。尽管存在这种巨大的潜在多样性，在研究的 120 个亲本 CEU 染色体中，只观察到七个单体型（其中五个观察到超过一次），这反映了自它们最近的共同祖先之后明显不相关的个体也有着共同的血统。

在这样的设定中，很容易解释这一对最常见的度量 LD 的方法：D' 和 r^2。（对于这些度量方法更全面的定义见补充信息。）D' 在不存在专性重组时被定义为 1，只由于重组或频发突变而减小 [27]。相反，r^2 就是简单的两个 SNP 间相关系数的平方。因而，当两个 SNP 在系统树的同一分支上产生并且不受重组的破坏时，r^2 为 1；但是当 SNP 在不同分支上产生，或者一个一开始很强的相关被互换所破坏时，r^2 值小于 1。

In this region, $D' = 1$ for all marker pairs, as there is no evidence of historical recombination. In contrast, and despite great simplicity of haplotype structure, r^2 values display a complex pattern, varying from 0.0003 to 1.0, with no relationship to physical distance. This makes sense, however, because without recombination, correlations among SNPs depend on the historical order in which they arose, not the physical order of SNPs on the chromosome.

Most importantly, the seeming complexity of r^2 values can be deconvolved in a simple manner: only seven different SNP configurations exist in this region, with all but two chromosomes matching five common haplotypes, which can be distinguished from each other by typing a specific set of four SNPs. That is, only a small minority of sites need be examined to capture fully the information in this region.

Variation in local recombination rates is a major determinant of LD. Recombination in the ancestors of the current population has typically disrupted the simple picture presented above. In the human genome, as in yeast[44], mouse[45] and other genomes, recombination rates typically vary dramatically on a fine scale, with hotspots of recombination explaining much crossing over in each region[28]. The generality of this model has recently been demonstrated through computational methods that allow estimation of recombination rates (including hotspots and coldspots) from genotype data[46,47].

The availability of nearly complete information about common DNA variation in the ENCODE regions allowed a more precise estimation of recombination rates across large regions than in any previous study. We estimated recombination rates and identified recombination hotspots in the ENCODE data, using methods previously described[46] (see Supplementary Information for details). Hotspots are short regions (typically spanning about 2 kb) over which recombination rates rise dramatically over local background rates.

Whereas the average recombination rate over 500 kb across the human genome is about 0.5 cM[48], the estimated recombination rate across the 500-kb ENCODE regions varied nearly tenfold, from a minimum of 0.19 cM (ENm013.7q21.13) to a maximum of 1.25 cM (ENr232.9q34.11). Even this tenfold variation obscures much more dramatic variation over a finer scale: 88 hotspots of recombination were identified (Fig. 8; see also Supplementary Fig. 7)—that is, one per 57 kb—with hotspots detected in each of the ten regions (from 4 in 12q12 to 14 in 2q37.1). Across the 5 Mb, we estimate that about 80% of all recombination has taken place in about 15% of the sequence (Fig. 9, see also refs 46, 49).

在这个区域，由于没有证据表明历史上有过重组，对于所有的标志物对而言，$D'=1$。相反，尽管单体型结构特别简单，r^2 值却呈现复杂的模式，在 0.0003 到 1.0 之间变化，与物理距离无关。这一点可以得到解释，因为在没有重组的情况下，SNP 间的相关依赖于它们出现的历史顺序，而不是它们在染色体上的物理顺序。

最重要的是，这种看起来复杂的 r^2 值能够以一种简单的方式去卷积：只有七种不同的 SNP 组合存在于这个区域，除了两条染色体外所有的染色体与五种常见的单体型相匹配，这些单体型能够通过输入特异的一组四个 SNP 而互相区分开来。也就是说，要完全捕获这个区域的信息，只需要检查少量的位点。

局部重组率的差异是 LD 的一个主要决定因素 现有种群的祖先中的重组通常已经破坏了上面呈现的简单图景。人类基因组与酵母 [44]、小鼠 [45] 和其他基因组一样，重组率一般在精细尺度上差异巨大，而重组热点解释了各个区域的很多互换 [28]。这个模型的普遍性最近已经通过计算方法得到证实，这种计算方法能够从基因型数据估计重组率（包括热点和冷点）[46,47]。

ENCODE 区域中常见 DNA 变异的接近完整的信息使我们能够比任何之前的研究在大的区域中更加精确地估计重组率。利用之前描述的方法 [46]，我们在 ENCODE 数据中估计了重组率，并且鉴定了重组热点（详见补充信息）。热点是短区域（一般长约 2 kb），其上的重组率比局部背景率高得多。

人类基因组中 500 kb 尺度上的平均重组率大约是 0.5 cM[48]。在 500 kb 的 ENCODE 区域中，重组率的变化幅度估计将近 10 倍，从最小的 0.19 cM（ENm013.7q21.13）到最大的 1.25 cM（ENr232.9q34.11）。然而这种 10 倍的差别仍然模糊了在更精细的尺度上更加巨大的差别：鉴定到 88 个重组热点（图 8；也可见补充信息图 7）——也就是每 57 kb 一个，并且存在于所有 10 个区域（从 12q12 区域的 4 个到 2q37.1 的 14 个）。在这个 5 Mb 的区域中，我们估计大约所有重组的 80% 发生在约 15% 的序列中（图 9，也可见参考文献 46、49）。

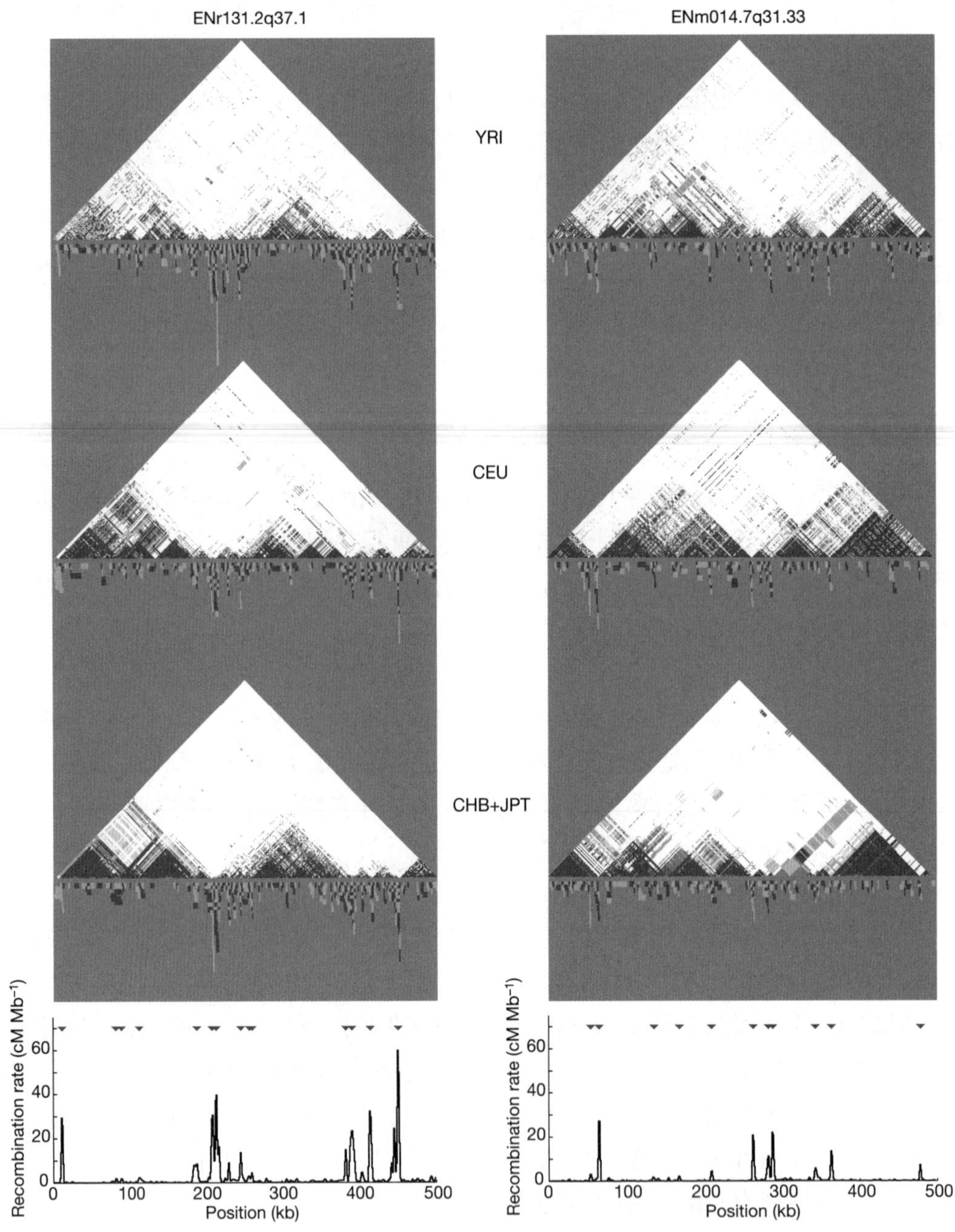

Fig. 8. Comparison of linkage disequilibrium and recombination for two ENCODE regions. For each region (ENr131.2q37.1 and ENm014.7q31.33), D' plots for the YRI, CEU and CHB+JPT analysis panels are shown: white, $D' < 1$ and LOD < 2; blue, $D' = 1$ and LOD < 2; pink, $D' < 1$ and LOD $\geqslant$ 2; red, $D' = 1$ and LOD $\geqslant$ 2. Below each of these plots is shown the intervals where distinct obligate recombination events must have occurred (blue and green indicate adjacent intervals). Stacked intervals represent regions where there are multiple recombination events in the sample history. The bottom plot shows estimated recombination rates, with hotspots shown as red triangles[46].

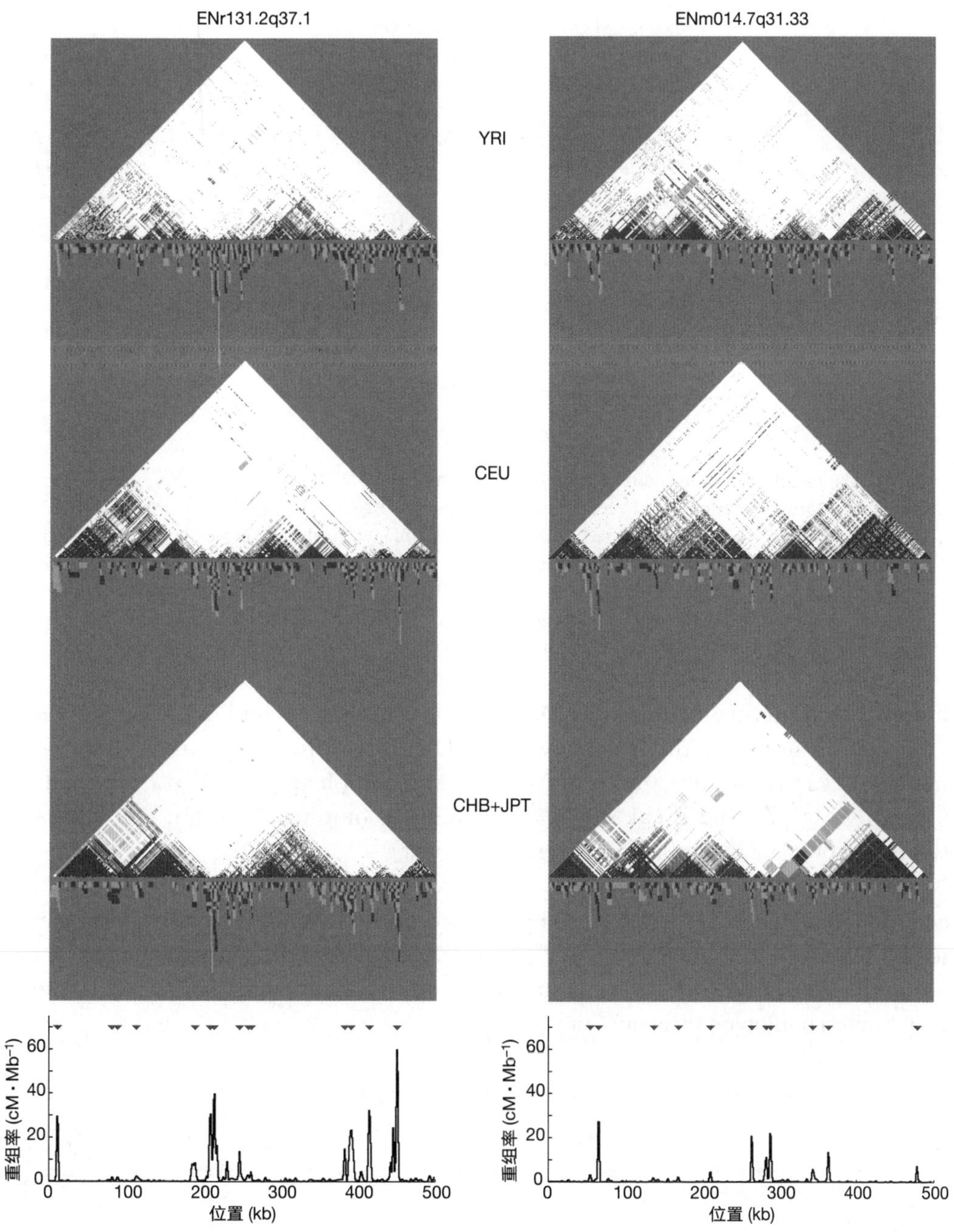

图 8. 两个 ENCODE 区域的 LD 和重组的比较。每个区域(ENr131.2q37.1 和 ENm014.7q31.33)的 YRI、CEU 和 CHB+JPT 分析小组的 D' 图：白色，$D'<1$ 且 LOD <2；蓝色，$D'=1$ 且 LOD <2；粉色，$D'<1$ 且 LOD $\geqslant 2$；红色，$D'=1$ 且 LOD $\geqslant 2$。各个图的下方展示了不同专性重组事件发生的区间(蓝色和绿色表示邻近的区间)。堆积的区间代表在样本历史中有多个重组事件发生的区域。底部的图展示估计的重组率，热点以红色三角形表示 [46]。

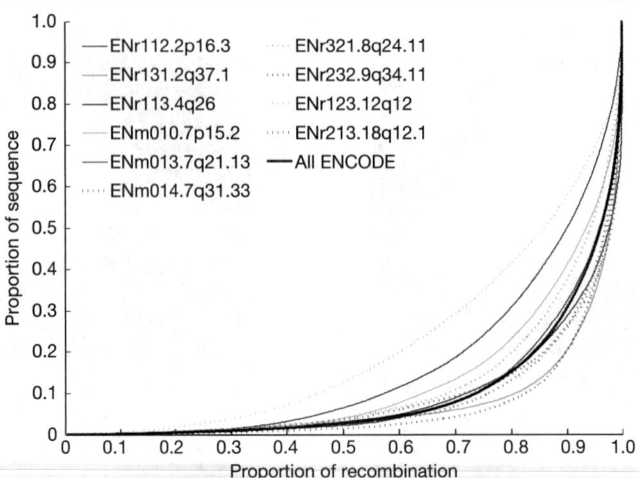

Fig. 9. The distribution of recombination events over the ENCODE regions. Proportion of sequence containing a given fraction of all recombination for the ten ENCODE regions (coloured lines) and combined (black line). For each line, SNP intervals are placed in decreasing order of estimated recombination rate[46], combined across analysis panels, and the cumulative recombination fraction is plotted against the cumulative proportion of sequence. If recombination rates were constant, each line would lie exactly along the diagonal, and so lines further to the right reveal the fraction of regions where recombination is more strongly locally concentrated.

A block-like structure of human LD. With most human recombination occurring in recombination hotspots, the breakdown of LD is often discontinuous. A "block-like" structure of LD is visually apparent in Fig. 8 and Supplementary Fig. 7: segments of consistently high D' that break down where high recombination rates, recombination hotspots and obligate recombination events[50] all cluster.

When haplotype blocks are more formally defined in the ENCODE data (using a method based on a composite of local D' values[30], or another based on the four gamete test[51]), most of the sequence falls into long segments of strong LD that contain many SNPs and yet display limited haplotype diversity (Table 5).

Table 5. Haplotype blocks in ENCODE regions, according to two methods

Parameter	YRI	CEU	CHB+JPT
Method based on a composite of local D' values[30]			
Average number of SNPs per block	30.3	70.1	54.4
Average length per block (kb)	7.3	16.3	13.2
Fraction of genome spanned by blocks (%)	67	87	81
Average number of haplotypes (MAF $\geqslant$ 0.05) per block	5.57	4.66	4.01
Fraction of chromosomes due to haplotypes with MAF $\geqslant$ 0.05 (%)	94	93	95
Method based on the four gamete test[51]			
Average number of SNPs per block	19.9	24.3	24.3

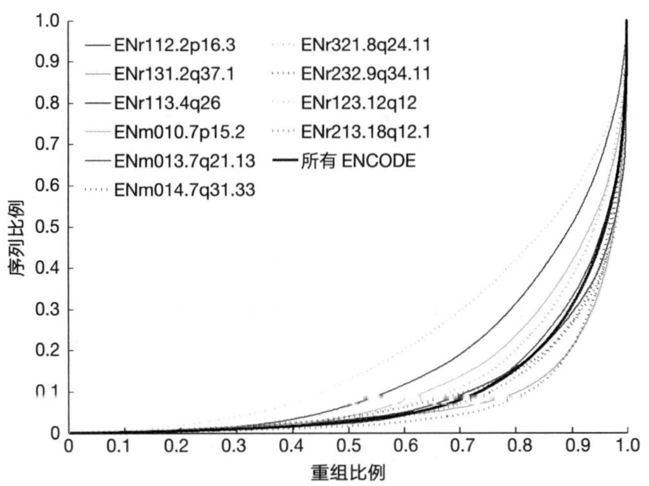

图 9. ENCODE 区域重组事件的分布。对于 10 个 ENCODE 区域（彩色线）和合并的全部区域（黑色线），含有特定比例重组的序列所占的比例。对于各条线，SNP 区间按估计的重组率逐渐减小的次序放置[46]，该重组率是各分析小组合计的，x 轴是累积的重组比例，y 轴是累积的序列比例。如果重组率是恒定的，各条线将会正好沿着对角线，所以偏向右边的线反映的是重组更加局部集中的区域所占的比例。

人类 LD 的区块样结构 大部分人类重组发生在重组热点处，LD 的分解常常是不连续的。LD 的“区块样”结构在图 8 和补充信息图 7 中看起来很明显：在高重组率、重组热点和专性重组事件[50]聚集的区域分成一段一段的高 D' 的片段。

当单体型区块在 ENCODE 数据中更正规地定义时（利用基于局部 D' 值组合的方法[30]，或者基于四配子检验的另一方法[51]），大部分的序列落入强 LD 的长片段，这些片段含有很多 SNP 但是表现出有限的单体型多样性（表 5）。

表 5. 根据两种方法得到的 ENCODE 区域的单体型块

参数	YRI	CEU	CHB+JPT
基于局部 D' 值组合的方法[30]			
每个区块的 SNP 平均数量	30.3	70.1	54.4
每个区块平均长度（kb）	7.3	16.3	13.2
被区块横跨的基因组的比例（%）	67	87	81
每个区块的单体型（MAF ≥ 0.05）平均数量	5.57	4.66	4.01
MAF ≥ 0.05 的单体型占染色体的比例（%）	94	93	95
基于四配子检验的方法[51]			
每个区块的 SNP 平均数量	19.9	24.3	24.3

Continued

Parameter	YRI	CEU	CHB+JPT
Average length per block (kb)	4.8	5.9	5.9
Fraction of genome spanned by blocks (%)	86	84	84
Average number of haplotypes (MAF ⩾ 0.05) per block	5.12	3.63	3.63
Fraction of chromosomes due to haplotypes with MAF ⩾ 0.05 (%)	91	95	95

Specifically, addressing concerns that blocks might be an artefact of low marker density[52], in these nearly complete data most of the sequence falls into blocks of four or more SNPs (67% in YRI to 87% in CEU) and the average sizes of such blocks are similar to initial estimates[30]. Although the average block spans many SNPs (30–70), the average number of common haplotypes in each block ranged only from 4.0 (CHB+JPT) to 5.6 (YRI), with nearly all haplotypes in each block matching one of these few common haplotypes. These results confirm the generality of inferences drawn from disease-mapping studies[27] and genomic surveys with smaller sample sizes[29] and less complete data[30].

Long-range haplotypes and local patterns of recombination. Although haplotypes often break at recombination hotspots (and block boundaries), this tendency is not invariant. We identified all unique haplotypes with frequency more than 0.05 across the 269 individuals in the phased data, and compared them to the fine-scale recombination map. Figure 10 shows a region of chromosome 19 over which many such haplotypes break at identified recombination hotspots, but others continue. Thus, the tendency towards colocalization of recombination sites does not imply that all haplotypes break at each recombination site.

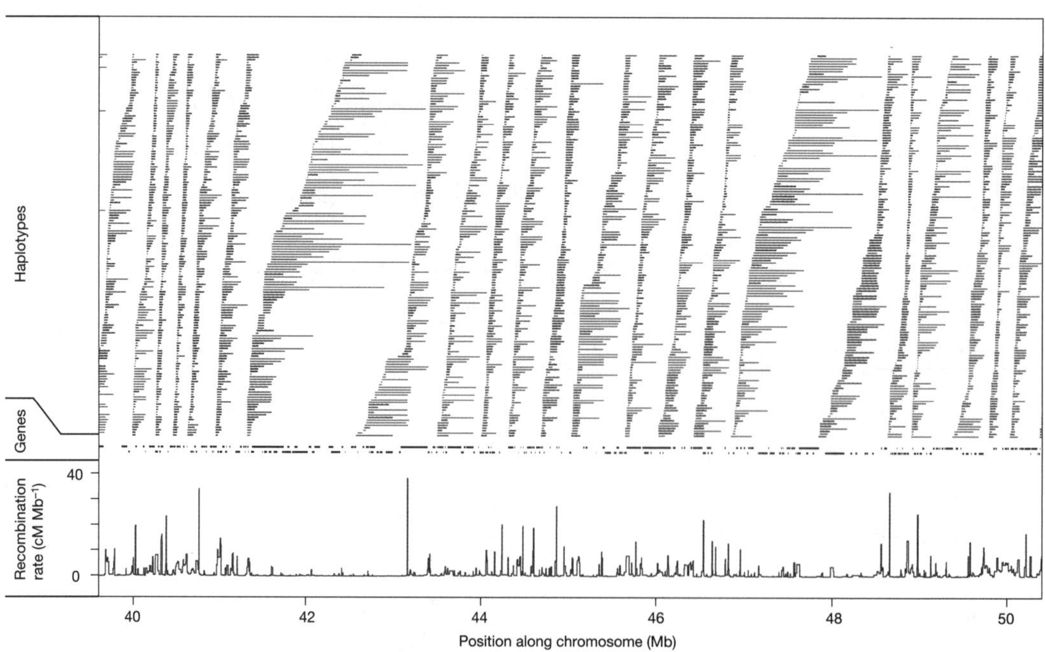

Fig. 10. The relationship among recombination rates, haplotype lengths and gene locations. Recombination

续表

参数	YRI	CEU	CHB+JPT
每个区块的平均长度(kb)	4.8	5.9	5.9
被区块横跨的基因组的比例(%)	86	84	84
每个区块的单体型(MAF ≥ 0.05)平均数量	5.12	3.63	3.63
MAF ≥ 0.05 的单体型占染色体的比例(%)	91	95	95

特别地，在这些接近完整的数据中，大部分的序列落入含有四个或更多 SNP 的区块中(从 YRI 中的 67% 到 CEU 中的 87%)，而且这些区块的平均大小与之前的估计类似 [30]，这解决了关于区块可能是低标签密度的假象的担忧 [57]。尽管一般的区块横跨很多 SNP(30 ~ 70 个)，但是在各个区块中，常见单体型的平均数量只有 4.0(CHB+JPT)到 5.6 个(YRI)，几乎所有的单体型都是不常见的单体型。这些结果证实了从样本量更小 [29]、数据更不完整 [30] 的疾病遗传定位研究 [27] 以及基因组调查所得出的推断的普遍性。

远距离单体型和局部重组模式　尽管单体型经常在重组热点(和区块的边界)中断，这个趋势并不是一成不变的。我们鉴定了阶段性数据中所有在 269 个个体中频率均超过 0.05 的独特单体型，并将它们与精细尺度的重组图谱相比较。图 10 展示了 19 号染色体的一个区域，这个区域上很多这样的单体型在鉴定的重组热点位置中断，但是其他的则是连续的。因而，重组位点共定位的趋势并不意味着所有单体型在各个重组位点中断。

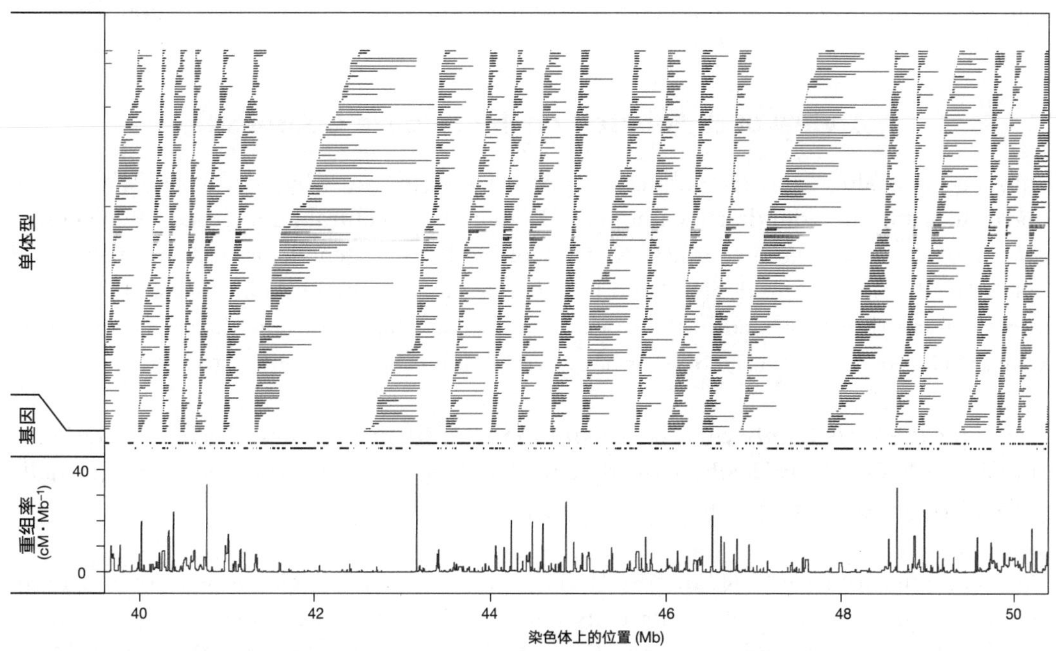

图 10. 重组率、单体型长度和基因位置之间的关系。重组率单位为 cM · Mb^{-1}(蓝色)。在 19 号染色体的

rates in cM Mb^{-1} (blue). Non-redundant haplotypes with frequency of at least 5% in the combined sample (bars) and genes (black segments) are shown in an example gene-dense region of chromosome 19 (19q13). Haplotypes are coloured by the number of detectable recombination events they span, with red indicating many events and blue few.

Some regions display remarkably extended haplotype structure based on a lack of recombination (Supplementary Fig. 8a, b). Most striking, if unsurprising, are centromeric regions, which lack recombination: haplotypes defined by more than 100 SNPs span several megabases across the centromeres. The X chromosome has multiple regions with very extensive haplotypes, whereas other chromosomes typically have a few such domains.

Most global measures of LD become more consistent when measured in genetic rather than physical distance. For example, when plotted against physical distance, the extent of pairwise LD varies by chromosome; when plotted against average recombination rate on each chromosome (estimated from pedigree-based genetic maps) these differences largely disappear (Supplementary Fig. 6). Similarly, the distribution of haplotype length across chromosomes is less variable when measured in genetic rather than physical distance. For example, the median length of haplotypes is 54.4 kb on chromosome 1 compared to 34.8 kb on chromosome 21. When measured in genetic distance, however, haplotype length is much more similar: 0.104 cM on chromosome 1 compared to 0.111 cM on chromosome 21 (Supplementary Fig. 9).

The exception is again the X chromosome, which has more extensive haplotype structure after accounting for recombination rate (median haplotype length = 0.135 cM). Multiple factors could explain different patterns on the X chromosome: lower SNP density, smaller sample size, restriction of recombination to females and lower effective population size.

A View of LD Focused on the Putative Causal SNP

Although genealogy and recombination provide insight into why nearby SNPs are often correlated, it is the redundancies among SNPs that are of central importance for the design and analysis of association studies. A truly comprehensive genetic association study must consider all putative causal alleles and test each for its potential role in disease. If a causal variant is not directly tested in the disease sample, its effect can nonetheless be indirectly tested if it is correlated with a SNP or haplotype that has been directly tested.

The typical SNP is highly correlated with many of its neighbours. The ENCODE data reveal that SNPs are typically perfectly correlated to several nearby SNPs, and partially correlated to many others.

We use the term proxy to mean a SNP that shows a strong correlation with one or more others. When two variants are perfectly correlated, testing one is exactly equivalent to testing the other; we refer to such collections of SNPs (with pairwise $r^2 = 1.0$ in the HapMap samples) as "perfect proxy sets".

一个高密度基因示例区域（19q13）内展示了在合并样本中频率至少为 5% 的非冗余单体型（棒状）和基因（黑色片段）。根据单体型区域内可检测的重组事件的数量，单体型标记为不同的颜色，红色代表很多的重组事件，而蓝色代表很少的重组事件。

由于缺少重组，一些区域表现出明显扩展的单体型结构（补充信息图 8a、8b）。最值得注意的是缺少重组的着丝粒区域：超过 100 个 SNP 组成的单体型横跨着丝粒前后数 Mb 的范围。X 染色体的多个区域含有范围非常大的单体型，而其他染色体则通常仅含有少量这样的区域。

当以遗传学距离而不是物理距离测量时，大多数对 LD 的全局性测量变得更加一致。例如，当针对物理距离作图时，成对 LD 的程度随染色体的不同而变化；当针对各条染色体上的平均重组率（估计自基于家系的遗传图谱）作图时，这些差异大部分消失了（补充信息图 6）。类似地，当以遗传学距离而不是物理距离测量时，单体型长度在染色体上的分布差异也更小。例如，单体型长度的中位数在 1 号染色体上是 54.4 kb，而在 21 号染色体上则是 34.8 kb。但是，当以遗传学距离测量时，单体型长度则接近得多：1 号染色体上是 0.104 cM，21 号染色体上是 0.111 cM（补充信息图 9）。

X 染色体同样是例外，扣除重组率的影响后，它含有范围更广的单体型结构（单体型长度中位数 ＝0.135 cM）。有多个因素可以解释 X 染色体上呈现的不同模式：更低的 SNP 密度、更小的样本量、限于女性的重组以及更小的有效种群尺寸。

集中于潜在致病 SNP 的 LD 概览

尽管家系和重组为解释为什么邻近的 SNP 常常关联提供了线索，但是对于关联研究的设计和分析，SNP 中的冗余才是重中之重。一个真正全面的遗传关联研究必须考虑所有可能致病的等位基因，并且检验它们在疾病中潜在的作用。如果一个导致疾病的变异没有在疾病样本中被直接检验，但是它与一个已经被直接检验的 SNP 或单体型相关联，则它的作用就可以被间接检验。

一个典型的 SNP 与很多邻近 SNP 高度相关 ENCODE 数据反映了 SNP 通常与多个邻近的 SNP 完美相关，与很多其他的 SNP 部分相关。

我们使用术语“代理者”来表示与一个或者更多其他 SNP 高度相关的一个 SNP。当两个变异完美相关时，检验一个就完全等同于检验另一个；我们称这样的 SNP 集（在 HapMap 样本中成对的 $r^2 = 1.0$）为“完美的代理者集”。

Considering only common SNPs (the target of study for the HapMap Project) in CEU in the ENCODE data, one in five SNPs has 20 or more perfect proxies, and three in five have five or more. In contrast, one in five has no perfect proxies. As expected, perfect proxy sets are smaller in YRI, with twice as many SNPs (two in five) having no perfect proxy, and a quarter as many (5%) having 20 or more (Figs 11 and 12). These patterns are largely consistent across the range of frequencies studied by the project, with a trend towards fewer proxies at MAF < 0.10 (Fig. 11). Put another way, the average common SNP in ENCODE is perfectly redundant with three other SNPs in the YRI samples, and nine to ten other SNPs in the other sample sets (Fig. 13).

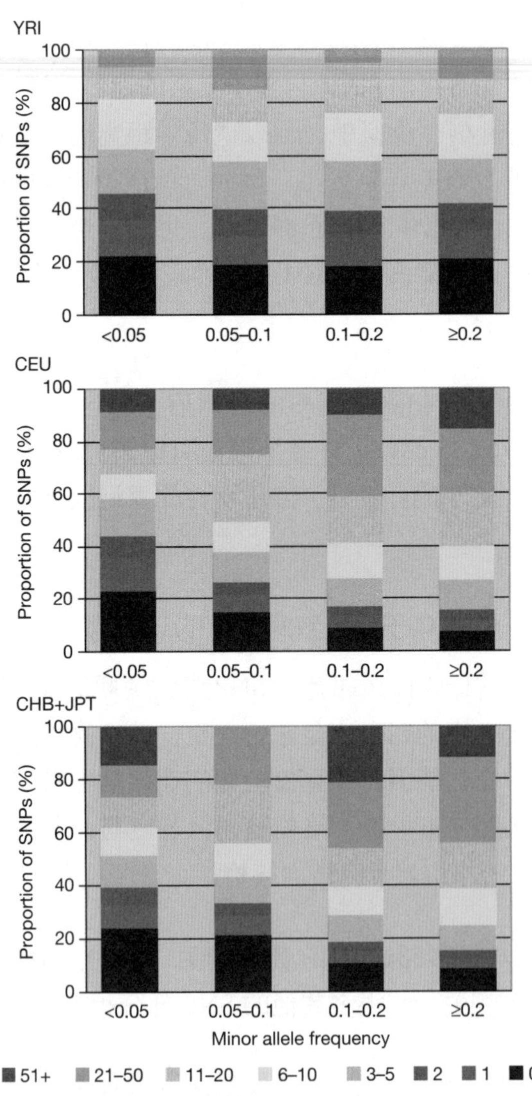

Fig. 11. The number of proxy SNPs ($r^2 \geqslant 0.8$) as a function of MAF in the ENCODE data.

只考虑ENCODE数据CEU中常见的SNP(HapMap计划的研究目标)，则每五个SNP中有一个有20个或更多的完美代理者，有三个含有五个或更多的完美代理者，只有一个没有完美代理者。正如预期的，完美代理者集在YRI中更小，没有完美代理者的SNP(每五个中有两个)是另两个小组的两倍，含有20个或更多代理者的SNP(5%)是另两个小组的四分之一(图11和12)。在计划所研究的频率范围内，这些模式大部分是一致的，并且表现出MAF < 0.10时代理者更少的趋势(图11)。换句话说，ENCODE中常见的SNP，在YRI样本中平均跟三个其他SNP完全冗余，在其他样本集中平均跟9到10个SNP完全冗余(图13)。

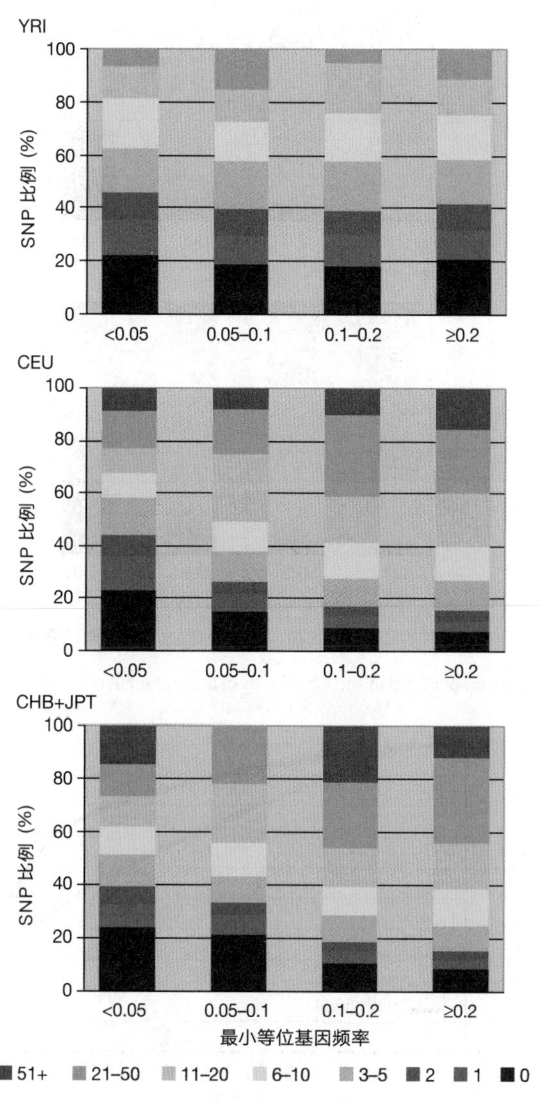

图11. ENCODE数据中不同MAF的SNP的代理者SNP($r^2 \geqslant 0.8$)数量

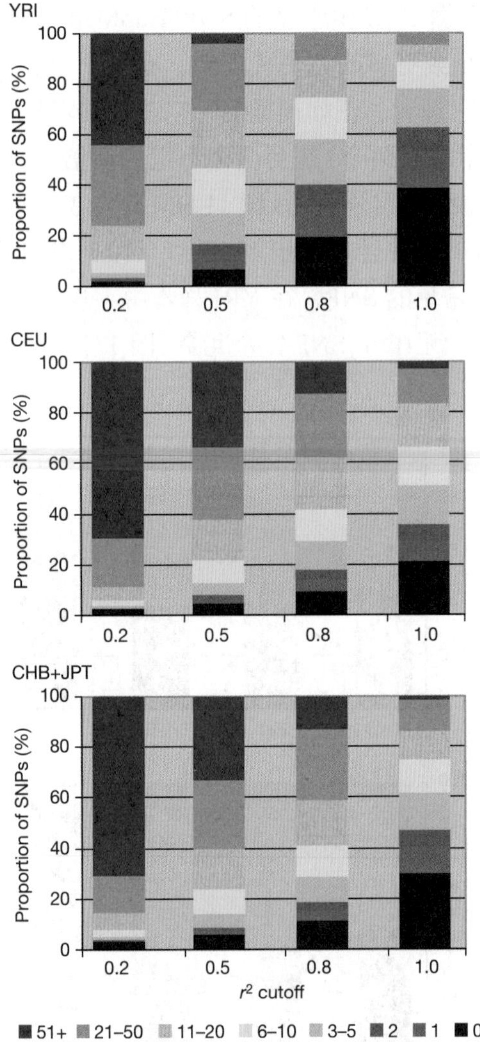

Fig. 12. The number of proxies per SNP in the ENCODE data as a function of the threshold for correlation (r^2).

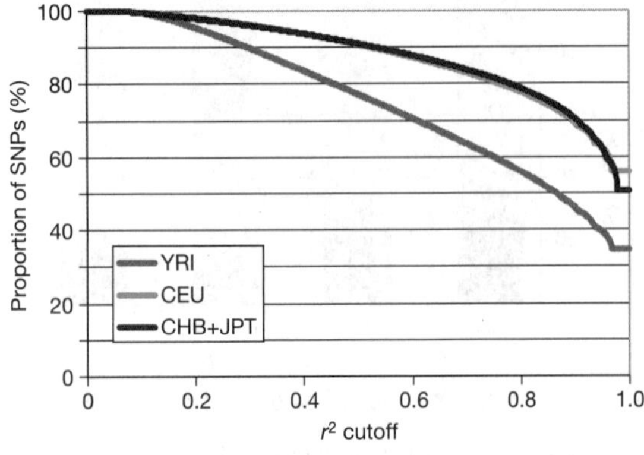

Fig. 13. Relationship in the Phase I HapMap between the threshold for declaring correlation between proxies and the proportion of all SNPs captured.

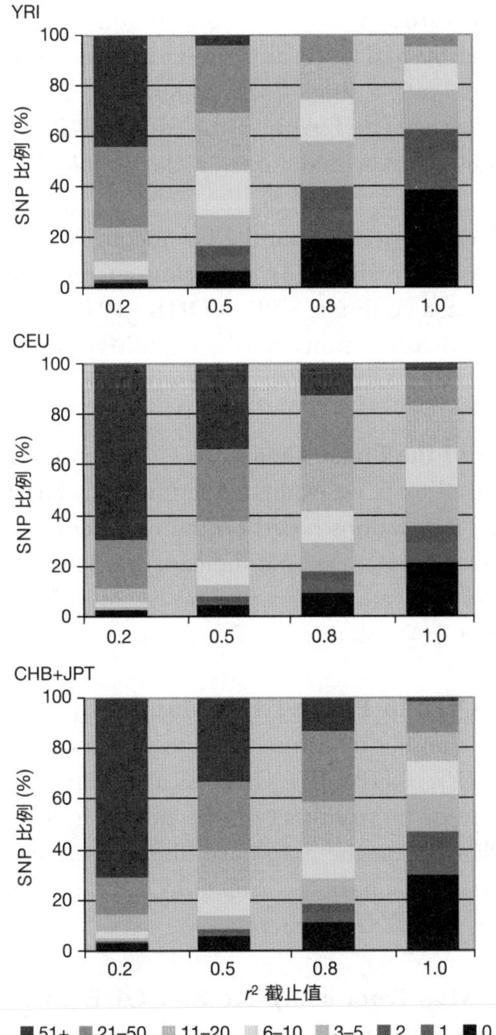

图 12. ENCODE 数据中不同相关性阈值(r^2)的每个 SNP 的代理者数量

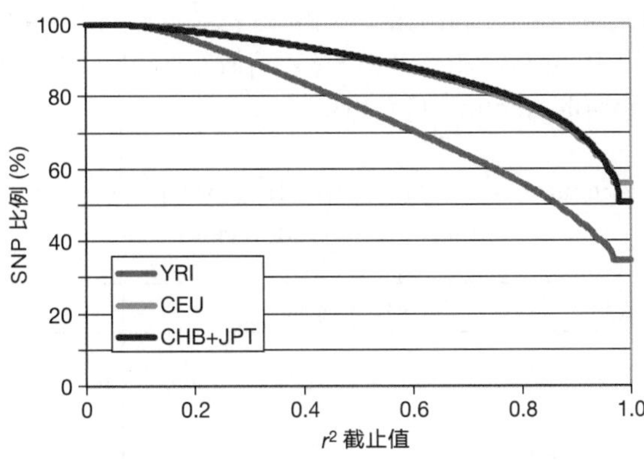

图 13. 第 I 阶段 HapMap 中用于判定代理者间相关的阈值和所有捕获到的 SNP 的比例之间的关系

Of course, to be detected through LD in an association study, correlation need not be complete between the genotyped SNP and the causal variant. For example, under a multiplicative disease model and a single-locus χ^2 test, the sample size required to detect association to an allele scales as $1/r^2$. That is, if the causal SNP has an $r^2 = 0.5$ to one tested in the disease study, full power can be maintained if the sample size is doubled.

The number of SNPs showing such substantial but incomplete correlation is much larger. For example, using a looser threshold for declaring correlation ($r^2 \geqslant 0.5$), the average number of proxies found for a common SNP in CHB+JPT is 43, and the average in YRI is 16 (Fig. 12). These partial correlations can be exploited through haplotype analysis to increase power to detect putative causal alleles, as discussed below.

Evaluating performance of the Phase I map. To estimate the proportion of all common SNPs captured by the Phase I map, we evaluated redundancy among SNPs on the genome-wide map, and performed simulations based on the more complete ENCODE data. The two methods give highly similar answers, and indicate that Phase I should provide excellent power for CEU, CHB and JPT, and substantial power for YRI. Phase II, moreover, will provide nearly complete power for all three analysis panels.

Redundancies among SNPs in Phase I HapMap. Redundancy offers one measure that Phase I has sampled densely in comparison to the underlying scale of correlation. Specifically, 50% (YRI) to 75% (CHB+JPT, CEU) of all SNPs on the Phase I map are highly correlated ($r^2 \geqslant 0.8$) to one or more others on the map (Fig. 13; see also Supplementary Fig. 10). Over 90% of all SNPs on the map have highly statistically significant correlation to one or more neighbours. These partial correlations can be combined to form haplotypes that are even better proxies for a SNP of interest.

Modelling Phase I HapMap from complete ENCODE data. A second approach to evaluating the completeness of the Phase I data involves thinning the more complete ENCODE data to match Phase I for allele frequency and SNP density. Simulated Phase I HapMaps were used to evaluate coverage in relation to the full set of common SNPs (Table 6), and provided nearly identical estimates to those above: 45% (YRI) to 74% (CHB+JPT, CEU) of all common SNPs are predicted to have a proxy with $r^2 \geqslant 0.8$ to a SNP included in the Phase I HapMap (Supplementary Fig. 11).

Table 6. Coverage of simulated Phase I and Phase II HapMap to capture all common SNPs in the ten ENCODE regions

Analysis panel	Per cent maximum $r^2 \geqslant 0.8$	Mean maximum r^2
Phase I HapMap		
YRI	45	0.67
CEU	74	0.85
CHB+JPT	72	0.83

当然，为了能在关联研究中通过 LD 被检测到，基因分型的 SNP 和导致表型的变异间的相关性不需要是完全的。例如，在乘性疾病模型和单基因座 χ^2 检验下，检测一个等位基因的相关性所要求的样本大小是 $1/r^2$。也就是说，当在疾病研究中导致表型的 SNP 与被检验的 SNP 的 $r^2 = 0.5$ 时，如果样本大小翻倍就能够保持完全的效能。

表现出这种相当程度但不完全相关的 SNP 的数量大得多。例如，利用一个宽松的阈值来判定相关性（$r^2 \geqslant 0.5$），在 CHB+JPT 中一个常见 SNP 的代理者的平均数量是 43 个，在 YRI 中是 16 个（图 12）。正如下文讨论的，这些部分相关可以通过单体型分析而被挖掘，进而来增加检测可能导致表型的等位基因的能力。

第 I 阶段图谱质量的评估　为了估计第 I 阶段图谱捕获的所有常见 SNP 的比例，我们评估了全基因组范围图谱上 SNP 间的冗余，并且基于更完整的 ENCODE 数据进行了模拟。这两种方法给出了高度相似的答案，说明第 I 阶段图谱应该为 CEU、CHB 和 JPT 提供了非常好的效能，为 YRI 提供了相当程度的效能。并且，第 II 阶段将会为所有三个分析小组提供接近完全的效能。

第 I 阶段 HapMap 中 SNP 间的冗余　与潜在的相关性指标相比，冗余提供了一种测量手段，第 I 阶段图谱已经对其进行了密集取样。特别地，第 I 阶段图谱上所有 SNP 中的 50%（YRI）到 75%（CHB+JPT、CEU）与一个或更多其他 SNP 高度相关（$r^2 \geqslant 0.8$）（图 13；也可见补充信息图 10）。图谱上所有 SNP 中超过 90% 与一个或更多邻近 SNP 具有统计学意义上的高度相关。这些部分相关能被联合起来形成单体型，这些单体型对于感兴趣的 SNP 甚至是更好的代理者。

基于完整 ENCODE 数据的第 I 阶段 HapMap 建模　评估第 I 阶段数据完整性的第二个方法包括缩减更加完整的 ENCODE 数据来匹配第 I 阶段数据的等位基因频率和 SNP 密度。模拟的第 I 阶段 HapMap 被用于评估其相对全套常见 SNP 的覆盖度（表 6），它提供了跟上面几乎一致的估计值：所有常见 SNP 的 45%（YRI）到 74%（CHB+JPT、CEU）被预测在第 I 阶段的 HapMap 中有一个 $r^2 \geqslant 0.8$ 的代理者（补充信息图 11）。

表 6. 模拟的 HapMap 第 I 和第 II 阶段捕获 10 个 ENCODE 区域中所有常见 SNP 的覆盖度

分析小组	最大 r^2 值 $\geqslant 0.8$ 的比例	平均最大 r^2 值
第 I 阶段 HapMap		
YRI	45	0.67
CEU	74	0.85
CHB+JPT	72	0.83

Continued

Analysis panel	Per cent maximum $r^2 \geqslant 0.8$	Mean maximum r^2
Phase II HapMap		
YRI	81	0.90
CEU	94	0.97
CHB+JPT	94	0.97

Simulated Phase I HapMaps were generated from the phased ENCODE data (release 16c1) by randomly picking SNPs that appear in dbSNP build 121 (excluding "non-rs" SNPs in release 16a) for every 5-kb bin until a common SNP was picked (allowing up to three attempts per bin). The Phase II HapMap was simulated by picking SNPs at random to achieve an overall density of 1 SNP per 1 kb. These numbers are averages over 20 independent iterations for all ENCODE regions in all three analysis panels.

Statistical power in association studies may be more closely approximated by the average (maximal) correlation value between a SNP and its best proxy on the map, rather than by the proportion exceeding an arbitrary (and stringent) threshold. The average values for maximal r^2 to a nearby SNP range from 0.67 (YRI) to 0.85 (CEU and CHB+JPT).

Modelling Phase II HapMap from complete ENCODE data. A similar procedure was used to generate simulated Phase II HapMaps from ENCODE data (Table 6). Phase II is predicted to capture the majority of common variation in YRI: 81% of all common SNPs should have a near perfect proxy ($r^2 \geqslant 0.8$) to a SNP on the map, with the mean maximal r^2 value of 0.90. Unsurprisingly, the CEU, CHB and JPT samples, already well served by Phase I, are nearly perfectly captured: 94% of all common sites have a proxy on the map with $r^2 \geqslant 0.8$, with an average maximal r^2 value of 0.97.

These analyses indicate that the Phase I and Phase II HapMap resources should provide excellent coverage for common variation in these population samples.

Selection of Tag SNPs for Association Studies

A major impetus for developing the HapMap was to guide the design and prioritization of SNP genotyping assays for disease association studies. We refer to the set of SNPs genotyped in a disease study as tags. A given set of tags can be analysed for association with a phenotype using a variety of statistical methods which we term tests, based either on the genotypes of single SNPs or combinations of multiple SNPs.

The shared goal of all tag selection methods is to exploit redundancy among SNPs, maximizing efficiency in the laboratory while minimizing loss of information[24,27]. This literature is extensive and varied, despite its youth. Some methods require that a single SNP serve as a proxy for other, untyped variants, whereas other methods allow combinations of alleles (haplotypes) to serve as proxies; some make explicit use of LD blocks whereas others are agnostic to such descriptions. Although it is not practical to implement all such

续表

分析小组	最大 r^2 值 ≥ 0.8 的比例	平均最大 r^2 值
第 II 阶段 HapMap		
YRI	81	0.90
CEU	94	0.97
CHB+JPT	94	0.97

模拟的第 I 阶段 HapMap 是从区分染色体的 ENCODE 数据（版本 16c1）产生的，方法是每 5 kb 随机选取出现在 dbSNP 版本 121（不包括版本 16a 中的“non-rs”SNP）中的 SNP，直到取到一个常见 SNP（每个 5 kb 区间允许三次尝试）。第 II 阶段 HapMap 的模拟是通过随机取 SNP 以达到总体每 1 kb 1 个 SNP 的密度。这些数值是三个分析小组中所有 ENCODE 区域 20 次独立迭代的平均值。

通过 SNP 和它在图谱上的最佳代理者的平均（最大）相关值，而不是通过超过一个主观设定的（并且严格的）阈值的比例，可以对关联研究中的统计效能进行更接近的估算。SNP 与附近 SNP 的最大 r^2 的平均值在 0.67（YRI）到 0.85（CEU 和 CHB+JPT）之间变化。

基于完整 ENCODE 数据的第 II 阶段 HapMap 建模 类似的步骤被用于从 ENCODE 数据产生模拟的第 II 阶段 HapMap（表 6）。第 II 阶段预计能捕获 YRI 中绝大多数的常见变异：所有常见 SNP 的 81% 应该在图谱上有一个接近完美的代理者（$r^2 \geq 0.8$），平均最大 r^2 值是 0.90。与预期相符，CEU、CHB 和 JPT 样本已经在第 I 阶段满足要求，被近乎完美地捕获：所有常见位点的 94% 在图谱上有一个 $r^2 \geq 0.8$ 的代理者，平均最大 r^2 值为 0.97。

这些分析说明第 I 和第 II 阶段 HapMap 的资源应该能够非常好地覆盖这些种群样本中的常见变异。

用于关联研究的标签 SNP 的选择

开展 HapMap 研究的一个主要的动力是指导疾病关联研究中 SNP 基因分型实验的设计和优先次序。我们称疾病研究中被基因分型的 SNP 集为标签。基于单个 SNP 或多个 SNP 的组合的基因型，利用我们称之为检验的各种统计学方法，可以分析一组给定的标签与某种表型的关联。

所有标签选择方法的共同目标是利用 SNP 间的冗余，最大化实验效率，同时最小化信息的丢失[24,27]。尽管刚起步，但这样的文献广泛而多样。一些方法要求单个 SNP 作为其他未分型变异的代理者，而其他方法允许等位基因的组合（单体型）作为代理者；一些人明确使用 LD 区块，而其他人则没有这样的描述。尽管在本计划的

methods at the project website, the HapMap genotypes are freely available and investigators can apply their method of choice to the data. To assist users, both a single-marker tagging method and a more efficient multimarker method have been implemented at http://www.hapmap.org.

Tagging using a simple pairwise method. To illustrate general principles of tagging, we first applied a simple and widely used pairwise algorithm[53,54]: SNPs are selected for genotyping until all common SNPs are highly correlated ($r^2 \geqslant 0.8$) to one or more members of the tag set.

Starting from the substantially complete ENCODE data, the density of common SNPs can be reduced by 75–90% with essentially no loss of information (Fig. 14). That is, the genotyping burden can be reduced from one common SNP every 500 bp to one SNP every 2 kb (YRI) to 5 kb (CEU and CHB+JPT). Because LD often extends for long distances, studies of short gene segments tend to underestimate the redundancy across the genome[43].

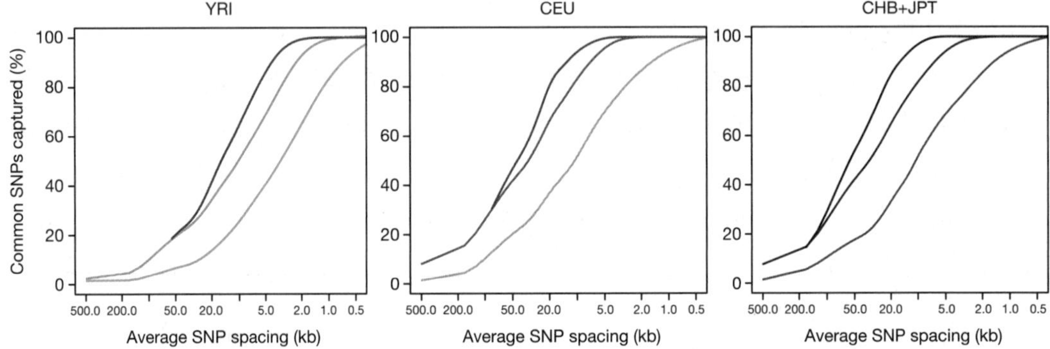

Fig. 14. Tag SNP information capture. The proportion of common SNPs captured with $r^2 \geqslant 0.8$ as a function of the average tag SNP spacing is shown for the phased ENCODE data, plotted (left to right) for tag SNPs prioritized by Tagger (multimarker and pairwise) and for tag SNPs picked at random. Results were averaged over all the ENCODE regions.

Although tags selected based on LD offer the greatest improvements in efficiency and information capture, even randomly chosen subsets of SNPs offer considerable efficiencies (Fig. 14).

The data also reveal a rule of diminishing returns: a small set of highly informative tags captures a large fraction of all variation, with additional tags each capturing only one or a few proxies. For example, in CHB+JPT the most informative 1% of all SNPs (one per 50 kb) is able to proxy (at $r^2 \geqslant 0.8$) for 40% of all common SNPs, whereas a substantial proportion of SNPs have no proxies at all.

These observations are encouraging with respect to genome-wide association studies. A set of SNPs typed every 5–10 kb across the genome (within the range of current technology) can capture nearly all common variation in the genome in the CEU and CHB+JPT samples, with more SNPs required in the YRI samples.

网站上实现所有这些方法并不现实，但是 HapMap 的基因型数据是自由获取的，研究者可以用他们选择的方法分析这些数据。为了帮助使用者，单个标志物标签法和更有效的多标志物法都已经在 http://www.hapmap.org 网站上实现了。

使用简单成对方法进行的标记 为了说明标记的一般原则，我们首先应用了一个简单而广泛使用的成对算法[53,54]：选择 SNP 进行基因分型，直到所有常见 SNP 与一个或更多的标签集成员高度连锁（$r^2 \geqslant 0.8$）。

先以相当完整的 ENCODE 数据为例，常见 SNP 的密度能被减少 75%～90% 而基本不丢失信息（图 14）。也就是说，基因分型的负担能从每 500 bp 一个常见 SNP 减少到每 2 kb（YRI）或每 5 kb（CEU 和 CHB+JPT）一个 SNP。因为 LD 常常延伸很长距离，短基因片段的研究通常低估基因组中的冗余度[43]。

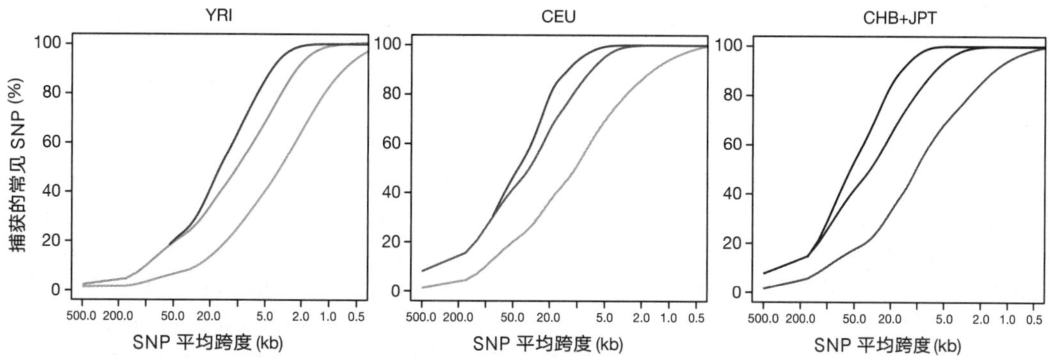

图 14. 标签 SNP 信息捕获。图中展示的是 ENCODE 数据中，被不同平均跨度的标签 SNP 捕获（$r^2 \geqslant 0.8$）的常见 SNP 的比例，从左到右依次为由 Tagger 软件（多标志物和成对标志物）定为优先级的标签 SNP 和随机选取的标签 SNP。结果是所有 ENCODE 区域的平均值。

基于 LD 选择的标签在效率和信息的捕获上提供了最大的改进，而即使是随机选择的 SNP 子集也提供了相当高的效率（图 14）。

这个数据还反映了一个回报不断减少的规律：一个小的高度富含信息的标签集可以捕获所有变异中的大部分，而增加的标签仅捕获一个或很少的代理者。例如，在 CHB+JPT 中，所有 SNP（每 50 kb 一个）中最富有信息的 1% 能够代理（$r^2 \geqslant 0.8$）所有常见 SNP 的 40%，而相当大比例的 SNP 完全没有代理者。

对于全基因组关联研究，这些现象是鼓舞人心的。对于 CEU 和 CHB+JPT 样本，在全基因组中每 5～10 kb 被分型的一组 SNP（现有技术的能力范围之内）就能够捕获基因组中几乎所有的常见变异，对于 YRI 样本则需要更多的 SNP。

Tagging from the genome-wide map. Whereas analysis of the complete ENCODE data set reveals the maximal efficiency likely to be possible with this tag selection strategy, analysis of the Phase I map illuminates the extent to which the current resource can be used for near-term studies. Specifically, using the same pairwise tagging approach above, 260,000 (CHB+JPT) to 474,000 (YRI) SNPs are required to capture all common SNPs in the Phase I data set (Table 7). That is, being incomplete and thus less redundant, the Phase I data are much less compressible by tag SNP selection than are the ENCODE data. Nevertheless, even at this level a half to a third of all SNPs can be selected as proxies for the remainder (and, by inference, the bulk of other common SNPs in the genome).

Table 7. Number of selected tag SNPs to capture all observed common SNPs in the Phase I HapMap

r^2 threshold*	YRI	CEU	CHB+JPT
$r^2 \geqslant 0.5$	324,865	178,501	159,029
$r^2 \geqslant 0.8$	474.409	293,835	259,779
$r^2 = 1.0$	604,886	447,579	434,476

Tag SNPs were picked to capture common SNPs in HapMap release 16c1 using the software program Haploview.
* Pairwise tagging at different r^2 thresholds.

Increasing the efficiency of tag SNPs. Although the pairwise method is simple, complete and straightforward, efficiency can be improved with a number of simple changes. First, relaxing the threshold on r^2 for tag SNP selection substantially reduces the number of tag SNPs selected, with only a modest decrease in the correlations among SNPs (Table 7). For example, reducing the r^2 threshold from 0.8 to 0.5 decreases the number of tag SNPs selected from the HapMap by 39% in CHB+JPT (260,000 to 159,000) and 32% in YRI (474,000 to 325,000). The average r^2 value between tags and other (unselected) SNPs falls much less dramatically than the number of tags selected, increasing efficiency. Whether such a loss of power is justified by the disproportionate reduction in work is a choice each investigator will need to make.

A second enhancement exploits multimarker haplotypes. Many investigators have discussed using multiple SNPs (in haplotypes and regression models) to serve as proxies for untyped sites[55-58], which may reduce the number of tags required and increase the power of analyses performed. Figure 14 illustrates the point with one such method[55], showing that a multimarker method allows greater coverage for a fixed set of markers (or, alternatively, fewer markers to achieve the same coverage). Although a full consideration of this issue is beyond the scope of this paper, the availability of these and other data should allow the comparison and application of such methods.

A third approach to increasing efficiency is to prioritize tags based on the number of other SNPs captured. Whereas 260,000 SNPs are required to provide $r^2 \geqslant 0.8$ for all SNPs in the Phase I HapMap (CHB+JPT), the best 10,000 such SNPs (4%) capture 22% of all common

从全基因组图谱进行的标签选取 尽管对完整 ENCODE 数据集的分析反映了利用这种标签选择策略可能具有的最大效率，对第 I 阶段图谱的分析则阐明了现有资源能够在多大程度上被用于近期的研究。特别地，利用相同的上述成对标签方法，为捕获第 I 阶段数据集中所有常见 SNP，需要 260,000(CHB+JPT) 到 474,000(YRI) 个 SNP(表 7)。也就是说，第 I 阶段数据不够完整因而冗余更少，与 ENCODE 数据相比，能够通过标签 SNP 选择而被压缩的程度要小很多。但是，即使在这种水平下，SNP 中的一半或三分之一可以被选作剩余 SNP(推断可知也是基因组中其他常见 SNP 的大部分) 的代理者。

表 7. 为捕获第 I 阶段 HapMap 中观察到的所有常见 SNP 所选择的标签 SNP 的数量

r^2 阈值 *	YRI	CEU	CHB+JPT
$r^2 \geqslant 0.5$	324,865	178,501	159,029
$r^2 \geqslant 0.8$	474,409	293,835	259,779
$r^2 = 1.0$	604,886	447,579	434,476

标签 SNP 是利用软件 Haploview 选取的，以捕获 HapMap 版本 16c1 中常见的 SNP。
* 在不同 r^2 阈值下取成对的标签。

增加标签 SNP 的效率 尽管成对的方法简单、完整而直接，我们还是可以通过一些简单的改变提高效率。首先，放宽用于标签 SNP 选择的 r^2 的阈值，可以大大减少选择的标签 SNP 的数量，而在 SNP 间的相关性上仅稍有下降(表 7)。例如，r^2 阈值从 0.8 减少到 0.5，在 CHB+JPT 中减少了 39% 从 HapMap 选择的标签 SNP 的数量(从 260,000 到 159,000)，而在 YRI 中减少了 32%(从 474,000 到 325,000)。标签与其他(未选择) SNP 间平均 r^2 值下降的水平要比选择的标签数量的下降少得多，因而提高了效率。是否这样一种效能的损失会因为不成比例地减少工作量变得合理，是每个研究者需要作出的一个选择。

第二个提高的方法是利用多标志物单体型。很多研究者已经讨论了利用多个 SNP(在单体型和回归模型中) 作为未分型位点的代理者 [55-58]，这可能能够减少需要的标签的数量，并且增加分析的效能。图 14 利用一个这样的方法 [55] 说明了这一点，它表明，对于固定的一组标志物，多标志物方法能够达到更高的覆盖度(换言之，更少的标志物可以达到相同的覆盖度)。尽管全面考虑这个问题超出了本文的范畴，已有的这些以及其他数据足以对这样的方法进行比较和应用。

第三个增加效率的方法是基于捕获的其他 SNP 的数量对标签划分优先级。尽管要求 260,000 个 SNP 才能以 $r^2 \geqslant 0.8$ 捕获第 I 期 HapMap 里所有的 SNP (CHB+JPT)，其中最佳的 10,000 个 (4%) 就能以 $r^2 \geqslant 0.8$ 捕获所有常见变异位点的

variable sites with $r^2 \geqslant 0.8$ (Table 8). Such prioritization can be applied using different weights for SNPs based on genomic annotation (for example, non-synonymous coding SNPs, SNPs in conserved non-coding sequence, and candidate genes of biological interest).

Table 8. Proportion of common SNPs in Phase I captured by sets of tag SNPs

Tag SNP set size	Common SNPs captured (%)		
	YRI	CEU	CHB+JPT
10,000	12.3	20.4	21.9
20,000	19.1	30.9	33.2
50,000	32.7	50.4	53.6
100,000	47.2	68.5	72.2
250,000	70.1	94.1	98.5

As in Table 7, tag SNPs were picked to capture common SNPs in HapMap release 16c1 using Haploview, selecting SNPs in order of the fraction of sites captured. Common SNPs were captured by fixed-size sets of pairwise tags at $r^2 \geqslant 0.8$.

Tag transferability across populations. The most complete set of tags would be those based on all 269 samples; however, many studies may be performed in individuals more closely related to one particular HapMap population, and efficiency may be gained by selecting tags only from that population sample. (Selecting tags in a HapMap population sample that is known to be more distantly related than is another, for example, using CEU to pick tags for a study of Japanese, seems inefficient.)

An important question is how tags selected in one or more analysis panels will transfer to disease studies performed in these or other populations. Our data do not address this question directly, although the known similarity of allele and haplotype frequencies across populations within continents[41] is encouraging. More data are clearly needed, however.

Tag selection based on initial genotyping. Whereas the discussions above assume *de novo* selection of SNPs, many investigators will have already performed initial studies, and wish to design follow-on experiments. The HapMap data can be used to highlight SNPs that might potentially explain a positive association signal, or those that were poorly captured (and thus still need to be tested) after a negative scan. In cases where multiple SNPs are both associated with the trait and with each other, the HapMap data can be queried to identify whether samples from any other analysis panel show a breakdown of LD in that region, and thus the possibility of narrowing the span over which the causal variant may reside.

Applications to the Analysis of Association Data

Beyond guiding selection of tag SNPs, HapMap data can inform the subsequent analysis and interpretation in disease association studies.

22%(表 8)。这样的划分优先级法可以通过根据基因组注释对 SNP 使用不同的权重来应用(例如，非同义编码 SNP、在保守的非编码序列中的 SNP，以及有生物学意义的候选基因)。

表 8. 通过不同标签 SNP 集捕获的第 I 阶段中常见 SNP 的比例

标签 SNP 集大小	捕获的常见 SNP(%)		
	YRI	CEU	CHB+JPT
10,000	12.3	20.4	21.9
20,000	19.1	30.9	33.2
50,000	32.7	50.4	53.6
100,000	47.2	68.5	72.2
250,000	70.1	94.1	98.5

如表 7 一样，标签 SNP 是使用 Haploview 选取的，以捕获 HapMap 版本 16c1 中常见的 SNP，SNP 按捕获的位点的比例顺序选择。常见 SNP 被固定大小的一组成对标签以 $r^2 \geqslant 0.8$ 捕获。

标签在不同种群间的可转移性　最完整的一组标签是基于所有 269 个样本的标签；但是，很多研究可能是在与一个特定的 HapMap 种群亲缘关系更近的个体中进行的，因此效率仅在从那个种群样本中选择标签时可能会增加。(在亲缘关系更远的 HapMap 种群样本中选择标签看起来是低效的，例如使用 CEU 种群选取用于研究日本人的标签。)

一个重要的问题是：在一个或更多个分析小组中选择的标签如何转移到在这些或者其他种群中所开展的疾病研究。尽管已知的同一大陆上各个种群中等位基因和单体型频率的相似性[41]是令人鼓舞的，但是我们的数据并没有直接回答这个问题。很明显，要回答这个问题还需要更多的数据。

基于初步基因分型的标签选择　上述的讨论是假定从头选择 SNP 的，但是很多研究者已经进行了初步研究，希望在此基础上设计后续实验。HapMap 数据可以用来突出有望解释阳性关联信号的 SNP，或者经过阴性扫描不容易被捕获的(因而仍需要被检验)SNP。在多个 SNP 都跟性状关联并且 SNP 也彼此关联的情况下，可以查询 HapMap 数据来鉴定任一其他分析小组的样本是否在那个区域表现为 LD 的断裂，以及缩小导致性状的变异可能存在的区域范围的可能性。

在关联数据分析中的应用

在疾病关联研究中，HapMap 数据除了指导标签 SNP 的选择，还能够影响后续的分析和解释。

Analysis of an existing genotype data set. The HapMap can be used to inform association testing, regardless of how tags were selected. Specifically, as long as the SNPs genotyped in a disease study have also been typed in the HapMap samples, it is possible to identify which SNPs are well captured by the genotyped SNPs (either singly, or in haplotype combinations), and which are not[55].

This is of particular importance for genome-wide association studies performed using array-based, standardized genotyping reagents, which do not allow investigators to choose their own sets of tag SNPs. The Affymetrix 120K SNP array data included in Phase I of the HapMap provides a simple example: in CEU 48% of HapMap SNPs have substantial pairwise correlation ($r^2 \geqslant 0.5$) to one or more of the 120K SNPs on the array. An additional 13%, however, are not correlated to a single SNP, but are to a specific haplotype of two members of the 120K panel. By identifying such haplotype predictors in the HapMap, and testing them (in addition to the single SNPs) in a disease study, it is likely that power will be increased (I. Pe'er *et al.*, manuscript in preparation).

Evaluating statistical significance and interpreting results. An important challenge in genome-wide association testing is to develop statistical procedures that minimize false positives without greatly sacrificing true positives. The challenge is amplified by the correlated nature of polymorphism data, which makes simple frequentist approaches that assume independence (such as Bonferonni correction) highly conservative. To illustrate this point, we used the ENCODE data to estimate the "effective number of independent tests" (the statistical burden of testing all common (MAF $\geqslant 0.05$) variation) across large genomic regions. Specifically, we re-sampled from the phased ENCODE chromosomes to create mock case-control panels in which all common SNPs were observed, but there was not a causal allele. The resulting χ^2 distribution for association indicates that complete testing of common variation in each 500-kb region is equivalent to performing about 150 independent statistical tests (in CEU and CHB+JPT) and about 350 tests (in YRI). Although it will probably be desirable to perform such empirical estimates of significance within each disease study, these results illustrate how Bonferonni correction overestimates the statistical penalty of performing many correlated tests.

Study of less common alleles. We have focused primarily on the hypothesis that a single, common causal allele exists, and needs to be tested for association to disease. Of course, in many cases the causal allele(s) will be less common, and might be missed by such an approach.

It is possible to perform additional haplotype tests, beyond those that capture known polymorphisms, in the hope of capturing less common or unrepresented alleles[56]. Such haplotype analysis has a long history and proven value in mendelian genetics; the causal mutation is generally rare and unexamined during initial genotyping, but is frequently recognized by its presence on a long, unique haplotype of common alleles[18,19,59-62].

Admixture mapping. Although not designed specifically to enable admixture mapping[63],

对一个已有的基因型数据集的分析　HapMap 能够用于关联检验，而不受标签如何选择的影响。特别地，只要在疾病研究中被基因分型的 SNP 也在 HapMap 样本中被分型，就有可能鉴定哪些 SNP 被已进行基因分型的 SNP 较好地捕获(单一地或者在单体型组合中)，哪些没有被捕获[55]。

这对于使用基于芯片的、标准化的基因分型试剂所进行的全基因组关联研究特别重要，这些研究不允许研究者选择他们自己的标签 SNP 集。包含在第 I 阶段的 HapMap 的 Affymetrix 120K SNP 芯片数据提供了一个简单的例子：在 CEU 中，48% 的 HapMap SNP 与芯片上的 120K SNP 中的一个或者更多有着相当程度的成对相关性($r^2 \geqslant 0.5$)。另外还有 13% 的 SNP 不与单个 SNP 相关，但是与 120K 组中两个成员组成的特定单体型相关。通过鉴定 HapMap 中这样的单体型预测者，并且在疾病研究中对它们(以及单个 SNP)进行检验，效能有可能会被提高(佩尔等，稿件准备中)。

统计学意义评估和结果解释　在全基因组关联检验中一个重要的挑战是开发最小化假阳性且不大幅牺牲真正阳性数据的统计学方法。多态性数据的关联特性将这个挑战进一步放大，使得假定独立的简单频率论方法(如邦费罗尼校正)高度保守。为了阐释这一点，我们使用 ENCODE 数据来估计在大的基因组区域内“独立检验的有效数量”(检验所有常见(MAF ⩾ 0.05)变异的统计学负担)。特别地，我们从区分染色体的 ENCODE 染色体中重新取样，进而产生模拟的样本–对照组，其中可观察到所有常见 SNP，但是没有引起疾病的等位基因。对于关联分析，得到的 χ^2 分布说明对各个 500 kb 区域内常见变异的完整检验等同于进行大约 150 次独立统计学检验(在 CEU 和 CHB+JPT 中)和大约 350 次检验(在 YRI 中)。尽管可能需要在各个疾病研究中都进行这样的对统计学意义的经验估计，这些结果阐释了邦费罗尼校正是如何高估了进行大量相关性检验的统计惩罚的。

不常见等位基因的研究　我们之前主要聚焦于这样的假说，即单个常见的引起疾病的等位基因是存在的，需要检验其与疾病的关联。当然，在很多情况下，导致疾病的等位基因是不常见的，可能被这样的方法所遗漏。

除了捕获已知的多态性的检验，还可以进行额外的单体型检验，以尝试捕获不常见或者无代表的等位基因[56]。这种单体型分析有着很长的历史，已经证明在孟德尔遗传研究中具有价值；引起疾病的突变通常是罕见的，并且在一开始的基因分型中未被检测到，但是经常由于其存在于常见等位基因的长的唯一的单体型上而被识别[18,19,59-62]。

混合作图　尽管没有特别设计来实现混合作图[63]，但是 HapMap 已经为这种方

the HapMap has helped lay the groundwork for this approach. Admixture mapping requires a map of SNPs that are highly differentiated in frequency across population groups. By typing many SNPs in samples from multiple geographical regions, the data have helped to identify such SNPs for the design of genome-wide admixture mapping panels[64,65] and can be further used to identify candidate SNPs with large allele frequency differences for follow-up of positive admixture scan results[66].

Loss of heterozygosity in tumours. Loss of heterozygosity (LOH) in tumour tissue can be a powerful indicator of the location of tumour suppressor genes, and genome-wide, fine-scale LOH analysis has been empowered by genome-wide SNP arrays[67]. Germline DNA is not always available from the same subjects, however, and even if available, typing of germline DNA doubles project costs. In lower density scans for LOH (with markers far apart relative to the scale of LD), long runs of homozygosity in tumours are nearly always indicative of LOH. However, at higher densities runs of homozygosity can be due to haplotype homozygosity in the inherited germline DNA, rather than LOH.

The HapMap data can help minimize this difficulty; previous probabilities for homozygosity based on known frequencies of haplotypes in the HapMap data can be used to distinguish homozygosity due to haplotype sharing rather than LOH[68].

Identifying Structural Variants in HapMap Data

Structural variations—segments where DNA is deleted, duplicated, or rearranged—are common[69,70] and have an important role in diseases[71-73]. The HapMap can provide some insight into structural variation because, in many cases, structural variants reveal themselves through signatures in SNP genotype data. In particular, polymorphic deletions are important to discover, because loss of genetic material is of obvious functional relevance, and results in aberrant patterns of SNP genotypes. These include apparent non-mendelian inheritance of SNP alleles, null genotypes and deviations from Hardy–Weinberg equilibrium. However, such SNPs are routinely discarded as technical failures of genotyping.

Thus, we scanned the unfiltered Phase I HapMap data using an approach developed and validated to identify polymorphic deletions from clusters of SNPs with aberrant genotype patterns (calibrated across the multiple centres and genotyping platforms[74]). In total, 541 candidate deletion polymorphisms were identified, of which 150 were common enough to be observed as homozygotes.

The properties of these candidate deletions, including experimental validation of 90 candidates, are described in ref. 74. Validated polymorphisms include 10 that remove coding exons of genes, such that in many cases individuals are homozygous null for the encoded transcript. Analysis of confirmed deletions often shows strong LD with nearby SNPs, indicating that LD-based approaches can be useful for detecting disease associations due to structural (as well as SNP) variants.

法打下了基础。混合作图需要在不同种群中频率高度不同的一群 SNP 的图谱。通过对来自多个地理区域的样本的很多 SNP 进行分型，得到的数据已经帮助鉴定到可用于设计全基因组混合作图的 SNP[64,65]，并且能被进一步用于鉴定后续阳性混合扫描结果所需的存在较大等位基因频率差异的候选 SNP[66]。

肿瘤中的杂合性缺失 肿瘤组织中的杂合性缺失（LOH）可以用作抑癌基因位置的强有力的指示物，全基因组 SNP 芯片使得全基因组范围的精细 LOH 分析成为可能[67]。然而，对于同一个个体，种系 DNA 并不总是能够获得的，而且即使获得了，对种系 DNA 的基因型分型也会使项目的花费加倍。在低密度扫描 LOH 时（相对于 LD 的尺度，标志物之间相距较远），肿瘤中纯合性的大量出现几乎总是提示 LOH 的存在。不过，在高密度扫描 LOH 时，纯合性的大量出现可能是由于遗传的种系 DNA 中单体型的纯合性，而不是 LOH。

HapMap 数据能够帮助将这种困难最小化；前期基于 HapMap 数据中单体型的已知频率得到的纯合性可能性可以用于区分单体型共享而非 LOH 导致的纯合性[68]。

HapMap 数据中结构变异的鉴定

结构变异——片段上发生 DNA 缺失、重复或重排——是常见的[69,70]，并且在疾病中有着重要的作用[71-73]。HapMap 能够提供关于结构变异的信息，因为在很多情况下，结构变异可通过 SNP 基因型数据中的特征反映出来。发现多态性的缺失尤其重要，因为遗传材料的缺失是明显与功能相关的，并且会导致 SNP 基因型的异常模式。这包括 SNP 等位基因明显的非孟德尔遗传、频率为零的基因型和哈迪-温伯格平衡的偏离。然而，这样的 SNP 通常作为基因分型的技术问题而被弃掉。

因而，在扫描未过滤的第 I 阶段 HapMap 数据时，我们采用了一种为从带有异常基因型模式的 SNP 簇（在多个中心和基因分型平台进行了校准[74]）中鉴定多态性缺失而开发和验证的方法。总共 541 个候选的缺失多态性被鉴定出来，其中 150 个较为常见，可观察到是纯合的。

这些候选缺失的组成，包括对 90 个候选者的实验验证，都在参考文献 74 中有所描述。得到验证的多态性中有 10 个使基因缺失外显子，在很多情况下，这种个体纯合缺少编码的转录本。对已证实的缺失的分析经常发现这种缺失与邻近 SNP 存在强 LD，表明基于 LD 的方法可以用于检测结构（以及 SNP）变异导致的疾病关联。

Polymorphic inversions may also be reflected in the HapMap data as long regions where multiple SNPs are perfectly correlated: because recombination between an inverted and non-inverted copy is lethal, the inverted and non-inverted copies of the region evolve independently. A striking example corresponds to the known inversion polymorphism on chromosome 17, present in 20% of the CEU chromosomes, that has been associated with fertility and total recombination rate in females among Icelanders[75]. Long LD may also arise, however, due to a low recombination rate or certain forms of natural selection, as discussed below.

Insights into Recombination and Natural Selection

In addition to its intended function as a resource for disease studies, the HapMap data provide clues about the biology of recombination and history of natural selection.

A genome-wide map of recombination rates at a fine scale. On the basis of the HapMap data, we created a fine-scale genetic map spanning the human genome (Supplementary Fig. 12), including 21,617 identified recombination hotspots (one per 122 kb).

Both the number and intensity of hotspots contribute to overall variation in recombination rate. For example, we selected 25 regions of 5 Mb as having the highest (>2.75 cM Mb^{-1}) and lowest (<0.5 cM Mb^{-1}) rates of recombination in the deCODE (pedigree-based) genetic map[48]. We detected recombination hotspots in all regions, even the lowest. But in the high cM Mb^{-1} regions hotspots are more closely spaced (one per 84 kb) and have a higher average intensity (0.124 cM) as compared to the low cM Mb^{-1} regions (one every 208 kb, and 0.051 cM, respectively).

Estimates of recombination rates and identified hotspots are robust to the specific markers and samples studied. Specifically, we compared these results to a similar analysis[76] of the data of ref. 77 (with about 1.6 million SNPs genotyped in 71 individuals). We find nearly complete correlation in rate estimates at a coarse scale (5 Mb) between these two surveys ($r^2 = 0.99$) and to the pedigree map ($r^2 = 0.95$). Very substantial correlation is found at finer scales: $r^2 = 0.8$ at 50 kb and $r^2 = 0.59$ at 5 kb. Moreover, of the 21,617 hotspots identified using the HapMap data, 78% (16,923) were also identified using the data of ref. 77.

The ability to detect events depends on marker density, with the larger number of SNPs studied by ref. 77 increasing power to detect hotspots, and presumably precision of rate estimates. There are, however, substantial genomic regions where the HapMap data have a higher SNP density. For example, more hotspots are detected on chromosomes 9 and 19 from the HapMap data. We expect that Phase II of HapMap will provide a genome-wide recombination map of substantially greater precision than either ref. 77, or Phase I, at fine scales.

多态性的倒位可能也在 HapMap 数据中有所反映，反映为多个 SNP 完美相关的长区域：因为倒位和非倒位拷贝间的重组是致死的，所以区域的倒位和非倒位拷贝各自独立进化。一个明显的例子是 17 号染色体上的一个已知的倒位多态性，它存在于 20% 的 CEU 染色体中，在冰岛女性中与生育力和总重组率相关 [75]。不过，长 LD 也可能由于低重组率或者特定形式的自然选择而增加，详见下文的讨论。

对重组和自然选择的认识

除了原本预期的作为疾病研究的资源这一功能外，HapMap 数据还为研究重组的机理和自然选择的历史提供了线索。

重组率的全基因组精细图谱 基于 HapMap 数据，我们绘制了整个人类基因组的精细遗传学图谱（补充信息图 12），包括了鉴定到的 21,617 个重组热点（每 122 kb 一个）。

重组热点的数量和强度都对重组率的整体变化有所贡献。例如，我们选择了 deCODE（基于家系）遗传图谱 [48] 中含有最高重组率（ > 2.75 cM · Mb^{-1}）和最低重组率（ < 0.5 cM · Mb^{-1}）的 25 个 5 Mb 的区域。我们检测了所有区域中的重组热点，甚至是最低重组率的区域。不过，与低 cM · Mb^{-1} 区域相比，在高 cM · Mb^{-1} 区域中重组热点靠得更近（每 84 kb 一个），平均强度更高（0.124 cM）（低 cM · Mb^{-1} 区域分别是每 208 kb 一个和 0.051 cM）。

对重组率和鉴定到的重组热点的估计，对于特定标志物和研究样本而言是稳健的。特别地，我们将这些结果与参考文献 77 的数据（含有 71 个个体的大约 160 万个基因分型的 SNP）的一个类似的分析 [76] 进行了比较。我们发现，在粗略的尺度（5 Mb）上，这两项研究之间相比（$r^2 = 0.99$）以及与家系图谱相比（$r^2 = 0.95$），重组率的估计几乎完全相关。在更加精细的尺度上，我们也发现了相当程度的相关性：在 50 kb 上 $r^2 = 0.8$，在 5 kb 上 $r^2 = 0.59$。并且，利用 HapMap 数据鉴定到的 21,617 个重组热点中，有 78%（16,923 个）也在参考文献 77 的数据中鉴定到。

检测重组事件的能力取决于标志物的密度，参考文献 77 所研究的 SNP 数量更多，这增加了其检测重组热点的效能，并且很可能提高了重组率估计的准确性。不过，也有很大一部分基因组区域，其中 HapMap 数据有着更高的 SNP 密度。例如，在 9 号和 19 号染色体上，在 HapMap 数据中检测到了更多的重组热点。我们预期，第 II 阶段的 HapMap 将会提供一个在精细尺度上与参考文献 77 或者第 I 阶段 HapMap 相比准确度大大提高的全基因组范围的重组图谱。

Little is yet known about the molecular determinants of recombination hotspots. In an analysis of the data of ref. 77, another study (ref. 76) found significant evidence for an excess of the THE1A/B retrotransposon-like elements within recombination hotspots, and more strikingly for a sixfold increase of a particular motif (CCTCCCT) within copies of the element in hotspots, compared to copies of the element outside hotspots. In analysing the HapMap data, we confirmed these findings (Supplementary Fig. 13). Furthermore, THE1B elements with the motif are particularly enriched within 1.5 kb of the centre of the hotspots compared to flanking sequence ($P < 10^{-16}$).

Correlations of LD with genomic features. Variation in recombination rate is important, in large part, because of its impact on LD. We thus examined genome-wide LD for correlation to recombination rates, sequence composition and gene features.

We confirmed previous observations that LD is generally low near telomeres, elevated near centromeres, and correlated with chromosome length (Fig. 15; see also Supplementary Figs 8b and 14)[48,78-80]. These patterns are due to recombination rate variation as discussed above. We also confirmed previously described relationships between LD and G+C content[78,81,82], sequence polymorphism[83] and repeat composition[78,82].

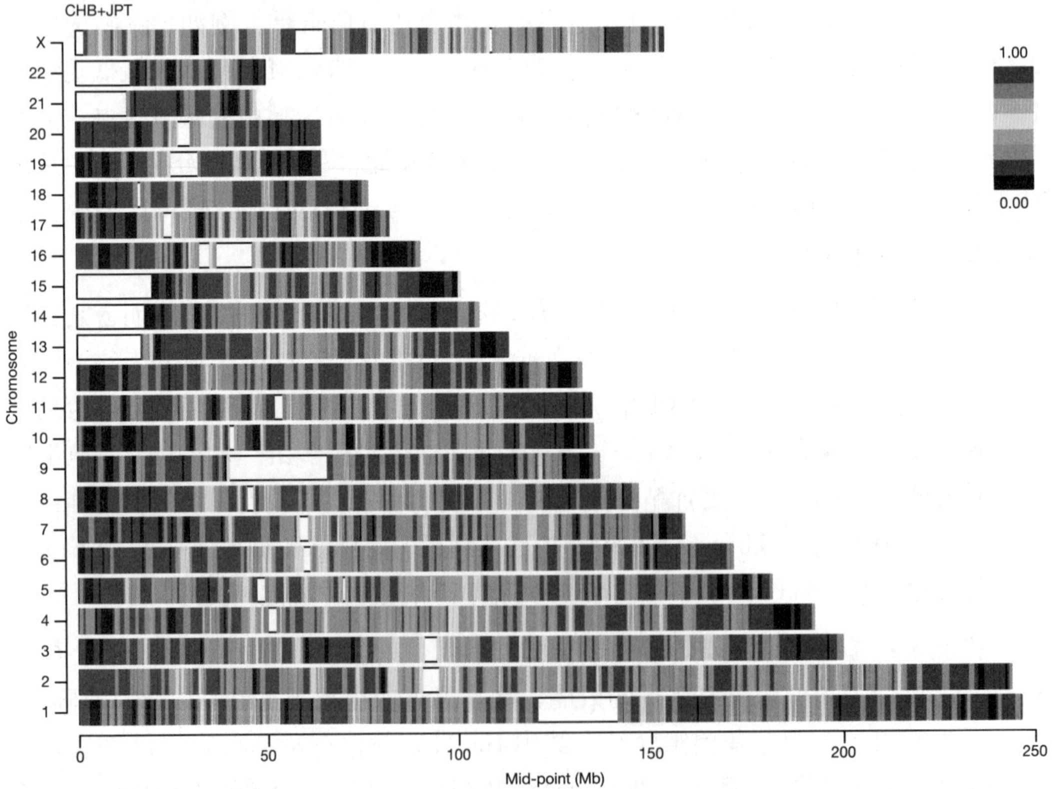

Fig. 15. Length of LD spans. We fitted a simple model for the decay of linkage disequilibrium[103] to windows of 1 million bases distributed throughout the genome. The results of model fitting are summarized for the CHB+JPT analysis panel, by plotting the fitted r^2 value for SNPs separated by 30 kb. The overall pattern

关于重组热点的分子机制目前还知之甚少。在分析参考文献 77 的数据时，另一项研究（参考文献 76）发现在重组热点中有明显的 THE1A/B 逆转录转座子样元件的富集，并且更明显地是，相对于重组热点外的元件，重组热点中的元件中一个特殊的模体（CCTCCCT）有六倍的增加。在分析 HapMap 数据的过程中，我们证实了这些发现（补充信息图 13）。并且，与旁侧序列相比，含有这个模体的 THE1B 元件在重组热点中心的 1.5 kb 范围内特别富集（$P < 10^{-16}$）。

LD 与基因组特征的相关性 重组率的变化的重要性很大程度上是由于它对 LD 的影响。因此我们研究了全基因组的 LD 与重组率、序列组成和基因特征的相关性。

我们证实了之前的发现，即 LD 普遍在端粒附近低，在着丝粒附近较高，并且与染色体的长度相关（图 15；也可见补充信息图 8b 和 14）[48,78-80]。正如上文的讨论，这种模式是由于重组率的变化。我们也证实了之前描述的 LD 与 G+C 含量 [78,81,82]、序列多态性 [83] 和重复序列组成 [78,82] 的关系。

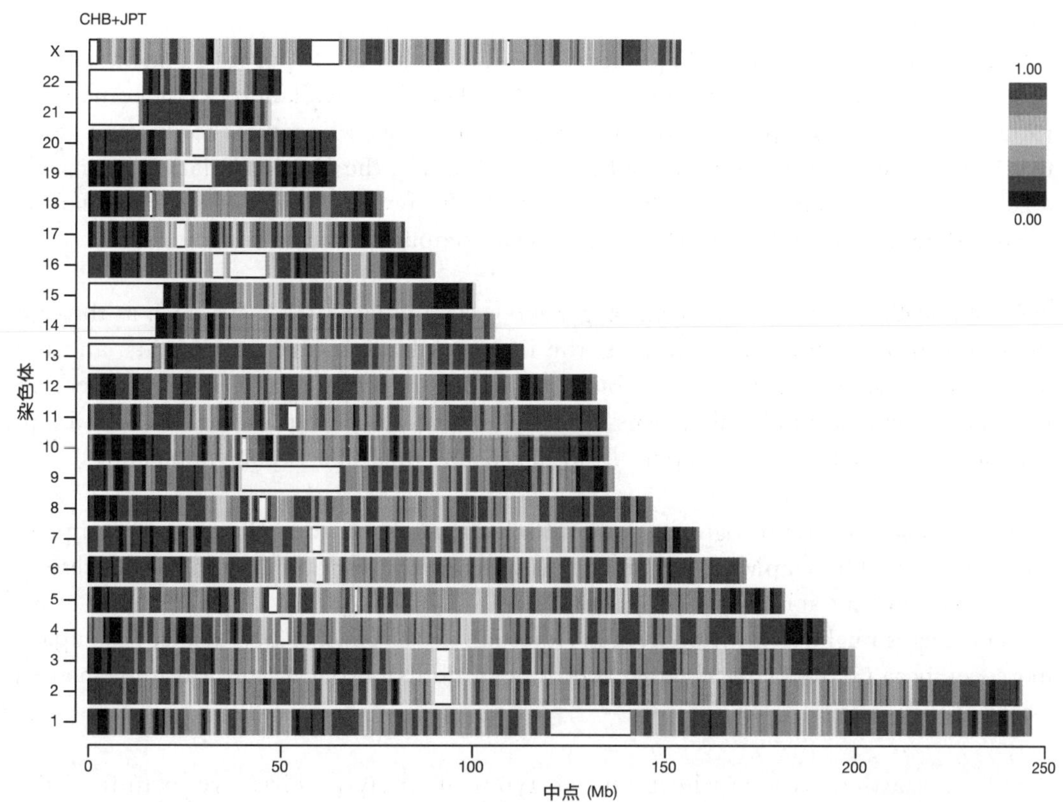

图 15. LD 横跨的长度。对于连锁不平衡的减弱 [103]，我们以全基因组范围内的 100 万碱基为窗口拟合了一个简单的模型。通过画出相距 30 kb 的 SNP 的拟合 r^2 值，我们总结了 CHB+JPT 分析小组模型拟合

of variation was very similar in the other analysis panels[84] (see Supplementary Information).

We observe, for the first time, that LD tracks with both the density and functional classification of genes. We examined quartiles of the genome based on extent of LD, and looked for correlations to gene density. Surprisingly, we find that both the top and bottom quartiles of the genome have greater gene density as compared to the middle quartiles (6.7 as compared to 6.1 genes per Mb), as well as percentage of bases in codons (1.24% as compared to 1.08%). We have no explanation for this observation.

Although the majority of gene classes are equally divided between these two extreme quartiles of the genome, some classes of genes show a marked skew in their distribution[64,84,85]. Genes involved in immune responses and neurophysiological processes are more often located in regions of low LD, whereas genes involved in DNA and RNA metabolism, response to DNA damage and the cell cycle are preferentially located in regions of strong linkage disequilibrium. It is intriguing to speculate that the extent of LD (and sequence diversity) might track with gene function due to natural selection, with increased diversity being favoured in genes involved in interface with the environment such as the immune response[86], and disadvantageous for core cell biological processes such as DNA repair and packaging[87,88].

Natural selection. The preceding observation highlights the hypothesis that signatures of natural selection are present in the HapMap data. The availability of genome-wide variation data makes it possible to scan the genome for such signatures to discover genes that were subject to selection during human evolution[89]; the HapMap data also provide a genome-wide empirical distribution against which previous claims of selection can be evaluated (rather than relying solely on theoretical computer simulations).

Natural selection influences patterns of genetic variation in various ways, such as through the removal of deleterious mutations, the fixation of advantageous variants, and the maintenance of multiple alleles through balancing selection. Each form of selection may have occurred uniformly across the world (and thus be represented in all human populations) or have been geographically localized (and thus differ among populations).

Nearly all methods for recognizing natural selection rely on the collection of complete sequence data. The HapMap Project's focus on common variation—and the process of SNP selection that achieved a preponderance of high-frequency alleles (Fig. 5)—thus prevents their straightforward application. Adjusting for the effect of SNP choice is complex, moreover, because SNP choice varied over time as dbSNP evolved, and was implemented locally at each centre.

For these reasons, we focus here on two types of analysis. First, we examined the distributions of signatures of selection across the genome. Although the absolute value of these measures is difficult to interpret (owing to SNP ascertainment), the most extreme cases in a genome-wide distribution are important candidates to evaluate for selection. Second,

的结果。变异的总体模式在其他分析小组中是非常类似的[84](详见补充信息)。

我们第一次观察到LD与基因的密度和功能分类均同步。我们基于LD的含量研究了基因组的四分位数，寻找其与基因密度的相关性。令人惊奇的是，我们发现与中间四分位数相比，基因组的上四分位数和下四分位数均含有更高的基因密度(中间四分位数每Mb 6.1个基因，上、下四分位数每Mb 6.7个)，位于编码区的碱基比例也是如此(前者为1.08%，后者为1.24%)。我们对这种现象还没有解释。

尽管大部分基因分类在基因组的两个极端四分位间是均等分布的，也有一些基因分类在分布上呈现明显的倾向性[64,84,85]。参与免疫反应和神经生理过程的基因更多位于低LD区域，而参与DNA和RNA代谢、DNA损伤反应和细胞周期的基因更多位于强连锁不平衡的区域。一种有趣的猜测是，由于自然选择，LD的含量(以及序列的多样性)可能与基因的功能同步，基因多样性的增加在环境界面如免疫反应相关的基因中是有利的[86]，而在核心细胞生物学过程如DNA修复和包装相关的基因中是不利的[87,88]。

自然选择　前面的发现突出了自然选择的特征存在于HapMap数据中这一假说。全基因组变异数据的获得，使得对这些特征进行全基因组扫描以发现在人类进化过程中被选择的基因成为可能[89]；HapMap数据还提供了一个全基因组的经验分布，以此可以评估之前声称的自然选择(而不是仅仅依赖于理论上的计算机模拟)。

自然选择通过多种方式对遗传变异的模式产生影响，这些方式如去除有害突变，固定有利变异，以及通过平衡选择对多个等位基因进行保持。各种形式的选择可能在全世界一致地发生(因而在所有人类种群中都表现出来)或者局限于特定的地理位置(因而在不同种群中有差异)。

几乎所有识别自然选择的方法都依赖于收集完整的序列数据。HapMap计划聚焦于常见变异——以及实现了高频等位基因的数量优势的SNP选择过程(图5)，因而使这些遗传变异不能直接应用。并且，调整SNP选择的影响是复杂的，因为随着dbSNP的演化，SNP选择也随时间变化，并在各个中心中局部执行。

因为这些原因，我们这里聚焦于两种类型的分析。首先，我们研究了自然选择的特征在全基因组的分布。尽管这些测量的绝对值难以解释(原因在于SNP检测方法)，但是对于自然选择，在基因组范围的分布中最极端的例子是重要的待评估的候

we compared across functional categories, because SNP choice was largely agnostic to such features, and thus systematic differences may be a sign of selection.

The outcomes of these analyses confirm a number of previous hypotheses about selection and identify new loci as candidates for selection.

Evidence for selective sweeps in particular genomic regions. First we consider population differentiation, generally accepted as a clue to past selection in one of the populations. The HapMap data reveal 926 SNPs with allele frequencies that differ across the analysis panels in a manner as extreme as the well-accepted example of selection at the Duffy (*FY*) locus (Supplementary Fig. 8c). Of these 926 SNPs, 32 are non-synonymous coding SNPs and many others occur in transcribed regions, making them strong candidates for functional polymorphisms that have experienced geographically restricted selection pressures (see Table 9 and Supplementary Information for details). In particular, the *ALMS1* gene on chromosome 2 has six amino acid polymorphisms that show very strong population differentiation.

Table 9. High-differentiation non-synonymous SNPs

Chromosome	Position (base number)	Gene*	SNP
1	54,772,383	*THEA*	rs1702003
1	156,000,000	*FY*	rs12075
1	244,000,000	Q8NGY8_human†	rs7555046
2	3,184,917	*COLEC11*	rs7567833
2	73,563,622	*ALMS1*	rs3813227
2	73,589,553	*ALMS1*	rs6546837
2	73,591,645	*ALMS1*	rs6724782
2	73,592,163	*ALMS1*	rs6546839
2	73,629,222	*ALMS1*	rs2056486
2	73,629,311	*ALMS1*	rs10193972
2	109,000,000	*EDAR*	rs3827760
3	182,000,000	*FXR1*	rs11499
3	185,000,000	*MCF2L2*	rs7639705
4	41,844,599	*SLC30A9*	rs1047626
4	46,567,077	ENSG00000172895.1	rs5825
4	101,000,000	*ADH1B*	rs1229984
8	10,517,787	*RP1L1*	rs6601495
8	146,000,000	*SLC39A4*	rs1871534
10	50,402,145	*ERCC6*	rs4253047
10	71,002,210	*NEUROG3*	rs4536103
11	46,701,579	*F2*	rs5896

选者。其次，我们比较了各个功能分类，因为 SNP 选择在很大程度上对于这些特征是不可知的，因而系统的差异可能是选择的一个标志。

这些分析的结果证实了大量的之前关于自然选择的假说，并且鉴定到了新的自然选择的候选基因座。

特定基因组区域内选择性清理的证据 首先，我们考虑了种群差异，它通常被认为是其中一个种群中以前的自然选择的线索。HapMap 数据发现了 926 个 SNP，其等位基因频率在各个分析小组中存在差异，且这种差异跟公认的在 Duffy(*FY*) 基因座的自然选择的例子一样极端(补充信息图 8c)。在这 926 个 SNP 中，有 32 个是非同义编码 SNP，并且很多其他的 SNP 位于转录区域中，使得它们成为经历过地理位置限制的选择压力的功能多态性的有力候选者(详见表 9 和补充信息)。特别地，2 号染色体上的 *ALMS1* 基因有六个表现出非常强的种群差异的氨基酸多态性。

表 9. 高度分化的非同义 SNP

染色体	位置(碱基数量)	基因 *	SNP
1	54,772,383	*THEA*	rs1702003
1	156,000,000	*FY*	rs12075
1	244,000,000	Q8NGY8_human†	rs7555046
2	3,184,917	*COLEC11*	rs7567833
2	73,563,622	*ALMS1*	rs3813227
2	73,589,553	*ALMS1*	rs6546837
2	73,591,645	*ALMS1*	rs6724782
2	73,592,163	*ALMS1*	rs6546839
2	73,629,222	*ALMS1*	rs2056486
2	73,629,311	*ALMS1*	rs10193972
2	109,000,000	*EDAR*	rs3827760
3	182,000,000	*FXR1*	rs11499
3	185,000,000	*MCF2L2*	rs7639705
4	41,844,599	*SLC30A9*	rs1047626
4	46,567,077	ENSG00000172895.1	rs5825
4	101,000,000	*ADH1B*	rs1229984
8	10,517,787	*RP1L1*	rs6601495
8	146,000,000	*SLC39A4*	rs1871534
10	50,402,145	*ERCC6*	rs4253047
10	71,002,210	*NEUROG3*	rs4536103
11	46,701,579	*F2*	rs5896

Continued

Chromosome	Position (base number)	Gene*	SNP
15	46,213,776	*SLC24A5*	rs1426654
15	61,724,262	*HERC1*	rs7162473
16	30,996,126	*ZNF646*	rs749670
16	46,815,699	*ABCC11*	rs17822931
17	26,322,430	*RNF135*	rs7225888
17	26,399,303	ENSG00000184253.2	rs6505228
18	66,022,323	*RTTN*	rs3911730
19	5,782,891	*FUT6*	rs364637
19	47,723,209	*CEACAM1*	rs8110904
22	18,164,095	*GNB1L*	rs2073770
X	65,608,007	*EDA2R*	rs1385699

* Where no standard gene abbreviation exists, the ENSEMBL gene ID has been given.
† It is unclear from current annotations whether this is a pseudogene.

Another signature of an allele having risen to fixation through selection is that all other diversity in the region is eliminated (known as a selective sweep). We identified extreme outliers in the joint distribution of heterozygosity (as assessed from shotgun sequencing SNP discovery projects) and either population differentiation or skewing of allele frequency towards rare alleles in each analysis panel (Supplementary Fig. 15). We identified 19 such genomic regions (13 on autosomes, 6 on the X chromosome) as candidates for future study (Supplementary Table 4); these include candidates for population-specific sweeps and sweeps in the ancestral population. Encouragingly, this analysis includes among its top-scoring results the *LCT* gene, which influences the ability to digest dairy products[90] and has been shown to be subject to past natural selection[91].

Long haplotypes as candidates for natural selection. Selective sweeps that fail to fix in the population, as well as balancing selection, lead to haplotypes that are relatively high in frequency and long in duration. In the *HLA* region (which is widely believed to have been influenced by balancing selection) multiple haplotypes of 500 SNPs that extend more than 1 cM in length are observed with a frequency in the HapMap samples of more than 1%. We identified other such occurrences of long haplotypes across the genome (Supplementary Fig. 8 and Supplementary Tables 5 and 6).

An approach to long haplotypes designed specifically to identify regions having undergone partial selective sweeps is the long range haplotype (LRH) test[91,92], which compares the length of each haplotype to that of others at the locus, matched across the genome based on frequency. Previously identified outliers to the genome-wide distribution for the LRH test (Fig. 16) that have been identified as candidates for selection include the *LCT* gene in the CEU sample (empirical P-value = 1.3×10^{-9}), which was an outlier for the

续表

染色体	位置（碱基数量）	基因 *	SNP
15	46,213,776	*SLC24A5*	rs1426654
15	61,724,262	*HERC1*	rs7162473
16	30,996,126	*ZNF646*	rs749670
16	46,815,699	*ABCC11*	rs17822931
17	26,322,430	*RNF135*	rs7225888
17	26,399,303	ENSG00000184253.2	rs6505228
18	66,022,323	*RTTN*	rs3911730
19	5,782,891	*FUT6*	rs364637
19	47,723,209	*CEACAM1*	rs8110904
22	18,164,095	*GNB1L*	rs2073770
X	65,608,007	*EDA2R*	rs1385699

* 没有标准基因缩写的使用 ENSEMBL 基因 ID。
† 根据现有的注释不清楚它是否是一个假基因。

一个等位基因通过自然选择而被固定的另一个特征是区域内所有其他的多样性都被消除（这被称为选择性清除）。我们在杂合性（SNP 发现计划进行鸟枪法测序所检测的）与种群差异或各分析小组中等位基因频率向罕见等位基因的偏斜的联合分布中鉴定到了极端的离群值（补充信息图 15）。我们鉴定到了 19 个这样的基因组区域（13 个位于常染色体上，6 个位于 X 染色体上），将其作为下一步研究的候选者（补充信息表 4）；它们包括种群特异性的清理以及古老种群中的清理的候选者。鼓舞人心的是，这个分析在高分结果中包括了 *LCT* 基因，该基因影响消化奶制品的能力 [90]，并且已经被发现经历过以前的自然选择 [91]。

作为自然选择候选者的长单体型 没有在种群中固定下来的选择性清理，以及平衡选择，产生了频率相对较高且跨度相对较长的单体型。在 *HLA* 区域（普遍认为它被平衡选择所影响过）中，多个单体型含有 500 个 SNP，长度上超过 1 cM，在 HapMap 样本中频率超过 1%。我们还在全基因组中鉴定到了其他这样的长单体型情况（补充信息图 8 和补充信息表 5 和 6）。

长距离单体型（LRH）检验 [91,92] 是一种专门设计的用来鉴定经历过部分选择性清理的长单体型区域的方法，它是将各个单体型的长度与全基因中频率匹配的其他单体型的长度相比较。之前通过 LRH 检验鉴定到的相对于全基因组分布的（图 16），已经被认为是选择候选者的离群值，包括 CEU 样本中的 *LCT* 基因（经验 P 值 $=1.3\times10^{-9}$）——上面提到的杂合性/等位基因频率检验的一个离群值，以及 YRI 样

heterozygosity/allele frequency test above, and the *HBB* gene (empirical *P*-value = 1.39×10^{-5}) in the YRI sample. However, most of the strongest signals in the LRH test (Table 10) were not previously hypothesized as undergoing selection.

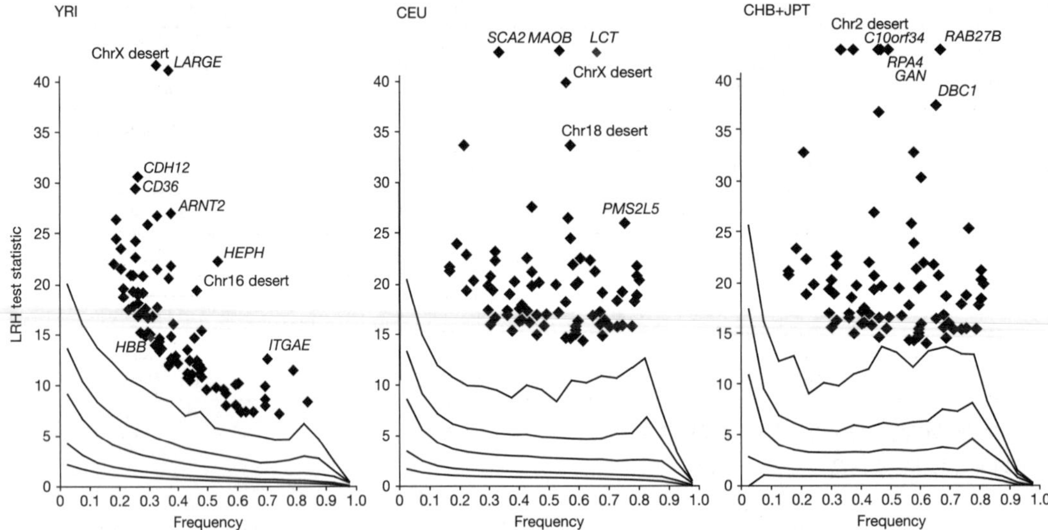

Fig. 16. The distribution of the long range haplotype (LRH[92]) test statistic for natural selection. In the YRI analysis panel, diversity around the *HBB* gene is highlighted by the red point. In the CEU analysis panel, diversity within the LCT gene region is similarly highlighted.

Table 10. Candidate loci in which selection occurred

Chromosome	Position (base number) at centre	Genes in region	Population	Haplotype frequency	Empirical *P*-value
2	137,224,699	*LCT*	CEU	0.65	1.25×10^{-9}
5	22,296,347	*CDH12, PMCHL1*	YRI	0.25	5.77×10^{-8}
7	79,904,387	*CD36*	YRI	0.24	2.72×10^{-6}
7	73,747,934	*PMS2L5, WBSCR16*	CEU	0.76	3.37×10^{-6}
12	109,892,896	*CUTL2*	CEU	0.36	7.95×10^{-9}
15	78,558,508	*ARNT2*	YRI	0.32	6.92×10^{-7}
16	75,661,011	Desert	YRI	0.46	5.01×10^{-7}
17	3,945,580	*ITGAE, GSG2,* HSA277841, *CAMKK1, P2RX1*	YRI	0.70	9.26×10^{-7}
18	24,502,756	Desert	CEU	0.57	2.23×10^{-7}
22	32,459,471	*LARGE*	YRI	0.36	7.82×10^{-9}
X	20,171,291	Desert	YRI	0.33	5.02×10^{-9}
X	64,323,320	*HEPH*	YRI	0.55	3.02×10^{-8}
X	42,763,073	*MAOB*	CEU	0.53	4.21×10^{-9}
X	34,399,948	Desert	CEU	0.57	8.85×10^{-8}

本中的 *HBB* 基因（经验 P 值 $=1.39\times10^{-5}$）。不过，LRH 检验中大部分的最强信号（表 10）之前并不认为经历过自然选择。

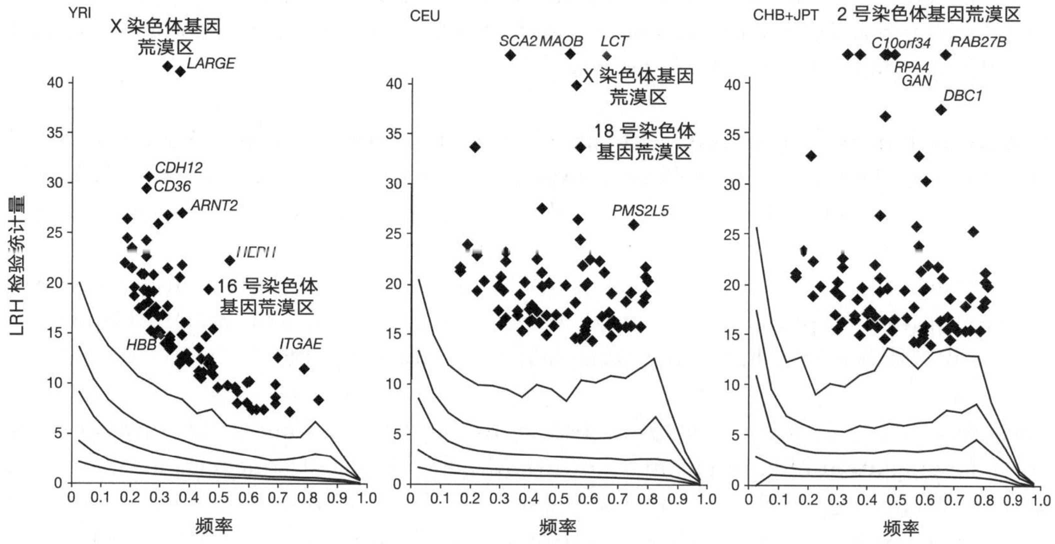

图 16. 自然选择的长距离单体型（LRH[92]）检验统计量的分布。在 YRI 分析小组，*HBB* 基因周围的多样性通过红点突出显示。在 CEU 分析小组中，*LCT* 基因区域内的多样性以类似的方式突出显示。

表 10. 发生选择的候选基因座

染色体	中心位置（碱基数量）	区域内基因	种群	单体型频率	经验 P 值
2	137,224,699	*LCT*	CEU	0.65	1.25×10^{-9}
5	22,296,347	*CDH12*、*PMCHL1*	YRI	0.25	5.77×10^{-8}
7	79,904,387	*CD36*	YRI	0.24	2.72×10^{-6}
7	73,747,934	*PMS2L5*、*WBSCR16*	CEU	0.76	3.37×10^{-6}
12	109,892,896	*CUTL2*	CEU	0.36	7.95×10^{-9}
15	78,558,508	*ARNT2*	YRI	0.32	6.92×10^{-7}
16	75,661,011	Desert	YRI	0.46	5.01×10^{-7}
17	3,945,580	*ITGAE*、*GSG2*、HSA277841、*CAMKK1*、*P2RX1*	YRI	0.70	9.26×10^{-7}
18	24,502,756	Desert	CEU	0.57	2.23×10^{-7}
22	32,459,471	*LARGE*	YRI	0.36	7.82×10^{-9}
X	20,171,291	Desert	YRI	0.33	5.02×10^{-9}
X	64,323,320	*HEPH*	YRI	0.55	3.02×10^{-8}
X	42,763,073	*MAOB*	CEU	0.53	4.21×10^{-9}
X	34,399,948	Desert	CEU	0.57	8.85×10^{-8}

These four tests overlap only partially in the hypotheses they address—heterozygosity, for example, is sensitive to older sweeps, whereas the haplotype tests are most powerful for partial sweeps—but encouragingly some candidate regions are found by more than one test. In particular, six regions are identified both by long haplotypes and by low heterozygosity, and three regions (*LCT* on chromosome 2, and two regions on the X chromosome at 20 and 65 Mb) are identified by three different tests.

Confirming purifying selection at conserved non-coding elements. Finally, we used the HapMap data to test an important hypothesis from comparative genomics. Genomic sequencing has shown that about 5% of the human sequence is highly conserved across species, yet less than half of this sequence spans known functional elements such as exons[45]. It is widely assumed that conserved non-genic sequences lack diversity because of selective constraint (that is, purifying selection), but such regions may simply be coldspots for mutation, and thus be of little value as candidates for functional study.

Analysis of allele frequencies helps to resolve this uncertainty. Functional constraint, but not a low mutation rate, results in a downward skew in allele frequencies for conserved sequences as compared to neutral sequences[93,94]. We find that conserved non-genic sequences display a greater skew towards rare alleles than do intergenic regions, as predicted under purifying selection. This skew is less extreme than that observed for exons (Supplementary Fig. 16), reflecting either stronger purifying selection or the prioritization of coding SNPs for genotyping by the HapMap centres regardless of validation status. This novel evidence for ongoing constraint shows that conserved non-genic sequences are not mutational coldspots, and thus remain of high interest for functional study.

Conclusions

The International HapMap Project set out to create a resource that would accelerate the identification of genetic factors that influence medical traits. Analyses reported here confirm the generality of hotspots of recombination, long segments of strong LD, and limited haplotype diversity. Most important is the extensive redundancy among nearby SNPs, providing (1) the potential to extract extensive information about genomic variation without complete resequencing, and (2) efficiencies through selection of tag SNPs and optimized association analyses. Beyond the biomedical context, these data have made it possible to identify deletion variants in the genome, explore the nature of fine-scale recombination and identify regions that may have been subject to natural selection.

The HapMap Project (along with a previous genome-wide assessment of LD[77]) is a natural extension of the Human Genome Project. Where the reference sequence constructed by the Human Genome Project is informative about the vast majority of bases that are invariant across individuals, the HapMap focuses on DNA sequence differences among individuals. Our understanding of SNP variation and LD around common variants in the sampled populations is reasonably complete; the current picture is unlikely to change with additional

这四种检验提出的假设仅仅部分重合——例如，杂合性对更古老的清理敏感，而单体型检验对于部分清理最有效；但是令人鼓舞的是，一些候选区域在超过一种检验中发现。特别地，有六个区域同时被长单体型和低杂合性鉴定出来，有三个区域(2 号染色体上的 *LCT*，以及 X 染色体在 20 和 65 Mb 处的两个区域)被三种不同的检验同时鉴定出来。

在保守的非编码元件中验证纯化选择　最后，我们使用 HapMap 数据来检验一个来自比较基因组学的重要假说。基因组测序已经发现，大约 5% 的人类序列是在物种间高度保守的，但是其中横跨已知的功能元件如外显子[43]的只有不到一半。普遍认为，保守的非基因序列由于选择性限制(即纯化选择)而缺少多样性，但是，这样的区域也可能单纯是突变的冷点而几乎没有作为功能研究候选者的价值。

分析等位基因频率能帮助解决这种不确定性。功能限制，而非低突变率，会导致相比于中性序列，保守序列的等位基因频率向下偏斜[93,94]。我们发现，保守的非基因序列比基因间区域表现出更大的向罕见等位基因的偏斜，这与在纯化选择压力下预测的一致。这种偏斜没有外显子中观察到的那么极端(补充信息图 16)，不管验证状态如何，这反映了更强的纯化选择，或是 HapMap 中心对基因分型的编码 SNP 的优先选择。这个正在进行的限制的新证据表明，保守的非基因序列并不是突变冷点，因而在功能研究中仍然很有价值。

结　论

国际 HapMap 计划的建立是为了创造一种能够加快对影响医学特性的遗传因子的鉴定的资源。本文报道的分析结果验证了重组热点、强 LD 的长片段和有限的单体型多样性的普遍性。最重要的是邻近 SNP 间广泛的冗余性，这提供了(1)不需要全基因组重测序就可以提取大量基因组变异信息的可能，以及(2)通过标签 SNP 的选择和优化后的关联分析而提高的效率。除了生物医学方面，这些数据还使得鉴定基因组中的缺失突变，探索精细尺度上重组的本质以及鉴定可能经历过自然选择的区域成为可能。

HapMap 计划(以及之前的一项全基因组范围的对 LD 的评估[77])是人类基因组计划的自然的延伸。人类基因组计划搭建的参考序列表明绝大多数碱基在人与人之间是不变的，而 HapMap 聚焦于个体间存在差异的 DNA 序列。我们对于取样种群中的 SNP 变异和常见变异附近的 LD 的了解是相当完整的；即使再增加数据，目前

data. In other aspects—such as the fine details of local correlation among SNPs, rarer alleles, structural variants, and interpopulation differences—these resources are only a first step on the path towards a complete characterization of genetic variation of the human population. Planned extensions of the Phase I map include Phase II of HapMap, with genotyping of another 4.6 million SNPs attempted in the HapMap samples, and detailed genotyping of the HapMap ENCODE regions in additional members of each HapMap population sampled, as well as in samples from additional populations. These results should guide understanding of the robustness and transferability of LD inferences and tag SNPs selected from the current set of HapMap samples.

An important application of the HapMap data is to help make possible comprehensive, genome-wide association studies. There are now laboratory tools that make it practical to undertake such studies, and initial results are encouraging[13]. Given the low prior probability of causality for each SNP in the genome, however, rigorous standards of statistical significance will be needed to avoid a flood of false-positive results. Multiple replications in large samples provide the most straightforward path to identifying robust and broadly relevant associations. Given the potential for confusion if associations of uncertain validity are widely reported (and a persistent tendency towards genetic determinism in public discourse), we urge conservatism and restraint in the public dissemination and interpretation of such studies, especially if non-medical phenotypes are explored. It is time to create mechanisms by which all results of association studies, positive and negative, are reported and discussed without bias.

The success of the HapMap will be measured in terms of the genetic discoveries enabled, and improved knowledge of disease aetiology. Specifically, identifying which genes and pathways are causal in humans has the potential to provide a new and solid foundation for biomedical research. This is equally true whether the variants that lead to the discovery of those genes are themselves rare or common, or of large or small effect. The impact on diagnostics and targeted prevention, however, will depend on how predictive each given allele may be. Where genetic mechanisms underlie treatment responses, both more efficient trials and individualized preventive and treatment strategies may become practical[95].

Success identifying alleles conferring susceptibility or resistance to common diseases will also provide a deeper understanding of the architecture of disease: how many genes are involved in each case, whether and how alleles interact with one another[96] and with environmental exposures to shape clinical phenotypes. In this regard, it will be important to invest heavily in the discovery and characterization of relevant lifestyle factors, environmental exposures, detailed characterization of clinical phenotypes, and the ability to obtain such information in longitudinal studies of adequate size. Where environmental and behavioural factors vary across studies, replication will be hard to come by (as will clinical utility) unless we can learn to capture these variables with the same precision and completeness as genotypic variation. Technological innovation and international collaboration in these realms will probably be required (as they have been in the Human Genome Project and the HapMap) to advance the shared goal of understanding, and ultimately preventing, common human diseases.

的图景也基本不会改变。对于其他方面——比如SNP间局部相关性、更罕见的等位基因、结构变异和种群间差异等细节，这些资源仅仅是通往完整表征人类种群遗传变异的道路的第一步。第I阶段图谱的延伸计划包括HapMap的第II阶段，即在HapMap样本中对另外460万个SNP进行基因分型，以及在已取样的各HapMap种群的更多成员中和另外的种群的样本中，对HapMap ENCODE区域进行详细的基因分型。这些结果会对理解LD推断和从现有HapMap样本集中选择的标签SNP的稳健性和可转移性提供指导。

HapMap数据的一个重要应用是帮助我们实现全面的全基因组关联研究。现有的实验工具已经使进行这样的研究切实可行，并且初步结果是令人鼓舞的[13]。但是，鉴于基因组中各个SNP导致疾病的先验概率偏低，统计显著性将需要严格的标准来避免大量的假阳性结果。大量样本中的多个重复提供了最直接的鉴定稳健且广泛相关的关联的方法。鉴于可能存在的对不确定有效的关联是否被广泛报道的困惑(以及公共讨论中朝向遗传决定论的持续倾向)，我们呼吁在公开宣传和说明这样的研究时保持保守和克制，特别是研究非医学表型时。现在需要建立一种机制，以使所有关联研究的结果，无论阳性或阴性，都得到无偏倚的报道和讨论。

HapMap的成功会通过它带来的遗传学发现和提高的病因知识来衡量。特别地，对人类中导致疾病的基因和通路的鉴定有望为生物医学研究提供新的坚实基础。不管使我们发现这些致病基因的遗传变异是罕见的还是常见的，效应是大还是小，这一点是始终成立的。不过，对诊断和靶向预防的影响将取决于各个给定的等位基因可以在多大程度上预测疾病发生。一旦清楚了治疗响应下的遗传机制，更加有效的试验和个性化预防与治疗策略都可能成为现实[95]。

成功鉴定影响常见疾病易感性或抵抗性的等位基因还将会帮助我们更加深入地理解疾病的机制：多少基因参与其中，等位基因间[96]以及等位基因与环境是否和如何相互作用进而影响临床表型。在这一点上，大力投入以发现和表征相关生活方式因子、环境暴露、详细的临床表型，以及在足够大的纵向研究中获取这种信息，将会十分重要。各个研究的环境和行为因子不同，除非我们能学会以与遗传学变异相同的准确度和完整性捕获这些变化，否则将难以进行重复(临床应用也是如此)。这些领域有可能需要技术的革新和国际合作(像人类基因组计划和HapMap那样)，以推进理解和最终预防常见人类疾病这一共同的目标。

Methods

The project was undertaken by investigators from Japan, the United Kingdom, Canada, China, Nigeria and the United States, and from multiple disciplines: sample collection, sequencing and genotyping, bioinformatics, population genetics, statistics, and the ethical, legal, and social implications of genetic research. The Supplementary Information contains information about project participants and organization.

Choice of DNA samples. Any choice of DNA samples represents a compromise: a single population offers simplicity, but cannot be representative, whereas grid-sampling is representative of the current worldwide population, but is neither practical nor captures historical genetic diversity. The project chose to include DNA samples based on well-known patterns of allele frequencies across populations[41], reflecting historical genetic diversity[31,32].

For practical reasons, the project focused on SNPs present at a minor allele frequency (MAF) $\geqslant 0.05$ in each analysis panel, and thus studied a sufficient number of individuals to provide good power for this frequency range[31]. Cell lines and DNA are available at the Coriell Institute for Medical Research (http://locus.umdnj.edu/nigms/products/hapmap.html).

Community engagement was employed to explain the project, and to learn how the project was viewed, in the communities where samples were collected[31,32]. Papers describing the community engagement processes are being prepared.

One JPT sample was replaced for technical reasons, but not in time for inclusion in this report. We surveyed cryptic relatedness among the study participants, and identified a small number of pairs with unexpectedly high allele sharing (Supplementary Information). As the total level of sharing is not great, and as a subset of analyses performed without these individuals were unchanged, we include these individuals in the data and analyses presented here.

Genome-wide SNP discovery. The project required a dense map of SNPs, ideally containing information about validation and frequency of each candidate SNP. When the project started, the public SNP database (dbSNP) contained 2.6 million candidate SNPs, few of which were annotated with the required information.

To generate more SNPs and obtain validation information, shotgun sequencing of DNA from whole-genome libraries and flow-sorted chromosomes was performed[31], augmented by analysis of sequence traces produced by Applied Biosystems[97,98], and information on 1.6 million SNPs genotyped by Perlegen Sciences[77], including 425,000 not in dbSNP when released (Supplementary Table 7). The HapMap Project contributed about 6 million new SNPs to dbSNP.

At the time of writing (October 2005) dbSNP (http://www.ncbi.nlm.nih.gov/projects/SNP/) contains 9.2 million candidate human SNPs, of which 3.6 million have been validated by both alleles having been seen two or more times during discovery ("double-hit" SNPs), and 2.4 million have genotype

方 法

本计划是由来自日本、英国、加拿大、中国、尼日利亚和美国的研究者实施的，这些研究者来自多个方向：样本收集，测序和基因分型，生物信息学，种群遗传学，统计学，以及遗传学研究的伦理、法律和社会影响。补充信息包括了参与本计划的人员和组织的信息。

DNA 样本的选择 任何一种 DNA 样本选择方法都代表着一种妥协：单一种群简单，但是不够有代表性，而网格采样代表了目前全球的种群，但是既不现实也不能捕获历史上的遗传多样性。本计划选择基于种群间等位基因频率的已知模式来纳入 DNA 样本[41]，进而反映历史上的遗传多样性[31,32]。

出于实际的原因，本计划聚焦于在各个分析小组中最小等位基因频率（MAF）≥0.05 的 SNP，因而研究了足量的个体来为这个频率段提供较好的效能[31]。细胞系和 DNA 可在科里尔医学研究所获取（http://locus.umdnj.edu/nigms/products/hapmap.html）。

本研究在样本采集的社区开展了参与活动，包括解释本计划，以及了解本计划是如何被看待的[31,32]。介绍社区参与过程的文章正在准备中。

只有一个 JPT 样本由于技术原因被替换掉，但是没有及时包含进本报道中。我们还调查了样本之间的隐性亲缘关系，鉴定到少量几对含有超预期的高等位基因共享性（详见补充信息）。由于总体的共享性水平并不高，以及不含这些个体的分析结果不变，我们在本文的数据和分析中还是包含了这些个体。

全基因组范围的 SNP 发现 本计划要求高密度的 SNP 图谱，最好包含各个候选 SNP 的验证和频率的信息。在本计划启动时，公共的 SNP 数据库（dbSNP）包含 260 万个候选 SNP，其中很少有所要求的注释信息。

为了产生更多的 SNP 并获得验证信息，我们对来自全基因组文库和流式筛选的染色体的 DNA 进行了鸟枪法测序[31]，并辅以 Applied Biosystems 公司的序列踪迹分析[97,98]，以及 Perlegen Sciences 公司基因分型的 160 万个 SNP 的信息[77]，这其中有 425,000 个在数据释放时并不在 dbSNP 中（补充信息表格 7）。HapMap 计划为 dbSNP 贡献了大约 600 万个新的 SNP。

在写作本文时（2005 年 10 月），dbSNP（http://www.ncbi.nlm.nih.gov/projects/SNP/）包含 920 万个候选人类 SNP，其中 360 万个 SNP 的两个等位基因都在发现过程中被观察到两次或更多

data (Fig. 1).

Comprehensive study of common variation across 5 Mb of DNA. To study patterns of genetic variation as comprehensively as possible, we selected ten 500-kb regions from the ENCODE Project[33]. These ten regions were chosen in aggregate to approximate the genome-wide average for G+C content, recombination rate, percentage of sequence conserved relative to mouse sequence, and gene density (Table 2).

In each such region additional sequencing and genotyping were performed to obtain a much more complete inventory of common variation. Specifically, bidirectional PCR-based sequencing was performed across each 500-kb region in 48 individuals (16 YRI, 16 CEU, 8 CHB, 8 JPT). Although the intent was for these same DNA samples to be included in Phase I, eight Yoruba and one Han Chinese sample used in sequencing were not among the 269 samples genotyped. (The nine samples are available from Coriell.)

All variants found by sequencing, and any others in dbSNP (build 121) not found by sequencing, were genotyped in all 269 HapMap samples. If the first attempt at genotyping was unsuccessful, a second platform was tried for each SNP.

False-positive and false-negative rates in PCR-based SNP discovery. The false-positive rate of SNP discovery by PCR-based resequencing was estimated at 7–11% (for the two sequencing centres), based on genotyping of each candidate SNP in the same samples used for discovery.

The false-negative rate of SNP discovery by PCR resequencing was estimated at 6%, using as the denominator a set of SNPs previously in dbSNP and confirmed by genotyping in the specific individuals sequenced. The false-negative rate was considerably higher, however, for singletons (SNPs seen only as a single heterozygote): 15% of singletons covered by high-quality sequence data were not detected by the trace analysis, and another 25% were missed due to a failure to obtain a high-quality sequence over the relevant base in the one heterozygous individual (D. J. Richter *et al.*, personal communication).

False-positive and false-negative rates in dbSNP. The false positive rate (candidate SNPs that cannot be confirmed as variable sites) estimated for dbSNP was 17%. This represents an upper bound, because dbSNP entries that are monomorphic in the 269 HapMap samples could be rare variants, or polymorphic in other samples. We note that as the catalogue of dbSNP gets deeper, the rate at which candidate SNPs are monomorphic in any given sample is observed to rise (Supplementary Table 8). This is expected because the number of rare SNPs and false positives scales with depth of sequencing, whereas the number of true common variants will plateau.

SNP genotyping for the genome-wide map. Genotyping assays were designed from dbSNP, with priority given to SNPs validated by previous genotyping data or both alleles having been seen more than once in discovery. Data from the Chimpanzee Genome Sequencing Project[99] were used in SNP validation if they confirmed the ancestral status of a human allele seen only once in discovery (Supplementary Information). Non-synonymous coding SNPs were also prioritized for genotyping.

次而得到验证（“双打击”的 SNP），240 万个含有基因型数据（图 1）。

对 5 Mb DNA 区域内常见变异的全面研究 为了尽可能全面地研究遗传变异的模式，我们从 ENCODE 计划中选择了 10 个 500 kb 区域[33]。这 10 个区域合起来可近似代表全基因组平均的 G+C 含量、重组率、相对于小鼠基因组保守序列的百分比，以及基因密度（表 2）。

在各个这样的区域中，我们进行了额外的测序和基因分型，以获得更加完整的常见变异清单。特别地，对 48 个个体（16 个 YRI、16 个 CEU、8 个 CHB、8 个 JPT）在各个 500 kb 区域上进行了基于双向 PCR 的测序。尽管目的是将这些用在测序中的 DNA 样本都包含进第 I 阶段 HapMap，但是其中有 8 个约鲁巴人和 1 个汉族中国人样本并没有包含在基因分型的 269 个样本当中。（这 9 个样本可从科里尔医学研究所获得。）

通过测序发现的所有变异，以及任何其他未被测序发现但在 dbSNP（版本 121）中的变异，都在所有 269 个 HapMap 样本中进行了基因分型。对于各个 SNP，如果第一次基因分型尝试不成功，则尝试第二个基因分型平台。

基于 PCR 的 SNP 发现中的假阳性率和假阴性率 通过基于 PCR 的重测序进行 SNP 发现，其假阳性率估计为 7%～11%（对于两个测序中心），这是基于在相同的样本中对各个候选 SNP 进行基因分型得出来的。

通过 PCR 重测序进行 SNP 发现，其假阴性率估计为 6%，这是使用之前在 dbSNP 中的并在被测序的特定个体中通过基因分型而被验证的 SNP 集作为分母得出来的。不过，对于单体 SNP（仅作为单个杂合子而被发现），假阴性率则高得多：被高质量的测序数据所覆盖的单体 SNP 中有 15% 没有被踪迹分析检测到，另外还有 25% 由于不能在唯一的杂合个体中获得相关碱基的高质量测序而被漏掉（里克特等，个人交流）。

dbSNP 中的假阳性率和假阴性率 对于 dbSNP，假阳性率（不能被证实在人群中存在变异的候选 SNP）估计为 17%。这代表着上限，因为在 269 个 HapMap 样本中呈单态性的 dbSNP 条目有可能是罕见变异，或者在其他样本中是多态性的。我们注意到，随着 dbSNP 目录的深入，候选 SNP 在任一给定样本中是单态性的比率是增加的（补充信息表 8）。这是符合预期的，因为罕见 SNP 和假阳性的数量与测序的深度成比例，而真正的常见变异的数量则将会保持稳定。

全基因组图谱的 SNP 基因分型 基因分型实验是根据 dbSNP 而设计的，首先针对由之前的基因分型数据验证过的 SNP，或者两个等位基因在发现过程中都被观察到超过一次的 SNP。对于在发现过程中仅被看到一次的人类等位基因，如果来自黑猩猩基因组测序计划[99]的数据确认了该等位基因的祖先状态，这些数据就被用于 SNP 验证。非同义编码 SNP 也被

Two whole-genome, array-based genotyping reagents were used efficiently to increase SNP density: 40,000 SNPs from Illumina, and 120,000 SNPs from Affymetrix[100].

To monitor progress, the genome was partitioned into 5-kb bins, with genotyping continuing through iterative rounds until a set of predetermined "stopping rules" was satisfied in each analysis panel. (1) Minor allele frequency: in each analysis panel a common SNP (MAF $\geqslant 0.05$) was obtained in each 5-kb bin. (2) Spacing: the distance between adjacent SNPs was 2–8 kb, with at least 9 SNPs across 50 kb. (3) "HapMappable" genome: with available technologies it is challenging to study centromeres, telomeres, gaps in genome sequence, and segmental duplications. The project identified such regions[101] (Supplementary Table 9), spanning 4.4% of the finished human genome sequence, in which only a single attempt to develop a genotyping assay was required. (4) Three strikes, you're out: if the above rules were not satisfied after three attempts to develop an assay in a given 5-kb region, or if all available SNPs in dbSNP had been tried, genotyping was considered complete for Phase I. Two attempts were considered sufficient if one attempt was of a SNP previously shown to have MAF $\geqslant 0.05$ in the appropriate population sample in a previous genome-wide survey[77]. (5) Quality control: ongoing and standardized quality control (QC) filters and three rounds of quality assessment (QA) were used to ensure and document the high quality of the genotype data.

QC filters were systematically performed, with each SNP tested for completeness ($>80\%$), consistency across five duplicate genotypes ($\leqslant 1$ discrepancy), mendelian inheritance in 60 trios ($\leqslant 1$ discrepancy in each of YRI and CEU), and Hardy–Weinberg equilibrium ($P > 0.001$[102]). SNPs in the Phase I data set passed all the QC filters in all the analysis panels and were polymorphic in the HapMap samples. Failing SNPs were released (with a special flag), as they can help to identify polymorphisms under primers, insertions/deletions, paralogous loci and natural selection.

Three QA exercises were carried out. First, a calibration exercise to "benchmark" each platform and laboratory protocol. Second, a mid-project evaluation of each genotyping centre. Third, a blind analysis of a random sample of the complete Phase I data set. A number of SNPs were genotyped more than once during the project, or by other investigators, providing additional information about data quality. See the Supplementary Information for full information about the QA exercises.

An exhaustive approach was taken to mtDNA. Alignment of more than 1,000 publicly available mtDNA sequences of African (n = 87), European (n = 928) and Asian (n = 238) geographical origin[34] was used to identify 210 common variants (MAF $\geqslant 0.05$ in at least one continental region) that were attempted in the samples.

Data release. Data deposited at the Data Coordination Center and released at http://www.hapmap.org, a Japanese mirror site http://hapmap.jst.go.jp/ and dbSNP include ascertainment status of each SNP at the time of selection, primer sequences, protocols for genotyping, genotypes for each sample, allele frequencies, and, for SNPs that failed QC filters, a code indicating the mode(s) of failure.

Initially, because of concern that third parties might seek patents on HapMap data, users were required to agree to a web-based "click-wrap license", assenting that they would not prevent others from using the data (http://www.hapmap.org/cgi-perl/registration). In December 2004 this license

优先进行基因分型。两个全基因组的基于芯片的基因分型试剂被高效用于增加SNP密度：40,000个SNP来自Illumina，120,000个SNP来自Affymetrix[100]。

为了监控进度，基因组被分成5 kb的块，基因分型通过迭代循环持续进行，直到满足在各个分析小组中预先确定的一套“停止规则”。(1)最小等位基因频率：在各个分析小组中每个5 kb块均获得一个常见SNP(MAF ≥ 0.05)。(2)间隔：相邻SNP间的距离是2～8 kb，50 kb区域内至少有9个SNP。(3)“可被HapMap的”基因组：使用现有技术研究着丝粒、端粒、基因组序列缺口和片段重复还比较有挑战性。本计划鉴定了占已完成的人类基因组序列的4.4%的区域[101](补充信息表9)，在这些区域内设计基因分型实验只需要尝试一次。(4)三振出局：如果在一个给定的5 kb区域内设计一个基因分型实验时经过三次尝试而上述规则都不满足，或者如果dbSNP中所有SNP都已经被尝试了，则对于第I阶段HapMap，基因分型被认为是完整的。如果其中一次尝试针对的是一个在之前的全基因组调查中在合适的种群样本中MAF ≥ 0.05的SNP[77]，则两次尝试也被认为是足够的。(5)质量控制：不间断的标准化质量控制(QC)过滤和三轮质量评估(QA)被用来确保和记录基因型数据的高质量。

我们系统实施了QC过滤，检验各个SNP的完整性(＞80%)、五个重复基因型间的一致性(≤1个不一致)、60个三联家系中的孟德尔遗传(在YRI和CEU中均≤1个不一致)，以及哈迪–温伯格平衡($P > 0.001$[102])。第I阶段数据集中的SNP在所有分析小组中都通过了所有的QC过滤，并且在HapMap样本中是多态性的。失败的SNP也被释放(含有一个特殊的标记)，因为它们能够帮助鉴定在引物、插入/缺失、同源基因座和自然选择下的多态性。

我们实施了三项QA。第一是“基准”各个平台和实验室操作手册的一项校准。第二是在计划的中期对各个基因分型中心进行的评估。第三是对完整的第I阶段数据集的随机样本进行的盲分析。大量的SNP在计划中被基因分型超过一次，或者被其他的研究者进行过分型，这提供了额外的关于数据质量的信息。关于QA的完整信息见补充信息。

对于mtDNA，我们采取了穷举的方法。超过1,000个可公开获得的来自非洲(n = 87)、欧洲(n = 928)和亚洲(n = 238)的mtDNA序列[34]被用来鉴定在样本中尝试过的210个常见变异(在至少一个大陆区域中MAF ≥ 0.05)。

数据释放 储存在数据协调中心，并在http://www.hapmap.org、一个日本镜像网站http://hapmap.jst.go.jp/和dbSNP上释放的数据包括各个SNP在选择时的确定状态、引物序列、基因分型的操作手册、各个样本的基因型、等位基因频率，此外，对于未通过QC过滤的SNP，数据还包含一个表示失败模式的编码。

最初，由于考虑到第三方可能会寻求HapMap数据的专利，所有的使用者都被要求同意一项基于网络的“点击同意许可”，表示其同意不会阻止其他人使用数据(http://www.hapmap.

was dropped, and all data were released without restriction into the public domain.

(**437**, 1299-1320; 2005)

Received 11 August; accepted 12 September 2005.

References:

1. Lechler, R. & Warrens, A. *HLA in Health and Disease* 2nd edn (Academic Press, San Diego, California, 2005).
2. Strittmatter, W. J. & Roses, A. D. Apolipoprotein E and Alzheimer's disease. *Annu. Rev. Neurosci.* **19**, 53-77 (1996).
3. Dahlbäck, B. Resistance to activated protein C caused by the factor V $R^{506}Q$ mutation is a common risk factor for venous thrombosis. *Thromb. Haemost.* **78**, 483-488 (1997).
4. Altshuler, D. *et al.* The common PPARγ Pro12Ala polymorphism is associated with decreased risk of type 2 diabetes. *Nature Genet.* **26**, 76-80 (2000).
5. Deeb, S. S. *et al.* A Pro12Ala substitution in PPARγ 2 associated with decreased receptor activity, lower body mass index and improved insulin sensitivity. *Nature Genet.* **20**, 284-287 (1998).
6. Florez, J. C., Hirschhorn, J. & Altshuler, D. The inherited basis of diabetes mellitus: implications for the genetic analysis of complex traits. *Annu. Rev. Genomics Hum. Genet.* **4**, 257-291 (2003).
7. Begovich, A. B. *et al.* A missense single-nucleotide polymorphism in a gene encoding a protein tyrosine phosphatase (*PTPN22*) is associated with rheumatoid arthritis. *Am. J. Hum. Genet.* **75**, 330-337 (2004).
8. Bottini, N. *et al.* A functional variant of lymphoid tyrosine phosphatase is associated with type I diabetes. *Nature Genet.* **36**, 337-338 (2004).
9. Bell, G. I., Horita, S. & Karam, J. H. A polymorphic locus near the human insulin gene is associated with insulin-dependent diabetes mellitus. *Diabetes* **33**, 176-183 (1984).
10. Ueda, H. *et al.* Assocation of the T-cell regulatory gene *CTLA4* with susceptibility to autoimmune disease. *Nature* **423**, 506-511 (2003).
11. Ogura, Y. *et al.* A frameshift mutation in *NOD2* associated with susceptibility to Crohn's disease. *Nature* **411**, 603-606 (2001).
12. Hugot, J. P. *et al.* Association of NOD2 leucine-rich repeat variants with susceptibility to Crohn's disease. *Nature* **411**, 599-603 (2001).
13. Klein, R. J. *et al.* Complement factor H polymorphism in age-related macular degeneration. *Science* **308**, 385-389 (2005).
14. Haines, J. L. *et al.* Complement factor H variant increases the risk of age-related macular degeneration. *Science* **308**, 419-421 (2005).
15. Edwards, A. O. *et al.* Complement factor H polymorphism and age-related macular degeneration. *Science* **308**, 421-424 (2005).
16. Puffenberger, E. G. *et al.* A missense mutation of the endothelin-B receptor gene in multigenic Hirschsprung's disease. *Cell* **79**, 1257-1266 (1994).
17. Emison, E. S. *et al.* A common sex-dependent mutation in a *RET* enhancer underlies Hirschsprung disease risk. *Nature* **434**, 857-863 (2005).
18. Kerem, B. *et al.* Identification of the cystic fibrosis gene: genetic analysis. *Science* **245**, 1073-1080 (1989).
19. Hästbacka, J. *et al.* Linkage disequilibrium mapping in isolated founder populations: diastrophic dysplasia in Finland. *Nature Genet.* **2**, 204-211 (1992).
20. Pritchard, J. K. & Przeworski, M. Linkage disequilibrium in humans: models and data. *Am. J. Hum. Genet.* **69**, 1-14 (2001).
21. Jorde, L. B. Linkage disequilibrium and the search for complex disease genes. *Genome Res.* **10**, 1435-1444 (2000).
22. Reich, D. E. *et al.* Linkage disequilibrium in the human genome. *Nature* **411**, 199-204 (2001).
23. Kruglyak, L. Prospects for whole-genome linkage disequilibrium mapping of common disease genes. *Nature Genet.* **22**, 139-144 (1999).
24. Johnson, G. C. *et al.* Haplotype tagging for the identification of common disease genes. *Nature Genet.* **29**, 233-237 (2001).
25. Nickerson, D. A. *et al.* DNA sequence diversity in a 9.7-kb region of the human lipoprotein lipase gene. *Nature Genet.* **19**, 233-240 (1998).
26. Zhu, X. *et al.* Localization of a small genomic region associated with elevated ACE. *Am. J. Hum. Genet.* **67**, 1144-1153 (2000).
27. Daly, M. J., Rioux, J. D., Schaffner, S. F., Hudson, T. J. & Lander, E. S. High-resolution haplotype structure in the human genome. *Nature Genet.* **29**, 229-232 (2001).
28. Jeffreys, A. J., Kauppi, L. & Neumann, R. Intensely punctate meiotic recombination in the class II region of the major histocompatibility complex. *Nature Genet.* **29**, 217-222 (2001).
29. Patil, N. *et al.* Blocks of limited haplotype diversity revealed by high-resolution scanning of human chromosome 21. *Science* **294**, 1719-1723 (2001).
30. Gabriel, S. B. *et al.* The structure of haplotype blocks in the human genome. *Science* **296**, 2225-2229 (2002).
31. The International HapMap Consortium. The International HapMap Project. *Nature* **426**, 789-796 (2003).
32. The International HapMap Consortium. Integrating ethics and science in the International HapMap Project. *Nature Rev. Genet.* **5**, 467-475 (2004).
33. The ENCODE Project Consortium. The ENCODE (ENCyclopedia Of DNA Elements) Project. *Science* **306**, 636-640 (2004).
34. Herrnstadt, C. *et al.* Reduced-median-network analysis of complete mitochondrial DNA coding-region sequences for the major African, Asian, and European haplogroups. *Am. J. Hum. Genet.* **70**, 1152-1171 (2002).
35. Jobling, M. A. & Tyler-Smith, C. The human Y chromosome: an evolutionary marker comes of age. *Nature Rev. Genet.* **4**, 598-612 (2003).
36. The Y Chromosome Consortium. A nomenclature system for the tree of human Y-chromosomal binary haplogroups. *Genome Res.* **12**, 339-348 (2002).
37. Underhill, P. A. *et al.* The phylogeography of Y chromosome binary haplotypes and the origins of modern human populations. *Ann. Hum. Genet.* **65**, 43-62 (2001).
38. Marchini, J. *et al.* A comparison of phasing algorithms for trios and unrelated individuals. *Am. J. Hum. Genet.* (in the press).

org/cgi-perl/registration）。2004 年 12 月，这一许可被终止，所有数据面向公众无限制释放。

（李平 翻译；常江 审稿）

39. Stephens, M. & Donnelly, P. A comparison of Bayesian methods for haplotype reconstruction from population genotype data. *Am. J. Hum. Genet.* **73**, 1162-1169 (2003).

40. Wright, S. *Evolution and the Genetics of Populations Volume 2: the Theory of Gene Frequencies* 294-295 (Univ. of Chicago Press, Chicago, 1969).

41. Rosenberg, N. A. *et al.* Genetic structure of human populations. *Science* **298**, 2381-2385 (2002).

42. Li, N. & Stephens, M. Modeling linkage disequilibrium and identifying recombination hotspots using single-nucleotide polymorphism data. *Genetics* **165**, 2213-2233 (2003).

43. Pe'er, I. *et al.* Reconciling estimates of linkage disequilibrium in the human genome. *Genome Res.* (submitted).

44. Lichten, M. & Goldman, A. S. Meiotic recombination hotspots. *Annu. Rev. Genet.* **29**, 423-444 (1995).

45. Mouse Genome Sequencing Consortium. Initial sequencing and comparative analysis of the mouse genome. *Nature* **420**, 520-562 (2002).

46. McVean, G. A. *et al.* The fine-scale structure of recombination rate variation in the human genome. *Science* **304**, 581-584 (2004).

47. Crawford, D. C. *et al.* Evidence for substantial fine-scale variation in recombination rates across the human genome. *Nature Genet.* **36**, 700-706 (2004).

48. Kong, A. *et al.* A high-resolution recombination map of the human genome. *Nature Genet.* **31**, 241-247 (2002).

49. Winckler, W. *et al.* Comparison of fine-scale recombination rates in humans and chimpanzees. *Science* **308**, 107-111 (2005).

50. Myers, S. R. & Griffiths, R. C. Bounds on the minimum number of recombination events in a sample history. *Genetics* **163**, 375-394 (2003).

51. Hudson, R. R. & Kaplan, N. L. Statistical properties of the number of recombination events in the history of a sample of DNA sequences. *Genetics* **111**, 147-164 (1985).

52. Phillips, M. S. *et al.* Chromosome-wide distribution of haplotype blocks and the role of recombination hot spots. *Nature Genet.* **33**, 382-387 (2003).

53. Chapman, J. M., Cooper, J. D., Todd, J. A. & Clayton, D. G. Detecting disease associations due to linkage disequilibrium using haplotype tags: a class of tests and the determinants of statistical power. *Hum. Hered.* **56**, 18-31 (2003).

54. Carlson, C. S. *et al.* Selecting a maximally informative set of single-nucleotide polymorphisms for association analyses using linkage disequilibrium. *Am. J. Hum. Genet.* **74**, 106-120 (2004).

55. de Bakker, P. I. W. *et al.* Efficiency and power in genetic association studies. *Nature Genet.* Advance online publication, 23 October 2005 (doi:10.1038/ng1669).

56. Lin, S., Chakravarti, A. & Cutler, D. J. Exhaustive allelic transmission disequilibrium tests as a new approach to genome-wide association studies. *Nature Genet.* **36**, 1181-1188 (2004).

57. Weale, M. E. *et al.* Selection and evaluation of tagging SNPs in the neuronal-sodium-channel gene *SCN1A*: implications for linkage-disequilibrium gene mapping. *Am. J. Hum. Genet.* **73**, 551-565 (2003).

58. Stram, D. O. *et al.* Choosing haplotype-tagging SNPs based on unphased genotype data using a preliminary sample of unrelated subjects with an example from the Multiethnic Cohort Study. *Hum. Hered.* **55**, 27-36 (2003).

59. de la Chapelle, A. & Wright, F. A. Linkage disequilibrium mapping in isolated populations: the example of Finland revisited. *Proc. Natl Acad. Sci. USA* **95**, 12416-12423 (1998).

60. Mootha, V. K. *et al.* Identification of a gene causing human cytochrome c oxidase deficiency by integrative genomics. *Proc. Natl Acad. Sci. USA* **100**, 605-610 (2003).

61. Engert, J. C. *et al.* ARSACS, a spastic ataxia common in northeastern Québec, is caused by mutations in a new gene encoding an 11.5-kb ORF. *Nature Genet.* **24**, 120-125 (2000).

62. Richter, A. *et al.* Location score and haplotype analyses of the locus for autosomal recessive spastic ataxia of Charlevoix-Saguenay, in chromosome region 13q11. *Am. J. Hum. Genet.* **64**, 768-775 (1999).

63. Chakraborty, R. & Weiss, K. M. Admixture as a tool for finding linked genes and detecting that difference from allelic association between loci. *Proc. Natl Acad. Sci. USA* **85**, 9119-9123 (1988).

64. Smith, M. W. & O'Brien, S. J. Mapping by admixture linkage disequilibrium: advances, limitations and guidelines. *Nature Rev. Genet.* **6**, 623-632 (2005).

65. Smith, M. W. *et al.* A high-density admixture map for disease gene discovery in African Americans. *Am. J. Hum. Genet.* **74**, 1001-1013 (2004).

66. Zhu, X. *et al.* Admixture mapping for hypertension loci with genome-scan markers. *Nature Genet.* **37**, 177-181 (2005).

67. Zhao, X. *et al.* An integrated view of copy number and allelic alterations in the cancer genome using single nucleotide polymorphism arrays. *Cancer Res.* **64**, 3060-3071 (2004).

68. Huang, J. *et al.* Whole genome DNA copy number changes identified by high density oligonucleotide arrays. *Hum. Genomics* **1**, 287-299 (2004).

69. Iafrate, A. J. *et al.* Detection of large-scale variation in the human genome. *Nature Genet.* **36**, 949-951 (2004).

70. Sebat, J. *et al.* Large-scale copy number polymorphism in the human genome. *Science* **305**, 525-528 (2004).

71. Stankiewicz, P. & Lupski, J. R. Genome architecture, rearrangements and genomic disorders. *Trends Genet.* **18**, 74-82 (2002).

72. Gonzalez, E. *et al.* The influence of *CCL3L1* gene-containing segmental duplications on HIV-1/AIDS susceptibility. *Science* **307**, 1434-1440 (2005).

73. Singleton, A. B. *et al.* α-Synuclein locus triplication causes Parkinson's disease. *Science* **302**, 841 (2003).

74. McCarroll, S. *et al.* Common deletion variants in the human genome. *Nature Genet.* (in the press).

75. Stefansson, H. *et al.* A common inversion under selection in Europeans. *Nature Genet.* **37**, 129-137 (2005).

76. Myers, S., Bottolo, L., Freeman, C., McVean, G. & Donnelly, P. A fine-scale map of recombination rates and recombination hotspots in the human genome. *Science* **310**, 321-324 (2005).

77. Hinds, D. A. *et al.* Whole-genome patterns of common DNA variation in three human populations. *Science* **307**, 1072-1079 (2005).

78. Yu, A. *et al.* Comparison of human genetic and sequence-based physical maps. *Nature* **409**, 951-953 (2001).

79. Broman, K. W., Murray, J. C., Sheffield, V. C., White, R. L. & Weber, J. L. Comprehensive human genetic maps: individual and sex-specific variation in

recombination. *Am. J. Hum. Genet.* **63**, 861-869 (1998).

80. Weissenbach, J. *et al.* A second-generation linkage map of the human genome. *Nature* **359**, 794-801 (1992).

81. Fullerton, S. M., Bernardo Carvalho, A. & Clark, A. G. Local rates of recombination are positively correlated with GC content in the human genome. *Mol. Biol. Evol.* **18**, 1139-1142 (2001).

82. Dawson, E. *et al.* A first-generation linkage disequilibrium map of human chromosome 22. *Nature* **418**, 544-548 (2002).

83. Begun, D. J. & Aquadro, C. F. Levels of naturally occurring DNA polymorphism correlate with recombination rates in *D. melanogaster*. *Nature* **356**, 519-520 (1992).

84. Smith, A. V., Thomas, D. J., Munro, H. M. & Abecasis, G. R. Sequence features in regions of weak and strong linkage disequilibrium. *Genome Res.* **15**, 1519-1534 (2005).

85. The Gene Ontology Consortium. Gene ontology: tool for the unification of biology. *Nature Genet.* **25**, 25-29 (2000).

86. Trachtenberg, E. *et al.* Advantage of rare HLA supertype in HIV disease progression. *Nature Med.* **9**, 928-935 (2003).

87. Pehrson, J. R. & Fuji, R. N. Evolutionary conservation of histone macroH2A subtypes and domains. *Nucleic Acids Res.* **26**, 2837-2842 (1998).

88. Modrich, P. & Lahue, R. Mismatch repair in replication fidelity, genetic recombination, and cancer biology. *Annu. Rev. Biochem.* **65**, 101-133 (1996).

89. Nielsen, R. Human genomics: disclosure of variation. *Nature* **434**, 288-289 (2005).

90. Enattah, N. S. *et al.* Identification of a variant associated with adult-type hypolactasia. *Nature Genet.* **30**, 233-237 (2002).

91. Bersaglieri, T. *et al.* Genetic signatures of strong recent positive selection at the lactase gene. *Am. J. Hum. Genet.* **74**, 1111-1120 (2004).

92. Sabeti, P. C. *et al.* Detecting recent positive selection in the human genome from haplotype structure. *Nature* **419**, 832-837 (2002).

93. Dermitzakis, E. T. *et al.* Numerous potentially functional but non-genic conserved sequences on human chromosome 21. *Nature* **420**, 578-582 (2002).

94. Margulies, E. H., Blanchette, M., Haussler, D. & Green, E. D. Identification and characterization of multi-species conserved sequences. *Genome Res.* **13**, 2507-2518 (2003).

95. Need, A. C., Motulsky, A. G. & Goldstein, D. B. Priorities and standards in pharmacogenetic research. *Nature Genet.* **37**, 671-681 (2005).

96. Brem, R. B., Storey, J. D., Whittle, J. & Kruglyak, L. Genetic interactions between polymorphisms that affect gene expression in yeast. *Nature* **436**, 701-703 (2005).

97. Istrail, S. *et al.* Whole-genome shotgun assembly and comparison of human genome assemblies. *Proc. Natl Acad. Sci. USA* **101**, 1916-1921 (2004).

98. Venter, J. C. *et al.* The sequence of the human genome. *Science* **291**, 1304-1351 (2001).

99. The Chimpanzee Sequencing and Analysis Consortium. Initial sequence of the chimpanzee genome and comparison with the human genome. *Nature* **437**, 69-87 (2005).

100. Matsuzaki, H. *et al.* Genotyping over 100,000 SNPs on a pair of oligonucleotide arrays. *Nature Methods* **1**, 109-111 (2004).

101. Bailey, J. A. *et al.* Recent segmental duplications in the human genome. *Science* **297**, 1003-1007 (2002).

102. Wigginton, J. E., Cutler, D. J. & Abecasis, G. R. A note on exact tests of Hardy–Weinberg equilibrium. *Am. J. Hum. Genet.* **76**, 887-893 (2005).

103. Hill, W. G. & Weir, B. S. Maximum-likelihood estimation of gene location by linkage disequilibrium. *Am. J. Hum. Genet.* **54**, 705-714 (1994).

Supplementary Information is linked to the online version of the paper at www.nature.com/nature.

Acknowledgements. We thank many people who contributed to this project: J. Beck, C. Beiswanger, D. Coppock, A. Leach, J. Mintzer and L. Toji (Coriell Institute for Medical Research) for transforming the Yoruba, Japanese and Han Chinese samples, distributing the DNA and cell lines, storing the samples for use in future research, and producing the community newsletters and reports; J. Greenberg and R. Anderson (NIH National Institute of General Medical Sciences) for providing funding and support for cell line transformation and storage in the NIGMS Human Genetic Cell Repository at the Coriell Institute; T. Dibling, T. Ishikura, S. Kanazawa, S. Mizusawa and S. Saito (SNP Research Center, RIKEN) for help with genotyping; C. Hind and A. Moghadam for technical support in genotyping and all members of the subcloning and sequencing teams at the Wellcome Trust Sanger Institute; X. Ke (Wellcome Trust Centre for Human Genetics at the University of Oxford) for help with data analysis; Oxford E-Science Centre for provision of high-performance computing resources; H. Chen, W. Chen, L. Deng, Y. Dong, C. Fu, L. Gao, H. Geng, J. Geng, M. He, H. Li, H. Li, S. Li, X. Li, B. Liu, Z. Liu, F. Lu, F. Lu, G. Lu, C. Luo, X. Wang, Z. Wang, C. Ye and X. Yu (Beijing Genomics Institute) for help with genotyping and sample collection; X. Feng, Y. Li, J. Ren and X. Zhou (Beijing Normal University) for help with sample collection; J. Fan, W. Gu, W. Guan, S. Hu, H. Jiang, R. Lei, Y. Lin, Z. Niu, B. Wang, L. Yang, W. Yang, Y. Wang, Z. Wang, S. Xu, W. Yan, H. Yang, W. Yuan, C. Zhang, J. Zhang, K. Zhang and G. Zhao (Chinese National Human Genome Center at Shanghai) for help with genotyping; P. Fong, C. Lai, C. Lau, T. Leung, L. Luk and W. Tong (University of Hong Kong, Genome Research Centre) for help with genotyping; C. Pang (Chinese University of Hong Kong) for help with genotyping; K. Ding, B. Qiang, J. Zhang, X. Zhang and K. Zhou (Chinese National Human Genome Center at Beijing) for help with genotyping; Q. Fu, S. Ghose, X. Lu, D. Nelson, A. Perez, S. Poole, R. Vega and H. Yonath (Baylor College of Medicine); C. Bruckner, T. Brundage, S. Chow, O. Iartchouk, M. Jain, M. Moorhead and K. Tran (ParAllele Bioscience Inc.); N. Addleman, J. Atilano, T. Chan, C. Chu, C. Ha, T. Nguyen, M. Minton and A. Phong (UCSF) for help with genotyping, and D. Lind (UCSF) for help with quality control and experimental design; R. Donaldson and S. Duan (Washington University) for help with genotyping, and J. Rice and N. Saccone (Washington University) for help with experimental design; J. Wigginton (University of Michigan) for help with implementing and testing QA/QC software; A. Clark, B. Keats, R. Myers, D. Nickerson and A. Williamson for providing advice to NIH; J. Melone, M. Weiss and E. DeHaut-Combs (NHGRI) for help with project management; M. Gray for organizing phone calls and meetings; D. Leja for help with figures; the Yoruba people of Ibadan, Nigeria, the people of Tokyo, Japan, and the community at Beijing Normal University, who participated in public consultations and community engagements; the people in these communities who were generous in donating their blood samples; and the people in the Utah CEPH community who allowed the samples they donated earlier to be used for the Project. We also thank A. Clark, E. Lander, C. Langley and R. Lifton for comments on earlier drafts of the manuscript. This work was supported by the Japanese Ministry of Education, Culture,

Sports, Science, and Technology, the Wellcome Trust, Nuffield Trust, Wolfson Foundation, UK EPSRC, Genome Canada, Génome Québec, the Chinese Academy of Sciences, the Ministry of Science and Technology of the People's Republic of China, the National Natural Science Foundation of China, the Hong Kong Innovation and Technology Commission, the University Grants Committee of Hong Kong, the SNP Consortium, the US National Institutes of Health (FIC, NCI, NCRR, NEI, NHGRI, NIA, NIAAA, NIAID, NIAMS, NIBIB, NIDA, NIDCD, NIDCR, NIDDK, NIEHS, NIGMS, NIMH, NINDS, NLM, OD), the W.M. Keck Foundation, and the Delores Dore Eccles Foundation.

Author Contributions. David Altshuler, Lisa D. Brooks, Aravinda Chakravarti, Francis S. Collins, Mark J. Daly and Peter Donnelly are members of the writing group responsible for this manuscript.

Author Information. Reprints and permissions information is available at npg.nature.com/reprintsandpermissions. The authors declare competing financial interests: details accompany the paper at www.nature.com/nature. Correspondence and requests for materials should be addressed to D.A. (altshuler@molbio.mgh.harvard.edu) or P.D. (donnelly@stats.ox.ac.uk).

Rain, Winds and Haze during the Huygens Probe's Descent to Titan's Surface

M. G. Tomasko *et al.*

Editor's Note

The Cassini spacecraft carried the Huygens mission, whose goal was to land on the surface of Saturn's largest moon, Titan. During Huygens' parachute descent, an array of instruments studied the moon's atmosphere, winds and surface. The lander survived on the surface for about three hours. This paper is one of two presenting the key findings of the mission. It reports that the moon's surface contains features resembling those made by flowing liquid on Earth, presumably due to liquid hydrocarbons. The landing site looks like a dry river or lake bed, scattered with small rounded "rocks", which are probably hydrocarbon-coated water ice. The atmospheric haze that obscures the surface from telescopes is made of small particles, most probably of complex and as yet poorly characterized organic matter.

The irreversible conversion of methane into higher hydrocarbons in Titan's stratosphere implies a surface or subsurface methane reservoir. Recent measurements from the cameras aboard the Cassini orbiter fail to see a global reservoir, but the methane and smog in Titan's atmosphere impedes the search for hydrocarbons on the surface. Here we report spectra and high-resolution images obtained by the Huygens Probe Descent Imager/Spectral Radiometer instrument in Titan's atmosphere. Although these images do not show liquid hydrocarbon pools on the surface, they do reveal the traces of once flowing liquid. Surprisingly like Earth, the brighter highland regions show complex systems draining into flat, dark lowlands. Images taken after landing are of a dry riverbed. The infrared reflectance spectrum measured for the surface is unlike any other in the Solar System; there is a red slope in the optical range that is consistent with an organic material such as tholins, and absorption from water ice is seen. However, a blue slope in the near-infrared suggests another, unknown constituent. The number density of haze particles increases by a factor of just a few from an altitude of 150 km to the surface, with no clear space below the tropopause. The methane relative humidity near the surface is 50 per cent.

THE surface of Titan has long been studied with various instruments, including those on the Hubble Space Telescope (HST) and ground-based adaptive optics systems[1]. More recently, Cassini investigations using the charge-coupled device (CCD) camera[2], the

惠更斯空间探测器下落至土卫六表面过程中的雨、风和雾霾

托马斯科等

编者按

卡西尼号土星探测器承担着惠更斯计划，惠更斯号探测器的目标是在土星最大的卫星土卫六的表面着陆。在惠更斯号打开降落伞的降落阶段，一系列探测仪器研究了土卫六的大气、风以及地表。着陆器在卫星表面工作了约3个小时。本文是报道本次探测任务主要发现的两篇文章之一，文章指出卫星的表面存在一些地貌特征，类似地球表面上液体流动留下的痕迹，推测可能和液态的烃有关。着陆点看起来像干涸的河床或者湖床，遍布圆球形的小“岩石”，这些很有可能是烃包被的水冰。由于大气中雾霾的存在，使得从望远镜只能观察到模糊的地表影像。雾霾是由细小粒子构成的，很可能成分复杂且存在迄今为止不为人知的有机物质。

在土卫六平流层探测到甲烷转变为高级烃的不可逆过程，这意味着土卫六的表面或地下存在甲烷储备。最近卡西尼轨道飞行器上搭载的照相机进行的观测并没能发现全球性的甲烷储备，但是土卫六大气中的甲烷和浓雾阻碍了在其表面寻找烃的工作。本文中，我们将报道惠更斯探测器在土卫六大气降落过程中，搭载的惠更斯降落成像仪 / 光谱辐射计测得的光谱和高分辨率图像。尽管这些图像并没有直接展现土卫六表面的液态烃湖，但是确实存在液态物质在土卫六表面流过留下的痕迹。土卫六上的地貌与地球表面有惊人的相似性，比较明亮的高地区域存在复杂交错的沟渠系统，它们汇入平坦而阴暗的低地。惠更斯号着陆后拍摄的照片显示的环境是一处干涸的河床。土卫六表面测得的红外反射光谱，不同于太阳系内任何其他已知的天体，在其光学波段的红端（较大波长）存在倾斜谱结构（红坡），这与一类有机物质（比如索林斯）的光谱特征相一致，水冰的吸收峰是可见的。尽管如此，近红外区域的蓝端倾斜结构（蓝坡）则预示另外一种未知的成分的存在。雾霾颗粒的数量密度从 150 km 的海拔到表面增大了数倍，并且在对流顶层以下没有明显的净空区域。甲烷在地表附近的相对湿度为 50%。

借助包括哈勃空间望远镜（HST）搭载的仪器和地基自适应光学系统等在内的各种仪器，人们已经对土卫六的表面进行了长期的研究[1]。最近，卡西尼号飞船使用

Visible and Infrared Mapping Spectrometer (VIMS) instrument[3], and the Radio Detection and Ranging (RADAR) imaging system[4] have provided more detailed views of Titan's surface in the hope of revealing how the methane in Titan's atmosphere is replenished from the surface or interior of Titan. Of the Cassini imagers, the Imaging Science Subsystem (ISS) camera is potentially capable of the greatest spatial resolution, but Titan's obscuring haze limits its resolution on the surface to about 1 km, a value roughly similar to that available from VIMS and the radar imaging system. At this resolution, the bright and dark regions observed on the surface of Titan have proved difficult to interpret. Owing to its proximity to the surface, the Descent Imager/Spectral Radiometer (DISR) camera on the Huygens probe was capable of a linear resolution of some metres from a height of 10 km. In addition, the lower the probe descended, the less haze lay between the camera and the ground. The DISR was capable of linear resolution orders of magnitude better than has been available from orbit, although of a much smaller portion of Titan's surface. Also, a lamp was used at low altitude to measure the continuous reflectance spectrum of the surface without the complications introduced by observations through large amounts of methane and aerosol haze[5].

In addition to studying the surface of Titan, the DISR took measurements of solar radiation in the atmosphere. Spectrometers looking upward at continuum wavelengths (between the major methane absorptions) as well as downward measured the vertical distribution and wavelength dependence of the aerosol haze opacity. Measurements of the polarization of light at a scattering angle of 90° constrained the small dimension of the haze particles. Measurements of the brightness in the solar aureole around the Sun determined the projected area of the haze particles. Observations in the methane bands determined the methane mole fraction profile.

Data collection during the descent proceeded mostly, although not exactly, as planned. Turbulence during the first half of the descent tipped the probe more rapidly than expected, causing the Sun sensor to remain locked on the azimuth of the Sun for only a few successive rotations at a time. Below about 35 km, the signal from the direct solar beam was lost by the Sun sensor owing to the unexpectedly low temperature of this detector. These effects caused data from each of the DISR sub-instruments to be collected at mostly random, instead of specific, azimuths. Additionally, the probe rotated in the intended direction for only the first ten minutes before rotating in the opposite sense for the remainder of the descent. This resulted in ineffective baffling of the direct solar beam for the upward-looking visible spectrometer and the solar aureole camera. Consequently, some measurements made by the solar aureole camera are saturated, and the separation of the direct and diffuse solar beams in the visible spectral measurements must be postponed until a good model of the probe attitude versus time is available. Finally, the loss of one of the two radio communication channels in the probe receiver aboard the orbiter resulted in the loss of half the images as well as several other low-altitude spectrometer measurements.

Despite these misfortunes, the DISR instrument collected a unique and very useful data set. Images of the surface with unprecedented resolution were collected over the boundary

自身搭载的电荷耦合器件（CCD）照相机[2]、可见光和红外测绘光谱仪（VIMS）[3]以及雷达成像系统[4]，对土卫六表面进行了更加仔细的观测，并期望能够揭示土卫六上的甲烷是如何从地表或地下重新回到大气中的。在卡西尼号所有的成像仪中，成像科学子系统（ISS）有能力可以获得最高的空间分辨率，但是在土卫六表面雾霾的影响下，其分辨率被限制在与 VIMS 以及雷达成像系统大致相当的水平，仅为 1 km 左右。在这一分辨率下，土卫六表面上被观测到的明亮和阴暗的区域被证明是难以解译的。由于惠更斯空间探测器接近土卫六表面，它搭载的降落成像仪 / 光谱辐射计（DISR）相机在 10 km 的高度可以获得米量级的线性分辨率。此外，探测器高度越低，则相机与地面之间的雾霾的厚度越小。这样，DISR 可以获得比在轨道上高数个量级的线性分辨率，但观测到的土卫六地表范围减小很多。同时，惠更斯号在较低的高度借助照明灯测量了表面的连续反射光谱，这样避免了透过大量的甲烷和气溶胶薄雾进行观测所引起的光谱复杂性[5]。

除了研究土卫六表面，DISR 系统还测量了大气中的太阳辐射。采用连续波段（甲烷主吸收峰之间的波段），光谱仪分别朝上和朝下测量了薄雾在垂直方向上的分布及其不同波长的不透明度。在 90° 散射角处测量光的偏振可以约束雾霾颗粒是小颗粒的，而对太阳周围的日晕亮度的测量可以确定雾霾颗粒的投影面积。甲烷波段的观测可以确定甲烷的摩尔分数。

在惠更斯号降落过程中，尽管数据的采集工作基本按预定计划进行，但也出现了一些意外情况。在降落过程的前半段，湍流造成的探测器的倾斜比预期要快很多，这导致探测器每次保持太阳传感器锁定在太阳所在的方向角的姿态只能持续几次探测器旋转周期。降至约 35 km 以下，由于探测器的温度出乎意料的低，太阳传感器无法感应来自太阳光直射的信号。以上意外导致 DISR 子设备得到的数据并没能像预计的一样从特定的方位角方向获取，大部分都有随机的方位角。另一件意料之外的事情是，探测器仅仅在前 10 分钟内按照预定的方向旋转，然后在剩余的下降时间就一直以相反的方向转动。这导致与朝上观测的可见光谱分析仪和太阳光晕照相机有关的直射太阳光没有被有效地遮挡。因此，太阳光晕照相机进行的某些测量过度曝光，可见光谱测量中的直射太阳光束与弥散太阳光束暂时也难以有效分离，除非找到关于探测器姿态与时间关系的合理模型。最后，轨道器搭载的探测器接收天线损失掉了两个无线电通信频道中的一个，造成一半图像丢失，同时也造成光谱仪部分低空测量数据的丢失。

尽管遭遇了这些不幸，DISR 中的设备还是采集到了一个独一无二且非常有用的数据集。轨道器拍摄到了明亮区域和暗区域之间的边界地带史无前例的高分辨率图

between bright and dark terrain seen from the orbiter. Owing to redundant transmission over both communication channels during most of the descent, almost all of the spectral and solar aureole observations were received. A very large set of high-quality spectra were obtained with good altitude resolution and with good coverage in azimuth both away from and towards the Sun. The images, the spectra, the Sun sensor pulses, the recording of the gain in the Cassini radio receiver, and information from Very Long Baseline Interferometry (VLBI) observations from Earth together will permit reconstruction of the probe attitude relative to the Sun as a function of time during the descent, enabling a full analysis of the spectral data. The large number of solar aureole measurements included several acquired near the Sun and many polarization measurements opposite to the Sun. The surface science lamp worked exactly as planned, permitting surface reflection measurements even in strong methane absorption bands. Operations after landing included the collection of successive images as well as spectral reflectance measurements of the surface illuminated by the lamp from an assumed height of roughly 30 cm.

Taken together, the new observations shed substantial light on the role played by methane in forming the surface of Titan and how it is recycled into the atmosphere. The substantial relative humidity of methane at the surface and the obvious evidence of fluid flow on the surface provide evidence for precipitation of methane onto the surface and subsequent evaporation. Some indications of cryovolcanic flows are also seen. The vertical distribution and optical properties of Titan's haze have been characterized to aid the interpretation of remote measurements of the spectral reflection of the surface. The speed and direction of Titan's winds has also been measured for comparison with future dynamical models that include the radiative heating and cooling rates implied by the haze.

Physical Processes That Form the Surface

The imagers provided views of Titan's previously unseen surface, thus allowing a deeper understanding of the moon's geology. The three DISR cameras were designed to provide overlapping coverage for an unbroken 360°-wide swath stretching from nadir angles between 6° and 96°. Some 20 sets of such images were planned during the descent. Because of the opacity of the haze in the passband of our imager, surface features could be discerned in the images only below about 50 km, limiting the number of independent panoramic mosaics that can be made of the surface. The loss of half of the images meant that Titan's surface was not covered by systematic overlapping triplets, as expected. Three different views of Titan's surface are shown in Figs 1 to 3. A view of Titan's surface after the Huygens landing is shown in Fig. 4.

The highest view (Fig. 1), projected from an altitude of 34 km, displays an albedo variation very similar to the highest-resolution images provided by the ISS or VIMS cameras on the Cassini orbiter. It shows Titan's surface to consist of brighter regions separated by lanes or lineaments of darker material. No obvious impact features are visible, although craters less than roughly 10 km should not be abundant as a result of atmospheric shielding[6]. In the

像。在下降过程的大部分时间，两个通信频道都进行了冗余传输，这使得我们可以接收到大部分光谱以及太阳光晕的观测结果。惠更斯探测器获得了大量高质量的光谱，并且无论是朝向太阳还是背向太阳都具有极佳的高度分辨率和方位覆盖。通过获得的图像、光谱、太阳传感器脉冲、卡西尼无线电接收机的增益记录以及地球上甚长基线干涉测量（VLBI）信息，我们可以重建出下降阶段中探测器相对太阳的姿态随时间的变化函数，从而使我们能够全面地分析光谱数据。所获得的大量的太阳光晕测量包括在指向太阳方向附近进行的一些测量和大量的背向太阳方向的偏振测量。用于土卫六表面测量的科学照明灯完全按预期进行工作，使得即使是在强甲烷吸收波段中，也可以进行表面反射光谱的相关测量。探测器着陆后的操作包括采集连续图像以及测量地表的光谱反射率，其中反射率测量中照明灯在高度约为 30 cm 提供照明。

综上，惠更斯探测器进行的最新观测在相当程度上揭示了甲烷在土卫六地表形成过程中的作用，以及甲烷是如何通过循环进入大气的。地表附近甲烷有较高的相对湿度，而且地表明显存在液体流动的痕迹，这些证据表明甲烷经历了凝结降落以及再蒸发的循环过程。同时，也发现了冰火山作用相关的流动构造的迹象。土卫六上雾霾的垂直分布和光学属性也被确定，有助于解释地表谱反射率的遥测结果。另外，探测器对土卫六上的风速和风向也进行了测量，可以用于和将来的动力学模型的对比，模型将包含由雾霾得出的辐射加热速率与冷却速率。

形成土卫六表面的物理过程

惠更斯探测器上的成像仪提供了大量前所未有的关于土卫六地表的信息，使得更加深入地了解这颗卫星的地质情况成为可能。DISR 系统的三台照相机可以提供 360° 水平无缝覆盖全景图像，其天底角介于 6° 到 96° 之间。探测器原本预计在下降过程中拍摄大约 20 组这样的照片。由于雾霾对于成像仪的滤波通带的不透明性，只有在约 50 km 以下高度拍摄的照片中才可以分辨出地表特征，从而也限制了不同的表面全景拼接图像的数量。同时，前面提到的图像丢失一半意味着土卫六的表面不能够按预期那样被三台照相机所拍摄的照片系统的交叠全部覆盖。图 1 到图 3 给出了土卫六表面的三幅图片。图 4 给出了惠更斯探测器着陆后拍摄的一幅地表图片。

视角最高的视图（图 1）是以 34 km 高度作为基点进行的投影，表明星体反照率的变化，这种变化与卡西尼轨道器上的 ISS 或 VIMS 照相机提供的最高分辨率图像非常相似。从图像中可知，土卫六表面由明亮区域以及分割这些区域的较暗的水道或线性构造构成。尽管由于大气的遮挡可以观察到的直径小于 10 km 的陨石坑应该

rightmost (eastern) part of the mosaic the images become sharper as the lower altitude of the camera causes the scale to decrease and the contrast to increase. More than a dozen brighter areas in that region seem to be elongated along a direction parallel with the main bright/dark boundary of that region. At the limit of resolution, narrow dark channels cut the bright terrain.

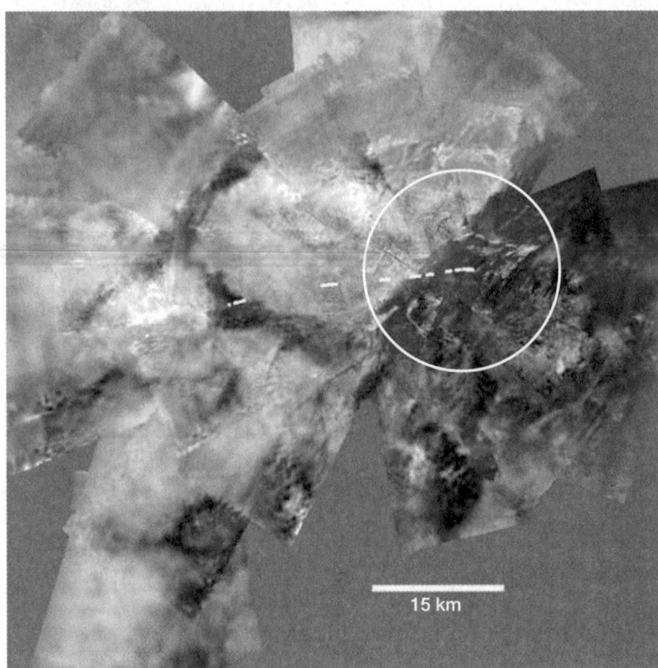

Fig. 1. View of Titan from 34 km above its surface. High-altitude (49 to 20 km) panoramic mosaic constructed from the DISR High and Medium Resolution Imagers (HRI and MRI) as projected from 34 km. The preliminary ground-track solution (indicated as small white points on gnomonic ground projection) represents the location of the probe when data were collected; north points up; scale indicated (although subsequent analysis indicates that north lies some 5–10° to the left of straight up in this and the two subsequent figures). Starting from the first surface image at 49 km, the probe moves in an east-northeastwardly direction at an initial speed of 20 m s^{-1}. Brighter regions separated by lanes or lineaments of darker material are seen. No obvious crater-like features are visible. The circle indicates the outline of the next-lowest pan, in Fig. 2. The method used for construction of panoramic mosaics incorporates knowledge of the probe's spatial location (longitude, latitude and altitude) and attitude (roll, pitch and yaw) at each image. With the exception of altitude, provided by the Huygens Atmospheric Structure Instrument[12] pressure sensor, none of these variables was directly measured. They are found through an iterative process in which a panorama is created, providing an improved ground-track and azimuth model, which results in an upgraded trajectory, which can improve the panorama, and so on. The current lack of pitch and roll knowledge constitutes the main source of error in the current composition and quality of the panoramas as well as the ground-track and wind speed determination reported below. Vigorous contrast-stretching in the images is required to reveal details washed out by the haze particle density at all altitudes in Titan's atmosphere. This contrast-stretching also displays the occasional ringing of the Discrete Cosine Transform data compressor, which appear as regular lines of bright and dark patterns, particularly in the MRI images.

The next view, Fig. 2, projected from an altitude of 8 km, reveals a large number of these channels (detailed in Fig. 5). The channel networks have two distinct patterns: short stubby

不会很多，图像中没有看到明显的陨石坑地貌[6]。在拼接图像的最右侧(东侧)，图像变得锐度增加，这是由于照相机拍摄时的高度降低，从而使拍摄范围变小，对比度增加。图中十多块明亮区域看似沿主要的明暗边界方向被拉伸了。在现有的最高分辨率下，图中的较为阴暗的狭谷截切了明亮区域。

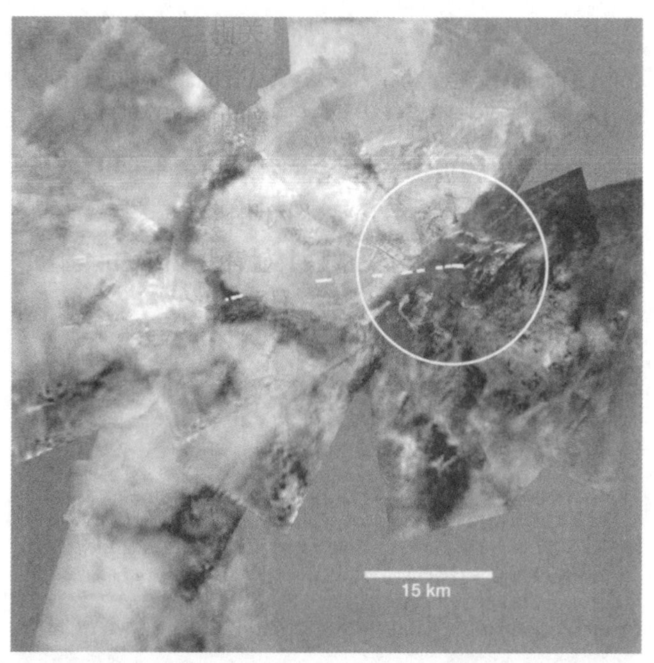

图 1. 土卫六上空 34 km 高度拍摄的图像。通过 DISR 高分辨率成像仪和中分辨率成像仪(即 HRI 和 MRI)构成的高空(49～20 km)全景拼接图，以 34 km 高度处作为基点投影。图中用小白点标出了探测器飞行轨迹在地面上的球心投影，这一初步的探测器飞行轨迹地表投影给出了探测器采集数据时的位置。图中正上方为北，比例尺标于图中(后续分析表明，图中实际的北方应该相对竖直向上方向向左偏移约 5°～10°，后面的两幅图也是如此)。探测器从 49 km 的高度开始拍摄第一幅照片，然后以 20 m/s 的初始速度向东北偏东方向飞行。由图可知，较亮的区域被较暗的水道或线性构造分开，但没有发现明显的类似环形山的地貌。图中的圆圈表示下一个全景镜头的轮廓，如图 2 所示。全景图像拼接的构建方法将探测器在拍摄每张照片时的空间定位(经度、纬度和高度)与姿态(滚动、俯仰和偏航)两方面的数据纳入考量。对于以上变量，除了高度可以通过惠更斯大气构造探测仪上的压强传感器直接测量[12]，其他变量都不能被直接测量。以上参数可以通过构建全景图像的迭代过程得到，该迭代过程可以改进探测器轨道的地表投影和方位模型，而模型又可以反过来校正探测器飞行轨道，进一步改善全景图像质量等。俯仰和滚动等姿态信息的缺乏，是目前全景图合成及其图像品质，以及后文轨迹的地面投影和风速确定中误差的主要来源。土卫六大气中的雾霾颗粒将使图像变得模糊不清，因此需要进行足够的对比度增强才能够显示更多的细节。对比度增强的操作也会偶尔表现出离散余弦变换数据压缩器的振荡效应，表现为规律性重复出现的明暗相间的条纹，这种效应对于中分辨率图像的影响特别严重。

下一幅拼接图如图 2 所示，以 8 km 的高度作为基点作投影，展现了大量的沟渠结构(具体细节如图 5 所示)。图中的沟渠网络具有两种明显不同的样式：一种是短

features and dendritic features with many branches. The region of the stubby network towards the west is significantly brighter than the dendritic region. The stubby channels are shorter, wider and relatively straight. They are associated with and often begin or end in dark circular areas, which suggest ponds or perhaps pits. The morphology of rectilinear networks with stubby heads is consistent with spring-fed channels or arroyos.

Fig. 2. View from 8 km. Medium-altitude (17 to 8 km) panoramic mosaic projected from 8 km. As in Fig. 1, the preliminary ground-track solution is indicated as points; north is up; scale indicated. At 11 km, the wind direction is at 0° (eastward), reaching −20° (southeastward) at an altitude of 8.5 km. The narrow dark lineaments, interpreted as channels, cut brighter terrain. The circle indicates the outline of the low-altitude pan in Fig. 3.

The dendritic network is consistent with rainfall drainage channels, implying a distributed rather than a localized source of a low-viscosity liquid. Stereo analysis of the dendritic region indicates an elevation of 50–200 m relative to the large darker plain to the south. It suggests that the brighter areas within the darker terrain are higher as well. The topographic differences are evident in Figs 6 and 7, which are three-dimensional renderings of the area just north of the landing site produced from the DISR images. They include the major bright–dark interface seen from above in Fig. 5. Figure 2 depicts many examples of these darker lanes of material between topographically higher, brighter areas. In fact the low contrast of the lowland plane argues that the entire dark region floods, and as the liquid drains the local topography drives flows as seen in the images. If the darker region is interpreted as a dried lakebed, it is too large to have been caused by the creeks and channels visible in the images. It may have been created by other larger river systems or some large-scale catastrophic event, which predates deposition by the rivers seen in these images.

而粗的断株状结构，另一种是具有很多分支的树枝状结构。前者相对更短、更宽和更直。图中指向西侧的断株样网络所在的区域明显比树枝状区域更加明亮。这些断株样沟渠通常起始或终止于深色的圆形区域，这一区域可能是池塘或深坑。以短粗结构为端点的直线网络地貌符合源于泉水的河道或河谷地貌特征。

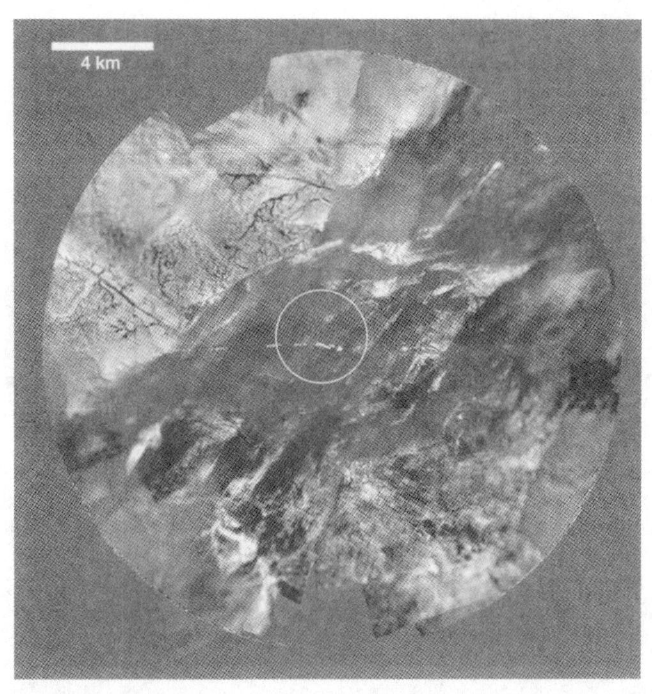

图 2. 8 km 高度的视图，是中空（17 ~ 8 km）所拍摄照片的全景拼接图。就像图 1 那样，探测器轨迹地面投影用点表示，图片正上方为北，比例尺标于图中。在 11 km 高度，探测到的风向位于 0°（朝正东方向）；在 8.5 km 高度，探测到的风向变为 −20°（朝东南方向）。图中细窄而暗的线性构造被解译为沟渠，截切了明亮区域。图中圆圈标记出了图 3 中的全景图像轮廓范围。

树枝状网络的地貌特征与降雨控制的水系河道相符，这意味着存在广泛分布的而非局域性的低黏度的液态物质。关于树枝状区域的立体信息分析显示，该区域相对南部的较暗的大型平原地区抬高了 50 ~ 200 m；这说明暗色区域内部的亮色区块同样会比周围的暗色区域更高。图 6 和图 7 是通过 DISR 图像生成的着陆点北部区域的三维效果图，它们包含了图 5 俯视图中主要的明暗分界，显然明暗区域的地貌特征存在很大区别。图 2 显示了多个存在于地形较高、较明亮的区域之间的较暗的带状分布的物质的实例。事实上，低地平原的较小的起伏说明整个暗色区域被液态物质湮没，而且地势会驱动液态物质的流动，表现出图像中所见的地貌。如果认为较暗的区域是干涸的湖床，显然面积太大，图像中所见到的小溪和河道不可能汇成这样大的湖泊。因此，暗色区域可能是由其他较大的河流系统汇成或者由某种大尺度的灾变造成，但这种灾变要早于图像中所见的河流沉积地貌。

The interpretation of the dark lanes within the brighter highlands as drainage features is so compelling as to dominate subsequent interpretation of other areas of images such as Figs 2 and 3. The prevailing bright–dark boundary of the region becomes a coastline, the bright areas separated from this boundary become islands. Bright streaks running parallel to the albedo boundary may be drift deposits or splays fractured off the bright highlands owing to faulting along the shoreline.

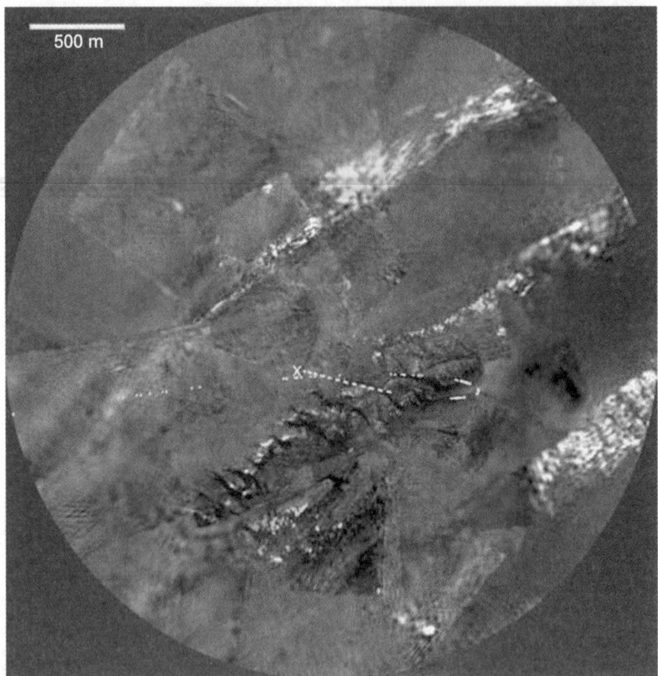

Fig. 3. View from 1.2 km. Low-altitude (7 to 0.5 km) panoramic mosaic projected from 1,200 m. As in Figs 1 and 2, the preliminary ground track is indicated as points; north is up; scale indicated. The probe's steady east-northeast drift halts altogether at an altitude of 7 km and reverses, moving west-northwest for some 1 km during the last 15 min of descent. Note the ridge near the centre, cut by a dozen darker lanes or channels. The projected landing site is marked with an "X" near the continuation of one of the channels, whose direction matches the orientation of the stream-like clearing in the near-foreground of the southward-looking surface image, Fig. 4.

When coupled with Fig. 4, which is an image of a typical offshore dark region, it is clear that the analogy has a limit. At present there is no liquid in the large dark lakebed imaged in Figs 1 to 5. The bright lobate feature, split by an apparently straight dark lane in the western part of the mosaic in Fig. 2, is a possible fissure-fed cryovolcanic flow. However, Fig. 4 also reveals rocks which—whether made of silicates or, more probably hydrocarbon-coated water-ice—appear to be rounded, size-selected and size-layered as though located in the bed of a stream within the large dark lakebed. Rounded stones approximately 15 cm in diameter and probably composed of water-ice, lie on top of a darker, finer-grained surface.

将明亮高地区域内的暗色条带看作用于排水的河道是非常可信的一种解释，以至于可以直接主导了图 2 和图 3 等其他图像的解译。这样的话，这一区域内主要的明暗边界可以被认为是海岸线，而被这些边界围限的明亮区域可以解释成岛屿。平行于反照率差异边界延伸的明亮条纹可能是冰川沉积物或者是沿海岸线发生断层活动由明亮高地断裂形成的斜面。

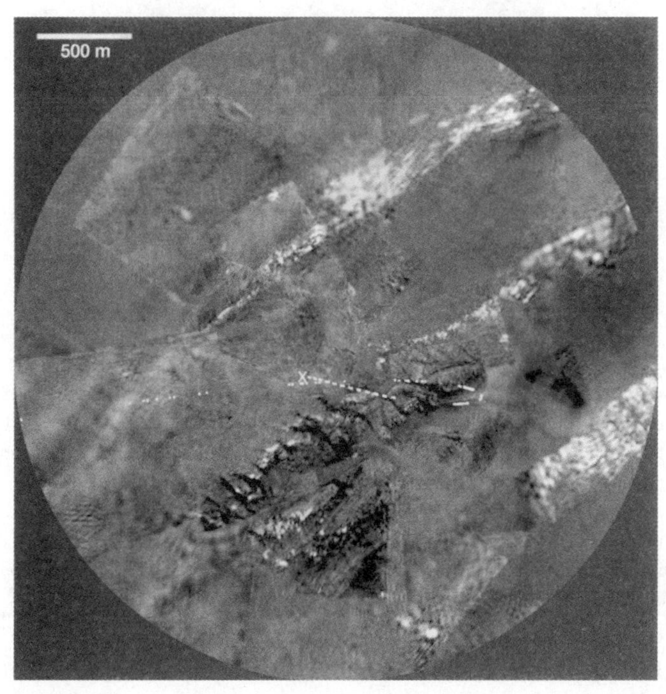

图 3. 1.2 km 高度处的视图。在 7 km ~ 0.5 km 范围内拍摄的图像，被用于制作以 1,200 m 为基点的投影全景图像。如图 1 和图 2，初步推测的探测器轨迹地面投影用点表示，图片正上方为北，比例尺标于图中。探测器首先朝东北偏东方向稳定飘移，到达 7 km 高度时完全停止，后反向运动，在下降阶段的最后 15 分钟探测器向西北偏西移动了大约 1 km。值得注意的是，图中心附近的山脊被若干暗色河道切割。探测器的着陆地点在图中的投影位置用“X”标记出，该位置靠近一条渠道的延长段，渠道与图 4 中显示的视角朝南的地表图像靠近前景的类似溪流的净空地带的延伸方向相一致。

图 4 是一片典型的离岸的暗色区域的图像，结合该图可知以上类比显然存在局限性。目前，图 1 至图 5 中的巨型暗色湖床区域并不存在液体。图 2 中西侧被一条直的暗色河道分隔的明亮叶状结构，可能是发源于裂缝的冰岩浆流。然而，图 4 展现的石块就好像是处在大型暗色湖床上的河床内，有较好的磨圆度、分选度和粒序层理，石块的成分可能是硅酸盐，或者更有可能是烃包被的水冰。这些直径大约为 15 cm、可能是由水冰构成的圆形石块位于由更细粒物质构成的更暗的表面之上。

Fig. 4. The view from Titan's surface. Merged MRI and SLI images acquired after the Huygens probe soft-landing. Horizon position implies a pitch of the DISR nose upward by $1.7 \pm 0.2°$ with no measurable roll. "Stones" 10–15 cm in size lie above darker, finer-grained substrate in a variable spatial distribution. Brightening of the upper left side of several rocks suggests solar illumination from that direction, implying a southerly view, which agrees with preliminary evidence from other data sets. A region with a relatively low number of rocks lies between clusters of rocks in the foreground and the background and matches the general orientation of channel-like features in the low-altitude pan of Fig. 3. The bright spot in the lower right corner is the illumination of the DISR surface science lamp.

It is interesting to compare the brightness and colour of the scene shown in Fig. 4 with that of a similar scene on the Earth. The brightness of the surface of the Earth illuminated by full sunlight is about half a million times greater than when illuminated by a full moon. The brightness of the surface of Titan is about a thousand times dimmer than full solar illumination on the Earth (or 500 times brighter than illumination by full moonlight). That is, the illumination level is about that experienced about 10 min after sunset on the Earth. The colour of the sky and the scene on Titan is rather orange due to the much greater attenuation of blue light by Titan's haze relative to red light. If the Sun is high in the sky, it is visible as a small, bright spot, ten times smaller than the solar disk seen from Earth, comparable in size and brightness to a car headlight seen from about 150 m away. The Sun casts sharp shadows, but of low contrast, because some 90% of the illumination comes from the sky. If the Sun is low in the sky, it is not visible.

The sizes of the more than 50 stones in the image in Fig. 4 vary between 3 mm in diameter, the resolution limit of the imager, and 15 cm. No rocks larger than 15 cm are seen. The resolution of the last images taken before landing from a height of 200–300 m would be sufficient to identify metre-sized objects, and none are seen in the 40×35 m field of view. Figure 8 shows the R value, a measure of the fraction of the surface covered by rocks of

图 4. 土卫六表面的图像。该图像是由惠更斯探测器软着陆后拍摄的 MRI 与 SLI 图像拼接而成。由地平线的位置可知 DISR 的仰角为 1.7° ±0.2°，但没有可见的左右倾斜。粒径 10 ~ 15 cm 的“石块”不均匀地散布在更暗更细粒的基底上。几块岩石的左上部分被照亮，指示了太阳光的照射位于这一方向，即来自南方，这与其他数据得到的初步结论相符。图中前景石块群和远景石块群之间存在石块数量相对较少的区域，其延伸方向与图 3 低空拼接图中沟渠地貌的大体指向相匹配。图中右下角的亮斑是 DISR 地表科学照明灯的照明。

将图 4 中场景的亮度、颜色与地球上类似的场景进行比较会是非常有趣的。在地球表面上，太阳直射亮度是满月照射亮度的约 50 万倍。而土卫六表面的亮度是地球上太阳直射亮度的约 1,000 分之一，或者约为地球满月照射亮度的 500 倍。即，这一照明水平相当于地球上日落 10 分钟后的情形。土卫六天空和景观呈现橙色，这是由于土卫六上的雾霾对蓝光的衰减作用远大于对红光的作用。因此，如果太阳高悬于天空，看起来就像一个非常小的亮点，是地球上看到的日面的 10 分之一，其尺寸和亮度与在约 150 m 远的地方看一辆小汽车的前灯的情况相当。太阳在土卫六表面投射下清晰的阴影，但是对比度较低，因为大约 90% 的照明来源于天空。如果太阳在天空的位置较低，则无法看到太阳。

对图 4 中 50 多块石头的尺寸进行统计，发现它们的直径尺寸介于 3 mm（成像仪分辨率的极限）到 15 cm 之间，没有发现大于 15 cm 的石块。探测器着陆前的最后一组图像拍摄于 200 ~ 300 m 高度，其分辨率足以识别出米量级的物体，然而在 40 m × 35 m 的视场中并没有发现任何这一尺寸的物体。*R* 值通常被用于表现陨石坑及其周围的溅射物的尺寸–密度分布，在这里表征了土卫六地表被给定尺寸的岩石

a given size frequently used to describe the size distribution of impact craters or crater ejecta. A larger fraction of the surface is covered with rocks greater than 5 cm as opposed to smaller pebbles. The dominance of the cobbles 5–15 cm in size suggests that rocks larger than ~15 cm cannot be transported to the lakebed, while small pebbles (< 5 cm) are quickly removed from the surface. Figure 8 confirms the visual impression given by Fig. 4 that the surface coverage of rocks in the foreground of the image (< 80 cm horizontal distance from the probe) is higher than in the region beyond (about 80–160 cm). However, this trend is not seen for the pebbles less than 5 cm in size.

Elongated dark trails aligned with the general trend of the possible stream-bed visible in the centre of Fig. 4 extend from several of the distant boulders. The direction of the trails agrees with the general northwest–southeast alignment of the stream-like features shown in Fig. 3, because the last upward-looking spectra indicate that the probe settled with DISR facing southward. Images taken from the surface show no traces of the landing of the probe. The viewing direction is probably generally not downwind (the parachute is not visible).

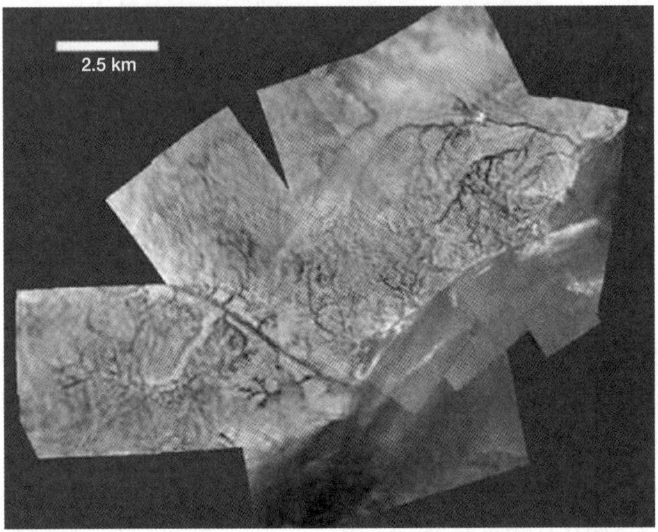

Fig. 5. View of "shoreline" and channels. Panoramic mosaic projected from 6.5 km. showing expanded view of the highlands and bright-dark interface. As in previous figures, north is up; scale indicated. Branching and rectilinear channel networks of dark lanes are shown along an albedo boundary approximately 12 km long.

When coupled with the shapes, size selection and layering of the stones in Fig. 4, the elongated islands and their orientation parallel to the coastline in Fig. 1, the stubby and dendritic channel networks, as well as the ponds in Fig. 2 and Fig. 5, the major elements of the Titan surface albedo variations can be interpreted to be controlled by flow of low-viscosity fluids driven by topographic variation, whether caused by precipitation (the dendritic networks) or spring-fed flows (the stubby networks). We thus interpret the bright–dark albedo difference as follows: irrigation of the bright terrain results in darker material being removed and carried into the channels, which discharge it into the region offshore,

覆盖的比例，图 8 给出了 R 值。土卫六表面大部分被尺寸大于 5 cm 的岩石覆盖而不是更小的卵石，5 ~ 15 cm 的鹅卵石占主体，表明尺寸大于 ~15 cm 的石块不能够被运移到湖床，而尺寸较小的卵石（< 5 cm）则会很快地被从地表运移走。图 8 中 R 值的分布进一步证实了图 4 给人的视觉印象，即图像前景区域（与探测器的水平距离小于 80 cm）的石块覆盖率高于更远的区域（大约在 80 ~ 160 cm）。然而，尺寸小于 5 cm 的石子并没有表现出这种趋势。

从远处几块巨砾延伸出拉长的暗色尾迹，其延伸方向与图 4 中心可辨的疑似河床地形的总体走势平行。同时，尾迹的方向与图 3 中河流状地貌西北–东南的大体走向也相符。因为最后一批朝上拍摄的光谱显示探测器着陆后 DISR 是面向南方的，DISR 在土卫六表面拍摄的图像没有显示探测器着陆的痕迹。视野方向可能不是大体上的下风向（视野内并没有看到降落伞）。

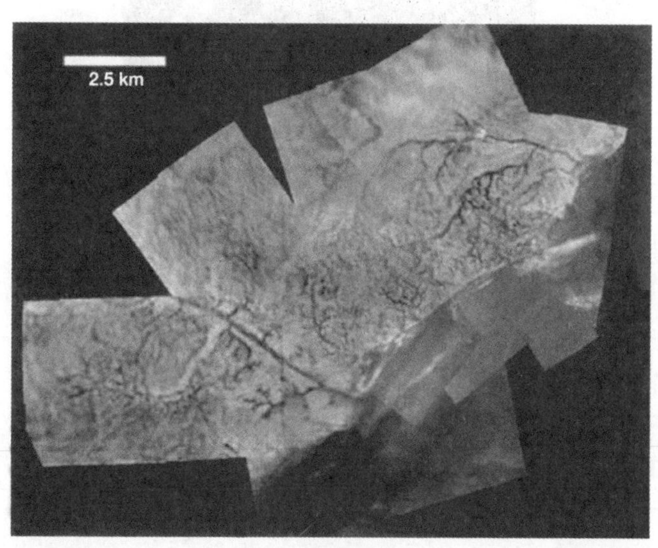

图 5."海岸线"和河道的图像。6.5 km 高度投影得到的全景拼接图。该图对高地区域和明暗分界进行了放大。与前面的图像一样，其正上方为北，比例尺标于图中。图上展现了沿大约 12 km 长的反照率边界分布的暗色河道的分支和直线网络。

综合图 4 中石块的形状、尺寸筛选以及层状分布，图 1 中和海岸线平行的拉长的岛屿，以及图 2 和图 5 中的断株状、树枝状河道网络及池塘，可以认为形成土卫六表面反照率变化的主要因素是地形起伏驱动的低黏度流体的流动，这些流体可能是源于降雨（对应树枝状河道网络）或源于泉（对应断株状河道网络）。因此，我们对明暗反照率的不同作以下解释：对明亮地带的冲刷将导致较暗物质被带离并被搬运进河道。河流在离岸区域将这些暗色物质卸载，因此这些区域变暗。如阵风等风成

thereby darkening it. Eolian processes such as wind gusts coupled with Titan's low gravity (compared to Earth) may aid this migration.

The dark channels visible in the lowest panorama (Fig. 3) seem to suggest south-easterly fluid flow across the lower plane, depositing or exposing the brighter materials (water ice?) along the upstream faces of the ridges.

Stereographic rendering of the dendritic channels just north of the probe landing site (Figs 6 and 7) shows that the slopes in bright terrain being dissected by the putative methane river channels are extremely rugged; slopes of the order of 30° are common. This suggests relatively rapid erosion by flows in the river beds, resulting in the deeply incised valleys. Erosion by steep landslides on slopes approaching the angle of repose is probably the primary mechanism by which the rugged topography is formed. Figure 7 shows two stereographic views of the shoreline and hillside north of the landing site.

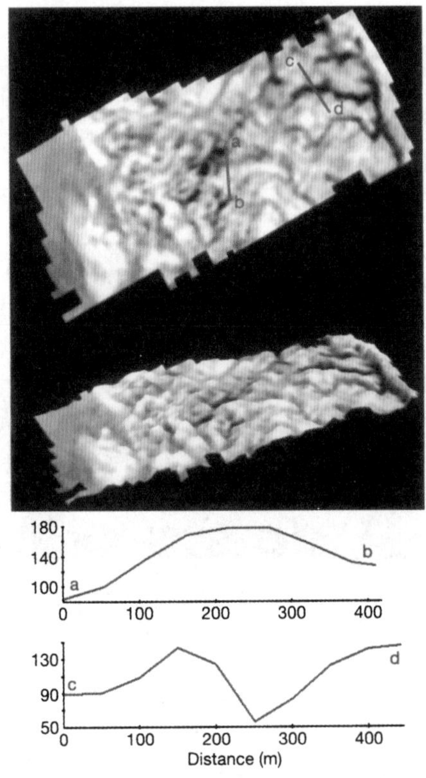

Fig. 6. Topographic model of highland region ~5 km north of the Huygens landing site. The top panel shows an orthorectified HRI image from stereo pair (vertical view). The middle panel shows a perspective view of the topographic model with ~50° tilt angle. No vertical exaggeration was applied (it is 1:1). The bottom panel shows profiles (a–b and c–d from the top panel) that illustrate the extremely steep topography in the region dissected by the drainages. All dimensions are in metres. A DISR stereo pair (using HRI frame 450 and MRI frame 601) was photogrammetrically analysed using a digital stereo workstation. The overlapping area of stereo coverage is about 1 × 3 km; the convergence angle is ~25°. The coincidence of the drainage patterns with the valley floors gives confidence in the reality of the topographic model;

过程和土卫六(相比地球)较低的引力相结合可能有助于这一物质运移过程。

从图 3 全景图像中的暗色河道看，流体是流向东南方向，穿过低地平原，沿山脊面向上游的坡面沉积下或暴露出明亮的物质(水冰?)。

图 6 和图 7 给出了探测器着陆点北侧的树枝状河道的立体影像，显示被假定的甲烷河流分隔的明亮地带中的斜坡是非常崎岖的，其坡度一般在 30° 的量级。这意味着河床遭受了流体相对快速的侵蚀，从而形成了深切河谷。河岸两侧的斜坡经过滑坡作用的侵蚀接近休止角，这种侵蚀过程可能是形成崎岖地形的主要机制。图 7 给出了着陆点北侧海岸线和山坡的立体影像。

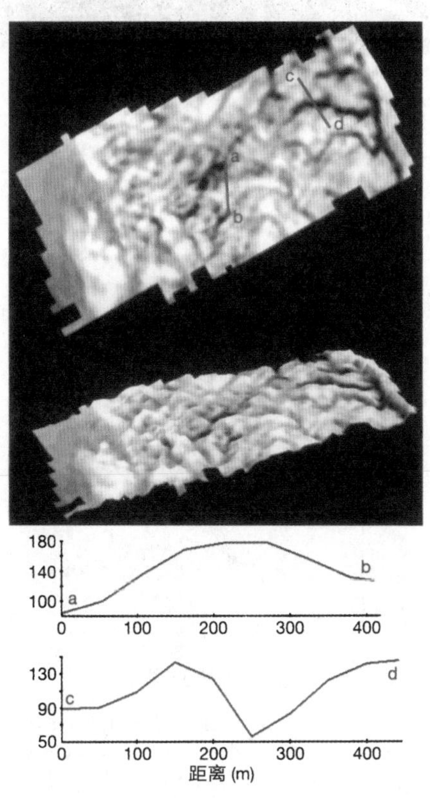

图 6. 惠更斯号着陆点北侧约 5 km 处高地区域的地貌模型。其中，上图表示一对立体图(垂直视图)经过正射校正合成的 HRI 图像；中图表示以约 50° 倾角观察地貌模型的透视图，其垂直方向比例尺没有放大(1∶1)；下图给出了上图中从 a 到 b 以及从 c 到 d 点的地形剖面，展示了被河流系统所分割区域的极为陡峭的地形，三维坐标的单位都是米。我们借助数字立体影像工作站，对两幅 DISR 立体图像(HRI 第 450 帧和 MRI 第 601 帧)进行了摄影测绘分析。两幅立体图像的交叠面积大约为 1 km × 3 km，会聚角约为 25°。河流系统的分布与谷底位置的相符使我们相信该地貌模型的真实性，其中高度的精确

the height accuracy is ~20 m. This preliminary model has been arbitrarily levelled so that the elevation differences are only relative.

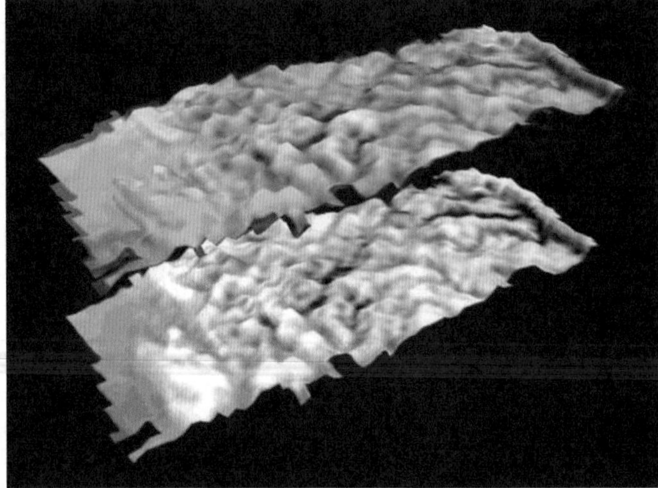

Figure 7. Titan's surface. Perspective view of Titan's surface using a topographic model of the highland region ~5 km north of the Huygens probe landing site derived from the DISR images. The model in greyscale and false colour shows the elevation (pale white highest). The lowland plane or lakebed is to the left side of the display (in blue); the northern highlands (with the dendritic channels) is to the right.

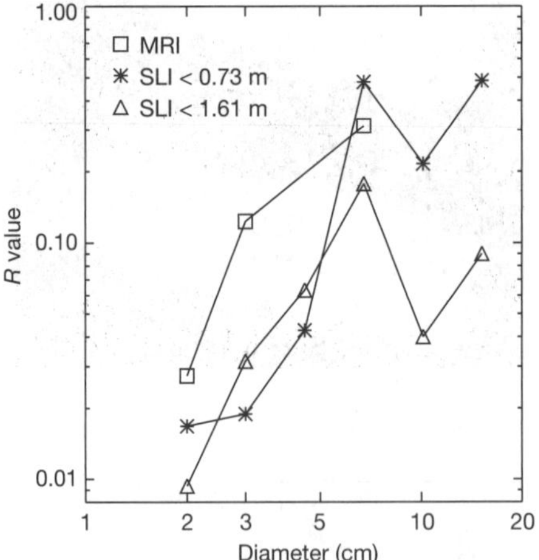

Fig. 8. Distribution of rock on the surface. Rocks larger than 1.63 cm as an R-plot, frequently used to describe size distribution of impact craters or crater ejecta. If N is the number of rocks per centimetre increment of rock size, the fraction of the surface area A covered by rocks with diameters between d and $d+\Delta d$ is approximately $N \times \Delta d \times d^2/A$. By keeping the size bin Δd proportional to the diameter d, the quantity $N \times d^3/A$ (the R value) is also proportional to the surface fraction covered by rocks of diameter d (with a proportionality constant of ~3). The plot shows R values derived from rock counts from the SLI and MRI surface images. For the SLI, R values from counts up to a distance from the probe of 73 cm and up to 161 cm are presented in separate curves. The comparison between the two curves suggests that the count is complete in the displayed size range. The increase of the R value with size corresponds to a higher fraction of the surface covered with large rocks than with smaller ones.

度约为 20 m。值得注意的是，这一初步模型的基准面高度是随意选取的，所以高程的变化是相对的。

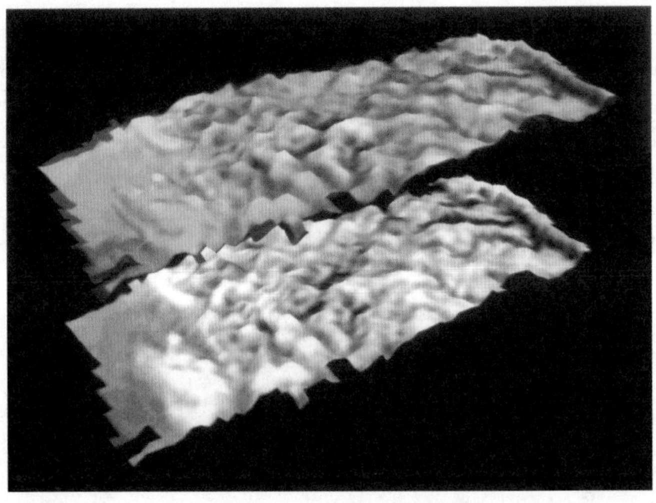

图 7. 土卫六表面的图像。由 DISR 图像得到的、惠更斯号探测器着陆点北侧约 5 km 处高地区域的地貌模型的透视图。其中，模型中的灰度和假彩色表示地形的高度（浅白色代表最高）。图中左侧的蓝色区域代表低地或者湖床，而右侧是布满树枝状沟渠的北方高地。

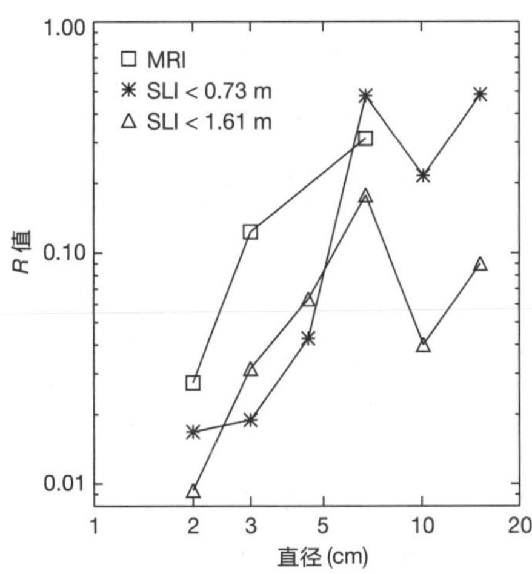

图 8. 地表岩石分布。图中给出了尺寸大于 1.63 cm 的石块的 R 值图，其中 R 通常被用于描述陨石坑及陨石坑周围溅射物的尺寸分布。如果用 N 表示石块尺寸每增加一厘米间隔范围内的石块数量，那么直径介于 d 和 $d+\Delta d$ 之间的石块在大小为 A 的表面积内的覆盖比例为 $N\times\Delta d\times d^2/A$。如果令尺寸间隔 Δd 正比于石块直径 d，则 R 值的大小 $N\times d^3/A$，也正比于直径为 d 的石块覆盖的面积比例，比例常数约为 3。图中 R 值是通过统计 SLI 和 MRI 表面图像中的石块得到的。对于 SLI，距离探测器 73 cm ~ 161 cm 的石块统计结果由单独的一条曲线表现。通过比较这两条曲线可知，计数统计在所示的尺寸范围内都是完整的。图中 R 值随石块尺寸增加而增加，符合大石块的表面覆盖率高于较小石块的现象。

The Wind Profile

Assembly of the panoramic mosaics leads to the construction of a descent trajectory as part of an iterative process of image reconstruction. The trajectory can be used to derive the probe ground track and extract the implied wind velocity as a function of altitude. Correlation of the roughly 200 usable images acquired by DISR during its descent resulted in longitude and latitude values versus time, displayed in Fig. 9. Fitted by polynomials, these ground tracks were differentiated with respect to time and scaled to derive the horizontal wind speed and direction as functions of altitude.

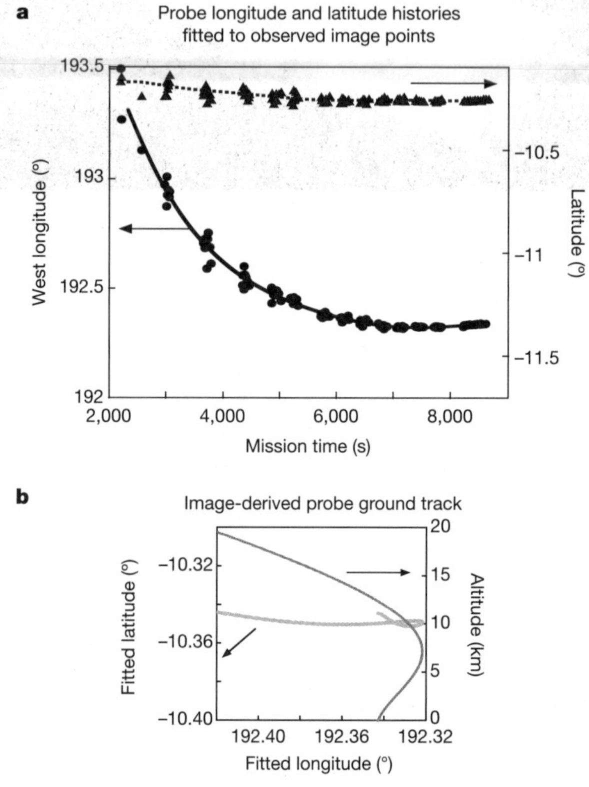

Fig. 9. Probe ground track. **a**, Sub-probe west longitude and latitude histories of the Huygens probe derived from panoramic image reconstructions. Arrows indicate the appropriate vertical axis. The image data points from which the latitude and longitude were derived are shown as triangles and dots respectively. The dotted and solid lines show polynomial fits to the data. Results adjusted to agree with the Descent Trajectory Working Group (DTWG-3) values at 2,200 s after T_0 (mission time). **b**, Probe longitude versus latitude (thicker line) and versus altitude (thinner line). The altitude axis needs to be expanded by a factor of almost six to recover one-to-one correspondence on a linear scale because the total longitudinal variation is less than 4 km. Using the Doppler Wind Experiment's (DWE) high-altitude references[33], the touchdown point (the predicted landing site) was extrapolated to west longitude 192.34°, latitude −10.34°. Using DTWG-3 high-altitude references, the touchdown point was extrapolated to 192.36°, latitude −10.36°.

The results indicate that the probe's steadily increasing eastward drift caused by Titan's

风况描述

构建全景拼接图像的同时可以构造出质量较好的探测器降落过程的空间运动轨迹，这也是图像重建迭代过程的重要部分。通过轨迹可以推导出探测器轨迹在地面上的投影，并且可以提取出不同高度的风速信息。我们将 DISR 在下降阶段获取的大约 200 幅有用图像相互关联，以此计算出探测器的经度和纬度随时间的变化，如图 9 所示。我们对轨迹地面投影进行多项式拟合，然后对时间做微分并换算成实际空间距离，可以得出水平的风速和风向随高度变化的函数。

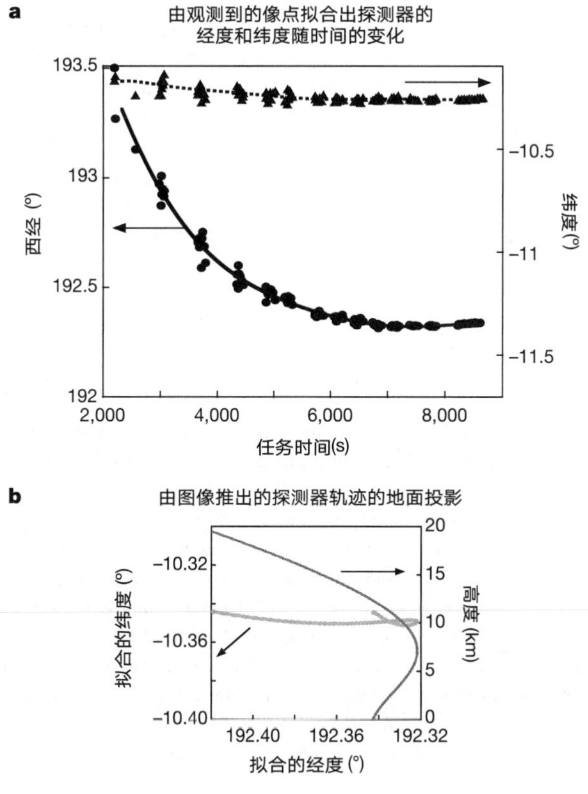

图 9. 探测器轨迹地面投影。**a**, 根据全景图像重建获得的惠更斯着陆探测器的西经和纬度随时间的变化。图中的箭头指向各自对应的纵轴，三角和圆点分别表示用于推出纬度和经度的图像数据点。图中的点线和实线是对以上数据的拟合曲线。结果已经经过校正以符合下降轨迹工作组（DTWG-3）在 T_0（任务时间）之后 2,200 s 的值。**b**, 探测器经度与纬度（粗线）以及经度与高度（细线）的关系。由于经向距离总变化小于 4 km，与高度对应的轴需要放大接近 6 倍才能实现比例尺上的一一对应。使用多普勒风实验仪（DWE）中的高空数据作为参照 [33]，推算出降落点（预计的着陆点）为西经 192.34°，纬度为 −10.34°。采用 DTWG-3 的高空数据作为参照，则推算出着地点位于西经 192.36°，纬度 −10.36°。

以上结果表明，探测器从 50 km 高度下降至 30 km 高度过程中，由于土卫六风

prograde winds, as shown in Fig. 10, slowed from near 30 to 10 m s^{-1} between altitudes of 50 and 30 km and slowed more rapidly (from 10 to 4 m s^{-1}) between altitudes of 30 and 20 km. The winds drop to zero and reverse at around 7 km, near the expected top of the planetary boundary layer, producing a west-northwestwardly motion extending for about 1 km during the last 15 min of the descent (see Fig. 3). The generally prograde nature of the winds between 50 and 10 km agrees with models of Titan's zonal winds available before the arrival of Cassini or the Huygens probe[7], although the wind speed is somewhat less than the average predicted before entry.

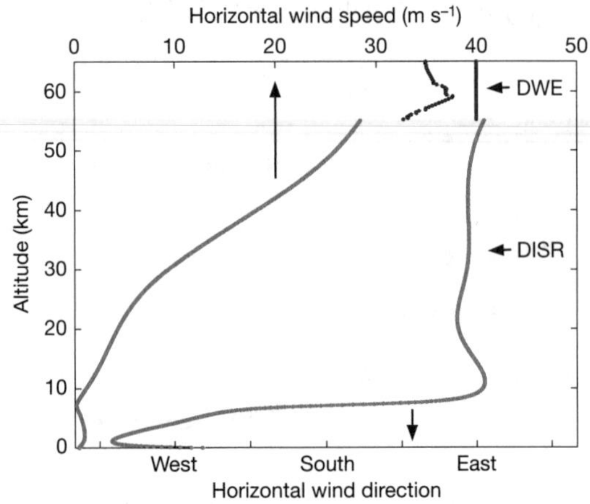

Fig. 10. Observed winds. Horizontal wind speed and direction (counter-clockwise from east) as a function of altitude. The green lines are the DISR data and the blue lines are the high altitude DWE data[33] (showing reasonable consistency between the two). The lines on the left show the wind-speed profile, and on the right is the wind direction. The wind is computed from the combined longitude and latitude reconstructions displayed in Fig. 9. Titan's prograde winds slow from about 28 m s^{-1} at 50 km to 10 m s^{-1} near 30 km altitude, then decrease more rapidly from 30 km (10 m s^{-1}) down to 7 km where they drop to zero. Below 7 km (which is near the expected top of the planetary boundary layer) the winds reverse and become retrograde, and the speed increases to about 1 m s^{-1} around 2–3 km before dropping to almost zero (~0.3 m s^{-1}) near the surface. The direction begins as due east, and then turns through south (beginning between 9–7 km) to the west-northwest between 7–5 km. The winds are extrapolated to be retrograde at the surface, but the two-sigma error bars (not shown) of 1 m s^{-1} at the surface could include surface prograde winds. The error bars at 55 km altitude (4 m s^{-1}) are consistent with continuity from the DWE measurements.

The planetary boundary layer is calculated to have a thickness of between 4 and 8 km, based on scaling the Earth's near-equatorial planetary-boundary-layer thickness of 1–2 km by the inverse square-rooted ratio of the planetary rotation rates. The minimum horizontal wind speed at 7 km can thus be an indication of entry into the boundary layer[8]. The reversal of wind direction at this altitude is also consistent with the Voyager-derived equatorial temperature profile[9], wherein the temperature gradient changes from dry adiabatic to a sub-adiabatic temperature gradient above 4 km altitude, indicating the top of the boundary layer.

The current ground-track and wind-speed analysis predicts winds of about 0.3 to 1 m s^{-1} near the surface. This velocity can be produced by any number of sources including pressure and temperature gradients and tides[10].

的顺行影响平稳地向东飘移，如图 10 所示，但速度从大约 30 m/s 降至 10 m/s；在从 30 km 高度下降至 20 km 高度过程中，速度迅速地从 10 m/s 降至 4 m/s。当探测器降至大约 7 km 高度（靠近预测的行星边界层顶部）时，风速减为零，然后风向改为相反的方向，造成探测器在降落阶段的最后 15 分钟朝西北偏西的方向飘移了约 1 km，如图 3 所示。50 km 到 10 km 高度的风向和土卫六转动的方向相同，这与卡西尼或惠更斯探测器到达前就已提出的土卫六纬向风模型相一致 [7]，但风速比探测器进入前的预测平均值略小。

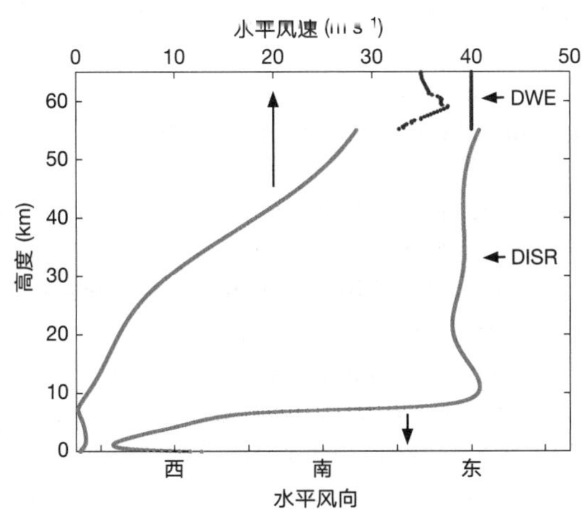

图 10. 观测到的风况。图中给出了水平风速和风向（自东沿逆时针方向）与高度的关系。其中，绿线代表 DISR 数据，而蓝线代表高空 DWE 数据 [33]，二者之间显示出合理的一致性。图中左侧曲线给出了风速剖面，而右侧曲线给出了风向，这两条曲线由图 9 中的经纬度信息重建得到。土卫六上的顺行风的风速由 50 km 高度处大约 28 km/s 下降到 30 km 高度附近约 10 m/s，最后在 7 km 高度迅速降为 0。在 7 km（靠近预期的行星边界层顶部）以下，风向则变为逆行方向，并且风速在 2 ~ 3 km 高度增加到 1 m/s 左右，在到达表面附近时则几乎为 0（~0.3 m/s）。风向首先向正东，然后转向南（起始于 9 ~ 7 km），并在 7 ~ 5 km 转为西北偏西。尽管地表附近风向的推断结果是反向风，但地表附近 1 m/s 风速的 2σ 误差范围（未在图中显示）包含了一部分顺向风的区间。55 km 高度处（对应风速 4 m/s）的误差范围与 DWE 测量得到的连续性相一致。

对土卫六和地球的旋转速率比值的平方根取倒数，并乘以地球赤道附近的行星边界层厚度（1 ~ 2 km），就可以计算出土卫六的行星边界层厚度为 4 ~ 8 km。7 km 高度处的最小水平风速，可以被认为是进入行星边界层的信号 [8]。而在 7 km 高度处的风向改变也与旅行者号飞船得出的赤道温度剖面相符 [9]。旅行者号的测量表明，在大于 4 km 的高度，温度梯度从干绝热变化到亚绝热，这意味着边界层的顶部。

目前，通过对探测器轨迹地面投影和风速分析，预测土卫六地表附近的风速大约为 0.3 ~ 1 m/s，这一速度可能是由气压、温度梯度和潮汐等因素引起 [10]。

Migration of Surface Material

The acquisition of visible spectra at known locations in the images allowed correlation of the reflectance spectra with different types of terrain. The Downward-Looking Visible Spectrometer (DLVS) was an imaging spectrometer measuring light between 480 and 960 nm as it projected the image of the slit onto the ground into up to 20 spatial resolution elements for nadir angles from 10° to 50°. Spectra were collected at nominally the same azimuths as the images, though often at slightly different altitudes (on different probe rotations). Interpolation between the times at which the spectral and image data were obtained located the spectra within the images. Determination of the surface reflectivity was hindered by scattering from the haze between the camera and the surface as well as by methane absorption. Correlation of the spectra with images was therefore best performed on measurements during which the altitude changed only slightly.

The centre of the image in Fig. 11 is displayed in true colour (that is, as the human eye would see it under Titan's atmosphere) using actual spectral data from one panorama. The area between the spectra is interpolated in azimuth. The coverage with spectra is similar to that shown in Fig. 12. The orange colour is due mainly to the illumination of the surface. Scattering and absorption (which dominate in the blue) cause the perceived true colour of the surface to change from yellow to orange with decreasing altitude. Note that the passband of the cameras peaked in the near infrared (at 750 nm), and therefore the brightness variations in the images would not necessarily be seen by the human eye.

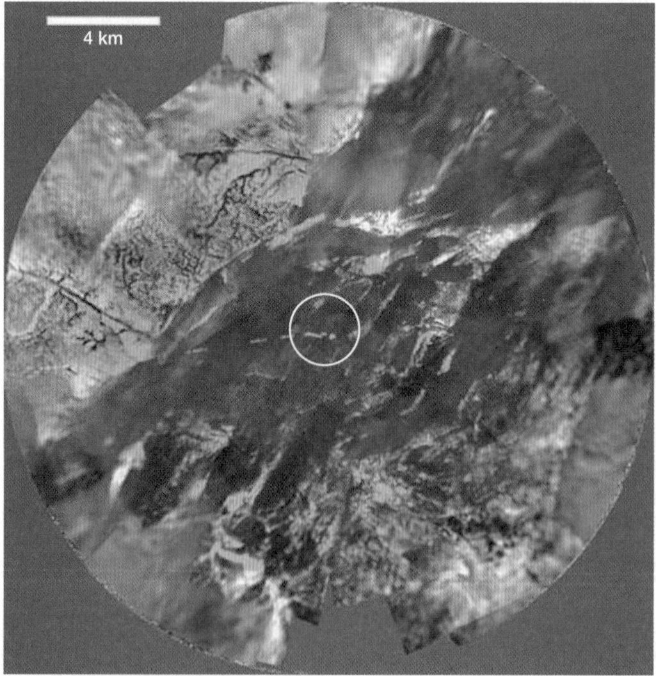

Fig. 11. The surface of Titan displayed in true colour. As seen from an altitude of 8 km. See Fig. 12 for the location of the spectrometer's footprints. Some bright features appear to be overexposed because they are

地表物质的运移

通过采集图像中已知位置的可见光谱，可以将反射光谱与不同类型的地形建立联系。探测器采用俯视可见光谱分析仪（DLVS），是一种用于测量 480 ~ 960 nm 波段光线的成像光谱分析仪，它将分析仪狭缝的像投影到地面上，在 10°到 50°的天底角最多可以覆盖 20 个空间分辨单元。虽然标称可以测量与图像相同方位角的光谱，但光谱与对应图像的采集高度往往略有不同（探测器未发生明显的偏转）。通过对光谱和图像数据的采集时刻进行内插分析，可以在图像中确定光谱数据对应的覆盖区域。照相机和土卫六地表之间的雾霾的散射作用和甲烷吸收的影响阻碍了地表反射率的确定。因此，测量过程中探测器高度改变较轻微的情况下，光谱和图像之间会有较好的相关性。

借助全景照片的真实光谱数据，图 11 中心附近的图像可以以真实彩色呈现（即和人眼在土卫六大气下实际看到的颜色一样）。光谱之间的区域按方位进行内插。光谱覆盖范围类似于图 12 所示。橙色的色调主要是由于地面被照亮。随着高度的降低，散射和吸收（主要作用在蓝色波段）将使地面的真实颜色从黄色变为橙色。需要注意的是，照相机的滤波通带峰值处在近红外 750 nm，因此，人眼实际观察不一定看到图像中那样的亮度变化。

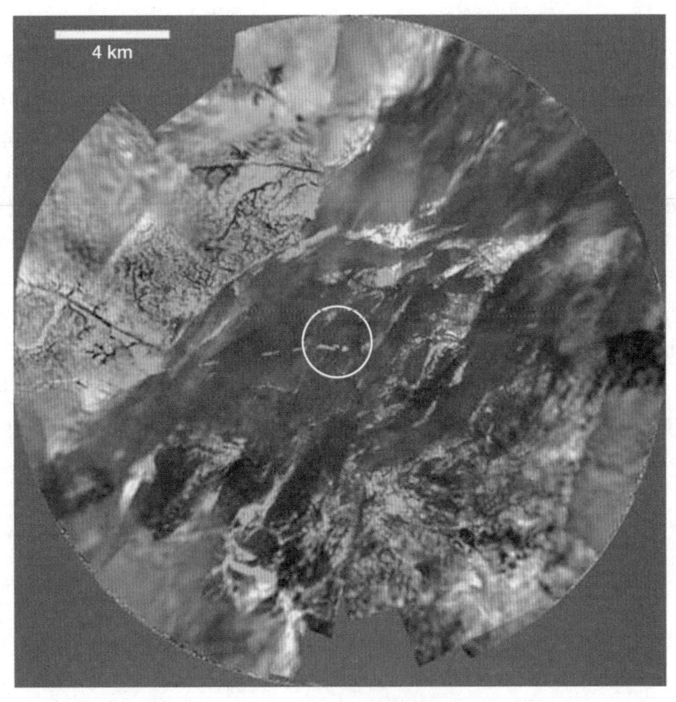

图 11. 真实颜色下的土卫六表面图像。从 8 km 高度看到的图像，图 12 标注了光谱仪覆盖区域的位置。图中亮度主要是根据近红外光谱得出的，部分地貌单元由于颜色过亮而产生过曝。真实彩色表达为红绿

too bright for their colour (the brightness in this image mainly derives from the near-infrared spectrum). True colour is expressed in red-green-blue (RGB) values that are derived by multiplying the spectra with the Commission Internationale de l'Eclairage colour-matching functions (with the 6,500-K correlated colour temperature, the D65 white point). The circle shown is the extent of the lowest panorama (Fig. 3).

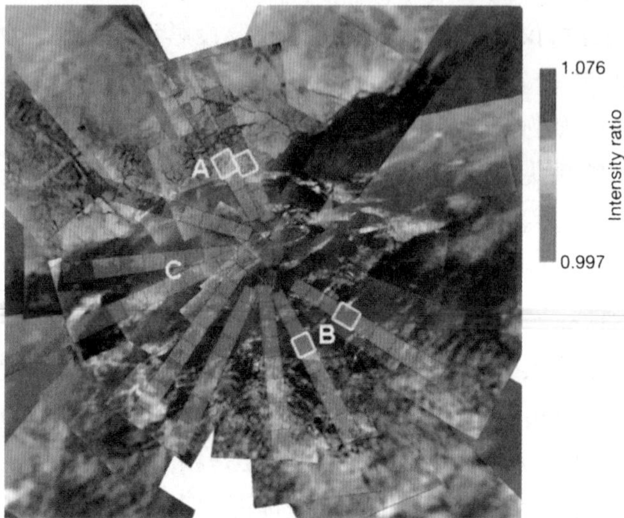

Fig. 12. Reflectivity samples of Titan's surface. A panorama of Titan's surface overlaid with DLVS footprints coloured according to the 827 nm/751 nm intensity ratio, coded from red (high) to green (low). Spectral footprints (the small rectangles) selected for analysis in Fig. 13 are outlined in white. The panorama shows an area of 23 by 23 km. Areas A, B and C are referred to in Fig. 13.

In Fig. 12 the images are correlated with the ratio of the intensity in two methane windows (827 nm/751 nm) located in the infrared part of the spectrum where scattering is minimal and the systematic variability with nadir angle can be ignored. Reddening (high 827 nm/751 nm ratio) is concentrated at the area covered with drainage channels (north and northwest in the pan of Fig. 12) and to a lesser degree to the lake area adjacent to the coastline. The lake area in the southeast is not reddened. A preliminary analysis of spectra recorded in other panoramic cycles indicates that the land area in the northeast, which is not covered by drainage channels, is only moderately reddened compared to the river area. The reddening is not restricted to these two methane windows. Figure 13 shows that, in fact, it is present over the whole visual range, amounting to about 6% per 100 nm (note that atmospheric backscatter dominates over surface reflection at wavelengths below 600 nm).

蓝(RGB)颜色值，这些值是由光谱乘以国际照明委员会颜色匹配函数(具有 6,500 K 色温，D65 白点)得到的。图中圆圈给出了最低全景图像(如图 3)的范围。

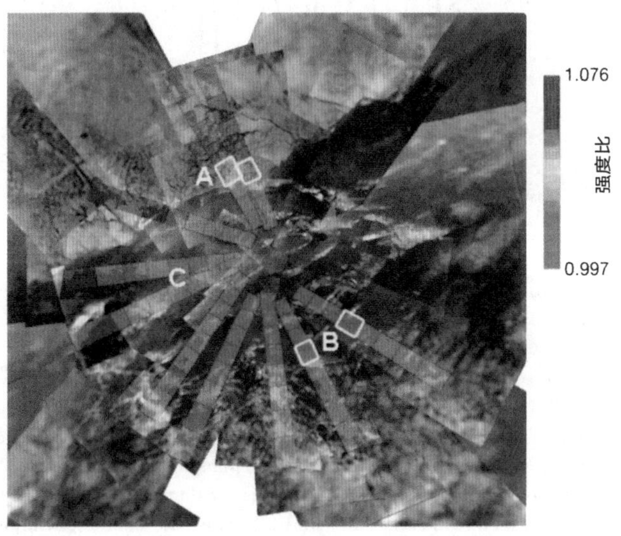

图 12. 土卫六地表的反射率样例。图中将土卫六表面的全景图像与 DLVS 的覆盖区域叠加在一起，其中 DLVS 覆盖区域按照 827 nm/751 nm 两波段强度之比的大小着色，即比值越大则为红色，比值较小则为绿色。图 13 分析的光谱覆盖区用白色矩形框标记出。全景图像的面积为 23 km × 23 km。而区域 A、B、C 光谱具体情况可以参考图 13。

在图 12 中，图像与两个甲烷窗口信号(827 nm/751 nm)的强度之比关联起来，这两个窗口位于红外波段，在该波段散射效应最小，并且天底角的系统性变化可以忽略。红化(即高 827 nm/751 nm 比值)集中在排水河道分布的区域(图 12 中的北侧和西北侧)；在靠近海岸线的湖区也存在相对较弱的红化。位于东南侧的湖区没有红化。我们对其他全景拍摄周期记录的光谱数据进行了初步分析，发现东北方向的陆地区域(未分布排水河道)相对于河流区域只是被中等红化。红化现象并不是局限在两个甲烷观测窗口。如图 13 所示，实际上红化存在于整个可见光波段，相当于每 100 nm 大约为 6% 的变化(值得注意的是对于小于 600 nm 的波段，大气背散射的影响显著地强于表面反射)。

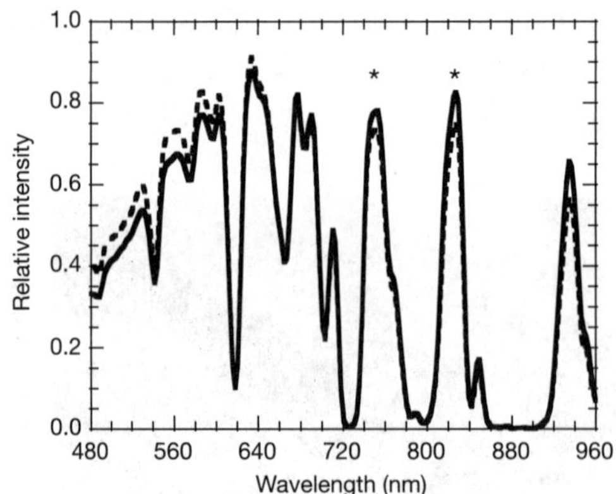

Fig. 13. Spectral comparison of bright highlands and dark lowlands. Spectra of the dendritic river highlands and lakebed lowlands areas are compared. To restrict the influence of the atmosphere, only spectra recorded at the same nadir angle are selected. The solid and dashed curves are the average spectra associated with the two spectral pixels outlined in white near the areas marked "A" (dendritic highlands, solid) and "B" (dark lakebed, dashed), respectively in Fig. 12. The two pixels bordering "C" in Fig. 12 yield a spectrum intermediate to the "A" and "B" spectra. The spectra have been corrected for albedo by dividing by the total intensity to emphasize the difference in slope. Not shown here is that the reflectivity of the dendritic area ("A") is higher than that of the lakebed area ("B") at all wavelengths by roughly a factor of two. The asterisks denote the methane windows taken for the reddening ratio in Fig. 12.

The DLVS data clearly show that the highlands (high-albedo area) are redder than the lakebed (low-albedo area). Spectra of the lakebed just south of the coastline are less red than the highlands but clearly more red than the lakebed further away (that is, to the southeast). The data suggests that the brighter (redder) material of the hilly area may be of local origin, and is corrugated by rivers and drainage channels, and that the darker material (less red) is a substance that seems to be washed from the hills into the lakebed. It could be connected to the alteration of the highland terrain, either by precipitation, wind and/or cryoactivity. Additionally, it could indicate that the surface of the lowland area may be covered by different materials in regions that exhibit diverse morphology.

Surface Reflectivity and Methane Mole Fraction

Spectra taken near Titan's surface allow measurement of its reflectance and determination of the local methane mole fraction. These measurements provide clues as to the make-up of Titan's crust. The Downward-Looking and Upward-Looking Infrared Spectrometers (DLIS and ULIS) cover the region from 840 to 1,700 nm with a resolution of 15 to 20 nm. The ULIS looks up at a half-hemisphere through a diffuser while the DLIS projects its slit into a 3 by 9° field centred at a 20° nadir angle. Below 700 m altitude, a 20-W lamp was turned on to illuminate the surface at wavelengths where solar light had been completely absorbed by methane in Titan's atmosphere. At low altitudes we took repeated DLIS and ULIS spectra at short integration times (1 s). Nine DLIS and seven ULIS spectra were received between 734 and 21 m altitude. The DLIS continued to measure surface spectra free of atmospheric

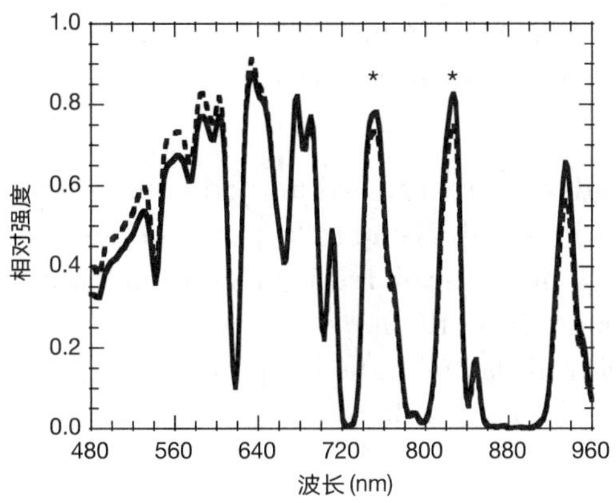

图 13. 较亮高地和较暗低地的光谱对比。我们在图中将分布有树枝状河流的高地的光谱与湖床低地区域的光谱进行了对比。为了限制大气的影响，我们只挑选了以相同天底角采集的光谱。图中实线（虚线）分别表示图 12 中 A(B) 区域附近白框标示的两个光谱像素的平均光谱，A 区域是树枝状高地，用实线表示；B 区域是暗的湖床，用虚线表示。而图 12 中 C 区周围的两个像素的平均光谱介于 A 和 B 光谱之间。图中光谱已经通过除以总亮度进行了反照率校正，从而可以突出光谱斜率的不同。图中未显示的信息是，在整个波段范围，树枝状区域（A 区域）的反射率是湖床区域（B 区域）的 2 倍左右。图中星号表示两个甲烷吸收窗波段，这两个波段用于表征图 12 中的红化程度。

DLVS 数据清楚地表明，具有高反照率的高地比低反照率的湖床红化程度更高，紧邻海岸线南侧的湖床的光谱比高地的红化要弱，但比离岸更远些（即东南方向）的湖床红化明显更强。数据说明丘陵地带的更亮（更红）物质可能是在原位形成的，并在河流的作用下形成波状褶皱结构，而较暗（更不红）的物质则看起来像是被从山地冲洗进湖床的。这可能与高原地形的改造有关，可能是降雨、风和（或者）冰冻等机制造成的。此外，低地地区的一些地区展现出地貌的多样性，这些区域的表面可能覆盖着不同的物质。

表面反射率与甲烷摩尔分数

在接近土卫六表面得到的光谱使我们可以测量其反射率，并确定局部的甲烷摩尔分数。这些测量计算可以为确定土卫六地壳组成提供线索。探测器采用的俯视红外光谱分析仪（DLIS）和仰视红外光谱分析仪（ULIS）可以覆盖从 840 nm 到 1,700 nm 的波段，其分辨率为 15 ~ 20 nm。ULIS 可以通过漫射器向上观测整个半球，而 DLIS 则将其狭缝投影到以 20° 天底角为中心、大小为 3° × 9° 的视场。在小于 700 m 高度，一盏功率为 20 W 的照明灯被开启用来照射土卫六表面，其发光波长恰是太阳光被土卫六大气中的甲烷完全吸收的波段。在低空，我们以较短的积分时间（1 s）不断重复测量 DLIS 和 ULIS 光谱。在 734 m 到 21 m 的下降高度之间，探测器总共

methane absorption after landing. About 20 such identical spectra were acquired from a distance of a few tens of centimetres of the surface.

DLIS spectra at all altitudes clearly showed an additional signal when the lamp was on. However, at the highest altitudes, the lamp reflection from the surface was negligible, so the additional signal was solely due to scattered light from the lamp into the instrument. This scattered light was estimated from the intensity level in the strong methane bands in the highest-altitude spectrum recorded with the lamp on, and removed from all DLIS spectra. After this correction, only spectra at 36 m and especially 21 m showed significant signal from the lamp. This signal dominated the upward intensity due to solar illumination in all regions of strong and moderate methane absorption, while the latter dominated in the methane windows.

This spectrum, which represents the product of the ground reflectivity and the two-way methane transmission, is shown in Fig. 14 and compared to synthetic spectra with methane mole fractions of 3%, 5% and 7%. The ground reflectivity assumed in these model calculations is shown in the inset. Four methane bands are seen in the lamp-only spectrum. The good correlation with the models, notably in the weak structures at 1,140, 1,370 and 1,470 nm, indicates a high signal-to-noise ratio of about 50. The best fit is achieved with a methane mole fraction of 5%, which is in firm agreement with the 4.9% *in situ* measurements made by the Gas Chromatograph Mass Spectrometer[11]. Most structures are well reproduced; notable exceptions are the detailed shape of the 1,000 nm band and the absorption shoulder near 1,320 nm.

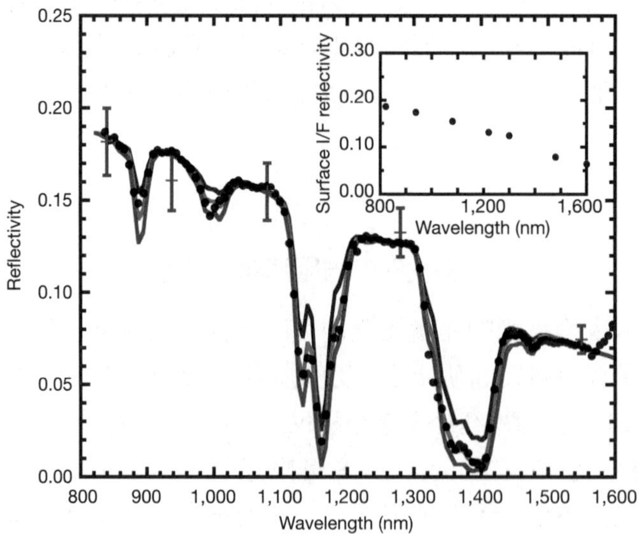

Fig. 14. Derivation of methane mole fraction. Lamp-only downward-looking spectrum from altitude of 21 m (black data points). The red line with three-sigma error bars indicate absolute reflectivity in methane windows estimated from infrared measurements. This spectrum is compared to three models: 3% (blue), 5% (green), and 7% (red) methane mole fractions. These models make use of surface reflectivity at seven

采集了 9 张 DLIS 光谱和 7 张 ULIS 光谱。探测器着陆后，DLIS 在没有大气甲烷吸收影响的情况下继续测量土卫六表面的光谱；在距离表面大约数十厘米的位置获得了大约 20 张这样的光谱。

所有高度的 DLIS 光谱都清楚地显示，当照明灯打开后，出现了一个额外的信号。然而当探测器位于最高的高度时，地面对照明灯光的反射则可以忽略，因此光谱中的额外信号应该仅仅是来自照明灯光的散射。根据照明灯打开后记录的最高高度光谱中的强甲烷吸收带的强度，可以估算出照明灯散射光的强度值，并将其从所有的 DLIS 光谱中剔除掉。经过此校正之后，只有在 36 m，尤其是 21 m 高度处的光谱还存在明显的照明灯信号。该信号决定了向上的光强，这是由于太阳光的所有波段都有强–中等的甲烷吸收，而后者只是在甲烷窗口才起决定作用。

图 14 中给出的光谱是地面反射率与来回两次通过甲烷的穿透率的乘积所给出的结果，并与甲烷摩尔分数分别为 3%、5% 以及 7% 的合成光谱相比较。图 14 右上角的插图给出了计算模型中采用的反射率。在只有照明灯情况下的光谱中可以观察到四个甲烷吸收带。测量数据与模型符合得很好，在 1,140 nm、1,370 nm 和 1,470 nm 位置均出现了较弱的吸收结构，这表明了约为 50 的高信噪比。甲烷摩尔分数为 5% 时，可以实现最佳拟合，这与使用气相色谱–质谱联用仪进行原位测量得到的 4.9% 的甲烷摩尔分数严格一致 [11]。光谱中大多数结构都可以被很好地重现，但 1,000 nm 处吸收带以及 1,320 nm 附近的吸收肩的细节存在明显的差异。

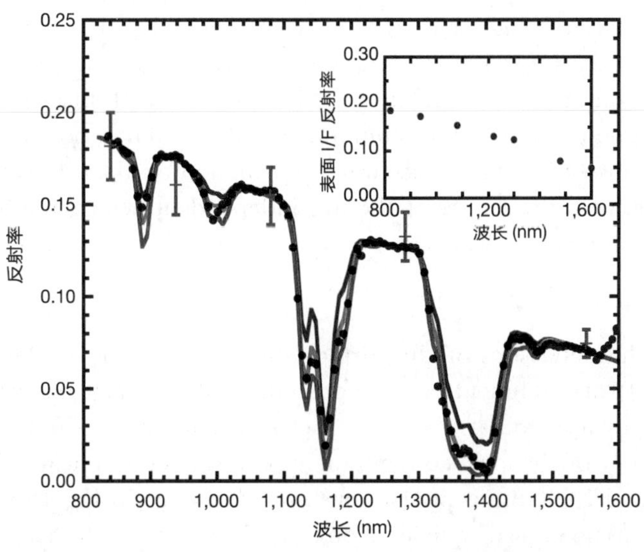

图 14. 甲烷摩尔分数的推导。图中给出了开启照明灯后，在 21 m 高度测量得到的俯视光谱（黑点表示数据点）。红线表示通过红外测量估计得到的甲烷窗口绝对反射率，同时给出了 3σ 误差区间。作为对比，图中也给出了甲烷摩尔分数分别为 3%、5% 和 7% 的三种模型结果，分别用蓝线、绿线和红线表示。这些模型结果通过对 7 个波长处的表面反射率（如图中右上小图中蓝点所示；其中 I/F 表示谱强度与太阳

wavelengths (shown in inset by the blue points; I/F is the ratio of the intensity to the solar flux divided by π) and linearly interpolated between. From the lamp-on infrared spectra, a lamp-only spectrum at 21 m (representing the spectrum observed by DLIS in the absence of solar illumination) was obtained as follows. First, the reflectivity in regions of negligible methane absorption (at 840, 940, 1,070, 1,280 and 1,500–1,600 nm) was estimated by the ratio of mean upward intensity (measured by DLIS) to mean downward intensity (measured by ULIS). The mean upward intensity is the average measured over the seven low-altitude DLIS spectra showing no contribution from the lamp (734 to 53 m). The mean downward intensity was obtained by averaging the strongest intensity with the weakest intensity. This average gives reasonable approximation of the downward flux divided by π. The ratio of the mean upward intensity to the mean downward intensity gives reflectivity. Two corrections were required in this analysis: correction for the spatial response of the ULIS diffuser and correction for the solar illumination in the DLIS 21-m spectrum. The correction for diffuser response ranged from 15% (840 nm) to 25% (1,550 nm), assuming a haze optical depth of ~2 at 938 nm. The contribution of solar illumination in the DLIS spectrum at 21 m was eliminated by subtracting the average of the DLIS spectra at 85 and 109 m where the lamp contribution was negligible. The difference spectrum was then divided by the spectral response of the lamp and scaled by a constant to match the continuum reflectivities inferred previously, producing the lamp-only spectrum at 21 m.

We conclude that the methane abundance is $5 \pm 1\%$ in the atmosphere near the surface. The corresponding (two-way, and including the $20°$ lamp inclination) methane column abundance in the spectrum is 9.6 m-amagat (or a column 9.6 m high at the standard temperature and pressure of 273 K and 1 atmosphere). With a temperature of 93.8 K and pressure of 1467.6 mbar (ref. 12), the relative humidity of methane is about 50% using Brown and Ziegler's saturation law[13]. Therefore, methane near the surface is not near saturation, and ground fogs caused by methane in the neighbourhood of the landing site are unlikely.

The ratio of the observed spectrum to the methane transmission, restricted to spectral regions where the latter is higher than 90%, is shown in Fig. 15a ("plus" symbols). It is compared to one of the DLIS spectra after landing, divided by the lamp spectral response and rescaled by a constant reflectivity factor. Note that although this spectrum shows signs of methane absorption at 1,140–1,160 nm and 1,300–1,400 nm, no attempt of correction was made, in the absence so far of accurate information on the absorption path lengths. The agreement between the shapes of the two independent determinations of the ground reflectivity adds confidence to the result.

The four major characteristics of the surface spectrum are: (1) a relatively low albedo, peaking around 0.18 at 830 nm; (2) a red slope in the visible range; (3) a quasilinear decrease of the reflectivity by a factor of about two between 830 and 1,420 nm; and (4) a broad absorption, by ~30% of the local continuum, apparently centred near 1,540 nm (although its behaviour beyond 1,600 nm is poorly constrained) as seen in Fig. 15b. This spectrum is very unusual and has no known equivalent on any other object in the Solar System.

通量的比值除以 π）进行线性插值得到。根据开启照明灯后的红外光谱可以推出 21 m 高度处只有照明灯信号的光谱（代表了没有太阳光照环境下的 DLIS 观测光谱），具体步骤如下。首先，在某些甲烷吸收可以忽略的波段（如 840 nm、940 nm、1,070 nm、1,280 nm 以及 1,500 ~ 1,600 nm），反射率可以通过向上的平均谱强度（DLIS 测量得到）与向下的平均谱强度（ULIS 测量得到）的比值估计出来。其中，向上的平均谱强度由不存在照明灯光影响的 7 个低高度 DLIS 光谱测量值（734 m 到 53 m）的平均后得到。通过对最大强度和最小强度取平均可以求出向下的平均谱强度；该平均值是向下光通量除以 π 的结果的合理近似。向上的平均谱强度与向下的平均谱强度的比值给出了反射率。在该分析中，需要考虑两种校准：ULIS 漫射器的空间响应的校准和 21 m 高度处 DLIS 光谱的太阳光照的校准。假设雾霾在 938 nm 处的光学深度约为 2，则散射器响应的校准范围介于 15%（对应 840 nm）和 25%（对应 1,550 nm）之间。由于在 85 m 和 109 m 高度处照明灯对 DLIS 光谱的贡献可以忽略，因此通过减去这两个高度的 DLIS 光谱平均值，从而消除 21 m 高度处 DLIS 光谱中太阳光照的贡献。然后用相减后得到的光谱除以照明灯的光谱响应，并乘以一个常数，从而与前面推论的连续谱反射率相符；这样就得到了只考虑照明灯情况的 21 m 高度处的反射率光谱。

因此我们得出以下结论：近地表大气中甲烷的丰度为 5% ± 1%。相应的（双程的且包括了 20 度照明灯倾角的）谱中甲烷柱丰度为 9.6 米阿马加（即相当于 273 K 的标准温度和 1 个标准大气压下 9.6 米高度柱）。在实际温度为 93.8 K 和气压为 1467.6 mbar 的条件下[12]，利用布朗和齐格勒的饱和定律[13]，可以计算出甲烷的相对湿度大约为 50%。因此，地表附近的甲烷并未接近饱和，着陆点附近的雾不太可能是由甲烷形成的。

图 15a 给出了光谱区域内观测到的光谱数据与甲烷穿透率的比值（用加号表示），限制在后者高于 90% 的区域内。图中也从着陆后的 DLIS 光谱中选择了一个与之进行比较，其中该光谱曲线是除以照明灯的光谱响应并利用常量反射率因子重新调整后的结果。值得注意的是，尽管该光谱在 1,140 ~ 1,160 nm 和 1,300 ~ 1,400 nm 波段出现了甲烷吸收的迹象，由于迄今还未获得关于吸收波程长度的精确信息，因此我们并没有尝试对光谱进行校正。由该图可知，通过两种独立方法确定的地面反射率符合得很好，从而进一步提高了结果的可信度。

表面光谱主要有以下四个特征：（1）相对较低的反照率，在 830 nm 达到峰值约 0.18；（2）在可见光波段存在红坡；（3）反射率从 830 nm 到 1,420 nm 准线性地降低了约一半；（4）如图 15b 所示，存在以 1,540 nm 附近为中心的较宽的吸收峰（波段宽度约占局域连续谱的 30%），但是在大于 1,600 nm 的情况未得到很好的约束。该光谱特别不同寻常，在太阳系的其他天体上还未发现这样的情况。

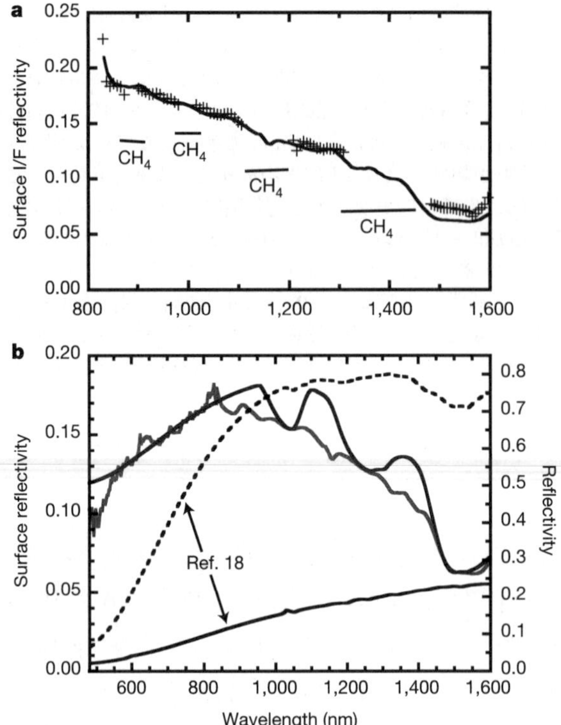

Fig. 15. Reflectance of Titan's surface. **a**, Reflectivity measured from 21 m altitude ("plus" symbols) compared to the reflectivity after landing (solid curve). The methane absorption bands are indicated by the CH_4 symbol. **b**, Surface reflectivity as measured after landing (red line). It is compared with a simulation (blue line) of a mixture of large-grained (750 μm) low-temperature water ice, yellow tholins, and an unknown component with featureless blue slope between 850 and 1,500 nm. Spectra of two different organic tholins: a yellow tholin (dashed line) and a dark tholin (solid black line) from ref. 18 are also shown for comparison (reflectance scale reduced by a factor of 4). We are attempting to identify or synthesize the missing blue material in our laboratory.

Ground-based spectroscopic observations have provided strong evidence, although spectrally restricted to the methane windows, for the presence of water ice on Titan's surface[14], coexisting in variable proportions with a dark component, presumably of organic nature[15,16]. Water ice may explain the 1,540 nm band, as illustrated in Fig. 15b by a simulation of the reflectance spectrum of a mixture of low-temperature water ice[17], yellow tholins[18] and a spectrally neutral dark component. This identification is reasonable in the context of the light-coloured rocks present at the landing site (Fig. 4), but not conclusive, because some organics do show absorption at a similar wavelength. This is the case, notably, for bright yellow-orange tholins produced in laboratory experiments[19,20] (shown in Fig. 15b), which partly contribute to this band in the simulation and which may account for the red slope in the visible range of the surface spectrum. It is probably this material, existing as aerosol particles, that absorbs the blue wavelengths, which would explain the yellow-orange colour of Titan's atmosphere as seen from space or from the surface.

We note the remarkable absence of other absorption features in the surface spectrum along with the 1,540 nm band. This is at odds with predictions that some specific chemical bonds,

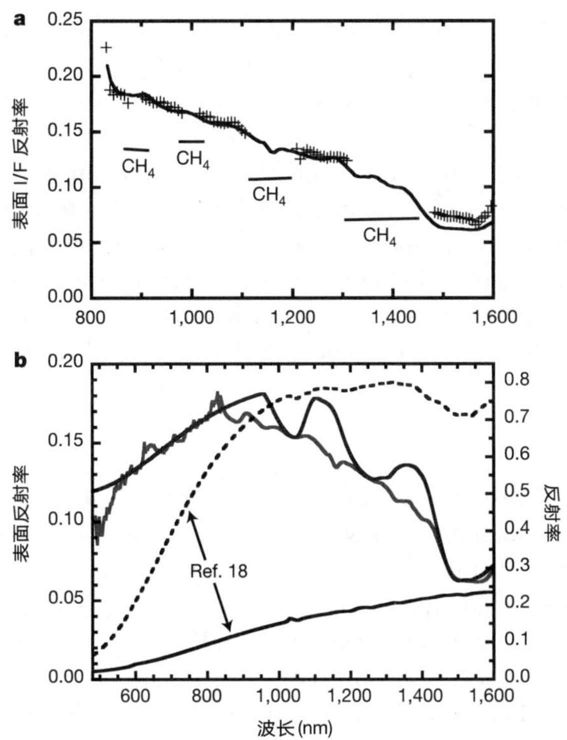

图 15. 土卫六地表的反射率。**a**, 在 21 m 高度测量得到的反射率（加号表示）与着陆后测量的反射率（实线表示）的对比。图中的 CH_4 标出了甲烷吸收波段的位置。**b**, 探测器着陆后测量得到的表面反射率（红线）与模拟结果（蓝线）的对比。其中，模拟曲线采用的是包含大颗粒（750 μm）的低温水冰、黄色的索林斯以及一种在 850 ~ 1,500 nm 之间存在无特征蓝坡的未知成分的混合物。同时图中还给出了黄色索林斯（虚线）和暗色索林斯（黑色实线）两种有机物的谱线用于对比（反射强度的纵坐标比例尺缩小为四分之一）[18]。目前，我们正在实验室尝试识别或合成这种缺失的蓝色物质。

尽管基于地表的光谱观测结果限制在几个甲烷窗口内，但该结果也提供了强有力的证据表明土卫六表面存在水冰 [14]；这些水冰与某种可能具有有机物特性的较暗的成分以变化的比例共存 [15,16]。水冰也许可以解释 1,540 nm 处的吸收峰，图 15b 中给出了低温水冰 [17]、黄色索林斯 [18] 以及一种光谱呈中性的暗色组分的混合物的反射率模拟结果。由于着陆点附近的岩石颜色较淡（见图 4），前文所作的鉴别具有合理性但不是结论性的，因为一些有机物在相似的波段也具有吸收峰。实验室制备的亮橘黄色的索林斯尤其如此 [19,20]（如图 15b 所示），它对于模拟中出现的吸收峰具有一定的贡献，同时可能用于解释表面光谱可见光范围内的红端斜坡。或许就是这种物质以气溶胶粒子的形式存在并吸收了蓝色光波段的光，造成了从太空或从土卫六表面观察到的土卫六大气呈现橘黄色。

我们注意到表面光谱中并没有与 1,540 nm 吸收带共存的其他吸收结构，这与以往研究得出的预测不符。根据先前的研究，在表面光谱中，将会出现某些化学键

in particular C–H or C≡N, and possibly the individual bands of atmospherically abundant species, such as ethane (C_2H_6), acetylene (C_2H_2), propane (C_3H_8), ethylene (C_2H_4), hydrogen cyanide (HCN) and their polymers, would show up as signatures in the surface spectrum.

The most intriguing feature in the surface spectrum is its quasilinear featureless "blue slope" between 830 and 1,420 nm. As briefly illustrated in Fig. 15b, a featureless blue slope is not matched by any combination of laboratory spectra of ices and complex organics, including various types of tholins. Depending on their composition and structural state (for example, abundance, extension and/or clustering of sp^2 carbon bonds), organic materials in the near-infrared exhibit either distinct absorption bands (for example, bright yellow-orange tholins, Fig. 15b), or a feature-poor red slope (for medium to low-albedo organics), or a very dark and flat spectrum[18,21].

Assessing the material responsible for the blue slope is a major challenge and also a prerequisite for a secure identification of the 1,540 nm band. If this band is indeed mostly due to water ice, an intimate mixing of this ice with a material displaying a strong "infrared-blue" absorption would explain the absence of the weaker H_2O bands at 1.04 and 1.25 μm in the surface spectrum, as demonstrated for several dark icy satellites, where these bands are hidden by the presence of an organic component (but neutral or reddish). Decreasing the water-ice grain size alone cannot suppress the 1.04- and 1.25-μm bands and at the same time maintain the apparent blue slope that is produced by large-grained water ice (considering only the continuum absorption between the infrared bands). To hide these weak water bands efficiently, the mixture would need to be ice and a material having a stronger and decreasing-with-wavelength infrared absorption.

Haze Particle Size

The haze particles in Titan's atmosphere have long been known to produce both high linear polarization and strong forward scattering. This has been taken to imply that the particles are aggregates of small "monomers" in open structures. The amount of linear polarization constrains the size of the small dimension (monomer radius) while the forward scattering or wavelength dependence of extinction optical depth determines the overall size of the particle or the number of monomers constituting the aggregate.

The DISR instrument measured the degree of linear polarization of scattered sunlight by measuring a vertical strip of sky in two bands centred at 492 and 934 nm. Some 50 measurements of this type were collected during the Titan descent. For the small monomer sizes expected, the direction of polarization would be perpendicular to the scattering plane and reach a maximum near 90° scattering angle at an azimuth opposite to the Sun, and would have a maximum electric field vector in the horizontal direction. We eliminated any polarization measurements made by the DISR that did not have this character, assuming that they were not made at the desired azimuth.

（尤其是 C–H 或 C≡N 键）的特征谱结构，也会出现大气中丰度较高的化合物的特征吸收带，例如乙烷（C_2H_6）、乙炔（C_2H_2）、丙烷（C_3H_8）、乙烯（C_2H_4）、氢氰酸（HCN）以及它们的聚合物。

表面光谱中最有趣的特征是位于 830 nm ~ 1,420 nm 之间的准线性无特征“蓝坡”结构。该斜坡结构还不能与水冰和复杂有机物（包括各种索林斯）的实验室光谱的任何组合相匹配。基于以上物质的不同组成和结构状态（比如，sp^2 碳化学键的丰度、延展和（或）成簇），得到的有机物质在近红外波段或者显示出明显的吸收峰（比如明亮的橘黄色索林斯，如图 15b 所示），或者显示出无特征的红坡（对于中低反照率的有机物），或者显示出非常暗的平坦谱线 [18,21]。

确定“蓝坡”光谱结构产生于何种物质是一项主要的挑战，这也是正确识别 1,540 nm 吸收带的前提。如果这一波段确实是主要由水冰造成的，这种冰与具有强“红外–蓝”吸收特征的物质的均匀混合将解释表面光谱 1.04 μm 和 1.25 μm 处更弱的 H_2O 吸收峰的缺失。这已经在数颗暗色的冰卫星上得到证实，这是因为这些吸收峰被有机物的谱结构（但是呈现出中性或红色的特征）所掩盖。仅仅减小水冰的晶粒尺寸并不能既抑制 1.04 μm 和 1.25 μm 处的吸收峰，又保持明显的蓝坡，因为蓝坡结构是由较大的水冰晶粒产生的（只考虑了红外波段吸收峰之间的连续吸收结构）。为了能够有效遮盖这些较弱的水的吸收峰，这种混合物质应该包含水冰和另一种具有更强的、但随波长增加而减弱的红外波段吸收特征的物质。

雾霾颗粒的尺寸

人们早就知道土卫六大气中的雾霾颗粒可以产生高度线偏振以及很强的前向散射。这意味着这些颗粒是较小的“单体”通过开放式结构构成的聚集体。线偏振的量约束了颗粒具有小维度的尺寸（单体的半径），而通过前向散射或消光光学深度与波长的关系可以确定颗粒的整体尺寸或构成聚集体的单体数目。

DISR 设备在以 492 nm 和 934 nm 为中心的两个波段，通过测量天空的一段垂直条带，得到太阳散射光的线偏振度。类似的测量在探测器降落过程中大约进行了 50 次。对于预期的较小单体尺寸，偏振方向将垂直于散射面并在背向太阳的方向达到最大偏振度（接近 90°），同时，在水平方向上具有最大电场矢量。我们假定不满足以上特征的偏振测量结果是由于 DISR 没有朝向预期的方位上，因此我们将这些结果进行了剔除。

Several polarization measurements showing the expected behaviour in Titan's atmosphere were obtained. A gradual rise to a maximum near a scattering angle of 90° was observed, followed by a decrease on the other side of this peak. The solar aureole camera made several of these measurements at different times through the descent that show a smooth decrease in polarization with increasing optical depth into the atmosphere (Fig. 16). Figure 16a shows a maximum degree of linear polarization of about 60% at altitudes above 120 km in the 934-nm channel. Below, we show that the optical depth at 934 nm is a few tenths at this location in the descent. Comparisons of this degree of polarization with model computations for different-sized fractal aggregate particles produced by binary cluster collision aggregation indicate that the radii of the monomers comprising the aggregate particles is near 0.05 μm, almost independent of the number of monomers in the particle.

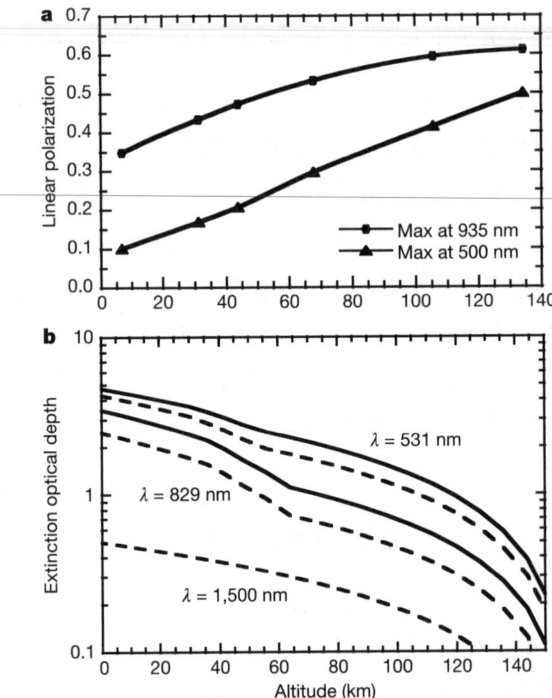

Fig. 16. Haze properties. **a**, The maximum degree of linear polarization measured opposite to the Sun as a function of altitude in our 500-nm channel (triangles) and in the 935-nm channel (dots). **b**, Extinction optical depth versus altitude for three wavelengths, 531 (top), 829 (middle) and 1,500 nm (bottom). The dashed curves correspond to $N = 256$ monomers, and the solid curves correspond to $N = 512$ monomers of the aggregate particles that make up Titan's haze. Note that the 531 (top) curve was constrained above 40 km and extrapolated to the ground. More explicit constraints from the infrared spectrometer will be available after the probe azimuth with time is determined.

Haze Optical Depth and Vertical Distribution

Before the Huygens probe descent, several workers considered the possibility that the haze in Titan's atmosphere clears below an altitude of some 50 or 70 km (ref. 22) owing to condensation of hydrocarbon gases produced at high altitudes that diffuse to lower,

我们获得了数个与预期的土卫六大气特征相符的偏振测量的结果。我们观察到偏振在 90° 散射角附近逐渐增至最大值，然后在此峰值另一侧逐渐减小。日晕照相机在下降阶段的不同时间对偏振进行了数次测量，结果表明偏振随着大气光学深度的增加而平稳减小(图 16)。其中，由图 16a 可知，934 nm 观测通道的线偏振度在大于 120 km 的高度达到最大值，约为 60%。而图 16b 则表明在下降过程中这一地点 934 nm 通道测量的光深大小仅为十分之几。我们把二元团簇碰撞聚集产生的不同尺寸的分形聚集颗粒带入模型进行计算，并与偏振度的测量结果进行比较，发现构成聚集体颗粒的单体的半径大约为 0.05 μm，且单体的半径几乎与颗粒中单体的数量无关。

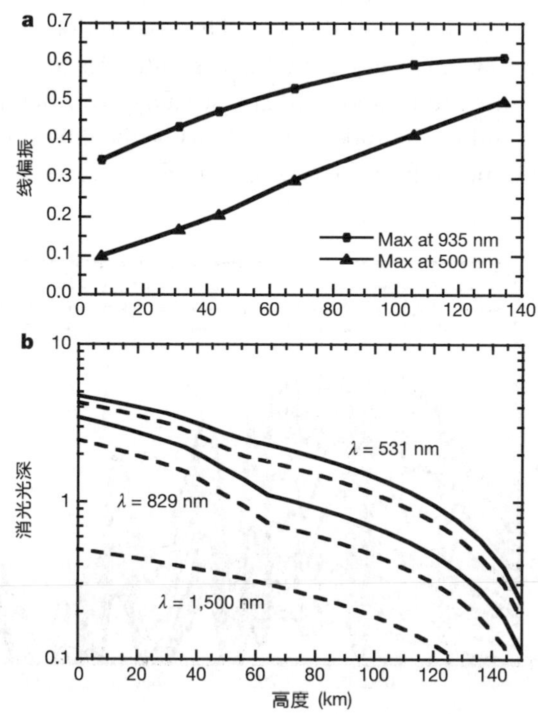

图 16. 雾霾的特性。**a**, 在背向太阳方向，500 nm 和 935 nm 两个通道测量得到的最大线偏振度与高度的关系，分别用三角和圆点表示出。**b**, 图中给出了在三个不同的波长(即 531 nm、829 nm 以及 1,500 nm)处，消光光深与高度的关系，对应曲线分别位于图上部、中部、下部。图中虚线对应组成土卫六雾霾的聚集颗粒的单体数量 $N = 256$ 的情况，实线则对应单体数量 $N = 512$ 的情况。值得注意的是，531 nm(上部)的曲线仅仅在 40 km 以上被约束，40 km 高度到地面的曲线由外推获得。在确定探测器方位随时间变化的关系后，还能进一步从红外光谱分析仪获得更精确的约束。

雾霾的光深和垂直分布

在惠更斯探测器降落之前，若干研究者认为当低于某一高度(约 50 km 或 70 km)时 [22]，土卫六大气中的雾霾可能会消失，这是因为生成于高层大气的碳氢化

colder, levels of the stratosphere. If such a clearing were to occur in Titan's atmosphere, the intensity seen by the downward-looking DISR spectrometer would be relatively constant below the altitude at which the clearing began.

The brightness looking downward averaged along the DLVS slit and averaged over azimuth increases by a factor of two from the surface to 30 km altitude at a wavelength of 830 nm as shown in Fig. 17a. The increase at 830 nm is due almost solely to scattering by haze between 30 km and the surface. These observations demonstrate that there is significant haze opacity at all altitudes throughout the descent, extending all the way down to the surface.

The brightness of the visible spectra looking upward depends on the azimuth relative to the Sun. Although the probe attitude is not yet well known, it is clear that the minimum intensities are found looking away from the Sun. The upward-looking spectra looking away from the Sun start with low intensities at the highest altitudes and increase in intensity as the altitude decreases from 140 to about 50 km (see Fig. 17b). Below 50 km the intensity decreases at short wavelengths as altitude decreases, while the intensity in continuum regions longer than 700 nm continues to increase, as shown in Fig. 17c.

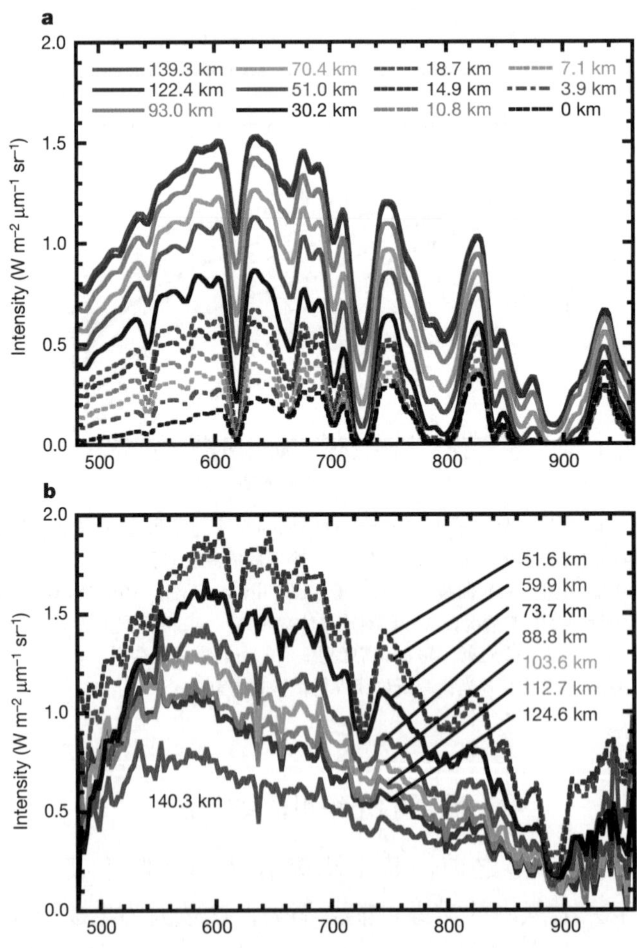

合物气体向下扩散到平流层较冷较低的平面，从而发生凝聚。如果确实如此的话，则当惠更斯探测器降至某一高度后，由于雾霾开始消失，俯视 DISR 光谱分析仪观察到的强度将保持相对固定。

俯视观测到的沿 DLVS 狭缝亮度的以及对方位平均的 830 nm 波长处的平均亮度，从地表到 30 km 高度增加了一倍，如图 17a 所示。这种增加主要是由于 30 km 高度与地表之间的雾霾产生的散射效应。这些观测证实了在整个下降的不同高度，直至土卫六表面都存在显著的由雾霾引起的不透明度。

仰视观测到的可见光谱的亮度则与相对太阳的方位角相关。尽管探测器的姿态尚未完全确定，但探测到的最小强度显然位于背向太阳的方向。在探测器位于最高高度时，背离太阳方向的仰视光谱测量得到的光谱强度较低；随着高度从 140 km 逐渐降低到 50 km，光谱强度逐渐增加，如图 17b 所示。在 50 km 高度以下，短波长波段的强度随着高度降低而减小，但大于 700 nm 的连续波段的强度则继续增加，如图 17c 所示。

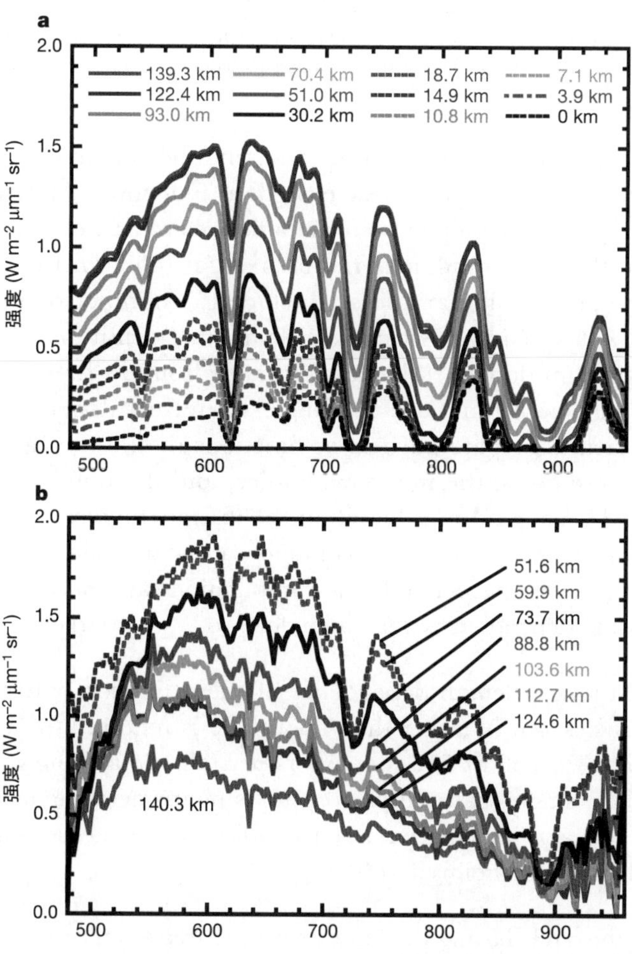

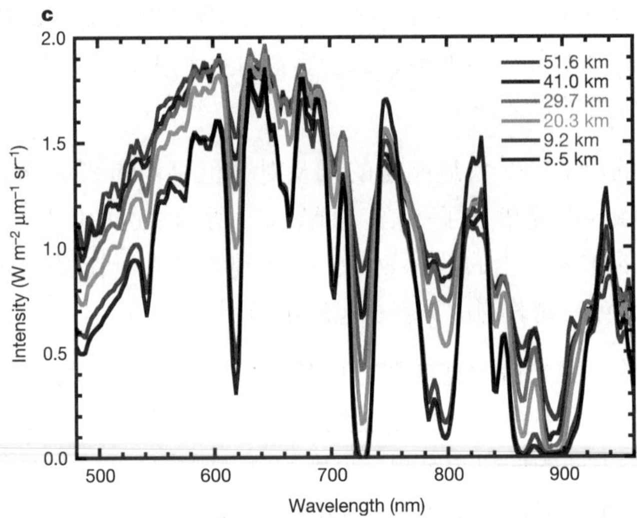

Fig. 17. Atmospheric spectra. **a**, The average intensity looking downward averaged over azimuth and over the length of the slit (10° to 50° nadir angle) as a function of wavelength for several altitudes as labelled. **b**, The intensity measured by the Upward-Looking Visible Spectrometer in the direction opposite the Sun as a function of wavelength for several altitudes as labelled. Note that the brightness begins at a low level at 140 km, and increases as altitude decreases. **c**, Same as **b** but for altitudes below 50 km. Note that the brightness away from the Sun decreases with decreasing altitude at short wavelengths, but increases in continuum regions longward of 700 nm.

The intensity looking upward away from the Sun, and the azimuthally averaged intensity looking downward at each continuum wavelength as functions of altitude, constrain the vertical distribution of aerosol opacity in the atmosphere as well as the aerosol single-scattering albedo. With the monomer radius fixed at 0.05 μm from the polarization measurements, the adjustable parameters include the number of monomers in each aggregate particle, N, as well as the local particle number density, n, in cm^{-3} as a function of altitude. An algorithm developed by (and available from) M.L. was used to determine the single-scattering phase function, the single-scattering albedo, and the extinction cross-section for each aggregate particle as functions of the wavelength, the real and imaginary refractive indices, the monomer radius, and the number of monomers per aggregate particle. This algorithm is based on the discrete dipole approximation and the T-matrix method (M. Lemmon, personal communication) to evaluate the single-scattering properties of the aggregate particles. These computations are most accurate at relatively small particle sizes and depend on extrapolation for N of 256 or larger.

For large particles the wavelength dependence of the extinction optical depth is smaller than for small particles. An N larger than about 100 is required to fit the observations. Hence, models with $N = 256$ or 512 monomers per particle are shown, even though for these values of N the single-scattering algorithm is not as accurate as desired. For these initial models, the real and imaginary refractive indices for the aerosols are taken from the measurements of laboratory tholins in ref. 23.

The radius, R_p, of the circle having the same projected area as an aggregate particle is given

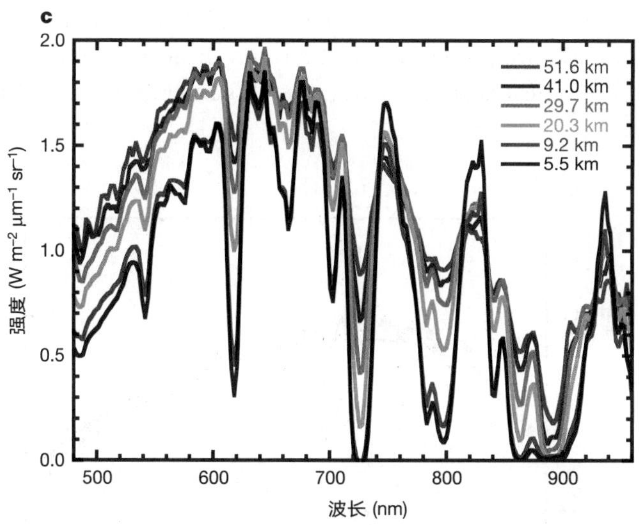

图 17. 大气光谱。**a**, 图中给出了几个不同高度下俯视光谱平均强度与波长的关系（见图中标注），其中平均强度是基于方位角和狭缝长度方向（天底角 10° 到 50° 之间）的平均。**b**, 在背向太阳的方向，仰视可见光波段光谱仪测量得到的不同高度下的强度与波长的关系（见图中标注）。值得注意的是，140 km 高度对应的起始亮度较低，但亮度随着高度的下降而增加。**c** 与 **b** 图表示的关系相同，但曲线对应的 50 km 以下的高度。值得注意的是，背向太阳方向的光亮度在短波长波段随着高度的减小而减小，但是在大于 700 nm 的波段则随高度减小而增加。

背向太阳方向的仰视测量强度和在各个连续波段得到的对方位平均的俯视强度相对于高度的变化函数，可以用于确定大气中气溶胶不透明度的垂直分布以及气溶胶单散射反照率。根据偏振测量，单体半径固定值为 0.05 μm，那么可调参量包括每个聚集体中颗粒的单体数量 N 以及粒子数局部密度 n（单位为个/cm^3）的高度变化函数。利用莱蒙提出的算法（可联系莱蒙获取），可以确定单散射相函数、单散射反照率、每个聚集颗粒的消光截面（波长的函数）、折射指数的实部和虚部、单体的半径以及每个聚集体中的单体数量。该算法基于离散偶极子近似和 T 矩阵方法（莱蒙，个人交流），可以评估聚集颗粒的单散射属性。以上计算对于相对较小的颗粒尺寸具有最高的精确度，而且依托于 $N \geqslant 256$ 的外推。

对于较大颗粒而言，较小颗粒的消光光深与波长的关系更加密切。为了拟合观测结果，N 的取值需要大于 100，因此我们在模型中设定每个颗粒中的单体数为 256 或 512。尽管对于这样的 N 取值，单散射算法不如想要的那样精确。对于以上初始模型，气溶胶折射指数的实部和虚部取自实验室中索林斯的测量结果 [23]。

设某一圆面积与单一的聚集颗粒的投影面积相同，则该圆的半径 R_p 可以由公式

by $R_p = r\sqrt{(N^{0.925})}$, where r is the monomer radius and N is the number of monomers. Particles with 256 or 512 monomers have the same projected areas (which control their forward scattering properties) as circles with radii 0.65 and 0.9 μm, respectively.

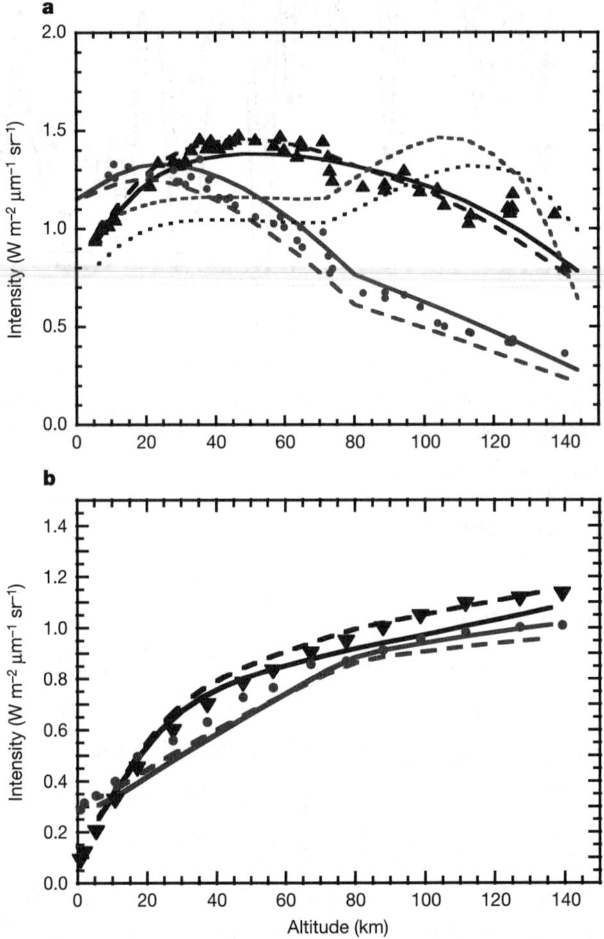

Fig. 18. Haze models versus observations. **a**, Measured upward-looking intensity (points) away from the Sun versus altitude for 531 (blue) and 829 nm (red). Three models are shown compared to the observations at each wavelength. The solid curves are for 512 monomers in each aggregate particle. The model at 531 nm has 12 particles cm^{-3} above 80 km and 18 particles cm^{-3} below that altitude. The corresponding model at 829 nm has 20 particles cm^{-3} above 80 km and 60 cm^{-3} below. The models indicated by long-dashed lines have 256 monomers per particle and at 531 nm the number density is 20 particles cm^{-3} above 80 km and 40 cm^{-3} below. At 829 nm the number density is 30 cm^{-3} above 80 km and 100 cm^{-3} below. The number density of particles differs slightly with wavelength because the model of fractal aggregate particles does not yet reproduce the wavelength dependence of the cross-section to high accuracy. The models indicated by short-dashed lines have 256 monomers per particle, and have the same number of total particles as the models indicated by long-dashed lines, but all the particles are concentrated above 72 km with a clear space below. Such models with clear spaces are clearly not in agreement with the observations. **b**, Downward-looking measured intensities versus altitude (plotted as points) for 531 (blue points) and 829 nm (red points). The two models (plotted as curves) are the same models as those shown by long-dashed lines and solid curves in **a**.

$R_p = r\sqrt{N^{0.925}}$ 计算得到，其中 r 表示单体的半径，N 表示单体的数量。不妨设两种颗粒分别由 256 个单体和 512 个单体组成，则与二者投影面积（决定着前向散射性质）相等的圆的半径分别为 0.65 μm 和 0.9 μm。

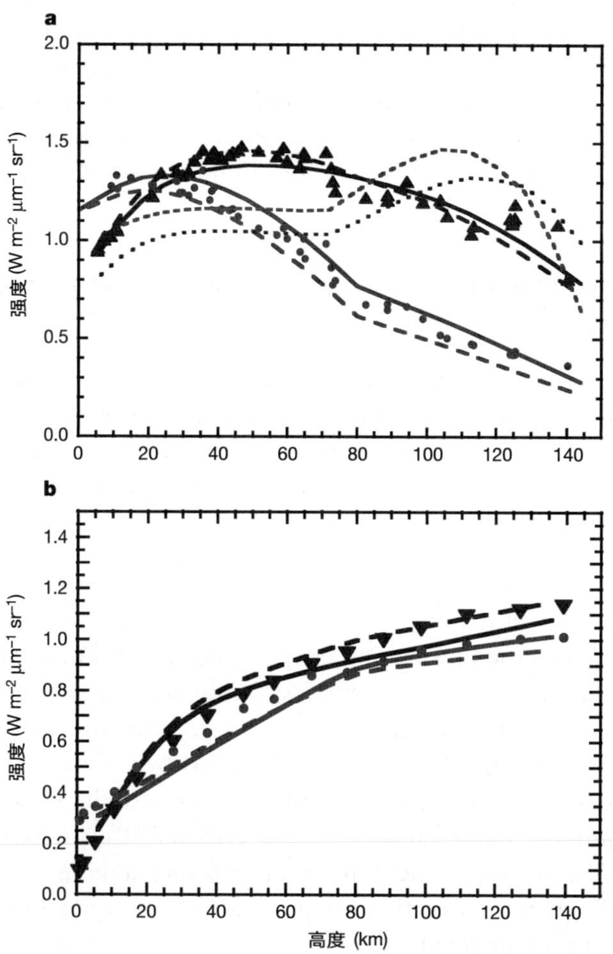

图 18. 雾霾模型与观测结果的对比。**a**, 图中给出了在背向太阳的方向仰视强度（用点表示）与高度的关系，蓝色与红色分别对应 531 nm 和 829 nm 观测通道。同时作为对比，对于每个通道还给出了三个模型的结果。其中，实线代表每个聚集颗粒包含 512 个单体的情况。对于 531 nm 通道，该模型在大于 80 km 高度的粒子数密度为 12 个/cm³，在低于 80 km 的高度则为 18 个/cm³。对于 829 nm 通道，该模型大于 80 km 高度的粒子数密度为 20 个/cm³，在低于 80 km 的高度则为 60 个/cm³。图中用长虚线标示的模型代表每个聚集颗粒包含 256 个单体，对于 531 nm 通道，大于 80 km 高度的粒子数密度为 20 个/cm³，在低于 80 km 的高度则为 40 个/cm³；对于 829 nm 通道，大于 80 km 高度的粒子数密度为 30 个/cm³，在低于 80 km 的高度则为 100 个/cm³。由以上数据可知，粒子数密度随波长变化而有些许变化，这是因为我们根据分形颗粒模型得出的截面积与波长的关系准确度并不高。图中短虚线表示每个颗粒由 256 个单体构成的模型结果，粒子（颗粒）总数与长虚线相同，但是所有的粒子都集中在 72 km 高度以上，低于 72 km 的空间则没有粒子。显然这样一种存在粒子净空层的模型与观测结果不符。**b**, 图中数据点标示了 531 nm 和 829 nm 两个通道测量得到的俯视测量的强度与高度的关系，分别用蓝色点和红色点表示。拟合曲线采用的两种模型与图 **a** 中长虚线和实线所代表的模型一致。

Comparison of the observed downward-streaming intensity looking away from the Sun at wavelengths of 531 and 829 nm with plane-parallel radiative transfer models constrains the vertical distribution of optical depth in Titan's atmosphere. The vertical distribution of particles can be adjusted to fit these curves arbitrarily well. In this preliminary work, only one constant number density above an altitude of 80 km and a second constant number density below 80 km were considered for models with $N = 256$ and $N = 512$, as shown in Fig. 18a. The number densities are larger in the lower half of the atmosphere than the upper half, but only by modest factors of two to three. The number densities are not exactly equal in the models at different wavelengths, but this is probably due to the extrapolation in the wavelength dependence of the cross-sections in the models for these relatively large N values at the shortest wavelengths (and largest size parameters). Average number densities in the entire atmosphere between 30 and 65 cm^{-3} are required if the number of monomers per particle is 256. Average number densities between 15 and 40 cm^{-3} are required if N is 512.

Models with a clear space below an altitude of 72 km are also shown in Fig. 18a. It is apparent that no such clear space exists in the region of the probe's entry. It will be interesting to examine the range of parameters in cloud physics models needed to reproduce the continuous variation of haze opacity throughout Titan's atmosphere.

Haze models must reproduce the upward-streaming intensity observed in the atmosphere as well as the downward-streaming intensity. Models with $N = 256$ and 512 at 531 and 829 nm are compared to the upward intensity averaged in azimuth and along the slit in Fig. 18b. While the fit is not exact, it is clear that the models that fit the downward intensity away from the Sun are also in reasonable agreement with the measured upward intensities. It is interesting to note the ground reflectivity implied by the measurements at 531 and 829 nm. These values include the true shape of the diffuser in the Upward-Looking Visible Spectrometer and produce ground reflectivities of 0.13 at 531 nm and 0.19 at 829 nm. The corrected value at 829 nm is in good agreement with the value measured by the Infrared Spectrometer (0.18) when a correction for the diffuser's non-ideal shape is made to the infrared spectrometer measurements.

Haze Structure and Methane Absorptions

Haze models must fit the observations at all wavelengths. How well do the models derived from the visible spectrometer fit the DISR observations in the infrared? The ULIS spectra in Fig. 19 clearly show absorption by the methane bands around 890, 1,000, 1,160 and 1,380 nm. The depths of these bands increase with decreasing altitude as a result of increasing methane column density. They are correctly reproduced by radiative transfer calculations based on an exponential sum formulation for the methane absorption and a stratospheric methane mole fraction of 1.6% (ref. 24). The agreement is worse at low altitudes in the troposphere, probably owing to inaccuracies in the methane absorption

将 531 nm 和 829 nm 两个波长观测到的背向太阳的下行辐射流强度，与平行平面的辐射传输模型进行比较，可以约束土卫六大气中光深的垂直分布。通过调整颗粒的垂直分布，可以对这些曲线进行很好的拟合。在前期工作中，对于 $N = 256$ 和 $N = 512$ 两种模型，我们只假设 80 km 以上的高度具有某一恒定粒子数密度 $N = 256$，而 80 km 以下高度具有另一恒定粒子数密度 $N = 512$，如图 18a 所示。下半层大气层中的粒子数密度比上半层大气层中的粒子数密度大，但只是上半层的 2 到 3 倍。但是在不同波长的模型中，粒子数密度并不是完全相同的，这可能是因为模型引入的截面与波长的关系在短波长和较大 N 取值条件（参数达到最大）下的外推造成的。如果每个颗粒的单体数量为 256，则整个大气层中的平均粒子数密度需要介于 30 个/cm^3 到 65 个/cm^3 之间。如果每颗粒的单体数量为 512，则整个大气层中的平均粒子数密度需要介于 15 个/cm^3 到 40 个/cm^3 之间。

同时，图 18a 还给出了假定 72 km 以下高度不存在雾霾的模型结果。显然，在探测器进入的区域并不存在这样的情况。检验用于重现土卫六大气中薄雾不透明度连续变化的云层物理模型参量的范围将是一项非常有意思的工作。

雾霾模型必须能够重现大气中的仰视方向辐射流强度和俯视方向辐射流强度。我们在 531 nm 和 829 nm 分别将 $N = 256$ 和 $N = 512$ 代入模型，并将模型结果与在不同方位上做平均并沿狭缝做平均后的向上辐射流强度进行比较，如图 18b 所示。尽管拟合还不够精确，但是可以清晰地看出用于拟合背向太阳的向下辐射流强度的模型也可以在一定程度上拟合向上辐射流强度。值得注意的是，在 531 nm 和 829 nm 测量可以得到包括仰视可见光谱仪中漫射器的真实形状等值，并以此获得在 531 nm 和 829 nm 测得的地面反射率（分别对应 0.13 和 0.19）。以上校正后的 829 nm 反射率与经过校正（考虑了红外光谱仪测量过程中漫射器的非理想形状）后的红外光谱仪的测量结果（0.18）相吻合。

雾霾的结构和甲烷的吸收

雾霾模型必须在所有波长都能拟合观测结果。那么根据可见光谱分析仪得到的模型在多大程度上可以拟合 DISR 在红外波段的观测结果呢？图 19 中的 ULIS 光谱清楚地显示，在 890 nm、1,000 nm、1,160 nm 以及 1,380 nm 附近存在甲烷的吸收峰。由于甲烷柱密度随着高度的降低而增加，所以这些吸收峰的深度也随着高度的降低而增加。根据甲烷吸收的指数求和公式以及平流层甲烷摩尔分数 1.6%，以上吸收带可以由辐射传输理论准确地计算出来[24]。但是在对流层的较低高度，理论计算和实测结果的吻合度较差，这可能是由于采用的甲烷吸收系数不够准确。在甲烷窗

coefficients. In the methane windows the downward average intensity varies by 25% or less between 104 and 10 km, indicating relatively low aerosol absorption in the infrared range.

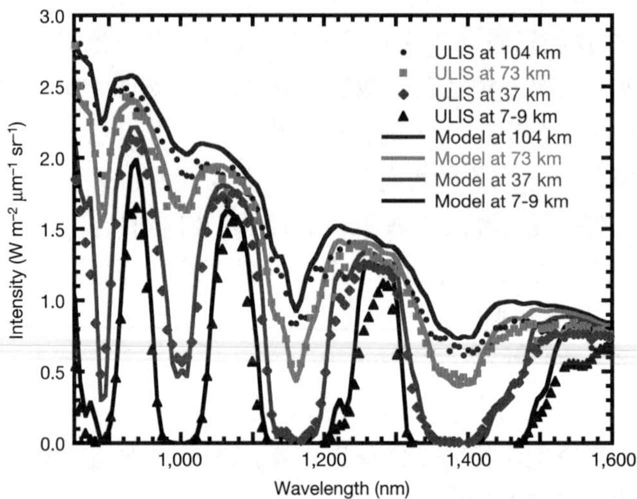

Fig. 19. Vanishing sunlight. ULIS spectra (points) recorded at various altitudes and showing the growth of the methane-band absorption with depth in the atmosphere. The spectra at altitudes greater than 3.5 km have been integrated over several probe rotations and correspond approximately (but not exactly) to azimuth-averaged intensities. The full analysis of these observations must await refinement of the attitude of the probe as a function of time, a task still in progress. Models, with a methane mole fraction of 1.6% in the stratosphere increasing to 5% at the surface, are shown for comparison (lines). At low altitudes (< 20 km), the mismatch between model and observations in the methane windows, specifically around 1,280 and 1,520–1,600 nm, is probably due to errors in the methane absorption coefficients at long path-lengths. The model slightly overestimates the intensity in the 104-km spectrum, because the latter does not correspond to an exact azimuth average. Radiative transfer calculations were based on a 16-term exponential-sum formulation of the methane absorption properties. In the near-infrared ($\lambda > 1{,}050$ nm), these absorption coefficients were calculated from a band model with a modified temperature dependence designed to better match the low-temperature observations. In the visible ($\lambda < 1{,}050$ nm), the absorption coefficients of Karkoschka[34] were used. In practice, for 30 pressure–temperature conditions representative of 30 levels in Titan's atmosphere[12], methane transmissions were calculated for 60 different paths, convolved to the resolution of the DISR spectrometers, and this ensemble of convolved transmissions was fitted each time with an exponential-sum model.

At altitudes less than 3.5 km, we used single short exposures for the infrared spectra rather than long time averages. The 940 nm intensity in the last three ULIS spectra is about four times larger that in the first three spectra, indicating that the Sun is located in their field of view (see Fig. 20). The contrast between the most intense spectrum (with the Sun in the field of view) and the weakest one (with the Sun out of the field of view) increases with wavelength, reaching 17 at 1,550 nm, a consequence of the decreasing haze optical depth. This contrast can be used to constrain the haze optical depth, assuming that the spectra correspond approximately to solar azimuths of 0 and 180°. A satisfactory model, using aggregate particles of 256 monomers, a 0.05-μm monomer radius, and a uniform concentration of 52 particles cm^{-3}, indicates an optical depth of about 2 at 940 nm, decreasing to 0.5 at 1,550 nm. Models with one-half and twice the particle density (and

口，向下平均强度在 104 km 到 10 km 高度之间变化了 25% 或者更小，这说明气溶胶在红外波段的吸收相对较弱。

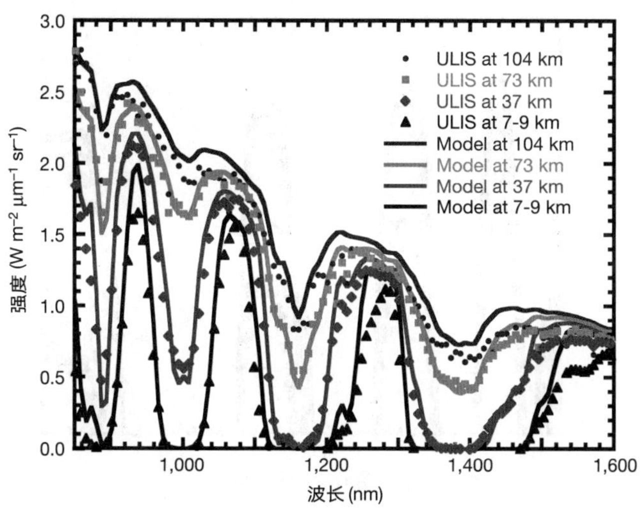

图 19. 逐渐消失的阳光。不同高度处的 ULIS 光谱记录（图中标示为数据点），表明甲烷吸收峰的强度随着进入大气层深度的增加而增加。大于 3.5 km 高度的光谱是对若干个探测器旋转周期进行积分的结果，大体上（但不是完全地）对应了方位平均强度。关于这些观测结果的全面分析尚需精确确定探测器高度与时间的关系，这一工作目前正在进行中。图中的曲线给出了我们采用的模型结果（用实线表示）作为对比，其中采用的甲烷摩尔分数在平流层为 1.6%，然后在表面处增加到 5%。对于较低的高度（< 20 km），模型与观测结果在甲烷吸收窗口，特别是在 1,280 nm、1,520 ~ 1,600 nm 附近，存在较大差异；这可能是由于甲烷吸收系数存在误差经过长光程的放大导致的。模型稍稍高估了 104 km 高度处的光谱强度，因为该光谱并不精确地匹配方位平均。辐射传输计算主要基于甲烷吸收特性的 16 项多项式求和公式。在近红外区域（λ > 1,050 nm），通过吸收波段模型可以计算出吸收系数，其中，为更好地拟合低温观测结果，我们对温度的依赖关系进行了修正。在可见光波段（λ < 1,050 nm），则采用了卡尔科施卡计算的吸收系数 [34]。实际操作中，对于代表土卫六大气 30 个高度的 30 个气压–温度条件 [12]，为 60 条不同的路径计算甲烷的穿透率，并与 DISR 光谱分析仪的分辨率作卷积，作卷积后的穿透率每次用一个指数求和模型进行拟合。

在低于 3.5 km 的高度，我们采用单次短时曝光（而不是长时曝光取平均）测量红外光谱。最后的三张 ULIS 光谱中 940 nm 的强度大约是最先三张 ULIS 光谱中 940 nm 强度的 4 倍，这说明太阳位于后三张光谱的拍摄视场中，参见图 20。当太阳位于视野之内时光谱强度显然最高，当太阳位于视野之外时光谱强度则最低；最强与最弱光谱之间对比度随着波长增加而增加，并在 1,550 nm 处达到 17，这是由于雾霾的光深随波长减小。假设以上光谱对应的太阳方位角分别约为 0° 和 180°，则以上对比度可以用来确定雾霾的光深。假设聚集颗粒由 256 个单体组成，每个单体的半径为 0.05 μm，粒子的均匀密度为 52 个/cm^3，这一可以被接受的模型可以计算得到：940 nm 处对应的光深大约为 2，在 1,550 nm 处降为 0.5。如果采用粒子数密度

hence optical depth) yield a contrast between spectra with the Sun in and out of the field about twice as large and half as large (respectively) as observed.

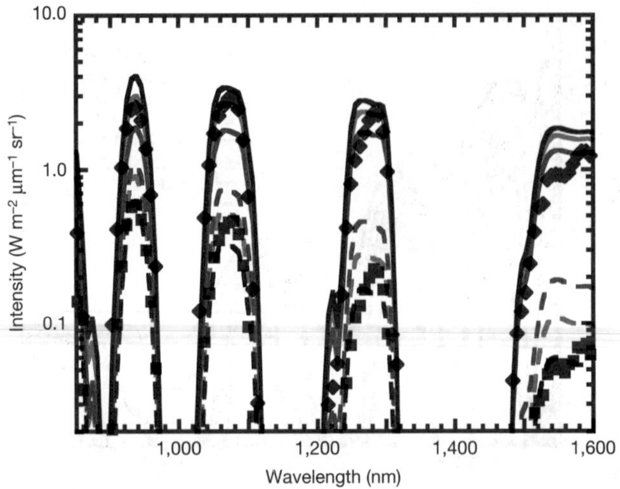

Fig. 20. Determination of total haze optical depth. ULIS spectra (black points) recorded at 734 m (diamonds) and 53 m (squares) above the surface with integration times of 1 s. The one with the highest intensity (734 m) has the Sun in its field of view; the lower one (53 m) does not. The contrast between the two in the methane windows increases with wavelength and is a sensitive function of the haze optical depth. The nominal model, shown for comparison (green line), has an optical depth of 2 at 940 nm decreasing to 0.5 at 1,550 nm. Calculations correspond to intensities averaged over the field of view and azimuths of 0 and 180 degrees with respect to the Sun. Other models show the effect of doubling (red) and halving (blue) the particle concentration. Solid lines show model intensity towards the Sun while the dashed lines show the intensity with the instrument facing away from the Sun.

The methane bands are prominent in the DLIS spectra at all altitudes (see Fig. 21a). The residual intensity in the cores of these bands is due to scattering by aerosol particles between the probe and the surface. Its variation provides a constraint of the vertical profile of the haze particles between approximately 150 and 40 km, as illustrated in Fig. 21a. The method is not sensitive at low altitudes because of absorption of the downward solar flux in the methane bands. A model with a constant particle concentration with altitude provides a good fit of the methane bands. Moderate variations of the particle concentration with height are also acceptable, but a model with a clear space in the lower stratosphere is inconsistent with the data.

The optical depths in the models that fit the visible and infrared spectral observations with $N = 256$ and 512 are shown as functions of altitude in Fig. 16b. The number density was assumed to be 52 cm^{-3}, independent of altitude for the infrared models with $N = 256$. The models computed for comparison with the visible spectrometer contained constant and different number densities above and below 80 km. The average number density above and below 80 km for the models with $N = 256$ (30 and 65) are in reasonable agreement with the single value used in the models derived from the infrared spectrometer. The same particle number densities give the required optical depths from 900 to 1,550 nm, indicating that the algorithm for generating cross-sections from particle sizes is working in a consistent manner. At shorter wave-lengths (531 nm), the size parameter is sufficiently large that the cross-

分别另取为以上粒子数密度（也是光学深度）的 1/2 和两倍的模型，则相应的太阳在视场内外两种光谱的对比度分别为观测值的两倍和一半。

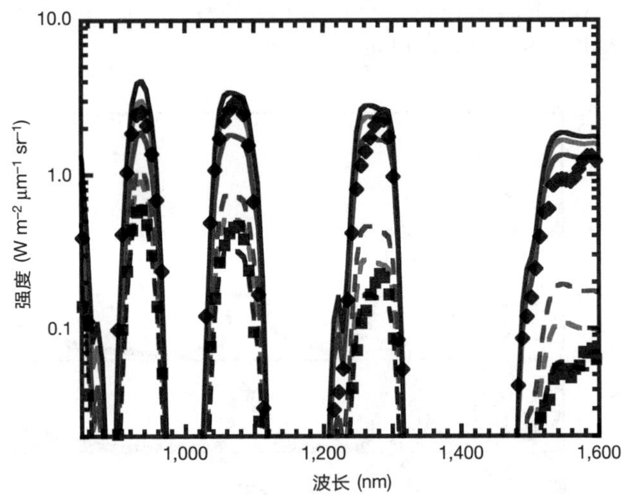

图 20. 雾霾总光深的确定。图中给出了土卫六地表之上 734 m（标示为菱形）和 53 m（标示为方形）处的 ULIS 光谱，积分时间为 1 s。其中，734 m 高度的光谱强度最大，视场中包含了太阳，而强度较弱的 53 m 光谱视场中没有太阳。两个甲烷窗口的对比度随着波长的增加而增加，并且对雾霾光深的变化很敏感。图中给出了标准模型作对比（绿线），其雾霾光深在 940 nm 处为 2，然后逐渐降至 1,550 nm 处的 0.5。计算得到的强度是对视场和相对太阳的 0 和 180° 方位角进行平均后的结果。其他模型表明粒子浓度加倍（红）或减半（蓝）的效应。图中实线表示朝向太阳的光谱强度，而虚线表示仪器背向太阳的光谱强度。

如图 21a 所示，在所有高度的 DLIS 光谱中，甲烷对应的吸收峰都是非常明显的。图中吸收峰核心区内存在强度残差，这是探测器与地表之间的气溶胶粒子散射造成的。这一强度的变化可以用于确定约 150 ~ 40 km 高度之间的雾霾颗粒的垂直分布，如图 21a 所示。但该方法在较低高度并不灵敏，因为向下照射的太阳光通量在甲烷吸收峰内被吸收。如果假定粒子浓度不随高度变化，那么相应的模型可以较好地拟合甲烷吸收峰。粒子浓度随高度有适度的变化的模型也是可接受的；但平流层下部有净空层的模型与数据不符。

图 16b 给出了与可见光和红外观测结果吻合的模型中光深与高度的关系，其中 N 的取值分别为 256 和 512。在红外模型（$N = 256$）中假设粒子数密度为 52 个/cm^3，并且与高度变化无关。用于与可见光谱分析仪对比的模型中，以 80 km 高度为界，上下分别采用了恒定但彼此不同的粒子数密度。$N = 256$ 的模型中 80 km 上下的平均粒子数密度（分别为 30 与 65），和基于红外光谱分析仪的模型采用的单值粒子数密度相符。相同的粒子数密度给出了 900 ~ 1,500 nm 范围的光深，这说明根据颗粒尺寸推出截面积的算法始终是适用的。在较短的波长（531 nm），粒子尺寸参量过大，

section algorithm is not as accurate, and the number density is decreased slightly to give models that fit the observations. The variation of optical depth with wavelength is modest, decreasing by only about a factor of 2.8 from 500 to 1,000 nm. If 512 monomers are used for the particles, the wavelength dependence is even less steep. The haze optical depth as a function of wavelength is presented in Fig. 21b.

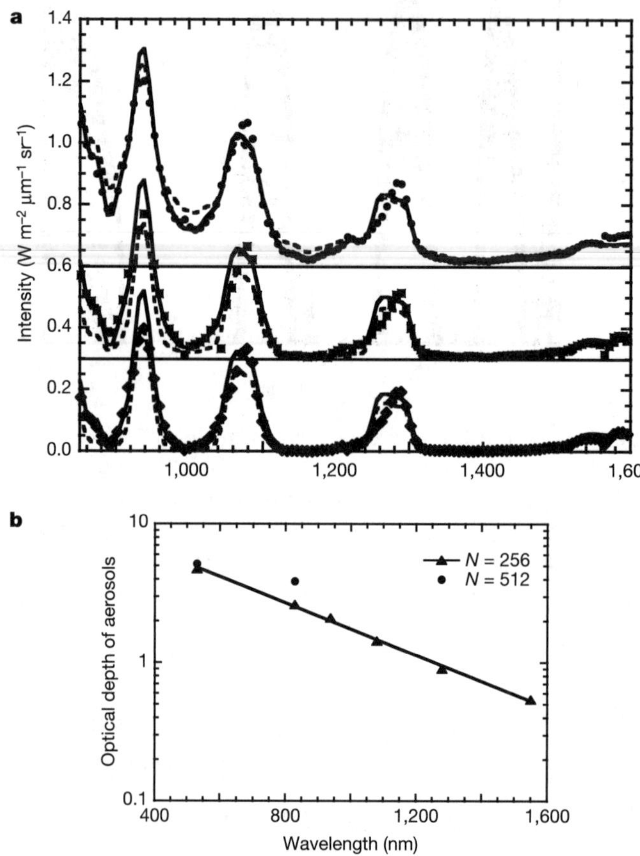

Fig. 21. Haze vertical structure and total optical depth. **a**, DLIS spectra recorded at three altitudes: 104 km (top, dots), 82 km (middle, squares), and 57 km (bottom, diamonds). The intensities for the higher altitudes have been displaced in 0.3 increments for clarity. The points are measured data; the lines are the nominal model; the dashed lines are a modified model with the same optical depth as the nominal model, but with all the haze particles concentrated above 72 km (clearing below). The residual intensity in the core of the methane bands is a sensitive indicator of the presence of scattering particles beneath the probe. The model with the clearing produces too much emission in the core of the CH_4 bands at high altitude and not enough at low altitude. **b**, Total extinction optical depth of the haze alone versus wavelength. The triangles are for models with 256 monomers per particle. The points at the two shortest wavelengths are from models that fit the visible spectrometer measurements. The other four points are from models that fit the infrared spectrometer measurements. The dots are for models with 512 monomers per particle that fit the visible spectrometer measurements.

导致截面算法不够准确，粒子数密度需要略微减小才能使模型与观测结果相符。当波长从 500 nm 增加到 1,000 nm，光深减小为原光深的 2.8 分之一，这一变化并不很大。如果聚集颗粒为 512 个单体构成，则这种变化会更小。图 21b 给出了雾霾光深与波长的关系。

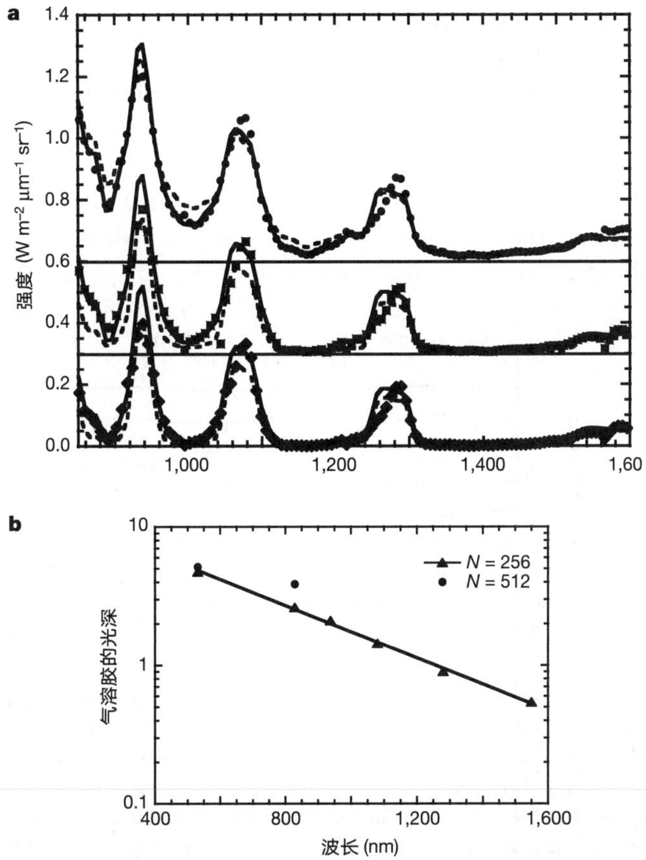

图 21. 雾霾的垂直结构和总光深。**a**, 图中分别给出了高度为 104 km（上部圆点）、82 km（中部方块）以及 57 km（下部菱形）的 DLIS 光谱测量数据。为更清楚地显示在一幅图中，较高高度的光谱分别相对较低高度的光谱向上移动 0.3 个单位。图中点表示测得的数据，实线表示标称模型，虚线表示与标称模型具有相同光深的修正模型，其所有的雾霾颗粒都集中在大约 72 km 高度以上（下面没有雾霾颗粒）。图中甲烷吸收峰核心区域存在的强度残差是探测器下面存在散射粒子的很敏感的指示物。对于较高的高度，修正模型在甲烷吸收峰核心附近的辐射强度过大，而对于较低的高度又过小。**b**, 图中给出了雾霾的总消光光深与波长的关系。其中三角表示每个聚集颗粒由 256 个单体构成的模型，最短的两个波长处的数据点源自可见光谱仪测量结果的拟合模型，而另外四个数据点则源自红外光谱仪测量结果的拟合模型。图中圆点代表了与可见光谱仪测量结果拟合的表示每个聚集颗粒由 512 个单体构成的模型。

A Thin Layer of Haze near 21 km Altitude

Many workers have suggested that hydrocarbons produced at very high altitudes could diffuse downward to cooler levels where they could condense on haze particles. Do our intensity profiles looking towards the horizon detect any thin haze layers at specific altitudes that might be due to this mechanism?

Figure 22 shows the normalized profile of intensity measured by the Side-Looking Imager (SLI) compared to a model. The left-hand side of the plot shows normalized intensity as a function of nadir angle for the observations at altitudes ranging from 20.4 to 22.3 km. The observations at 22.1 km and above and at 20.4 km and below show smooth functions of nadir angle. However, for the measurement at 20.8, and the two measurements at 20.9 km altitude, a dip of about 2% is seen near a nadir angle of 90°.

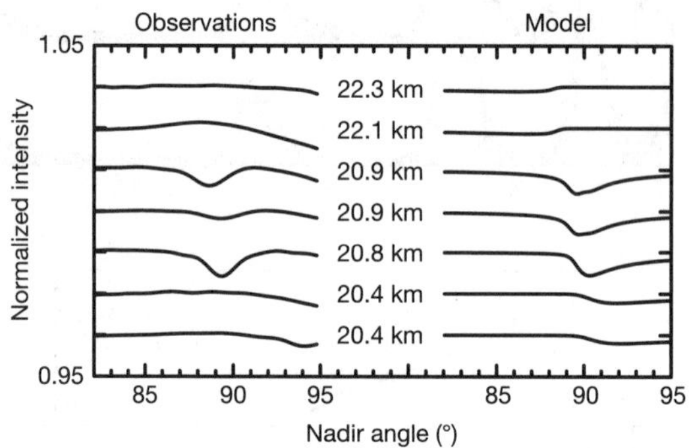

Fig. 22. Thin cloud layer observation at 21 km. In the left-hand side are the intensity profiles as a function of nadir angle divided by the average intensity profile measured by the SLI at the altitudes indicated. The right-hand side of the figure shows the model intensity profiles computed for a cloud layer of absorption optical depth 0.001 which is 1 km thick at an altitude of 21 km. The model is able to reproduce the 2% contrast feature seen in the observations at altitudes of 20.8 and 20.9 km. If the layer is mostly illuminated by diffuse light, the absorption optical depth is equal to total optical depth times the difference between the single-scattering albedos of the material in the layer and the albedo of the background haze. If the layer is primarily illuminated by direct sunlight, the absorption optical depth of the haze is proportional to the total optical depth of the haze times the difference between the phase functions of the material in the layer and the background haze at the scattering angles for any observation.

The curves in the right-hand side of Fig. 22 show the intensities of a model having a thin additional layer of haze at an altitude of 20.9 km. The haze layer of vertical absorption optical depth within a factor of two of 0.001 with a gaussian profile between 1 and 2 km thick can reproduce the depth of the feature. The location of the layer is at 21.0 ± 0.5 km, where the local temperature is 76 K and the pressure is 450 mbar (ref. 12).

This feature at 21 km occurs in the troposphere and may be an indication of methane condensation. It is the only indication of a thin layer seen in the set of SLI images taken from 150 km to the surface. Evidence of condensation of hydrocarbons in the lower stratosphere, where several hydrocarbons might be expected to condense[25], has not yet been

21 km 高度附近的雾霾薄层

许多研究者认为，非常高的大气中产生的碳氢化合物，可以向下扩散到较冷的层位，并在那里凝结成雾霾颗粒。那么我们朝向地平线方向的光谱强度剖面，是否会探测到某一高度存在由于以上机制形成的雾霾薄层呢？

图 22 给出了侧视成像仪（SLI）测量得到的归一化的强度分布曲线，以及与相应模型的比较。图中左侧表示归一化强度与天底角的关系曲线，对应的观测高度从 20.4 km 到 22.3 km。测量曲线在 22.1 km 及以上的区间和 20.4 km 以下的区间内都平滑连续，但是对于位于 20.8 km 高度的一条测量曲线和位于 20.9 km 高度的两条测量曲线，在天底角 90° 附近均存在一个大约 2% 的凹陷。

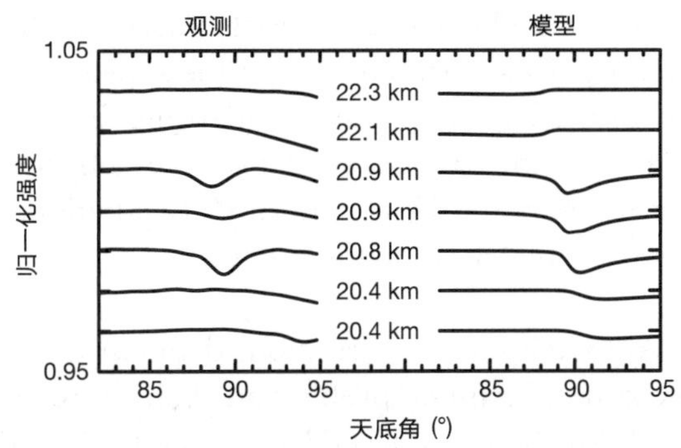

图 22. 21 km 高度处薄云层的观测结果。图中左侧给出了 SLI 在不同高度测量得到的强度与天底角的关系曲线，其中强度是实际强度除以平均强度后的结果。右侧则给出了对厚度为 1 km 的云层的模型计算结果，其中该云层的高度为 21 km，吸收光深为 0.001。该模型可以重现观测结果中 20.8 km 和 20.9 km 曲线 2% 的下陷结构。如果该云层主要由漫射光照亮，则吸收光深等于总光深乘以云层物质的单散射反照率与背景雾霾反照率之差。如果云层主要受直射阳光照亮，对于任何观测，雾霾的吸收光深正比于雾霾的总光深乘以云层物质与散射角方向处的背景雾霾之间的相位函数之差。

图 22 右侧给出了模型的强度曲线，其中该模型在 20.9 km 的高度附加了一层较薄的雾霾。如果这层薄雾霾的垂直吸收光深在 1 ~ 2 km 的厚度范围内为高斯分布，并且其值在 0.001 的两倍范围内变化，则该模型也可以产生类似左图的凹陷结构。该层薄雾霾的高度为 21.0 ± 0.5 km，局域温度为 76 K，压强为 450 mbar[12]。

21 km 处的雾霾薄层位于对流层内，可能意味着甲烷的凝结。SLI 在探测器从 150 km 高度到着陆过程中拍摄的图像里，这是唯一预示雾霾薄层存在的观测结果。但是目前尚未发现人们预期的碳氢化合物在平流层底层的凝结现象（人们预测有几

found, but the search is continuing.

Unravelling Titan's Mysteries

Some of the major questions about Titan concern the nature of the source of methane that replaces the irreversible loss at high altitudes by photochemistry that produces a host of complex organic compounds. Open pools of liquid hydrocarbons on the surface have been suggested, as well as cryovolcanism. Also, if methane photochemistry has been occurring over the lifetime of the Solar System the organic products of these processes should have accumulated to significant depths on the surface and should be seen in images and spectra of the surface.

Although no such liquid bodies were directly imaged by DISR, there is compelling evidence for fluid flow on the surface of Titan, including the dendritic and stubby drainage channel networks, the rounded and size-graded "rocks" at the surface landing site, and the morphology of the shoreline, offshore structures, and the appearance of the darker lakebed region. The stubby networks may imply sapping or spring-fed flows, as the existence of liquid pools on the surface and the frequency of precipitation that could cause the deep dendritic drainage channels are both still unconfirmed. In addition, there are at least a few structures that suggest cryovolcanic flows on the surface.

The ground track derived by the image correlations demonstrates a zonal wind field that is mostly prograde. The general altitude profile and shape agree with predicted average models of the zonal wind flow between 50 and 10 km altitude[7], although with a reduced intensity. Below 10 km, falling wind speeds and an abrupt change of wind direction indicate a planetary boundary layer some 7–8 km thick, scaling nicely from near-equatorial terrestrial boundary layers.

Spatially resolved spectral reflectance measurements of different regions on the surface suggest that the uplands are redder than the lowland lakebed regions. The regions near the mouths of the rivers are also redder than the lake regions. A host of questions about the sequence of flooding and formation of these structures is suggested by these observations.

The reflectivity of the surface at the landing site was measured from 480 nm to 1,600 nm without the interference of methane absorption bands or haze opacity. The peak reflectivity in the dark regions is about 0.18 at 830 nm and decreases towards longer and shorter wavelengths. The red slope in the visible is consistent with organic material, such as tholins, but the blue infrared slope is still unexplained. Between 1,500 and 1,600 nm the reflectivity is low (0.06) and flat, consistent with water ice. Nevertheless, the decrease in reflectivity from 900 to 1,500 nm does not show the expected weak absorption bands of water ice near 1,000 and 1,200 nm, and the identity of the surface component responsible for this blue slope remains unknown.

种碳氢化合物应该会在那里凝结[25]），相关搜寻正在进行中。

揭秘土卫六

关于土卫六的一些主要问题都涉及了甲烷的来源问题。在高层大气中，不可逆光化学反应通过消耗甲烷不断产生大量有机化合物，而源源不断进入大气、填补这一消耗的甲烷来自哪里呢？人们认为土卫六表面存在着液态烃构成的露天水池以及冰火山活动。同时，如果甲烷光化学反应在太阳系整个寿命周期中都不断进行，那么该过程中产生的有机产物就会在土卫六表面积累并达到明显的厚度，从而可以在图像和光谱中被观测到。

然而，DISR 图像并没有直接观测到这样的液体，不过还是有令人信服的证据支持土卫六表面存在液体的流动。这些证据包括树枝状和断株状沟渠网络，着陆点附近有分选性的圆“石”，海岸线地形结构，离岸的各种结构以及较暗的湖床区域的外观。其中，这些断株状网络可能意味着液体下切侵蚀的河道或者源自泉眼的液体流；而表面存在液体湖泊以及形成较深的树枝状排水沟道的频繁降雨，这两者尚未得到证实。此外，至少有几处结构表明地表存在冰火山作用相关的流动构造。

根据图像关联性推出的探测器轨迹地面投影说明了该处的风场是一个总体上与土卫六自转同向的纬向风场。该风场的大体高度剖面和速度曲线的形状与 50 ~ 10 km 高度之间的纬向风模型的预测相一致[7]，虽然风的强度相比于模型有所减小。在 10 km 高度以下，风速的降低和风向的骤然改变表明，行星边界层厚度大约为 7 ~ 8 km，与地球赤道附近的行星边界层有很好的对应关系。

土卫六表面的空间分辨光谱反射率测量表明，高地比湖床低地区域红化程度更高，河口附近的区域也比湖区更加红化。这些观测结果向我们提出了与洪水活动的顺序以及这些地形结构如何形成等相关的许多问题。

在不受甲烷吸收峰和雾霾不透明度干扰的情况下，探测器在着陆点附近测量了 480 ~ 1,600 nm 波段的表面反射率。暗色区域的反射率在 830 nm 处到达峰值，大约为 0.18，然后随着波长的增加和减小都降低。其中，可见光波段的红坡与有机物质（如索林斯）的结果相符，但是红外波段的蓝坡目前尚未得到解释。从 1,500 nm 到 1,600 nm 的范围内，反射率较低（0.06）且比较平坦，与水冰的特征相符。不过，从 900 nm 到 1,500 nm 反射率的下降过程中并未发现 1,000 nm 和 1,200 nm 附近预期存在的较弱的水冰吸收峰。是什么表面物质导致了这一蓝坡结构目前还不得而知。

The nature of the haze aerosols measured by DISR is different in significant ways from the view before the Huygens mission. Before the Huygens probe, cloud physics models with sedimentation and coagulation predicted a strong increase in haze density with decreasing altitude[9]. In addition, measurements of the high degree of linear polarization in light scattered from Titan near a phase angle of 90° by the Pioneer and Voyager spacecraft could only be matched by spherical particles having radii less than or equal to 0.1 μm. Such small particles produced a strong increase in optical depth with decreasing wavelength shortward of 1,000 nm. Fitting the strong methane band at 890 nm constrained the amount of haze at high altitudes. This haze became optically much thicker at the wavelength of the weaker methane band at 619 nm. To fit the observed strength of this band it was necessary to remove the haze permitted by the cloud physics calculations at altitudes below about 70 km by invoking condensation of organic gases produced at very high altitudes as they diffused down to colder levels[26]. The condensation of many organic gases produced by photochemistry at high altitudes on Titan was suggested by Sagan and Thompson in ref. 25, and seemed consistent with this view.

The next development in Titan haze models (pioneered by R.W., P.S., Cabane, M. and M.L.) included the use of fractal aggregate haze particles that had a small component (monomer) with a radius of about 0.06 μm to produce strong linear polarization[26-30]. These monomers stuck together in an aggregation of many tens (or more) monomers. The large size of the aggregation could produce the strong forward scattering required from the Titan haze aerosols while preserving the high degree of linear polarization. However, it was quite laborious to compute the single-scattering properties of such aggregate particles for more than about 100 monomers at visible wavelengths. Particles with an effective radius of about 0.35 μm were required to produce the degree of forward scattering observed by Voyager[31]. This required the number of monomers in an aggregate particle to be about 45, and permitted single-scattering computations of the cross-section and phase function of the particles over the visible range. Of course, even larger numbers of monomers per particle would have matched the observations at high phase angles on Voyager, but these were difficult to perform and have largely gone unexplored. If larger particles had been used, however, the optical depth of the aerosols at shorter wavelengths would not have been nearly so large, and the clear space below 70 km may well not have been necessary.

The new DISR observations give a measurement of the monomer radius of 0.05 μm, in good agreement with previous estimates. Significantly, however, they show that the haze optical depth varies from about 2 at 935 nm to only about 4.5 at 531 nm, and the number of monomers in a haze particle is therefore probably several hundred. A value of 256 for N gives a projected area equal to that of a sphere of radius 0.65 μm, about twice as large as previously assumed. With $N = 512$, the equivalent sphere with the same projected area has a radius of 0.9 μm, nearly three times the size previously used. In any case, it seems that the size of the aggregate particles is several times as large as in some of the older models. A better estimate of the particle size will be available after the analysis of the solar aureole measurements of the variation in brightness near the Sun. In addition, measurements by the DISR violet photometer will extend the optical measurements of the haze to

由 DISR 测量得到的土卫六上的雾霾气溶胶的特性与惠更斯任务之前的观点是非常不同的。在惠更斯探测器之前，考虑了凝固和沉积作用的云层物理模型预测随着高度的降低，雾霾的密度将有较大的增幅[9]。此外，先驱者号和旅行者号飞船在 90° 相位角测量了土卫六的散射光，发现存在高度线偏振，两个飞船的测量结果只有在球形颗粒的半径小于或等于 0.1 μm 的情况下才能吻合。这么小的颗粒随着波长（小于 1,000 nm 范围内）的减小，其光深大幅增加。通过拟合 890 nm 的甲烷强吸收峰，可以约束高空雾霾的总量。但在 619 nm 的甲烷弱吸收峰，这些雾霾的光学厚度则要大得多。为了拟合该吸收峰的观测强度，有必要援引产生于高层大气的有机气体扩散到下面较冷的大气层后将发生凝结的观点[26]，以此来移除云层物理模型中允许的 70 km 以下高度的雾霾。萨根和汤普森在参考文献 25 中提出了土卫六高层大气中的许多有机气体通过光化学发生凝结的观点，看似与上述观点相符。

通过引入雾霾颗粒分形聚集，进一步发展了土卫六雾霾模型（由韦斯特、史密斯、卡班和莱蒙率先提出），其采用的小成分（单体）的半径约为 0.06 μm，可以产生很强的线偏振[26-30]。聚集颗粒由数十个（或更多）单体彼此吸附而成。这么大的聚集颗粒不但可以产生土卫六雾霾气溶胶需要的非常强的前向散射，而且能够保持高度线偏振。然而，如果构成颗粒的单体超过 100 个，则计算这么大颗粒在可见光波段的单散射特性是非常艰难的。为了解释旅行者号观测到的前向散射度，聚集颗粒的有效半径需要大约为 0.35 μm[31]，则相应的单体数量大约为 45，这样就可以在可见光范围进行粒子的截面积和相函数的单散射计算。当然，每个颗粒有更大的单体数量也能匹配旅行者号在高相位角的观测结果，但是这是很难计算的，并且在很大程度上也未探索过。然而，如果采用较大的聚集颗粒，这些气溶胶在短波长波段的光深并不一定这么大，而且也没必要假设 70 km 以下存在净空层。

最近的 DISR 观测结果表明，构成雾霾颗粒的单体的半径为 0.05 μm，这与先前的估计相符。但是，明显地，雾霾的光深从 935 nm 处大约为 2 变到 531 nm 处的 4.5，而雾霾颗粒中的单体数量可能高达数百个。如果假设单体数量 $N = 256$，则其投影面积相当于一个半径为 0.65 μm 球体，相当于先前假定值的两倍。如果假设单体数量 $N = 512$，则相同投影面积的等价球体的半径为 0.9 μm，该尺寸几乎是先前值的三倍。无论哪种情况，聚集颗粒的尺寸看起来都是原先的模型中对应值的数倍。通过对太阳光晕进行太阳周围亮度的变化测量并进行分析，可以更好地估计出粒子的尺寸。此外，DISR 紫外光度计的测量可以将雾霾的光学测量范围扩展到更短的波

wavelengths as short as the band from 350 to 480 nm, also helping to constrain the size of the haze particles.

The number density of the haze particles does not increase with depth nearly as dramatically as predicted by the older cloud physics models. In fact, the number density increases by only a factor of a few over the altitude range from 150 km to the surface. This implies that vertical mixing is much less than had been assumed in the older models where the particles are distributed approximately as the gas is with altitude. In any case, no clear space at low altitudes, which was suggested earlier[32], was seen.

The methane mole fraction of 1.6% measured in the stratosphere by the Composite Infrared Spectrometer (CIRS) and the Gas Chromatograph Mass Spectrometer is consistent with the DISR spectral measurements. At very low altitudes (20 m) DISR measured $5 \pm 1\%$ for the methane mole fraction.

Finally, the entire set of DISR observations gives a new view of Titan, and reinforces the view that processes on Titan's surface are more similar to those on the surface of the Earth than anywhere else in the Solar System.

(**438**, 765-778; 2005)

M. G. Tomasko[1], B. Archinal[2], T. Becker[2], B. Bézard[3], M. Bushroe[1], M. Combes[3], D. Cook[2], A. Coustenis[3],C. de Bergh[3], L. E. Dafoe[1], L. Doose[1], S. Douté[4], A. Eibl[1], S. Engel[1], F. Gliem[5], B. Grieger[6], K. Holso[1], E. Howington-Kraus[2], E. Karkoschka[1], H. U. Keller[6], R. Kirk[2], R. Kramm[6], M. Küppers[6], P. Lanagan[1], E. Lellouch[3], M. Lemmon[7], J. Lunine[1,8], E. McFarlane[1], J. Moores[1], G. M. Prout[1], B. Rizk[1], M. Rosiek[2], P. Rueffer[5], S. E. Schröder[6], B. Schmitt[4], C. See[1], P. Smith[1], L. Soderblom[2], N. Thomas[9] & R. West[10]

[1] Lunar and Planetary Laboratory, University of Arizona, 1629 E. University Blvd, Tucson, Arizona 85721-0092, USA

[2] US Geological Survey, Astrogeology, 2225 N. Gemini Drive, Flagstaff, Arizona 86001, USA

[3] LESIA, Observatoire de Paris, 5 place Janssen, 92195 Meudon, France

[4] Laboratoire de Planétologie de Grenoble, CNRS-UJF, BP 53, 38041 Grenoble, France

[5] Technical University of Braunschweig, Hans-Sommer-Str. 66, D-38106 Braunschweig, Germany

[6] Max Planck Institute for Solar System Research, Max-Planck-Str. 2, D-37191 Katlenburg-Lindau, Germany

[7] Department of Physics, Texas A&M University, College Station, Texas 77843-3150, USA

[8] Istituto Nazionale di Astrofisica — Istituto di Fisica dello Spazio Interplanetario (INAF-IFSI ARTOV), Via del Cavaliere, 100, 00133 Roma, Italia

[9] Department of Physics, University of Bern, Sidlerstr. 5, CH-3012 Bern, Switzerland

[10] Jet Propulsion Laboratory, 4800 Oak Grove Drive, Pasadena, California 91109, USA

Received 26 May; accepted 8 August 2005. Published online 30 November 2005.

References:

1. Coustenis, A. *et al.* Maps of Titan's surface from 1 to 2.5 μm. *Icarus* **177**, 89-105 (2005).
2. Porco, C. C. *et al.* Imaging of Titan from the Cassini spacecraft. *Nature* **434**, 159-168 (2005).
3. Sotin, C. *et al.* Infrared images of Titan. *Nature* **435**, 786-789 (2005).
4. Elachi, C. *et al.* Cassini radar views the surface of Titan. *Science* **308**, 970-974 (2005).
5. Tomasko, M. G. *et al.* The Descent Imager/Spectral Radiometer (DISR) experiment on the Huygens entry probe of Titan. *Space Sci. Rev.* **104**, 469-551 (2002).
6. Ivanov, B. A., Basilevski, A. T. & Neukem, G. Atmospheric entry of large meteoroids: implication to Titan. *Planet. Space Sci.* **45**, 993-1007 (1997).
7. Flasar, F. M., Allison, M. D. & Lunine, J. I. Titan zonal wind model. *ESA Publ.* **SP-1177**, 287-298 (1997).

段(350 ~ 480 nm)，这也有助于约束雾霾颗粒的尺寸大小。

雾霾颗粒的粒子数密度，并不像早期的云层物理模型预测的那样随着深度的增加而急剧增加。事实上，粒子数密度从 150 km 高度到表面仅仅增加了几倍。这意味着垂直混合程度要比早期模型的假设小得多，后者认为粒子随高度的分布大体和气体的分布是一样的。总之，和先前的研究一样[32]，并没有在较低高度发现雾霾净空层。

合成红外光谱仪(CIRS)和气相色谱–质谱联用分析仪测得的甲烷摩尔分数为 1.6%，这与 DISR 光谱测量结果相一致。在非常低的高度(20 m)，DISR 给出的甲烷摩尔分数为 5% ± 1%。

最后，DISR 的整个观测使我们对土卫六有了全新的认识，使我们进一步意识到土卫六表面上发生的各种过程比太阳系内其他任何天体都更类似于地球表面上发生的过程。

(金世超 翻译；陈含章 审稿)

8. Bond, N. A. Observations of planetary boundary-layer structure in the eastern equatorial Pacific. *J. Atmos. Sci.* **5**, 699-706 (1992).

9. Lindal, G. F., Wood, G. E., Hotz, H. B. & Sweetnam, D. N. The atmosphere of Titan: An analysis of the Voyager 1 radio occultation measurements. *Icarus* **53**, 348-363 (1983).

10. Tokano, T. & Neubauer, F. M. Tidal winds on Titan caused by Saturn. *Icarus* **158**, 499-515 (2002).

11. Niemann, H. B. *et al.* The abundances of constituents of Titan's atmosphere from the GCMS instrument on the Huygens probe. *Nature* doi:10.1038/nature04122 (this issue).

12. Fulchignoni, M. *et al.* *In situ* measurements of the physical characteristics of Titan's environment. *Nature* doi:10.1038/nature04314 (this issue).

13. Brown, G. N. Jr & Ziegler, W. T. in *Advances in Cryogenetic Engineering* (ed. Timmerhaus, K. D.) Vol. 25, 662-670 (Plenum, New York, 1980).

14. Griffith, C. A., Owen, T., Geballe, T. R., Rayner, J. & Rannou, P. Evidence for the exposure of water ice on Titan's surface. *Science* **300**, 628-630 (2003).

15. Coustenis, A., Lellouch, E., Maillard, J.-P. & McKay, C. P. Titan's surface: composition and variability from the near-infrared albedo. *Icarus* **118**, 87-104 (1995).

16. Lellouch, E., Schmitt, B., Coustenis, A. & Cuby, J.-G. Titan's 5-μm lightcurve. *Icarus* **168**, 204-209 (2004).

17. Grundy, W. & Schmitt, B. The temperature-dependent near-infrared absorption spectrum of hexagonal H_2O ice. *J. Geophys. Res. E* **103**, 25809-25822 (1998).

18. Bernard, J.-M. *et al.* Evidence for chemical variations at the micrometric scale of Titan's tholins: Implications for analysing Cassini-Huygens data. *Icarus* (submitted).

19. Coll, P. *et al.* Experimental laboratory simulation of Titan's atmosphere: aerosols and gas phase. *Planet. Space Sci.* **47**, 1331-1340 (1999).

20. Bernard, J.-M. *et al.* Experimental simulation of Titan's atmosphere: detection of ammonia and ethylene oxide. *Planet. Space Sci.* **51**, 1003-1011 (2003).

21. Moroz, L. V., Arnold, G., Korochantsev, A. V. & Wäsch, R. Natural solid bitumens as possible analogs for cometary and asteroid organics. 1. Reflectance spectroscopy of pure bitumens. *Icarus* **134**, 253-268 (1998).

22. Toon, O. B., McKay, C. P., Griffith, C. A. & Turco, R. P. A physical model of Titan's aerosols. *Icarus* **95**, 24-53 (1992).

23. Khare, B. N. *et al.* Optical constants of organic tholins produced in a simulated Titanian atmosphere—From soft X-ray to microwave frequencies. *Icarus* **60**, 127-137 (1984).

24. Flasar, F. M. *et al.* Titan's atmospheric temperatures, winds, and composition. *Science* **308**, 975-978 (2005).

25. Sagan, C. & Thompson, W. R. Production and condensation of organic gases in the atmosphere of Titan. *Icarus* **59**, 133-161 (1984).

26. Lemmon, M. T. *Properties of Titan's Haze and Surface.* PhD dissertation, (Univ. Arizona, 1994).

27. West, R. A. & Smith, P. H. Evidence for aggregate particles in the atmospheres of Titan and Jupiter. *Icarus* **90**, 330-333 (1991).

28. Cabane, M., Chassefière, E. & Israel, G. Formation and growth of photochemical aerosols in Titan's atmosphere. *Icarus* **96**, 176-189 (1992).

29. Cabane, M., Rannou, P., Chassefière, E. & Israel, G. Fractal aggregates in Titan's atmosphere. *Planet. Space Sci.* **41**, 257-267 (1993).

30. West, R. A. Optical properties of aggregate particles whose outer diameter is comparable to the wavelength. *Appl. Opt.* **30**, 5316-5324 (1991).

31. Rages, K. B. & Pollack, J. Vertical distribution of scattering hazes in Titan's upper atmosphere. *Icarus* **55**, 50-62 (1983).

32. McKay, C. P. *et al.* Physical properties of the organic aerosols and clouds on Titan. *Planet. Space Sci.* **49**, 79-99 (2001).

33. Bird, M. K. *et al.* The vertical profile of winds on Titan. *Nature* doi:10.1038/nature04060 (this issue).

34. Karkoschka, E. Methane, ammonia, and temperature measurements of the Jovian planets and Titan from CCD-spectrophotometry. *Icarus* **133**, 134-146 (1998).

Acknowledgements. We thank the people from the following organizations whose dedication and effort have made this project successful: AETA (Fontenay-aux-Roses, France), Alcatel Space (Cannes, France), Collimated Holes Inc., EADS Deutschland GmbH (formerly Deutsche Aerospace AG, Munich, Germany), ETEL Motion Technology (Mortiers, Switzerland), The European Space Agency's (ESA) European Space and Technology Centre (ESTEC), The European Space Operations Centre (ESOC), The Jet Propulsion Laboratory (JPL), Laboratoire de Planétologie de Grenoble (CNRS-UJF), Loral Fairchild (Tustin, California, USA), Martin Marietta Corporation (Denver, Colorado, USA), Max-Planck-Institut für Sonnensystemforschung (Katlenburg-Lindau, Germany), Observatoire de Paris (Meudon, France), Technische Universität Braunschweig (TUB), Thomson-CSF (Grenoble, France), University of Arizona's Kuiper Lunar and Planetary Laboratory (LPL), and the US Geological Survey (Flagstaff, Arizona, USA).

Author Information. Reprints and permissions information is available at npg.nature.com/reprintsandpermissions. The authors declare no competing financial interests. Correspondence and requests for materials should be addressed to C.S. (csee@lpl.arizona.edu).

Discovery of a Cool Planet of 5.5 Earth Masses through Gravitational Microlensing

J.-P Beaulieu *et al.*

Editor's Note

The search for extrasolar planets continues apace, with nearly 4,000 known in mid-2019. It is difficult from the surface of the Earth to find planets of about the mass of the Earth using the traditional Doppler technique of looking for the wobble in the parent star's position. Here Jean-Philippe Beaulieu and his colleagues report a planet of only 5.5 Earth masses, found using gravitational microlensing (the bending and focusing of light rays by gravity). The planet is about 2.6 astronomical units (the Earth–Sun distance) from its parent star—about where the asteroid belt is in our solar system. The planet orbits the star acting as a lens, which passed between our solar system and a more distant, lensed star.

In the favoured core-accretion model of formation of planetary systems, solid planetesimals accumulate to build up planetary cores, which then accrete nebular gas if they are sufficiently massive. Around M-dwarf stars (the most common stars in our Galaxy), this model favours the formation of Earth-mass ($M_\oplus$) to Neptune-mass planets with orbital radii of 1 to 10 astronomical units (AU), which is consistent with the small number of gas giant planets known to orbit M-dwarf host stars[1-4]. More than 170 extrasolar planets have been discovered with a wide range of masses and orbital periods, but planets of Neptune's mass or less have not hitherto been detected at separations of more than 0.15 AU from normal stars. Here we report the discovery of a $5.5^{+5.5}_{-2.7}M_\oplus$ planetary companion at a separation of $2.6^{+1.5}_{-0.6}$ AU from a $0.22^{+0.21}_{-0.11}M_\odot$ M-dwarf star, where $M_\odot$ refers to a solar mass. (We propose to name it OGLE-2005-BLG-390Lb, indicating a planetary mass companion to the lens star of the microlensing event.) The mass is lower than that of GJ876d (ref. 5), although the error bars overlap. Our detection suggests that such cool, sub-Neptune-mass planets may be more common than gas giant planets, as predicted by the core accretion theory.

GRAVITATIONAL microlensing events can reveal extrasolar planets orbiting the foreground lens stars if the light curves are measured frequently enough to characterize planetary light curve deviations with features lasting a few hours[6-9]. Microlensing is most sensitive to planets in Earth-to-Jupiter-like orbits with semi-major axes in the range 1–5 AU. The sensitivity of the microlensing method to low-mass planets is restricted by the finite angular size of the source stars[10,11], limiting detections to planets of a few $M_\oplus$ for giant

通过微引力透镜发现一颗5.5倍地球质量的冷行星

博利厄等

编者按

搜寻系外行星的工作进展迅速，截至2019年年中，已有将近4,000颗系外行星被发现。在地表利用传统的多普勒技术探测母恒星位置移动，进而发现与地球质量相当的行星是十分困难的。本文中，让–菲利普·博利厄与他的同事报道了通过微引力透镜（光线被引力弯曲和聚焦）找到的一颗仅有地球质量5.5倍的行星。这颗行星大致距离其母恒星2.6个天文单位（日地距离），大约对应于太阳系中小行星带所处的位置。围绕母星转动的行星与其母星，在经过我们的太阳系与更遥远的、被透镜的恒星之间时，表现为一个“透镜”。

核吸积模型是当前较为流行的行星系统形成模型，该模型认为微行星会聚集形成一个致密的行星核。如果它们的质量足够大，那么它们将逐渐吸积星云气体。M型矮星是银河系中最常见的恒星，根据核吸积模型，在它们附近形成的行星质量通常介于地球质量（$M_{\oplus}$）和海王星质量之间，相应的轨道半径为1～10个天文单位（AU）。这个预言与实际观测结果相符合，因为已知围绕M型矮星运动的气体巨行星的数目很少[1-4]。目前，人们已经发现了超过170颗的系外行星，其质量和轨道周期的范围跨度很大，但是迄今尚未发现与正常恒星相距超过0.15AU、质量与海王星相当或略小的行星。在本文中，我们发现了一颗质量为$5.5^{+5.5}_{-2.7}M_{\oplus}$的类行星伴天体，其宿主恒星是质量为$0.22^{+0.21}_{-0.11}M_{\odot}$的M型矮星（$M_{\odot}$表示太阳的质量），二者相距$2.6^{+1.5}_{-0.6}$ AU。我们建议将这颗星体命名为OGLE-2005-BLG-390Lb，用以表示是微引力透镜事件中与恒星一起组成透镜系统的行星伴星。相比GJ8764d，虽然两者的误差棒相互重叠，但是这颗行星的质量更小[5]。我们的探测研究结果表明，同核吸积模型预测的一样，这些温度偏低的亚海王星质量的行星可能比气体巨行星更为普遍。

如果光变曲线的采样频率足够高，以致能够识别时标为几个小时的行星光变曲线偏移特征，微引力透镜事件可以揭示围绕前景透镜恒星转动的系外行星[6-9]。微引力透镜对于探测那些轨道半长轴介于1～5 AU之间（对应从地球到木星之间的轨道范围）的行星最为灵敏。该方法在探测低质量行星方面的灵敏度主要受限于背景源恒星有限的角尺度[10,11]，因此，对于巨型源恒星，行星质量的探测下限大致为几个

source stars, but allowing the detection of planets as small as $0.1M_{\oplus}$ for main-sequence source stars in the Galactic Bulge. The PLANET collaboration[12] maintains the high sampling rate required to detect low-mass planets while monitoring the most promising of the > 500 microlensing events discovered annually by the OGLE collaboration, as well as events discovered by MOA. A decade of pioneering microlensing searches has resulted in the recent detections of two Jupiter-mass extrasolar planets[13,14] with orbital separations of a few AU by the combined observations of the OGLE, MOA, MicroFUN and PLANET collaborations. The absence of perturbations to stellar microlensing events can be used to constrain the presence of planetary lens companions. With large samples of events, upper limits on the frequency of Jupiter-mass planets have been placed over an orbital range of 1–10 AU, down to $M_{\oplus}$ planets[15-17] for the most common stars of our galaxy.

On 11 July 2005, the OGLE Early Warning System[18] announced the microlensing event OGLE-2005-BLG-390 (right ascension α = 17 h 54 min 19.2 s, declination $\delta = -30° 22' 38''$, J2000) with a relatively bright clump giant as a source star. Subsequently, PLANET, OGLE and MOA monitored it with their different telescopes. After peaking at a maximum magnification of $A_{max} = 3.0$ on 31 July 2005, a short-duration deviation from a single lens light curve was detected on 9 August 2005 by PLANET. As described below, this deviation was due to a low-mass planet orbiting the lens star.

From analysis of colour-magnitude diagrams, we derive the following reddening-corrected colours and magnitudes for the source star: $(V-I)_0 = 0.85$, $I_0 = 14.25$ and $(V-K)_0 = 1.9$. We used the surface brightness relation[20] linking the emerging flux per solid angle of a light-emitting body to its colour, calibrated by interferometric observations, to derive an angular radius of 5.25 ± 0.73 μas, which corresponds to a source radius of $9.6 \pm 1.3R_{\odot}$ (where $R_{\odot}$ is the radius of the Sun) if the source star is at a distance of 8.5 kpc. The source star colours indicate that it is a 5,200 K giant, which corresponds to a G4 III spectral type.

Figure 1 shows our photometric data for microlensing event OGLE-2005-BLG-390 and the best planetary binary lens model. The best-fit model has $\chi^2 = 562.26$ for 650 data points, seven lens parameters, and 12 flux normalization parameters, for a total of 631 degrees of freedom. Model length parameters in Table 1 are expressed in units of the Einstein ring radius R_E (typically ~2 AU for a Galactic Bulge system), the size of the ring image that would be seen in the case of perfect lens–source alignment. In modelling the light curve, we adopted linear limb darkening laws[21] with $\Gamma_I = 0.538$ and $\Gamma_R = 0.626$, appropriate for this G4 III giant source star, to describe the centre-to-limb variation of the intensity profile in the I and R bands. Four different binary lens modelling codes were used to confirm that the model we present is the only acceptable model for the observed light curve. The best alternative model is one with a large-flux-ratio binary source with a single lens, which has gross features that are similar to a planetary microlensing event[22]. However, as shown in Fig. 1, this model fails to account for the PLANET-Perth, PLANET-Danish and OGLE

太阳质量。但对于星系核球中的主序源恒星，则允许探测到 0.1 个太阳质量大小的行星。PLANET 合作组织 [12] 每年对光学引力透镜实验（OGLE）合作组织以及天文物理微透镜观测（MOA）合作组织发现的超过 500 个最有可能的微引力透镜事件进行监控，其采样频率足够高，可以满足低质量行星探测的要求。结合过去十年 OGLE、MOA、MicroFUN 和 PLANET 合作组织对微引力透镜早期搜寻的观测结果，人们最近发现了两颗质量与木星相当、轨道半径相差几个 AU 的系外行星 [13,14]。无扰动的恒星微引力透镜事件可以用来限制透镜中行星伴星的存在。基于大样本的微引力透镜事件，人们不仅对轨道半径在 1 ~ 10 AU 范围内木星质量的行星出现频率给出了上限，还对于银河系中最常见的恒星周围低至 $M_{\oplus}$的行星出现频率给出了限制 [15-17]。

2005 年 7 月 11 日，OGLE 的预警系统 [18] 报告发现了微引力透镜事件 OGLE-2005-BLG-390（赤经 $\alpha = 17$ h 54 min 19.2 s，赤纬 $\delta = -30°22'38''$，J2000），其源恒星为一颗相对明亮的团簇巨星。随后，PLANET、OGLE 和 MOA 分别利用望远镜对其进行了监控观测。在 2005 年 7 月 31 日该引力透镜事件的放大率达到峰值 $A_{max} = 3.0$ 之后，2005 年 8 月 9 日 PLANET 探测到一个短时偏离单透镜光变曲线的现象。如下所述，这种偏离是由一颗围绕透镜恒星运动的低质量行星引起的。

通过分析颜色–星等图，我们可以推知这颗源恒星经过红化校正后的颜色和星等，结果如下：$(V-I)_0 = 0.85$，$I_0 = 14.25$，$(V-K)_0 = 1.9$。我们利用面亮度关系 [20] 将发光天体单位立体角内的辐射通量与颜色联系起来。经过干涉观测校正，我们推导出源恒星的角半径为 5.25 ± 0.73 μas。如果这颗源恒星的距离为 8.5 kpc，那么上述角半径值对应的物理半径为 $9.6 \pm 1.3R_{\odot}$，其中 $R_{\odot}$表示太阳的半径。这颗源恒星的颜色表明，它是一颗温度为 5,200 K 的巨星，对应于 G4 III 光谱类型。

图 1 给出了微引力透镜事件 OGLE-2005-BLG-390 的测光数据以及最佳的行星双透镜模型。最优拟合模型的 $\chi^2 = 562.26$，该拟合具有 650 个数据点、7 个透镜参数、12 个通量归一化参数，总自由度为 631。表 1 给出的模型长度参数以爱因斯坦环半径 R_E（对于一个银河系核球系统而言，其典型值约为 2 AU）为单位，这个爱因斯坦环的半径对应当透镜、源恒星以及观测者精确排列在一条直线上时所能观测到的环形图像的尺寸。在建模光变曲线时，我们采用线性临边昏暗定律 [21] 来描述 I 和 R 波段从中心到边缘的发光强度轮廓，参数取值为 $\Gamma_I = 0.538$，$\Gamma_R = 0.626$，适用于 G4 III 巨型源恒星。四套不同的双透镜建模代码被用来验证我们这里所展示的模型，是针对观测到的光变曲线的唯一可接受模型。最好的替代模型包含单一的前景透镜恒星，背景源为具有大流量比的双星，该模型可以粗略地产生类似于行星微引力透镜事件的特征 [22]。然而，如图 1 所示，这一模型无法解释由 PLANET-Perth 天文台

measurements near the end of the planetary deviation, and it is formally excluded by $\Delta\chi^2 = 46.25$ with one less model parameter.

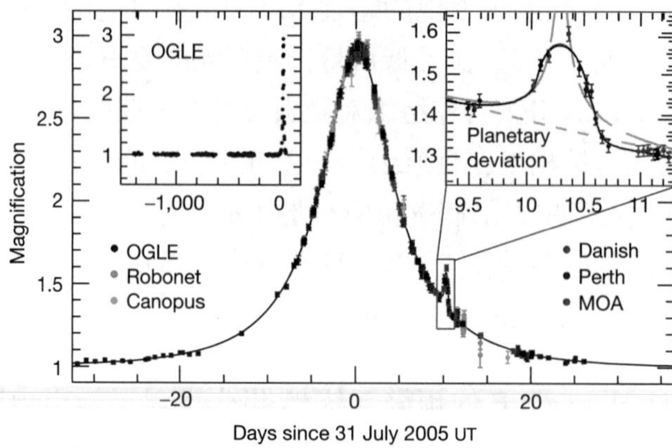

Fig. 1. The observed light curve of the OGLE-2005-BLG-390 microlensing event and best-fit model plotted as a function of time. Error bars are 1σ. The data set consists of 650 data points from PLANET Danish (ESO La Silla, red points), PLANET Perth (blue), PLANET Canopus (Hobart, cyan), RoboNet Faulkes North (Hawaii, green), OGLE (Las Campanas, black), MOA (Mt John Observatory, brown). This photometric monitoring was done in the I band (with the exception of the Faulkes R-band data and the MOA custom red passband) and real-time data reduction was performed with the different OGLE, PLANET and MOA data reduction pipelines. Danish and Perth data were finally reduced by the image subtraction technique[19] with the OGLE pipeline. The top left inset shows the OGLE light curve extending over the previous 4 years, whereas the top right one shows a zoom of the planetary deviation, covering a time interval of 1.5 days. The solid curve is the best binary lens model described in the text with $q = 7.6 \pm 0.7 \times 10^{-5}$, and a projected separation of $d = 1.610 \pm 0.008 R_E$. The dashed grey curve is the best binary source model that is rejected by the data, and the dashed orange line is the best single lens model.

Table 1 Microlensing fit parameters

d	$1.610 \pm 0.008 R_E$
q	$(7.6 \pm 0.7) \times 10^{-5}$
Closest approach	$0.359 \pm 0.005 R_E$
Einstein ring radius crossing time	11.03 ± 0.11 days
Time of closest approach	31.231 ± 0.005 July 2005 UT
Source star radius crossing time	0.282 ± 0.010 days
θ	2.756 ± 0.003 rad

The parameters for the best binary lens model for the OGLE 2005-BLG-390 microlensing event light curve are shown with their 1σ uncertainties. Some of these parameters are scaled to the Einstein ring radius, which is given by $R_E = 2\sqrt{GMD_L(D_S - D_L)/(c^2 D_S)}$, where M is the mass of the lens, G is the newtonian constant of gravitation, c is the speed of light in vacuum, and D_L and D_S are the lens and source distances, respectively.

The planet is designated OGLE-2005-BLG-390Lb, where the "Lb" suffix indicates the secondary component of the lens system with a planetary mass ratio. The microlensing fit

望远镜、PLANET-Danish 天文台望远镜以及 OGLE 观测到的行星偏离末段附近的数据，该模型缺少一个模型参数，因 $\Delta\chi^2 = 46.25$ 从严格意义上被排除。

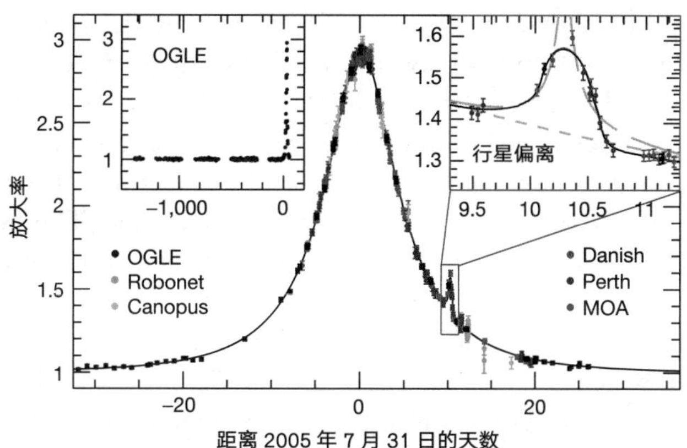

图 1. 观测到的 OGLE-2005-BLG-390 微引力透镜事件的光变曲线以及最佳拟合模型(横轴为时间)。误差棒为 1σ。图中共有 650 个数据点，分别来源于 PLANET-Danish 天文台望远镜(拉西亚欧南台(ESO)，红色)、PLANET-Perth 天文台望远镜(蓝色)、PLANET-Canopus 天文台望远镜(霍巴特，青色)、RoboNet 北福克斯望远镜(夏威夷，绿色)、OGLE(拉斯坎帕纳斯，黑色)以及 MOA(约翰山天文台，棕色)。其中，除福克斯望远镜采用 R 波段数据，MOA 采用特制的红色通带外，其他测光监测均在 I 波段进行。我们采用 OGLE、PLANET 以及 MOA 不同的数据处理软件对数据进行实时处理。PLANET-Danish 和 PLANET-Perth 天文台望远镜的数据最终通过 OGLE 数据处理中的图像扣除技术 [19] 进行处理。左上方的插图给出了前四年里的 OGLE 测得的光变曲线，右上方插图给出了行星偏离部分的放大图，其时间跨度为 1.5 天。实线表示文中所述的最佳双透镜模型的理论预言，$q = 7.6 \pm 0.7 \times 10^{-5}$，投影间隔 $d = 1.610 \pm 0.008R_E$。图中灰色虚线表示不被数据支持的最佳双源模型的理论预言。图中橙色虚线表示最佳单透镜模型的理论预言。

表 1. 微引力透镜的拟合参数

参数	值
d	$1.610 \pm 0.008R_E$
q	$(7.6 \pm 0.7) \times 10^{-5}$
最接近点	$0.359 \pm 0.005R_E$
爱因斯坦环半径的穿越时间	11.03 ± 0.11 天
最接近点时间	31.231 ± 0.005 2005 年 7 月 UT
源恒星半径的穿越时间	0.282 ± 0.010 天
θ	2.756 ± 0.003 rad

上表给出了用于拟合 OGLE 2005-BLG-390 透镜事件光变曲线的最佳双透镜模型的参数值，同时给出了各参数 1σ 不确定度。表中部分参数以爱因斯坦半径 R_E 作为单位，$R_E = 2\sqrt{GMD_L(D_S - D_L)/(c^2D_S)}$，其中 M 代表透镜的质量，G 表示牛顿万有引力常数，c 表示真空中的光速，D_S 和 D_L 分别表示透镜距离和源距离。

这颗行星被命名为 OGLE-2005-BLG-390Lb，其中后缀“Lb”表示透镜系统包含具有行星质量的次等组分。微引力透镜拟合仅直接确定行星与恒星的质量比

only directly determines the planet–star mass ratio, $q = 7.6 \pm 0.7 \times 10^{-5}$, and the projected planet–star separation, $d = 1.610 \pm 0.008 R_E$. Although the planet and star masses are not directly determined for planetary microlensing events, we can derive their probability densities. We have performed a bayesian analysis[23] employing the Galactic models and mass functions described in refs 11 and 23. We averaged over the distances and velocities of the lens and source stars, subject to the constraints due to the angular diameter of the source and the measured parameters given in Table 1. This analysis gives a 95% probability that the planetary host star is a main-sequence star, a 4% probability that it is a white dwarf, and a probability of < 1% that it is a neutron star or black hole. The host star and planet parameter probability densities for a main sequence lens star are shown in Fig. 2 for the Galactic model used in ref. 23. The medians of the lens parameter probability distributions yield a companion mass of $5.5^{+5.5}_{-2.7} M_{\oplus}$ and an orbital separation of $2.6^{+1.5}_{-0.6}$ AU from the $0.22^{+0.21}_{-0.11} M_{\odot}$ lens star, which is located at a distance of $D_L = 6.6 \pm 1.0$ kpc. These error bars indicate the central 68% confidence interval. These median parameters imply that the planet receives radiation from its host star that is only 0.1% of the radiation that the Earth receives from the Sun, so the probable surface temperature of the planet is ~50 K, similar to the temperatures of Neptune and Pluto.

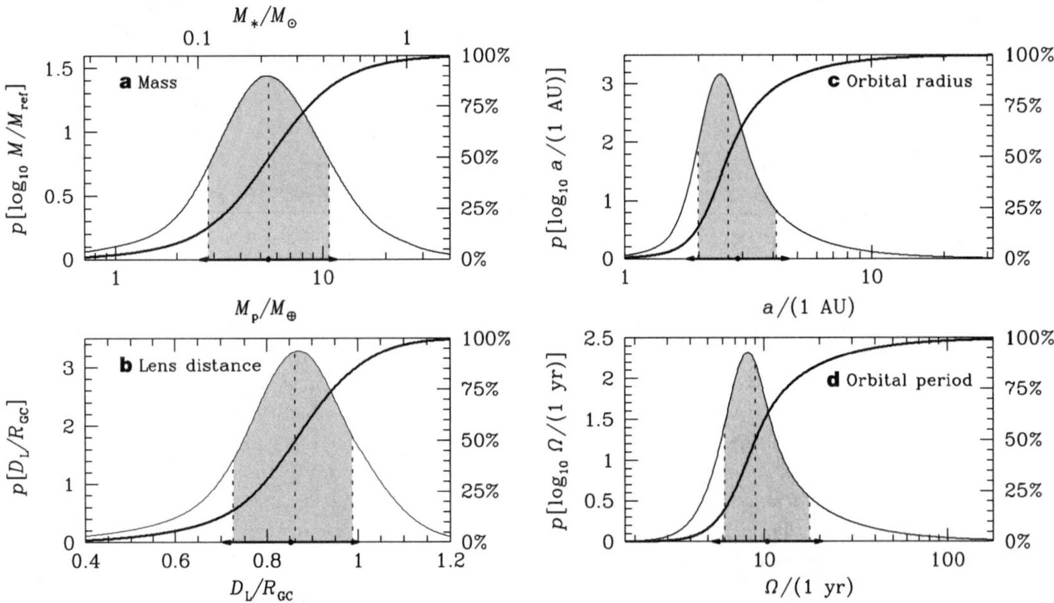

Fig. 2. Bayesian probability densities for the properties of the planet and its host star. **a**, The masses of the lens star and its planet (M_* and M_p respectively), **b**, their distance from the observer (D_L), **c**, the three-dimensional separation or semi-major axis *a* of an assumed circular planetary orbit; and **d**, the orbital period Ω of the planet. (In **a**, M_{ref} refers to $M_{\oplus}$ on the upper x axis and $M_{\odot}$ on the lower x axis.) The bold, curved line in each panel is the cumulative distribution, with the percentiles listed on the right. The dashed vertical lines indicate the medians, and the shading indicates the central 68.3% confidence intervals, while dots and arrows on the abscissa mark the expectation value and standard deviation. All estimates follow from a bayesian analysis assuming a standard model for the disk and bulge population of the Milky Way, the stellar mass function of ref. 23, and a gaussian prior distribution for $D_S = 1.05 \pm 0.25 R_{GC}$ (where $R_{GC} = 7.62 \pm 0.32$ kpc for the Galactic Centre distance). The medians of these distributions yield a $5.5^{+5.5}_{-2.7} M_{\oplus}$

$q = 7.6 \pm 0.7 \times 10^{-5}$，以及行星与恒星的投影距离，$d = 1.610 \pm 0.008 R_E$。尽管我们不能通过行星微引力透镜事件直接确定恒星和行星的质量，但是却可以推导出它们的概率密度。我们通过引入银河系模型和质量函数[11,23]，进行了贝叶斯分析[23]。基于源的角直径以及表 1 中测量参数的限制，我们对透镜和源恒星在距离和速度上进行了平均。该分析结果表明，该行星宿主恒星是一颗主序恒星的概率为 95%，是一颗白矮星的概率为 4%，是一颗中子星或黑洞的概率小于 1%。图 2 给出了银河系模型[23]下透镜系统中以主序星作为宿主恒星及其行星各参数的概率密度。透镜参数概率分布的中值表明伴星的质量为 $5.5^{+5.5}_{-2.7} M_\oplus$，围绕质量为 $0.22^{+0.21}_{-0.11} M_\odot$ 的透镜恒星运动，轨道半径为 $2.6^{+1.5}_{-0.6}$ AU，整个透镜系统与我们的距离为 $D_L = 6.6 \pm 1.0$ kpc。图中误差棒给出了中心附近 68% 的置信区间。这些中值参数表明该行星从宿主恒星接收到的辐射，仅为地球从太阳接收到的辐射的 0.1%，因此这颗行星的表面温度很可能仅为 50 K 左右，类似于海王星和冥王星。

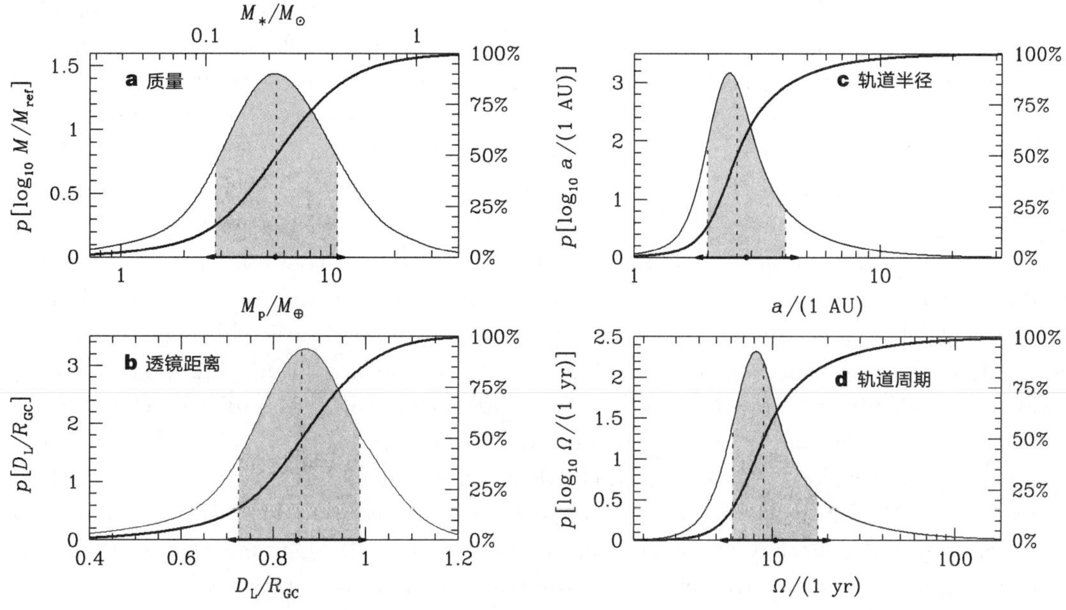

图 2. 行星及其宿主恒星属性的贝叶斯概率密度。**a**，透镜恒星及其行星的质量，分别用 M_* 和 M_p 表示；**b**，透镜恒星及其行星与观测者的距离（D_L）；**c**，假定圆形行星轨道的三维间隔或半长轴 a；**d**，行星的轨道周期 Ω。（在图 **a** 中，M_{ref} 表示在顶部 x 轴以 $M_\oplus$ 为单位，在底部 x 轴以 $M_\odot$ 为单位）在每幅图中，加粗曲线表示累计分布，右侧坐标轴给出了相应的百分比值。图中垂直虚线表示中值，阴影部分给出了中心附近 68.3% 的置信区间，横坐标轴上的点和箭头分别标记出期望值和标准差。以上所有估算均是在假定一个标准的银河系盘和核球模型，文献 23 中的恒星质量函数以及 $D_S = 1.05 \pm 0.25 R_{GC}$ 满足高斯先验分布（其中 R_{GC} 代表银河中心的距离，$R_{GC} = 7.62 \pm 0.32$ kpc）的情况下利用贝叶斯分析得到的。这些分布的中值表明行星的质量为 $5.5^{+5.5}_{-2.7} M_\oplus$，在银河系核球区域的 M 型矮星的质量为 $0.22^{+0.21}_{-0.11} M_\odot$，二者相距 $2.6^{+1.5}_{-0.6}$ AU，而整个透镜系统与太阳的距离为 $D_L = 6.6 \pm 1.0$ kpc。行星的中值周期为 9^{+9}_{-3} 年。对概率分布取对数平均（遵循开普勒第三定律），可知这颗行星的轨道周期为 10.4 年，与宿主恒星的距离为 2.9 AU，恒星和行星的质量分别为 $0.22 M_\odot$ 和 $5.5 M_\oplus$。每幅图中的纵坐标的方括号中列出了该概率密度的独

planetary companion at a separation of $2.6^{+1.5}_{-0.6}$ AU from a $0.22^{+0.21}_{-0.11}M_{\odot}$ Galactic Bulge M-dwarf at a distance of 6.6 ± 1.0 kpc from the Sun. The median planetary period is 9^{+9}_{-3} years. The logarithmic means of these probability distributions (which obey Kepler's third law) are a separation of 2.9 AU, a period of 10.4 years, and masses of $0.22M_{\odot}$ and $5.5M_{\oplus}$ for the star and planet, respectively. In each plot, the independent variable for the probability density is listed within square brackets. The distribution of the planet–star mass ratio was taken to be independent of the stellar mass, and a uniform prior distribution was assumed for the planet–star separation distribution.

The parameters of this event are near the limits of microlensing planet detectability for a giant source star. The separation of $d = 1.61$ is near the outer edge of the so-called lensing zone[7], which has the highest planet detection probability, and the planet's mass is about a factor of two above the detection limit set by the finite size of the source star. Planets with $q > 10^{-3}$ and $d \approx 1$ are much easier to detect, and so it may be that the parameters of OGLE-2005-BLG-390Lb represent a more common type of planet. This can be quantified by simulating planetary light curves with different values of q and θ (where θ is the angle of source motion with respect to the lens axis) but the remaining parameters are fixed to the values for the three known microlensing planets. We find that the probability of detecting a $q \approx 4-7 \times 10^{-3}$ planet, like the first two microlens planets[13,14], is ~50 times larger than the probability of detecting a $q = 7.6 \times 10^{-5}$ planet like OGLE-2005-BLG-390Lb. This suggests that, at the orbital separations probed by microlensing, sub-Neptune-mass planets are significantly more common than large gas giants around the most common stars in our Galaxy. Similarly, the first detection of a sub-Neptune-mass planet at the outer edge of the "lensing zone" provides a hint that these sub-Neptune-mass planets may tend to reside in orbits with semi-major axes $a > 2$ AU.

The core-accretion model of planet formation predicts that rocky/icy $5-15M_{\oplus}$ planets orbiting their host stars at 1–10 AU are much more common than Jupiter-mass planets, and this prediction is consistent with the small fraction of M-dwarfs with planets detected by radial velocities[3,5] and with previous limits from microlensing[15]. Our discovery of such a low-mass planet by gravitational microlensing lends further support to this model, but more detections of similar and lower-mass planets over a wide range of orbits are clearly needed. Planets with separations of ~0.1 AU will be detected routinely by the radial velocity method or space observations of planetary transits in the coming years[24-27], but the best chance to increase our understanding of such planets over orbits of 1–10 AU in the next 5–10 years is by future interferometer programs[28] and more advanced microlensing surveys[11,29,30].

(**439**, 437-440; 2006)

立变量。这里认为行星–恒星质量比的分布与恒星的质量无关，并假定行星与宿主恒星之间的距离分布服从均匀的先验分布。

这次事件的各参数值接近巨型源恒星的微引力透镜探测极限。恒星、行星相距 $d = 1.61$ 已接近所谓的“透镜区域”的外边界 [7]，从而具有最高的行星探测概率；同时行星的质量是由源恒星的有限尺度所设定的探测极限的 2 倍。显然，那些 $q>10^{-3}$ 且 $d\approx 1$ 的行星更加容易探测到，因此 OGLE-2005-BLG-390Lb 可能代表了一类更为普遍的行星。这可以在不同 q 和 θ 取值条件下通过对行星光变曲线进行模拟仿真而被量化(其中 θ 表示源恒星运动相对透镜主轴的角度)，而剩余参量则对应已知的三个微透镜行星取固定值。我们发现，探测到一颗 $q\approx 4\sim 7\times 10^{-3}$ 的行星(类似首先发现的两颗透镜行星)的概率是探测到一颗 $q = 7.6\times 10^{-5}$ 的行星(类似 OGLE-2005-BLG-390Lb)的概率的约 50 倍。这说明，在微引力透镜可探测的轨道间隔的范围内，对于我们星系中的众多恒星，其四周具有亚海王星质量的行星比大型气态巨行星明显更加普遍。类似地，在“透镜区域”外边缘首次探测到亚海王星质量的行星，暗示着这类行星的轨道半长轴倾向于 >2 AU。

行星形成的核吸积模型预测，在宿主恒星附近普遍存在的是质量为 $5\sim 15M_{\oplus}$、轨道半长轴在 1 ~ 10 AU 范围内的行星，而不是木星质量大小的行星。这与运用径向速度法很少探测到 M 型矮星附近的行星 [3,5] 的观测结果相吻合，同时也与过去的微引力透镜限制一致 [15]。我们通过微引力透镜方法发现的这样一颗低质量的行星，进一步支持了该模型，但是我们仍需探测更多类似的位于不同轨道上的低质量行星来验证这一模型。在未来几年中，通过径向速度法或空间观测掩食法探测轨道半长轴约为 0.1 AU 的行星 [24-27] 将成为一种常态。然而，未来 5 ~ 10 年里，借助将来的干涉仪探测项目 [29] 以及更为先进的微引力透镜巡天 [11,28,30] 是进一步了解那些轨道介于 1 ~ 10 AU 的行星性质的最好的方法。

（金世超 刘项琨 翻译；李然 审稿）

J.-P. Beaulieu[1,4], D. P. Bennett[1,3,5], P. Fouqué[1,6], A. Williams[1,7], M. Dominik[1,8], U. G. Jørgensen[1,9], D. Kubas[1,10], A. Cassan[1,4], C. Coutures[1,11], J. Greenhill[1,12], K. Hill[1,12], J. Menzies[1,13], P. D. Sackett[1,14], M. Albrow[1,15], S. Brillant[1,10], J. A. R. Caldwell[1,16], J. J. Calitz[1,17], K. H. Cook[1,18], E. Corrales[1,4], M. Desort[1,4], S. Dieters[1,12], D. Dominis[1,19], J. Donatowicz[1,20], M. Hoffman[1,19], S. Kane[1,21], J.-B. Marquette[1,4], R. Martin[1,7], P. Meintjes[1,17], K. Pollard[1,15], K. Sahu[1,22], C. Vinter[1,9], J. Wambsganss[1,23], K. Woller[1,9], K. Horne[1,8], I. Steele[1,24], D. M. Bramich[1,8,24], M. Burgdorf[1,24], C. Snodgrass[1,25], M. Bode[1,24], A. Udalski[2,26], M. K. Szymański[2,26], M. Kubiak[2,26], T. Więckowski[2,26], G. Pietrzyński[2,26,27], I. Soszyński[2,26,27], O. Szewczyk[2,26], Ł. Wyrzykowski[2,26,28], B. Paczyński[2,29], F. Abe[3,30], I. A. Bond[3,31], T. R. Britton[3,15,32], A. C. Gilmore[3,15], J. B. Hearnshaw[3,15], Y. Itow[3,30], K. Kamiya[3,30], P. M. Kilmartin[3,15], A. V. Korpela[3,33], K. Masuda[3,30], Y. Matsubara[3,30], M. Motomura[3,30], Y. Muraki[3,30], S. Nakamura[3,30], C. Okada[3,30], K. Ohnishi[3,34], N. J. Rattenbury[3,28], T. Sako[3,30], S. Sato[3,35], M. Sasaki[3,30], T. Sekiguchi[3,30], D. J. Sullivan[3,33], P. J. Tristram[3,32], P. C. M. Yock[3,32] & T. Yoshioka[3,30]

[1] PLANET/RoboNet Collaboration (http://planet.iap.fr and http://www.astro.livjm.ac.uk/RoboNet/)

[2] OGLE Collaboration (http://ogle.astrouw.edu.pl)

[3] MOA Collaboration (http://www.physics.auckland.ac.nz/moa)

[4] Institut d'Astrophysique de Paris, CNRS, Université Pierre et Marie Curie UMR7095, 98bis Boulevard Arago, 75014 Paris, France

[5] University of Notre Dame, Department of Physics, Notre Dame, Indiana 46556-5670, USA

[6] Observatoire Midi-Pyrénées, Laboratoire d'Astrophysique, UMR 5572, Université Paul Sabatier—Toulouse 3, 14 avenue Edouard Belin, 31400 Toulouse, France

[7] Perth Observatory, Walnut Road, Bickley, Perth, WA 6076, Australia

[8] Scottish Universities Physics Alliance, University of St Andrews, School of Physics and Astronomy, North Haugh, St Andrews KY16 9SS, UK

[9] Niels Bohr Institutet, Astronomisk Observatorium, Juliane Maries Vej 30, 2100 København Ø, Denmark

[10] European Southern Observatory, Casilla 19001, Santiago 19, Chile

[11] CEA DAPNIA/SPP Saclay, 91191 Gif-sur-Yvette cedex, France

[12] University of Tasmania, School of Mathematics and Physics, Private Bag 37, Hobart, TAS 7001, Australia

[13] South African Astronomical Observatory, PO Box 9, Observatory 7935, South Africa

[14] Research School of Astronomy and Astrophysics, Australian National University, Mt Stromlo Observatory, Weston Creek, ACT 2611, Australia

[15] University of Canterbury, Department of Physics and Astronomy, Private Bag 4800, Christchurch 8020, New Zealand

[16] McDonald Observatory, 16120 St Hwy Spur 78 #2, Fort Davis, Texas 79734, USA

[17] Boyden Observatory, University of the Free State, Department of Physics, PO Box 339, Bloemfontein 9300, South Africa

[18] Lawrence Livermore National Laboratory, IGPP, PO Box 808, Livermore, California 94551, USA

[19] Universität Potsdam, Institut für Physik, Am Neuen Palais 10, 14469 Potsdam, Astrophysikalisches Institut Potsdam, An der Sternwarte 16, D-14482, Potsdam, Germany

[20] Technische Universität Wien, Wiedner Hauptstrasse 8/020 B.A. 1040 Wien, Austria

[21] Department of Astronomy, University of Florida, 211 Bryant Space Science Center, Gainesville, Florida 32611-2055, USA

[22] Space Telescope Science Institute, 3700 San Martin Drive, Baltimore, Maryland 21218, USA

[23] Astronomisches Rechen-Institut (ARI), Zentrum für Astronomie, Universität Heidelberg, Mönchhofstrasse 12-14, 69120 Heidelberg, Germany

[24] Astrophysics Research Institute, Liverpool John Moores University, Twelve Quays House, Egerton Wharf, Birkenhead CH41 1LD, UK

[25] Astronomy and Planetary Science Division, Department of Physics, Queen's University Belfast, Belfast, UK

[26] Obserwatorium Astronomiczne Uniwersytetu Warszawskiego, Aleje Ujazdowskie 4, 00-478 Warszawa, Poland

[27] Universidad de Concepcion, Departamento de Fisica, Casilla 160-C, Concepcion, Chile

[28] Jodrell Bank Observatory, The University of Manchester, Macclesfield, Cheshire SK11 9DL, UK

[29] Princeton University Observatory, Peyton Hall, Princeton, New Jersey 08544, USA

[30] Solar-Terrestrial Environment Laboratory, Nagoya University, Nagoya 464-860, Japan

[31] Institute for Information and Mathematical Sciences, Massey University, Private Bag 102-904, Auckland, New Zealand

[32] Department of Physics, University of Auckland, Private Bag 92019, Auckland, New Zealand

[33] School of Chemical and Physical Sciences, Victoria University, PO Box 600, Wellington, New Zealand

[34] Nagano National College of Technology, Nagano 381-8550, Japan

[35] Department of Astrophysics, Faculty of Science, Nagoya University, Nagoya 464-860, Japan

Received 28 September; accepted 14 November 2005.

References:

1. Safronov, V. *Evolution of the Protoplanetary Cloud and Formation of the Earth and Planets* (Nauka, Moscow, 1969).
2. Wetherill, G. W. Formation of the terrestrial planets. *Annu. Rev. Astron. Astrophys* **18**, 77-113 (1980).
3. Laughlin, G., Bodenheimer, P. & Adams, F. C. The core accretion model predicts few jovian-mass planets orbiting red dwarfs. *Astrophys. J.* **612**, L73-L76 (2004).
4. Ida, S. & Lin, D. N. C. Toward a deterministic model of planetary formation. II. The formation and retention of gas giant planets around stars with a range of metallicities. *Astrophys. J.* **616**, 567-572 (2004).
5. Rivera, E. *et al.* A ~7.5 Earth-mass planet orbiting the nearby star, GJ 876. *Astrophys. J.* (in the press).
6. Mao, S. & Paczynski, B. Gravitational microlensing by double stars and planetary systems. *Astrophys. J.* **374**, L37-L40 (1991).
7. Gould, A. & Loeb, A. Discovering planetary systems through gravitational Microlenses. *Astrophys. J.* **396**, 104-114 (1992).
8. Wambsganss, J. Discovering Galactic planets by gravitational microlensing: magnification patterns and light curves. *Mon. Not. R. Astron. Soc.* **284**, 172-188 (1997).
9. Griest, K. & Safizadeh, N. The use of high-magnification microlensing events in discovering extrasolar planets. *Astrophys. J.* **500**, 37-50 (1998).
10. Bennett, D. P. & Rhie, S. H. Detecting Earth-mass planets with gravitational microlensing. *Astrophys. J.* **472**, 660-664 (1996).
11. Bennett, D. P. & Rhie, S. H. Simulation of a space-based microlensing survey for terrestrial extrasolar planets. *Astrophys. J.* **574**, 985-1003 (2002).
12. Albrow, M. *et al.* The 1995 pilot campaign of PLANET: searching for microlensing anomalies through precise, rapid, round-the-clock monitoring. *Astrophys. J.* **509**, 687-702 (1998).
13. Bond, I. A. *et al.* OGLE 2003-BLG-235/MOA 2003-BLG-53: A planetary microlensing event. *Astrophys. J.* **606**, L155-L158 (2004).
14. Udalski, A. *et al.* A jovian-mass planet in microlensing event OGLE-2005-BLG- 071. *Astrophys. J.* **628**, L109-L112 (2005).
15. Gaudi, B. S. *et al.* Microlensing constraints on the frequency of Jupiter-mass companions: analysis of 5 years of PLANET photometry. *Astrophys. J.* **566**, 463-499 (2002).
16. Abe, F. *et al.* Search for low-mass exoplanets by gravitational microlensing at high magnification. *Science* **305**, 1264-1267 (2004).
17. Dong, S. *et al.* Planetary detection efficiency of the magnification 3000 microlensing event OGLE-2004-BLG-343. *Astrophys. J.* (submitted); preprint at 〈http://arXiv.org/astro-ph/0507079〉 (2005).
18. Udalski, A. The optical gravitational lensing experiment. Real time data analysis systems in the OGLE-III survey. *Acta Astron.* **53**, 291-305 (2003).
19. Alard, C. Image subtraction using a space-varying kernel. *Astron. Astrophys. Suppl.* **144**, 363-370 (2000).
20. Kervella, P. *et al.* Cepheid distances from infrared long-baseline interferometry. III. Calibration of the surface brightness-color relations. *Astron. Astrophys.* **428**, 587-593 (2004).
21. Claret, A., Diaz-Cordoves, J. & Gimenez, A. Linear and non-linear limb-darkening coefficients for the photometric bands R I J H K. *Astron. Astrophys. Suppl.* **114**, 247-252 (1995).
22. Gaudi, B. S. Distinguishing between binary-source and planetary microlensing perturbations. *Astrophys. J.* **506**, 533-539 (1998).
23. Dominik, M. Stochastical distributions of lens and source properties for observed galactic microlensing events. *Mon. Not. R. Astron. Soc.* (submitted); preprint at 〈http://arXiv.org/astro-ph/0507540〉 (2005).
24. Vogt, S. S. *et al.* Five new multicomponent planetary systems. *Astrophys. J.* **632**, 638-658 (2005).
25. Mayor, M. *et al.* The CORALIE survey for southern extrasolar planets. XII. Orbital solutions for 16 extrasolar planets discovered with CORALIE. *Astron. Astrophys.* **415**, 391-402 (2004).
26. Borucki, W. *et al.* in *Second Eddington Workshop: Stellar Structure and Habitable Planet Finding* (eds Favata, F., Aigrain, S. & Wilson, A.) 177-182 (ESA SP-538, ESA Publications Division, Noordwijk, 2004).
27. Moutou, C. *et al.* Comparative blind test of five planetary transit detection algorithms on realistic synthetic light curves. *Astron. Astrophys.* **437**, 355-368 (2005).
28. Sozzeti, A. *et al.* Narrow-angle astrometry with the space interferometry mission: the search for extrasolar planets. I. Detection and characterization of single planets. *Pub. Astron. Soc. Pacif.* **114**, 1173-1196 (2002).
29. Bennett, D. P. in *ASP Conf. Ser. on Extrasolar Planets: Today and Tomorrow* (eds Beaulieu, J.-P., Lecavelier des Etangs, A. & Terquem, C.) Vol. 321, 59-68 (ASP, 2004).
30. Beaulieu, J. P. *et al.* PLANET III: searching for Earth-mass planets via microlensing from Dome C? *ESA Publ. Ser.* **14**, 297-302 (2005).

Acknowledgements. PLANET is grateful to the observatories that support our science (the European Southern Observatory, Canopus, Perth; and the South African Astronomical Observatory, Boyden, Faulkes North) and to the ESO team in La Silla for their help in maintaining and operating the Danish telescope. Support for the PLANET project was provided by CNRS, NASA, the NSF, the LLNL/NNSA/DOE, PNP, PICS France-Australia, D. Warren, the DFG, IDA and the SNF. RoboNet is funded by the UK PPARC and the FTN was supported by the Dill Faulkes Educational Trust. Support for the OGLE project, conducted at Las Campanas Observatory (operated by the Carnegie Institution of Washington), was provided by the Polish Ministry of Science, the Foundation for Polish Science, the NSF and NASA. The MOA collaboration is supported by MEXT and JSPS of Japan, and the Marsden Fund of New Zealand.

Author Information. The photometric data set is available at planet.iap.fr and ogle.astrouw.edu.pl. Reprints and permissions information is available at npg.nature.com/reprintsandpermissions. The authors declare no competing financial interests. Correspondence and requests for materials should be addressed to J.P.B. (beaulieu@iap.fr) or D.P.B. (bennett@nd.edu).

Folding DNA to Create Nanoscale Shapes and Patterns

P. W. K. Rothemund

Editor's Note

Organizing molecules into structures with dimensions of several nanometres is a key objective in nanotechnology. DNA has emerged as a particularly versatile material for such molecular engineering, because it can be programmed to self-assemble in highly specific and selective ways, governed by the rules of complementary base-pairing. Artificial DNA strands had previously been used to make topologically complex molecules. But the strategy described here by Paul Rothemund at the California Institute of Technology goes much further, enabling the design of DNA that can be folded into just about any arbitrary two-dimensional pattern. Rothemund's DNA map of the world at a scale of $1:2\times10^{14}$ is emblematic of the power of contemporary nanotechnology.

"Bottom-up fabrication", which exploits the intrinsic properties of atoms and molecules to direct their self-organization, is widely used to make relatively simple nanostructures. A key goal for this approach is to create nanostructures of high complexity, matching that routinely achieved by "top-down" methods. The self-assembly of DNA molecules provides an attractive route towards this goal. Here I describe a simple method for folding long, single-stranded DNA molecules into arbitrary two-dimensional shapes. The design for a desired shape is made by raster-filling the shape with a 7-kilobase single-stranded scaffold and by choosing over 200 short oligonucleotide "staple strands" to hold the scaffold in place. Once synthesized and mixed, the staple and scaffold strands self-assemble in a single step. The resulting DNA structures are roughly 100 nm in diameter and approximate desired shapes such as squares, disks and five-pointed stars with a spatial resolution of 6 nm. Because each oligonucleotide can serve as a 6-nm pixel, the structures can be programmed to bear complex patterns such as words and images on their surfaces. Finally, individual DNA structures can be programmed to form larger assemblies, including extended periodic lattices and a hexamer of triangles (which constitutes a 30-megadalton molecular complex).

IN 1959, Richard Feynman put forward the challenge of writing the *Encyclopaedia Britannica* on the head of a pin[1], a task which he calculated would require the use of dots 8 nm in size. Scanning probe techniques have essentially answered this challenge: atomic force microscopy[2] (AFM) and scanning tunnelling microscopy[3,4] (STM) allow us to manipulate individual atoms. But these techniques create patterns serially (one line or

折叠 DNA 以形成纳米尺度的形状和图案

罗特蒙德

编者按

将分子在几个纳米的尺度上组织起来是纳米技术的一个重要目标。DNA 已经成为分子工程中用途非常广泛的材料，因为它可以在遵循碱基互补配对的前提下，通过编程方法以高度特异性和选择性的方式自组装。之前人工合成 DNA 链已经被用来组装具有复杂拓扑结构的分子。但是本文中美国加州理工学院的保罗·罗特蒙德所描述的方法远远跨出一步，这种方法能够设计 DNA 使其折叠成几乎任意的二维图案。罗特蒙德设计的比例尺为 $1:2\times10^{14}$ 的 DNA 世界地图，标志着当代纳米技术的巨大力量。

“自下而上的组装方法”，即通过原子与分子的固有性质以指导其自组装过程，被广泛地用来制造相对简单的纳米结构。这一方法的核心目标之一在于创造出具有高度复杂性的纳米结构，其复杂性程度可与一般通过“自上而下”方法所得到的相媲美。DNA 分子的自组装为这一目标的实现提供了一个引人关注的途径。本文中我要描述一种简单的方法，它能够将长的单链 DNA 分子折叠成任意的二维形状。在设计所需形状时，先将一条 7,000 多个碱基的单链脚手架链像光栅一样填充起来，再选取 200 多条短链的低聚核苷酸“订书针链”将模板在合适的位置固定。合成并混合后，订书针链和脚手架链就会通过单步反应完成自组装。所得 DNA 结构直径大约为 100 nm，并且接近所需形状，如正方形、圆盘和五角星等，其空间分辨率为 6 nm。由于每个低聚核苷酸可以充当一个 6 nm 的像素，因此可以通过编程在该结构表面上设计诸如词汇和图像等复杂图案。最后，单个 DNA 结构可以通过编程形成更大的组装，包括扩展的周期性晶格和一个三角形六聚体（构成一个 30 兆道尔顿的分子复合物）等。

1959 年，理查德·费曼发起了在针头上写下《大英百科全书》的挑战 [1]，根据他的计算，实现这一任务需要使用大小为 8 nm 的像素点。扫描探针技术实际上已经能够应对这一挑战：原子力显微镜 [2]（AFM）和扫描隧道显微镜 [3,4]（STM）使我们能够对单个原子进行操纵。但是这些技术只能逐次创造图案（一次一条线或一个像素），

one pixel at a time) and tend to require ultrahigh vacuum or cryogenic temperatures. As a result, methods based on self-assembly are considered as promising alternatives that offer inexpensive, parallel synthesis of nanostructures under mild conditions[5]. Indeed, the power of these methods has been demonstrated in systems based on components ranging from porphyrins[6] to whole viral particles[7]. However, the ability of such systems to yield structures of high complexity remains to be demonstrated. In particular, the difficulty of engineering diverse yet specific binding interactions means that most self-assembled structures contain just a few unique positions that may be addressed as "pixels".

Nucleic acids can help overcome this problem: the exquisite specificity of Watson–Crick base pairing allows a combinatorially large set of nucleotide sequences to be used when designing binding interactions. The field of "DNA nanotechnology"[8,9] has exploited this property to create a number of more complex nanostructures, including two-dimensional arrays with 8–16 unique positions and less than 20 nm spacing[10,11], as well as three-dimensional shapes such as a cube[12] and truncated octahedron[13]. However, because the synthesis of such nanostructures involves interactions between a large number of short oligonucleotides, the yield of complete structures is highly sensitive to stoichiometry (the relative ratios of strands). The synthesis of relatively complex structures was thus thought to require multiple reaction steps and purifications, with the ultimate complexity of DNA nanostructures limited by necessarily low yields. Recently, the controlled folding of a long single DNA strand into an octahedron was reported[14], an approach that may be thought of as "single-stranded DNA origami". The success of this work suggested that the folding of long strands could, in principle, proceed without many misfoldings and avoid the problems of stoichiometry and purification associated with methods that use many short DNA strands.

I now present a versatile and simple "one-pot" method for using numerous short single strands of DNA to direct the folding of a long, single strand of DNA into desired shapes that are roughly 100 nm in diameter and have a spatial resolution of about 6 nm. I demonstrate the generality of this method, which I term "scaffolded DNA origami", by assembling six different shapes, such as squares, triangles and five-pointed stars. I show that the method not only provides access to structures that approximate the outline of any desired shape, but also enables the creation of structures with arbitrarily shaped holes or surface patterns composed of more than 200 individual pixels. The patterns on the 100-nm-sized DNA shapes thus have a complexity that is tenfold higher than that of any previously self-assembled arbitrary pattern and comparable to that achieved using AFM and STM surface manipulation[4].

Design of Scaffolded DNA Origami

The design of a DNA origami is performed in five steps, the first two by hand and the last three aided by computer (details in Supplementary Note S1). The first step is to build a geometric model of a DNA structure that will approximate the desired shape. Figure 1a shows an example shape (outlined in red) that is 33 nm wide and 35 nm tall. The shape

而且往往需要超高真空或者超低温。因此，基于自组装过程的方法被认为是最有发展前景的选择，它成本低，且能在温和条件下并行合成纳米结构[5]。事实上，这种方法的力量已经在基于从卟啉[6]到整个病毒颗粒[7]等组分的各种系统中得到证实。但是，利用这种方法来制备高度复杂性结构的能力有待于进一步的验证。尤其是设计多种具有特异性的相互作用这点非常困难，这使得大多数自组装结构只包含很少可以被称为"像素"的独特部位。

核酸有助于克服这一难题：沃森–克里克碱基配对结构的高度特异性使得我们在设计相互作用时可以使用组合学意义上的大量核苷酸序列。"DNA 纳米技术"[8,9]领域便是要探索这一性质以创造出多种更为复杂的纳米结构，其中包括具有 8～16 个独特部位和间隔不到 20 nm 的二维阵列[10,11]，以及诸如立方体[12]和截角八面体[13]等三维形状。但是，由于上述纳米结构的合成涉及大量短链低聚核苷酸之间的相互作用，于是完整结构的产率很容易受到化学计量比（各链的相对比例）的影响。因此一般认为相对复杂结构的合成需要多步反应与纯化，DNA 纳米结构的最终复杂性必定受到低产率的限制。最近，有人报道将长的单链 DNA 可控折叠成八面体[14]，这种方法被认为是"单链 DNA 折纸术"。这一研究的成功意味着，在原则上，长链的折叠可以准确无误地进行，而且避免了使用大量短链 DNA 组装时涉及的化学计量与纯化的问题。

现在我将提出一种通用的而又简单的"一锅法"，利用大量短的单链 DNA 来指导一条长的单链 DNA 折叠成所需形状，其直径约为 100 nm 且具有约 6 nm 的空间分辨率。我将通过组装诸如正方形、三角形和五角星等六种不同的形状来阐明这种被我称之为"脚手架式 DNA 折纸术"方法的普遍性。我要指出的是，这种方法不仅提供了构建所需形状的轮廓的途径，而且还能够创造出由 200 多个独立像素点构成的具有任意形状的孔洞或表面图案的结构。因此 100 nm 大小的 DNA 表面上的图案，其复杂度是此前所得到的任何一种自组装图案的复杂度的十倍，并且可以与利用 AFM 和 STM 表面操纵技术所得到的图案[4]相媲美。

脚手架式 DNA 折纸术的设计

DNA 折纸术的设计是通过五步来进行的，前两步由手工实现，后三步则需计算机辅助（细节见附注 S1）。第一步是搭建一个与所需形状相似的 DNA 结构的几何模型。图 1a 展示了一个示例形状（其外廓为红色），宽 33 nm，高 35 nm。用偶数个平

is filled from top to bottom by an even number of parallel double helices, idealized as cylinders. The helices are cut to fit the shape in sequential pairs and are constrained to be an integer number of turns in length. To hold the helices together, a periodic array of crossovers (indicated in Fig. 1a as small blue crosses) is incorporated; these crossovers designate positions at which strands running along one helix switch to an adjacent helix and continue there. The resulting model approximates the shape within one turn (3.6 nm) in the x-direction and roughly two helical widths (4 nm) in the y-direction. As noticed before in DNA lattices[15], parallel helices in such structures are not close-packed, perhaps owing to electrostatic repulsion. Thus the exact y-resolution depends on the gap between helices. The gap, in turn, appears to depend on the spacing of crossovers. In Fig. 1a crossovers occur every 1.5 turns along alternating sides of a helix, but any odd number of half-turns may be used. In this study, data are consistent with an inter-helix gap of 1 nm for 1.5-turn spacing and 1.5 nm for 2.5-turn spacing, yielding a y-resolution of 6 or 7 nm, respectively.

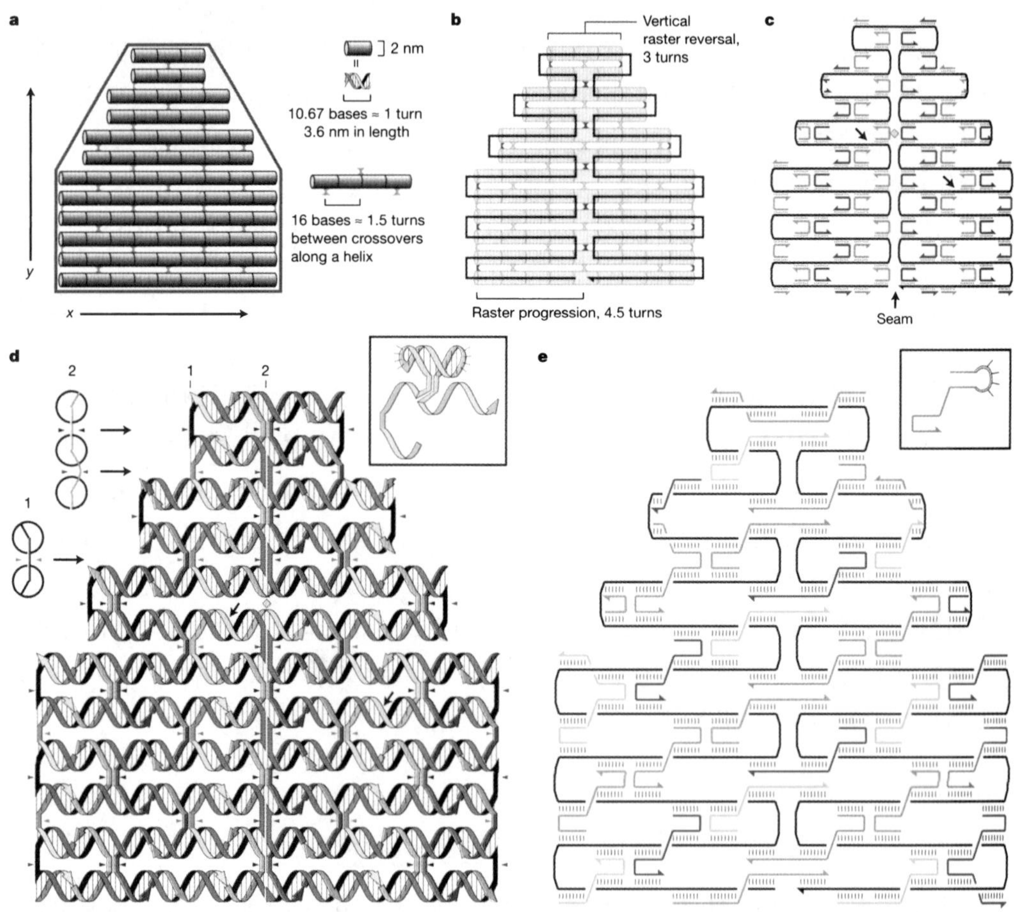

Fig. 1. **Design of DNA origami. a**, A shape (red) approximated by parallel double helices joined by periodic crossovers (blue). **b**, A scaffold (black) runs through every helix and forms more crossovers (red). **c**, As first designed, most staples bind two helices and are 16-mers. **d**, Similar to **c** with strands drawn

行的双螺旋从顶部至底部将该形状填满(理想化地表示为圆柱形)。将螺旋进行剪切形成连续碱基对并且将其长度限制在整数个螺旋。为将螺旋维持在一起，要建立一种周期性的交叉阵列(在图 1a 中用小的蓝色交叉表示)；这些交叉指明了沿着某一螺旋延伸的链转向一个邻近螺旋并继续延伸的位点。所得模型模拟的形状在 x 方向上 1 个螺距内(3.6 nm)，在 y 方向上则大致两个螺旋宽度(4 nm)。如同之前已在 DNA 晶格中注意到的那样 [15]，这种结构中的平行螺旋并不是密堆积的，这可能是由于静电斥力。因此 y 方向上精确的分辨率取决于螺旋之间的空隙。反过来，空隙则取决于交叉的间隔。在图 1a 中，交叉沿着螺旋的侧面交替出现，每 1.5 个螺距出现一次，但是半个螺距的奇数倍都是可以用的。在这项研究中，所有数据符合 1.5 个螺距有 1 nm 的交叉螺旋空隙，2.5 个螺距有 1.5 nm 空隙，因此 y 方向上的分辨率分别为 6 nm 或 7 nm。

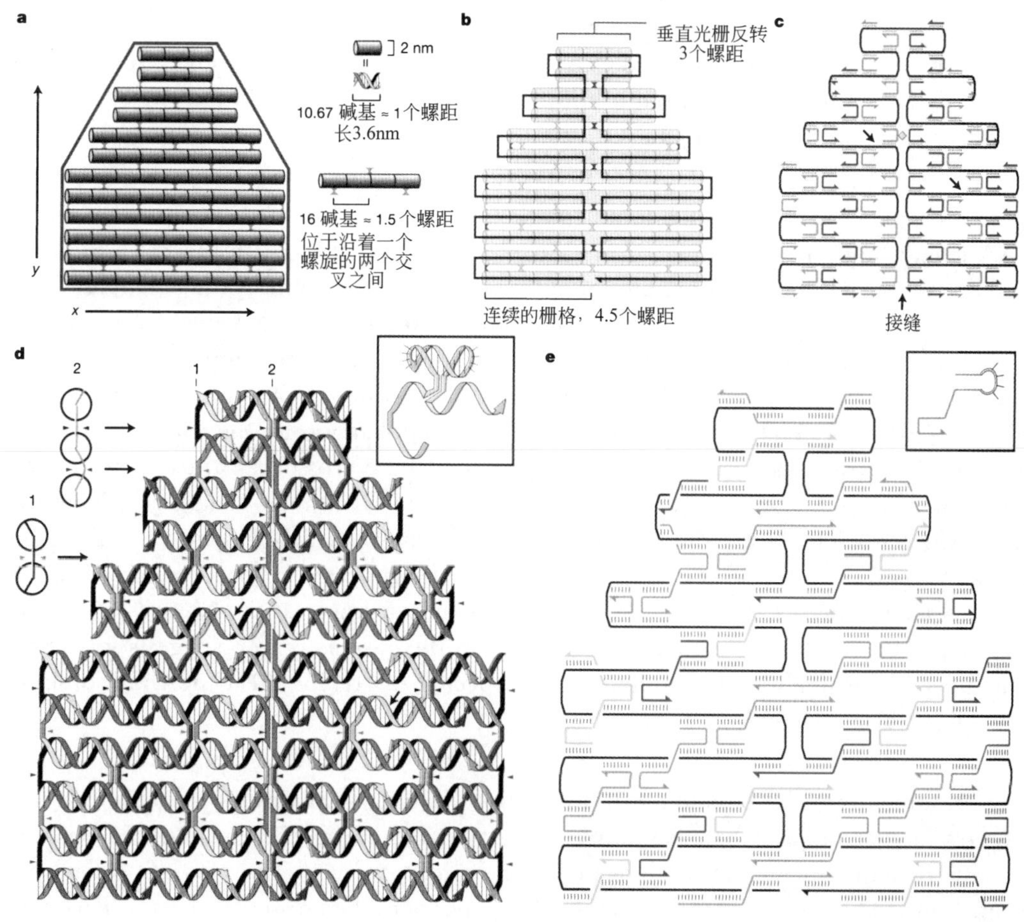

图 1. **DNA 折纸术的设计。a**，用周期性交叉(蓝色)连接起来的平行双螺旋所模拟而成的形状(红色)。**b**，一根遍及每一螺旋并且形成更多交叉(红色)的脚手架(黑色)。**c**，如同最初所设计的，大多数订书针链连接两个螺旋，长度为 16 个碱基。**d**，类似于 **c**，其中的订书针链画成螺旋。红色三角形指的是脚

as helices. Red triangles point to scaffold crossovers, black triangles to periodic crossovers with minor grooves on the top face of the shape, blue triangles to periodic crossovers with minor grooves on bottom. Cross-sections of crossovers (1, 2, viewed from left) indicate backbone positions with coloured lines, and major/minor grooves by large/small angles between them. Arrows in **c** point to nicks sealed to create green strands in **d**. Yellow diamonds in **c** and **d** indicate a position at which staples may be cut and resealed to bridge the seam. **e**, A finished design after merges and rearrangements along the seam. Most staples are 32-mers spanning three helices. Insets show a dumbbell hairpin (**d**) and a 4-T loop (**e**), modifications used in Fig. 3.

Conceptually, the second step (illustrated in Fig. 1b) proceeds by folding a single long scaffold strand (900 nucleotides (nt) in Fig. 1b) back and forth in a raster fill pattern so that it comprises one of the two strands in every helix; progression of the scaffold from one helix to another creates an additional set of crossovers, the "scaffold crossovers" (indicated by small red crosses in Fig. 1b). The fundamental constraint on a folding path is that the scaffold can form a crossover only at those locations where the DNA twist places it at a tangent point between helices. Thus for the scaffold to raster progressively from one helix to another and onto a third, the distance between successive scaffold crossovers must be an odd number of half turns. Conversely, where the raster reverses direction vertically and returns to a previously visited helix, the distance between scaffold crossovers must be an even number of half-turns. Note that the folding path shown in Fig. 1b is compatible with a circular scaffold and leaves a "seam" (a contour which the path does not cross).

Once the geometric model and a folding path are designed, they are represented as lists of DNA lengths and offsets in units of half-turns. These lists, along with the DNA sequence of the actual scaffold to be used, are input to a computer program. Rather than assuming 10.5 base pairs (bp) per turn (which corresponds to standard B-DNA twist), the program uses an integer number of bases between periodic crossovers (for example, 16 bp for 1.5 turns). It then performs the third step, the design of a set of "staple strands" (the coloured DNA strands in Fig. 1c) that provide Watson–Crick complements for the scaffold and create the periodic crossovers. Staples reverse direction at these crossovers; thus crossovers are antiparallel, a stable configuration well characterized in DNA nanostructures[16]. Note that the crossovers in Fig. 1c are drawn somewhat misleadingly, in that single-stranded regions appear to span the inter-helix gap even though the design leaves no bases unpaired. In the assembled structures, helices are likely to bend gently to meet at crossovers so that only a single phosphate from each backbone occurs in the gap (as ref. 16 suggests for similar structures). Such small-angle bending is not expected to greatly affect the width of DNA origami (see also Supplementary Note S2).

The minimization and balancing of twist strain between crossovers is complicated by the non-integer number of base pairs per half-turn (5.25 in standard B-DNA) and the asymmetric nature of the helix (it has major and minor grooves). Therefore, to balance the strain[15] caused by representing 1.5 turns with 16 bp, periodic crossovers are arranged with a glide symmetry, namely that the minor groove faces alternating directions in alternating columns of periodic crossovers (see Fig. 1d, especially cross-sections 1 and 2). Scaffold crossovers are not balanced in this way. Thus in the fourth step, the twist of scaffold

手架交叉，黑色三角形对应的周期性交叉处小沟在底面，蓝色三角形对应的交叉处小沟在顶面。交叉的横截面（1，2，从左面看去）上以有色线段显示出骨架的位置，而大 / 小沟则用它们之间的大 / 小角度来表现。**c** 中的箭头表示缺口，它们闭合起来以产生 **d** 中的绿色订书针链。**c** 和 **d** 中的黄色菱形指出订书针链可以在此处被截短并重新闭合而将接缝连接起来。**e**，沿接缝经过合并重排后的设计。大多数订书针链长度为 32 个碱基，跨越 3 个螺旋。小图显示了一个哑铃形发卡结构（**d**）和一个 4–T 环（**e**），图 3 中用它们进行了修饰。

从概念上讲，第二步（如图 1b 所示）是通过将一条长的脚手架单链（图 1b 中所示为 900 个核苷酸（nt））在光栅中来回折叠以填充图形，从而使其构成每个双螺旋中的一条链；脚手架从一个螺旋到另一个螺旋的推进产生一组附加的交叉，即“脚手架交叉”（在图 1b 中用小的红色交叉表示）。对于折叠路径的基本限制是，只有在那些 DNA 扭转出现在两个螺旋间切点的位置时，脚手架才能形成交叉。因此当脚手架在光栅中从一个螺旋渐次移向下一个以及第三个时，连续的脚手架交叉点之间的距离必定是半螺距的奇数倍。反过来，当光栅在垂直方向上反转过来并返回之前曾到达的螺旋时，脚手架交叉点之间的距离必定是半螺距的偶数倍。注意图 1b 中所显示的折叠路径是与环形脚手架相一致的，并且留下了一条“接缝”（一条折叠路径未曾穿过的周线）。

一旦设计好几何模型和折叠路径，便可以将 DNA 的长度以及补偿以半螺旋为单位进行列表。将这些列表连同实际应用的脚手架的 DNA 序列，输入到计算机程序之中。程序并未假定每个螺距有 10.5 个碱基对（bp）（相当于标准 B-DNA 扭转），而是在周期性交叉点之间使用整数个碱基（例如每 1.5 个螺距有 16 bp）。接着便实施第三步，用沃森–克里克互补准则为脚手架设计一系列“订书针链”（图 1c 中彩色 DNA 链）产生周期性的交叉。订书针链在这些交叉处倒转方向；因此交叉是反平行的，这是一种很好地表征过的 DNA 纳米结构的稳定构型[16]。注意图 1c 中的交叉的画法是有些误导性的，因为单链区域似乎跨越了螺旋间空隙，然而设计中没有留下未配对碱基。在组装结构中，螺旋应该微微有些弯曲使其能在交叉处相遇，因而每一骨架中只有一个磷酸酯出现在空隙中（如同参考文献 16 所提出的类似结构那样）。预期这种小角度弯曲不会显著地影响 DNA 折叠结构的宽度（另见附注 S2）。

由于每半个螺距之中的非整数碱基对（标准 B-DNA 中是 5.25）以及螺旋的不对称性（其中有大沟和小沟），最小化和平衡交叉之间的扭曲张力是非常复杂的。因此，为了平衡采用 1.5 个螺距中 16 bp 产生的张力[15]，周期性交叉的排列具有滑移对称性，即小沟在交替的周期性交叉列中面朝交替的方向（见图 1d，尤其是横截面 1 和 2）。脚手架交叉不是以这种方式来平衡的。于是在第四步中，计算出脚手架交

crossovers is calculated and their position is changed (typically by a single bp) to minimize strain; staple sequences are recomputed accordingly. Along seams and some edges the minor groove angle (150°) places scaffold crossovers in tension with adjacent periodic crossovers (Fig. 1d, cross-section 2); such situations are left unchanged.

Wherever two staples meet there is a nick in the backbone. Nicks occur on the top and bottom faces of the helices, as depicted in Fig. 1d. In the final step, to give the staples larger binding domains with the scaffold (in order to achieve higher binding specificity and higher binding energy which results in higher melting temperatures), pairs of adjacent staples are merged across nicks to yield fewer, longer, staples (Fig. 1e). To strengthen a seam, an additional pattern of breaks and merges may be imposed to yield staples that cross the seam; a seam spanned by staples is termed "bridged". The pattern of merges is not unique; different choices yield different final patterns of nicks and staples. All merge patterns create the same shape but, as shown later, the merge pattern dictates the type of grid underlying any pixel pattern later applied to the shape.

Folding M13mp18 Genomic DNA into Shapes

To test the method, circular genomic DNA from the virus M13mp18 was chosen as the scaffold. Its naturally single-stranded 7,249-nt sequence was examined for secondary structure, and a hairpin with a 20-bp stem was found. Whether staples could bind at this hairpin was unknown, so a 73-nt region containing it was avoided. When a linear scaffold was required, M13mp18 was cut (in the 73-nt region) by digestion with *Bsr*BI restriction enzyme. While 7,176 nt remained available for folding, most designs did not fold all 7,176 nt; short ($\leqslant$ 25 nt) "remainder strands" were added to complement unused sequence. In general, a 100-fold excess of 200–250 staple and remainder strands were mixed with scaffold and annealed from 95 °C to 20 °C in < 2 h. When samples were deposited on mica, only folded DNA structures stuck to the surface while excess staples remained in solution; AFM imaging thus proceeded under buffer without prior purification. Six different folds were explored; Fig. 2 gives their folding paths and their predicted and experimentally observed DNA structures. (Models and staple sequences are given in Supplementary Note S3, final designs appear in Supplementary Note S12. Experimental methods are given in Supplementary Note S4, results described here but not shown are in Supplementary Note S5.) Of the products imaged by AFM, a particular structure was considered qualitatively "well-formed" if it had no defect (hole or indentation in the expected outline) greater than 15 nm in diameter. For each fold the fraction of well-formed structures, as a percentage of all distinguishable structures in one or more AFM fields, was calculated as a rough estimate of yield. I note that while some structures classified as well-formed had 15-nm defects, most had no defects greater than 10 nm in diameter.

叉的扭曲并改变其位置（典型情况下是变动一个 bp）以使张力最小化；相应地，订书针链的序列被重新指定。沿着接缝和某些棱处，小沟角度（150°）使脚手架交叉与其毗邻的周期性交叉产生张力（图 1d，横截面 2）；这种情况并未被改变。

无论两个订书针链在哪里相遇，骨架中都会有一个缺口。缺口出现在螺旋的顶部和底部的面上，如图 1d 所示。在最后一步中，为了使订书针链与脚手架链具有更大的连接范围（目的是获得更高的连接特异性和更高的结合能，从而具有更高的熔点），将邻近的各对订书针链跨过缺口合并，以产生出更少、更长的订书针链（图 1e）。为了使接缝更坚固，可以额外打断并连接一些订书针链使其穿过接缝；被订书针所横跨的接缝称为“桥联的”。接合的图案并不唯一；不同的选择导致最终图案中的缺口和订书针链不同。全部接合图案制造出相同的形状，但是如同后面将会指出的，接合图案决定了后面将会应用于此形状的任何像素图案的背景光栅类型。

将 M13mp18 基因组 DNA 折叠成型

为了检验该方法，选取来自病毒 M13mp18 的圆形基因组 DNA 作为脚手架。对其天然的 7,249 个核苷酸序列进行二级结构的检测，发现其中具有一个茎长 20 bp 的发夹状结构。由于不知道订书针链能否在这个发夹结构上弯曲，因此避开包含它的 73 个核苷酸区域。在需要线性的脚手架链时，用 *Bsr*BI 限制性内切酶通过降解作用将 M13mp18（在 73 个核苷酸区域中）切开。尽管可供折叠用的部分还有 7,176 个核苷酸，大多数设计却并不折叠全部 7,176 个核苷酸；而是用短的（≤25 个核苷酸）“余链”以互补未用到的序列。一般而言，将有 100 倍过量的 200～250 个的订书针和余链与脚手架混合，并在 2 小时内从 95 ℃退火到 20 ℃。将样品沉积在云母上之后，只有折叠的 DNA 结构才能固定在表面上，而多余的订书针则留在溶液中；因此可以不经过预先提纯而在缓冲液中进行 AFM 成像。我们尝试了 6 种不同的折叠；图 2 给出了它们的折叠路径和预测的以及实验测得的 DNA 结构。（附注 S3 中给出了模型和订书针的序列，最终设计出现在附注 S12 中。附注 S4 中给出了实验方法，这里只描述而并没有显示的结果则见于附注 S5。）在用 AFM 成像的结果中，如果没有直径超过 15 nm 的缺陷（洞或者是预期外廓中的锯齿形缺口），可以从质量上认为“良好地形成”一种特定的结构。对每一种折叠，用良好形成的部分的比例来估算产率，并用在一个或多个 AFM 场中占所有可区分结构的百分数表示。我注意到，尽管某些划归到良好形成类的结构具有 15nm 的缺陷，但大多数结构没有直径超过 10 nm 的缺陷。

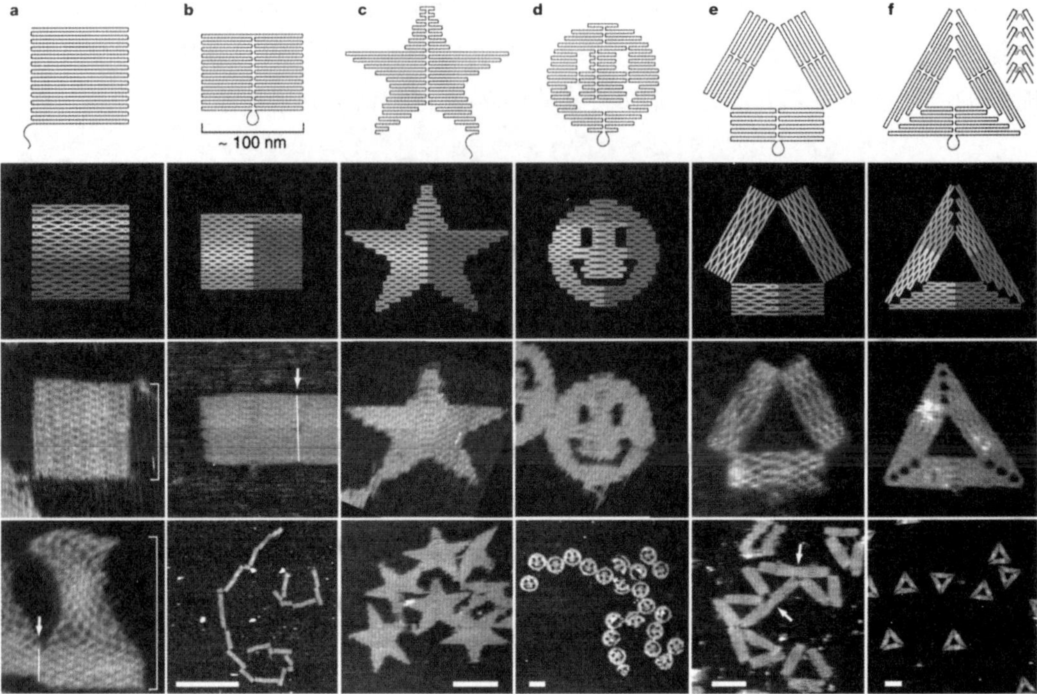

Fig. 2. **DNA origami shapes.** Top row, folding paths. **a**, square; **b**, rectangle; **c**, star; **d**, disk with three holes; **e**, triangle with rectangular domains; **f**, sharp triangle with trapezoidal domains and bridges between them (red lines in inset). Dangling curves and loops represent unfolded sequence. Second row from top, diagrams showing the bend of helices at crossovers (where helices touch) and away from crossovers (where helices bend apart). Colour indicates the base-pair index along the folding path; red is the 1st base, purple the 7,000th. Bottom two rows, AFM images. White lines and arrows indicate blunt-end stacking. White brackets in **a** mark the height of an unstretched square and that of a square stretched vertically (by a factor > 1.5) into an hourglass. White features in **f** are hairpins; the triangle is labelled as in Fig. 3k but lies face down. All images and panels without scale bars are the same size, 165 nm × 165 nm. Scale bars for lower AFM images: **b**, 1 μm; **c**–**f**, 100 nm.

First, a simple 26-helix square was designed (Fig. 2a). The square had no vertical reversals in raster direction, required a linear scaffold, and used 2.5-turn crossover spacing. Most staples were 26-mers that bound each of two adjacent helices as in Fig. 1c, but via 13 bases rather than 8. The design was made assuming a 1.5-nm inter-helix gap; an aspect ratio of 1.05 (93.9 nm × 89.5 nm) was expected. By AFM, 13% of structures were well-formed squares (out of $S = 45$ observed structures) with aspect ratios from 1.00 to 1.07 and bore the expected pattern of crossovers (Fig. 2a, upper AFM image). Of the remaining structures, ~25% were rectangular fragments, and ~25% had an hourglass shape that showed a continuous deformation of the crossover lattice (Fig. 2a, lower AFM image). Sequential imaging documented the stretching of a square into an hourglass, suggesting that hourglasses were originally squares that stretched upon deposition or interaction with the AFM tip. No subsequent designs exhibited stretching. Other designs had either a tighter 1.5-turn spacing with 32-mer staples spanning three helical domains (Fig. 2b–d, f) or smaller domains that appeared to slide rather than stretch (Fig. 2e).

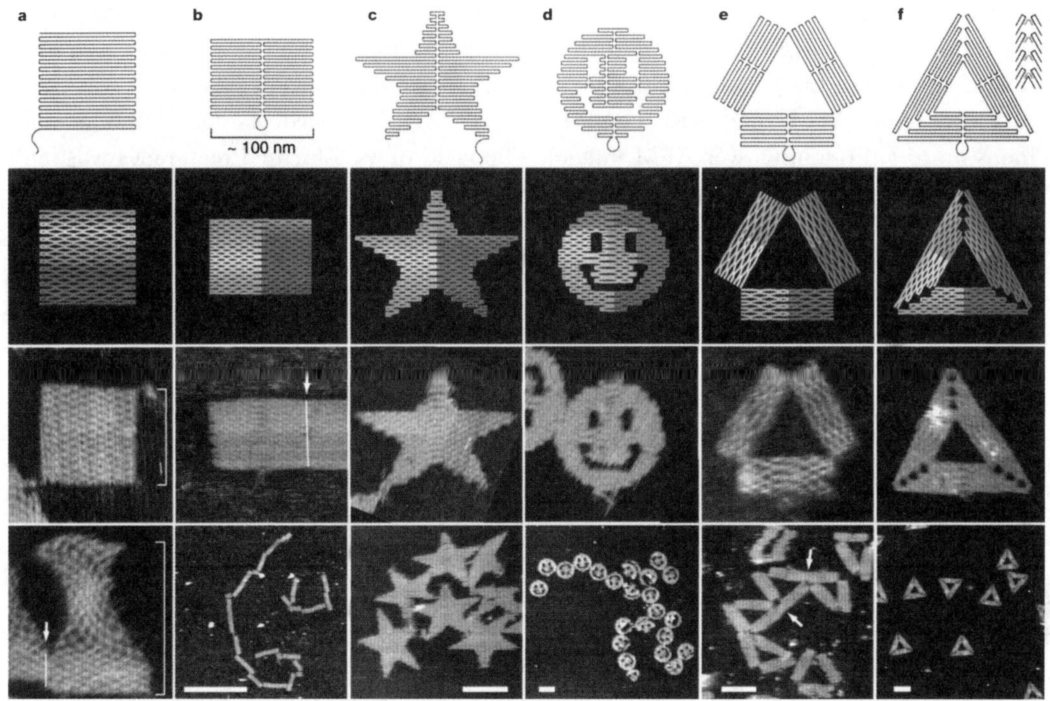

图 2. **DNA 折纸术造型**。顶行，折叠路径。**a**，正方形；**b**，矩形；**c**，五角星形；**d**，有三个洞的圆盘；**e**，由矩形区域构成的三角形；**f**，由梯形区域构成并且其间有桥联（小图中的红色线段）的锐角三角形。摇摆的曲线和环表示未折叠的序列。从顶部数第二行，图片显示螺旋在交叉处（螺旋发生接触）有弯曲而后远离交叉（螺旋分开）。颜色指示沿折叠路径的碱基对顺序；红色是第一个碱基，紫色是第 7,000 个。底部的两行是 AFM 图像。白线和箭头表示平端的堆积。**a** 中的白色括弧标示出一个未拉伸正方形的高度和一个在垂直方向上拉伸（> 1.5 倍）为沙漏形的正方形的高度。**f** 中的白色特征为发夹结构；三角形的标记如同图 3k 中那样，不过面朝下放置。所有没有比例尺的图像和画板都具有相同的尺寸，165 nm × 165 nm。下方 AFM 图像的比例尺：**b**，1 μm；**c** ~ **f**，100 nm。

首先设计一个简单的 26 螺旋正方形（图 2a）。该正方形在光栅方向上没有垂直的反转，它需要一根线性脚手架，并且要用到 2.5 螺距交叉间隔。大多数订书针长度是 26 个碱基，如图 1c 所示那样将两个邻近螺旋连接起来，不过是借助于 13 个碱基而不是 8 个碱基。设计时假定螺旋间的空隙有 1.5 nm；预期会有 1.05 的纵横比（93.9 nm × 89.5 nm）。根据 AFM 可知，13% 的结构是良好形成的正方形（在 $S = 45$ 观测到的结构中），其纵横比为 1.00 到 1.07，并且具有预期的交叉图案（图 2a，上方的 AFM 图像）。在其余结构中，约 25% 具有矩形形态，还有约 25% 具有沙漏形状并且显示出交叉晶格的连续形变（图 2a，下方的 AFM 图像）。连续成像证实了从正方形到沙漏形的拉伸过程，表明沙漏最初本是正方形，由于沉积作用或者与 AFM 探针的相互作用而拉伸。后续设计都没有呈现出拉伸。其他设计或者是具有更为紧凑的 1.5 螺距间隔，长度为 32 个碱基的订书针跨越 3 个螺旋区域（图 2b ~ d，f）；或者是更小些的区域，它们更倾向于滑移而不是拉伸（图 2e）。

To test the formation of a bridged seam, a rectangle was designed (Fig. 2b) according to the scheme outlined in Fig. 1e using 1.5-turn crossover spacing, 32-mer staples and a circular scaffold. As seen in Fig. 2b, the central seam and associated pattern of crossovers was easily visualized (upper AFM image). Rectangles stacked along their vertical edges, often forming chains up to 5 μm long (lower AFM image). The yield of well-formed rectangles was high (90%, $S = 40$), and so rectangles were used to answer basic questions concerning inter-helix gaps, base-stacking, defects and stoichiometry. AFM drift often distorts aspect ratios so that inter-helix gaps cannot be inferred from the aspect ratio of a single rectangle. A range of aspect ratios implied a gap size from 0.9 to 1.2 nm; later designs assume 1 nm. Whatever the exact value, it is consistent: aspect ratios were invariant along stacked chains with dozens of rectangles. Such stacking was almost completely abolished by omitting staples along vertical edges. On the other hand, stacking across the seam of an unbridged rectangle (as in Fig. 1c) kept 65% of structures ($S = 40$) well-formed; the rest showed some degree of dislocation at the seam. Other defects, such as the intentional omission of single staples, could be visualized as 5–10-nm holes. However, sharp tips and high tapping amplitudes were required; repeated scanning created holes difficult to distinguish from holes due to missing strands. This effect also increased uncertainty when stoichiometry was varied. When staple excesses of approximately 100:1 and 9:1 were used, the frequencies of 5–10-nm holes (a few per rectangle) were indistinguishable. At 2:1, rectangles were similar; perhaps a greater fraction were malformed. At 1.5:1, rectangles formed but had holes up to ~10% of their area in size. At a 1:1 ratio, < 1% of structures were rectangular.

To demonstrate the creation of arbitrary shapes, a five-pointed star was designed with 1.5-turn spacing, 32-mer staples and a linear rather than circular scaffold (Fig. 2c). Designed assuming a 1.5-nm inter-helix gap (the work was carried out before the gap for 1.5-turn spacing was measured), the stars are somewhat squat (Fig. 2c, upper AFM image). Still, the stars show that the width of a shape may be approximated to within one DNA turn. Many of the structures observed were star fragments (Fig. 2c, lower AFM image), and only 11% ($S = 70$) were well-formed. The low yield of stars (and squares, see above) may be due to strand breakage occurring during *Bsr*BI digestion or subsequent steps to remove the enzyme; when untreated circular scaffold was folded into stars, 63% ($S = 43$) were well-formed. To show that DNA origami need not be topological disks, and that scaffolds can be routed arbitrarily through shapes, a three-hole disk was designed (Fig. 2d). Although the shape approximated is symmetric, the folding path is highly asymmetric and has five distinct seams. Unlike the rectangles, which rarely break or fold, three-hole disks exhibit several characteristic deformations (Fig. 2d, lower AFM image); still, 70% ($S = 90$) were well-formed.

DNA origami is not limited to the approximation of shapes by raster fill: some shapes can be created more exactly by combining distinct raster fill domains in non-parallel arrangements. Figure 2e shows a triangle built from three separate, 2.5-turn spacing rectangular domains; only single covalent bonds along the scaffold hold the domains together. But the desired equiangular triangles (upper AFM image) were rarely observed (< 1%, $S = 199$). As seen in

为了检验桥联接缝的形成，根据图 1e 中所描绘的概图设计一个矩形（图 2b），其中用到 1.5 螺距交叉间隔，长度 32 个碱基的订书针和一根圆形脚手架。如同在图 2b 中所看到的，中心接缝和相关的交叉图案很容易看到（上方的 AFM 图像）。矩形沿着它们的垂直边堆叠，经常会形成长达 5 μm 的链（下方的 AFM 图像）。良好形成的矩形的产率很高（90%，$S = 40$），因此可以用矩形来研究与螺旋间空隙、碱基堆叠、缺陷和化学计量比等有关的基本问题。AFM 漂移经常会扭曲纵横比，使得螺旋间空隙不能利用单个矩形的纵横比推断出来。从纵横比范围可以推测出空隙大小为 0.9 ~ 1.2 nm；后面的设计假定为 1 nm。不管具体数值是多少，它是前后一致的：纵横比对于沿着堆叠链方向的数十个矩形是不变的。如果去掉沿垂直棱的订书针的话，这种堆叠几乎被完全破坏。另一方面，穿过非桥联矩形的接缝的堆叠（如图 1c 所示）则保持着 65%（$S = 40$）的良好形成的结构；其余部分在接缝处表现出不同程度的位错。其他缺陷，例如有意去掉某条订书针，可以看到一个 5 ~ 10 nm 的洞。但是，需要尖锐的针尖和高的敲击幅度；重复的扫描产生出的洞很难与归因于缺失链的洞相区分。这种效应增加了在改变化学计量比时的不确定性。当所用的订书针的过量值近似为 100 : 1 和 9 : 1 时，5 ~ 10 nm 洞的出现频率（每个矩形有若干个）是不可区分的。在过量值为 2 : 1 时，矩形是相似的；可能会有较大比例是畸形的。在 1.5 : 1 时，能够形成矩形，但是洞占总面积多达 10% 左右。在 1 : 1 的比例时，< 1% 的结构是矩形的。

为了证明这种方法能制备任意的形状，用 1.5 螺距间隔，长度为 32 个碱基的订书针和一根线性而不是圆形的脚手架设计了一个五角星（图 2c）。设计时假定螺旋间的空隙为 1.5 nm（这项工作是在测定 1.5 螺距间隔的空隙之前进行的），因此所得五角星似乎有些矮胖（图 2c，上方 AFM 图像）。五角星的情形再一次表明，形状的宽度可以近似地在一个 DNA 螺距之内。所观测到的很多结构是星形的碎片（图 2c，下方的 AFM 图像），只有 11%（$S = 70$）是良好形成的。星形的低产率（正方形的，见上）可以归结为在 *Bsr*BI 降解或随后移除酶的过程中发生的链断裂；在用未经处理的圆形脚手架折叠成星形时，有 63%（$S = 43$）是良好形成的。为了说明 DNA 折纸术并不一定需要拓扑意义上的圆盘，其脚手架可以转变为任意形状，我设计了一个有 3 个洞的圆盘（图 2d）。尽管所模拟的形状是对称的，折叠路径却是高度不对称的，而且具有 5 个不同的接缝。与几乎不会破裂或弯折的矩形不同，三洞圆盘呈现出几种特征性的变形（图 2d，下方 AFM 图像）；仍然有 70%（$S = 90$）是良好形成的。

DNA 折纸术用来模拟形状的方法并不限于光栅填充：某些形状能够通过将不同的光栅填充区域合并为非平行排布而更为准确地制造出来。图 2e 显示了一个三角形，它是用 3 个独立的、2.5 螺距间隔的矩形区域搭建起来的；只有沿着脚手架上的单个共价键将各区域维持在一起。但是所需的正三角形（上方的 AFM 图像）却很少

the lower AFM image, stacking caused rectangular domains of separate triangles to bind; this effect and the flexibility of the single-bond joints at the vertices may account for the ease with which these triangles deform. To solve these problems, "sharp triangles", built from trapezoidal domains with 1.5-turn spacing, were designed (Fig. 2f). The slanted edges of the trapezoids meet at the triangle vertices and allow the addition of bridging staples along these interfaces. Sharp triangles remained separated and equiangular (Fig. 2f, lower AFM image); 88% were well-formed ($S = 78$). Even when bridging staples at the vertices were not used, a large number of sharp triangles were well-formed (55%, $S = 22$). These "weakened" sharp triangles provided the most stringent test of the estimated inter-helix gap, because too high or low an estimate would have caused gaps or overlaps between trapezoids. Gaps of 10 nm occasionally appeared but overlaps were never observed, suggesting that 1 nm may be a slight underestimate of the inter-helix gap.

Patterning and Combining DNA Origami

In addition to binding the DNA scaffold and holding it in shape, staple strands provide a means for decorating shapes with arbitrary patterns of binary pixels. Given a shape, the original set of staples is taken to represent binary "0"s; a new set of labelled staples, one for each original staple, is used to represent binary "1"s. Patterns are created by mixing appropriate subsets of these strands. In this way, any desired pattern can be made.

In principle, a variety of DNA modifications—for example, biotin or fluorophores—could serve as labels. Here, "dumbbell hairpins" (Fig. 1d inset, Supplementary Note S6), designed to avoid dimerization at high concentration, were added to the middle of 32-mer staples at the position of merges made during design. Depending on the merge pattern, the resulting pixel pattern was either rectilinear, with adjacent columns of hairpins on alternate faces of the shape, or staggered and nearly hexagonally packed, with all hairpins on the same face. In AFM images labelled staples give greater height contrast (3 nm above the mica) than unlabelled staples (~1.5 nm), which results in a pattern of light "1" and dark "0" pixels. Several patterns (Fig. 3), each with ~200 pixels, illustrate the generality of this technique.

能被看到（< 1%，$S = 199$）。如同在下方 AFM 图像中所看到的那样，堆叠导致不同三角形中的矩形区域结合起来；这种效应加上顶点处单键结合的易变性，可以解释这些三角形为什么这么容易变形。为了解决这些问题，设计了“锐角三角形”（图 2f），它是用有 1.5 螺距间隔的梯形区域搭建起来的。梯形的斜边在三角形顶点处相遇，并且允许在界面处加入桥联的订书针。锐角三角形保持着等角并彼此独立（图 2f，下方 AFM 图像）；88%（$S = 78$）是良好形成的。即使在顶点处没有使用桥联订书针，仍然有大量锐角三角形是良好形成的（55%，$S = 22$）。这些“弱化的”锐角三角形为所估计的螺旋间空隙提供了最为严格的检验，因为过高或者过低的估计值都可能会导致梯形之间的空隙或重叠。偶尔会出现 10 nm 的空隙，但是未曾观测到重叠，这意味着 1 nm 对于螺旋间空隙来说可能是一个略微偏低的估值。

DNA 折纸术的绘图与合并

除去将 DNA 脚手架结合并保持其形状之外，订书针链还提供了一种以任意二元像素图案修饰该形状的方法。假定一个形状，最初的一组订书针用来表示二进制的“0”；另一组标记了的订书针，每条都对应于最初的每条订书针，用来表示二进制的“1”。通过混合这些订书针链的适当子集，即可创造出图案。利用这种方法，能得到任何预想的图案。

原则上讲，有很多种 DNA 修饰——例如生物素或荧光团——可以充当标记。本文中，将为避免高浓度时的二聚作用而设计了的“哑铃型发夹”（图 1d 小图，附注 S6），在设计订书针的接合时加入长度为 32 个碱基的订书针的中部。所得到的像素图案取决于接合图案：或者是直线形，其中发夹的相邻列位于该式样中交替的面上；或者是交错的近似于六方堆积，其中所有的发夹都位于相同的面上。在 AFM 图像中，标记的订书针表现出比未标记的订书针（约 1.5 nm）更大的高度对比（3 nm，在云母表面），结果是图案中的像素“1”明亮而“0”灰暗。各自具有大约 200 像素的几种图案（图 3）说明了这种技术的普适性。

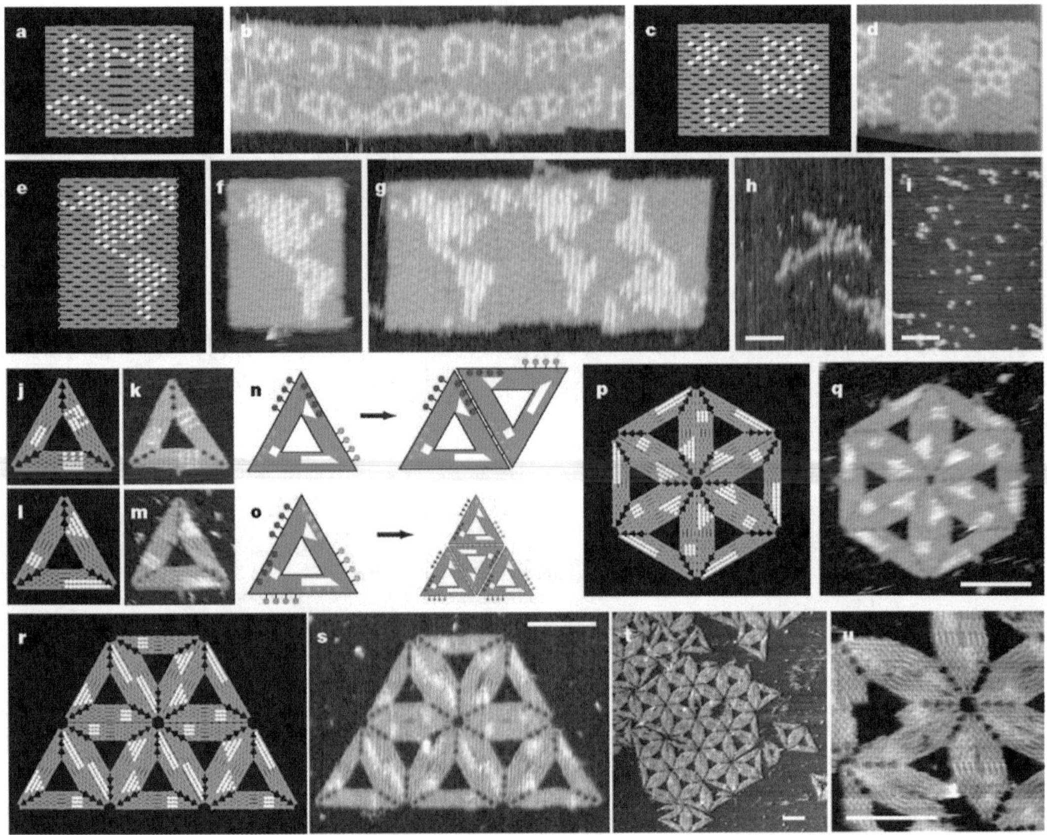

Fig. 3. **Patterning and combining DNA origami. a**, Model for a pattern representing DNA, rendered using hairpins on a rectangle (Fig. 2b). **b**, AFM image. One pixelated DNA turn (~100 nm) is 30 × the size of an actual DNA turn (~3.6 nm) and the helix appears continuous when rectangles stack appropriately. Letters are 30 nm high, only 6 × larger than those written using STM in ref. 3; 50 billion copies rather than 1 were formed. **c**, **d**, Model and AFM image, respectively, for a hexagonal pattern that highlights the nearly hexagonal pixel lattice used in **a–i**. **e–i**, Map of the western hemisphere, scale 1:2 × 10^{14}, on a rectangle of different aspect ratio. Normally such rectangles aggregate (**h**) but 4-T loops or tails on edges (white lines in e) greatly decrease stacking (**i**). **j–m**, Two labellings of the sharp triangle show that each edge may be distinguished. In **j–u**, pixels fall on a rectilinear lattice. **n–u**, Combination of sharp triangles into hexagons (**n, p, q**) or lattices (**o, r–u**). Diagrams (**n, o**) show positions at which staples are extended (coloured protrusions) to match complementary single-stranded regions of the scaffold (coloured holes). Models (**p, r**) permit comparison with data (**q, s**). The largest lattice observed comprises only 30 triangles (**t**). **u** shows close association of triangles (and some breakage). **d** and **f** were stretched and sheared to correct for AFM drift. Scale bars: **h**, **i**, 1 μm; **q**, **s–u**, 100 nm.

Yields of patterned origami were similar to those of unpatterned origami; for the pattern in Fig. 3a, 91% ($S = 85$) of rectangles were well-formed. Because rectilinear patterns imaged poorly, only staggered patterns were examined quantitatively. Distances measured between pairs of "1" pixels in alternating columns (two pixel widths: 11.5 ± 0.9 nm, mean ± s.d., $n = 26$) and adjacent rows (one pixel height: 6.6 ± 0.5 nm, $n = 24$) are consistent with the theoretically expected pixel size of 5.4 nm × 6 nm. Most defects take the form of "missing pixels"; that is, pixels that should image as "1"s but image as "0"s instead. 94% of "1" pixels (of 1,080 observed) were visualized. Whether missing pixels represent real defects or artefacts

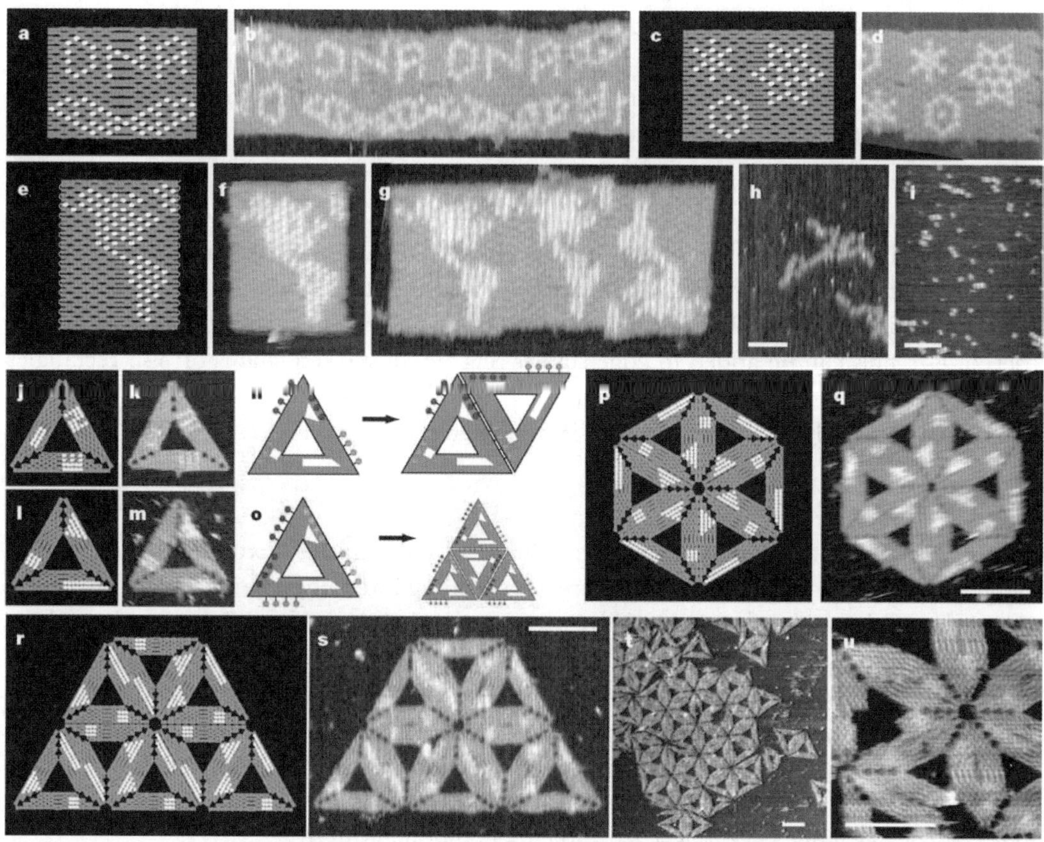

图 3. **DNA 折纸术的绘图与结合。a**，呈现 DNA 的图像的模型，利用矩形上的发夹来实现（图 2b）。**b**，AFM 图像。单像素的一个 DNA 螺距（约 100 nm）的大小是一个实际 DNA 螺距（约 3.6 nm）的 30 倍，并且矩形以适当方式堆积时螺旋看起来是连续的。字母高度为 30 nm，是参考文献 3 中用 STM 写成的那些字母的 6 倍；形成的副本是 500 亿而不是 1 份。**c**，**d**，分别为一个六边形图案的模型和 AFM 图像，它强调了 **a** ~ **i** 中所使用的近似六边形像素晶格。**e** ~ **i**，西半球的地图（比例尺为 1：2×10^{14}）画在不同纵横比的矩形之上。正常情况下这些矩形会聚集（**h**），但是棱上的 4-T 环或尾部（**e** 中的白色线段）有效地避免了堆积（**i**）。**j** ~ **m**，锐角三角形的两个标记表明各条边是可区分的。在 **j** ~ **u** 中，像素落在直线形晶格上。**n** ~ **u**，锐角三角形结合成六边形（**n**，**p**，**q**）或晶格（**o**，**r** ~ **u**）。图像（**n**，**o**）显示出订书针进行扩展（彩色的突起）以便与脚手架的单链区域（彩色的洞）互补的部位。模型（**p**，**r**）与数据（**q**，**s**）是可比的。所观测到的最大晶格仅由 30 个三角形组成（**t**）。**u** 显示出与三角形（以及某些断裂）的密切关联。**d** 和 **f** 被拉伸并切割，以修正 AFM 的漂移。比例尺：**h**，**i**，1 μm；**q**，**s** ~ **u**，100 nm。

有绘图的折纸术的产率与没有绘图的折纸术的产率相近；对于图 3a 中的图案而言，91%（$S=85$）的矩形是良好形成的。由于直线形图案成像不佳，因此只对交错型图案进行了定量检测。所测得的交替列（两个像素宽度：11.5 ± 0.9 nm，平均值 ± 标准差，$n=26$）与相邻行中（一个像素的高度：6.6 ± 0.5 nm，$n=24$）一对“1”像素之间的距离与理论预期的像素尺寸 5.4 nm × 6 nm 相吻合。大多数缺陷体现为“丢失像素”的形式；也就是说，这些像素应该描绘的是“1”但却描绘了“0”。其中 94%

of imaging is unknown; sequential AFM images occasionally showed "1" pixels that later converted irreversibly to "0" pixels, suggesting tip-induced damage. Stoichiometric errors, synthetic errors, or unwanted secondary structure are not implicated for any particular strand, as the position of missing pixels appeared random (Fig. 3b, f and g).

Stacking of shapes along blunt-ended helices provides an uncontrolled mechanism for the creation of larger structures (Fig. 3b). Instead of removing staples on the edge of a rectangle to avoid stacking (as described previously), 4-T hairpin loops (four thymines in a row, Fig. 1e, inset) or 4-T tails can be added to edge staples (Fig. 3e, f); stacked chains of 3–5 rectangles still formed (Fig. 3g), but 30% of rectangles ($S = 319$) occurred as monomers (Fig. 3i). Without hairpins, all rectangles occurred in aggregates (Fig. 3h).

Controlled combination of shapes was achieved by designing "extended staples" that connected shapes along their edges. To create a binding interaction between two particular edges, extended staples were designed by merging and breaking normal staples along these edges (Supplementary Note S7). Starting with sharp triangles, this approach was used to create finite (hexagons; Fig. 3n, p, q) as well as periodic structures (triangular lattice; Fig. 3o, r–u). I note that the successful combination of shapes (unlike the successful formation of individual shapes) is in principle very sensitive to the concentrations of extended staples (which should ideally be equal to that of the scaffold). Poor stoichiometry may play a role in the poor yield of hexagons ($< 2\%$, $S = 70$) and lattices (not measured).

Discussion

The scaffolded self-assembly of DNA strands has been used to create linear structures[17,18] and proposed as a method for creating arbitrary patterns[18,19]. But the widespread use of scaffolded self-assembly, and in particular the use of long DNA scaffolds in combination with hundreds of short strands, has been inhibited by several misconceptions: it was assumed that (1) sequences must be optimized[20] to avoid secondary structure or undesired binding interactions, (2) strands must be highly purified, and (3) strand concentrations must be precisely equimolar. These three criteria are important for the formation of many DNA nanostructures and yet all three are ignored in the present method. For example, M13mp18 is essentially a natural sequence that has a predicted secondary structure which is more stable (lower in energy) than similar random sequences (Supplementary Note S8). Further, stocks of staples each contained a few per cent truncation products, stock concentrations were measured with at least 10% error, and staples were used successfully at stoichiometries that varied over an order of magnitude.

I suggest that several factors contribute to the success of scaffolded DNA origami (even though the method ignores the normal, careful practices of DNA nanotechnology).

的“1”像素（在 1,080 个观测到的像素之中）能被观察到。还不知道丢失像素所表现的究竟是真实的缺陷还是成像过程中的假象；连续的 AFM 成像偶尔会出现一开始为“1”像素随后不可逆地转变为“0”像素的情况，这意味着探针造成的损伤。对任一特定的链而言，化学计量误差、合成误差或者是不需要的二级结构都并未牵扯其中，因为丢失像素的部位似乎是随机的（图 3b，f 和 g）。

形状沿着平端螺旋的堆积为产生较大结构（图 3b）提供了一种无法控制的机制。不同于移除矩形边上的订书针以避免堆积（如同前面所描述的那样），4-T 发夹环（一行中有 4 个胸腺嘧啶，图 1e 小图）或 4-T 尾部可以加入边上的订书针中（图 3c 和 f）；3 ~ 5 个矩形的堆叠链仍然会形成，但是 30% 的矩形（$S = 319$）是作为单体而出现的（图 3i）。没有发夹的话，所有矩形都以聚集体形式出现（图 3h）。

通过设计“扩展型订书针”以沿形状的边将其连接，可以实现形状的可控结合。把沿着这些边的订书针链重新接合和打断，可以得到具有特定相互作用的两条边（附注 S7）。从锐角三角形开始，用这种方式可以创造有限的（六边形；图 3n，p 和 q）和周期性的结构（三角形晶格；图 3o，r ~ u）。我注意到，形状的成功结合（与单个形状的成功形成不同）在原则上极容易受到扩展型订书针浓度（在理想情况下，它应该等于脚手架的浓度）的影响。不准确的化学计量会导致六边形（ < 2%，$S = 70$）和晶格（未检测）的低产率。

讨　论

DNA 链以脚手架方式进行的自组装已经被用来制造多种线形结构 [17,18]，而且被提出可作为一种能创造任意图案的方法 [18,19]。但是脚手架式自组装的广泛应用，特别是长 DNA 脚手架在与数百个短订书针链相结合时的应用，却一直由于若干误解而不得发挥：人们假定（1）必须要对序列进行优化 [20] 以避免二级结构或不需要的结合作用，（2）DNA 链必须是高度纯化的，以及（3）DNA 链的浓度必须精确为等物质的量。这三个准则对于很多种 DNA 纳米结构的形成是重要的，但在我目前提出的这种方法中它们都是被忽略的。例如，M13mp18 实际上是一种天然序列，具有预期的二级结构，比类似的随机序列（附注 S8）更为稳定（能量更低）。此外，每个订书针的原料都含有少量百分比的截头产物，原料浓度的测定至少有 10% 的误差，而且当化学计量在一个数量级的水平上变动时仍可成功地使用订书针。

我认为，有几个因素对脚手架式 DNA 折纸术的成功很重要（尽管该方法忽略了 DNA 纳米技术常规的、谨慎的做法）。它们是：（1）链的进入，（2）过量订书针，

These are (1) strand invasion, (2) an excess of staples, (3) cooperative effects and (4) design that intentionally does not rely on binding between staples. Briefly (details are given in Supplementary Note S9), strand invasion may allow correct binding of excess full-length staples to displace unwanted secondary structure, incorrect staples, or grossly truncated staples. Further, each correct addition of a staple organizes the scaffold for subsequent binding of adjacent staples and precludes a large set of undesired secondary structures. Last, because staples are not designed to bind one another, their relative concentrations do not matter.

The method presented here is easy to implement, high yield and relatively inexpensive. Three months of effort went into the design program. In addition, each structure required about one week to design and one week to synthesize (commercially); the mixing and annealing of strands required a few hours. The greatest experimental difficulty was acquiring high-resolution AFM images, typically taking two days per structure. For rigid designs using circular scaffolds (rectangles with patterns, three-hole disks, and sharp triangles), yields of qualitatively well-formed structures were at least 70%. A better understanding of folding will depend on less-destructive imaging and quantification of small (<15 nm) defects. A possible objection to the routine use of the method is the potential cost of staples; unlike the scaffold, staples cannot be cloned. However, unpurified strands are inexpensive so that the scaffold constitutes 80% of the cost, even when using a 100-fold excess of staples (Supplementary Note S10).

I believe that scaffolded DNA origami can be adapted to create more complex or larger structures. For example, the design of three-dimensional structures should be accessible using a straightforward adaptation of the raster fill method given here. If non-repetitive scaffolds of megabase length can be prepared, micrometre-size origami with 20,000 features may be possible. However, the requirement for unique sequence information means that the method cannot be scaled up arbitrarily; whenever structures above a critical size or level of complexity are desired, it will therefore be necessary to combine scaffolded DNA origami with hierarchical self-assembly[10,11], algorithmic self-assembly[22], or top-down fabrication techniques.

An obvious application of patterned DNA origami would be the creation of a "nanobreadboard", to which diverse components could be added. The attachment of proteins[23], for example, might allow novel biological experiments aimed at modelling complex protein assemblies and examining the effects of spatial organization, whereas molecular electronic or plasmonic circuits might be created by attaching nanowires, carbon nanotubes or gold nanoparticles[24]. These ideas suggest that scaffolded DNA origami could find use in fields as diverse as molecular biology and device physics.

(**440**, 297-302; 2006)

Paul W. K. Rothemund

Departments of Computer Science and Computation & Neural Systems, California Institute of Technology, Pasadena, California 91125, USA

(3)协同效应和(4)有意不依赖订书针间结合的设计。简单地说(附注 S9 中给出了细节)，链的进入使得过量的完整长度订书针正确结合，以取代那些不需要的二级结构、不正确的订书针或者是不准确的截头订书针。另外，每一个新的正确添加的订书针都会组织脚手架以使其适合与随后的相邻订书针的连接，从而排除了大量不需要的二级结构。最后，由于订书针并不是被设计成彼此连接的，因此它们的相对浓度并没有影响。

这里所介绍的方法容易施行，产率高而且成本相对较低。设计程序需要三个月的努力。此外，每一种结构大约需要一个星期来设计和一个星期来合成(商业化方式的)；订书针链的混合与退火需要几个小时。实验中最大的困难是获得高分辨率的 AFM 图像，通常情况下每个结构需要花两天时间。对于用圆形脚手架进行的稳定设计来说(有图案的矩形、三洞圆盘和锐角三角形)，从质量上看良好形成的结构的产率至少有 70%。对折叠过程更好的理解将取决于破坏更少的成像和对小(<15 nm)缺陷的定量。反对把这一方法投入常规使用的理由可能是订书针的潜在消耗；与脚手架不同，订书针是不可复制的。但是，未经纯化的链成本很低，因此，即使是在使用 100 倍过量的订书针时，脚手架占成本消耗的 80%(附注 S10)。

我相信脚手架式 DNA 折纸术可以适用于创造更复杂或者更大的结构。例如，三维结构的设计应该可以通过直接改进本文中所给出的光栅填充方法而达到。如果能够制备出长达兆碱基的非重复性脚手架，那么具有 20,000 个特征的微米尺寸折纸结构就有可能实现。但是，对于独特的序列信息的需求意味着这种方法无法任意地按比例扩大；一旦需要超过某一临界尺寸或者是复杂性水平的结构，就必须要将脚手架式 DNA 折纸术与分级自组装 [10,11]、算法式自组装 [22] 或者自顶向下的技术相结合。

绘制图案的 DNA 折纸术的一个明显的应用，是“纳米模拟电路板”的制造，板上可以添加各种不同的组件。例如，蛋白质的附着 [23] 可能会使新型的生物学实验得以进行，这些实验将会有助于对复杂的蛋白质组装过程建立模型以及检测空间组织化过程的影响，而附着纳米导线、碳纳米管或者是金纳米颗粒 [24] 则有可能制造出分子电路或等离子电路。这些想法意味着脚手架式 DNA 折纸术可以在与分子生物学和器件物理学这种多样化的领域中找到应用。

（王耀杨 翻译；刘冬生 审稿）

Received 7 September 2005; accepted 12 January 2006.

References:

1. Feynman, R. P. There's plenty of room at the bottom. *Engineering and Science* 23 (5), 22-36 (Caltech, February, 1960).
2. Junno, T., Deppert, K., Montelius, L. & Samuelson, L. Controlled manipulation of nanoparticles with an atomic force microscope. *Appl. Phys. Lett.* **66**, 3627-3629 (1995).
3. Eigler, D. M. & Schweizer, E. K. Positioning single atoms with a scanning tunnelling microscope. *Nature* **344**, 524-526 (1990).
4. Heinrich, A. J., Lutz, C. P., Gupta, J. A. & Eigler, D. M. Molecular cascades. *Science* **298**, 1381-1387 (2002).
5. Whitesides, G. M., Mathias, J. P. & Seto, C. T. Molecular self-assembly and nanochemistry: a chemical strategy for the synthesis of nanostructures. *Science* **254**, 1312-1319 (1991).
6. Yokoyama, T., Yokoyama, S., Kamikado, T., Okuno, Y. & Mashiko, S. Self-assembly on a surface of supramolecular aggregates with controlled size and shape. *Nature* **413**, 619-621 (2001).
7. Mao, C. B. *et al.* Virus-based toolkit for the directed synthesis of magnetic and semiconducting nanowires. *Science* **303**, 213-217 (2004).
8. Seeman, N. C. Nucleic-acid junctions and lattices. *J. Theor. Biol.* **99**, 237-247 (1982).
9. Seeman, N. C. & Lukeman, P. S. Nucleic acid nanostructures: bottom-up control of geometry on the nanoscale. *Rep. Prog. Phys.* **68**, 237-270 (2005).
10. Chworos, A. *et al.* Building programmable jigsaw puzzles with RNA. *Science* **306**, 2068-2072 (2004).
11. Park, S. H. *et al.* Finite-size, fully-addressable DNA tile lattices formed by hierarchical assembly procedures. *Angew. Chem.* **118**, 749-753 (2006).
12. Chen, J. & Seeman, N. C. The synthesis from DNA of a molecule with the connectivity of a cube. *Nature* **350**, 631-633 (1991).
13. Zhang, Y. & Seeman, N. C. The construction of a DNA truncated octahedron. *J. Am. Chem. Soc.* **116**, 1661-1669 (1994).
14. Shih, W. M., Quispe, J. D. & Joyce, G. F. A 1.7-kilobase single-stranded DNA that folds into a nanoscale octahedron. *Nature* **427**, 618-621 (2004).
15. Rothemund, P. W. K. *et al.* Design and characterization of programmable DNA nanotubes. *J. Am. Chem. Soc.* **26**, 16344-16353 (2004).
16. Fu, T.-J. & Seeman, N. C. DNA double-crossover molecules. *Biochemistry* **32**, 3211-3220 (1993).
17. LaBean, T. H., Winfree, E. & Reif, J. H. in *DNA Based Computers V* (eds Winfree, E. & Gifford, D. K.) 123-140 (Vol. 54 of DIMACS, AMS Press, Providence, Rhode Island, 1999).
18. Yan, H., LaBean, T. H., Feng, L. & Reif, J. H. Directed nucleation assembly of DNA tile complexes for barcode-patterned lattices. *Proc. Natl Acad. Sci, USA* **100**, 8103-8108 (2003).
19. Reif, J. H. in *Proc. 29th Int. Colloquium on Automata, Languages, and Programming* (*ICALP*) (eds Widmayer, P., Ruiz, F. T., Bueno, R. M., Hennessy, M., Eidenbenz, S. & Conejo, R.) 1-21 (Vol. 2380 of Lecture Notes in Computer Science, Springer, New York, 2002).
20. Seeman, N. C. *De novo* design of sequences for nucleic acid structural engineering. *J. Biomol. Struct. Dyn.* **8**, 573-581 (1990).
21. Winfree, E. in *DNA Based Computers* (eds Lipton, R. J. & Baum, E. B.) 199-221 (Vol. 27 of DIMACS, AMS Press, Providence, Rhode Island, 1996).
22. Rothemund, P. W. K., Papadakis, N. & Winfree, E. Algorithmic self-assembly of DNA Sierpinski triangles. *PloS Biol.* **2**, e424 (2004).
23. Yan, H., Park, S. H., Finkelstein, G., Reif, J. H. & LaBean, T. H. DNA-templated self-assembly of protein arrays and highly conductive nanowires. *Science* **301**, 1882-1884 (2003).
24. Le, J. D. *et al.* DNA-templated self-assembly of metallic nanocomponent arrays on a surface. *Nano Lett.* **4**, 2343-2347 (2004).

Supplementary Information is linked to the online version of the paper at www.nature.com/nature.

Acknowledgements. I thank E. Winfree for discussions and providing a stimulating laboratory environment; B. Yurke for the term "nanobreadboard"; N. Papadakis, L. Adleman, J. Goto, R. Barish, R. Schulman, R. Hariadi, M. Cook and M. Diehl for discussions; B. Shaw for a gift of AFM tips; A. Schmidt for coordinating DNA synthesis; and K. Yong, J. Crouch and L. Hein for administrative support. This work was supported by National Science Foundation Career and Nano grants to E. Winfree as well as fellowships from the Beckman Foundation and Caltech Center for the Physics of Information.

Author Information. Reprints and permissions information is available at npg.nature.com/reprintsandpermissions. The author declares competing financial interests: details accompany the paper on www. nature.com/nature. Correspondence and requests for materials should be addressed to P.W.K.R. (pwkr@dna.caltech.edu).

Long γ-ray Bursts and Core-collapse Supernovae Have Different Environments

A. S. Fruchter *et al.*

Editor's Note

Long gamma-ray bursts (GRBs) are astrophysical phenomena generally believed to arise in the collapse and subsequent explosion of a very massive star. They had been associated with somewhat anomalous supernovae, but the nature of the link was unclear. Here Andrew Fruchter and colleagues report a careful analysis of galaxies hosting long GRBs and find that the bursts are far more concentrated in the brightest regions of galaxies than are the supernovae. These bright regions have the most massive stars. The authors conclude that bursts are best associated with such stars, and speculate that there is a preference for these stars to occur in environments relatively poor in elements heavier than helium.

When massive stars exhaust their fuel, they collapse and often produce the extraordinarily bright explosions known as core-collapse supernovae. On occasion, this stellar collapse also powers an even more brilliant relativistic explosion known as a long-duration γ-ray burst. One would then expect that these long γ-ray bursts and core-collapse supernovae should be found in similar galactic environments. Here we show that this expectation is wrong. We find that the γ-ray bursts are far more concentrated in the very brightest regions of their host galaxies than are the core-collapse supernovae. Furthermore, the host galaxies of the long γ-ray bursts are significantly fainter and more irregular than the hosts of the core-collapse supernovae. Together these results suggest that long-duration γ-ray bursts are associated with the most extremely massive stars and may be restricted to galaxies of limited chemical evolution. Our results directly imply that long γ-ray bursts are relatively rare in galaxies such as our own Milky Way.

IT is an irony of astrophysics that stellar birth is most spectacularly marked by the deaths of massive stars. Massive stars burn brighter and hotter than smaller stars, and exhaust their fuel far more rapidly. Therefore a region of star formation filled with low mass stars still early in their lives, and in some cases still forming, may also host massive stars already collapsing and producing supernovae. Indeed, with the exception of the now famous type Ia supernovae, which have been so successfully used for cosmological studies[1,2] and which are thought to be formed by the uncontrolled nuclear burning of stellar remnants comparable in mass to the Sun[3], all supernovae are thought to be produced by the collapse of massive stars. The collapse of the most massive stars (tens of solar masses) is thought to leave behind either black holes or neutron stars, depending largely on the state of chemical evolution of the material that formed the star, whereas the demise of stars between approximately 8 and 20 solar masses produces only neutron stars[4].

长伽马射线暴与核坍缩超新星具有不同的环境

弗鲁赫特等

编者按：

长时标伽马射线暴（简称长暴）一般被认为由大质量恒星的核坍缩及后续爆发所产生。它们与某些超新星成协出现，但成协的本质尚不清楚。本文中，安德鲁·弗鲁赫特与其合作者对长暴和超新星的宿主星系进行了细致的对比分析。他们发现相比于超新星，长暴发生位置更多地集中在星系中最亮的区域，而这些亮区域恰恰存在有超大质量的恒星。作者得出结论，长暴最有可能与星系中的超大质量恒星有着物理关联，并由此推测长暴仅发生在低金属丰度的宿主星系环境中。

当大质量恒星耗尽燃料，它们的内核会发生坍缩，接着星体常常会产生明亮的爆炸，该现象被称为核坍缩超新星。有时这种恒星坍缩会产生更亮的相对论性的爆炸，爆炸会产生持续时标超过两秒的伽马射线辐射，并且辐射能被空间高能卫星探测到。我们简称其为长时标伽马射线暴（即长暴）。一般会认为这些长暴和核坍缩超新星应该处于类似的星系环境中。但我们揭示这一预想是错误的。我们发现长暴比核坍缩超新星更加集中在宿主星系中最亮的区域。而且，长暴的宿主星系明显比核坍缩超新星的宿主星系更暗、更不规则。这些结论表明长暴和超大质量恒星成协，而且可能仅发生在金属丰度较低的星系中。我们的结果直接表明长暴很少会发生在类似银河系样的宿主星系中。

恒星诞生最显著的标志是大质量恒星的死亡，这可以说是天体物理学的一个讽刺。大质量恒星比小质量恒星燃烧时更亮更热，也更快地消耗完自己的燃料。因此在恒星诞生的区域，那里一方面遍布着处于生命期早期的小质量恒星，某些情况下仍有恒星生成；另一方面也可能包含已经坍缩并产生超新星的大质量恒星。诚然，除了最近著名的成功用作宇宙学研究[1,2]的 Ia 型超新星，其他类型超新星都被认为是由大质量恒星的核坍缩形成的。Ia 型超新星一般被认为是类似太阳质量大小的恒星遗迹在不可控制的核燃烧[3]下形成的。普遍认为，大多数大质量恒星（几十个太阳质量）坍缩后会留有黑洞或中子星，这主要取决于构成此恒星物质的化学演化状态。不过质量约为 8 到 20 倍太阳质量的恒星死亡后仅会产生中子星[4]。

Gamma-ray bursts (GRBs), like supernovae, are a heterogeneous population. GRBs can be divided into two classes: short, hard bursts, which last between milliseconds and about two seconds and have hard high-energy spectra, and long, soft bursts, which last between two and tens of seconds, and have softer high-energy spectra[5]. Only very recently have a few of the short bursts been well localized, and initial studies of their apparent hosts suggest that these bursts may be formed by the binary merger of stellar remnants[6,7]. In contrast, the afterglows of over 80 long GRBs (LGRBs) have been detected in the optical and/or radio parts of the spectrum. And as a result of these detections, it has become clear that LGRBs, like core-collapse supernovae, are related to the deaths of young, massive stars. It is these objects, born of the deaths of massive stars, that we study here.

LGRBs are generally found in extremely blue host galaxies[8-11] that exhibit strong emission lines[12,13], suggesting a significant abundance of young, very massive stars. Furthermore, whereas the light curves of the optical transients associated with LGRBs are often dominated by radiation from the relativistic outflow of the GRB, numerous LGRBs have shown late-time "bumps" in their light curves consistent with the presence of an underlying supernova[14-16]. In several cases spectroscopic evidence has provided confirmation of the light of a supernova superposed on the optical transient[17-20]. Indeed, given the large variations in the brightnesses of optical transients and supernovae, and the limited observations on some GRBs, it seems plausible that all LGRBs have an underlying supernova[21]. And although the energy released in a LGRB often appears to the observer to be orders of magnitude larger than that of a supernova, there is now good evidence suggesting that most LGRBs are highly collimated and often illuminate only a few per cent of the sky[22,23]. When one takes this into account, the energy released in LGRBs more closely resembles that of energetic supernovae. However, not all core-collapse supernovae may be candidates for the production of GRBs. The supernovae with good spectroscopic identifications so far associated with GRBs have been type Ic—that is, core-collapse supernovae that show no evidence of hydrogen or helium in their spectra. (Type Ib supernovae, which are often studied together with type Ic, have spectra that are also largely devoid of hydrogen lines but show strong helium features.) A star may therefore need to lose its outer envelope if a GRB is to be able to burn its way through the stellar atmosphere[24]. Studies that have compared the locations of type Ib/c supernovae with the more numerous type II supernovae (core-collapse supernovae showing hydrogen lines) in local galaxies so far show no differences in either the type of host or the placement of the explosion on the host[25,26]. This result led the authors of ref. 25 to argue that core-collapse supernovae all come from the same mass range of progenitor stars, but that type Ib/c supernovae may have had their envelopes stripped by interaction with a binary stellar companion. Whether type Ic supernovae come from single stars, or binary stars, or both, it is very likely that only a small fraction of these supernovae produce GRBs[27].

Given the common massive stellar origins of core-collapse supernovae and LGRBs, one might expect that their hosts and local environments are quite similar. It has long been argued that core-collapse supernovae should track the blue light in the Universe (the light from massive stars is blue), both in their distribution among galaxies and within their host galaxies themselves. One would expect similar behaviour from LGRBs, and indeed

伽马射线暴，类似于超新星，也分有子类。伽马暴可以分为两类：短硬暴，持续时间大约几毫秒到两秒，具有硬的高能能谱；长软暴，持续时间在两秒到几百秒之间，具有较软的高能能谱[5]。数个短暴的位置直到最近才得到确认。初步研究其宿主星系可知这些短暴可能由两颗恒星各自演化成最后状态的星体然后并合形成[6,7]。相比之下，在光学或射电波段探测到超过80个长暴余辉。这些探测结果显示长暴和核坍缩超新星一样与年轻大质量恒星的死亡有关。我们这里研究的正是这些从大质量恒星死亡中诞生的天体。

长暴一般在呈现强发射线[12,13]的极端蓝宿主星系中被发现[8-11]，这明显说明有大量的年轻的特大质量恒星存在。而且，尽管与长暴成协的光学暂现源通常由来自伽马暴相对论性外向流的辐射主导，还是有很多长暴在它们光变曲线的后期出现“隆起”，与潜在的超新星的出现一致[14-16]。在多个实例中，有光谱证据显示超新星的光线叠加在光学暂现源上[17-20]。确实，考虑到光学暂现源和超新星巨大的亮度变化范围以及对一些伽马暴的有限观测，似乎每颗长暴都具有成协的超新星[21]。尽管长暴释放的各向同性能量通常看来比超新星大几个量级，但是现在有很好的证据显示大多数长暴都高度准直且仅仅照亮几个百分比的天空[22,23]。考虑到这个因素，长暴释放的真实能量更接近能量更强的超新星。不过，不是所有核坍缩超新星都可能是产生伽马暴的候选体。目前与伽马暴成协光谱证认良好的超新星一般都是Ic型超新星——即光谱中没有氢线和氦线的核坍缩超新星(Ib型超新星一般会与Ic型超新星一同被研究，其光谱大多也没有氢线，但表现出很强的氦线特征)。恒星需要失去外包层才能够使伽马暴冲破恒星外表面[24]。目前对邻近星系中Ib/c型超新星和数量更多的II型超新星(具有氢线的核坍缩超新星)位置的比较研究显示二者在宿主星系的类型或在宿主星系爆发的位置均没有差异[25,26]。这一结果使得文献25的作者得出结论，核坍缩超新星全部来自同一质量范围的前身星，但是Ib/c型超新星可能已经通过与双星中的伴星的相互作用被剥离了包层。不管Ic型超新星源自单星、双星抑或二者兼而有之，这些超新星中只有很少一部分会产生伽马暴[27]。

核坍缩超新星和长暴都起源于大质量恒星，我们预计它们的宿主星系与其当地环境都很相似。一直以来都有观点主张核坍缩超新星在星系里的分布和在宿主星系本身里的分布应该都可以示踪宇宙中的蓝光(来自大质量恒星的光是蓝色的)。对长暴也有类似的估计。事实上已经有研究报道了这一相关的粗略证据[28]。这里我们利

rough evidence for such a correlation has been reported[28]. Here we use the high resolution available from Hubble Space Telescope (HST) images, and an analytical technique developed by us that is independent of galaxy morphology, to study the correlation between these objects and the light of their hosts. We also compare the sizes, morphologies and brightnesses of the LGRB hosts with those of the supernovae. Our results reveal surprising and substantial differences between the birthplaces of these cosmic explosions. We find that whereas core-collapse supernovae trace the blue light of their hosts, GRBs are far more concentrated on the brightest regions of their hosts. Furthermore, while the hosts of core-collapse supernovae are approximately equally divided between spiral and irregular galaxies, the overwhelming majority of GRBs are on irregulars, even when we restrict the GRB sample to the same redshift range as the supernova sample. We argue that these results may be best understood if GRBs are formed from the collapse of extremely massive, low-metallicity stars.

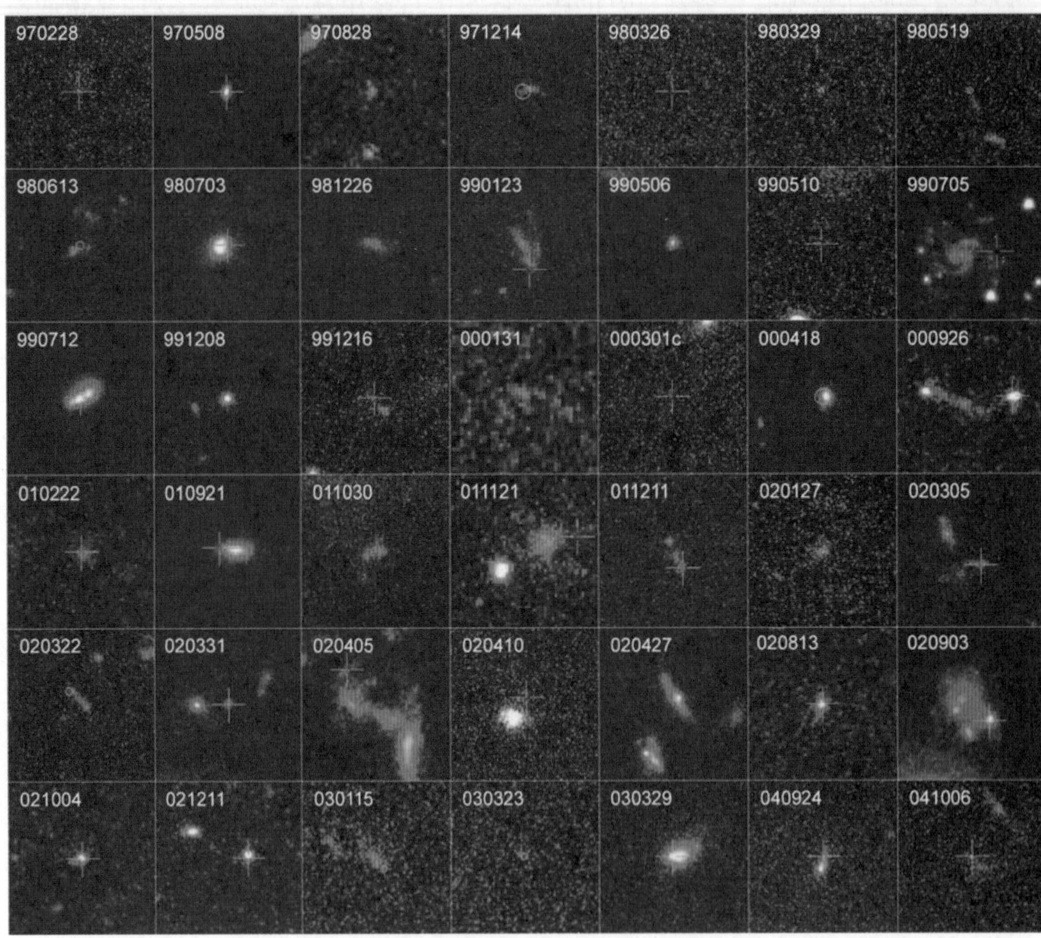

Fig. 1. A mosaic of GRB host galaxies imaged by HST. Each individual image corresponds to a square region on the sky 3.75″ on a side. These images were taken with the Space Telescope Imaging Spectrograph (STIS), the Wide-Field and Planetary Camera 2 (WFPC2) and the Advanced Camera for Surveys (ACS) on HST. In cases where the location of the GRB on the host is known to better than 0.15″ the position of the GRB is shown by a green mark. If the positional error is smaller than the point spread function of the image (0.07″ for STIS and ACS, 0.13″ for WFPC2) the position is marked by a cross-hair;

用哈勃太空望远镜(HST)的高分辨率图像，以及我们提出的独立于星系形态的分析技术来研究这些天体和它们宿主星系光线之间的相关。我们也比较了长暴宿主星系和超新星宿主星系的大小、形态和亮度。结果显示这些宇宙爆炸诞生地存在令人惊讶的重大差异。我们发现尽管核坍缩超新星示踪宿主星系的蓝光，但伽马暴会更加集中于宿主星系的最亮区域。而且，核坍缩超新星的宿主星系中旋涡和不规则星系大致各占一半；甚至将伽马暴样本的选取限定在与超新星样本一样的红移区间时亦是如此。如果伽马暴形成于金属丰度低的超大质量恒星的坍缩，那么也许就能够理解这些结果。

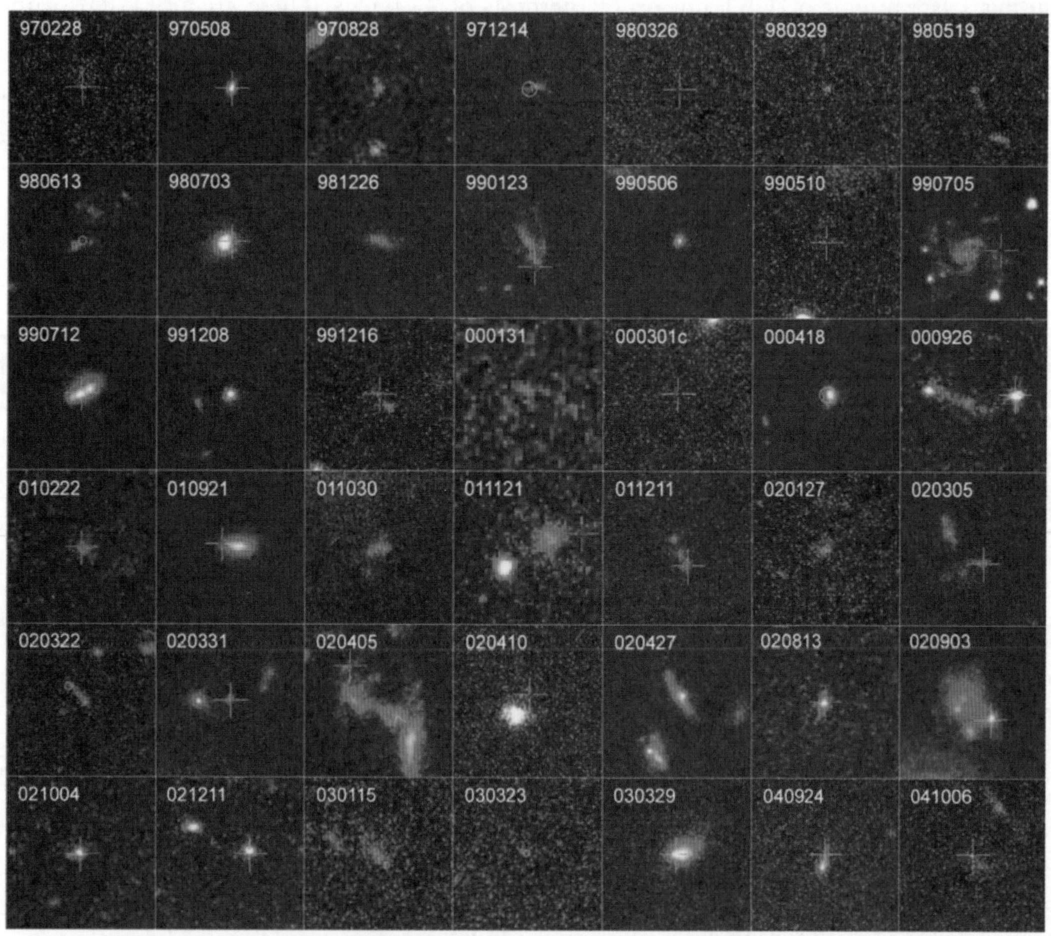

图 1. HST 获得的伽马暴宿主星系拼接组合图。每幅单独的图像对应天球上单边 3.7″的正方形天区。这些图像来自 HST 上的太空望远镜成像摄谱仪(STIS)、大视场行星相机 2(WFPC2)和高新巡天相机(ACS)。位置精度优于 0.15″的伽马暴用绿色标记出来。位置误差小于图像的点扩散函数(STIS 和 ACS 为 0.07″，WFPC2 的为 0.13″)的伽马暴用十字标记；其他位置误差用圆圈表示。STIS 图像都是白光下拍摄的(没有滤光片)，WFPC2 和 ACS 的图像大部分是在 F606W 滤光片下拍摄的(有几种情况此滤光

otherwise the positional error is indicated by a circle. The STIS images were all taken in white light (no filter), and in most cases the WFPC2 and ACS images are in the F606W filter (though in a few cases where images in this filter were not available we have used images in F555W or F775W). The STIS and F606W images can be thought of as broad "V" or visual images, and are, for galaxies exhibiting typical colours of GRB hosts, the single most sensitive settings for these cameras. F555W is close to the ground-based Johnson V band, and F775W corresponds to the ground-based Johnson I band. Owing to the redshifts of the hosts, these images generally correspond to blue or ultraviolet images of the hosts in their rest frame, and thus detect light largely produced by the massive stars in the hosts.

The Sample

Over 40 LGRBs have been observed with the HST at various times after outburst. The HST is unique in its capability easily to resolve the distant hosts of these objects. Shown in Fig. 1 is a mosaic of HST images of the hosts of 42 bursts. These are all LGRBs with public data that had an afterglow detected with better than 3σ significance and a position sufficiently well localized to determine a host galaxy. A list of all the GRBs used in this work can be found in Supplementary Tables 1–3.

The supernovae discussed here were all discovered as part of the Hubble Higher z Supernova Search[29,30], which was done in cooperation with the HST GOODS survey[31]. The GOODS survey observed two ~150 arcmin^2 patches of sky five times each, in epochs separated by 45 days. Supernovae were identified by image subtraction. Here we discuss only the core-collapse supernovae identified in this survey. A list of the supernovae used is presented in Supplementary Table 4, and images of the supernova hosts can be seen in Fig. 2.

Positions of GRBs and Supernovae on Their Hosts

If LGRBs do in fact trace massive star formation, then in the absence of strong extinction we should find a close correlation between their position on their host galaxies and the blue light of those galaxies. However, many of the GRB hosts and quite a few of the supernova hosts are irregular galaxies made up of more than one bright component. As a result, the common astronomical procedure of identifying the centroid of the galaxy's light, and then determining the distance of the object in question from the centroid, is not particularly appropriate for these galaxies—the centroid of light may in fact lie on a rather faint region of the host (examine GRBs 000926 and 020903 in Fig. 1 for excellent illustrations of this effect). We have therefore developed a method that is independent of galaxy morphology. We sort all of the pixels of the host galaxy image from faintest to brightest, and ask what fraction of the total light of the host is contained in pixels fainter than or equal to the pixel containing the explosion. If the explosions track the distribution of light, then the fraction determined by this method should be uniformly distributed between zero and one. (A detailed exposition of this method can be found in Supplementary Information).

片下的图像无法获取，我们用 F555W 或 F775W 中的图像代替）。STIS 和 F606W 的图像可以被认为是宽“V”或者目视图像，而且对于呈现基本色彩的伽马暴宿主，这些相机的设置都是最为敏感的。F555W 接近于地基 Johnson V 波段，F775W 对应于地基 Johnson I 波段。由于宿主星系的红移，这些图像总体上都对应各自静止系的宿主星系的蓝或紫外图像，因此探测到的大部分光线都源自宿主星系里的大质量恒星。

样　本

HST 在爆后不同时间内观测了超过 40 颗长暴。HST 分辨这些天体遥远的宿主星系的能力是独一无二的。图 1 展示的是 HST 测得的 42 颗伽马暴宿主星系的拼接组合图。这些皆是存在公开数据的长暴中测得的余辉，显著性优于 3σ，且具有足以定位到的宿主星系。这一工作中所有伽马暴都在补充表格 1 ~ 3 中列出。

本文讨论的超新星是哈勃高红移超新星搜寻项目中[29,30]发现的一部分超新星，该研究是与 HST GOODS 巡天合作完成的。GOODS 巡天观测了两块约 150 arcmin^2 的天区，每块天区观测 5 次，每次间隔 45 天。通过图像相减发现超新星。这里使用的超新星列在补充表格 4 中，超新星宿主星系的图像见图 2。

伽马暴和超新星在宿主星系的位置

如果长暴真的与大质量恒星密切相关，那么在没有强消光的情况下我们应该能够发现长暴在宿主星系的位置和星系蓝光具有很好的相关。不过，许多伽马暴宿主星系和一些超新星宿主星系是不规则星系，由多于 1 个亮成分组成。因此，尽管天文学普遍的处理步骤是先确定星系光的形心，然后确定要研究的天体离形心的距离，但对于这些星系不是特别适用——光的形心可能位于宿主星系的一个相当暗弱的区域（图 1 中伽马暴 000926 和 020903 很好地描述了这一效应）。我们因此创立了一种独立于星系形态的方法。我们把宿主星系的所有像素从暗到亮排序，然后分别找出像素中包含宿主全部光的部分与包含爆发的部分，再求得前者与后者相比同等亮或更为暗弱的像素的比例。如果爆炸示踪了光的分布，那么这种方法确定的比率应该从 0 到 1 均匀分布（此方法的具体说明见补充资料）。

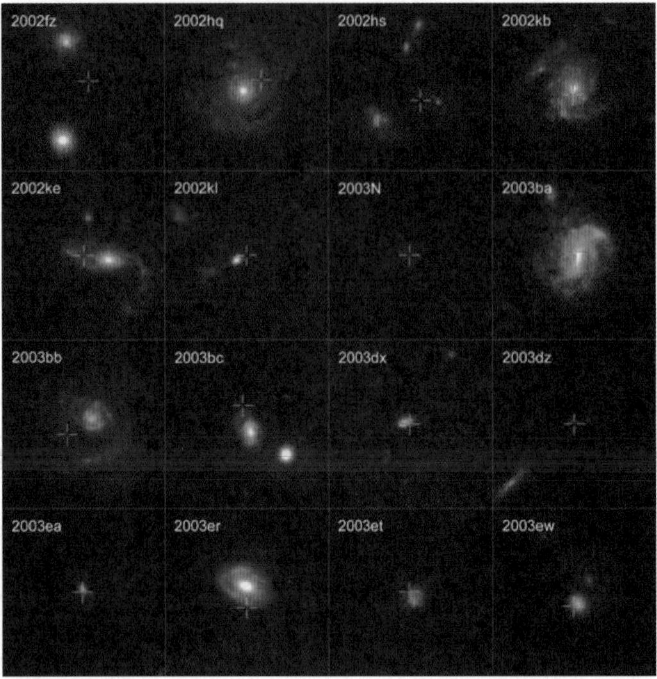

Fig. 2. A mosaic of core-collapse supernova host galaxies imaged with HST as part of the GOODS programme. Each image in the mosaic has a width of 7.5″ on the sky, and thus twice the field of view of each image in the GRB mosaic. The position of each supernova on its host galaxy is marked. In all cases, these positions are known to sub-pixel accuracy. Supernovae in the GOODS sample were identified by ref. 30 as either type Ia or core-collapse supernovae on the basis of their colours, luminosities and light curves, as data allowed (a supernova exploding near the beginning or end of one of the multi-epoch observing runs would have much less data, and sometimes poor colour information). Thus bright type Ib and Ic supernovae, which have colours and luminosities similar to type Ia supernovae, would probably have been classified as type Ia (unless a grism spectrum was taken—however, only a small fraction of objects were observed spectroscopically). On the other hand, fainter type Ib and Ic supernovae ($M_B \gtrsim -18$) could in principle be identified from photometric data; however, in practice the data were rarely sufficient for a clear separation from other core-collapse supernovae. On the basis of surveys of nearby galaxies, one might expect approximately 20% of the core-collapse supernovae to be type Ib or Ic[49,50].

As can be seen in Fig. 3, the core-collapse supernovae do track the light of their hosts as well as could be expected given their small number statistics. A Kolmogorov–Smirnov (KS) test finds that the distribution of the supernovae is indistinguishable from the distribution of the underlying light. The situation is clearly different for LGRBs. As can be seen in Fig. 3, the GRBs do not simply trace the blue light of the hosts; rather, they are far more concentrated on the peaks of light in the hosts than the light itself. A KS test rejects the hypothesis that GRBs are distributed as the light of their hosts with a probability greater than 99.98%. Furthermore, this result is robust: it shows no dependence on GRB host size or magnitude. And in spite of the relatively small number of supernova hosts for which a comparison can be made, the two populations are found by the KS test to be drawn from different distributions with ~99% certainty. In the next section, we show that the surprising differences in the locations of these objects on the underlying light of their hosts may be due not only to the nature of their progenitor stars but also that of their hosts.

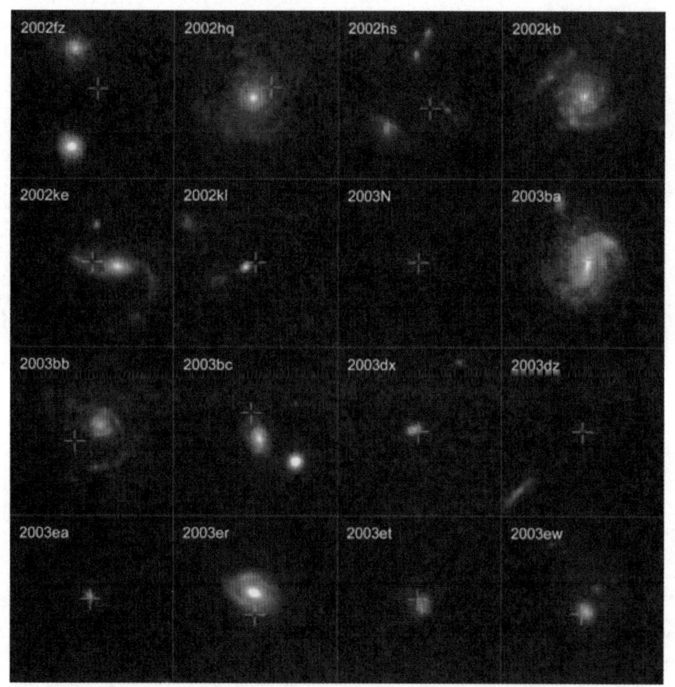

图 2. 作为 GOODS 项目一部分，HST 获得的核坍缩超新星宿主星系拼接组合图。拼接图中的每幅图像对应天球上单边 7.5″的天区，因此是伽马暴拼接图中每幅图像视场的两倍。宿主星系里超新星的位置都标记了出来。所有这些位置精度都是在亚像素级别的。GOODS 的超新星样本都在文献 30 中根据颜色、光度和光变曲线数据确定为 Ia 型或核坍缩超新星（如果超新星爆发时接近多次观测中某一次观测的开始或结束，则它的数据会比较少，有时颜色信息也很差）。因此具有类似 Ia 型超新星颜色和光度的亮 Ib 或 Ic 型超新星可能被分类作 Ia 型超新星（除非拍摄获得棱栅光谱——不过，只有对一小部分天体是可以通过光谱观测的）。另一方面，较暗弱的 Ib 或 Ic 型超新星（$M_B \geqslant -18$）原则上可以依据测光数据分类；不过，实际操作中，缺少充分的数据将其与另外的核坍缩超新星区分。在对附近星系的巡天基础上，估计大约有 20%的核坍缩超新星是 Ib 或 Ic 型超新星[49,50]。

如图 3 所示，在小数目统计下，核坍缩超新星确实如预想的一样很好地示踪了宿主星系光线。柯尔莫戈洛夫 - 斯米尔诺夫（KS）检验发现，超新星的分布和基底光的分布难以区分。而对于长暴，情况则明显不同。如图 3 所示，伽马暴不仅示踪了宿主星系的蓝光，而且它们比光本身更加集中在宿主星系的光的峰值处。KS 检验有大于 99.8% 的概率否定伽马暴与宿主星系光具有相同分布的假设。而且这一结论是很可靠的：它与伽马暴宿主星系的大小或星等无关。尽管超新星宿主星系的数目较少，不足以做出比较，KS 检测发现这两个星族在约 99%确定度下分布不同。在本文的下一节中，我们将展示这些天体在宿主星系基底光线位置的惊人差异可能不仅源自它们前身星的性质，同时也与其宿主星系的性质有关。

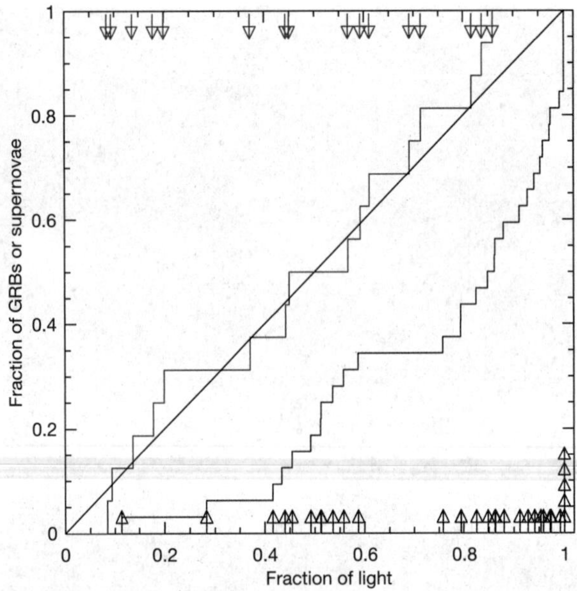

Fig. 3. The locations of the explosions in comparison to the host light. For each object, an arrow indicates the fraction of total host light in pixels fainter than or equal to the light in the pixel at the location of the transient. The cumulative fraction of GRBs or supernovae found at a given fraction of the total light is shown as a histogram. The blue arrows and histogram correspond to the GRBs, and the red arrows and histogram correspond to the supernovae. Were the GRBs and supernovae to track the light identically, their histograms would follow the diagonal line. Whereas the supernova positions do follow the light within the statistical error, the GRBs are far more concentrated on the brightest regions of their hosts. Thus although the probability of a supernova exploding in a particular pixel is roughly proportional to the surface brightness of the galaxy at that pixel, the probability of a GRB at a given location is effectively proportional to a higher power of the local surface brightness.

A Comparison of the Host Populations

An examination of the mosaics of the GRB and supernova hosts (Figs 1 and 2) immediately shows a remarkable contrast—only one GRB host in this set of 42 galaxies is a grand-design spiral, whereas nearly half of the supernova hosts are grand-design spirals. One might wonder if this effect is due to a difference in redshift distribution— the core-collapse supernovae discovered by the GOODS collaboration all lie at redshift $z < 1.2$, whereas LGRBs can be found at much larger redshifts where grand-design spirals are rare to non-existent. Yet if we restrict the GRB population to $z < 1.2$ (and thus produce a population with a nearly identical mean and standard deviation in redshift space compared to the GOODS core-collapse supernovae), the situation remains essentially unchanged: only one out of the eighteen GRB hosts is a grand-design spiral. (For a detailed comparison of GRB hosts to field galaxies, rather than the supernova selected galaxies shown here, see ref. 32.)

Were the difference in spiral fraction the only indication of a difference in the host populations, we could not rule out random chance—given the small number statistics, both populations are barely consistent with each other and a spiral fraction of ~25%. However,

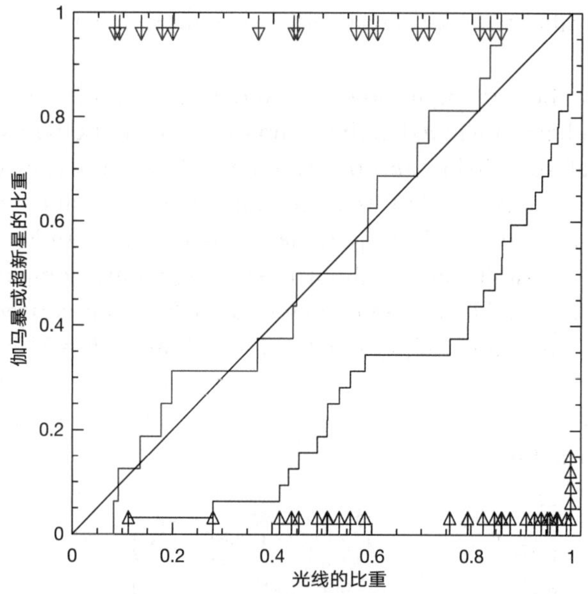

图 3. 爆炸位置和宿主星系光线位置的比较。对于每个天体，箭头代表总宿主光线较暂现位置处光线暗弱或相等的像素的比重。在总光线的给定比例下找到的伽马暴或超新星的累积比重用直方图表示。蓝色箭头和直方图代表伽马暴，红色箭头和直方图代表超新星。如果伽马暴或超新星能同样跟踪光线，那么它们的直方图将沿着对角线分布。超新星位置在统计误差范围的确示踪了光线，而伽马暴的位置则更加集中于宿主星系的极亮区域。因此，尽管超新星在某个像素爆发的概率约正比于在该像素的星系表面亮度，但是在某个位置出现伽马暴的概率实际上正比于能量更高的近域表面亮度。

宿主星族的比较

检查伽马暴和超新星宿主星系的拼接图像（图 1 和图 2）立刻会发现二者的显著差异——在本组 42 个星系中只有一个伽马暴宿主星系是宏大的旋涡结构，而接近一半的超新星宿主星系是宏大旋涡结构。这不禁让人想知道这一效应是否是由于红移分布的不同导致的——GOODS 巡天发现的核坍缩超新星红移全部低于 1.2；长暴可以在更大红移处被发现，但那里却很少甚至没有宏大旋涡星系存在。不过如果我们将伽马暴族限制在 $z < 1.2$（从而产生平均值和标准偏差在红移空间与 GOODS 核坍缩超新星相当的星族），情况基本没有变化：18 个伽马暴宿主星系中只有一个是宏大旋涡星系。（如果相较于此处根据超新星而选择星系进行比较，读者想要获取伽马暴宿主星系和场星系的详细对比，请见文献 32。）

假如旋涡比率的差异是宿主星族的唯一指标，那么我们就不能排除随机因素的影响——因为在小数目统计资料下，星族之间几乎不可能出现彼此一致，旋涡比率约为

the host populations differ strongly in ways other than morphology.

In Fig. 4 we compare the 80% light radius (r_{80}) and absolute magnitude distributions of the GRB and supernova hosts. Included in the comparison are all LGRBs with known redshifts $z < 1.2$ at the time of submission and the 16 core-collapse supernovae of GOODS with spectroscopic or photometric redshifts (see the Supplementary Tables for a complete list of the GRBs, supernovae and associated parameters used in this study). The small minority of GRB hosts in this redshift range without HST imaging are compared only in absolute magnitude and not in size. The absolute magnitudes have been derived from the observed photometry using a cosmology of $\Omega_m = 0.27$, $\Lambda = 0.73$ and $H_0 = 71$ km s^{-1}Mpc^{-1}, and the magnitudes have been corrected for foreground Galactic extinction[33]. For a technical discussion of the determination of the magnitude and size of individual objects, see Supplementary Information.

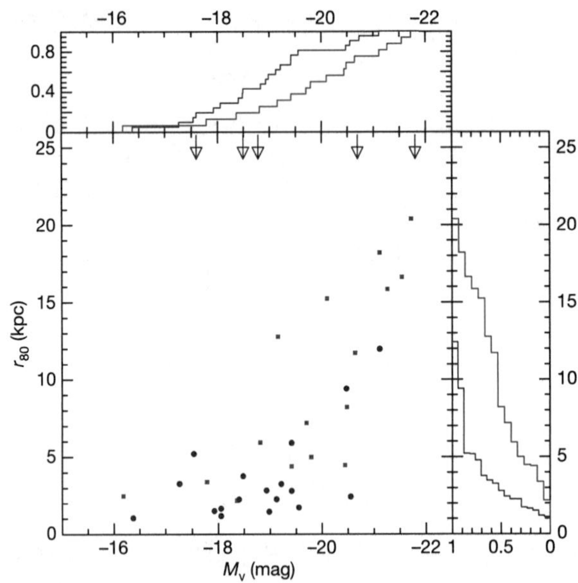

Fig. 4. A comparison of the absolute magnitude and size distributions of the GRB and supernova hosts. In the main panel, the core-collapse supernova hosts are represented as red squares, and the LGRB hosts as blue circles. The absolute magnitudes of the hosts are shown on the x axis, and the lengths of the semi-major axes of the hosts on the y axis. The plot is then projected onto the two side panels where a histogram is displayed for each host population in each of the dimensions—absolute magnitude and semi-major axis. Shown as blue arrows are the absolute magnitudes of GRB hosts with $z < 1.2$ that have been detected from the ground but have not yet been observed by HST. These hosts are only included in the absolute magnitude histogram. The hosts of GRBs are both smaller and fainter than those of supernovae.

As can be readily seen, the two host populations differ substantially both in their intrinsic magnitudes and sizes. The GRB hosts are fainter and smaller than the supernova hosts. Indeed, KS tests reject the hypothesis that these two populations are drawn from the same population with certainties greater than 98.6% and 99.7% for the magnitude and size distributions, respectively.

25%的情况。不过，不同的宿主星系族除了形态之外，在其他方面也有很大的差异。

图 4 中我们对伽马暴和超新星宿主星系的 80%光半径（r_{80}）和绝对星等分布进行了比较。包含的比较对象皆为截至投稿时间已知的红移 $z < 1.2$ 的长暴和 16 个具有光谱或测光红移 GOODS 的核坍缩超新星（伽马暴、超新星以及本研究相关参数的完整列表参见补充表格）。在此红移范围的小部分没有 HST 成像的伽马暴宿主星系仅比较绝对星等，不比较大小。绝对星等是在宇宙学参数 $\Omega_M = 0.27$，$\Lambda = 0.73$ 和 $H_0 = 71\ \mathrm{km \cdot s^{-1} \cdot Mpc^{-1}}$ 的条件下，经过银河系前景消光修正后的测光观测数据获得的[33]。获得单个天体星等和大小的技术讨论见补充资料。

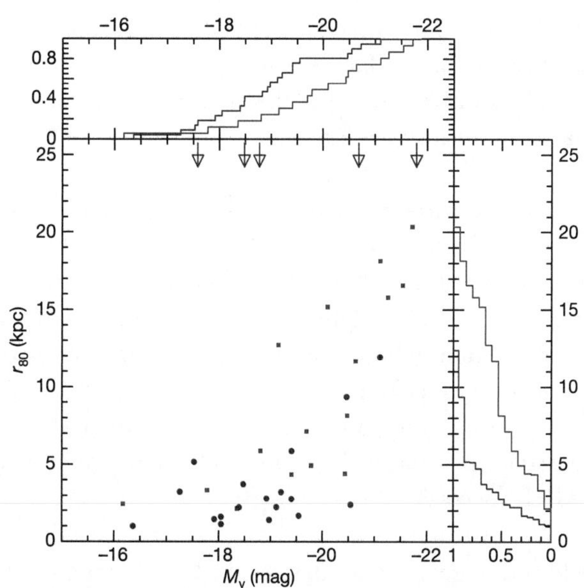

图 4. 伽马暴和超新星宿主星系绝对星等和大小分布的比较。在主图中，核坍缩超新星宿主星系用红色方块表示，长暴宿主星系用蓝色圆圈表示。x 轴表示宿主星系的绝对星等，y 轴表示半长轴的长度。图像投影到面板的两侧得到在每个纬度上——绝对星等和半长轴——各个宿主星系族的直方图。蓝色箭头表示的是在地面探测到，但是没有在 HST 观测到的 $z < 1.2$ 伽马暴宿主星系的绝对星等。这些宿主星系只包含在绝对星等直方图中。伽马暴的宿主星系比超新星的更小更暗弱。

如上所示，两个宿主星系星族在本征星等和大小上都存在巨大差异。伽马暴宿主星系比超新星宿主星系更暗弱且更小。事实上，KS 检验分别在星等和大小分布上以高于 98.6% 和 99.7% 的确定度否定了这两类星系是同一种星系的假设。

Discussion

Although the evidence is now overwhelming that both core-collapse supernovae and LGRBs are formed by the collapse of massive stars, our observations show that the distribution of these objects on their hosts, and the nature of the hosts themselves, are substantially different. How then can this be? We propose here that these surprising findings are the result of the dependence of the probability of GRB formation on the state of the chemical evolution of massive stars in a galaxy.

Even before the association of LGRBs with massive stars had been established, a number of theorists had suggested that these objects could be formed by the collapse of massive stars, which would leave behind rapidly spinning black holes. An accretion disk about the black hole would power the GRB jet. These models, sometimes referred to as "hypernovae" or "collapsar" models, implicitly require very massive stars, as only stars greater than about 18 solar masses form black holes. But in fact it was widely suspected that even more massive stars would be required—if only to provide the required large energies, and to limit the numbers of supernovae progressing to GRBs.

We conclude that LGRBs do indeed form from the most massive stars and that this is the reason that they are even more concentrated on the blue light of their hosts than the light itself. The most massive stars (O stars) are frequently found in large associations. These associations can be extremely bright, and can indeed provide the peak of the light of a galaxy—particularly if that galaxy is a faint, blue irregular, as are the GRB hosts in general. Indeed, a connection of LGRBs with O stars (and perhaps Wolf–Rayet stars) is a natural one— given the strong emission lines (including Ne [III]) seen in many of these hosts[12,13] and the evidence for possible strong winds off the progenitors of the GRBs seen in absorption in some LGRB spectra[34,35].

However, O stars are found in galaxies of all sizes. Indeed, studies of the Magellanic clouds suggest that the initial distribution of masses of stars at formation in these dwarf galaxies is essentially identical to that in our much larger spiral, the Milky Way[36]. Therefore, a difference in the initial mass function of stars is unlikely to be responsible for the differences between the hosts. We propose that the fundamental differences between the LGRB and supernova host populations are not their size or luminosity, but rather their metallicity, or chemical evolution. Some evidence of this already exists. The hosts of seven LGRBs (GRBs 980425 (P.M.V., personal communication), 990712[13], 020903[37], 030323[38], 030329[17], 031203[39] and 050730[40]) have measurements of, or limits on, their metallicity, and in all cases the metallicity is less than one-third solar. The small size and low luminosity of the GRB hosts is then a result of the well known correlation between galaxy mass and metallicity (see ref. 41 and references therein).

But why do LGRBs occur in low-metallicity galaxies? This may be a direct result of the evolution of the most massive stars. It has recently been proposed that metal-rich stars

讨　论

尽管现在有决定性的证据表明核坍缩超新星与长暴是由大质量恒星坍缩而成，但是我们的观测显示这两者在宿主星系里的分布与宿主星系本身的性质上明显不同。这又是为什么呢？我们推测这些令人惊讶的发现是由于伽马暴的形成取决于星系中大质量恒星化学演化状态。

在确定长暴和大质量恒星的成协之前，很多理论已经提出这些天体可能是由大质量恒星坍缩形成，并留下一个快速自转的黑洞。黑洞周围的吸积盘将为伽马暴喷流提供能量。这些模型，即被人们称为“极超新星”或“坍缩星”的模型，内含了形成黑洞需要质量非常大的恒星的要求，例如只有大于 18 个太阳质量的恒星才能形成黑洞。但是实际上人们很大程度上猜测需要更大质量的恒星——如果仅仅是为了提供需要的巨大能量，并限制超新星演化到伽马暴的数目。

我们的结论是，长暴确实由最大质量恒星形成，这就是为什么它们在宿主星系的蓝光区比自身光区更集中。最大质量恒星（O 型星）经常在大的星协里被发现。这些星协极其亮，事实上能够提供星系的光线峰值——特别是如果那个星系是一个微弱的蓝色不规则星系，就像一般的伽马暴宿主星系一样。事实上，长暴和 O 型星（也可能沃尔夫－拉叶星）的关联是很自然的——因为在这些宿主星系里，很多都发现了强发射线（包括 Ne[III]）[12,13]，同时有证据显示在某些长暴光谱吸收中发现可能源自伽马暴前身星的强风[34,35]。

不过，所有大小的星系中都有 O 型星。事实上，对麦哲伦云的研究显示在这些矮星系中恒星形成时的初始质量分布本质上与我们所生活的更大的旋涡星系——银河系——是一样的[36]。因此，恒星初始质量分布的差异不太可能导致两类宿主星系的差异。我们认为长暴和超新星宿主星系星族的根本差异不在于其大小或者光度，而是金属丰度或者说化学演化。目前已经有一些这方面的证据。具有测量或极限金属丰度的 7 个长暴的宿主星系（伽马暴 980425（弗雷埃斯维克，个人交流），990712[13]，020903[37]，030323[38]，030329[17]，031203[39] 和 050730[40]），它们的金属丰度全都小于太阳水平的 1/3。伽马暴宿主星系的小尺度和低光度则由著名的星系质量和金属丰度相关导致的（见参考文献 41 及其中的文献）。

但是为什么长暴存在于低金属丰度星系？这可能是最大质量恒星演化直接导致的。最近有研究提出具有几十个太阳质量的富金属恒星的表面风很强（光压作用在富

with masses of tens of solar masses have such large winds off their surfaces (due to the photon pressure on their metal-rich atmospheres) that they lose most of their mass before they collapse and produce supernovae[4]. As a result they leave behind neutron stars, not the black holes necessary for LGRB formation. Ironically, stars of 15–30 solar masses may still form black holes, as they do not possess radiation pressure sufficient to drive off their outer envelopes. Direct evidence for this scenario comes from recent work showing that the Galactic soft γ-ray repeater, SGR 1820–06, is in a cluster of extremely young stars of which the most massive have only started to collapse[42]—yet the progenitor of SGR 1820–06 collapsed to a neutron star, not a black hole. Recent observations of winds from very massive (Wolf–Rayet) stars provide further support for this scenario: outflows from the low-metallicity stars in the Large Magellanic Cloud are substantially smaller than those seen from more metal-rich Galactic stars[43]. The possible importance of metallicity in LGRB formation has therefore not escaped the notice of theorists[44,45].

A preference for low metallicity may also explain one of the most puzzling results of GRB host studies. None of the LGRB hosts is a red, sub-millimetre bright galaxy. These highly dust-enshrouded galaxies at redshifts of ~1–3 are believed to be the sites of a large fraction of the star formation in the distant Universe[46]. And although some LGRB hosts do show sub-millimetre emission, none have the red colours characteristic of the majority of this population. However, it is likely that these red dusty galaxies have substantial metallicities at all redshifts. The low metallicity of hosts may also help explain the fact that a substantial fraction of high-redshift LGRB hosts display strong Lyman-α emission[47].

All well-classified supernovae associated with LGRBs are of type Ic, presumably because the presence of a hydrogen envelope about the collapsing core can block the emergence of a GRB jet[24]. Thus only those supernovae whose progenitors have lost some, but not too much, mass appear to be candidates for the formation of a GRB. Given the large numbers of type Ic supernovae in comparison to the estimated numbers of LGRBs, however, it is likely that only a small fraction of type Ic supernovae produce LGRBs. Indeed, even the number of unusually energetic type Ib/Ic supernovae appears to dwarf the LGRB population[48]. Another process, perhaps the spin-up of the progenitor in a binary[27], may decide which type Ic supernovae produce LGRBs. Interestingly, it was the similar distribution of supernovae on their hosts, and particularly the fact that type Ib/Ic were no more correlated than type II with the UV bright regions of their hosts, that led ref. 25 to the conclusion that type Ib/Ic supernovae form from binaries. LGRBs clearly track light differently from the general type Ic population. However, the samples used by refs 25 and 26 were from supernovae largely discovered on nearby massive galaxies—dwarf irregular hosts are underrepresented in these samples. It will be particularly interesting to see whether large unbiased supernova surveys at present underway produce similar locations for their supernovae.

We do not know, however, what separates the small fraction of low-metallicity type Ic supernovae that turn into LGRBs from the rest of the population. Potentially, the answer is the amount of angular momentum available in the core to form the jet. In this case, the

金属大气上导致)，在它们坍缩形成超新星之前已经丢失了大部分质量[4]。因此它们留下的是中子星而不是形成长暴所需要的黑洞。具有讽刺意味的是，15～30倍太阳质量的恒星仍然能形成黑洞，因为它们没有足够的辐射压以驱动它们的外包层。这个图景的直接证据来自最近的研究，研究显示银河系的软伽马射线复现源SGR1820-06处于由极年轻恒星组成的星团中，其中最大质量的一批恒星刚刚开始坍缩[42]——不过SGR1820-06的前身星已然坍缩成中子星，而不是黑洞。最近对来自超大质量恒星(沃尔夫－拉叶星)星风的观测也进一步支持了这个图景：大麦哲伦云里的低金属度星的外向流比银河系中更为富金属的恒星的外向流明显要小[43]。在长暴形成过程中可能起重要作用的金属丰度因此也没能逃出理论学家的注意[44,45]。

低金属丰度也可能用来解释关于伽马暴宿主星系研究中最困难的一个谜题。长暴宿主星系中没有一个是红的、亚毫米波的亮星系。这些红移约为1～3处的被尘埃掩盖的星系被认为是遥远宇宙大部分恒星形成之地[46]。尽管一些长暴宿主星系有亚毫米辐射，但是没有一个具有大部分这类星族的红色特征。不过，很可能这些红色的富尘埃星系在所有红移处都具有充足的金属丰度。宿主星系低金属丰度或许也能帮助解释一部分高红移长暴宿主星系呈现的强莱曼α辐射[47]。

所有与长暴成协的，类型确定的超新星都是Ic型，这大概是由于坍缩核外的氢包层阻止了伽马暴喷流出现[24]。因此只有那些前身星已经丢失一些质量，但又不是很多的超新星才可能是伽马暴的候选。不过，考虑到Ic型超新星数量相比长暴的估计数目大得多，可能只有一小部分Ic型超新星会形成长暴。事实上，甚至非正常的富能量Ib/Ic型超新星的数目看起来都使长暴族相形见绌[48]。另一个过程，可能是双星前身星的加速[27]，可能决定了哪一种Ic型超新星产生长暴。有意思的是，正是超新星在其宿主星系的相似分布，尤其是Ib/Ic型超新星和II型超新星与其各自宿主星系的UV亮区域之间的相关都不紧密，导致文献25得出Ib/Ic型超新星来自双星的结论。长暴的清晰示踪的光线和一般Ic型超新星不同。不过，文献25和26使用的样本大部分来自附近大质量宿主星系——矮且不规则的星系中发现的超新星，因而是取样不足的。目前正在进行的大型、无偏超新星巡天发现的超新星是否也具有类似的位置将具有非凡的意义。

不过，我们不知道是什么机制把贫金属Ic型超新星从星族分离出来变成长暴。答案有可能是核中角动量是否足够大到形成喷流。在此情形下，贫金属的倾向可能表明双星中单星的演化相比双星相互作用在长暴的形成过程中占主导作用。通过研

preference for low metallicity may indicate that single star evolution dominates over binary interaction in forming LGRBs. Deep, high-spectral-resolution studies of LGRB afterglows may provide insight here, by allowing studies of the winds off the progenitor and any binary companion.

Only a small fraction of LGRBs are found in spiral galaxies, even for LGRBs with redshifts $z < 1$ where spirals are much more common. However, the local metallicity in spirals is known to be anti-correlated with distance from the centre of the galaxy. Thus one might expect LGRBs in spirals to violate the trend that we have seen for the general LGRB population, and avoid the bright central regions of their hosts. The present number of LGRBs known in spirals is still too small to test this prediction. But a sample size a few times larger should begin to allow such a test. Additionally, a survey of the metallicity of the hosts of the GOODS supernovae should find a higher average metallicity than that seen in GRB hosts. Finally, if low metallicity is indeed the primary variable in determining whether LGRBs are produced, then as we observe higher redshifts, where metallicities are lower than in most local galaxies, LGRBs should be more uniformly distributed among star-forming galaxies. Indeed, some evidence of this may already be present in the data[32]. LGRBs, however, are potentially visible to redshifts as high as $z \approx 10$. At significant redshifts, where the metallicities of even relatively large galaxies are expected to be low, we may find that LGRBs do become nearly unbiased tracers of star formation.

(**441**, 463-468; 2006)

A. S. Fruchter[1], A. J. Levan[1,2,3], L. Strolger[1,4], P. M. Vreeswijk[5], S. E. Thorsett[6], D. Bersier[1,7], I. Burud[1,8], J. M. Castro Cerón[1,9], A. J. Castro-Tirado[10], C. Conselice[11,12], T. Dahlen[13], H. C. Ferguson[1], J. P. U. Fynbo[9], P. M. Garnavich[14], R. A. Gibbons[1,15], J. Gorosabel[1,10], T. R. Gull[16], J. Hjorth[9], S. T. Holland[17], C. Kouveliotou[18], Z. Levay[1], M. Livio[1], M. R. Metzger[19], P. E. Nugent[20], L. Petro[1], E. Pian[21], J. E. Rhoads[1], A. G. Riess[1], K. C. Sahu[1], A. Smette[5], N. R. Tanvir[3], R. A. M. J. Wijers[22] & S. E. Woosley[6]

[1] Space Telescope Science Institute, 3700 San Martin Drive, Baltimore, Maryland 21218, USA

[2] Department of Physics and Astronomy, University of Leicester, University Road, Leicester, LE1 7RH, UK

[3] Centre for Astrophysics Research, University of Hertfordshire, College Lane, Hatfield, AL10 9AB, UK

[4] Physics & Astronomy, TCCW 246, Western Kentucky University, 1 Big Red Way, Bowling Green, Kentucky 42101, USA

[5] European Southern Observatory, Alonso de Córdova 3107, Casilla 19001, Santiago, Chile

[6] Department of Astronomy & Astrophysics, University of California, 1156 High Street, Santa Cruz, California 95064, USA

[7] Astrophysics Research Institute, Liverpool John Moores University, Twelve Quays House, Egerton Wharf, Birkenhead, CH41 1LD, UK

[8] Norwegian Meteorological Institute, PO Box 43, Blindern, N-0313 Oslo, Norway

[9] Dark Cosmology Centre, Niels Bohr Institute, University of Copenhagen, DK-2100 Copenhagen, Denmark

[10] Instituto de Astrofísica de Andalucía (CSIC), Camino Bajo de Huétor, 50, 18008 Granada, Spain

[11] California Institute of Technology, Mail Code 105-24, Pasadena, California 91125, USA

[12] School of Physics and Astronomy, University of Nottingham, University Park, Nottingham, NG7 2RD, UK

[13] Department of Physics, Stockholm University, SE-106 91 Stockholm, Sweden

[14] Physics Department, University of Notre Dame, 225 Nieuwland Hall, Notre Dame, Indiana 46556, USA

[15] Vanderbilt University, Department of Physics and Astronomy, 6301 Stevenson Center, Nashville, Tennessee 37235, USA

[16] Code 667, Extraterrestial Planets and Stellar Astrophysics, Exploration of the Universe Division, [17] Code 660.1, Goddard Space Flight Center, Greenbelt, Maryland 20771, USA

[18] NASA/Marshall Space Flight Center, VP-62, National Space Science & Technology Center, 320 Sparkman Drive, Huntsville, Alabama 35805, USA

究前身星风和双子星伴星，长暴余辉的深、高光谱分辨率研究可能会为此问题提供一些思路。

即便在红移 $z<1$ 里十分常见的旋涡星系，也只有小部分长暴被发现存在于其中。不过，旋涡星系的局部金属丰度与离星系中心的距离反相关。因此可以预测旋涡星系中的长暴与我们在一般长暴族观测到的趋势不同，它们将回避其宿主星系中心亮区域。目前已知的旋涡星系中的长暴数量还太少，不足以验证这个预测。等样本数量再增大几倍时就可以进行验证。另外，对 GOODS 超新星宿主星系金属丰度巡天发现的平均金属丰度应该超过伽马暴宿主星系的观测值。最后，如果贫金属丰度确实是决定是否形成长暴的主要参数，那么在更高红移处的金属丰度比大部分邻近星系低，应该看到更多的长暴不均匀地分布于正在形成恒星的星系中。确实，这个趋势可能已经在数据中有所展示[32]。不过，长暴可能在红移 $z\approx10$ 处都能看到。在显著的红移处，相对较大的星系的金属丰度预期会降低，因而我们可能会发现长暴成为恒星形成的无偏差示踪者。

（肖莉 翻译；徐栋 审稿）

[19] Renaissance Technologies Corporation, 600 Route 25A, East Setauket, New York 11733, USA
[20] Lawrence Berkeley National Laboratory, MS 50F-1650, 1 Cyclotron Road, Berkeley, California 94720, USA
[21] INAF, Osservatorio Astronomico di Trieste, Via G.B. Tiepolo 11, I-34131 Trieste, Italy
[22] Astronomical Institute "Anton Pannekoek", University of Amsterdam, Kruislaan 403, NL-1098 SJ Amsterdam, The Netherlands

Received 22 August 2005; accepted 5 April 2006. Published online 10 May 2006.

References:

1. Riess, A. G. *et al.* Observational evidence from supernovae for an accelerating universe and a cosmological constant. *Astron. J.* **116**, 1009-1038 (1998).
2. Perlmutter, S. *et al.* Measurements of Ω and Λ from 42 high-redshift supernovae. *Astrophys. J.* **517**, 565-586 (1999).
3. Branch, D., Livio, M., Yungelson, L. R., Boffi, F. R. & Baron, E. In search of the progenitors of type IA supernovae. *Publ. Astron. Soc. Pacif.* **107**, 1019-1028 (1995).
4. Heger, A., Fryer, C. L., Woosley, S. E., Langer, N. & Hartmann, D. H. How massive single stars end their life. *Astrophys. J.* **591**, 288-300 (2003).
5. Kouveliotou, C. *et al.* Identification of two classes of gamma-ray bursts. *Astrophys. J.* **413**, 101-104 (1993).
6. Gehrels, N. *et al.* A short γ-ray burst apparently associated with an elliptical galaxy at redshift $z = 0.225$. *Nature* **437**, 851-854 (2005).
7. Prochaska, J. X. *et al.* The galaxy hosts and large-scale environments of short-hard γ-ray bursts. *Astrophys. J. Lett.* (in the press): preprint at 〈http://arxiv.org/astro-ph/0510022〉 (2005).
8. Fruchter, A. S. *et al.* HST and Palomar imaging of GRB 990123: Implications for the nature of gamma-ray bursts and their hosts. *Astrophys. J.* **519**, 13-16 (1999).
9. Sokolov, V. V. *et al.* Host galaxies of gamma-ray bursts: Spectral energy distributions and internal extinction. *Astron. Astrophys.* **372**, 438-455 (2001).
10. Le Floc'h, E. *et al.* Are the hosts of gamma-ray bursts sub-luminous and blue galaxies? *Astron. Astrophys.* **400**, 499-510 (2003).
11. Christensen, L., Hjorth, J. & Gorosabel, J. UV star-formation rates of GRB host galaxies. *Astron. Astrophys.* **425**, 913-926 (2004).
12. Bloom, J. S., Djorgovski, S. G., Kulkarni, S. R. & Frail, D. A. The host galaxy of GRB 970508. *Astrophys. J.* **507**, L25-L28 (1998).
13. Vreeswijk, P. M. *et al.* VLT spectroscopy of GRB 990510 and GRB 990712: Probing the faint and bright ends of the gamma-ray burst host galaxy population. *Astrophys. J.* **546**, 672-680 (2001).
14. Bloom, J. S. *et al.* The unusual afterglow of the γ-ray burst of 26 March 1998 as evidence for a supernova connection. *Nature* **401**, 453-456 (1999).
15. Galama, T. J. *et al.* Evidence for a supernova in reanalyzed optical and near-infrared images of GRB 970228. *Astrophys. J.* **536**, 185-194 (2000).
16. Levan, A. *et al.* GRB 020410: A gamma-ray burst afterglow discovered by its supernova light. *Astrophys. J.* **624**, 880-888 (2005).
17. Hjorth, J. *et al.* A very energetic supernova associated with the γ-ray burst of 29 March 2003. *Nature* **423**, 847-850 (2003).
18. Stanek, K. Z. *et al.* Spectroscopic discovery of the supernova 2003dh associated with GRB 030329. *Astrophys. J.* 591, L17-L20 (2003).
19. Della Valle, M. *et al.* Evidence for supernova signatures in the spectrum of the late-time bump of the optical afterglow of GRB 021211. *Astron. Astrophys.* **406**, L33-L37 (2003).
20. Malesani, D. *et al.* SN 2003lw and GRB 031203: A bright supernova for a faint gamma-ray burst. *Astrophys. J.* **609**, L5-L8 (2004).
21. Zeh, A., Klose, S. & Hartmann, D. H. A systematic analysis of supernova light in gamma-ray burst afterglows. *Astrophys. J.* **609**, 952-961 (2004).
22. Panaitescu, A. & Kumar, P. Fundamental physical parameters of collimated gamma-ray burst afterglows. *Astrophys. J.* **560**, L49-L53 (2001).
23. Frail, D. A. *et al.* Beaming in gamma-ray bursts: Evidence for a standard energy reservoir. *Astrophys. J.* **562**, L55-L58 (2001).
24. MacFadyen, A. I., Woosley, S. E. & Heger, A. Supernovae, jets, and collapsars. *Astrophys. J.* **550**, 410-425 (2001).
25. van Dyk, S. D., Hamuy, M. & Filippenko, A. V. Supernovae and massive star formation regions. *Astron. J.* **111**, 2017-2027 (1996).
26. van den Bergh, S., Li, W. & Filippenko, A. V. Classifications of the host galaxies of supernovae, Set III. *Publ. Astron. Soc. Pacif.* **117**, 773-782 (2005).
27. Podsiadlowski, P., Mazzali, P. A., Nomoto, K., Lazzati, D. & Cappellaro, E. The rates of hypernovae and gamma-ray bursts: Implications for their progenitors. *Astrophys. J.* **607**, L17-L20 (2004).
28. Bloom, J. S., Kulkarni, S. R. & Djorgovski, S. G. The observed offset distribution of gamma-ray bursts from their host galaxies: A robust clue to the nature of the progenitors. *Astron. J.* **123**, 1111-1148 (2002).
29. Riess, A. G. *et al.* Identification of Type Ia supernovae at redshift 1.3 and beyond with the Advanced Camera for Surveys on the Hubble Space Telescope. *Astrophys. J.* **600**, L163-L166 (2004).
30. Strolger, L. G. *et al.* The Hubble Higher z Supernova Search: Supernovae to $z \sim 1.6$ and constraints on Type Ia progenitor models. *Astrophys. J.* **613**, 200-223 (2004).
31. Giavalisco, M. *et al.* The Great Observatories Origins Deep Survey: Initial results from optical and near-infrared imaging. *Astrophys. J.* **600**, L93-L98 (2004).
32. Conselice, C. J. *et al.* Gamma-ray burst selected high redshift galaxies: Comparison to field galaxy populations to $z \sim 3$. *Astrophys. J.* **633**, 29-40 (2005).
33. Schlegel, D. J., Finkbeiner, D. P. & Davis, M. Maps of dust infrared emission for use in estimation of reddening and cosmic microwave background radiation foregrounds. *Astrophys. J.* **500**, 525-553 (1998).
34. Mirabal, N. *et al.* GRB 021004: A possible shell nebula around a Wolf-Rayet star gamma-ray burst progenitor. *Astrophys. J.* **595**, 935-949 (2003).
35. Klose, S. *et al.* Probing a gamma-ray burst progenitor at a redshift of $z = 2$: A comprehensive observing campaign of the afterglow of GRB 030226. *Astron. J.* **128**, 1942-1954 (2004).
36. Weidner, C. & Kroupa, P. in *The Initial Mass Function 50 Years Later* (eds Corbelli, E., Plla, F. & Zinnecker, H.) 125-186 (Springer, Dordrecht, 2005).

37. Bersier, D. *et al.* Evidence for a supernova associated with the x-ray flash 020903. *Astrophys. J.* (submitted).

38. Vreeswijk, P. M. *et al.* The host of GRB 030323 at $z = 3.372$: A very high column density DLA system with a low metallicity. *Astron. Astrophys.* **419**, 927-940 (2004).

39. Prochaska, J. X. *et al.* The host galaxy of GRB 031203: Implications of its low metallicity, low redshift, and starburst nature. *Astrophys. J.* **611**, 200-207 (2004).

40. Chen, H.-W., Prochaska, J. X., Bloom, J. S. & Thompson, I. B. Echelle spectroscopy of a GRB afterglow at $z = 3.969$: A new probe of the interstellar and intergalactic media in the young Universe. *Astrophys. J.* **634**, L25-L28 (2005).

41. Kobulnicky, H. A. & Kewley, L. J. Metallicities of galaxies in the GOODS-North Field. *Astrophys. J.* **617**, 240-261 (2004).

42. Figer, D. F., Najarro, F., Geballe, T. R., Blum, R. D. & Kudritzki, R. P. Massive stars in the SGR 1806-20 cluster. *Astrophys. J.* **622**, L49-L52 (2005).

43. Crowther, P. A. & Hadfield, L. J. Reduced Wolf-Rayet line luminosities at low metallicity. *Astron. Astrophys.* **449**, 711-722 (2006).

44. Woosley, S. & Heger, A. The progenitor stars of gamma-ray Bursts. *Astrophys. J.* **637**, 914-921 (2005).

45. Yoon, S.-C. & Langer, N. Evolution of rapidly rotating metal-poor massive stars towards gamma-ray bursts. *Astron. Astrophys.* **443**, 643-648 (2005).

46. Chapman, S. C., Blain, A. W., Smail, I. & Ivison, R. J. A redshift survey of the submillimeter galaxy population. *Astrophys. J.* **622**, 772-796 (2005).

47. Fynbo, J. P. U. *et al.* On the Lyα emission from gamma-ray burst host galaxies: Evidence for low metallicities. *Astron. Astrophys.* **406**, L63-L66 (2003).

48. Soderberg, A. M., Nakar, E., Kulkarni, S. R. & Berger, E. Late-time radio observations of 68 Type Ibc supernovae: Strong constraints on off-axis gamma-ray bursts. *Astrophys. J.* **638**, 930-937 (2006).

49. van den Bergh, S. & Tammann, G. A. Galactic and extragalactic supernova rates. *Annu. Rev. Astron. Astrophys.* **29**, 363-407 (1991).

50. Mannucci, F. *et al.* The supernova rate per unit mass. *Astron. Astrophys.* **433**, 807-814 (2005).

Supplementary Information is linked to the online version of the paper at www.nature.com/nature.

Acknowledgements. Support for this research was provided by NASA through a grant from the Space Telescope Science Institute, which is operated by the Association of Universities for Research in Astronomy, Inc. Observations analysed in this work were taken by the NASA/ESA Hubble Space Telescope under programmes: 7785, 7863, 7966, 8189, 8588, 9074 and 9405 (Principal Investigator, A.S.F.); 7964, 8688, 9180 and 10135 (PI, S. R. Kulkarni); 8640 (PI, S.T.H.). We thank N. Panagia, N. Walborn and A. Soderberg for conversations; A. Filippenko and collaborators for early-time images of GRB 980326; and J. Bloom and collaborators for making public their early observations of GRB 020322.

Author Information. Reprints and permissions information is available at npg.nature.com/reprintsandpermissions. The authors declare no competing financial interests. Correspondence and requests for materials should be addressed to A.S.F. (fruchter@stsci.edu).

The Association of GRB 060218 with a Supernova and the Evolution of the Shock Wave

S. Campana *et al.*

Editor's Note

Although the association between gamma-ray bursts and a rare class of supernovae (exploding stars) is well established, how the jet that characterizes a burst emerges from the exploding star had been unclear. Here Sergio Campana and colleagues report observations of a fairly nearby burst, which they spotted very early and were thus able to study carefully its connection with the associated supernova. They conclude that the properties are best explained by the "break-out" of a shock wave formed when a fast-moving shell of detritus collided with a stellar wind surrounding the supernova's progenitor star. This means that they observed the supernova in the very act of exploding.

Although the link between long γ-ray bursts (GRBs) and supernovae has been established[1-4], hitherto there have been no observations of the beginning of a supernova explosion and its intimate link to a GRB. In particular, we do not know how the jet that defines a γ-ray burst emerges from the star's surface, nor how a GRB progenitor explodes. Here we report observations of the relatively nearby GRB 060218 (ref. 5) and its connection to supernova SN 2006aj (ref. 6). In addition to the classical nonthermal emission, GRB 060218 shows a thermal component in its X-ray spectrum, which cools and shifts into the optical/ultraviolet band as time passes. We interpret these features as arising from the break-out of a shock wave driven by a mildly relativistic shell into the dense wind surrounding the progenitor[7]. We have caught a supernova in the act of exploding, directly observing the shock break-out, which indicates that the GRB progenitor was a Wolf–Rayet star.

GRB 060218 was detected with the Burst Alert Telescope (BAT) instrument[8] onboard the Swift[9] space mission on 18.149 February 2006 Universal Time[5]. The burst profile is unusually long with a T_{90} (the time interval containing 90% of the flux) of $2,100 \pm 100$ s (Fig. 1). The flux slowly rose to the peak at 431 ± 60 s (90% containment; times are measured from the BAT trigger time). Swift slewed autonomously to the newly discovered burst. The X-ray Telescope (XRT)[10] found a bright source, which rose smoothly to a peak of ~ 100 counts s^{-1} (0.3–10 keV) at 985 ± 15 s. The X-ray flux then decayed exponentially with an *e*-folding time of $2,100 \pm 50$ s, followed around 10 ks later by a shallower power-law decay similar to that seen in typical GRB afterglows[11,12] (Fig. 2). The UltraViolet/Optical Telescope (UVOT)[13] found emission steadily brightening by a factor of 5–10 after the first detection, peaking in a broad plateau first in the ultraviolet (31.3 ± 1.8 ks at 188 nm) and later in the optical (39.6 ± 2.5 ks at 439 nm) parts of the spectrum. The light curves reached a minimum at about 200 ks, after which the ultraviolet light curves remained constant while

GRB 060218 与一个超新星成协以及激波的演化

坎帕纳等

编者按

尽管γ射线暴与一类稀有的超新星（即爆炸的恒星）之间的成协已经确立，作为γ射线暴特征的喷流如何从爆炸恒星出现却一直不为人所知。本文作者塞尔希奥·坎帕纳与他的同事报道了他们对一个非常近的γ射线暴的观测。他们很早便开始观察，因而可以细致地研究这一个γ射线暴与成协的超新星之间的联系。他们得出结论，对于所观测到的一系列特征最好的解释是快速运动的爆炸残余物壳层与超新星前身星周围的星风碰撞时形成的激波的"突破"。这意味着他们观测到的超新星正处于爆炸进行之中。

尽管长γ射线暴（GRB）和超新星之间的关联已经被确定[1-4]，迄今尚未观测到超新星爆发的开始时期及其与GRB之间的紧密联系。尤其是，我们不知道刻画γ射线暴的喷流是如何从恒星的表面突现出来的，也不知道GRB的前身星是怎样爆发的。这里我们报道就相对较近的GRB 060218（参考文献5）及其与超新星SN 2006aj（参考文献6）之间的联系的观测。除了典型的非热辐射，GRB 060218的X射线能谱还显示存在热成分，热成分随着时间推移逐渐变冷并转移到光学/紫外波段。我们将这些特征解释为由轻度相对论性壳层驱动的激波进入前身星周围的致密星风引发的激波暴[7]。我们正好捕捉到一颗超新星处于爆发之中，并直接观测到激波暴。此激波暴表明该GRB的前身星是一颗沃尔夫–拉叶星。

GRB 060218是于世界时2006年2月18.149日利用载于雨燕[9]天文台上的γ暴预警望远镜（BAT）[8]观测到的[5]。这个γ暴的轮廓不寻常的长，T_{90}（包含90%流量的时间区间）为2,100±100 s（图1）。流量缓慢上升，在431±60 s时达到峰值（包含90%；时间以BAT触发时间为起始）。雨燕自动转动到这个最新发现的γ暴。X射线望远镜（XRT）[10]发现了一颗亮源在平滑上升，并在985±15 s时到达峰值，约100次计数·s^{-1}（0.3～10 keV）。然后X射线流量指数衰减，e折时间为2,100±50 s，接下来约10 ks后是类似典型GRB余辉中看到的那种较缓的幂率衰减[11,12]（图2）。紫外/光学望远镜（UVOT）[13]发现在第一次探测后辐射稳定变亮为原来的5～10倍，在一个宽的平台中达到峰值，首次是出现于频谱的紫外波段（31.3±1.8 ks，位于188 nm），随后出现于光学波段（39.6±2.5 ks，位于439 nm）。光变曲线在约200 ks时达到最低值，之后紫外波段光变曲线一直稳定，光学波段再次变亮，约在

a rebrightening was seen in the optical bands, peaking again at about 700–800 ks (Fig. 2).

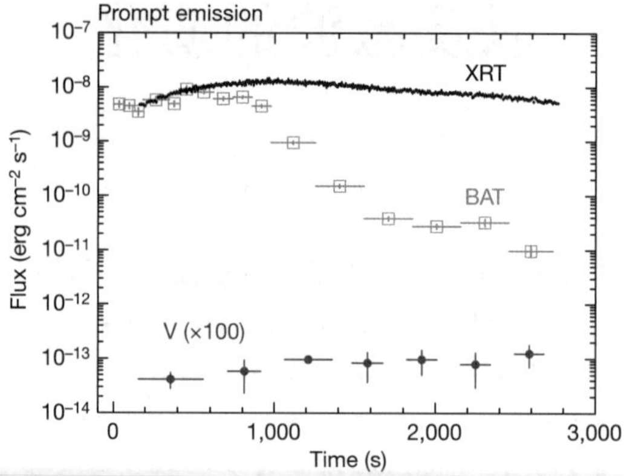

Fig. 1. Early Swift light curve of GRB 060218. GRB 060218 was discovered by the BAT when it came into the BAT field of view during a pre-planned slew. There is no emission at the GRB location up to −3,509 s. Swift slewed again to the burst position and the XRT and UVOT began observing GRB 060218 159 s later. For each BAT point we converted the observed count rate to flux (15–150 keV band) using the observed spectra. The combined BAT and XRT spectra were fitted with a cut-off power law plus a blackbody, absorbed by interstellar matter in our Galaxy and in the host galaxy at redshift $z = 0.033$. The host galaxy column density is $N_H^z = 5.0 \times 10^{21}$ cm^{-2} and that of our Galaxy is $(0.9–1.1) \times 10^{21}$ cm^{-2}. Errors are at 1σ significance. At a redshift $z = 0.033$ (corresponding to a distance of 145 Mpc with $H_0 = 70$ km s^{-1} Mpc^{-1}) the isotropic equivalent energy, extrapolated to the 1–10,000-keV rest-frame energy band, is $E_{iso} = (6.2 \pm 0.3) \times 10^{49}$ erg. The peak energy in the GRB spectrum is at $E_p = 4.9^{+0.4}_{-0.3}$ keV. These values are consistent with the Amati correlation, suggesting that GRB 060218 is not an off-axis event[26]. This conclusion is also supported by the lack of achromatic rise behaviour of the light curve in the three Swift observation bands. The BAT fluence is dominated by soft X-ray photons and this burst can be classified as an X-ray flash[27]. A *V*-band light curve is shown with red filled circles. For clarity the *V* flux has been multiplied by a factor of 100. Magnitudes have been converted to fluxes using standard UVOT zero points and multiplying the specific flux by the filter Full Width at Half Maximum (FWHM). Gaps in the light curve are due to the automated periodic change of filters during the first observation of the GRB.

Soon after the Swift discovery, low-resolution spectra of the optical afterglow and host galaxy revealed strong emission lines at a redshift of $z = 0.033$ (ref. 14). Spectroscopic indications of the presence of a rising supernova (designated SN 2006aj) were found three days after the burst[6,15] with broad emission features consistent with a type Ic supernova (owing to a lack of hydrogen and helium lines).

The Swift instruments provided valuable spectral information. The high-energy spectra soften with time and can be fitted with (cut-off) power laws. This power-law component can be ascribed to the usual GRB jet and afterglow. The most striking feature, however, is the presence of a soft component in the X-ray spectrum that is present in the XRT data up to ~10,000 s. The blackbody component shows a marginally decreasing temperature ($kT \approx 0.17$ keV, where k is the Boltzmann constant), and a clear increase in luminosity with time, corresponding to an increase in the apparent emission radius from $R^X_{BB} = (5.2 \pm 0.5) \times 10^{11}$ cm to $R^X_{BB} = (1.2 \pm 0.1) \times 10^{12}$ cm (Fig.3). During the rapid decay ($t \approx 7{,}000$ s), a blackbody component is still present in the data with a marginally cooler temperature ($kT = 0.10 \pm 0.05$ keV) and a comparable emission radius:

700 ~ 800 ks 再次达到峰值（图 2）。

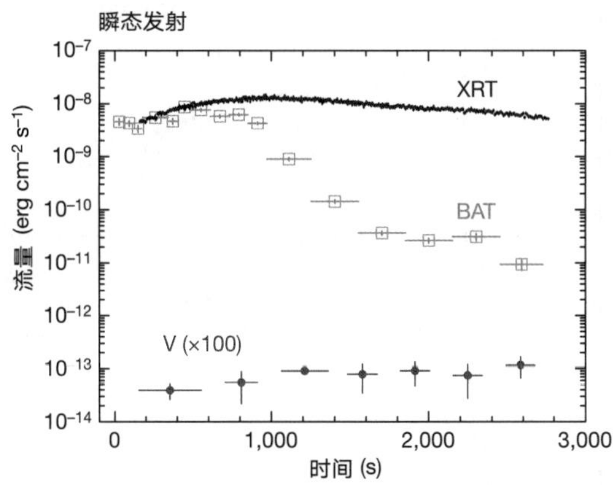

图 1. GRB 060218 的早期雨燕光变曲线。GRB 060218 是在 BAT 按预设回转时进入 BAT 的视野从而被发现的。直到 γ 暴前 3,509 s，在 GRB 的位置还没有辐射。雨燕重新回转到 γ 暴位置，XRT 和 UVOT 在 γ 暴后 159 s 开始观测 GRB 060218。对 BAT 的每个点，我们通过观测的能谱把观测计数率转化为流量（15 ~ 150 keV 波段）。我们使用一个截断幂率谱加一个黑体谱来拟合 BAT 和 XRT 组合能谱，其中考虑了银河系和红移 $z = 0.033$ 的宿主星系中星际介质的吸收。宿主星系的柱密度为 $N_H^z = 5.0 \times 10^{21}$ cm^{-2}，而银河系的柱密度为（0.9 ~ 1.1）$\times 10^{21}$ cm^{-2}。误差为 1σ。在红移 $z = 0.033$ 处（对应距离 145 Mpc，$H_0 = 70$ km · s^{-1} · Mpc^{-1}），外推到静止参照系 1 ~ 10,000 keV 能量区间的各向同性等效能量为 $E_{iso} = (6.2 \pm 0.3) \times 10^{49}$ erg。GRB 能谱的峰值能量为 $E_p = 4.9^{+0.4}_{-0.3}$ keV。这些值和阿马蒂关系一致，表明 GRB 060218 不是一个偏轴的事件 [26]。光变曲线在雨燕的三个观测波段没有出现无色的上升行为也支持了这一观点。BAT 的能流主要由软 X 射线光子主导，所以这一 γ 暴被分类为 X 射线闪变 [27]。V 波段的光变曲线用红色实心圆表示。为了更清晰地呈现，这里将 V 波段的流量乘以因子 100。星等转化成流量的过程中利用了标准 UVOT 零点，并将此流量乘以了滤波器的半高全宽（FWHM）。光变曲线中的间断是由于第一次观测该 GRB 时滤波器的自动周期性改变所造成的。

在雨燕的发现之后不久，光学余辉和宿主星系的低分辨率光谱显示在红移 $z = 0.033$ 处有很强的发射线（参考文献 14）。GRB 三天后 [6,15] 分光观测显示出现了一颗增亮中的超新星（命名为 SN 2006aj），宽线辐射特征表明其为 Ic 型超新星（由于缺乏氢线和氦线）。

雨燕搭载的设备提供了丰富的光谱信息。高能光谱随着时间软化，并可以用（截断的）幂率谱进行拟合。这个幂率谱成分通常来自 GRB 的喷流和余辉。不过最显著的特征是，在 XRT 数据中 X 射线谱存在直到约 10,000 s 的软成分。黑体成分呈现温度微弱的降低（$kT \approx 0.17$ keV，此处 k 为玻尔兹曼常数），而光度随时间显著地增大，这对应于视辐射半径从 $R_{BB}^X = (5.2 \pm 0.5) \times 10^{11}$ cm 增大到 $R_{BB}^X = (1.2 \pm 0.1) \times 10^{12}$ cm（图 3）。在快速衰减的过程（$t \approx 7,000$ s）中，黑体成分在数据中依然存在，具有一个稍微冷一点的温度（$kT = 0.10 \pm 0.05$ keV）和一个大小相当的辐射半径 $R_{BB}^X =$

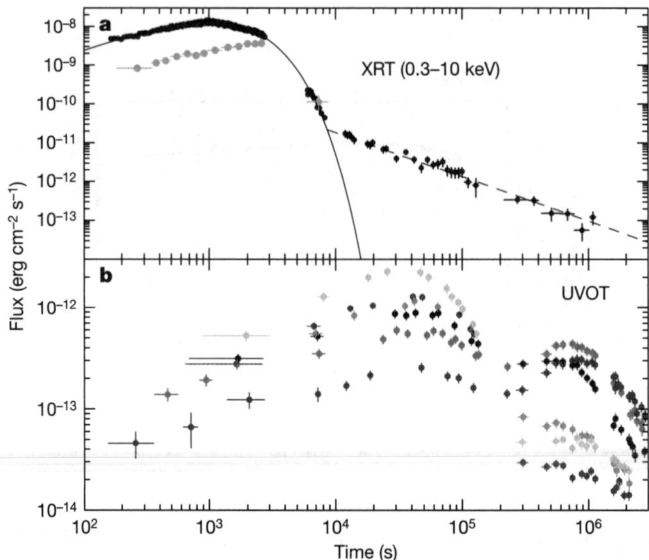

Fig. 2. Long-term Swift light curve of GRB 060218. **a**, The XRT light curve (0.3–10 keV) is shown with open black circles. Count-rate-to-flux conversion factors were derived from time-dependent spectral analysis. We also plotted (filled grey circles) the contribution to the 0.3–10-keV flux by the blackbody component. Its percentage contribution is increasing with time, becoming dominant at the end of the exponential decay. The X-ray light curve has a long, slow power-law rise followed by an exponential (or steep power-law) decay. At about 10,000 s the light curve breaks to a shallower power-law decay (dashed red line) with an index of -1.2 ± 0.1, characteristic of typical GRB afterglows. This classical afterglow can be naturally accounted for by a shock driven into the wind by a shell with kinetic energy $E_{\text{shell}} \approx 10^{49}$ erg. The t^{-1} flux decline is valid at the stage where the shell is being decelerated by the wind with the deceleration phase beginning at $t_{\text{dec}} \lesssim 10^4$ s for $\dot{M} \gtrsim 10^{-4}(v_{\text{wind}}/10^8)M_\odot\,\text{yr}^{-1}$ (where v_{wind} is in units of cm s^{-1}), consistent with the mass-loss rate inferred from the thermal X-ray component. Error bars are 1σ. **b**, The UVOT light curve. Filled circles of different colours represent different UVOT filters: red, V (centred at 544 nm); green, B (439 nm); dark blue, U (345 nm); light blue, UVW1 (251 nm); magenta, UVM1 (217 nm); and yellow, UVW2 (188 nm). Specific fluxes have been multiplied by their FWHM widths (75, 98, 88, 70, 51 and 76 nm, respectively). Data have been rebinned to increase the signal-to-noise ratio. The ultraviolet band light curve peaks at about 30 ks owing to the shock break-out from the outer stellar surface and the surrounding dense stellar wind, while the optical band peaks at about 800 ks owing to radioactive heating in the supernova ejecta.

$R_{\text{BB}}^{\text{X}} = (6.5^{+14}_{-4.4}) \times 10^{11}$ cm. In the optical/ultraviolet band at 9 hours (32 ks) the blackbody peak is still above the UVOT energy range. At 120 ks the peak of the blackbody emission is within the UVOT passband, and the inferred temperature and radius are $kT = 37^{+1.9}_{-0.9}$ eV and $R_{\text{BB}}^{\text{UV}} = 3.29^{+0.94}_{-0.93} \times 10^{14}$ cm, implying an expansion speed of $(2.7\pm0.8)\times10^9$ cm s^{-1}. This estimate is consistent with what we would expect for a supernova and it is also consistent with the line broadening observed in the optical spectra.

The thermal components are the key to interpreting this anomalous GRB. The high temperature (two million degrees) of the thermal X-ray component suggests that the radiation is emitted by a shock-heated plasma. The characteristic radius of the emitting region, $R_{\text{shell}} \approx (E/aT^4)^{1/3} \approx 5\times10^{12}$ cm (E is the GRB isotropic energy and a is the radiation density constant), corresponds to the radius of a blue supergiant progenitor. However, the

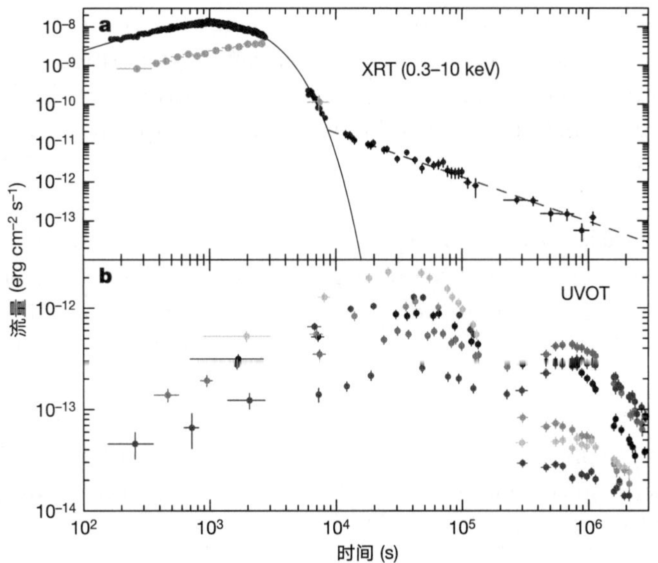

图 2. GRB 060218 的长期雨燕光变曲线。**a**，黑色空心圆代表 XRT(0.3 ~ 10 keV) 的光变曲线。计数率–流量转化因子来自依赖于时间的谱分析。我们也绘制了黑体成分对 0.3 ~ 10 keV 流量的贡献(灰色实心圆)。黑体成分的百分比随着时间而增大，在指数衰减的末期成为主导。X 射线光变曲线具有长而缓慢的幂率上升，然后指数(或以陡的幂率形式)衰减。在约 10,000 s 光变曲线偏折为较浅的幂率衰减(红色虚线)，指数为 -1.2 ± 0.1，具有典型的 GRB 余辉的特征。这个典型余辉可以很自然地用被壳层驱动进入星风的激波来描述，壳层动能为 $E_{\rm shell}\approx10^{49}$ erg。壳层被星风减速，减速相开始于 $t_{\rm dec}\lesssim10^4$ s，此时 $\dot{M}\gtrsim10^{-4}(v_{\rm wind}/10^8)M_\odot\ {\rm yr}^{-1}$(这里 $v_{\rm wind}$ 以 cm · s^{-1} 为单位)，流量下降符合 t^{-1}。这与从热 X 射线成分推导出的质量损失率一致。误差棒为 1σ。**b**，UVOT 光变曲线。不同颜色的实心圈代表不同的 UVOT 滤光片：红色，V(中心位于 544 nm)；绿色，B(439 nm)；深蓝色，U(345 nm)；浅蓝色，UVW1(251 nm)；品红，UVW1(217 nm)；黄色，UVW2(188 nm)。具体的流量已经乘以了各自的半高全宽(分别为 75 nm，98 nm，88 nm，70 nm，51 nm 和 76 nm)。数据被重新分组以增加信噪比。紫外波段的光变曲线的峰值大约在 30 ks 处，这是由于激波需要从恒星外表面和周围致密星风物质突破出来。然而光学波段的峰值大约在 800 ks 处，源自超新星喷出物中的放射性加热。

$(6.5^{+14}_{-4.4})\times10^{11}$ cm。在光学/紫外波段黑体的峰在 9 小时(32 ks)时仍然高于 UVOT 的能量范围。在 120 ks 时黑体的峰落入 UVOT 能量范围内，估计的温度和半径分别为 $kT=3.7^{+1.9}_{-0.9}$ eV 和 $R^{\rm X}_{\rm BB}=(3.29^{+0.94}_{-0.93})\times10^{14}$ cm，这表明膨胀速度为 $(2.7\pm0.8)\times10^9$ cm · s^{-1}。这一估计与我们对超新星的预期一致，也与光谱中观测到的谱线展宽一致。

热成分是解释这一反常 GRB 的关键。X 射线热成分的较高温度(两百万度)表明辐射是由激波加热的等离子体发出。辐射区的特征半径为 $R_{\rm shell}\approx(E/aT^4)^{1/3}\approx5\times10^{12}$ cm(E 是 GRB 各向同性的能量，a 是辐射密度常数)，对应于前身星为蓝巨星的半径。不过，超新星光谱缺少氢线表明它是一个更致密的源。大的辐射半径

lack of hydrogen lines in the supernova spectrum suggests a much more compact source. The large emission radius may be explained in this case by the existence of a massive stellar wind surrounding the progenitor, as is common for Wolf–Rayet stars. The thermal radiation is observed once the shock driven into the wind reaches a radius, $\sim R_{\rm shell}$, where the wind becomes optically thin.

The characteristic variability time is $R_{\rm shell}/c \approx 200$ s, consistent with the smoothness of the X-ray pulse and the rapid thermal X-ray flux decrease at the end of the pulse. We interpret this as providing, for the first time, a direct measurement of the shock break-out[16,17] of the stellar envelope and the stellar wind (first investigated by Colgate[18]). The fact that $R_{\rm shell}$ is larger than $R^{\rm X}_{\rm BB}$ suggests that the shock expands in a non-spherical manner, reaching different points on the $R_{\rm shell}$ sphere at different times. This may be due to a non-spherical explosion (such as the presence of a jet), or a non-spherical wind[19,20]. In addition, the shock break-out interpretation provides us with a delay between the supernova explosion and the GRB start of $\lesssim 4$ ks (ref. 21; see Fig. 1).

As the shock propagates into the wind, it compresses the wind plasma into a thin shell. The mass of this shell may be inferred from the requirement that its optical depth be close to unity, $M_{\rm shell} \approx 4\pi R^2_{\rm shell}/\kappa \approx 5\times10^{-7} M_\odot$ ($\kappa \approx 0.34$ cm^2 g^{-1} is the opacity). This implies that the wind mass-loss rate is $\dot{M} \approx M_{\rm shell} v_{\rm wind}/R_{\rm shell} \approx 3\times10^{-4} M_\odot$ yr^{-1}, for a wind velocity $v_{\rm wind} = 10^8$ cm s^{-1}, typical for Wolf–Rayet stars. Because the thermal energy density behind a radiation-dominated shock is $aT^4 \approx 3\rho v_s^2$ (ρ is the wind density at $R_{\rm shell}$ and v_s the shock velocity) we have $\rho \approx 10^{-12}$ g cm^{-3}, which implies that the shock must be (mildly) relativistic, $v_s \simeq c$. This is similar to GRB 980425/SN 1998bw, where the ejection of a mildly relativistic shell with energy of $\simeq 5\times10^{49}$ erg is believed to have powered radio[22-24] and X-ray emission[7].

The optical–ultraviolet emission observed at an early time of $t \lesssim 10^4$ s may be accounted for as the low-energy tail of the thermal X-ray emission produced by the (radiation) shock driven into the wind. At a later time, the optical–ultraviolet emission is well above that expected from the (collisionless) shock driven into the wind. This emission is most probably due to the expanding envelope of the star, which was heated by the shock passage to a much higher temperature. Initially, this envelope is hidden by the wind. As the star and wind expand, the photosphere propagates inward, revealing shocked stellar plasma. As the star expands, the radiation temperature decreases and the apparent radius increases (Fig. 3). The radius inferred at the peak of the ultraviolet emission, $R^{\rm UV}_{\rm BB} \approx 3\times10^{14}$ cm, implies that emission is arising from the outer $\sim 4\pi(R^{\rm UV}_{\rm BB})^2/\kappa \approx 10^{-3} M_\odot$ shell at the edge of the shocked star. As the photosphere rapidly cools, this component of the emission fades. The ultraviolet light continues to plummet as cooler temperatures allow elements to recombine and line blanketing to set in, while radioactive decay causes the optical light to begin rising to the primary maximum normally seen in supernova light curves (Fig. 2).

可以用存在于前身星周围的大质量星风来解释，这对于沃尔夫–拉叶星来说很常见。一旦被驱动进入星风的激波到达半径 $\sim R_{\rm shell}$，在此星风变得光薄，热辐射便会被观测到。

特征的变化时标为 $R_{\rm shell}/c \approx 200$ s，这与平滑的 X 射线脉冲以及在脉冲末端热 X 射线流量的快速减少一致。我们对此的解释是，这是首次直接观测到激波从恒星包层和星风中突破出来的激波暴[16,17]（科尔盖特[18]曾首次进行过研究）。$R_{\rm shell}$ 比 $R^{\rm X}_{\rm BB}$ 大的事实表明激波以非球形的方式膨胀，在不同的时间到达 $R_{\rm shell}$ 球上的不同点。这可能是由非球形的爆炸（例如存在喷流）或者非球形的星风[19,20]引起。另外，激波暴的解释为我们提供了超新星爆炸和 GRB 开始之间延迟时间的限制，为 $\lesssim 4$ ks。

激波传播到星风后压缩星风等离子体成为一个薄壳层。这个壳层的质量可以从光深满足接近于 1 这一条件推出，$M_{\rm shell} \approx 4\pi R^2_{\rm shell}/\kappa \approx 5\times10^{-7} M_\odot$（$\kappa \approx 0.34\ {\rm cm^2\cdot g^{-1}}$ 是不透明度）。这表明对星风速度为 $v_{\rm wind} = 10^8\ {\rm cm\cdot s^{-1}}$ 的典型沃尔夫–拉叶星而言，星风质量损失为 $\dot{M} \approx M_{\rm shell} v_{\rm wind}/R_{\rm shell} \approx 3\times10^{-4} M_\odot\ {\rm yr^{-1}}$。因为在辐射主导的激波之后的热能量密度为 $aT^4 \approx 3\rho v_{\rm s}^2$（$\rho$ 为 $R_{\rm shell}$ 处的星风密度，$v_{\rm s}$ 为激波速度），我们得到 $\rho \approx 10^{-12}\ {\rm g\cdot cm^{-3}}$，这意味着激波必须是（轻度）相对论性的，$v_{\rm s} \simeq c$。这与 GRB 980425/SN 1998bw 类似，我们相信其中是一个能量为 $\simeq 5\times10^{45}$ erg 的轻度相对论性壳层驱动了射电[22-24]和 X 射线辐射[7]。

早期 $t \lesssim 10^4$ s 时观测到的光学–紫外辐射可能是来自（辐射）激波进入星风产生的热 X 射线辐射的低能尾巴。之后，光学–紫外的辐射明显亮于对星风中的（无碰撞）激波的预期。这个辐射最有可能来自膨胀的恒星包层，恒星星包由于激波经过而加热到一个大大增高的温度。起初，这个包层隐埋在星风里。当恒星和星风膨胀，光球层向内传播，逐渐显示出激波加速后的恒星等离子体物质。当恒星膨胀时，辐射温度降低，视半径增大（图 3）。紫外辐射达到峰值时推算出的半径 $R^{\rm UV}_{\rm BB} \approx 3\times10^{14}$ cm，这表明辐射来自经激波扫过的恒星外层约 $4\pi(R^{\rm UV}_{\rm BB})^2/\kappa \approx 10^{-3} M_\odot$的壳层。当光球快速冷却时，辐射的这一成分衰退。温度降低导致元素发生重合，并开始出现谱线覆盖，因而紫外光继续急速下降。然而放射性衰变而导致可见光开始上升，并达到通常在超新星光变曲线中常见到的主极大值（图 2）。

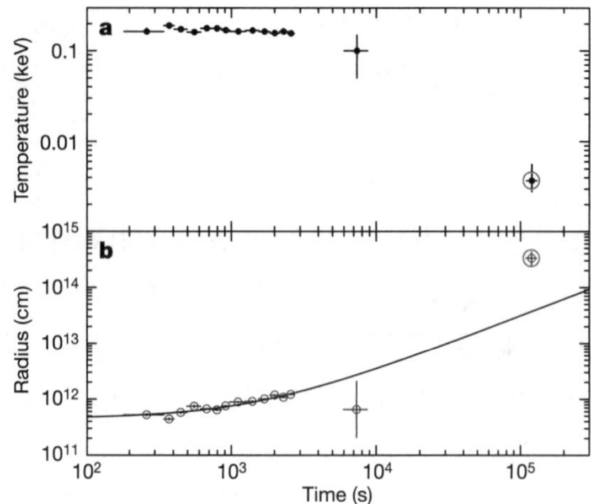

Fig. 3. Evolution of the soft thermal component temperature and radius. **a**, Evolution of the temperature of the soft thermal component. The joint BAT and XRT spectrum has been fitted with a blackbody component plus a (cut-off) power-law in the first ~3,000 s (see also the legend of Fig. 1). The last point (circled in green) comes from a fit to the six UVOT filters, assuming a blackbody model with Galactic reddening, $E(B-V)=0.14$, and host galaxy reddening. This reddening has been determined by fitting the Rayleigh–Jeans tail of the blackbody emission at 32 ks (9 hours). The data require an intrinsic $E(B-V)=0.20\pm0.03$ (assuming a Small Magellanic Cloud reddening law[28]). Error bars are 1σ. **b**, Evolution of the radius of the soft thermal component. The last point (circled in green) comes from the fitting of UVOT data. The continuous line represents a linear fit to the data.

Because the wind shell is clearly larger than the progenitor radius, we infer that the star radius is definitely smaller than 5×10^{12} cm. Assuming a linear expansion at the beginning (owing to light travel-time effects) we can estimate a star radius of $R_{\rm star}\approx(4\pm1)\times10^{11}$ cm. This is smaller than the radius of the progenitors of type II supernovae, like blue supergiants (4×10^{12} cm for SN 1987A, ref. 25) or red supergiants (3×10^{13} cm). Our results unambiguously indicate that the progenitor of GRB 060218/SN 2006aj was a compact massive star, most probably a Wolf–Rayet star.

(**442**, 1008-1010; 2006)

S. Campana[1], V. Mangano[2], A. J. Blustin[3], P. Brown[4], D. N. Burrows[4], G. Chincarini[1,5], J. R. Cummings[6,7], G. Cusumano[2], M. Della Valle[8,9], D. Malesani[10], P. Mészáros[4,11], J. A. Nousek[4], M. Page[3], T. Sakamoto[6,7], E. Waxman[12], B. Zhang[13], Z. G. Dai[13,14], N. Gehrels[6], S. Immler[6], F. E. Marshall[6], K. O. Mason[15], A. Moretti[1], P. T. O'Brien[16], J. P. Osborne[16], K. L. Page[16], P. Romano[1], P. W. A. Roming[4], G. Tagliaferri[1], L. R. Cominsky[17], P. Giommi[18], O. Godet[16], J. A. Kennea[4], H. Krimm[6,19], L. Angelini[6], S. D. Barthelmy[6], P. T. Boyd[6], D. M. Palmer[20], A. A. Wells[16] & N. E. White[6]

[1] INAF—Osservatorio Astronomico di Brera, via E. Bianchi 46, I-23807 Merate (LC), Italy
[2] INAF—Istituto di Astrofisica Spaziale e Fisica Cosmica di Palermo, via U. La Malfa 153, I-90146 Palermo, Italy
[3] UCL Mullard Space Science Laboratory, Holmbury St. Mary, Dorking, Surrey RH5 6NT, UK
[4] Department of Astronomy and Astrophysics, Pennsylvania State University, University Park, Pennsylvania 16802, USA
[5] Università degli studi di Milano Bicocca, piazza delle Scienze 3, I-20126 Milano, Italy
[6] NASA—Goddard Space Flight Center, Greenbelt, Maryland 20771, USA
[7] National Research Council, 2101 Constitution Avenue NW, Washington DC 20418, USA
[8] INAF—Osservatorio Astrofisico di Arcetri, largo E. Fermi 5, I-50125 Firenze, Italy
[9] Kavli Institute for Theoretical Physics, UC Santa Barbara, California 93106, USA

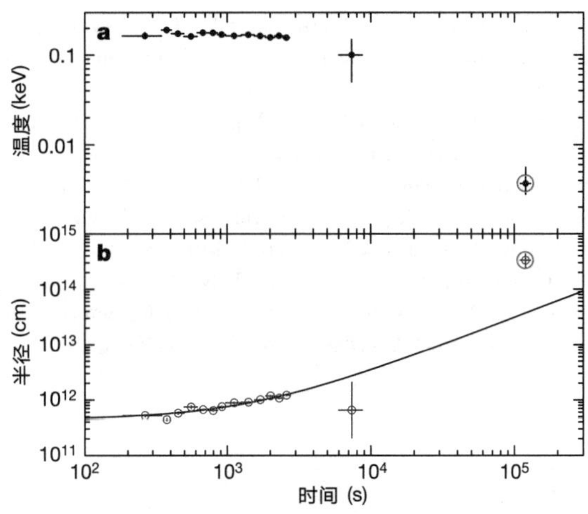

图 3. 软的热成分的温度和半径的演化。**a**，软的热成分温度的演化。BAT 和 XRT 的联合能谱在前约 3,000 s 用一个黑体成分加一个（截断的）幂率来拟合（见图 1 图注）。最后的点（绿色圆圈内）来自考虑了黑体模型在银河系红化 $E(B-V)=0.14$ 和宿主星系红化后对 UVOT 的 6 个滤光片的拟合。红化值的确定来自对在 32 ks（9 小时）黑体辐射的瑞利–金斯尾的拟合。数据要求本征红化 $E(B-V)=0.20\pm0.03$（假设为小麦哲伦云的红外定律 [28]）。误差棒为 1σ。**b**，软的热成分的半径的演化。最后的点（绿色圆圈内）来自对 UVOT 数据的拟合。连续线代表对数据的线性拟合。

因为星风壳层明显比前身星半径大，我们推断恒星半径必然小于 5×10^{12} cm。假设起初星体是线性膨胀的，（由光行时间效应）我们估计恒星半径为 $R_{\rm star}\approx(4\pm1)\times10^{11}$ cm。这比 II 型超新星的前身星半径，比如蓝巨星（对 SN 1987A 而言是 4×10^{12} cm，参考文献 25）或红巨星（3×10^{13} cm）要小。我们的结果毫无疑问地证明了 GRB 060218/SN 2006aj 的前身星为致密的大质量恒星，而且最可能是沃尔夫–拉叶星。

（肖莉 翻译；黎卓 审稿）

[10] International School for Advanced Studies (SISSA-ISAS), via Beirut 2-4, I-34014 Trieste, Italy
[11] Department of Physics, Pennsylvania State University, University Park, Pennsylvania 16802, USA
[12] Physics Faculty, Weizmann Institute, Rehovot 76100, Israel
[13] Department of Physics, University of Nevada, Box 454002, Las Vegas, Nevada 89154-4002, USA
[14] Department of Astronomy, Nanjing University, Nanjing, 210093, China
[15] PPARC, Polaris House, North Star Avenue, Swindon SN2 1SZ, UK
[16] Department of Physics and Astronomy, University of Leicester, University Road, Leicester LE1 7RH, UK
[17] Department of Physics and Astronomy, Sonoma State University, Rohnert Park, California 94928-3609, USA
[18] ASI Science Data Center, via G. Galilei, I-00044 Frascati (Roma), Italy
[19] Universities Space Research Association, 10211 Wincopin Circle Suite 500, Columbia, Maryland 21044-3431, USA
[20] Los Alamos National Laboratory, PO Box 1663, Los Alamos, New Mexico 87545, USA

Received 13 March; accepted 10 May 2005.

References:

1. Woosley, S. E. Gamma-ray bursts from stellar mass accretion disks around black holes. *Astrophys. J.* **405**, 273-277 (1993).
2. Paczyński, B. Are gamma-ray bursts in star-forming regions? *Astrophys. J.* **494**,L45-L48 (1993).
3. MacFadyen, A. I. & Woosley, S. E. Collapsars: gamma-ray bursts and explosions in "failed supernovae". *Astrophys. J.* **524**, 262-289 (1999).
4. Galama, T. J. *et al.* An unusual supernova in the error box of the γ-ray burst of 25 April 1998. *Nature* **395**, 670-672 (1998).
5. Cusumano, G. *et al.* GRB060218: Swift-BAT detection of a possible burst. *GCN Circ.* 4775 (2006).
6. Masetti, N. *et al.* GRB060218: VLT spectroscopy. *GCN Circ.* 4803 (2006).
7. Waxman, E. Does the detection of X-ray emission from SN1998bw support its association with GRB980425? *Astrophys. J.* **605**, L97-L100 (2004).
8. Barthelmy, S. D. *et al.* The Burst Alert Telescope (BAT) on the SWIFT Midex Mission. *Space Sci. Rev.* **120**, 143-164 (2005).
9. Gehrels, N. *et al.* The Swift gamma ray burst mission. *Astrophys. J.* **611**, 1005-1020 (2004).
10. Burrows, D. N. *et al.* The Swift X-Ray Telescope. *Space Sci. Rev.* **120**, 165-195 (2005).
11. Tagliaferri, G. *et al.* An unexpectedly rapid decline in the X-ray afterglow emission of long γ-ray bursts. *Nature* **436**, 985-988 (2005).
12. O'Brien, P. T. *et al.* The early X-ray emission from GRBs. *Astrophys. J.* (submitted); preprint at ⟨http://arXiv.org/astro-ph/0601125l⟩ (2006).
13. Roming, P. W. A. *et al.* The Swift Ultra-Violet/Optical Telescope. *Space Sci. Rev.* **120**, 95-142 (2005).
14. Mirabal, N. & Halpern, J. P. GRB060218: MDM Redshift. *GCN Circ.* 4792 (2006).
15. Pian, E. *et al.* An optical supernova associated with the X-ray flash XRF 060218. *Nature* doi:10.1038/nature05082 (this issue).
16. Ensman, L. & Burrows, A. Shock breakout in SN1987A. *Astrophys. J.* **393**, 742-755 (1992).
17. Tan, J. C., Matzner, C. D. & McKee, C. F. Trans-relativistic blast waves in supernovae as gamma-ray burst progenitors. *Astrophys. J.* **551**, 946-972 (2001).
18. Colgate, S. A. Prompt gamma rays and X-rays from supernovae. *Can. J. Phys.* **46**, 476 (1968).
19. Mazzali, P. A. *et al.* An asymmetric, energetic type Ic supernova viewed off-axis and a link to gamma-ray bursts. *Science* **308**, 1284-1287 (2005).
20. Leonard, D. C. *et al.* A non-spherical core in the explosion of supernova SN2004dj. *Nature* **440**, 505-507 (2006).
21. Norris, J. P. & Bonnell, J. T. How can the SN-GRB time delay be measured? *AIP Conf. Proc.* **727**, 412-415 (2004).
22. Kulkarni, S. R. *et al.* Radio emission from the unusual supernova 1998bw and its association with the gamma-ray burst of 25 April 1998. *Nature* **395**, 663-669 (1998).
23. Waxman, E. & Loeb, A. A subrelativistic shock model for the radio emission of SN1998bw. *Astrophys. J.* **515**, 721-725 (1999).
24. Li, Z.-Y. & Chevalier, R. A. Radio supernova SN1998bw and its relation to GRB980425. *Astrophys. J.* **526**, 716-726 (1999).
25. Arnett, W. D. *et al.* Supernova 1987A. *Annu. Rev. Astron. Astrophys.* **27**, 629-700 (1989).
26. Amati, L. *et al.* GRB060218: $E_{p,i}$–E_{iso} correlation. *GCN Circ.* 4846 (2006).
27. Heise, J., *et al.* in *Proceedings of "Gamma-Ray Bursts in the Afterglow Era"* (eds Costa, E., Frontera, F. & Hjorth, J.) 16-21 (Springer, Berlin/Heidelberg, 2001).
28. Pei, Y. C. Interstellar dust from the Milky Way to the Magellanic Clouds. *Astrophys. J.* **395**, 130-139 (1992).

Acknowledgements. We acknowledge support from ASI, NASA and PPARC.

Author Information. Reprints and permissions information is available at www.nature.com/reprints. The authors declare no competing financial interests.

Correspondence and requests for materials should be addressed to S.C. (sergio.campana@brera.inaf.it).

Analysis of One Million Base Pairs of Neanderthal DNA

R. E. Green *et al.*

Editor's Note

Neanderthals were the closest relatives of modern humans. They lived in Europe and Western Asia about 400,000–24,000 years ago. As modern humans were present in Europe at least 41,000 years ago, there is much speculation about whether Neanderthals and modern humans interbred. The only way to decide this is from genome analysis. In this paper Svante Pääbo and colleagues describe the sequencing of a million base pairs of Neanderthal nuclear DNA from a bone from Croatia. The full genome sequence took a further four years to complete, and the analysis of this and other Neanderthal genomes suggests that around 4% of the DNA of modern Europeans is of Neanderthal origin.

Neanderthals are the extinct hominid group most closely related to contemporary humans, so their genome offers a unique opportunity to identify genetic changes specific to anatomically fully modern humans. We have identified a 38,000-year-old Neanderthal fossil that is exceptionally free of contamination from modern human DNA. Direct high-throughput sequencing of a DNA extract from this fossil has thus far yielded over one million base pairs of hominoid nuclear DNA sequences. Comparison with the human and chimpanzee genomes reveals that modern human and Neanderthal DNA sequences diverged on average about 500,000 years ago. Existing technology and fossil resources are now sufficient to initiate a Neanderthal genome-sequencing effort.

NEANDERTHALS were first recognized as a distinct group of hominids from fossil remains discovered 150 years ago at Feldhofer in Neander Valley, outside Düsseldorf, Germany. Subsequent Neanderthal finds in Europe and western Asia showed that fossils with Neanderthal traits appear in the fossil record of Europe and western Asia about 400,000 years ago and vanish about 30,000 years ago. Over this period they evolved morphological traits that made them progressively more distinct from the ancestors of modern humans that were evolving in Africa[1,2]. For example, the crania of late Neanderthals have protruding mid-faces, brain cases that bulge outward at the sides, and features of the base of the skull, jaw and inner ears that set them apart from modern humans[3].

The nature of the interaction between Neanderthals and modern humans, who expanded out of Africa around 40,000–50,000 years ago and eventually replaced Neanderthals

对尼安德特人 DNA 的一百万碱基对的分析

格林等

编者按

尼安德特人是现代人类最近的亲戚。它们大约在 400,000 到 24,000 年前生活在欧洲和西亚。由于现代人类至少在 41,000 年前出现在欧洲，人们对尼安德特人和现代人类是否杂交有很多猜测。确定这一点的唯一方法是从基因组进行分析。在本文中，斯凡特·帕波及其同事描述了来自克罗地亚骨骼的尼安德特人核 DNA 的一百万个碱性对的测序。完整的基因组序列又用了四年时间才完成。对这个基因组序列和其他尼安德特人基因组的分析表明，现代欧洲人中约有 4%的 DNA 来自尼安德特人。

尼安德特人是与当代人类亲缘关系最近的已经灭绝的人类类群，所以他们的基因组能为我们提供一个独特的机遇来了解解剖学意义上完全的现代人所特有的遗传变化。我们已经甄别选定了一件罕见不受现代人 DNA 污染的有着 38,000 年历史的尼安德特人化石。对该化石提取出的 DNA 进行的直接高通量测序，目前已经得到了 100 多万碱基对的类人猿核 DNA 序列。该序列与人和黑猩猩的基因组对比表明，现代人和尼安德特人的 DNA 序列平均大约在 500,000 年前发生分歧。现有的技术和化石资源足以支撑着手开展尼安德特人基因组测序工作。

基于 150 年前发现于德国杜塞尔多夫外的尼安德谷费尔德霍费尔山洞的化石遗存，尼安德特人最初被认为是一个独特的人类类群。后来欧洲和西亚的尼安德特人的发现显示具有尼安德特人特征的化石在欧洲和西亚的化石记录中出现于约 400,000 年前，消失于约 30,000 年前。在这期间，他们进化出了与当时正在非洲进化的现代人祖先渐行渐远的形态特征 [1,2]。例如，晚期尼安德特人的头骨有着突出的中面部，两侧向外突出的脑颅以及将他们与现代人区别开来的颅底、颌骨和内耳特征 [3]。

现代人大约于 40,000 到 50,000 年前走出非洲，并最终取代了尼安德特人和其他生活在旧大陆的古老型人类。这些现代人与尼安德特人间相互交流的本质至今仍是

as well as other archaic hominids across the Old World is still a matter of some debate. Although there is no evidence of contemporaneous cohabitation at any single site, there is evidence of geographical and temporal overlap in their ranges before the disappearance of Neanderthals. Additionally, late in their history, some Neanderthal groups adopted cultural traits such as body decorations, potentially through cultural interactions with incoming modern humans[4].

In 1997, a segment of the hypervariable control region of the maternally inherited mitochondrial DNA (mtDNA) of the Neanderthal type specimen found at Feldhofer was sequenced. Phylogenetic analysis showed that it falls outside the variation of contemporary humans and shares a common ancestor with mtDNAs of present-day humans approximately half a million years ago[5,6]. Subsequently, mtDNA sequences have been retrieved from eleven additional Neanderthal specimens: Feldhofer 2 in Germany[7], Mezmaiskaya in Russia[8], Vindija 75, 77 and 80 in Croatia[9,10], Engis 2 in Belgium, La Chapelle-aux-Saints and Rochers de Villeneuve in France[10], Scladina in Belgium[11], Monte Lessini in Italy[12], and El Sidron 441 in Spain[13]. Although some of these sequences are extremely short, they are all more closely related to one another than to modern human mtDNAs[9,11].

This fact, in conjunction with the absence of any related mtDNA sequences in currently living humans or in a small number of early modern human fossils[5,10] strongly suggests that Neanderthals contributed no mtDNA to present-day humans. On the basis of various population models, it has been estimated that a maximal overall genetic contribution of Neanderthals to the contemporary human gene pool is between 25% and 0.1% (refs 10, 14). Because the latter conclusions are based on mtDNA, a single maternally inherited locus, they are limited in their ability to detect a Neanderthal contribution to the current human gene pool both by the vagaries of genetic drift and by the possibility of a sex bias in reproduction. However, both morphological evidence[4,15] and the variation in the modern human gene pool[16] support the conclusion that if any genetic contribution of Neanderthals to modern human occurred, it was of limited magnitude.

Neanderthals are the hominid group most closely related to currently living humans, so a Neanderthal nuclear genome sequence would be an invaluable resource for annotating the human genome. Roughly 35 million nucleotide differences exist between the genomes of humans and chimpanzees, our closest living relatives[17]. Soon, genome sequences from other primates such as the orang-utan and the macaque will allow such differences to be assigned to the human and chimpanzee lineages. However, temporal resolution of the genetic changes along the human lineage, where remarkable morphological, behavioural and cognitive changes occurred, are limited without a more closely related genome sequence for comparison. In particular, comparison to the Neanderthal would enable the identification of genetic changes that occurred during the last few hundred thousand years, when fully anatomically and behaviourally modern humans appeared.

一个存在争议的问题。尽管没有证据显示他们在任何一个地点同时生存过，但是有证据表明在尼安德特人消失之前，他们彼此存在地理和时代分布的交叠。此外，在他们的历史的晚期，有些尼安德特人群体接纳了身体装饰等文化特征，可能是他们通过与迁徙而来的现代人之间进行文化交流而实现的[4]。

1997 年，对发现于费尔德霍费尔的尼安德特人模式标本的母系遗传的线粒体 DNA（mtDNA）的高变控制区的一个片段进行了测序。系统发育分析的结果表明，该序列落于当代人的变异范围之外，并且于大概 500,000 年前与当代人的 mtDNA 共有一个祖先[5,6]。接着，从另外 11 个尼安德特人标本得到了 mtDNA 序列，包括德国的费尔德霍费尔 2[7]，俄罗斯的梅兹迈斯卡亚[8]，克罗地亚的温迪迦 75、77 和 80[9,10]，比利时的英格斯 2，法国的拉沙佩勒–欧赛恩茨和维伦纽夫岩[10]，比利时的斯卡拉迪纳[11]，意大利的蒙蒂莱西尼[12]，以及西班牙的埃尔西德隆 441[13]。尽管其中有些序列非常短，但是比起与现代人 mtDNA 的关系，它们彼此的亲缘关系要近得多[9,11]。

这一事实与现存的人类或少量早期现代人化石[5,10]中缺少任何相关的 mtDNA 序列一起强烈表明尼安德特人对当代人的 mtDNA 没有贡献。基于各种人类群体模型，已估算出尼安德特人对当代人基因库的最大整体遗传贡献在 25% 到 0.1% 之间（参考文献 10 和 14）。因为后面的推论是根据 mtDNA 推导出来的，而 mtDNA 只是一个单一的母系遗传位点，所以遗传漂移的不确定性及繁殖过程中性别差异的可能性导致它们在检测尼安德特人对当前人类基因库的贡献度方面的能力是有限的。然而，形态学证据[4,15]和现代人类基因库的变异[16]都支持如下结论，即如果尼安德特人对现代人有过任何的遗传贡献的话，那么这种贡献程度也是有限的。

尼安德特人是与现存的人类亲缘关系最密切的人类类群，所以尼安德特人的核基因组序列对于注释人类基因组来说将是无价的资源。我们现存的至亲黑猩猩与人类的基因组之间大概存在 35,000,000 个核苷酸的差异[17]。不久的将来，其他灵长类（如猩猩和猕猴）的基因组序列也会支持这种差异添加到人类和黑猩猩的演化谱系中。但是，由于所发生的形态、行为和认知变化非常显著，如果没有与人类亲缘关系更加接近的基因组序列能够用来做比较，发生在人类演化谱系中遗传变化的年代分辨率会受到局限。特别是，通过与尼安德特人进行比较，就可以对过去几十万年里出现的解剖学和行为学上完全的现代人发生的遗传变化进行鉴别。

Identification of a Neanderthal Fossil for DNA Sequencing

Although it is possible to recover mtDNA[18] and occasionally even nuclear DNA sequences[19-22] from well-preserved remains of organisms that are less than a few hundred thousand years old, determination of ancient hominid sequences is fraught with special difficulties and pitfalls[18]. In addition to degradation and chemical damage to the DNA that can cause any ancient DNA to be irretrievable or misread, contamination of specimens, laboratory reagents and instruments with traces of DNA from modern humans must be avoided. In fact, when sensitive polymerase chain reaction (PCR) is used, human mtDNA sequences can be retrieved from almost every ancient specimen[23,24]. This problem is especially severe when Neanderthal remains are studied because Neanderthal and human are so closely related that one expects to find few or no differences between Neanderthals and modern humans within many regions[25], making it impossible to rely on the sequence information itself to distinguish endogenous from contaminating DNA sequences. A necessary first step for sequencing nuclear DNA from Neanderthals is therefore to identify a Neanderthal specimen that is free or almost free of modern human DNA.

We tested more than 70 Neanderthal bone and tooth samples from different sites in Europe and western Asia for bio-molecular preservation by removing samples of a few milligrams for amino acid analysis. The vast majority of these samples had low overall contents of amino acids and/or high levels of amino acid racemization, a stereoisomeric structural change that affects amino acids in fossils, indicating that they are unlikely to contain retrievable endogenous DNA[26]. However, some of the samples are better preserved in that they contain high levels of amino acids (more than 20,000 p.p.m.), low levels of racemization of amino acids such as aspartate that racemize rapidly, as well as amino acid compositions that suggest that the majority of the preserved protein stems from collagen.

From 100–200 mg of bone from six of these specimens we extracted DNA and analysed the relative abundance of Neanderthal-like mtDNA sequences and modern human-like mtDNA sequences by performing PCR with primer pairs that amplify both human and Neanderthal mtDNA with equal efficiency. The amplification products span segments of the hypervariable region of the mtDNA in which all Neanderthals sequenced to date differ from all contemporary humans. From subsequent cloning into a plasmid vector and sequencing of more than a hundred clones from each product, we determined the ratio of Neanderthal-like to modern human-like mtDNA in each extract. We used two different primer pairs that amplify fragments of 63 base pairs and 119 base pair to gauge the contamination levels for different lengths of DNA molecules.

Figure 1 shows that the level of contamination differs drastically among the samples. Whereas only around 1% of the mtDNA present in three samples from France, Russia and Uzbekistan was Neanderthal-like, one sample from Croatia and one from Spain contained around 5% and 75% Neanderthal-like mtDNA, respectively. One bone (Vi-80) from Vindija Cave, Croatia, stood out in that ~99% of the 63-base-pair mtDNA segments and

用于 DNA 测序的尼安德特人化石的甄选

尽管从保存良好且年龄小于几十万年的生物遗骸中得到 mtDNA[18]，极少数情况下甚至得到核 DNA 序列 [19-22] 也是可能的，但是古老人类序列的确定充满特殊的困难和陷阱 [18]。除了导致古 DNA 不可获取或误读的 DNA 降解和化学损伤外，标本的污染、带有现代人 DNA 残留的实验室试剂和仪器也是必须要避免的。事实上，当使用敏感的聚合酶链式反应(PCR)时，几乎可以从每件古老的标本中得到人类 mtDNA 序列 [23,24]。当研究尼安德特人时，这一问题尤为严重，因为尼安德特人与现代人的亲缘关系如此之近，所以研究者会预期在许多区域上现代人和尼安德特人之间都很少有或根本没有差异 [25]，这使得仅仅依靠序列本身不可能将内源 DNA 与污染源 DNA 序列区别开。因此如果要对尼安德特人的核 DNA 进行测序，第一步必须要做的就是甄选一件没有或几乎没有受到现代人 DNA 污染的尼安德特人标本。

通过去除几毫克样品进行氨基酸分析，我们检测了欧洲和西亚的不同地点采集的 70 多个尼安德特人骨骼和牙齿样品以确认其生物分子保存状况。绝大部分样品的总氨基酸含量很低并且(或者)氨基酸外消旋化(一种对化石中的氨基酸产生影响的立体异构变化)水平很高，表明这些样本不可能含有可获取的内源 DNA[26]。然而，有些样本的保存情况较好，因为它们的氨基酸含量水平很高(大于 20,000 ppm)；而通常快速外消旋化的天冬氨酸等氨基酸的外消旋化水平很低；另外这些样本的氨基酸构成意味着大部分保存下来的蛋白质来源于胶原蛋白。

从其中六件标本的 100 mg 到 200 mg 骨骼样品中我们提取到了 DNA。通过执行等效引物对 PCR 扩增现代人和尼安德特人的 mtDNA，我们分析了似尼安德特人的 mtDNA 序列及似现代人的 mtDNA 序列的相对丰度。扩增产物涵盖了 mtDNA 高变区的片段。迄今为止，在这一区域中所有经过测序的尼安德特人与所有现代人都不一样。随后通过将 PCR 产物克隆到质粒载体中并对每个产物的一百多个克隆进行测序，我们确定了每个提取物中似尼安德特人 mtDNA 与似现代人 mtDNA 的比率。我们使用了两个不同引物对分别扩增 63 个碱基对(bp)和 119 个碱基对的片段，用以计量不同长度的 DNA 分子的污染水平。

图 1 显示样品间的污染水平显著不同。尽管来自法国、俄罗斯和乌兹别克斯坦的三个样品中只有 1% 左右的 mtDNA 是似尼安德特人的，而一个来自克罗地亚的样品和一个来自西班牙的样品分别含有 5% 和 75% 左右的似尼安德特人 mtDNA。来自克罗地亚温迪迦洞穴的一件骨骼标本(Vi-80)的数据异常突出，约 99% 的 63 bp

~94% of the 119-base pair segments are of Neanderthal origin. Assuming that the ratio of Neanderthal to contaminating modern human DNA is the same for mtDNA as it is for nuclear DNA, the Vi-80 bone therefore yields DNA fragments that are predominantly of Neanderthal origin and provided that the contamination rate was not increased during the downstream sequencing process, the extent of contamination in the final analyses is below ~6%.

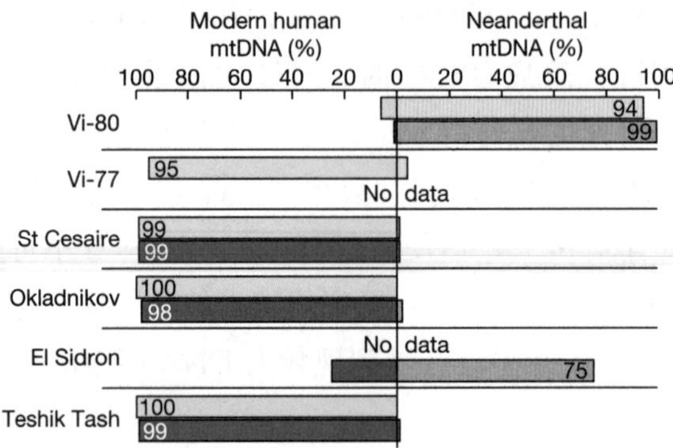

Fig. 1. Ratio of Neanderthal to modern human mtDNA in six hominid fossils. For each fossil, primer pairs that amplify a long (119 base pairs; upper lighter bars) and short (63 base pairs; lower darker bars) product were used to amplify segments of the mtDNA hypervariable region. The products were sequenced and determined to be either of Neanderthal (yellow) or modern human (blue) type.

The Vi-80 bone was discovered by M. Malez and co-workers in layer G3 of Vindija Cave in 1980. It has been dated by carbon-14 accelerator mass spectrometry to 38,310 ± 2,130 years before present and its entire mtDNA hypervariable region I has been sequenced[10]. Out of 14 Neanderthal remains from layer G3 that we have analysed, this bone is one of six samples that show good bio-molecular preservation, while the other eight bones show intermediate to bad states of preservation that do not suggest the presence of amplifiable DNA. Preservation conditions in Vindija Cave thus vary drastically from bone to bone, a situation that may be due to different extents of water percolation in different parts of the cave.

Direct Large-scale DNA Sequencing from the Vindija Neanderthal

Because the Vi-80 Neanderthal bone extract is largely free of contaminating modern human mtDNA, we chose this extract to perform large-scale parallel 454 sequencing[27]. In this technology, single-stranded libraries, flanked by common adapters, are created from the DNA sample and individual library molecules are amplified through bead-based emulsion PCR, resulting in beads carrying millions of clonal copies of the DNA fragments from the samples. These are subsequently sequenced by pyrosequencing on the GS20 454 sequencing system.

For several reasons, the 454 sequencing platform is extremely well suited for analyses of bulk

的 mtDNA 片段和约 94% 的 119 bp 的 mtDNA 片段源自尼安德特人。假设尼安德特人的核 DNA 与现代人污染源的核 DNA 的比率与 mtDNA 的情况一样，那么从该 Vi-80 骨骼标本得到的 DNA 片段应该主要来自尼安德特人。那么，如果在下游测序过程中没有增加污染率的话，那么最终分析时的污染率应该低于大约 6%。

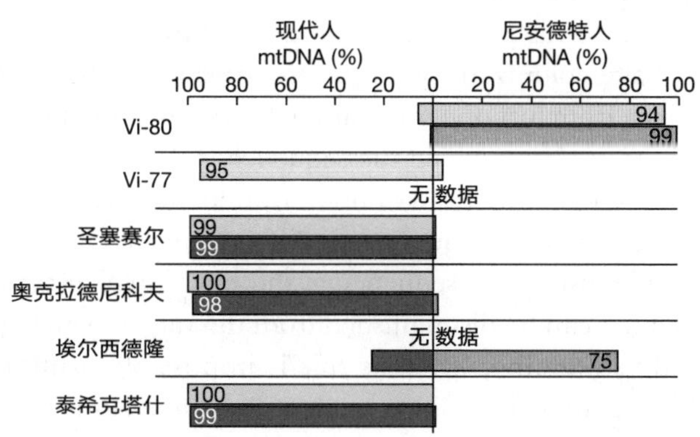

图 1. 六件人类化石标本中尼安德特人与现代人 mtDNA 的比例。对于每件化石标本，都使用了两个引物对来扩增 mtDNA 的高变区片段，其中一对扩增的产物较长（119 bp；上面的浅颜色横杠），另一对扩增的产物较短（63 bp；下面的深颜色横杠）。对产物进行了测序并确定其属于尼安德特人（黄色）类型还是现代人（蓝色）类型。

梅勒兹及其同事于 1980 年在温迪迦洞穴的 G3 层中发现了 Vi-80 骨骼标本。通过碳–14 加速器质谱法测定其年代为距今 38,310±2,130 年，其 mtDNA 高变区 I 已经完全测序[10]。我们已经对 G3 层发现的 14 个尼安德特人样本进行了分析，该骨骼标本是生物分子保存状况良好的六个样品之一，而其余八件骨骼标本则显示出中度至较差的保存状况，表明其不存在可扩增的 DNA。因此温迪迦洞穴中各骨骼标本彼此的保存状况非常不同，可能是由洞穴中不同部位的水渗透情况不同造成的。

温迪迦尼安德特人的直接大范围 DNA 测序

由于 Vi-80 尼安德特骨骼标本提取物基本排除了现代人 mtDNA 污染的可能，所以我们选择这一提取物来进行大范围平行 454 测序[27]。这一技术中，由 DNA 样本创建出两侧连有通用衔接子的单链 DNA 库，各个库中的分子通过磁珠乳液 PCR 进行扩增，于是就可以从样本中得到携带有数以百万计的 DNA 片段的克隆拷贝的磁珠。然后在 GS20 454 测序系统上通过焦磷酸测序法对这些 DNA 片段进行测序。

由于以下几个原因，454 测序平台非常适合于分析古代残骸中提取到的大量

DNA extracted from ancient remains[28]. First, it circumvents bacterial cloning, in which the vast majority of initial template molecules are lost during transformation and establishment of clones. Second, because each molecule is amplified in isolation from other molecules it also precludes template competition, which frequently occurs when large numbers of different DNA fragments are amplified together. Third, its current read length of 100–200 nucleotides covers the average length of the DNA preserved in most fossils[29]. Fourth, it generates hundreds of thousands of reads per run, which is crucial because the majority of the DNA recovered from fossils is generally not derived from the fossil species, but rather from organisms that have colonized the organism after its death[20,30]. Fifth, because each sequenced product stems from just one original single-stranded template molecule of known orientation, the DNA strand from which the sequence is derived is known[28]. This provides an advantage over traditional PCR from double-stranded templates, in which the template strand is not known, because the frequency of different nucleotide misincorporations can be deduced. For example, using 454 sequencing, the rate at which cytosine is converted to uracil and read as thymine can be distinguished from the rate at which guanine is converted to xanthine and read as adenine, whereas this is impossible using traditional PCR or bacterial cloning. This is important since nucleotide conversions and misincorporations in ancient DNA are caused by damage that affects different bases differently[28,31] and this pattern of false substitutions can be used to estimate the relative probability that a particular substitution (that is, the observation of a nucleotide difference between DNA sequences) represents the authentic DNA sequence of the organism versus an artefact from DNA degradation.

We recovered a total of 254,933 unique sequences from the Vi-80 bone (see Supplementary Methods). These were aligned to the human (build 36.1)[32], chimpanzee (build 1)[17] and mouse (build 34.1)[33] complete genome sequences, to environmental sample sequences in the GenBank *env* database (version 3, September 2005), and to the complete set of redundant nucleotide sequences in GenBank *nt* (version 3, September 2005, excluding EST, STS, GSS, environmental and HTGS sequences)[34] using the program BLASTN (NCBI version 2.2.12)[35]. The most similar database sequence for each query was identified and classified by its taxonomic order (Fig. 2) (see Supplementary Methods). No significant nucleotide sequence similarity in the databases was found for 79% of the fossil extract sequence reads. This is typical of large-scale sequencing both from other ancient bones[20,22,28] and from environmental samples[36,37], although some permafrost-preserved specimens can yield high amounts of endogenous DNA[22]. Sequences with similarity to a database sequence were classified by the taxonomic order of their most significant alignment. Actinomycetales, a bacterial order with many soil-living species, was the most populous order and accounted for 6.8% of the sequences. The second most populous order, to which 15,701 unique sequences or 6.2% of the sequence reads were most similar, was that of primates. All other individual orders were substantially less frequent. Notably, the average percentage identity for the primate sequence alignments was 98.8%, whereas it was 92–98% for the other frequently occurring orders. Thus, the primate reads, unlike many of the prokaryotic reads, are aligned to a very closely related species.

DNA[28]。首先，它绕过了细菌克隆，因为在转化和建立克隆的过程中会丢失绝大多数的初始模板分子。其次，由于每个分子都是同其他分子隔离开进行扩增的，所以可以预先排除模板竞争(模板竞争经常发生在同时扩增大量不同的 DNA 片段时)。第三，454 测序目前的读取长度是 100 到 200 个核苷酸，这一长度可以涵盖大部分化石中保存的 DNA 的平均长度 [29]。第四，454 测序技术每次运行可以产生数十万个读取，这一特点是很重要的，因为大部分从化石中复原的 DNA 都不是来自化石物种本身，而是来自那些在化石物种死后聚居其上的生物 [20,30]。第五，由于每个测序产物仅由一个已知方向的初始单链模板分子扩增而来，所以产生该序列的 DNA 链是已知的 [28]。传统 PCR 中，我们不知道模板链是哪条，所以 454 测序技术比传统的从双链模板得到 PCR 产物的技术更具优越性，因为同时可以推导出不同核苷酸错配率。举例而言，使用 454 测序技术，可以将胞嘧啶被转变为尿嘧啶而被读为胸腺嘧啶的比率与鸟嘌呤被转变为黄嘌呤而被读成腺嘌呤的比率区别开来，而这在使用传统 PCR 或者细菌克隆时是不可能办到的。由于古 DNA 中的损伤会不同程度地影响不同碱基而发生核苷酸转变和错配，而且这种虚假的碱基替换模式可以用来估计代表生物真实 DNA 序列的特定碱基替换(即观察到的 DNA 序列间的核苷酸差异)与 DNA 降解所导致人为替代的相对概率，所以该测序技术的这一特点非常重要 [28,31]。

我们总共从 Vi-80 骨骼标本中复原了 254,933 条不同序列(见补充方法)。使用 BLASTN 程序(NCBI 2.2.12 版本) [35] 将这些序列与人(版本 36.1) [32]、黑猩猩(版本 1) [17] 和小鼠(版本 34.1) [33] 的全基因组序列进行了对齐，与 GenBank *env* 数据库(2005 年 9 月第 3 版)中的环境样本序列进行了对齐，以及与 GenBank *nt*(2005 年 9 月第 3 版，不包括 EST、STS、GSS、环境和 HTGS 序列) [34] 中全套冗余核苷酸序列进行了对齐。根据“目”一级分类学单元将每次查询到的最相似的数据库序列进行确认并分类(图 2)(见补充方法)。其中 79% 的化石提取物的序列读取没有在数据库中找到有意义的核苷酸序列相似性。虽然一些永久冻土中保存的标本可以产生大量的内源 DNA[22]，但是对于其他古代骨骼 [20,22,28] 和环境样品 [36,37] 的大范围测序来说，这是很典型的现象。与数据库序列具有相似性的序列根据其最有意义的对齐的“目”一级分类学单元进行分类。放线菌目是一个含有许多生活在土壤中的种类的细菌目，是序列数目最多的细菌目，占序列的 6.8%。第二大目是灵长目，有 15,701 条不同序列或者说 6.2% 的序列读取与其最相似。所有其他的各目的频率都低得多。值得注意的是，灵长目序列对齐的平均百分率一致性达到了 98.8%，而其余经常出现的目是 92% 到 98%。因此，与许多原核生物的读取不同，灵长目的序列读取被对齐到了亲缘关系很近的物种上。

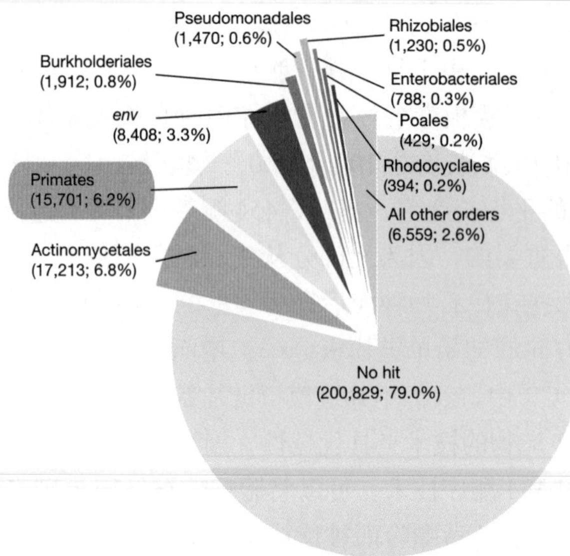

Fig. 2. Taxonomic distribution of DNA sequences from the Vi-80 extract. The taxonomic order of the database sequence giving the best alignment for each unique sequence read was determined. The most populous taxonomic orders are shown.

Neanderthal mtDNA Sequences

Among the 15,701 sequences of primate origin, we first identified all mtDNA in order to investigate whether their evolutionary relationship to the current human mtDNA pool is similar to what is known from previous analyses of Neanderthal mtDNA. A total of 41 unique DNA sequences from the Vi-80 fossil had their closest hits to different parts of the human mtDNA, and comprised, in total, 2,705 base pairs of unique mtDNA sequence. None of the putative Neanderthal mtDNA sequences map to the two hypervariable regions that have been previously sequenced in Neanderthals. We aligned these mtDNA sequences to the complete mtDNA sequences of 311 modern humans from different populations[38] as well as to the complete mtDNA sequences of three chimpanzees and two bonobos (Supplementary Information). A schematic neighbour-joining tree estimated from this alignment is shown in Fig. 3. In agreement with previous results, the Neanderthal mtDNA falls outside the variation among modern humans. However, the length of the branch leading to the Neanderthal mtDNA is 2.5 times as long as the branch leading to modern human mtDNAs. This is likely to be due to errors in our Neanderthal sequences derived from substitution artefacts from damaged, ancient DNA and from sequencing errors[28].

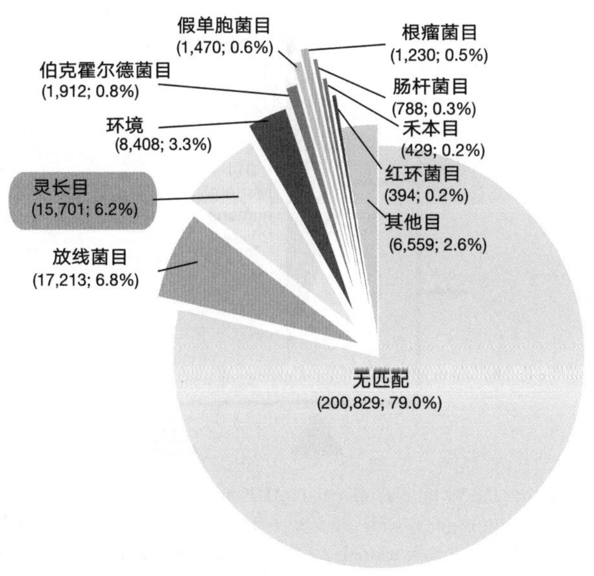

图 2. Vi-80 提取物的 DNA 序列的分类学分布情况。确定了与每条不同序列读取最佳对齐的数据库序列所属的分类学目。图中展示了序列数量最多的几个分类学目。

尼安德特人的 mtDNA 序列

在源于灵长目的 15,701 条序列中，我们首先鉴定了所有 mtDNA，以研究它们同现代人 mtDNA 库的演化关系是否与之前对尼安德特人 mtDNA 的分析结果相似。从 Vi-80 化石共得到 41 条与人类 mtDNA 的不同部位最为匹配的不同序列，总共包含 2,705 bp 的独特 mtDNA 序列。这些推定的尼安德特人 mtDNA 序列不能够映射到之前已测序的尼安德特人两个高变区。我们将这些 mtDNA 序列与来自不同人类群体的 311 个现代人的 mtDNA 全序列 [38] 及三只黑猩猩和两只倭黑猩猩的 mtDNA 全序列进行了对齐（见补充信息）。由该对齐估算出的邻接树的示意图如图 3 所示。得到的结果与以前的研究一致，尼安德特人 mtDNA 也落在现代人的变异范围之外。然而，尼安德特人 mtDNA 分支的长度是现代人 mtDNA 分支长度的 2.5 倍。导致这种情况的原因可能是尼安德特人序列中存在的由受损古 DNA 导致的人为碱基替换错误以及测序错误 [28]。

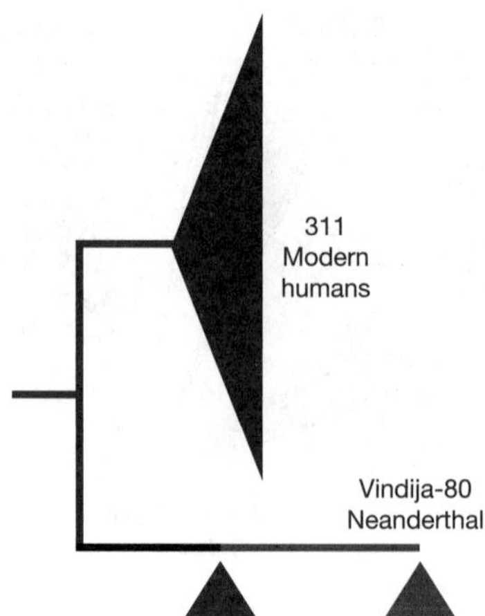

Fig. 3. Schematic tree relating the Vi-80 Neanderthal mtDNA sequences to 311 human mtDNA sequences. The Neanderthal branch length is given with uncorrected sequences (red triangle) and after correction of sequences via independent PCRs (black triangle). Chimpanzee and bonobo sequences (not shown) were used to root the neighbour-joining tree. Several substitution models (Kimura 2-parameter, Tajima-Nei, and Tamura 3-parameter with uniform or gamma-distributed ($\gamma = 0.5–1.1$) rates) yielded bootstrap support values for the human branch from 72–83%.

To analyse the extent to which errors occur in the Neanderthal mtDNA reads, we designed 29 primer pairs (Supplementary Methods) flanking all 39 positions at which the Vi-80 Neanderthal mtDNA sequences differed by substitutions from the consensus bases seen among the 311 human mtDNA sequences. These primer pairs, which are designed to yield amplification products that vary in length between 50 and 98 base pairs (including primers), were used in a multiplex two-step PCR[39] from the same Neanderthal extract that had been used for large-scale 454 sequencing. Twenty five of the PCR products, containing 34 of the positions where the Neanderthal differs from humans, were successfully amplified and cloned, and then six or more clones of each product were sequenced. The consensus sequence seen among these clones revealed the same nucleotides seen by the 454 sequencing at 20 of the 34 positions and no additional differences. Of the 14 positions found to represent errors in the sequence reads, seven were C to T transitions, four were G to A, two were G to T and one was T to C. This pattern of change is typical for ancient DNA, where deamination of cytosine residues[31] and, to a lesser extent, modifications of guanosine residues[28] have been found to account for the majority of nucleotide misincorporations during PCR.

These results also show that the likelihood of observing errors in the sequencing reads is drastically different depending on whether one considers nucleotide positions where a base in the Neanderthal mtDNA sequence differs from both the human and chimpanzee sequences, or positions where the Neanderthal differ from the humans but is identical to the chimpanzee mtDNA sequences. Among the mtDNA sequences analysed, there are 14 positions where the Neanderthal carries a base identical to the chimpanzee, and 13 of those were confirmed by PCR. In contrast, among the remaining 20 positions, where the Neanderthal sequences differed from both humans and chimpanzees, only seven were

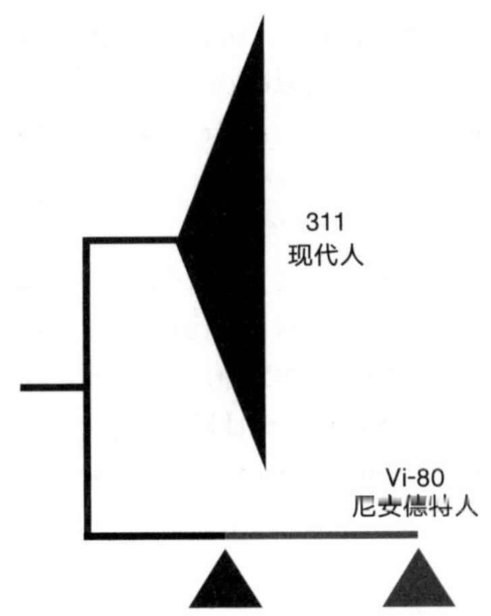

图 3. Vi-80 尼安德特人 mtDNA 序列和 311 个人类的 mtDNA 序列的关系树示意图。尼安德特人分支的长度根据未修正序列（红色三角形）及通过独立 PCR 修正过的序列（黑色三角形）分别给出。黑猩猩和倭黑猩猩的序列（图中未示出）用来作为邻接树的根部。几种替换模型（替代速率为均匀分布或伽马分布（$\gamma = 0.5 \sim 1.1$）的 Kimura 双参数模型、Tajima-Nei 模型以及 Tamura 三参数模型）对人类分支的自举支持值达到了 72% 到 83%。

为了分析在尼安德特人 mtDNA 序列读取中发生错误的程度，我们设计了 29 个引物对（见补充方法），从两侧囊括了 39 个 Vi-80 尼安德特人 mtDNA 序列由于替换而与 311 个人类 mtDNA 序列中所见的合意碱基不同的位置。这些被设计为获取 50 bp 到 98 bp（包括引物）长度不等的扩增产物的引物对被用于大范围 454 测序的尼安德特人相同提取物的多重两步 PCR[39]。最终成功扩增及克隆了 25 个包含尼安德特人不同于人类的 34 个位置的 PCR 产物。随后对每个产物挑选了六个或者更多克隆进行了测序。这些克隆的合意序列与 454 测序法测得的序列相比，这 34 个位置中的 20 个位置的核苷酸是相同的，而且不存在此外的差异。在这 14 个被发现是错误的序列读取中，其中七个是 C → T 转换，四个 G → A 转换，两个 G → T 转换及一个 T → C 转换。这种变化模式是古 DNA 的典型变化类型。这当中，胞嘧啶残基的脱氨基作用 [31] 以及程度较弱的鸟嘌呤核苷残基的修改 [28] 导致了 PCR 过程中发生的大多数核苷酸错参。

这些结果也表明在测序序列读取当中观察到错误发生的可能性的显著差异取决于一个人是否考虑尼安德特人 mtDNA 中某些核苷酸位置上一个碱基与人和黑猩猩序列都不同，或者在这些位置上尼安德特人与现代人不同但与黑猩猩 mtDNA 序列完全相同。在所分析的 mtDNA 序列中，在 14 个位置上，尼安德特人所带的碱基与黑猩猩相同，其中 13 个通过 PCR 得到了确认。相比之下，余下的 20 个位置上，尼安德特人与人类和黑猩猩都不相同，其中只有 7 处得到了确认。当只使用经过 PCR

confirmed. When only PCR-confirmed sequence data are used to estimate the mtDNA tree (Fig. 3), the Neanderthal branch has a length comparable to that of contemporary humans. This suggests that no large source of errors other than what is detected by the PCR analysis affects the sequences.

Using these PCR-confirmed substitutions and a divergence time between humans and chimpanzees of 4.7–8.4 million years[40-42], we estimate the divergence time for the mtDNA fragments determined here to be 461,000–825,000 years. This is in general agreement with previous estimates of Neanderthal–human mtDNA divergence of 317,000–741,000 years[6] based on mtDNA hypervariable region sequences and is compatible with our presumption that the mtDNA sequences determined from the Vi-80 extract are of Neanderthal origin.

Nuclear DNA Sequences

We next analysed the sequence reads whose closest matches are to the human or chimpanzee nuclear genomes and that are at least 30 base pairs long. Figure 4 shows where they map to the human karyotype (see Supplementary Methods). Overall, 0.04% of the autosomal genome sequence is covered by the Neanderthal reads—on average 3.61 bases per 10,000 bases. Both X and Y chromosomes are represented, with a lower coverage of 2.18 and 1.62 bases per 10,000, respectively, showing that the Vi-80 bone is derived from a male individual.

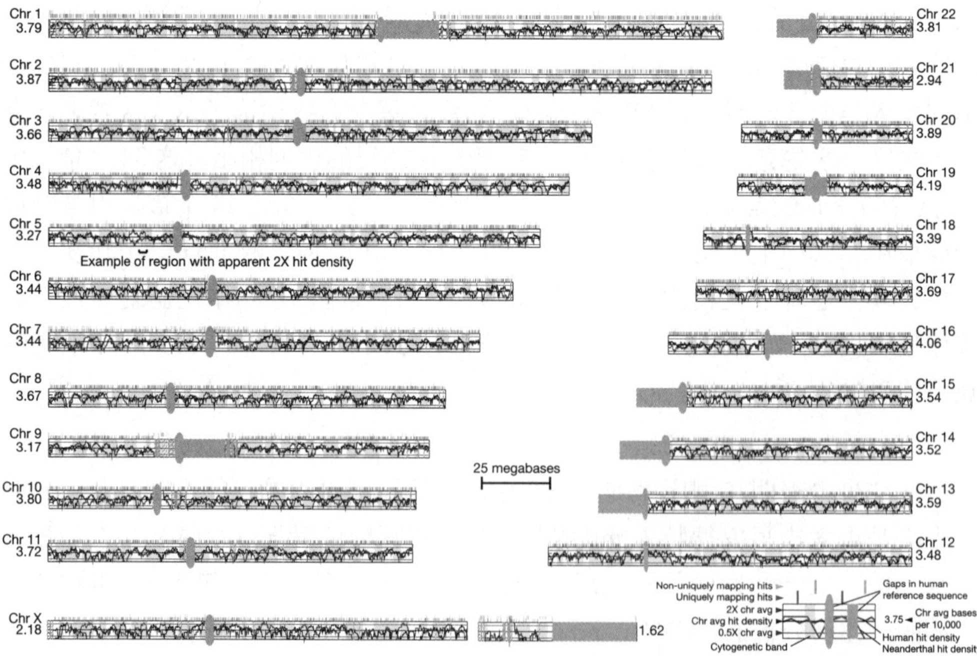

Fig. 4. Location on the human karyotype of Neanderthal DNA sequences. All sequences longer than 30 nucleotides whose best alignments were to the human genome are shown. The blue lines above each chromosome mark the position of all alignments that are unique in terms of bit-score within the human

确认过的序列数据来估算 mtDNA 树（图 3）时，尼安德特人分支的长度基本与现代人的分支长度相当。这表明除了 PCR 分析检测出的错误外，没有其他大的错误来源影响到序列的测定。

我们使用这些 PCR 确认过的碱基替换并采纳人类和黑猩猩间的分歧时间为距今 470 万年到 840 万年 [40-42]，由此估算出这里所确定的 mtDNA 片段的分歧时间是距今 461,000 年到 825,000 年。这与以前基于 mtDNA 高变区序列的尼安德特人–现代人 mtDNA 分歧时间为距今 317,000 年到 741,000 年 [6] 基本一致，也与我们的假定是相符的，即假定 Vi-80 提取物确定的 mtDNA 序列是源于尼安德特人的。

核 DNA 序列

接着我们分析了与现代人或黑猩猩核基因组最匹配的序列读取，这些序列至少 30 bp 长。图 4 展示了将它们映射到人类染色体核型上的情况（见补充方法）。总的来说，这些尼安德特人的序列读取涵盖了 0.04% 的常染色体基因组序列——平均每 10,000 碱基中有 3.61 个。其中也展示了 X 染色体和 Y 染色体的情况，涵盖范围比常染色体的稍低，分别是每 10,000 碱基有 2.18 个和 1.62 个，表明 Vi-80 骨骼标本是一个男性个体的。

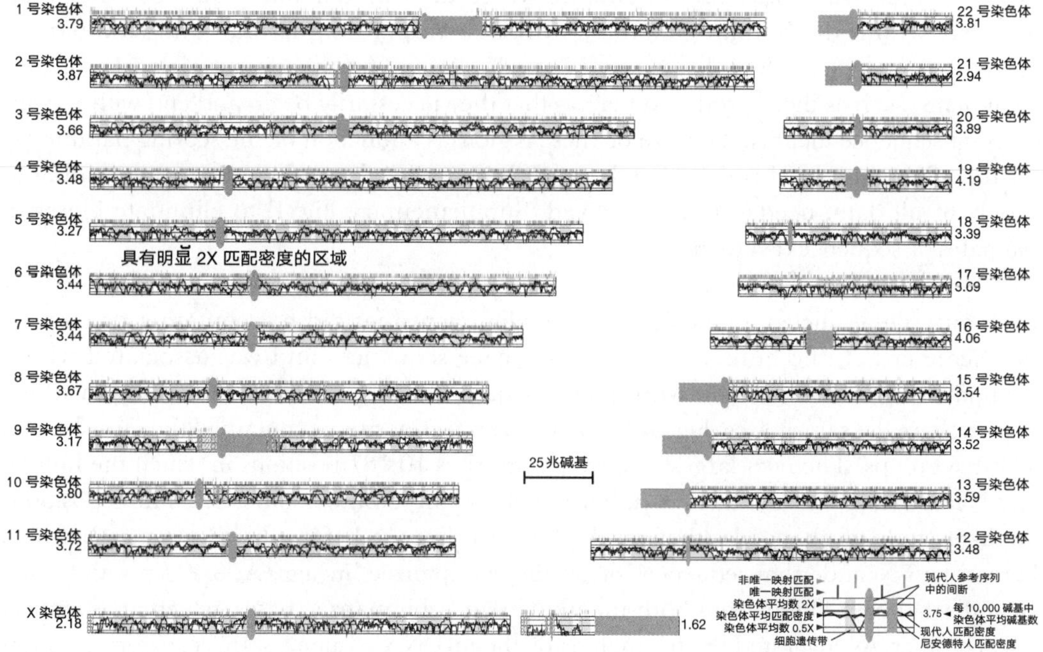

图 4. 尼安德特人 DNA 序列在人类染色体核型上的位置。所有长度大于 30 个核苷酸的序列在现代人基因组中的最佳对齐位置都在此示出。每条染色体上部的蓝线表示在现代人基因组中具有唯一位分值的最

genome. Orange lines are alignments that have more than one alignment of equal bit-score. To the left of each chromosome, the average number of Neanderthal bases per 10,000 is given. Lines (Neanderthal, blue; human, red) within each chromosome show the hit density, on a log-base 2 scale, within sliding windows of 3 megabases along each chromosome. The centre black lines indicate the average hit-density for the chromosomes. The purple lines above and below indicate hit densities of 2X and 1/2X the chromosome average, respectively. On chromosome 5, an example of a region of increased sequence density is highlighted. Sequence gaps in the human reference sequence are indicated by dark grey regions. Chromosomal banding pattern is indicated by light grey regions.

The data presented in Fig. 4 show that when the hit density for sequences that have a single best hit in the human genome is plotted along the chromosomes, several suggestive local deviations from the average hit density are seen, which may represent copy-number differences in the Neanderthal relative to the human reference genome. For comparison, we generated 454 sequence data from a DNA sample from a modern human. Interestingly, some of the deviations seen in the Neanderthal are present also in the modern human, whereas others are not. The latter group of sequences may indicate copy-number differences that are unique to the Neanderthal relative to the modern human genome sequence. Thus, when more Neanderthal sequence is generated in the future, it may be possible to determine copy number differences between the Neanderthal, the chimpanzee and the human genomes.

Patterns of Nucleotide Change on Lineages

We generated three-way alignments between all Neanderthal sequences that map uniquely within the human genome and the corresponding human and chimpanzee genome sequences (see Supplementary Methods). An important artefact of local sequence alignments, such as those produced here, is that they necessarily begin and end with regions of exact sequence identity. The size of these regions is a function of the scoring parameters for the alignment. In this case, five bases at both ends of the alignments, amounting to ~14% of all data, needed to be removed (Supplementary Fig. 1) to eliminate biases in estimates of sequence divergence.

Each autosomal nucleotide position in the alignment that did not contain a deletion in the Neanderthal, the human or the chimpanzee sequences and was associated with a chimpanzee genome position with quality score $\geqslant 30$ was classified according to which species share the same bases (Fig. 5). A total of 736,941 positions contained the same base in all three groups. The next largest category comprises 10,167 positions in which the human and Neanderthal base are identical, but the chimpanzee base is different. These positions are likely to have changed either on the hominid lineage before the divergence between human and Neanderthal sequences or on the chimpanzee lineage. At 3,447 positions, the Neanderthal base differs from both the human and chimpanzee bases, which are identical to each other. As suggested by the analysis of the mtDNA sequences, this category contains positions that have changed on the Neanderthal lineage, as well as a large proportion of errors that derive both from base damage that have accumulated in the ancient DNA and

佳对齐位置。橙色线表示具有相等位分值的超过一个的对齐位置。每条染色体的左侧给出了每 10,000 个碱基中尼安德特人碱基的平均数目。每条染色体内部的曲线(蓝色代表尼安德特人，红色代表现代人)表示沿每条染色体在三百万碱基滑动窗口内以 2 为底取对数的匹配密度。中间的黑线表示该染色体的平均匹配密度。上面和下面的紫色线分别表示相当于染色体平均数 2X 和 1/2X 的匹配密度。5 号染色体上，标出了一个序列密度增加的区域的例子。现代人参考序列中的序列间断用深灰色区域标出。染色体带型用浅灰色区域表示。

图 4 列出的数据表明当在现代人基因组中只有一个最佳匹配的序列的匹配密度沿着染色体被描绘的时候，能够看到几个平均匹配密度的提示性局部偏差，可能代表了尼安德特人相对于现代人参考基因组的拷贝数差异。为了进行比较，我们由一个现代人的 DNA 样本生成了 454 序列数据。有趣的是，某些在尼安德特人序列中看到的偏差同样也存在于现代人中，而其他却不存在。后一组序列可能暗示相对于现代人基因组序列而言，尼安德特人所特有的拷贝数差异。因此，如果将来可以生成更多的尼安德特人序列，那么就有可能确定尼安德特人、黑猩猩和现代人基因组间的拷贝数差异。

各谱系中的核苷酸变化模式

我们对所有在现代人基因组内有唯一映射的尼安德特人序列与相应的现代人和黑猩猩基因组序列进行了三重对齐(见补充方法)。就像这里的对齐一样，局部的序列对齐一个重要的人为因素就是它们必须以序列完全相同的区域开始和结束。这些区域的大小是对齐的评分参数的一个函数。这里，对齐两端各有五个碱基(补充图 1)，占所有数据的约 14%，需要予以剔除，以消除估算序列分化时的偏差。

尼安德特人、现代人或者黑猩猩序列中不包含缺失，且在黑猩猩基因组中位置质量得分 ≥30 那些对齐中，每个常染色体核苷酸位置都根据共有相同碱基的物种进行分类(图 5)。所有三组中含有相同碱基的总共有 736,941 个位置。第二大类中，现代人和尼安德特人具有相同碱基但与黑猩猩不同的位置有 10,167 个。这些位置可能在现代人和尼安德特人序列发生分歧之前或者是黑猩猩谱系中就发生了变化。尼安德特人在 3,447 个位置的碱基与人类和黑猩猩都不同，但后两者完全相同。正如 mtDNA 序列分析所表明的那样，这一类别包含了尼安德特人谱系中发生改变的位置，也包含了由古 DNA 中累积的碱基损伤以及测序错误导致的大部分错误。现代

from sequencing errors. At 434 positions, the human base differs from both the Neanderthal and chimpanzee bases, which are identical to each other. These positions are likely to have changed on the human lineage after the divergence from Neanderthal. Finally, a total of 51 positions contain different bases in all three groups.

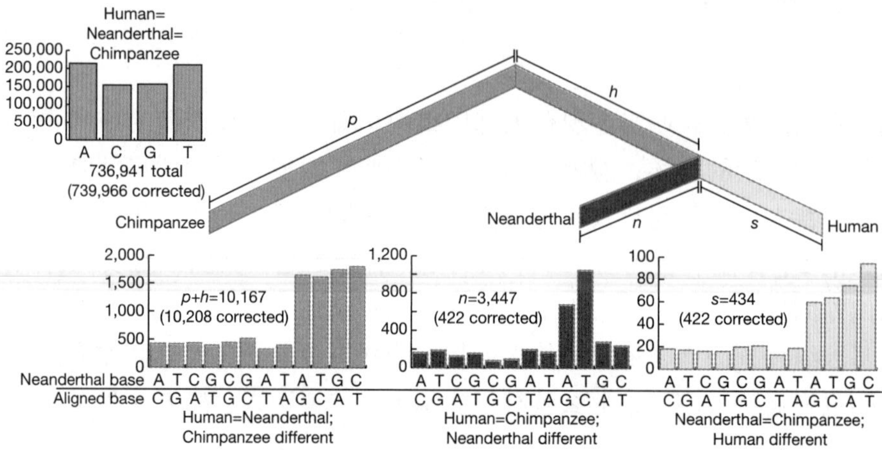

Fig. 5. Schematic tree illustrating the number of nucleotide changes inferred to have occurred on hominoid lineages. In blue is the distribution of all aligned positions that did not change on any lineage. In brown are the changes that occurred either on the chimpanzee lineage (p) or on the hominid lineage (h) before the human and Neanderthal lineages diverged. In red are the changes that are unique to the Neanderthal lineage (n), including all changes due to base-damage and base-calling errors. In yellow are changes unique to the human lineage. The distributions of types of changes in each category are also given. The numbers of changes in each category, corrected for base-calling errors in the Neanderthal sequence (see Supplementary Methods), are shown within parentheses.

Because the 454 sequencing technology allows the base in a base pair from which a sequence is derived to be determined, the relative frequencies of each of the 12 possible categories of base changes can be estimated for each evolutionary lineage. As seen in Fig. 5, the patterns of the chimpanzee-specific and human-specific changes are similar to each other in that the eight transversional changes are of approximately equal frequency and about fourfold less frequent than each of the four transitional changes, yielding a transition to transversion ratio of 2.04, typical of closely related mammalian genomes[43]. For the Neanderthal-specific changes the pattern is very different in that mismatches are dominated by C to T and G to A differences. Thus, the pattern of change seen among the Neanderthal-specific alignment mismatches is typical of the nucleotide substitution pattern observed in PCR of ancient DNA.

Consistent with this, modern human sequences determined by 454 sequencing show no excess amount of C to T or G to A differences (Supplementary Fig. 2), indicating that lesions in the ancient DNA rather than sequencing errors account for the majority of the errors in the Neanderthal sequences. Assuming that the evolutionary rate of DNA change was the same on the Neanderthal and human lineages, the majority of observed differences specific to the Neanderthal lineage are artefacts. All Neanderthal-specific changes were therefore

人在 434 个位置处的碱基与尼安德特人和黑猩猩都不同，而后两者是完全相同的。这些位置可能是人类与尼安德特人发生分歧之后才在人类谱系中发生了变化的。最后，有 51 个位置所有三组中的碱基均不同。

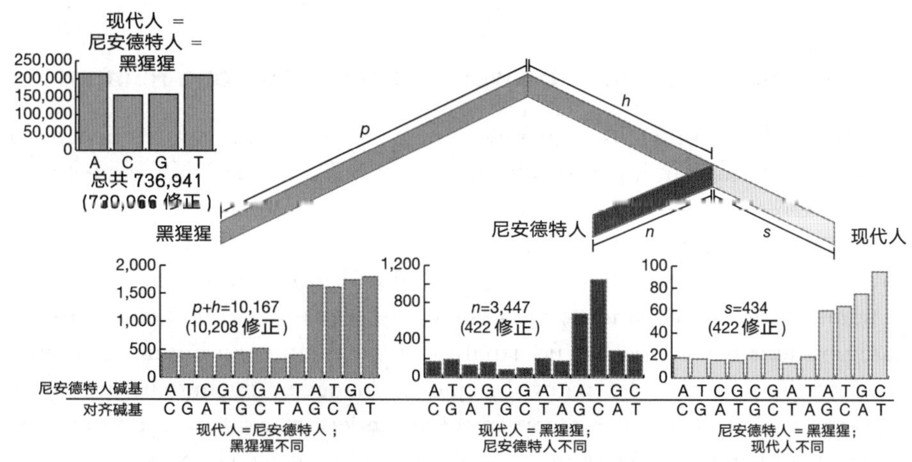

图 5. 人猿超科谱系上已发生的核苷酸变化数目的推测示意树。蓝色表示在任何谱系上都没有变化的对齐位置的分布情况。褐色表示在现代人和尼安德特人谱系分歧之前，黑猩猩谱系（p）或者人类谱系（h）上发生过的变化。红色表示尼安德特人谱系（n）独有的变化，包括碱基损伤和碱基识别错误等引起的所有变化。黄色是现代人谱系独有的变化。图中也给出了每个类别的变化类型的分布情况。圆括号中给出的是对尼安德特人序列碱基识别错误进行修正（见补充方法）后的每个类别的变化数目。

由于 454 测序技术能够确定序列来源的碱基对中的碱基，所以可以估算出每一演化谱系 12 个可能的碱基变化种类的每一种的相对发生频率。正如从图 5 中看到的，黑猩猩特有以及现代人特有的变化模式彼此很相似，因为八种颠换发生的频率基本相等，大概是四种转换中的每一种的发生频率的 1/4，产生了一个转换对颠换比率为 2.04 的结果，这对亲缘关系近的哺乳动物基因组而言是很具有代表性的 [43]。尼安德特人特有的变化主要是 C 到 T 和 G 到 A 错配的差异，所以与上述模式很不一样。因此，尼安德特人特有的对齐中的错配的变化模式，是在古 DNA 的 PCR 中典型的核苷酸替换。

与此一致，454 测序技术确定的现代人序列显示出没有过量的 C 到 T 或 G 到 A 的差异（补充图 2），这表明造成尼安德特人序列中出现的错误，大部分是由于古 DNA 中的损伤而非测序错误。假定 DNA 变化的演化速率在尼安德特人和现代人谱系中是相等的，那么大部分观察到的尼安德特谱系特有的差异就都是人为噪音了。因此在接下来的分析中我们忽略了所有尼安德特人特有的变化，尼安德特人序列只

disregarded in the subsequent analyses and the Neanderthal sequences were used solely to assign changes to the human or chimpanzee lineage where the human and chimpanzee genome sequences differ and the Neanderthal sequence carries either the human or the chimpanzee base.

Genomic Divergence between Neanderthals and Humans

Assuming that the rates of DNA sequence change along the chimpanzee lineage and the human lineage were similar, it can be estimated that 8.2% of the DNA sequence changes that have occurred on the human lineage since the divergence from the chimpanzee lineage occurred after the divergence of the Neanderthal lineage. However, although the Neanderthal-specific changes that are heavily influenced by errors are not used for this analysis, some errors in the single-pass sequencing reads from the Neanderthal extract will create positions where the Neanderthal is identical either to human or chimpanzee sequences, and thus affect the estimates of sequence change on the human and chimpanzee lineages. When the effects of such errors in the Neanderthal sequences are quantified and removed (see Supplementary Methods), ~7.9% of the sequence changes along the human lineage are estimated to have occurred after divergence from the Neanderthal. If the human–chimpanzee divergence time is set to 6,500,000 years (refs 40, 41, 44), this implies an average human–Neanderthal DNA sequence divergence time of ~516,000 years. A 95% confidence interval generated by bootstrap re-sampling of the alignment data gives a range of 465,000 to 569,000 years. Obviously, these divergence estimates are dependent on the human–chimpanzee divergence time, which is a much larger source of uncertainty.

We analysed the DNA sequences generated from a contemporary human using the same sequencing protocol as was used for the Neanderthal. Although ancient DNA is degraded and damaged, this comparison controls for many of the aspects of the analysis including sequencing and alignment methodology. In this case, ~7.1% of the divergence along the human lineage is assigned to the time subsequent to the divergence of the two human sequences. The average divergence time between alleles within humans is thus ~459,000 years with a 95% confidence interval between 419,000 and 498,000 years. As expected, this estimate of the average human diversity is less than the divergence seen between the human and the Neanderthal sequences, but constitutes a large fraction of it because much of the human sequence diversity is expected to predate the human–Neanderthal split[25]. Neanderthal genetic differences to humans must therefore be interpreted within the context of human diversity.

Ancestral Population Size

Humans differ from apes in that their effective population size is of the order of 10,000 while those of chimpanzees, gorillas and orangutans are two to four times larger[45-47]. Furthermore, the population size of the ancestor of humans and chimpanzees was found to

用来在现代人和黑猩猩基因组序列不同的位置并且在该位置尼安德特人序列或者携带现代人或者携带黑猩猩碱基的情况下，将变化分配到现代人或者黑猩猩谱系中去。

尼安德特人和现代人基因组之间的分歧

假定黑猩猩谱系和人类谱系中 DNA 序列变化速率是相似的，那么可以估算出从人类与黑猩猩谱系发生分歧后所发生的 DNA 序列变化中，有 8.2% 是在与尼安德特人谱系发生分歧之后发生的。但是，尽管本分析中并未使用那些受错误严重影响的尼安德特人特有的变化，但是有些来自尼安德特人提取物的单通道测序读取错误能告知尼安德特人与现代人或者黑猩猩序列相同的位置，因此会影响到对现代人和黑猩猩谱系中发生的序列变化的估算。当把尼安德特人序列中这种错误的影响量化并且去除后（见补充方法），可以估算出，现代人谱系中的 7.9% 的序列变化是发生在与尼安德特人发生分歧之后。如果人类–黑猩猩分歧时间被设定在 6,500,000 年（参考文献 40、41、44），那么就意味着现代人–尼安德特人 DNA 序列的平均分歧时间约为距今 516,000 年。通过对对齐数据进行自举重新抽样产生的 95% 置信区间给出的范围是距今 465,000 年至 569,000 年。很明显，这些分歧估值取决于人类–黑猩猩的分歧时间，这是一个更大的不确定性来源。

我们使用与尼安德特人同样的测序方案分析了一例现代人产生的 DNA 序列。尽管古 DNA 会被降解和损坏，但是这一比较在包括测序和对齐方法论在内的很多方面对分析进行了控制。这里，现代人谱系中 7.1% 的分歧被分配到两条人类序列发生分歧之后的时间里。于是，现代人中等位基因之间的平均分歧时间大约是距今 459,000 年，95% 置信区间的范围在距今 419,000 年到 498,000 年间。正如预期的那样，这一平均人类多样性的估值比现代人和尼安德特人序列间的分歧要晚，但是因为现代人序列许多多样性理应早于现代人–尼安德特人分歧时间，所以年代范围占据了其中很大一部分[25]。因此必须在现代人多样性这一背景下解读尼安德特人与现代人间的遗传差异。

祖先种群大小

人不同于猿，人的有效种群大小是 10,000，而黑猩猩、大猩猩和猩猩的有效种群大小是人类的两到四倍[45-47]。此外，研究发现人类和黑猩猩的祖先的种群大小与

be similar to those of apes, rather than to humans[42,48]. The Neanderthal sequence data now allow us to ask if the effective size of the population ancestral to humans and Neanderthals was large, as is the case for apes and the human–chimpanzee ancestor, or small, as for present-day humans.

We applied a method[42] that co-estimates the ancestral effective population size and the split time between Neanderthal and human populations (Fig. 6a; see Supplementary Methods). As seen in Fig. 6b, we recover a line describing combinations of population sizes and split times compatible with the data and lack power to be more precise (see Supplementary Methods and Results). Using this line we can estimate the ancestral population size, given estimates about the population split time from independent sources. If we use a split time of 400,000 years inferred from the fossil record (J. J. Hublin, personal communication), then our point estimate of the ancestral population size is ~ 3,000. Given uncertainty in both the sequence divergence time and the population split time, our estimate of the ancestral population size varies from 0 to 12,000.

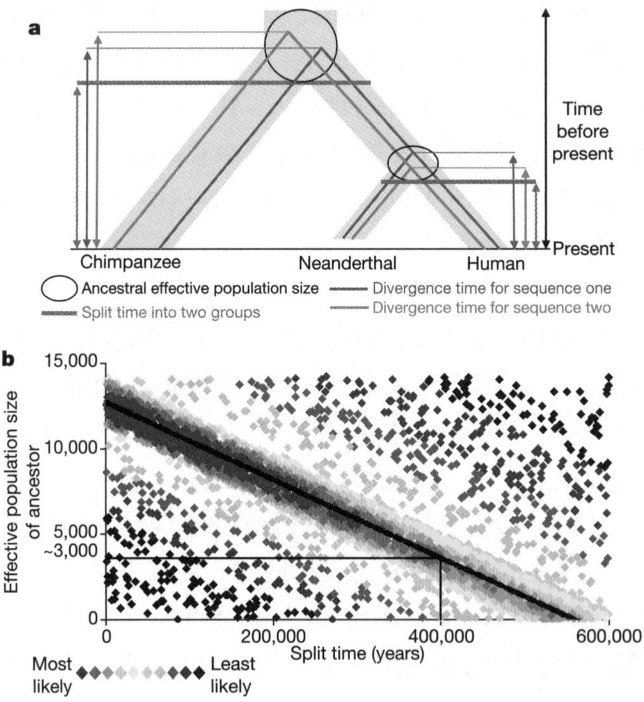

Fig. 6. Estimate of the effective population size of the ancestor of humans and Neanderthals. **a**, Schematic illustration of the model used to estimate ancestral effective population size. By split time, we mean the time, in the past, after which there was no more interbreeding between two groups. By divergence, we mean the time, in the past, at which two genetic regions separated and began to accumulate substitutions independently. Effective population size is the number of individuals needed under ideal conditions to produce the amount of observed genetic diversity within a population. **b**, The likelihood estimates of population split times and ancestral population sizes. The likelihoods are grouped by colour. The red–yellow points are statistically equivalent based on the likelihood ratio test approximation. The black line is the line of best fit to red–yellow points (see Supplementary Methods). This graph is scaled assuming a human–chimpanzee average sequence divergence time of 6,500,000 years.

猿类很相似，而非与人类相似[42,48]。现在，尼安德特人的序列数据允许我们提出如下问题，现代人和尼安德特人的祖先的有效种群大小是像猿类和人类–黑猩猩的祖先的种群一样大呢，还是像如今的现代人一样小？

我们运用了一种可以同时估算祖先有效种群大小和尼安德特人与现代人种群之间的分化时间的方法[42]（图 6a；见补充方法）。如图 6b 所示，我们复原了一条同时描述种群大小和分化时间组合且与数据相吻合的拟合线，但是统计学功效欠缺无法更加精确（见补充方法和结果）。使用这条拟合线，给定独立来源的种群的分化时间估值，我们可以估算出祖先种群的大小。如果我们使用根据化石记录推断的距今 400,000 年（胡布林，个人交流）作为分化时间的话，那么我们对祖先种群大小的点估计大约是 3,000。鉴于序列分歧时间和种群分化时间二者的不确定性，我们对祖先种群大小的估计从 0 到 12,000 不等。

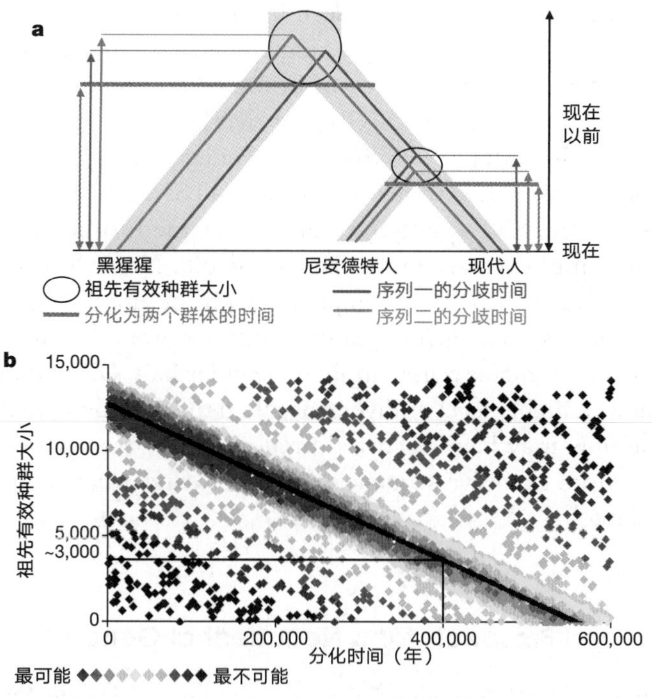

图 6. 现代人和尼安德特人的祖先的有效种群大小的估计。**a**，用来估算祖先有效种群大小的模型的示意图。我们所说的分化时间是指过去的某一时间，在该时间之后两个群体之间不再有杂交。我们所说的分歧是指过去某一时间，在该时间两个遗传区域发生分离并且开始独立积累替换。有效种群大小是指理想条件下能在一个种群内产生特定数量的可见遗传多样性所需的个体数目。**b**，种群分化时间和祖先种群大小的可能性估计。用不同颜色对可能性进行了分组。红黄点根据似然比检验逼近在统计意义上是等同的。黑线是红黄点的最佳拟合线（见补充方法）。该图比例设定的是在假定人类–黑猩猩的平均序列分歧时间为距今 6,500,000 年的前提下。

These results suggest that the population ancestral to present-day humans and Neanderthals was similar to present-day humans in having a small effective size and thus that the effective population size on the hominid lineage had already decreased before the split between humans and Neanderthals. Therefore, the small effective population size seen in present-day human samples may not be unique to modern humans, but was present also in the common ancestor of Neanderthals and modern humans. We speculate that a small effective size, perhaps associated with numerous expansions from small groups, was typical not only of modern humans but of many groups of the genus *Homo*. In fact, the origin of *Homo erectus* may have been associated with genetic or cultural adaptations that resulted in drastic population expansions as indicated by their appearance outside Africa around two million years ago.

Neanderthal Sequences and Human Polymorphisms

Another question that can be addressed with these data is how often the Neanderthal has the ancestral allele (that is, the same allele seen in the chimpanzee) versus the derived (or novel) allele at sites where humans carry a single nucleotide polymorphism (SNP). The latter case identifies SNPs that were present in the common ancestor of Neanderthals and present-day humans. Using the SNPs that overlap with our data from two large genome-wide data sets (HapMap[49], 786 SNPs and Perlegen[50], 318 SNPs), we find that the Neanderthal sample has the derived allele in ~30% of all SNPs. This number is presumably an overestimate since the SNPs analysed were ascertained to be of high frequency in present-day humans and hence are more likely to be old. Nevertheless, this high level of derived alleles in the Neanderthal is incompatible with the simple population split model estimated in the previous section, given split times inferred from the fossil record. This may suggest gene flow between modern humans and Neanderthals. Given that the Neanderthal X chromosome shows a higher level of divergence than the autosomes (R.E.G., unpublished observation), gene flow may have occurred predominantly from modern human males into Neanderthals. More extensive sequencing of the Neanderthal genome is necessary to address this possibility.

Rationale and Prospects for a Neanderthal Genome Sequence

We demonstrate here that DNA sequences can be generated from the Neanderthal nuclear genome by massive parallel sequencing on the 454 sequencing platform. It is thus feasible to determine large amounts of sequences from this extinct hominid. As a corollary, it is possible to envision the determination of a Neanderthal genome sequence. For several reasons, we believe that this would represent a valuable genomic resource.

First, a Neanderthal genome sequence would allow all nucleotide sequence differences as well as many copy-number differences between the human and chimpanzee genomes to be temporally resolved with respect to whether they occurred before the separation of

这些结果表明如今的现代人和尼安德特人的祖先种群与如今的现代人相似，有效种群大小都比较小，所以人类谱系的有效种群大小在现代人和尼安德特人发生分化之前已经有所减少。因此，我们看到的如今的现代人样本中有效种群小的现象，可能并不是现代人特有的，而是早在尼安德特人和现代人的共同祖先时就具有的。我们推测，小规模有效种群可能与小群体的扩张有关，不仅在现代人中很典型，在人属的许多其他类群中也很典型。事实上，直立人的起源可能与遗传和文化适应有关，这导致了明显的种群扩张，正如大约两百万年前他们就出现在非洲之外所表明的那样。

尼安德特人序列和人类的遗传多态性

基于这些数据可以探讨的另一个问题是尼安德特人在现代人携带的单核苷酸多态性（SNP）的位点处出现祖先型等位基因（即在黑猩猩中也见到的同样的等位基因）与衍生的（或新型）等位基因的频率。后一种情况可以识别出存在于尼安德特人和现代人的共同祖先中的 SNP。我们使用两大全基因组数据库中与我们的数据相重叠的那些 SNP（HapMap[49] 有 786 个 SNP 及 Perlegen[50] 有 318 个 SNP），发现尼安德特人样品具有的衍生等位基因约占到了所有 SNP 的 30%。这一数字很可能是过高估算，因为已经查明所有纳入分析的 SNP 在如今的现代人中都有很高的出现频率，故而其年代更有可能比较久远。然而，根据由化石记录推断的分化时间，尼安德特人中这种高水平衍生等位基因与前一部分估算的简单种群分化模型是相矛盾的。这可能表明现代人和尼安德特人之间存在基因流。考虑到尼安德特人的 X 染色体显示出比常染色体更高水平的分歧（理查德 · 格林，未发表观察结果），因此基因流可能主要是从现代人男性流向尼安德特人的。要确认这一可能性，须要对尼安德特人基因组进行更广泛的测序。

尼安德特人基因组序列的缘由和前景

我们在此证明可以通过在 454 测序平台上进行大规模平行测序从尼安德特人核基因组生成 DNA 序列。因此从这种灭绝的人类确定大量序列是可行的。作为必然的结果，确定尼安德特人基因组序列是可能的。由于如下原因，我们相信这将是一种非常有价值的基因组资源。

首先，尼安德特人基因组序列使得人类与黑猩猩基因组之间所有的核苷酸序列差异以及许多拷贝数差异在时间上得到解决，即差异究竟是发生在现代人与尼安德

humans from Neanderthals, or whether they occurred after or at the time of separation. The latter class of changes is of interest, because some of them will be associated with the emergence of modern humans. A Neanderthal genome sequence would therefore allow the research community to determine whether DNA sequence differences between humans and chimpanzees that are found to be functionally important represent recent changes on the human lineage. No data other than a Neanderthal genome sequence can provide this information.

Second, the fact that Neanderthals carry the derived allele for a substantial fraction of human SNPs suggests a method of identifying genomic regions that have experienced a selective sweep subsequent to the separation of human and Neanderthal populations. Such selective sweeps in the human genome will make the variation in these regions younger than the separation of humans and Neanderthals. As we show above, in regions not affected by sweeps a substantial proportion of polymorphic sites in humans will carry derived alleles in the Neanderthal genome sequence, whereas no sites will do so in regions affected by sweeps. This represents an approach to identifying selective sweeps in humans that is not possible from other data.

Third, once large amounts of Neanderthal genome sequence is generated, it will become possible to estimate the misincorporation probabilities for each class of nucleotide differences between the Neanderthal and chimpanzee genomes with high accuracy by analysing regions covered by many reads such as mtDNA, repeated genome regions of high sequence identity, as well as single-copy regions covered by multiple reads. Once this is done, the confidence that any particular nucleotide position where the Neanderthal differs from human as well as chimpanzee is correct can be reliably estimated. In combination with future knowledge about the function of genes and biological systems, comprehensive information from the Neanderthal genome will then allow aspects of Neanderthal biology to be deciphered that are unavailable by any other means.

Are fossil and technical resources today sufficient to imagine the determination of a Neanderthal genome sequence? The results presented here are derived from approximately one fifteenth of an extract prepared from ~100 mg of bone. To achieve one-fold coverage of the Neanderthal genome (3 gigabases) without any further improvement in technology, about twenty grams of bone and 6,000 runs on the current version of the 454 sequencing platform would be necessary. Although this is at present a daunting task, technical improvements in the procedures described here that would make the retrieval of DNA sequences of the order of ten times more efficient can easily be envisioned (our unpublished results). In view of that prospect, we have recently initiated a project that aims at achieving an initial draft version of the Neanderthal genome within two years.

(**444**, 330-336; 2006)

特人分离之前，还是发生在分离之后或分离之时。后一类变化更加应该被关注，因为其中有些与现代人的出现有关。因此尼安德特人的基因组序列将令学术界能够确定人和黑猩猩间功能上重要的 DNA 序列差异是否代表了人类谱系最近发生的变化。除了尼安德特人的基因组序列外，没有其他数据能够提供这一信息。

其次，尼安德特人携带着大量人类 SNP 的衍生等位基因这一事实指明了一种识别基因组中在现代人与尼安德特人种群分离之后经历了选择性清除的区域的方法。人类基因组中的这些选择性清除会使这些区域内的变异比现代人和尼安德特人的分离年轻。正如我们上面所陈述的，在未受选择性清除影响的区域中，对于人类的大部分多态位点来说，尼安德特人基因组序列中会携带存在的衍生等位基因，然而在受选择性清除影响的区域中，没有位点会出现这种情况。这代表一种识别人类中的选择性清除的方法，其他数据是不可能做到这点的。

再者，一旦生成了大量尼安德特人基因组序列，通过分析像 mtDNA 那样由多读取覆盖的区域、序列高度等同的重复基因组区域以及多读取覆盖的单拷贝区域，以高精度估算尼安德特人和黑猩猩基因组间每类核苷酸差异的错配概率将成为可能。一旦完成了这一步，就能够可靠地估计出尼安德特人不同于现代人及黑猩猩的任何特定核苷酸位置的正确性的可信度。加上将来对基因功能和生物系统的进一步了解，尼安德特人基因组提供的全面信息就可以对尼安德特人生物学的方方面面进行解读，这是其他方式无法做到的。

现在的化石和技术资源足够让我们奢想确定尼安德特人基因组序列吗？这里介绍的结果源自由约 100 mg 骨骼样品产生的提取物的大概 1/15。在无须任何技术改进的基础上，为了得到尼安德特人整个基因组（3 千兆碱基）的信息，在现有版本的 454 测序平台上，将需要约 20 g 骨骼，进行 6,000 次运行。尽管目前这是一项令人畏惧的任务，但是对这里描述的步骤进行技术改良之后，很轻易就能够更有效地获取十倍的 DNA 序列（我们尚未发表的结果）。鉴于这一前景，我们最近启动了一个旨在两年内完成尼安德特人基因组草图的项目。

（刘皓芳 翻译；张颖奇 审稿）

Richard E. Green[1], Johannes Krause[1], Susan E. Ptak[1], Adrian W. Briggs[1], Michael T. Ronan[2], Jan F. Simons[2], Lei Du[2], Michael Egholm[2], Jonathan M. Rothberg[2], Maja Paunovic[3‡] & Svante Pääbo[1]

[1] Max-Planck Institute for Evolutionary Anthropology, Deutscher Platz 6, D-04103 Leipzig, Germany

[2] 454 Life Sciences, 20 Commercial Street, Branford, Connecticut 06405, USA

[3] Institute of Quaternary Paleontology and Geology, Croatian Academy of Sciences and Arts, A. Kovacica 5/II, HR-10 000 Zagreb, Croatia

[‡] Deceased

Received 14 July; accepted 11 October 2006.

References:

1. Bischoff, J. L. *et al.* The Sima de los Huesos hominids date to beyond U/Th equilibrium (> 350 kyr) and perhaps to 400–500 kyr: New radiometric dates. *J. Archaeol. Sci.* **30**, 275-280 (2003).
2. Hublin, J-J. (ed.) *Climatic Changes, Paleogeography, and the Evolution of the Neandertals* (Plenum Press, New York, 1998).
3. Franciscus, R. G. (ed.) *Neanderthals* (Oxford Univ. Press, Oxford, 2002).
4. Hublin, J. J., Spoor, F., Braun, M., Zonneveld, F. & Condemi, S. A late Neanderthal associated with Upper Palaeolithic artefacts. *Nature* **381**, 224-226 (1996).
5. Krings, M. *et al.* Neandertal DNA sequences and the origin of modern humans. *Cell* **90**, 19-30 (1997).
6. Krings, M., Geisert, H., Schmitz, R. W., Krainitzki, H. & Pääbo, S. DNA sequence of the mitochondrial hypervariable region II from the Neandertal type specimen. *Proc. Natl Acad. Sci. USA* **96**, 5581-5585 (1999).
7. Schmitz, R. W. *et al.* The Neandertal type site revisited: Interdisciplinary investigations of skeletal remains from the Neander Valley, Germany. *Proc. Natl Acad. Sci. USA* **99**, 13342-13347 (2002).
8. Ovchinnikov, I. V. *et al.* Molecular analysis of Neanderthal DNA from the northern Caucasus. *Nature* **404**, 490-493 (2000).
9. Krings, M. *et al.* A view of Neandertal genetic diversity. *Nature Genet.* **26**, 144-146 (2000).
10. Serre, D. *et al.* No evidence of Neandertal mtDNA contribution to early modern humans. *PLoS Biol.* **2**, 313-317 (2004).
11. Orlando, L. *et al.* Revisiting Neandertal diversity with a 100,000 year old mtDNA sequence. *Curr. Biol.* **16**, R400-R402 (2006).
12. Caramelli, D. *et al.* A highly divergent mtDNA sequence in a Neandertal individual from Italy. *Curr. Biol.* **16**, R630-R632 (2006).
13. Lalueza-Fox, C. *et al.* Neandertal evolutionary genetics: mitochondrial DNA data from the Iberian peninsula. *Mol. Biol. Evol.* **22**, 1077-1081 (2005).
14. Currat, M. & Excoffier, L. Modern humans did not admix with Neanderthals during their range expansion into Europe. *PLoS Biol.* **2**, e421 (2004).
15. Stringer, C. Modern human origins: progress and prospects. *Phil. Trans. R. Soc. Lond. B* **357**, 563-579 (2002).
16. Takahata, N., Lee, S. H. & Satta, Y. Testing multiregionality of modern human origins. *Mol. Biol. Evol.* **18**, 172-183 (2001).
17. Chimpanzee Sequencing and Analysis Consortium. Initial sequence of the chimpanzee genome and comparison with the human genome. *Nature* **437**, 69-87 (2005).
18. Pääbo, S. *et al.* Genetic analyses from ancient DNA. *Annu. Rev. Genet.* **38**, 645-679 (2004).
19. Greenwood, A. D., Capelli, C., Possnert, G. & Pääbo, S. Nuclear DNA sequences from late Pleistocene megafauna. *Mol. Biol. Evol.* **16**, 1466-1473 (1999).
20. Noonan, J. P. *et al.* Genomic sequencing of Pleistocene cave bears. *Science* **309**, 597-599 (2005).
21. Rompler, H. *et al.* Nuclear gene indicates coat-color polymorphism in mammoths. *Science* **313**, 62 (2006).
22. Poinar, H. N. *et al.* Metagenomics to paleogenomics: large-scale sequencing of mammoth DNA. *Science* **311**, 392-394 (2006).
23. Hofreiter, M., Serre, D., Poinar, H. N., Kuch, M. & Pääbo, S. Ancient DNA. *Nature Rev. Genet.* **2**, 353-359 (2001).
24. Malmstrom, H., Stora, J., Dalen, L., Holmlund, G. & Gotherstrom, A. Extensive human DNA contamination in extracts from ancient dog bones and teeth. *Mol. Biol. Evol.* **22**, 2040-2047 (2005).
25. Pääbo, S. Human evolution. *Trends Cell Biol.* **9**, M13-M16 (1999).
26. Poinar, H. N., Höss, M., Bada, J. L. & Pääbo, S. Amino acid racemization and the preservation of ancient DNA. *Science* **272**, 864-866 (1996).
27. Margulies, M. *et al.* Genome sequencing in microfabricated high-density picolitre reactors. *Nature* **437**, 376-380 (2005).
28. Stiller, M. *et al.* Patterns of nucleotide misincorporations during enzymatic amplification and direct large-scale sequencing of ancient DNA. *Proc. Natl Acad. Sci. USA* **103**, 13578-13584 (2006).
29. Pääbo, S. Ancient DNA: extraction, characterization, molecular cloning, and enzymatic amplification. *Proc. Natl Acad. Sci. USA* **86**, 1939-1943 (1989).
30. Höss, M., Dilling, A., Currant, A. & Pääbo, S. Molecular phylogeny of the extinct ground sloth *Mylodon darwinii*. *Proc. Natl Acad. Sci. USA* **93**, 181-185 (1996).
31. Hofreiter, M., Jaenicke, V., Serre, D., Haeseler Av, A. & Pääbo, S. DNA sequences from multiple amplifications reveal artifacts induced by cytosine deamination in ancient DNA. *Nucleic Acids Res.* **29**, 4793-4799 (2001).
32. International Human Genome Sequencing Consortium. Initial sequencing and analysis of the human genome. *Nature* **409**, 860-921 (2001).
33. Mouse Genome Sequencing Consortium. Initial sequencing and comparative analysis of the mouse genome. *Nature* **420**, 520-562 (2002).
34. Benson, D. A., Karsch-Mizrachi, I., Lipman, D. J., Ostell, J. & Wheeler, D. L. GenBank. *Nucleic Acids Res.* **34**, D16-D20 (2006).
35. Altschul, S. F. *et al.* Gapped BLAST and PSI-BLAST: a new generation of protein database search programs. *Nucleic Acids Res.* **25**, 3389-3402 (1997).
36. Beja, O. *et al.* Construction and analysis of bacterial artificial chromosome libraries from a marine microbial assemblage. *Environ. Microbiol.* **2**, 516-529 (2000).

37. Venter, J. C. *et al.* Environmental genome shotgun sequencing of the Sargasso Sea. *Science* **304**, 66-74 (2004).

38. Ingman, M. & Gyllensten, U. mtDB: Human Mitochondrial Genome Database, a resource for population genetics and medical sciences. *Nucleic Acids Res.* **34**, D749-D751 (2006).

39. Krause, J. *et al.* Multiplex amplification of the mammoth mitochondrial genome and the evolution of Elephantidae. *Nature* **439**, 724-727 (2006).

40. Kumar, S., Filipski, A., Swarna, V., Walker, A. & Blair Hedges, S. Placing confidence limits on the molecular age of the human-chimpanzee divergence. *Proc. Natl Acad. Sci. USA* **102**, 18842-18847 (2005).

41. Patterson, N., Richter, D. J., Gnerre, S., Lander, E. S. & Reich, D. Genetic evidence for complex speciation of humans and chimpanzees. *Nature* **441**, 1103-1108 (2006).

42. Wall, J. D. Estimating ancestral population sizes and divergence times. *Genetics* **163**, 395-404 (2003).

43. Yang, Z. & Yoder, A. D. Estimation of the transition/transversion rate bias and species sampling. *J. Mol. Evol.* **48**, 274-283 (1999).

44. Innan, H. & Watanabe, H. The effect of gene flow on the coalescent time in the human-chimpanzee ancestral population. *Mol. Biol. Evol.* **23**, 1040-1047 (2006).

45. Kaessmann, H., Wiebe, V., Weiss, G. & Pääbo, S. Great ape DNA sequences reveal a reduced diversity and an expansion in humans. *Nature Genet.* **27**, 155-156 (2001).

46. Yu, N., Jensen-Seaman, M. I., Chemnick, L., Ryder, O. & Li, W.-H. Nucleotide diversity in gorillas. *Genetics* **166**, 1375-1383 (2004).

47. Fischer, A., Pollack, J., Thalmann, O., Nickel, B. & Pääbo, S. Demographic history and genetic differentiation in apes. *Curr. Biol.* **16**, 1133-1138 (2006).

48. Rannala, B. & Yang, Z. Bayes estimation of species divergence times and ancestral population sizes using DNA sequences from multiple loci. *Genetics* **164**, 1645-1656 (2003).

49. The International HapMap Consortium. A haplotype map of the human genome. *Nature* **437**, 1299-1320 (2005).

50. Hinds, D. A. *et al.* Whole genome patterns of common DNA variation in three human populations. *Science* **307**, 1072-1079 (2005).

Supplementary Information is linked to the online version of the paper at www.nature.com/nature.

Acknowledgements. We are indebted to G. Coop, W. Enard, I. Hellmann, A. Fischer, P. Johnson, S. Kudaravalli, M. Lachmann, T. Maricic, J. Pritchard, J. Noonan, D. Reich, E. Rubin, M. Slatkin, L. Vigilant and T. Weaver for discussions. We thank A. P. Derevianko, C. Lalueza-Fox, A. Rosas and B. Vandermeersch for fossil samples. We also thank the Croatian Academy of Sciences and Arts for support and the Innovation Fund of Max Planck Society for financial support. 454 Life Sciences thanks NHGRI for continued support for the development of this platform, as well as all of its employees who developed the sequencing system. R.E.G. is supported by an NSF postdoctoral fellowship in Biological Informatics.

Author Contributions. M.P. provided Neanderthal samples and palaeontological information; J.M.R. and S.P. conceived of and initiated the 454 Neanderthal sequencing approach; M.T.R. developed the library preparation method, and generated and processed the sequencing data; J.F.S. planned and coordinated library preparation and sequencing activities; L.D. processed and transferred data between 454 Life Sciences and the MPI; M.E. supervised, planned and coordinated research between MPI and 454 Life Sciences; J.K. and A.W.B. extracted ancient DNA and performed analyses in the "Identification of a Neanderthal fossil for DNA sequencing" section; J.K. and R.E.G. performed analyses in the "Neanderthal mtDNA sequences" section; R.E.G. performed the analyses in the sections "Direct large-scale DNA sequencing" to "Genomic divergence between Neanderthals and humans"; S.E.P. performed analyses in the sections "Ancestral population size" and "Neanderthal sequences and human polymorphisms"; S.P. conceived of the ideas presented in the section "Rationale and prospects for a Neanderthal genome sequence", and initiated, planned and coordinated the study; R.E.G., S.E.P., J.K. and S.P. wrote the paper.

Author Information. Neanderthal fossil extract sequences were deposited at EBI with accession numbers CAAN01000001-CAAN01369630. Reprints and permissions information is available at www.nature.com/reprints. The authors declare competing financial interests: details accompany the paper on www.nature.com/nature. Correspondence and requests for materials should be addressed to R.E.G. (green@eva.mpg.de).

Identification and Analysis of Functional Elements in 1% of the Human Genome by the ENCODE Pilot Project

The ENCODE Project Consortium*

Editor's Note

With the human genome sequence completed, the logical next step was to work out how cells interpret the genetic instructions. This paper reveals the first findings of the Encyclopedia of DNA Elements (ENCODE) project, an initiative aiming to identify the functional elements of the human genome. The pilot program, which scrutinizes just 1% of the genome, offers insights into the nature and evolution of DNA sequences. Around half the functional elements appear able to change sequence more freely than expected, challenging the idea that biologically relevant DNA resists change. The study also finds that most DNA is transcribed into RNA, undermining the common picture of our genome as a rather small number of discrete genes amidst a mass of inactive "junk DNA".

We report the generation and analysis of functional data from multiple, diverse experiments performed on a targeted 1% of the human genome as part of the pilot phase of the ENCODE Project. These data have been further integrated and augmented by a number of evolutionary and computational analyses. Together, our results advance the collective knowledge about human genome function in several major areas. First, our studies provide convincing evidence that the genome is pervasively transcribed, such that the majority of its bases can be found in primary transcripts, including non-protein-coding transcripts, and those that extensively overlap one another. Second, systematic examination of transcriptional regulation has yielded new understanding about transcription start sites, including their relationship to specific regulatory sequences and features of chromatin accessibility and histone modification. Third, a more sophisticated view of chromatin structure has emerged, including its inter-relationship with DNA replication and transcriptional regulation. Finally, integration of these new sources of information, in particular with respect to mammalian evolution based on inter- and intra-species sequence comparisons, has yielded new mechanistic and evolutionary insights concerning the functional landscape of the human genome. Together, these studies are defining a path for pursuit of a more comprehensive characterization of human genome function.

THE human genome is an elegant but cryptic store of information. The roughly three billion bases encode, either directly or indirectly, the instructions for synthesizing

---

*The full list of authors and affiliations has been removed. The original text is available in the *Nature* online archive.

通过 ENCODE 先导计划鉴定与分析人类基因组 1% 区域内的功能元件

ENCODE 计划协作组 *

编者按

人类基因组测序完成后的下一步必然是阐释细胞如何执行这个遗传指南。DNA 元件百科全书 (ENCODE) 计划是一个旨在鉴定人类基因组功能元件的倡议，本文介绍了该计划的早期发现。先导项目仅仔细研究了 1% 的基因组，使人们对 DNA 序列的本质和进化有了初步了解。大约半数的功能元件看起来比预期能够更自由地改变序列，这对生物学相关的 DNA 抵抗改变的观点提出了挑战。大多数的 DNA 被转录成 RNA，这个发现颠覆了我们把基因组看作是相当少量且分散的基因位于大量不活跃的“垃圾 DNA”之间的通俗认识。

作为 ENCODE 计划先导阶段的一部分，针对人类基因组 1% 的目标区域进行了多种不同的实验，得到了功能数据，我们对这些功能数据的产生和分析进行了报道。这些数据已经通过大量的进化和计算分析被进一步整合和扩充。总的来说，我们的结果在几个主要方面增加了对人类基因组功能的认识。首先，我们的研究提供了令人信服的证据，证明了基因组是广泛转录的，以至于它的大部分碱基都可以在原始转录本中找到，这包括非蛋白质编码的转录本和广泛的互相重合的转录本。第二，转录调控的系统研究对转录起始位点产生了新的认识，这包括其与特定调控序列的关系、染色可接近性和组蛋白修饰的特征。第三，一个关于染色质结构的更加复杂精致的图景已经浮现，这包括它与 DNA 复制和转录调控的相互关系。最后，这些新的信息的整合，特别是基于物种间和物种内序列比较对哺乳动物进化的研究，已经对人类基因组的功能图景产生了机制和进化方面新的认识。总而言之，这些研究为进一步阐明人类基因组功能指明了道路。

人类基因组以简洁而隐晦的方式储存信息。大约 30 亿个碱基通过直接或间接的方式编码合成分子的指令，指导合成几乎所有构成人类细胞、组织和器官的分子。

---

* 作者和其他附加信息已经移除，原文可以从《自然》在线数据库中获得。

nearly all the molecules that form each human cell, tissue and organ. Sequencing the human genome[1-3] provided highly accurate DNA sequences for each of the 24 chromosomes. However, at present, we have an incomplete understanding of the protein-coding portions of the genome, and markedly less understanding of both non-protein-coding transcripts and genomic elements that temporally and spatially regulate gene expression. To understand the human genome, and by extension the biological processes it orchestrates and the ways in which its defects can give rise to disease, we need a more transparent view of the information it encodes.

The molecular mechanisms by which genomic information directs the synthesis of different biomolecules has been the focus of much of molecular biology research over the last three decades. Previous studies have typically concentrated on individual genes, with the resulting general principles then providing insights into transcription, chromatin remodelling, messenger RNA splicing, DNA replication and numerous other genomic processes. Although many such principles seem valid as additional genes are investigated, they generally have not provided genome-wide insights about biological function.

The first genome-wide analyses that shed light on human genome function made use of observing the actions of evolution. The ever-growing set of vertebrate genome sequences[4-8] is providing increasing power to reveal the genomic regions that have been most and least acted on by the forces of evolution. However, although these studies convincingly indicate the presence of numerous genomic regions under strong evolutionary constraint, they have less power in identifying the precise bases that are constrained and provide little, if any, insight into why those bases are biologically important. Furthermore, although we have good models for how protein-coding regions evolve, our present understanding about the evolution of other functional genomic regions is poorly developed. Experimental studies that augment what we learn from evolutionary analyses are key for solidifying our insights regarding genome function.

The Encyclopedia of DNA Elements (ENCODE) Project[9] aims to provide a more biologically informative representation of the human genome by using high-throughput methods to identify and catalogue the functional elements encoded. In its pilot phase, 35 groups provided more than 200 experimental and computational data sets that examined in unprecedented detail a targeted 29,998 kilobases (kb) of the human genome. These roughly 30 Mb—equivalent to ~1% of the human genome—are sufficiently large and diverse to allow for rigorous pilot testing of multiple experimental and computational methods. These 30 Mb are divided among 44 genomic regions; approximately 15 Mb reside in 14 regions for which there is already substantial biological knowledge, whereas the other 15 Mb reside in 30 regions chosen by a stratified random-sampling method (see http://www.genome.gov/10506161). The highlights of our findings to date include:

- The human genome is pervasively transcribed, such that the majority of its bases are associated with at least one primary transcript and many transcripts link distal regions to established protein-coding loci.

对人类基因组的测序[1-3]为 24 条染色体提供了高精度的 DNA 序列。但是目前，我们还不能完整理解基因组的蛋白质编码部分，而对于非蛋白质编码的转录本和在时空水平上调控基因表达的基因组元件，更是知之甚少。为了了解人类基因组，以及它调控的生物学过程和它的缺陷导致疾病的机制，我们需要对它所编码的信息有一个更清楚的认知。

在过去的 30 多年里，基因组信息指导合成不同生物分子的分子机制一直是很多分子生物学研究的焦点。之前的研究通常聚焦于单个基因，获得一般规律，进而促进对转录、染色质重塑、信使 RNA 剪接、DNA 复制和众多其他基因组学过程的认识。尽管许多这样的规律也适用于其他研究的基因，但是它们通常还未提供在全基因组范围内认识生物学功能的信息。

第一个在基因组层面阐明人类基因组功能的研究是关于进化的行为。随着越来越多的脊椎动物基因组被测序[4-8]，人们越来越能够发现被进化的力量影响最多和最少的基因组区域。不过，尽管这些研究有力地表明，存在大量的基因组在进化上高度保守，但是这些研究在鉴定精确的保守碱基上还不够有力，对于为什么那些碱基在生物学上是重要的也几乎不能提供有用的信息。并且，尽管我们对于蛋白质编码区域如何进化有很好的模型，但是我们目前对其他有功能的基因组区域的了解还很欠缺。为了巩固我们对基因组功能的理解，通过实验研究扩充我们进化分析所得是关键。

DNA 元件百科全书（ENCODE）计划[9]旨在利用高通量方法对人类基因组所编码的功能元件进行鉴定和分类，进而提供人类基因组更加具有生物学信息的表征。在该计划的先导阶段，35 个课题组贡献了超过 200 个实验和计算数据集，以前所未有的详细度研究了人类基因组特定的 29,998 千碱基（kb）区域。这大约 30 Mb——相当于人类基因组的 ~ 1%——已经足够大和足够多样，可以严格地用来先期测试多种实验和计算方法。这 30 Mb 分布在 44 个基因组区域；大约 15 Mb 位于 14 个已经有大量的生物学知识的区域，而另外 15 Mb 则位于 30 个通过分层随机抽样方法选择的区域（见 http://www.genome.gov/10506161）。目前我们研究发现的亮点包括：

- 人类基因组是广泛转录的，以至于它的大部分碱基至少和一个原始转录本相关联，并且许多转录本将远端区域与已知的蛋白质编码基因座相关联。

- Many novel non-protein-coding transcripts have been identified, with many of these overlapping protein-coding loci and others located in regions of the genome previously thought to be transcriptionally silent.

- Numerous previously unrecognized transcription start sites have been identified, many of which show chromatin structure and sequence-specific protein-binding properties similar to well-understood promoters.

- Regulatory sequences that surround transcription start sites are symmetrically distributed, with no bias towards upstream regions.

- Chromatin accessibility and histone modification patterns are highly predictive of both the presence and activity of transcription start sites.

- Distal DNaseI hypersensitive sites have characteristic histone modification patterns that reliably distinguish them from promoters; some of these distal sites show marks consistent with insulator function.

- DNA replication timing is correlated with chromatin structure.

- A total of 5% of the bases in the genome can be confidently identified as being under evolutionary constraint in mammals; for approximately 60% of these constrained bases, there is evidence of function on the basis of the results of the experimental assays performed to date.

- Although there is general overlap between genomic regions identified as functional by experimental assays and those under evolutionary constraint, not all bases within these experimentally defined regions show evidence of constraint.

- Different functional elements vary greatly in their sequence variability across the human population and in their likelihood of residing within a structurally variable region of the genome.

- Surprisingly, many functional elements are seemingly unconstrained across mammalian evolution. This suggests the possibility of a large pool of neutral elements that are biochemically active but provide no specific benefit to the organism. This pool may serve as a "warehouse" for natural selection, potentially acting as the source of lineage-specific elements and functionally conserved but non-orthologous elements between species.

Below, we first provide an overview of the experimental techniques used for our studies, after which we describe the insights gained from analysing and integrating the generated data sets. We conclude with a perspective of what we have learned to date about this 1% of the human genome and what we believe the prospects are for a broader and deeper

- 许多新的非蛋白质编码转录本被鉴定，其中很多与蛋白质编码基因座重合，其他的则位于以前被认为是不转录的基因组区域。

- 大量的以前未被识别的转录起始位点被鉴定，其中很多具有与已知的启动子类似的染色质结构和序列特异的蛋白质结合特性。

- 转录起始位点附近的调控序列是对称分布的，并没有偏向上游区域。

- 染色质可接近性和组蛋白修饰特征不仅可以准确预测转录起始位点的存在，而且可以预测转录起始位点的活性。

- 远端的 DNA 酶 I 超敏感位点含有与启动子截然不同的组蛋白修饰特征；其中一些远端位点含有的标志物与绝缘子功能一致。

- DNA 复制时相与染色质的结构相关。

- 基因组总计 5% 的碱基可以明确地鉴定为在哺乳动物中进化保守；基于目前已有的实验结果，大约 60% 的保守碱基，有证据表明是有功能的。

- 尽管实验鉴定到的功能性基因组区域与进化保守的区域有广泛的重叠，但是这些实验确定的区域内，并非所有的碱基都有进化保守的证据。

- 不同的功能元件在人群中序列变异性方面以及位于基因组结构可变区的可能性方面差别很大。

- 出人意料的是，很多功能元件看起来在哺乳动物进化中并不保守。这暗示可能存在一个巨大的中性元件库，这些中性元件在生化上活跃，但是对机体并没有特别的益处。这个中性元件库可以作为一个用于自然选择的“仓库”，作为谱系特异元件以及种间功能保守但非同源元件的潜在来源。

接下来，我们首先概述我们研究中所用到的实验技术，然后描述由分析和整合这些实验产生的数据集所得到的见解。最后我们总结了目前 1% 的人类基因组中已有的发现，并对更广更深地研究人类基因组中的功能元件进行了展望。为了帮助读

investigation of the functional elements in the human genome. To aid the reader, Box 1 provides a glossary for many of the abbreviations used throughout this paper.

Box 1. Frequently used abbreviations in this paper

AR Ancient repeat: a repeat that was inserted into the early mammalian lineage and has since become dormant; the majority of ancient repeats are thought to be neutrally evolving

CAGE tag A short sequence from the 5′ end of a transcript

CDS Coding sequence: a region of a cDNA or genome that encodes proteins

ChIP-chip Chromatin immunoprecipitation followed by detection of the products using a genomic tiling array

CNV Copy number variants: regions of the genome that have large duplications in some individuals in the human population

CS Constrained sequence: a genomic region associated with evidence of negative selection (that is, rejection of mutations relative to neutral regions)

DHS DNaseI hypersensitive site: a region of the genome showing a sharply different sensitivity to DNaseI compared with its immediate locale

EST Expressed sequence tag: a short sequence of a cDNA indicative of expression at this point

FAIRE Formaldehyde-assisted isolation of regulatory elements: a method to assay open chromatin using formaldehyde crosslinking followed by detection of the products using a genomic tiling array

FDR False discovery rate: a statistical method for setting thresholds on statistical tests to correct for multiple testing

GENCODE Integrated annotation of existing cDNA and protein resources to define transcripts with both manual review and experimental testing procedures

GSC Genome structure correction: a method to adapt statistical tests to make fewer assumptions about the distribution of features on the genome sequence. This provides a conservative correction to standard tests

HMM Hidden Markov model: a machine-learning technique that can establish optimal parameters for a given model to explain the observed data

Indel An insertion or deletion; two sequences often show a length difference within alignments, but it is not always clear whether this reflects a previous insertion or a deletion

者阅读，框 1 提供了一个词汇表，里面有本论文所用的很多缩写词。

框 1. 本文常用的缩写词

AR 古老的重复：一种在早期就插入到哺乳动物谱系但自此休眠的重复；大部分古老的重复被认为正在进行中性进化

CAGE tag 转录本 5′端的短序列

CDS 编码序列：cDNA 区域或编码蛋白质的基因组区域

ChIP-chip 染色质免疫沉淀后利用基因组叠瓦式阵列检测产物

CNV 拷贝数变异：在人群中一些个体中发生大片段重复的基因组区域

CS 保守序列：有证据表明跟负选择相关的基因组区域（也就是，跟中性区域相比更排斥突变）

DHS DNA 酶 I 超敏感位点：跟附近区域相比，对 DNA 酶 I 非常敏感的基因组区域

EST 表达序列标签：表示在该点表达的 cDNA 短序列

FAIRE 基于甲醛的调控元件分离：一种检测开放染色质的方法，甲醛交联后利用基因组叠瓦式阵列检测产物

FDR 错误发现率：一种为统计学检验设定阈值以校正多重检验的统计学方法

GENCODE 通过人工审查和实验检测的方法，整合注释已知的 cDNA 和蛋白质资源，以定义转录本

GSC 基因组结构校正：一种适合统计学检验的方法，使其对基因组序列上的分布特征做出更少假设。这为标准检验提供了一种保守的校正

HMM 隐马尔可夫模型：一种机器学习技术，能为一个给定的模型建立最优的变量以解释观察到的数据

Indel 插入或缺失；在比对两个序列时经常出现长度不一的情况，但是有时并不清楚这反映的是一个插入还是缺失

PET A short sequence that contains both the 5′ and 3′ ends of a transcript

RACE Rapid amplification of cDNA ends: a technique for amplifying cDNA sequences between a known internal position in a transcript and its 5′ end

RFBR Regulatory factor binding region: a genomic region found by a ChIP-chip assay to be bound by a protein factor

RFBR-Seqsp Regulatory factor binding regions that are from sequence-specific binding factors

RT–PCR Reverse transcriptase polymerase chain reaction: a technique for amplifying a specific region of a transcript

RxFrag Fragment of a RACE reaction: a genomic region found to be present in a RACE product by an unbiased tiling-array assay

SNP Single nucleotide polymorphism: a single base pair change between two individuals in the human population

STAGE Sequence tag analysis of genomic enrichment: a method similar to ChIP-chip for detecting protein factor binding regions but using extensive short sequence determination rather than genomic tiling arrays

SVM Support vector machine: a machine-learning technique that can establish an optimal classifier on the basis of labelled training data

TR50 A measure of replication timing corresponding to the time in the cell cycle when 50% of the cells have replicated their DNA at a specific genomic position

TSS Transcription start site

TxFrag Fragment of a transcript: a genomic region found to be present in a transcript by an unbiased tiling-array assay

Un.TxFrag A TxFrag that is not associated with any other functional annotation

UTR Untranslated region: part of a cDNA either at the 5′ or 3′ end that does not encode a protein sequence

Experimental Techniques

Table 1 (expanded in Supplementary Information section 1.1) lists the major experimental techniques used for the studies reported here, relevant acronyms, and references reporting the generated data sets. These data sets reflect over 400 million experimental data points (603 million data points if one includes comparative sequencing bases). In describing the major

PET 同时包含一个转录本 5′和 3′端的短序列

RACE cDNA 末端快速扩增：一种扩增技术，扩增一个转录本中间的已知位置和它的 5′端之间的 cDNA 序列

RFBR 调控因子结合区域：ChIP-chip 实验发现的与蛋白质因子结合的基因组区域

RFBR-Seqsp 被序列特异的结合因子所结合的调控因子结合区域

RT–PCR 反转录酶聚合酶链式反应：扩增转录本特定区域的技术

RxFrag RACE 反应的片段：通过无偏性叠瓦式阵列实验在 RACE 产物中发现的基因组区域

SNP 单核苷酸多态性：人群中两个个体间单个碱基对的改变

STAGE 基因组丰度的序列标签分析：类似于 ChIP-chip 用于检测蛋白质因子结合区域的方法，不过是利用大量短序列确定，而不是基因组叠瓦式阵列实验

SVM 支持向量机：一种机器学习技术，能基于标记的训练数据建立最优的分类器

TR50 一种衡量复制时相的方法，对应于细胞周期中 50% 的细胞在特定的基因组位置复制 DNA 所用的时间

TSS 转录起始位点

TxFrag 转录本片段：通过无偏性叠瓦式阵列实验在转录本中发现基因组片段

Un.TxFrag 与任何功能注释均不关联的 TxFrag

UTR 非翻译区：位于 5′或 3′端的不编码蛋白质序列的部分 cDNA

实验技术

表 1(在补充信息 1.1 中有扩充)列出了本文报道的研究所用的主要实验技术、相关的首字母缩略词以及报道这些实验技术所涉及的参考文献。这些数据集反映了超过 4 亿个实验数据点(如果包含比较测序的碱基则是 6.03 亿个数据点)。在描述主

results and initial conclusions, we seek to distinguish "biochemical function" from "biological role". Biochemical function reflects the direct behaviour of a molecule(s), whereas biological role is used to describe the consequence(s) of this function for the organism. Genome-analysis techniques nearly always focus on biochemical function but not necessarily on biological role. This is because the former is more amenable to large-scale data-generation methods, whereas the latter is more difficult to assay on a large scale.

Table 1. Summary of types of experimental techniques used in ENCODE

Feature class	Experimental technique(s)	Abbreviations	References	Number of experimental data points
Transcription	Tiling array, integrated annotation	TxFrag, RxFrag, GENCODE	117 118 19 119	63,348,656
5′ ends of transcripts*	Tag sequencing	PET, CAGE	121 13	864,964
Histone modifications	Tiling array	Histone nomenclature†, RFBR	46	4,401,291
Chromatin‡ structure	QT-PCR, tiling array	DHS, FAIRE	42 43 44 122	15,318,324
Sequence-specific factors	Tiling array, tag sequencing, promoter assays	STAGE, ChIP-Chip, ChIP-PET, RFBR	41, 52 11, 120 123 81 34, 51 124 49 33 40	324,846,018
Replication	Tiling array	TR50	59 75	14,735,740
Computational analysis	Computational methods	CCI, RFBR cluster	80 125 10 16 126 127	NA
Comparative sequence analysis*	Genomic sequencing, multi-sequence alignments, computational analysis	CS	87 86 26	NA
Polymorphisms*	Resequencing, copy number variation	CNV	103 128	NA

* Not all data generated by the ENCODE Project.

† Histone code nomenclature follows the Brno nomenclature as described in ref. 129.

‡ Also contains histone modification.

要的结果和初始的结论上，我们试着区分“生化功能”和“生物学功能”。生化功能反映的是一个或多个分子的直接行为，而生物学功能则用来描述这个功能对机体的影响。全基因分析技术几乎都是聚焦于生化功能，并不一定关心生物学功能。这是因为前者更容易被大规模数据产生方法所检验，而后者在大通量下却更难分析。

表 1. ENCODE 中使用的实验技术类型总结

特征分类	实验技术	缩写	参考文献	实验数据点的数量
转录	叠瓦式阵列、整合注释	TxFrag、RxFrag、GENCODE	117 118 19 119	63,348,656
转录本 5′端 *	标签测序	PET、CAGE	121 13	864,964
组蛋白修饰	叠瓦式阵列	组蛋白系统命名法 †、RFBR	46	4,401,291
染色质结构 ‡	QT-PCR、叠瓦式阵列	DHS、FAIRE	42 43 44 122	15,318,324
序列特异因子	叠瓦式阵列、标签测序、启动子实验	STAGE、ChIP-chip、ChIP-PET、RFBR	41、52 11、120 123 81 34、51 124 49 33 40	324,846,018
复制	叠瓦式阵列	TR50	59 75	14,735,740
计算分析	计算方法	CCI、RFBR 簇	80 125 10 16 126 127	NA
序列比较分析 *	基因组测序、多重序列比对、计算分析	CS	87 86 26	NA
多态性 *	重测序、拷贝数目变异	CNV	103 128	NA

* 并非所有的数据都由 ENCODE 计划产生。

† 组蛋白密码系统命名法遵循参考文献 129 中描述的布尔诺系统命名法。

‡ 也包括组蛋白修饰。

The ENCODE pilot project aimed to establish redundancy with respect to the findings represented by different data sets. In some instances, this involved the intentional use of different assays that were based on a similar technique, whereas in other situations, different techniques assayed the same biochemical function. Such redundancy has allowed methods to be compared and consensus data sets to be generated, much of which is discussed in companion papers, such as the ChIP-chip platform comparison[10,11]. All ENCODE data have been released after verification but before this publication, as befits a "community resource" project (see http://www.wellcome.ac.uk/doc_wtd003208.html). Verification is defined as when the experiment is reproducibly confirmed (see Supplementary Information section 1.2). The main portal for ENCODE data is provided by the UCSC Genome Browser (http://genome.ucsc.edu/ENCODE/); this is augmented by multiple other websites (see Supplementary Information section 1.1).

A common feature of genomic analyses is the need to assess the significance of the co-occurrence of features or of other statistical tests. One confounding factor is the heterogeneity of the genome, which can produce uninteresting correlations of variables distributed across the genome. We have developed and used a statistical framework that mitigates many of these hidden correlations by adjusting the appropriate null distribution of the test statistics. We term this correction procedure genome structure correction (GSC) (see Supplementary Information section 1.3).

In the next five sections, we detail the various biological insights of the pilot phase of the ENCODE Project.

Transcription

Overview. RNA transcripts are involved in many cellular functions, either directly as biologically active molecules or indirectly by encoding other active molecules. In the conventional view of genome organization, sets of RNA transcripts (for example, messenger RNAs) are encoded by distinct loci, with each usually dedicated to a single biological role (for example, encoding a specific protein). However, this picture has substantially grown in complexity in recent years[12]. Other forms of RNA molecules (such as small nucleolar RNAs and micro (mi)RNAs) are known to exist, and often these are encoded by regions that intercalate with protein-coding genes. These observations are consistent with the well-known discrepancy between the levels of observable mRNAs and large structural RNAs compared with the total RNA in a cell, suggesting that there are numerous RNA species yet to be classified[13-15]. In addition, studies of specific loci have indicated the presence of RNA transcripts that have a role in chromatin maintenance and other regulatory control. We sought to assay and analyse transcription comprehensively across the 44 ENCODE regions in an effort to understand the repertoire of encoded RNA molecules.

Transcript maps. We used three methods to identify transcripts emanating from the ENCODE regions: hybridization of RNA (either total or polyA-selected) to unbiased tiling

ENCODE 先导计划旨在建立不同数据集发现结果的冗余。在一些情形下，这包括专门使用基于相似技术的不同实验结果，而在其他情形下，则是不同的技术测定同一生化功能。这种冗余使得不同方法可以进行比较，产生一致的数据集，这部分在关于比较的文章中有很多讨论，比如 ChIP-chip 平台的比较 [10,11]。作为一个“共享资源”计划，所有经过验证的 ENCODE 数据在本文发表前就已经发布了（详见 http://www.wellcome.ac.uk/doc_wtd003208.html）。验证的定义是实验被重复确认（详见补充信息 1.2）。UCSC 基因组浏览器提供获取 ENCODE 数据的主要入口（http://genome.ucsc.edu/ENCODE/）；除此之外还有多个其他的网站（详见补充信息 1.1）。

基因组分析的一个常见的特征是需要评估特征共发生的显著性，或者其他统计学检验的显著性。一个混淆因素是基因组的异质性，这会使得分布于基因组上的变量产生无意义的相关。我们已经开发和使用了一种统计学框架，它通过检验统计调整合适的零分布，以减少很多这样的隐含相关。我们将这种校正方法称为基因组结构校正（GSC）（详见补充信息 1.3）。

在接下来的五部分中，我们详细介绍 ENCODE 先导计划的各种生物学发现。

转　录

概况　RNA 转录本通过直接作为生物学活性分子，或者间接编码其他活性分子，参与很多细胞功能。关于基因组结构，传统上认为，不同的 RNA 转录本（如信使 RNA）由不同的基因座编码，每一种 RNA 转录本通常具有单独一种生物学功能（如编码一个特定的蛋白质）。但是，近年来这种描述逐渐变得复杂起来 [12]，我们已经发现其他形式的 RNA 分子（如小核仁 RNA 和微 RNA（miRNA））的存在，它们大多间插在编码蛋白质的基因区域。众所周知，目前观察到的 mRNA 和大的结构性 RNA 水平与细胞总 RNA 水平不符，这与上述发现是一致的，说明还有大量的 RNA 类型未被归类 [13-15]。另外，特定基因座的研究已经表明，RNA 转录本的存在在染色质维持和其他调节控制中有重要作用。为了理解作为一种分子库的编码 RNA，我们全面地检测和分析了 44 个 ENCODE 区域的转录。

转录本图谱　我们使用了三种方法来鉴定 ENCODE 区域产生的转录本：RNA 杂交（总的或多聚 A 筛选的）到无偏性叠瓦式阵列（详见补充信息 2.1），在 5′端或

arrays (see Supplementary Information section 2.1), tag sequencing of cap-selected RNA at the 5′ or joint 5′/3′ ends (see Supplementary Information sections 2.2 and S2.3), and integrated annotation of available complementary DNA and EST sequences involving computational, manual, and experimental approaches[16] (see Supplementary Information section 2.4). We abbreviate the regions identified by unbiased tiling arrays as TxFrags, the cap-selected RNAs as CAGE or PET tags (see Box 1), and the integrated annotation as GENCODE transcripts. When a TxFrag does not overlap a GENCODE annotation, we call it an Un.TxFrag. Validation of these various studies is described in papers reporting these data sets[17] (see Supplementary Information sections 2.1.4 and 2.1.5).

These methods recapitulate previous findings, but provide enhanced resolution owing to the larger number of tissues sampled and the integration of results across the three approaches (see Table 2). To begin with, our studies show that 14.7% of the bases represented in the unbiased tiling arrays are transcribed in at least one tissue sample. Consistent with previous work[14,15], many (63%) TxFrags reside outside of GENCODE annotations, both in intronic (40.9%) and intergenic (22.6%) regions. GENCODE annotations are richer than the more-conservative RefSeq or Ensembl annotations, with 2,608 transcripts clustered into 487 loci, leading to an average of 5.4 transcripts per locus. Finally, extensive testing of predicted protein-coding sequences outside of GENCODE annotations was positive in only 2% of cases[16], suggesting that GENCODE annotations cover nearly all protein-coding sequences. The GENCODE annotations are categorized both by likely function (mainly, the presence of an open reading frame) and by classification evidence (for example, transcripts based solely on ESTs are distinguished from other scenarios); this classification is not strongly correlated with expression levels (see Supplementary Information sections 2.4.2 and 2.4.3).

Table 2. Bases detected in processed transcripts either as a GENCODE exon, a TxFrag, or as either a GENCODE exon or a TxFrag

	GENCODE exon	TxFrag	Either GENCODE exon or TxFrag
Total detectable transcripts (bases)	1,776,157 (5.9%)	1,369,611 (4.6%)	2,519,280 (8.4%)
Transcripts detected in tiled regions of arrays (bases)	1,447,192 (9.8%)	1,369,611 (9.3%)	2,163,303 (14.7%)

Percentages are of total bases in ENCODE in the first row and bases tiled in arrays in the second row.

Analyses of more biological samples have allowed a richer description of the transcription specificity (see Fig. 1 and Supplementary Information section 2.5). We found that 40% of TxFrags are present in only one sample, whereas only 2% are present in all samples. Although exon-containing TxFrags are more likely (74%) to be expressed in more than one sample, 45% of unannotated TxFrags are also expressed in multiple samples. GENCODE annotations of separate loci often (42%) overlap with respect to their genomic coordinates, in particular on opposite strands (33% of loci). Further analysis of GENCODE-annotated sequences with respect to the positions of open reading frames revealed that some component exons do not have the expected synonymous versus non-synonymous substitution patterns of protein-coding sequence (see Supplement Information section 2.6)

同时在 5′/3′ 端采用标签测序经帽子筛选的 RNA（详见补充信息 2.2 和 S2.3），以及通过计算、手工和实验方法整合注释已知互补的 DNA 和 EST 序列 [16]（详见补充信息 2.4）。我们把经无偏性叠瓦式阵列鉴定到的区域简称为 TxFrag，将经帽子筛选的 RNA 称为 CAGE 或 PET 标签（详见框 1），将整合的注释称为 GENCODE 转录本。对于不与 GENCODE 注释重合的 TxFrag，我们称之为 Un.TxFrag。这些不同研究的验证在报道这些数据集的文章中有描述 [17]（详见补充信息 2.1.4 和 2.1.5）。

这些方法虽然再现了之前的发现，但这次不仅有更多组织取样，还整合了三种不同的方法，从而提供了更高的分辨率（见表 2）。首先，我们的研究发现，无偏性叠瓦式阵列上的碱基有 14.7% 在至少一种组织样本中转录。与之前的研究一致 [14,15]，很多（63%）TxFrag 位于 GENCODE 注释之外，内含子区有（40.9%），基因间区有（22.6%）。与相对保守的 RefSeq 或 Ensembl 注释相比，GENCODE 注释更丰富，2,608 个转录本聚集成 487 个基因座，平均每个基因座 5.4 个转录本。最后，对位于 GENCODE 注释之外的预测蛋白编码序列进行了广泛的检测，发现只有 2% 是阳性的 [16]，这说明 GENCODE 注释涵盖了几乎所有的蛋白质编码序列。既可通过可能的功能（主要是开放阅读框的存在）也可通过类别证据（例如，仅基于 EST 的转录本跟其他类型是不同的）对 GENCODE 注释进行分类；这种分类跟表达水平并没有很强的相关性（详见补充信息 2.4.2 和 2.4.3）。

表 2. 在处理过的转录本中检测到的是 GENCODE 外显子、TxFrag 或两者其一的碱基

	GENCODE 外显子	TxFrag	GENCODE 外显子或 TxFrag
总的检测到的转录本（碱基）	1,776,157 (5.9%)	1,369,611 (4.6%)	2,519,280 (8.4%)
在阵列的叠瓦区域检测到的转录本（碱基）	1,447,192 (9.8%)	1,369,611 (9.3%)	2,163,303 (14.7%)

比例是第一行中 ENCODE 中总的碱基和第二行阵列中叠瓦的碱基。

对更多生物学样本的分析更加丰富了对转录本特异性的描述（详见图 1 和补充信息 2.5）。我们发现 40% 的 TxFrag 仅在一种样本中存在，只有 2% 的 TxFrag 在所有样本中都存在。尽管含有外显子的 TxFrag 更倾向于（74%）在超过一种样本中表达，45% 未被注释的 TxFrag 也在多种样本中表达。不同基因座 42% 的 GENCODE 注释大多与它们的基因组位置重合，特别是与另一条链（33% 的基因座）。根据开放阅读框的位置进一步分析 GENCODE 注释的序列，发现一些组成的外显子没有预期的蛋白质编码序列中同义和非同义替换情况（详见补充信息 2.6），一些外显子含

and some have deletions incompatible with protein structure[18]. Such exons are on average less expressed (25% versus 87% by RT–PCR; see Supplementary Information section 2.7) than exons involved in more than one transcript (see Supplementary Information section 2.4.3), but when expressed have a tissue distribution comparable to well-established genes.

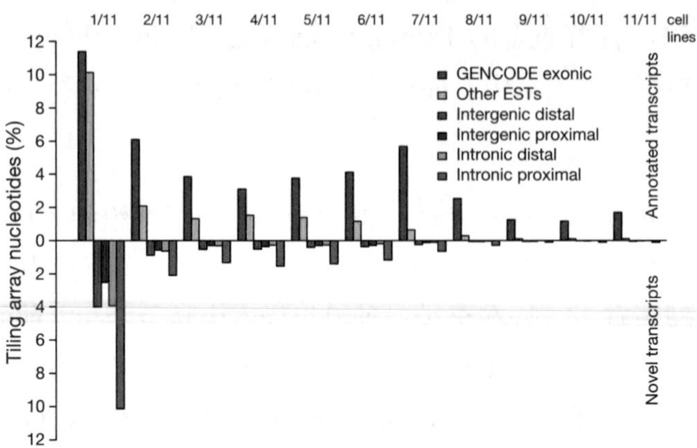

Fig. 1. Annotated and unannotated TxFrags detected in different cell lines. The proportion of different types of transcripts detected in the indicated number of cell lines (from 1/11 at the far left to 11/11 at the far right) is shown. The data for annotated and unannotated TxFrags are indicated separately, and also split into different categories based on GENCODE classification: exonic, intergenic (proximal being within 5 kb of a gene and distal being otherwise), intronic (proximal being within 5 kb of an intron and distal being otherwise), and matching other ESTs not used in the GENCODE annotation (principally because they were unspliced). The *y* axis indicates the per cent of tiling array nucleotides present in that class for that number of samples (combination of cell lines and tissues).

Critical questions are raised by the presence of a large amount of unannotated transcription with respect to how the corresponding sequences are organized in the genome—do these reflect longer transcripts that include known loci, do they link known loci, or are they completely separate from known loci? We further investigated these issues using both computational and new experimental techniques.

Unannotated transcription. Consistent with previous findings, the Un.TxFrags did not show evidence of encoding proteins (see Supplementary Information section 2.8). One might expect Un.TxFrags to be linked within transcripts that exhibit coordinated expression and have similar conservation profiles across species. To test this, we clustered Un.TxFrags using two methods. The first method[19] used expression levels in 11 cell lines or conditions, dinucleotide composition, location relative to annotated genes, and evolutionary conservation profiles to cluster TxFrags (both unannotated and annotated). By this method, 14% of Un.TxFrags could be assigned to annotated loci, and 21% could be clustered into 200 novel loci (with an average of ~7 TxFrags per locus). We experimentally examined these novel loci to study the connectivity of transcripts amongst Un.TxFrags and between Un.TxFrags and known exons. Overall, about 40% of the connections (18 out of 46) were validated by RT–PCR. The second clustering method involved analysing a time course (0, 2, 8 and 32 h) of expression changes in human HL60 cells following retinoic-acid

有与蛋白质结构不相容的缺失[18]。这种外显子(详见补充信息 2.7)比参与一种以上转录本的外显子(详见补充信息 2.4.3)平均表达水平更低(根据 RT–PCR 是 25% 比 87%)，不过，当它们表达时，其组织分布与已确定的基因相比差别不大。

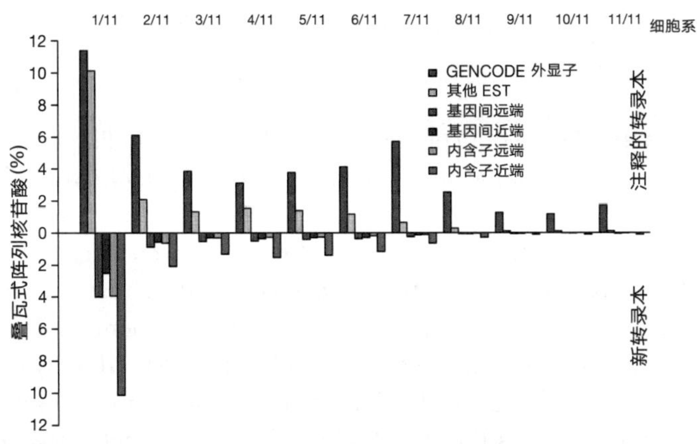

图 1. 不同细胞系中检测到的注释和未注释的 TxFrag。在指明数量的细胞系中检测到的不同类型的转录本的比例(从最左边的 1/11 到最右边的 11/11)。注释和未注释的 TxFrag 的数据分开展示，并且还基于 GENCODE 分类分成了不同的类别：外显子、基因间(距基因 5 kb 以内是近端，大于 5 kb 则是远端)、内含子(距内含子 5 kb 以内是近端，大于 5 kb 则是远端)和未在 GENCODE 注释中使用的匹配其他 EST(主要是因为它们没有剪接)。*y* 轴表示特定数量的样本(包括了细胞系和组织)特定分类下检测到的叠瓦式阵列核苷酸的百分比。

大量未注释转录本的存在带来了一个重要问题：它们相应的序列在基因组中是如何组织的——它们是否反映了包含已知基因座的更长转录本，它们是否与已知基因座有联系，或者它们是否完全独立于已知基因座之外？我们利用计算和新实验技术进一步探究了这些问题。

未被注释的转录 与之前的发现一致，没有证据表明 Un.TxFrag 编码蛋白质(详见补充信息 2.8)。我们期待 Un.TxFrag 在协调表达且在物种间有相似保守性的转录本中可能有关联。为检验这种猜测，我们利用两种方法将 Un.TxFrag 聚类。第一种方法[19]利用 11 种细胞系或条件下表达水平、二核苷酸构成、相对注释基因的位置和进化保守谱来聚类 TxFrag(未注释的和已注释的)。通过这种方法，14% 的 Un.TxFrag 能被分配到注释的基因座，21% 能被聚类成 200 个新的基因座(平均一个基因座大约 7 个 TxFrag)。我们利用实验检查了这些新的基因座，以研究 Un.TxFrag 内部以及 Un.TxFrag 跟已知的外显子间的转录本的关联性。总体上，大约 40% 的关联(46 个中有 18 个)被 RT–PCR 验证。第二种聚类方法是分析视黄酸刺激人 HL60 细胞后一段时间内(0、2、8 和 32 小时)的表达变化。这是一个描述注释基因座表达

stimulation. There is a coordinated program of expression changes from annotated loci, which can be shown by plotting Pearson correlation values of the expression levels of exons inside annotated loci versus unrelated exons (see Supplementary Information section 2.8.2). Similarly, there is coordinated expression of nearby Un.TxFrags, albeit lower, though still significantly different from randomized sets. Both clustering methods indicate that there is coordinated behaviour of many Un.TxFrags, consistent with them residing in connected transcripts.

Transcript connectivity. We used a combination of RACE and tiling arrays[20] to investigate the diversity of transcripts emanating from protein-coding loci. Analogous to TxFrags, we refer to transcripts detected using RACE followed by hybridization to tiling arrays as RxFrags. We performed RACE to examine 399 protein-coding loci (those loci found entirely in ENCODE regions) using RNA derived from 12 tissues, and were able to unambiguously detect 4,573 RxFrags for 359 loci (see Supplementary Information section 2.9). Almost half of these RxFrags (2,324) do not overlap a GENCODE exon, and most (90%) loci have at least one novel RxFrag, which often extends a considerable distance beyond the 5′ end of the locus. Figure 2 shows the distribution of distances between these new RACE-detected ends and the previously annotated TSS of each locus. The average distance of the extensions is between 50 kb and 100 kb, with many extensions (> 20%) being more than 200 kb. Consistent with the known presence of overlapping genes in the human genome, our findings reveal evidence for an overlapping gene at 224 loci, with transcripts from 180 of these loci (~50% of the RACE-positive loci) appearing to have incorporated at least one exon from an upstream gene.

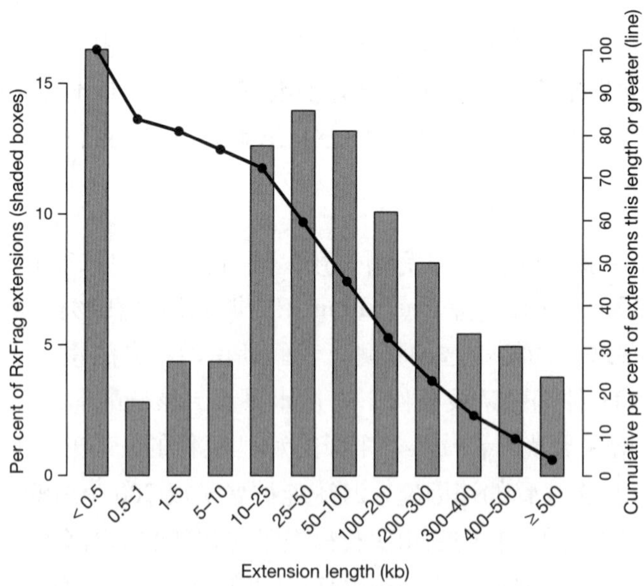

Fig. 2. Length of genomic extensions to GENCODE-annotated genes on the basis of RACE experiments followed by array hybridizations (RxFrags). The indicated bars reflect the frequency of extension lengths among different length classes. The solid line shows the cumulative frequency of extensions of that length or greater. Most of the extensions are greater than 50 kb from the annotated gene (see text for details).

变化的协调性项目，通过跟无关外显子相比较描绘注释基因座外显子表达水平的皮尔逊相关值(详见补充信息 2.8.2)。同样，靠近的 Un.TxFrag 协调表达，虽然水平更低一些，但仍然与随机组显著不同。这两种聚类方法说明很多 Un.TxFrag 存在协调关系，这与它们位于相关联转录本上一致。

转录本的关联性　我们联合使用 RACE 和叠瓦式阵列 [20] 来研究蛋白质编码基因座产生的转录本的多样性。类似于 TxFrag，我们称 RACE 后杂交到叠瓦式阵列所检测到的转录本为 RxFrag。利用 12 种组织的 RNA，我们通过 RACE 检测了 399 个蛋白质编码基因座(这些基因座全部位于 ENCODE 区域)，能够确定地在 359 个基因座检测到 4,573 个 RxFrag(详见补充信息 2.9)。这些 RxFrag 近乎半数(2,324 个)不与 GENCODE 外显子重合，并且大部分基因座(90%)至少有一个全新的 RxFrag，这些全新的 RxFrag 大多从基因座的 5′端延伸相当长的距离。图 2 展示了各个基因座中这些 RACE 检测到的全新的末端与已注释的 TSS 之间的距离的分布。延伸的平均距离在 50 kb 到 100 kb 之间，很多(> 20%)超过了 200 kb。与已知人类基因组中存在重合基因一致，我们发现 224 个基因座存在重合基因的证据，其中 180 个基因座(~ 50% 的 RACE 阳性基因座)的转录本看起来包含了上游基因的至少一个外显子。

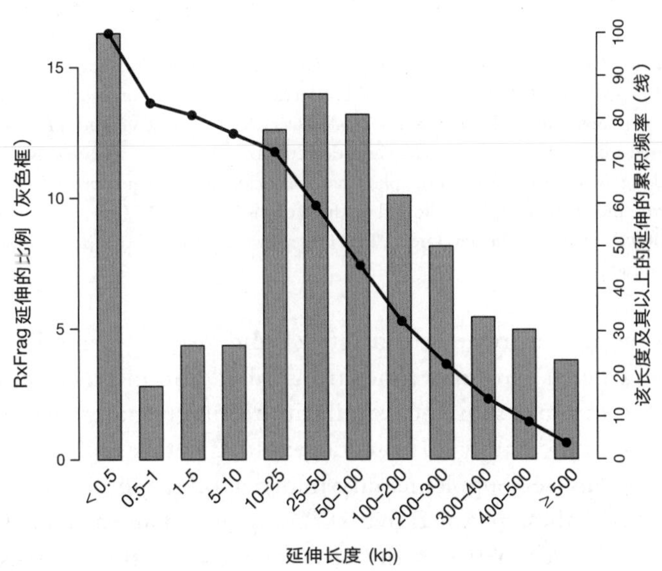

图 2. 基于 RACE 实验结合微阵列杂交(RxFrag)得到 GENCODE 注释基因的延伸长度。各个条表示在不同长度类别下延伸长度的频率。实线表示特定长度及其以上的延伸的累积频率。大部分延伸距离注释基因超过 50 kb(详见正文)。

To characterize further the 5′ RxFrag extensions, we performed RT–PCR followed by cloning and sequencing for 550 of the 5′ RxFrags (including the 261 longest extensions identified for each locus). The approach of mapping RACE products using microarrays is a combination method previously described and validated in several studies[14,17,20]. Hybridization of the RT–PCR products to tiling arrays confirmed connectivity in almost 60% of the cases. Sequenced clones confirmed transcript extensions. Longer extensions were harder to clone and sequence, but 5 out of 18 RT–PCR-positive extensions over 100 kb were verified by sequencing (see Supplementary Information section 2.9.7 and ref. 17). The detection of numerous RxFrag extensions coupled with evidence of considerable intronic transcription indicates that protein-coding loci are more transcriptionally complex than previously thought. Instead of the traditional view that many genes have one or more alternative transcripts that code for alternative proteins, our data suggest that a given gene may both encode multiple protein products and produce other transcripts that include sequences from both strands and from neighbouring loci (often without encoding a different protein). Figure 3 illustrates such a case, in which a new fusion transcript is expressed in the small intestine, and consists of at least three coding exons from the *ATP5O* gene and at least two coding exons from the *DONSON* gene, with no evidence of sequences from two intervening protein-coding genes (*ITSN1* and *CRYZL1*).

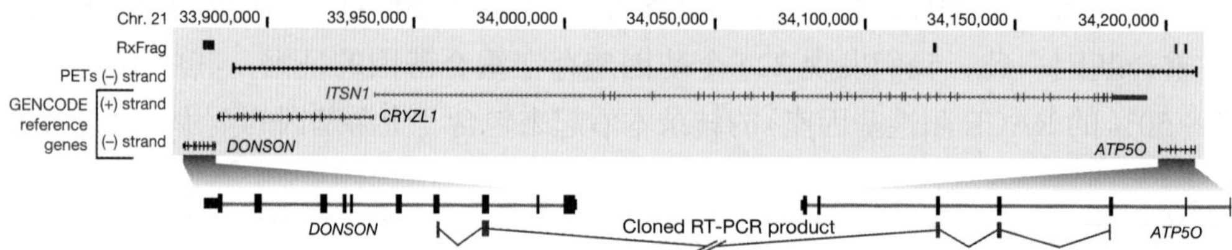

Fig. 3. Overview of RACE experiments showing a gene fusion. Transcripts emanating from the region between the *DONSON* and *ATP5O* genes. A 330-kb interval of human chromosome 21 (within ENm005) is shown, which contains four annotated genes: *DONSON*, *CRYZL1*, *ITSN1* and *ATP5O*. The 5′ RACE products generated from small intestine RNA and detected by tiling-array analyses (RxFrags) are shown along the top. Along the bottom is shown the placement of a cloned and sequenced RT–PCR product that has two exons from the *DONSON* gene followed by three exons from the *ATP5O* gene; these sequences are separated by a 300 kb intron in the genome. A PET tag shows the termini of a transcript consistent with this RT–PCR product.

Pseudogenes. Pseudogenes, reviewed in refs 21 and 22, are generally considered non-functional copies of genes, are sometimes transcribed and often complicate analysis of transcription owing to close sequence similarity to functional genes. We used various computational methods to identify 201 pseudogenes (124 processed and 77 non-processed) in the ENCODE regions (see Supplementary Information section 2.10 and ref. 23). Tiling-array analysis of 189 of these revealed that 56% overlapped at least one TxFrag. However, possible cross-hybridization between the pseudogenes and their corresponding parent genes may have confounded such analyses. To assess better the extent of pseudogene transcription, 160 pseudogenes (111 processed and 49 non-processed) were examined for expression using RACE/tiling-array analysis (see Supplementary Information section 2.9.2). Transcripts were detected for 14 pseudogenes (8 processed and 6 non-processed) in at least

为了进一步研究 5′ RxFrag 延伸的特征，我们对 550 个 5′ RxFrag（包括各个基因座延伸最长的 261 个）进行了 RT–PCR，然后进行克隆和测序。通过微阵列检测 RACE 产物的方法是一种之前报道的并且被很多研究验证过的联合方法[14,17,20]。将 RT–PCR 产物杂交到叠瓦式阵列验证了几乎 60% 的连通情况。被测序的克隆验证了转录本的延伸。越长的延伸越难克隆测序，但是 18 个超过 100 kb 的 RT–PCR 阳性延伸中，有 5 个经测序验证（详见补充信息 2.9.7 和参考文献 17）。大量的 RxFrag 延伸被检测到，并且有证据表明有相当数量的内含子转录本存在，都说明蛋白质编码基因座的转录要比之前想象的更加复杂。传统的观点认为很多基因含有一个或多个选择性转录本，编码选择性蛋白质，而我们的数据则暗示一个特定的基因可能既编码多个蛋白质产物，又产生包含两条链以及附近基因座序列的其他转录本（大多不编码其他蛋白质）。图 3 展示了这样一个例子来加以说明，在这个例子中，一个新的融合转录本在小肠中表达，它含有 *ATP5O* 基因的至少三个编码外显子和 *DONSON* 基因的至少两个编码外显子，但是并没有证据显示它含有位于这两个干扰蛋白编码基因（*ITSN1* 和 *CRYZL1*）的序列。

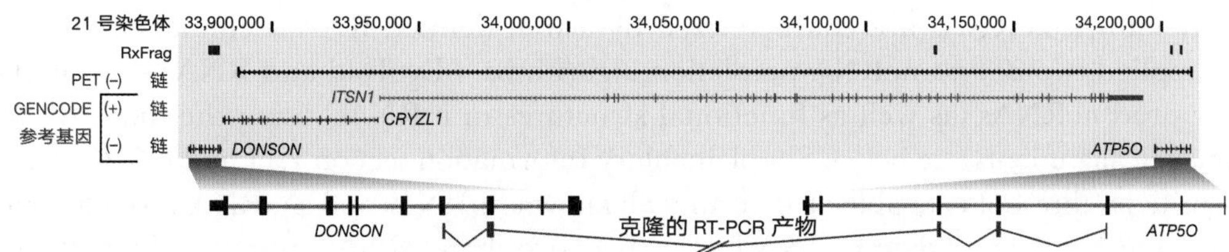

图 3. 概述 RACE 实验，展示基因融合现象。转录本产生于 *DONSON* 和 *ARP5O* 基因间区域。图中展示了人类 21 号染色体的一段 330 kb 的区域（在 ENm005 中），它包含四个注释基因：*DONSON*、*CRYZL1*、*ITSN1* 和 *ATP5O*。顶端展示的是叠瓦式阵列检测到的来自小肠 RNA 的 5′ RACE 产物（RxFrag）。底端展示的是克隆测序的 RT–PCR 产物的位置，该 PCR 产物含有 *DONSON* 基因的两个外显子和 *ATP5O* 基因的三个外显子；这些序列在基因组上被一个 300 kb 的内含子分开。PET 标签展示的与这个 RT–PCR 产物一致的一个转录本的末端。

假基因 假基因通常被认为是没有功能的基因拷贝，有时候会被转录，因为与正常基因序列的高度相似性常使转录分析变得复杂，在参考文献 21 和 22 中有对其的综述。利用多种计算方法，我们在 ENCODE 区域鉴定到 201 个假基因（124 个加工，77 个未加工）（见补充信息 2.10 和参考文献 23）。对其中的 189 个进行叠瓦式阵列分析，发现 56% 与至少一个 TxFrag 重合。不过，假基因可能和它们对应的亲本基因交叉杂交，进而可能使得上述分析产生混乱。为了更好地检测假基因转录的范围，利用 RACE/叠瓦式阵列检测 160 个假基因（111 个加工，49 个未加工）的表达（详见补充信息 2.9.2）。14 个假基因（8 个加工，6 个未加工）在 12 个 RNA 样本中的

one of the 12 tested RNA sources, the majority (9) being in testis (see ref. 23). Additionally, there was evidence for the transcription of 25 pseudogenes on the basis of their proximity (within 100 bp of a pseudogene end) to CAGE tags (8), PETs (2), or cDNAs/ESTs (21). Overall, we estimate that at least 19% of the pseudogenes in the ENCODE regions are transcribed, which is consistent with previous estimates[24,25].

Non-protein-coding RNA. Non-protein-coding RNAs (ncRNAs) include structural RNAs (for example, transfer RNAs, ribosomal RNAs, and small nuclear RNAs) and more recently discovered regulatory RNAs (for example, miRNAs). There are only 8 well-characterized ncRNA genes within the ENCODE regions (*U70*, *ACA36*, *ACA56*, *mir-192*, *mir-194-2*, *mir-196*, *mir-483* and *H19*), whereas representatives of other classes, (for example, box C/D snoRNAs, tRNAs, and functional snRNAs) seem to be completely absent in the ENCODE regions. Tiling-array data provided evidence for transcription in at least one of the assayed RNA samples for all of these ncRNAs, with the exception of mir-483 (expression of mir-483 might be specific to fetal liver, which was not tested). There is also evidence for the transcription of 6 out of 8 pseudogenes of ncRNAs (mainly snoRNA-derived). Similar to the analysis of protein-pseudogenes, the hybridization results could also originate from the known snoRNA gene elsewhere in the genome.

Many known ncRNAs are characterized by a well-defined RNA secondary structure. We applied two *de novo* ncRNA prediction algorithms—EvoFold and RNAz—to predict structured ncRNAs (as well as functional structures in mRNAs) using the multi-species sequence alignments (see below, Supplementary Information section 2.11 and ref. 26). Using a sensitivity threshold capable of detecting all known miRNAs and snoRNAs, we identified 4,986 and 3,707 candidate ncRNA loci with EvoFold and RNAz, respectively. Only 268 loci (5% and 7%, respectively) were found with both programs, representing a 1.6-fold enrichment over that expected by chance; the lack of more extensive overlap is due to the two programs having optimal sensitivity at different levels of GC content and conservation. We experimentally examined 50 of these targets using RACE/tiling-array analysis for brain and testis tissues (see Supplementary Information sections 2.11 and 2.9.3); the predictions were validated at a 56%, 65%, and 63% rate for Evofold, RNAz and dual predictions, respectively.

Primary transcripts. The detection of numerous unannotated transcripts coupled with increasing knowledge of the general complexity of transcription prompted us to examine the extent of primary (that is, unspliced) transcripts across the ENCODE regions. Three data sources provide insight about these primary transcripts: the GENCODE annotation, PETs, and RxFrag extensions. Figure 4 summarizes the fraction of bases in the ENCODE regions that overlap transcripts identified by these technologies. Remarkably, 93% of bases are represented in a primary transcript identified by at least two independent observations (but potentially using the same technology); this figure is reduced to 74% in the case of primary transcripts detected by at least two different technologies. These increased spans are not mainly due to cell line rearrangements because they were present in multiple tissue experiments that confirmed the spans (see Supplementary Information section 2.12).

至少 1 个样本中检测到转录本，大部分（9 个）在睾丸中被检测到（见参考文献 23）。另外，有证据表明 25 个假基因在转录，这是基于它们靠近（假基因末端距离 100 bp 以内）CAGE 标签（8 个）、PET（2 个）或 cDNA/EST（21 个）。总体来说，我们估计 ENCODE 区域内至少 19% 的假基因被转录，这与之前的估计是一致的[24,25]。

非蛋白质编码 RNA 非蛋白质编码 RNA（ncRNA）包括结构 RNA（如转运 RNA、核糖体 RNA 和核内小 RNA）和最近发现的调控 RNA（如 miRNA）。在 ENCODE 区域内只有 8 种熟知的 ncRNA 基因（*U70*、*ACA36*、*ACA56*、*mir-192*、*mir-194-2*、*mir-196*、*mir-483* 和 *H19*），其他类型（如 C/D 核仁小 RNA、tRNA 和功能性 snRNA）在 ENCODE 区域似乎完全不存在。叠瓦式阵列数据为所有这些 ncRNA 在至少一种被分析 RNA 样本中的转录提供了证据，这其中 mir-483 除外（mir-483 可能在胎肝中特异表达，而胎肝没有被检测）。也有证据表明 8 个 ncRNAs 假基因（主要是 snoRNA 产生的）中有 6 个转录。类似于蛋白质假基因的分析，杂交结果也可能是来源于基因组中其他地方的已知 snoRNA 基因。

很多已知的 ncRNA 具有明确的 RNA 二级结构。我们通过两种从头预测 ncRNA 的算法——EvoFold 和 RNAz，利用多物种序列比对来预测结构 ncRNA（详见下文、补充信息 2.11 和参考文献 26）。使用能检测所有已知的 miRNA 和 snoRNA 的敏感性阈值，我们通过 EvoFold 和 RNAz 分别鉴定到 4,986 和 3,707 个候选 ncRNA 基因座。只有 268 个基因座（分别是 5% 和 7%）在两个程序中都被发现，相对偶然概率有 1.6 倍富集；没有更高的重合是因为两个程序的最佳敏感性所要求的 GC 含量和保守性水平不同。我们使用 RACE/叠瓦式阵列实验在大脑和睾丸组织中检查了其中的 50 个（详见补充信息 2.11 和 2.9.3）；Evofold、RNAz 和两种方法联合预测的被验证率分别为 56%、65% 和 63%。

初级转录本 由于大量未被注释的转录本被检测到，加上我们对转录的基本复杂性认识的深入，使我们不得不检查初级（也就是未被剪接的）转录本在 ENCODE 区域内的分布范围。三个数据源为认识这些初级转录本提供了线索：GENCODE 注释、PET 和 RxFrag 延伸。图 4 总结了 ENCODE 区域内碱基与这些技术鉴定到的转录本重合的比例。值得注意的是，93% 的碱基位于被至少两种独立观察（但是有可能使用同一种技术）鉴定到的初级转录本中；如果是被至少两种不同的技术鉴定到的初级转录本，则这个数字下降到 74%。这些增加的跨度并不主要是由于细胞系的重排，因为在多种组织的实验中都验证了这种跨度的存在（详见补充信息 2.12）。这些

These estimates assume that the presence of PETs or RxFrags defining the terminal ends of a transcript imply that the entire intervening DNA is transcribed and then processed. Other mechanisms, thought to be unlikely in the human genome, such as *trans*-splicing or polymerase jumping would also produce these long termini and potentially should be reconsidered in more detail.

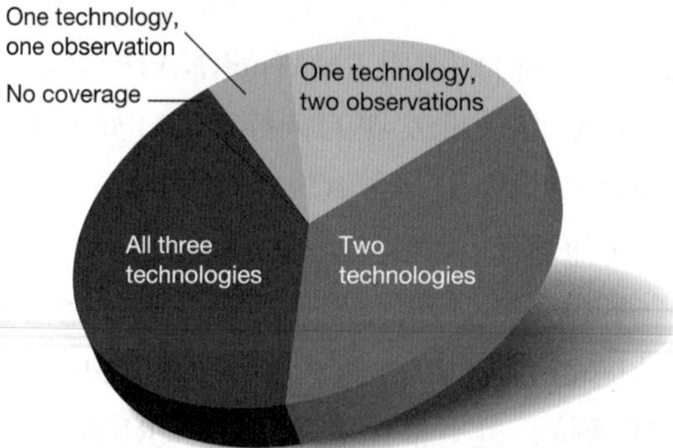

Fig. 4. Coverage of primary transcripts across ENCODE regions. Three different technologies (integrated annotation from GENCODE, RACE-array experiments (RxFrags) and PET tags) were used to assess the presence of a nucleotide in a primary transcript. Use of these technologies provided the opportunity to have multiple observations of each finding. The proportion of genomic bases detected in the ENCODE regions associated with each of the following scenarios is depicted: detected by all three technologies, by two of the three technologies, by one technology but with multiple observations, and by one technology with only one observation. Also indicated are genomic bases without any detectable coverage of primary transcripts.

Previous studies have suggested a similar broad amount of transcription across the human[14,15] and mouse[27] genomes. Our studies confirm these results, and have investigated the genesis of these transcripts in greater detail, confirming the presence of substantial intragenic and intergenic transcription. At the same time, many of the resulting transcripts are neither traditional protein-coding transcripts nor easily explained as structural non-coding RNAs. Other studies have noted complex transcription around specific loci or chimaeric-gene structures (for example refs 28–30), but these have often been considered exceptions; our data show that complex intercalated transcription is common at many loci. The results presented in the next section show extensive amounts of regulatory factors around novel TSSs, which is consistent with this extensive transcription. The biological relevance of these unannotated transcripts remains unanswered by these studies. Evolutionary information (detailed below) is mixed in this regard; for example, it indicates that unannotated transcripts show weaker evolutionary conservation than many other annotated features. As with other ENCODE-detected elements, it is difficult to identify clear biological roles for the majority of these transcripts; such experiments are challenging to perform on a large scale and, furthermore, it seems likely that many of the corresponding biochemical events may be evolutionarily neutral (see below).

估计假定如果存在定义转录本两个末端的 PET 或 RxFrag，就意味着整个中间 DNA 首先被转录，然后再被加工。被认为不可能存在于人类基因组的其他机制，例如反式剪接或者聚合酶跳跃，也会产生这种长末端，可能应该被更仔细地重新思考。

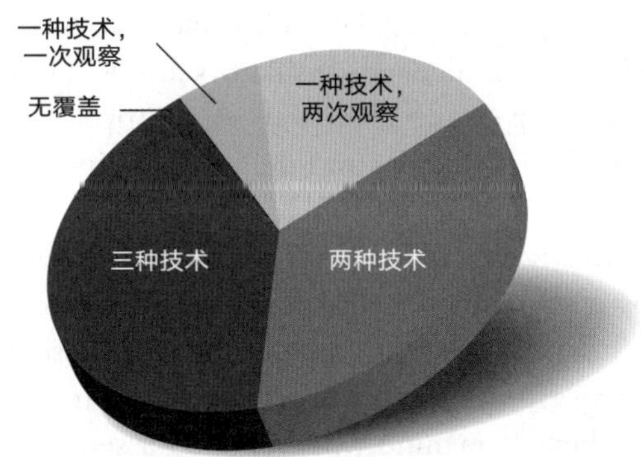

图 4. 原始转录本在 ENCODE 区域的覆盖度。三种不同的技术（来自 GENCODE、RACE 阵列实验(RxFrag) 和 PET 标签的整合注释）被用来评估初级转录本中碱基的存在。这些技术的应用使人们有机会对每一个发现进行多次观察。图中展示了在 ENCODE 区域中被检测的基因组碱基与下列情形相关的各自所占的比例：被三种技术都检测到，被三种技术中的两种检测到，被一种技术但是多次观察到，以及被一种技术一次观察到。图中还展示了没有检测到初级转录本覆盖的基因组碱基所占的比例。

之前的研究已经表明人类 [14,15] 和小鼠 [27] 基因组都存在类似的大量转录。我们的研究验证了这些结果，并且还更详细地研究了这些转录本的产生，验证了大量基因内和基因间转录的存在。同时，得到的很多转录本既不是传统的蛋白质编码转录本，也不能简单地解释为结构性非编码 RNA。其他的研究已经注意到在特定基因座或者嵌合基因结构存在复杂的转录（如参考文献 28 ~ 30），但是这些常常被认为是特例，我们的数据说明复杂的有插入的转录在很多基因座是常见的。下一部分呈现的结果展示了在新的 TSS 周围有大量的调控因子，与这种广泛转录一致。这些研究还不能回答这些未注释转录本的生物学意义。进化的信息（下文有详细介绍）在这方面是混杂的，例如，它暗示未注释的转录本比很多其他注释特征表现为更弱的进化保守性。和其他 ENCODE 检测到的元件一样，这些转录本中的大多数难以鉴定明确的生物学功能。大规模地进行这种实验很有挑战性，再者，有可能很多对应的生化事件在进化上是中性的（见下文）。

Regulation of Transcription

Overview. A significant challenge in biology is to identify the transcriptional regulatory elements that control the expression of each transcript and to understand how the function of these elements is coordinated to execute complex cellular processes. A simple, commonplace view of transcriptional regulation involves five types of *cis*-acting regulatory sequences—promoters, enhancers, silencers, insulators and locus control regions[31]. Overall, transcriptional regulation involves the interplay of multiple components, whereby the availability of specific transcription factors and the accessibility of specific genomic regions determine whether a transcript is generated[31]. However, the current view of transcriptional regulation is known to be overly simplified, with many details remaining to be established. For example, the consensus sequences of transcription factor binding sites (typically 6 to 10 bases) have relatively little information content and are present numerous times in the genome, with the great majority of these not participating in transcriptional regulation. Does chromatin structure then determine whether such a sequence has a regulatory role? Are there complex inter-factor interactions that integrate the signals from multiple sites? How are signals from different distal regulatory elements coupled without affecting all neighbouring genes? Meanwhile, our understanding of the repertoire of transcriptional events is becoming more complex, with an increasing appreciation of alternative TSSs[32,33] and the presence of non-coding[27,34] and anti-sense transcripts[35,36].

To better understand transcriptional regulation, we sought to begin cataloguing the regulatory elements residing within the 44 ENCODE regions. For this pilot project, we mainly focused on the binding of regulatory proteins and chromatin structure involved in transcriptional regulation. We analysed over 150 data sets, mainly from ChIP-chip[37-39], ChIP-PET and STAGE[40,41] studies (see Supplementary Information section 3.1 and 3.2). These methods use chromatin immunoprecipitation with specific antibodies to enrich for DNA in physical contact with the targeted epitope. This enriched DNA can then be analysed using either microarrays (ChIP-chip) or high-throughput sequencing (ChIP-PET and STAGE). The assays included 18 sequence-specific transcription factors and components of the general transcription machinery (for example, RNA polymerase II (Pol II), TAF1 and TFIIB/GTF2B). In addition, we tested more than 600 potential promoter fragments for transcriptional activity by transient-transfection reporter assays that used 16 human cell lines[33]. We also examined chromatin structure by studying the ENCODE regions for DNaseI sensitivity (by quantitative PCR[42] and tiling arrays[43,44], see Supplementary Information section 3.3), histone composition[45], histone modifications (using ChIP-chip assays)[37,46], and histone displacement (using FAIRE, see Supplementary Information section 3.4). Below, we detail these analyses, starting with the efforts to define and classify the 5′ ends of transcripts with respect to their associated regulatory signals. Following that are summaries of generated data about sequence-specific transcription factor binding and clusters of regulatory elements. Finally, we describe how this information can be integrated to make predictions about transcriptional regulation.

转录的调控

概况 生物学中一个重大的挑战是鉴定控制各个转录本表达的转录调控元件，以及理解这些元件的功能是如何协调以完成复杂的细胞过程。对转录调控一个简单朴素的认识包括五种类型的顺式调控元件——启动子、增强子、沉默子、绝缘子和基因座控制区 [31]。总体上，转录调控包括多种组分的相互作用，特定转录因子的可用性和特定基因组区域的开放性决定一个转录本的产生与否 [31]。不过，目前对转录调控的认识被认为过于简单，很多细节还未完善。例如，转录因子结合位点的共有序列（通常 6 到 10 个碱基）含有相对很少的信息量，在基因组中大量存在，其中大部分并不参与转录调控。接下来，是否染色质结构决定这样一个序列是否具有调控作用？是否有复杂的因子间相互作用整合来自多个位点的信号？来自不同的远端调控元件的信号如何联系起来并且还不影响所有邻近的基因？同时，随着对可变 TSS[32,33] 认识的更加深入以及非编码 [27,34] 和反义 [35,36] 转录本的存在，我们对转录事件库的理解正变得更加复杂。

为更好地理解转录调控，我们开始为位于 44 个 ENCODE 区域内的调控元件编制目录。对于这个先导计划，我们主要关注调控蛋白的结合和参与转录调控的染色质结构。我们分析了超过 150 个数据集，这些数据集主要来自 ChIP-chip[37-39]，ChIP-PET 和 STAGE[40,41] 研究（详见补充信息 3.1 和 3.2）。这些方法使用染色质免疫沉淀结合特定的抗体从而物理靠近目标表位富集 DNA。这些富集的 DNA 可以利用芯片（ChIP-chip）或高通量测序（ChIP-PET 和 STAGE）分析。这样的实验包括 18 个序列特异的转录因子和基本转录机制的组分（如 RNA 聚合酶 II(Pol II)、TAF1 和 TFIIB/GTF2B）。另外，我们利用 16 个人细胞系，通过瞬时转染报告实验检测了超过 600 个潜在的启动子片段的转录活性 [33]。我们还通过研究 ENCODE 区域的 DNaseI 敏感性（通过定量 PCR[42] 和叠瓦式阵列 [43,44]，详见补充信息 3.3）、组蛋白构成 [45]、组蛋白修饰（利用 ChIP-chip 实验）[37,46] 和组蛋白置换（使用 FAIRE，详见补充信息 3.4）检测了染色质的结构。下面，我们详细介绍这些信息，首先根据相关的调控信号来定义和分类转录本的 5′ 端；然后总结产生的关于序列特异转录因子结合和调控元件簇的数据；最后我们描述这些信息如何被整合起来，从而对转录调控做出预测。

Transcription start site catalogue. We analysed two data sets to catalogue TSSs in the ENCODE regions: the 5′ ends of GENCODE-annotated transcripts and the combined results of two 5′-end-capture technologies—CAGE and PET-tagging. The initial results suggested the potential presence of 16,051 unique TSSs. However, in many cases, multiple TSSs resided within a single small segment (up to ~200 bases); this was due to some promoters containing TSSs with many very close precise initiation sites[47]. To normalize for this effect, we grouped TSSs that were 60 or fewer bases apart into a single cluster, and in each case considered the most frequent CAGE or PET tag (or the 5′-most TSS in the case of TSSs identified only from GENCODE data) as representative of that cluster for downstream analyses.

The above effort yielded 7,157 TSS clusters in the ENCODE regions. We classified these TSSs into three categories: known (present at the end of GENCODE-defined transcripts), novel (supported by other evidence) and unsupported. The novel TSSs were further subdivided on the basis of the nature of the supporting evidence (see Table 3 and Supplementary Information section 3.5), with all four of the resulting subtypes showing significant overlap with experimental evidence using the GSC statistic. Although there is a larger relative proportion of singleton tags in the novel category, when analysis is restricted to only singleton tags, the novel TSSs continue to have highly significant overlap with supporting evidence (see Supplementary Information section 3.5.1).

Table 3. Different categories of TSSs defined on the basis of support from different transcript-survey methods

Category	Transcript survey method	Number of TSS clusters (non-redundant)*	*P* value†	Singleton clusters‡ (%)
Known	GENCODE 5′ ends	1,730	2×10^{-70}	25 (74 overall)
Novel	GENCODE sense exons	1,437	6×10^{-39}	64
	GENCODE antisense exons	521	3×10^{-8}	65
	Unbiased transcription survey	639	7×10^{-63}	71
	CpG island	164	4×10^{-90}	60
Unsupported	None	2,666	–	83.4

* Number of TSS clusters with this support, excluding TSSs from higher categories.

† Probability of overlap between the transcript support and the PET/CAGE tags, as calculated by the Genome Structure Correction statistic (see Supplementary Information section 1.3).

‡ Per cent of clusters with only one tag. For the "known" category this was calculated as the per cent of GENCODE 5′ ends with tag support (25%) or overall (74%).

Correlating genomic features with chromatin structure and transcription factor binding. By measuring relative sensitivity to DNaseI digestion (see Supplementary Information section 3.3), we identified DNaseI hypersensitive sites throughout the ENCODE regions. DHSs and TSSs both reflect genomic regions thought to be enriched for regulatory information and many DHSs reside at or near TSSs. We partitioned DHSs into those within 2.5 kb of a TSS (958; 46.5%) and the remaining ones, which were classified

转录起始位点的分类 我们分析了两个数据集，进而对 ENCODE 区域的 TSS 进行分类：GENCODE 注释的转录本的 5′端和两个 5′端捕获技术——CAGE 和 PET 标签的综合结果。起初的结果表明有 16,051 个唯一的 TSS 的潜在存在。但是，在很多情况下，多个 TSS 位于同一个小的片段内（最多 ~ 200 碱基），这是因为一些含有 TSS 的启动子含有很多非常靠近的精确的起始位点 [47]。为了使这种影响正常化，我们将相距 60 或者更少碱基的 TSS 归为一个簇，并且在这种情况下把频率最高的 CAGE 或 PET 标签（或者对于仅在 GENCODE 数据鉴定到 TSS 则是最 5′的 TSS）作为这个簇的代表用于下游分析。

上述分析在 ENCODE 区域产生了 7,157 个 TSS 簇。我们将这些 TSS 分为三类：已知的（在 GENCODE 确定的转录本的末端存在），全新的（被其他证据支持的）和不被支持的。基于支持的证据的性质，全新的 TSS 被进一步分类（详见表 3 和补充信息 3.5），利用 GSC 统计发现，产生的 4 个亚类都与实验证据高度重合。当仅分析单一标签时，尽管单一标签的相对比例在新类里更大，但是全新的 TSS 仍然与支持证据高度重合（详见补充信息 3.5.1）。

表 3. 基于支持不同转录本鉴定方法而定义的不同类的 TSS

类别	转录本鉴定方法	TSS 簇的数量（非冗余的）*	*P* 值†	单一簇‡(%)
已知的	GENCODE 5′端	1,730	2×10^{-70}	25（总共 74）
全新的	GENCODE 正义链外显子	1,437	6×10^{-39}	64
	GENCODE 反义链外显子	521	3×10^{-8}	65
	无偏性的转录鉴定	639	7×10^{-63}	71
	CpG 岛	164	4×10^{-90}	60
不被支持的	无	2,666	–	83.4

* 基于本文 TSS 簇的数量，不包括更高层级分类的 TSS。

† 转录本支持和 PET/CAGE 标签重合的概率，通过基因组结构校正统计计算（详见补充信息 1.3）。

‡ 只有一个标签的簇的比例。对于“已知的”这一类，计算的是含有标签支持的（25%）或所有的（74%）GENCODE 5′端的比例。

将基因组特征与染色质结构和转录因子结合相关联 通过检测对 DNaseI 切割的相对敏感度（详见补充信息 3.3），我们在整个 ENCODE 区域鉴定了 DNaseI 超敏感位点。DHS 和 TSS 都反映了被认为是富集调控信息的基因组区域，并且很多 DHS 位于 TSS 或其附近。我们将 DHS 分为在 TSS 2.5 kb 以内的（958；46.5%）和其他位于远端的（1,102；53.5%）。然后我们通过根据相对 TSS 或 DHS 的距离将信号聚集，

as distal (1,102; 53.5%). We then cross-analysed the TSSs and DHSs with data sets relating to histone modifications, chromatin accessibility and sequence-specific transcription factor binding by summarizing these signals in aggregate relative to the distance from TSSs or DHSs. Figure 5 shows representative profiles of specific histone modifications, Pol II and selected transcription factor binding for the different categories of TSSs. Further profiles and statistical analysis of these studies can be found in Supplementary Information 3.6.

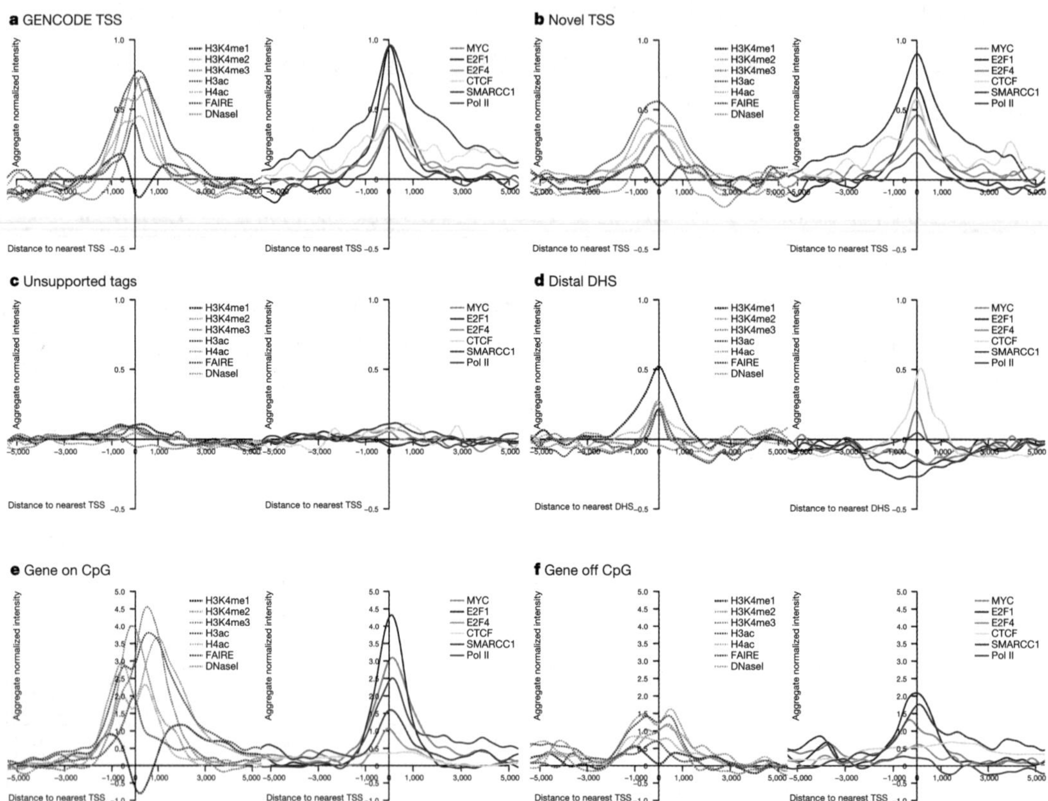

Fig. 5. Aggregate signals of tiling-array experiments from either ChIP-chip or chromatin structure assays, represented for different classes of TSSs and DHS. For each plot, the signal was first normalized with a mean of 0 and standard deviation of 1, and then the normalized scores were summed at each position for that class of TSS or DHS and smoothed using a kernel density method (see Supplementary Information section 3.6). For each class of sites there are two adjacent plots. The left plot depicts the data for general factors: FAIRE and DNaseI sensitivity as assays of chromatin accessibility and H3K4me1, H3K4me2, H3K4me3, H3ac and H4ac histone modifications (as indicated); the right plot shows the data for additional factors, namely MYC, E2F1, E2F4, CTCF, SMARCC1 and Pol II. The columns provide data for the different classes of TSS or DHS (unsmoothed data and statistical analysis shown in Supplementary Information section 3.6).

In the case of the three TSS categories (known, novel and unsupported), known and novel TSSs are both associated with similar signals for multiple factors (ranging from histone modifications through DNaseI accessibility), whereas unsupported TSSs are not. The enrichments seen with chromatin modifications and sequence-specific factors, along with the significant clustering of this evidence, indicate that the novel TSSs do not reflect false

对 TSS 和 DHS 与组蛋白修饰、染色质开放性和序列特异的转录因子结合相关的数据集进行了交叉分析。图 5 展示了特定的组蛋白修饰、Pol II 和特定转录因子结合不同类的 TSS 的代表性特征。这些研究更深入的特征和统计学分析可以在补充信息 3.6 中找到。

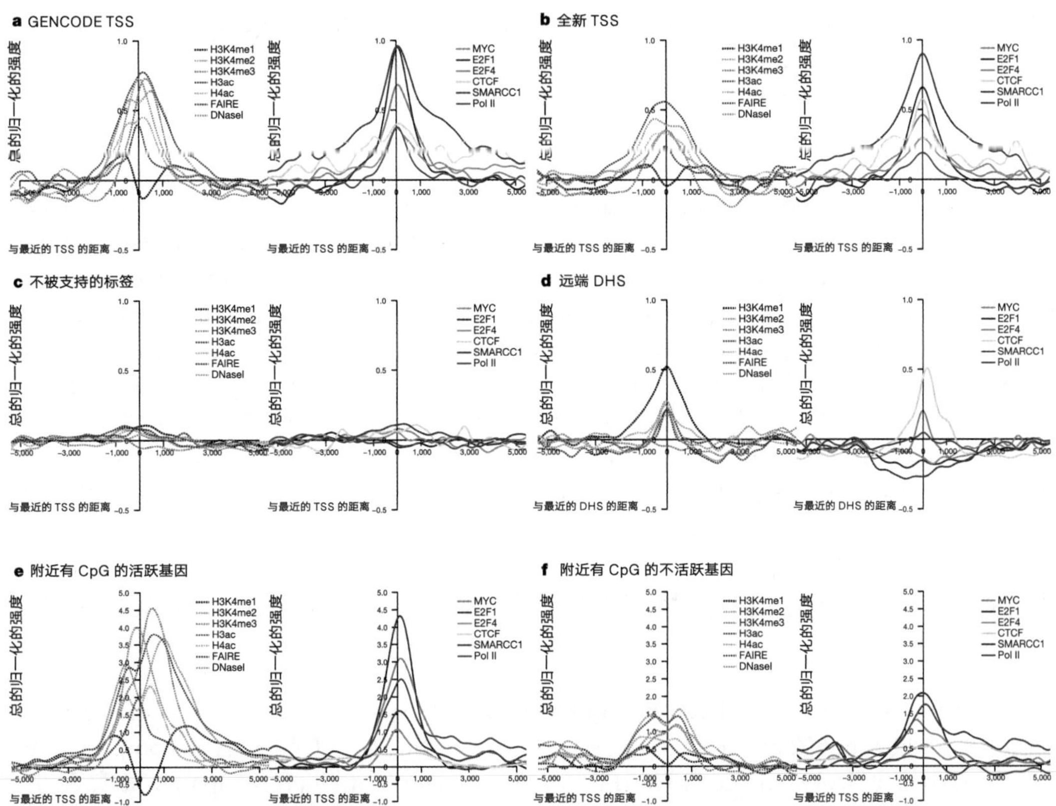

图 5. ChIP-chip 或染色质结构实验中的叠瓦式阵列实验的聚集信号，展示的是不同类型的 TSS 和 DHS 的情况。对于每一个图，信号先根据均值为 0、标准差为 1 被归一化，然后归一化后的值在相应类型的 TSS 或 DHS 的每一个位置求和，并利用核密度方法平滑化（详见补充信息 3.6）。对于各类位点有两个相邻的图。左图展示的是基本因子的数据：FAIRE 和 DNaseI 敏感性作为染色质开放性的检测方法，以及 H3K4me1、H3K4me2、H3K4me3、H3ac 和 H4ac 组蛋白修饰（如图所示）；右图展示的是另外的因子的数据，即 MYC、E2F1、E2F4、CTCF、SMARCC1 和 Pol II。各栏提供了不同类型的 TSS 或 DHS 的数据（未平滑的数据和统计学分析见补充信息 3.6）。

对于这三类 TSS（已知的、全新的和不被支持的），已知的和全新的 TSS 与多个因子类似的信号（从组蛋白修饰到 DNaseI 开放性）都有关联，而不被支持的 TSS 则不是这样。染色质修饰和序列特异的因子的富集，以及这些证据的明显聚集，说明全新的 TSS 并不是假阳性，可能使用和其他启动子一样的生物机制。序列特异的转

positives and probably use the same biological machinery as other promoters. Sequence-specific transcription factors show a marked increase in binding across the broad region that encompasses each TSS. This increase is notably symmetric, with binding equally likely upstream or downstream of a TSS (see Supplementary Information section 3.7 for an explanation of why this symmetrical signal is not an artefact of the analysis of the signals). Furthermore, there is enrichment of SMARCC1 binding (a member of the SWI/SNF chromatin-modifying complex), which persists across a broader extent than other factors. The broad signals with this factor indicate that the ChIP-chip results reflect both specific enrichment at the TSS and broader enrichments across ~5-kb regions (this is not due to technical issues, see Supplementary Information section 3.8).

We selected 577 GENCODE-defined TSSs at the 5′ ends of a protein-coding transcript with over 3 exons, to assess expression status. Each transcript was classified as: (1) "active" (gene on) or "inactive" (gene off) on the basis of the unbiased transcript surveys, and (2) residing near a "CpG island" or not ("non-CpG island") (see Supplementary Information section 3.17). As expected, the aggregate signal of histone modifications is mainly attributable to active TSSs (Fig. 5), in particular those near CpG islands. Pronounced doublet peaks at the TSS can be seen with these large signals (similar to previous work in yeast[48]) owing to the chromatin accessibility at the TSS. Many of the histone marks and Pol II signals are now clearly asymmetrical, with a persistent level of Pol II into the genic region, as expected. However, the sequence-specific factors remain largely symmetrically distributed. TSSs near CpG islands show a broader distribution of histone marks than those not near CpG islands (see Supplementary Information section 3.6). The binding of some transcription factors (E2F1, E2F4 and MYC) is extensive in the case of active genes, and is lower (or absent) in the case of inactive genes.

Chromatin signature of distal elements. Distal DHSs show characteristic patterns of histone modification that are the inverse of TSSs, with high H3K4me1 accompanied by lower levels of H3K4Me3 and H3Ac (Fig. 5). Many factors with high occupancy at TSSs (for example, E2F4) show little enrichment at distal DHSs, whereas other factors (for example, MYC) are enriched at both TSSs and distal DHSs[49]. A particularly interesting observation is the relative enrichment of the insulator-associated factor CTCF[50] at both distal DHSs and TSSs; this contrasts with SWI/SNF components SMARCC2 and SMARCC1, which are TSS-centric. Such differential behaviour of sequence-specific factors points to distinct biological differences, mediated by transcription factors, between distal regulatory sites and TSSs.

Unbiased maps of sequence-specific regulatory factor binding. The previous section focused on specific positions defined by TSSs or DHSs. We then analysed sequence-specific transcription factor binding data in an unbiased fashion. We refer to regions with enriched binding of regulatory factors as RFBRs. RFBRs were identified on the basis of ChIP-chip data in two ways: first, each investigator developed and used their own analysis method(s) to define high-enrichment regions, and second (and independently), a stringent false discovery rate (FDR) method was applied to analyse all data using three cut-offs (1%, 5% and 10%).

录因子在包含 TSS 的较宽区域内的结合明显增加。这种增加是明显对称的，在 TSS 上游或下游结合的可能性一样（详见补充信息 3.7，其中有解释为什么这种对称的信号不是一种信号分析的假象）。另外，还有 SMARCC1 结合的富集（SWI/SNF 染色质修饰复合物的一个组分），它比其他因子分布的范围更宽。这个因子较宽的信号暗示，ChIP-chip 结果不仅反映在 TSS 特异的富集，也反映在更宽的约 5 kb 区域的富集（这不是因为技术的原因，详见补充信息 3.8）。

我们选择 577 个 GENCODE 定义的 TSS 来评价表达水平，这些 TSS 都位于蛋白质编码且含有超过三个外显子的转录本的 5′端。各个转录本被分为：(1) 基于无偏性的转录本检测确定的“活跃的”（基因开启）或者“不活跃的”（基因关闭）和 (2) 位于“CpG 岛”附近或者不在附近的（“非 CpG 岛”）（详见补充信息 3.17）。正如预期的，组蛋白修饰的聚集信号主要位于活跃的 TSS（图 5），特别是那些 CpG 岛附近的。由于在 TSS 染色质的开放性，在 TSS 位置这些大的信号能看到明显的双峰（类似于之前在酵母中的发现 [48]）。正如预期的，很多组蛋白标志和 Pol II 信号现在明显是不对称的，Pol II 信号延伸到了基因区。但是，序列特异的因子仍然大部分是对称分布的。与不靠近 CpG 岛的 TSS 相比，CpG 岛附近的 TSS 呈现更宽的组蛋白标志分布（详见补充信息 3.6）。对于活跃基因，一些转录因子（E2F1、E2F4 和 MYC）广泛结合，而对于不活跃的基因，则结合水平较低（或不结合）。

远端元件的染色质标志 远端 DHS 呈现的特征性组蛋白修饰模式与 TSS 相反，高水平的 H3K4me1 伴随着低水平的 H3K4Me3 和 H3Ac（图 5）。很多在 TSS 结合水平较高的因子（如 E2F4）在远端 DHS 几乎没有富集，而其他因子（如 MYC）则在 TSS 和远端 DHS 都有富集 [49]。一个特别有意思的现象是，绝缘子相关因子 CTCF[50] 在远端 DHS 和 TSS 均有相对富集，这与 SWI/SNF 组分 SMARCC2 和 SMARCC1 明显不同，后者以 TSS 为中心。序列特异的因子的这种差异的行为表明，在远端调控位点和 TSS 之间，存在由转录因子介导的明显的生物学差异。

序列特异的调控因子结合的无偏性图谱 前面的部分聚焦于 TSS 或 DHS 定义的特定的位置。接下来我们以无偏性的方式分析序列特异的转录因子结合数据。我们称有调控因子富集结合的区域为 RFBR。基于 ChIP-chip 数据，RFBR 通过两种方式被鉴定：第一，每个研究者开发和使用他们自己的分析方法定义高富集区；第二（单独地），一种严格的错误发现率（FDR）方法被应用于分析所有的数据，它使用三个阈值（1%、5% 和 10%）。实验室特异的和基于 FDR 的方法高度相关，特别是对于

The laboratory-specific and FDR-based methods were highly correlated, particularly for regions with strong signals[10,11]. For consistency, we used the results obtained with the FDR-based method (see Supplementary Information section 3.10). These RFBRs can be used to find sequence motifs (see Supplementary Information section S3.11).

RFBRs are associated with the 5′ ends of transcripts. The distribution of RFBRs is non-random (see ref. 10) and correlates with the positions of TSSs. We examined the distribution of specific RFBRs relative to the known TSSs. Different transcription factors and histone modifications vary with respect to their association with TSSs (Fig. 6; see Supplementary Information section 3.12 for modelling of random expectation). Factors for which binding sites are most enriched at the 5′ ends of genes include histone modifications, TAF1 and RNA Pol II with a hypo-phosphorylated carboxy-terminal domain[51]—confirming previous expectations. Surprisingly, we found that E2F1, a sequence-specific factor that regulates the expression of many genes at the G1 to S transition[52], is also tightly associated with TSSs[52]; this association is as strong as that of TAF1, the well-known TATA box-binding protein associated factor 1 (ref. 53). These results suggest that E2F1 has a more general role in transcription than previously suspected, similar to that for MYC[54-56]. In contrast, the large-scale assays did not support the promoter binding that was found in smaller-scale studies (for example, on SIRT1 and SPI1 (PU1)).

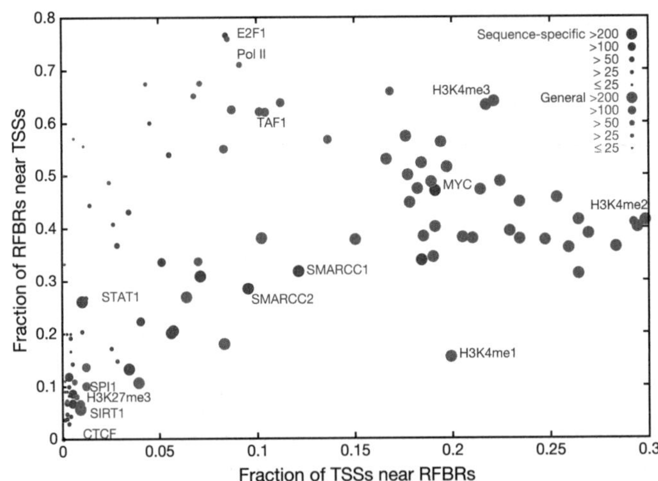

Fig 6. Distribution of RFBRs relative to GENCODE TSSs. Different RFBRs from sequence-specific factors (red) or general factors (blue) are plotted showing their relative distribution near TSSs. The *x* axis indicates the proportion of TSSs close (within 2.5 kb) to the specified factor. The *y* axis indicates the proportion of RFBRs close to TSSs. The size of the circle provides an indication of the number of RFBRs for each factor. A handful of representative factors are labelled.

Integration of data on sequence-specific factors. We expect that regulatory information is not dispersed independently across the genome, but rather is clustered into distinct regions[57]. We refer to regions that contain multiple regulatory elements as "regulatory clusters". We sought to predict the location of regulatory clusters by cross-integrating data generated using all transcription factor and histone modification assays, including

强信号区域[10,11]。为了统一性，我们使用基于 FDR 方法获得的结果（详见补充信息 3.10），这些 RFBR 可被用于发现序列模序（详见补充信息 S3.11）。

RFBR 与转录本的 5′端关联 RFBR 的分布不是随机的（详见参考文献 10），并且与 TSS 的位置相关。我们考察了特定 RFBR 相对已知的 TSS 的分布。不同的转录因子和组蛋白修饰在它们与 TSS 关联方面各不相同（图 6；随机期望的建模详见补充信息 3.12）。在基因 5′端最富集的与位点结合的因子包括组蛋白修饰、TAF1 和含有低磷酸化羧基端结构域的 RNA Pol II[51] 验证之前的预期。令人意外的是，我们发现 E2F1，一个在 G1 到 S 转换阶段调控很多基因表达的序列特异的因子[52]，也与 TSS 紧密关联[52]；这种关联跟 TAF1 与 TSS 的关联一样强，TAF1 即人们熟知的 TATA 框结合蛋白相关因子 1（参考文献 53）。这些结果说明相比于之前的猜测，E2F1 在转录中有着更基本的功能，这类似于 MYC[54-56] 的功能。相反，大规模的实验并不支持之前小规模研究发现的启动子的结合（如 SIRT1 和 SPI1（PU1））。

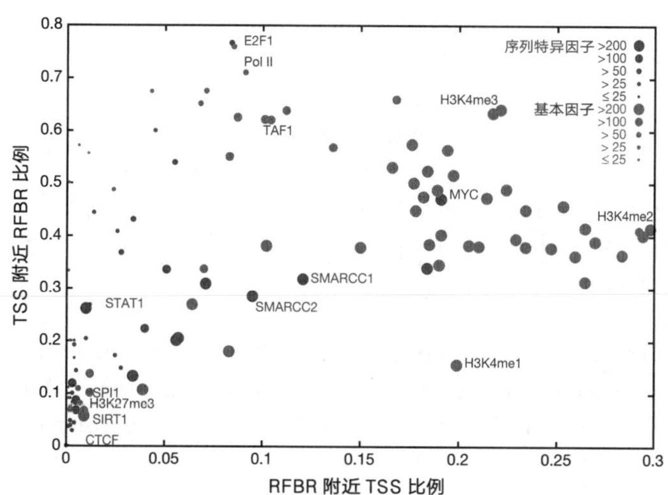

图 6. RFBR 相对 GENCODE TSS 的分布。图中展示的是源自序列特异的因子（红色）或基本因子（蓝色）的不同 RFBR 在 TSS 附近的相对分布。x 轴表示靠近（2.5 kb 以内）特定因子的 TSS 的比例。y 轴表示靠近 TSS 的 RFBR 的比例。圆圈的大小表示各个因子的 RFBR 的数量。少数几个代表性的因子被标记出来了。

序列特异的因子的数据的整合 我们预期调控信息并不是各自分散在基因组上，而是聚集在特定的区域[57]。我们称含有多个调控元件的区域为“调控簇”。利用所有转录因子和组蛋白修饰实验所产生的交叉整合的数据，包括在单个实验中低于主观阈值的结果，我们来预测调控簇的位置。特别地，我们利用了四种互为补充的

results falling below an arbitrary threshold in individual experiments. Specifically, we used four complementary methods to integrate the data from 129 ChIP-chip data sets (see Supplementary Information section 3.13 and ref. 58). These four methods detect different classes of regulatory clusters and as a whole identified 1,393 clusters. Of these, 344 were identified by all four methods, with another 500 found by three methods (see Supplementary Information section 3.13.5). 67% of the 344 regulatory clusters identified by all four methods (or 65% of the full set of 1,393) reside within 2.5 kb of a known or novel TSS (as defined above; see Table 3 and Supplementary Information section 3.14 for a breakdown by category). Restricting this analysis to previously annotated TSSs (for example, RefSeq or Ensembl) reveals that roughly 25% of the regulatory clusters are close to a previously identified TSS. These results suggest that many of the regulatory clusters identified by integrating the ChIP-chip data sets are undiscovered promoters or are somehow associated with transcription in another fashion. To test these possibilities, sets of 126 and 28 non-GENCODE-based regulatory clusters were tested for promoter activity (see Supplementary Information section 3.15) and by RACE, respectively. These studies revealed that 24.6% of the 126 tested regulatory clusters had promoter activity and that 78.6% of the 28 regulatory clusters analysed by RACE yielded products consistent with a TSS[58]. The ChIP-chip data sets were generated on a mixture of cell lines, predominantly HeLa and GM06990, and were different from the CAGE/ PET data, meaning that tissue specificity contributes to the presence of unique TSSs and regulatory clusters. The large increase in promoter proximal regulatory clusters identified by including the additional novel TSSs coupled with the positive promoter and RACE assays suggests that most of the regulatory regions identifiable by these clustering methods represent bona fide promoters (see Supplementary Information 3.16). Although the regulatory factor assays were more biased towards regions associated with promoters, many of the sites from these experiments would have previously been described as distal to promoters. This suggests that commonplace use of RefSeq- or Ensembl-based gene definition to define promoter proximity will dramatically overestimate the number of distal sites.

Predicting TSSs and transcriptional activity on the basis of chromatin structure. The strong association between TSSs and both histone modifications and DHSs prompted us to investigate whether the location and activity of TSSs could be predicted solely on the basis of chromatin structure information. We trained a support vector machine (SVM) by using histone modification data anchored around DHSs to discriminate between DHSs near TSSs and those distant from TSSs. We used a selected 2,573 DHSs, split roughly between TSS-proximal DHSs and TSS-distal DHSs, as a training set. The SVM performed well, with an accuracy of 83% (see Supplementary Information section 3.17). Using this SVM, we then predicted new TSSs using information about DHSs and histone modifications—of 110 high-scoring predicted TSSs, 81 resided within 2.5 kb of a novel TSS. As expected, these show a significant overlap to the novel TSS groups (defined above) but without a strong bias towards any particular category (see Supplementary Information section 3.17.1.5).

To investigate the relationship between chromatin structure and gene expression, we examined transcript levels in two cell lines using a transcript-tiling array. We compared

方法整合来自 129 个 ChIP-chip 数据集的数据（详见补充信息 3.13 和参考文献 58）。这四种方法检测不同类型的调控簇，合起来鉴定到 1,393 个簇。其中 344 个被四种方法都鉴定到，另外 500 个被三种方法鉴定到（详见补充信息 3.13.5）。被四种方法都鉴定到的 344 个调控簇中（或 1,393 个簇中的 65%），有 67% 位于已知的或全新的 TSS 2.5 kb 范围内（上文所定义的；按类分解情况详见表 3 和补充信息 3.14）。仅分析之前已经注释的 TSS（如 RefSeq 或者 Ensembl）发现大约 25% 的调控簇靠近之前鉴定的 TSS。这些结果说明通过整合 ChIP-chip 数据集鉴定到的调控簇中，有很多是还未被发现的启动子或者以其他的某种方式与转录存在着关联。为了检查这些可能性，启动子活性（详见补充信息 3.15）和 RACE 实验分别检测了 126 和 28 个非基于 GENCODE 的调控簇。这些研究发现 126 个被检测的调控簇中有 24.6% 具有启动子活性，28 个被 RACE 分析的调控簇中有 78.6% 产生产物，这与 TSS 一致[58]。与 CAGE/PET 数据不同，ChIP-chip 数据集产生于混合的细胞系，主要是 HeLa 和 GM06990，这意味着组织特异性也是存在独特的 TSS 和调控簇的一个原因。包括了额外全新的 TSS 以及阳性启动子和 RACE 实验后，鉴定到的启动子附近的调控簇大大增加了，这说明大部分通过这些聚类方法鉴定的调控区域代表的是真实的启动子（详见补充信息 3.16）。尽管调控因子实验更偏向启动子相关区域，很多这些实验得到的位点如果按以前的描述是远离启动子的。这说明简单地使用基于 RefSeq 或 Ensembl 定义的基因来确定是否靠近启动子会极大地高估远距离位点的数量。

基于染色质结构预测 TSS 和转录活性　TSS 与组蛋白修饰和 DHS 的高度的相关性促使我们探究是否仅基于染色质结构信息就能预测 TSS 的位置和活性。我们使用 DHS 周围的组蛋白修饰数据训练了一个支持向量机（SVM），用来区分靠近 TSS 的 DHS 和远离 TSS 的 DHS。我们选择了 2,573 个 DHS，大体分为靠近 TSS 的 DHS 和远离 TSS 的 DHS，作为一个训练集。SVM 表现良好，准确度为 83%（详见补充信息 3.17）。使用这个 SVM，然后我们利用 DHS 和组蛋白修饰信息预测新的 TSS——110 个高分预测的 TSS 中，81 个位于全新的 TSS 的 2.5 kb 范围内。正如预期的，它们与全新 TSS 组（上文定义的）有高度的重合，不过也没有特别偏向某一类（详见补充信息 3.17.1.5）。

为了研究染色质结构和基因表达的关系，我们利用转录本阵列实验在两种细胞系中检测了转录本的表达水平。我们将这个转录本数据与测定 ENCODE 区域组蛋白

this transcript data with the results of ChIP-chip experiments that measured histone modifications across the ENCODE regions. From this, we developed a variety of predictors of expression status using chromatin modifications as variables; these were derived using both decision trees and SVMs (see Supplementary Information section 3.17). The best of these correctly predicts expression status (transcribed versus non-transcribed) in 91% of cases. This success rate did not decrease dramatically when the predicting algorithm incorporated the results from one cell line to predict the expression status of another cell line. Interestingly, despite the striking difference in histone modification enrichments in TSSs residing near versus those more distal to CpG islands (see Fig. 5 and Supplementary Information section 3.6), including information about the proximity to CpG islands did not improve the predictors. This suggests that despite the marked differences in histone modifications among these TSS classes, a single predictor can be made, using the interactions between the different histone modification levels.

In summary, we have integrated many data sets to provide a more complete view of regulatory information, both around specific sites (TSSs and DHSs) and in an unbiased manner. From analysing multiple data sets, we find 4,491 known and novel TSSs in the ENCODE regions, almost tenfold more than the number of established genes. This large number of TSSs might explain the extensive transcription described above; it also begins to change our perspective about regulatory information—without such a large TSS catalogue, many of the regulatory clusters would have been classified as residing distal to promoters. In addition to this revelation about the abundance of promoter-proximal regulatory elements, we also identified a considerable number of putative distal regulatory elements, particularly on the basis of the presence of DHSs. Our study of distal regulatory elements was probably most hindered by the paucity of data generated using distal-element-associated transcription factors; nevertheless, we clearly detected a set of distal-DHS-associated segments bound by CTCF or MYC. Finally, we showed that information about chromatin structure alone could be used to make effective predictions about both the location and activity of TSSs.

Replication

Overview. DNA replication must be carefully coordinated, both across the genome and with respect to development. On a larger scale, early replication in S phase is broadly correlated with gene density and transcriptional activity[59-66]; however, this relationship is not universal, as some actively transcribed genes replicate late and vice versa[61,64-68]. Importantly, the relationship between transcription and DNA replication emerges only when the signal of transcription is averaged over a large window ($>$ 100 kb)[63], suggesting that larger-scale chromosomal architecture may be more important than the activity of specific genes[69].

The ENCODE Project provided a unique opportunity to examine whether individual histone modifications on human chromatin can be correlated with the time of replication and whether such correlations support the general relationship of active, open chromatin with early replication. Our studies also tested whether segments showing interallelic

修饰的 ChIP-chip 实验结果进行了比较。从中，我们以染色质修饰为变量开发了很多表达状态的预测器；这些是利用决策树和 SVM 派生的（详见补充信息 3.17）。其中最好的预测器正确预测了 91% 的表达状态（表达对不表达）。当预测算法吸收一个细胞系的结果去预测另一个细胞系的表达状态时，成功率并没有大幅降低。有趣的是，尽管组蛋白修饰富集情况在靠近 CpG 岛的 TSS 和和远离 CpG 岛的 TSS 上有很大不同（详见图 5 和补充信息 3.6），包含靠近 CpG 岛的信息后并没有对预测器有所改观。这说明尽管在这些 TSS 种类中组蛋白修饰有很大不同，利用不同组蛋白修饰水平的相互作用，仅需一个预测器即可。

总结一下，我们研究了很多数据集，目的是更全面地了解调控信息，不仅是特定位点（TSS 和 DHS）附近的，并且是以无偏性的方式。从多个数据集的分析中，我们在 ENCODE 区域发现了 4,491 个已知的和全新的 TSS，这几乎比已确定的基因数量多了十倍。这么大量的 TSS 可能解释了上文提到的广泛的转录，也开始改变我们关于调控信息的认识——如果没有这么一大类 TSS，很多调控簇将会被分类为远离启动子。除了揭露了靠近启动子的调控元件的丰富性外，我们还鉴定到相当数量的假定远端调控元件，特别是基于 DHS 的存在。我们关于远端调控元件的研究最可能被远端元件相关转录因子的数据的缺乏所影响，但是，我们还是清楚地检测到了一组被 CTCF 或 MYC 结合的远端 DHS 相关片段。最后，我们表明仅靠染色质结构的信息就能对 TSS 的位置和活性做出有效的预测。

复　制

概况　无论是在基因组范围内，还是在发育过程中，DNA 复制必须被精确地协调。在更大的尺度下，在 S 期的早期复制与基因密度和转录活性有着广泛的相关性 [59-66]；但是，这种关系并不是普遍适用的，因为一些活跃转录的基因复制较晚，反之亦然 [61,64-68]。重要的是，只有当转录信号在一个较大的窗口（ > 100 kb）中被平均时，转录与 DNA 复制之间的联系才会出现 [63]，这说明大尺度的染色体构建可能比特定基因的活性更重要 [69]。

ENCODE 计划提供了一个独特的机会来检验是否人类染色质上单一的组蛋白修饰可以与复制的时间相关联，以及是否这种关联支持活跃的开放染色质与早期复制

variation in the time of replication have two different types of histone modifications consistent with an interallelic variation in chromatin state.

DNA replication data set. We mapped replication timing across the ENCODE regions by analysing Brd-U-labelled fractions from synchronized HeLa cells (collected at 2 h intervals throughout S phase) on tiling arrays (see Supplementary Information section 4.1). Although the HeLa cell line has a considerably altered karyotype, correlation of these data with other cell line data (see below) suggests the results are relevant to other cell types. The results are expressed as the time at which 50% of any given genomic position is replicated (TR50), with higher values signifying later replication times. In addition to the five "activating" histone marks, we also correlated the TR50 with H3K27me3, a modification associated with polycomb-mediated transcriptional repression[70-74]. To provide a consistent comparison framework, the histone data were smoothed to 100-kb resolution, and then correlated with the TR50 data by a sliding window correlation analysis (see Supplementary Information section 4.2). The continuous profiles of the activating marks, histone H3K4 mono-, di-, and tri-methylation and histone H3 and H4 acetylation, are generally anti-correlated with the TR50 signal (Fig. 7a and Supplementary Information section 4.3). In contrast, H3K27me3 marks show a predominantly positive correlation with late-replicating segments (Fig. 7a; see Supplementary Information section 4.3 for additional analysis).

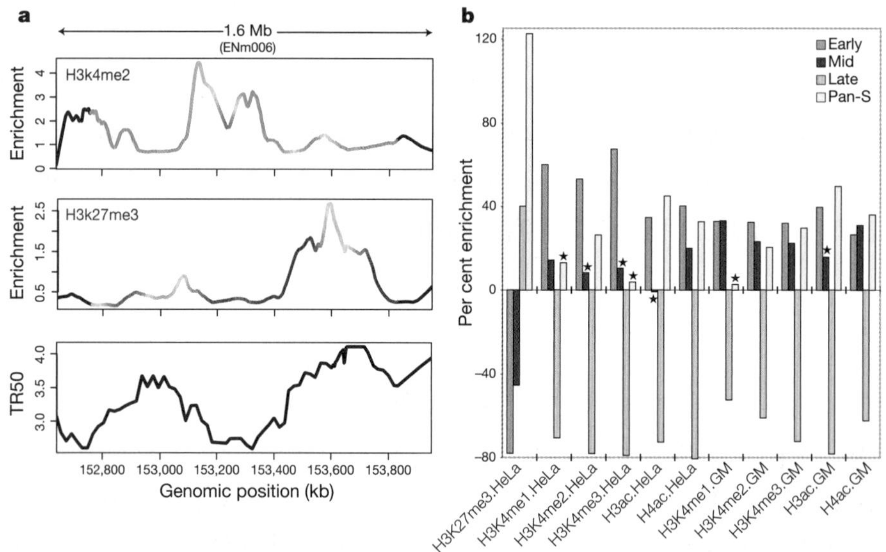

Fig. 7. Correlation between replication timing and histone modifications. **a**, Comparison of two histone modifications (H3K4me2 and H3K27me3), plotted as enrichment ratio from the Chip-chip experiments and the time for 50% of the DNA to replicate (TR50), indicated for ENCODE region ENm006. The colours on the curves reflect the correlation strength in a sliding 250-kb window. **b**, Differing levels of histone modification for different TR50 partitions. The amounts of enrichment or depletion of different histone modifications in various cell lines are depicted (indicated along the bottom as "histone mark.cell line"; GM = GM06990). Asterisks indicate enrichments/depletions that are not significant on the basis of multiple tests. Each set has four partitions on the basis of replication timing: early, mid, late and Pan-S.

的基本关系。我们的研究还检验了表现为等位基因间复制时间不同的片段是否有两种不同的组蛋白修饰，即染色质状态在等位基因间不同。

DNA 复制数据集 我们利用叠瓦式阵列分析 Brd-U 标记的同步化的 HeLa 细胞(在 S 期每隔 2h 收集)，绘制 ENCODE 区域的复制时间图谱(详见补充信息 4.1)。尽管 HeLa 细胞系核型变化较大，但是利用其他细胞系数据(见下文)进行校正发现，结果跟其他细胞系是相近的。结果以任一给定基因组位置在 50% 的细胞中被复制时的时间(TR50)表示，数值越大代表复制时间越晚。除了五个“激活的”组蛋白标志，我们还将 TR50 与 H3K27me3 关联了起来，H3K27me3 是一个与多梳蛋白介导的转录抑制相关的修饰 [70-74]。为了统一比较，组蛋白数据被平滑化到 100 kb 的分辨率，然后通过滑动窗口相关性分析将其与 TR50 数据关联(详见补充信息 4.2)。激活标志物组蛋白 H3K4 单、二和三甲基化以及组蛋白 H3 和 H4 乙酰化的连续图谱通常与 TR50 信号负相关(图 7a 和补充信息 4.3)。相反，H3K27me3 标志物则明显与晚期复制片段正相关(图 7a；额外的分析详见补充信息 4.3)。

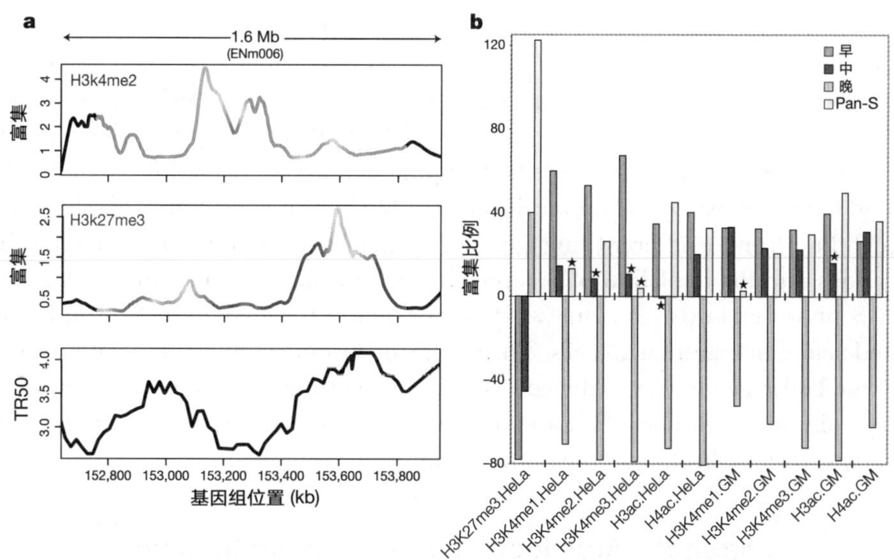

图 7. 复制时间和组蛋白修饰的相关性。**a**，比较两种组蛋白修饰(H3K4m2 和 H3K27me3)，以 ChIP-chip 实验得到的富集率和 50% 的 DNA 复制的时间(TR50)画图，展示的是 ENCODE 区域 ENm006。曲线的颜色表示在一个 250 kb 的滑动窗口内的相关强度。**b**，对于不同 TR50 段组蛋白修饰的不同水平。图中描绘的是不同组蛋白修饰在不同细胞系中富集或缺失的量(在底部以“组蛋白标志 . 细胞系”表示；GM = GM06990)。星号表示基于多重检验不显著的富集/缺失。基于复制时间各组含有四段：早、中、晚和 Pan-S。

Although most genomic regions replicate in a temporally specific window in S phase, other regions demonstrate an atypical pattern of replication (Pan-S) where replication signals are seen in multiple parts of S phase. We have suggested that such a pattern of replication stems from interallelic variation in the chromatin structure[59,75]. If one allele is in active chromatin and the other in repressed chromatin, both types of modified histones are expected to be enriched in the Pan-S segments. An ENCODE region was classified as non-specific (or Pan-S) regions when > 60% of the probes in a 10-kb window replicated in multiple intervals in S phase. The remaining regions were sub-classified into early-, mid- or late-replicating based on the average TR50 of the temporally specific probes within a 10-kb window[75]. For regions of each class of replication timing, we determined the relative enrichment of various histone modification peaks in HeLa cells (Fig. 7b; Supplementary Information section 4.4). The correlations of activating and repressing histone modification peaks with TR50 are confirmed by this analysis (Fig. 7b). Intriguingly, the Pan-S segments are unique in being enriched for both activating (H3K4me2, H3ac and H4ac) and repressing (H3K27me3) histones, consistent with the suggestion that the Pan-S replication pattern arises from interallelic variation in chromatin structure and time of replication[75]. This observation is also consistent with the Pan-S replication pattern seen for the H19/IGF2 locus, a known imprinted region with differential epigenetic modifications across the two alleles[76].

The extensive rearrangements in the genome of HeLa cells led us to ask whether the detected correlations between TR50 and chromatin state are seen with other cell lines. The histone modification data with GM06990 cells allowed us to test whether the time of replication of genomic segments in HeLa cells correlated with the chromatin state in GM06990 cells. Early- and late-replicating segments in HeLa cells are enriched and depleted, respectively, for activating marks in GM06990 cells (Fig. 7b). Thus, despite the presence of genomic rearrangements (see Supplementary Information section 2.12), the TR50 and chromatin state in HeLa cells are not far from a constitutive baseline also seen with a cell line from a different lineage. The enrichment of multiple activating histone modifications and the depletion of a repressive modification from segments that replicate early in S phase extends previous work in the field at a level of detail and scale not attempted before in mammalian cells. The duality of histone modification patterns in Pan-S areas of the HeLa genome, and the concordance of chromatin marks and replication time across two disparate cell lines (HeLa and GM06990) confirm the coordination of histone modifications with replication in the human genome.

Chromatin Architecture and Genomic Domains

Overview. The packaging of genomic DNA into chromatin is intimately connected with the control of gene expression and other chromosomal processes. We next examined chromatin structure over a larger scale to ascertain its relation to transcription and other processes. Large domains (50 to > 200 kb) of generalized DNaseI sensitivity have been detected around developmentally regulated gene clusters[77], prompting speculation that the genome is organized into "open" and "closed" chromatin territories that represent higher-order functional domains. We explored how different chromatin features, particularly histone

尽管大多数的基因组区域在 S 期以一种时间特异窗口复制，也有一些区域呈现非典型的复制模式（Pan-S），即复制信号出现在 S 期的多个阶段。我们已经表明这样一种复制模式源于染色质结构在等位基因间的不同[59,75]。如果一个等位基因位于活跃的染色质而另一个位于被抑制的染色质，则这两种修饰的组蛋白都会在 Pan-S 区间富集。当一个 ENCODE 区域在 10 kb 窗口里有 >60% 的探针在 S 期的多个阶段被复制，则这个区域被归类为非特异的（或者 Pan-S）区域。基于在 10 kb 窗口内的时间特异探针的平均 TR50 值，剩下的区域被进一步分类为早、中或晚期复制[75]。对于每一类复制时间的区域，我们确定了各种组蛋白修饰峰在 HeLa 细胞中的相对富集（图 7b；补充信息 4.4）。这个分析证实了激活和抑制的组蛋白修饰峰与 TR50 的相关性（图 7b）。有趣的是，Pan-S 区间是独特的，激活（H3K4me2、H3ac 和 H4ac）和抑制（H3K27me3）的组蛋白都有富集，这也说明 Pan-S 复制模式是由于在染色质结构和复制时间方面等位基因的差异[75]。这种现象也与 H19/IGF2 基因座的 Pan-S 复制模式一致，该基因座是一个已知的两个等位基因不同的表观遗传学修饰印记区域。

HeLa 细胞广泛的基因组重排使我们不禁要问检测到的 TR50 和染色质状态之间的相关性是否能在其他细胞系中看到。GM06990 细胞的组蛋白修饰数据使我们能够检测是否 HeLa 细胞中基因组片段的复制时间与 GM06990 细胞中的染色质状态相关。HeLa 细胞中的早期和晚期复制片段分别在 GM06990 细胞中富集和缺失激活的标志物（图 7b）。因而，尽管存在基因组的重排（详见补充信息 2.12），HeLa 细胞中 TR50 和染色质状态差不多就是一个组成性的基准，也能在源于另一谱系的细胞系中看到。S 期早期复制片段的这种对多种激活性组蛋白修饰的富集和对抑制性修饰的缺失，以前所未有的详细度和尺度拓展了之前在哺乳动物细胞中这一领域的工作。HeLa 基因组 Pan-S 区域的组蛋白修饰模式的二元性，以及在两种不同的细胞系（HeLa 和 GM06990）中染色质标志物和复制时间的一致性，验证了人类基因组中组蛋白修饰和复制的相关性。

染色质结构和基因组结构域

概况　基因组 DNA 包装成染色质与基因表达的调控以及其他染色质过程紧密相关。接下来，我们探究更大尺度的染色质结构来确定它与转录和其他过程的关系。在发育调控的基因簇附近已经检测到广泛的 DNaseI 敏感性的大结构域（50 到 >200 kb）[77]，这促使我们猜测基因组被组成“开放”和“闭合”的染色质区域，代表着更高级的功能结构域。我们探索了不同的染色质特征，特别是组蛋白修饰，在短

modifications, correlate with chromatin structure, both over short and long distances.

Chromatin accessibility and histone modifications. We used histone modification studies and DNaseI sensitivity data sets (introduced above) to examine general chromatin accessibility without focusing on the specific DHS sites (see Supplementary Information sections 3.1, 3.3 and 3.4). A fundamental difficulty in analysing continuous data across large genomic regions is determining the appropriate scale for analysis (for example, 2 kb, 5 kb, 20 kb, and so on). To address this problem, we developed an approach based on wavelet analysis, a mathematical tool pioneered in the field of signal processing that has recently been applied to continuous-value genomic analyses. Wavelet analysis provides a means for consistently transforming continuous signals into different scales, enabling the correlation of different phenomena independently at differing scales in a consistent manner.

Global correlations of chromatin accessibility and histone modifications. We computed the regional correlation between DNaseI sensitivity and each histone modification at multiple scales using a wavelet approach (Fig. 8 and Supplementary Information section 4.2). To make quantitative comparisons between different histone modifications, we computed histograms of correlation values between DNaseI sensitivity and each histone modification at several scales and then tested these for significance at specific scales. Figure 8c shows the distribution of correlation values at a 16-kb scale, which is considerably larger than individual *cis*-acting regulatory elements. At this scale, H3K4me2, H3K4me3 and H3ac show similarly high correlation. However, they are significantly distinguished from H3K4me1 and H4ac modifications ($P < 1.5 \times 10^{-33}$; see Supplementary Information section 4.5), which show lower correlation with DNaseI sensitivity. These results suggest that larger-scale relationships between chromatin accessibility and histone modifications are dominated by sub-regions in which higher average DNaseI sensitivity is accompanied by high levels of H3K4me2, H3K4me3 and H3ac modifications.

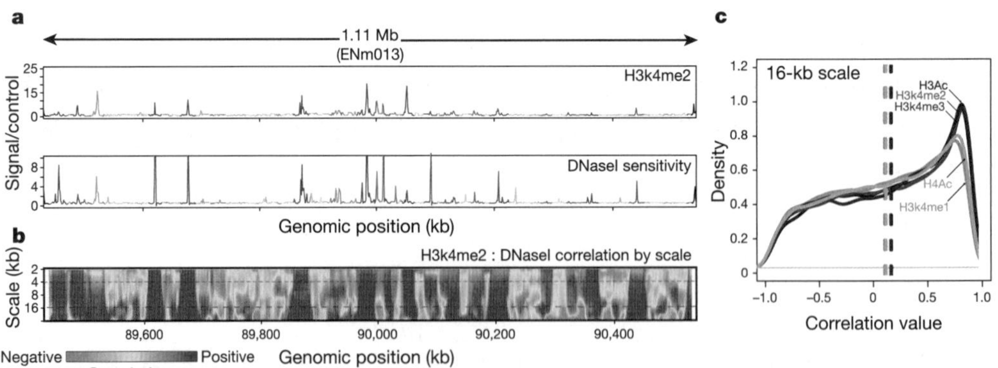

Fig. 8. Wavelet correlations of histone marks and DNaseI sensitivity. As an example, correlations between DNaseI sensitivity and H3K4me2 (both in the GM06990 cell line) over a 1.1-Mb region on chromosome 7 (ENCODE region ENm013) are shown. **a**, The relationship between histone modification H3K4me2 (upper plot) and DNaseI sensitivity (lower plot) is shown for ENCODE region ENm013. The curves are coloured with the strength of the local correlation at the 4-kb scale (top dashed line in panel **b**). **b**, The same data as in **a** are represented as a wavelet correlation. The *y* axis shows the differing scales decomposed

距离以及长距离方面，如何与染色质结构关联。

染色质开放性和组蛋白修饰 我们使用了组蛋白修饰研究和 DNaseI 敏感性数据集（上文介绍的）来探究整体的染色质开放性，而不是聚焦于特定的 DHS 位点（详见补充信息 3.1、3.3 和 3.4）。在大的基因组区域范围内的连续数据方面，一个主要的困难是确定合适的分析尺度（如 2 kb、5 kb、20 kb 等等）。为了解决这个问题，我们基于小波分析开发了一种方法，而小波分析是主要应用于信号处理领域的一种数学工具，最近已经被应用到连续值的基因组分析当中。小波分析提供了一种可以一致地将连续信号转换为不同尺度的方法，从而能以一种一致的方式将不同的现象在不同尺度下独立地关联。

染色质开放性与组蛋白修饰的整体相关性 我们利用小波的方法，在多种尺度下计算了 DNaseI 敏感性与每一种组蛋白修饰区域的相关性（图 8 和补充信息 4.2）。为了定量比较不同的组蛋白修饰，我们计算了 DNaseI 敏感性和每一种组蛋白修饰在多个尺度下的相关性值并画了直方图，然后对其进行检验以寻找在特定的尺度下的显著性。图 8c 展示了在 16 kb 尺度下相关性值的分布，它的值明显比单个顺式调控元件的大。在这个尺度下，H3K4me2、H3K4me3 和 H3ac 表现类似，都是高度相关。但是，它们跟 H3K4me1 和 H4ac 修饰显著不同（$P < 1.5 \times 10^{-33}$；详见补充信息 4.5），后者与 DNaseI 敏感性的相关性更低。这些结果表明染色质开放性和组蛋白修饰在大尺度下的关系，主要取决于更高 DNaseI 敏感性平均值伴随着较高水平 H3K4me2、H3K4me3 和 H3ac 修饰的亚区域。

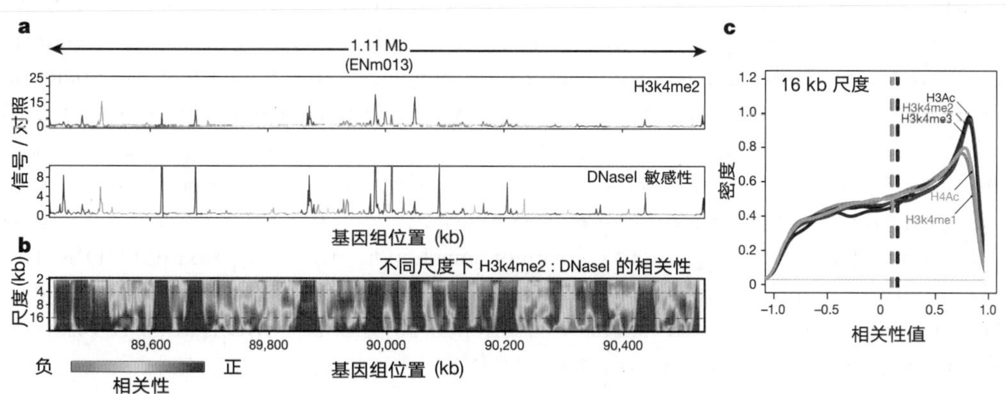

图 8. 组蛋白标志物和 DNaseI 敏感性的小波关联。作为一个例子，展示了在 7 号染色体超过 1.1 Mb 的区域内（ENCODE 区域 ENm013）DNaseI 敏感性和 H3K4me2（都是在 GM06990 细胞系）之间的相关性。**a**，展示的是 ENCODE 区域 ENm013 内组蛋白修饰 H3K4me2（上图）和 DNaseI 敏感性（下图）之间的关系。曲线的颜色是根据在 4 kb 尺度下局部相关性的强度（**b** 图中上方的虚线）。**b**，与 **a** 中同样的数据作

by the wavelet analysis from large to small scale (in kb); the colour at each point in the heatmap represents the level of correlation at the given scale, measured in a 20 kb window centred at the given position. **c**, Distribution of correlation values at the 16 kb scale between the indicated histone marks. The *y* axis is the density of these correlation values across ENCODE; all modifications show a peak at a positive-correlation value.

Local correlations of chromatin accessibility and histone modifications. Narrowing to a scale of ~2 kb revealed a more complex situation, in which H3K4me2 is the histone modification that is best correlated with DNaseI sensitivity. However, there is no clear combination of marks that correlate with DNaseI sensitivity in a way that is analogous to that seen at a larger scale (see Supplementary Information section 4.3). One explanation for the increased complexity at smaller scales is that there is a mixture of different classes of accessible chromatin regions, each having a different pattern of histone modifications. To examine this, we computed the degree to which local peaks in histone methylation or acetylation occur at DHSs (see Supplementary Information section 4.5.1). We found that 84%, 91% and 93% of significant peaks in H3K4 mono-, di- and tri-methylation, respectively, and 93% and 81% of significant peaks in H3ac and H4ac acetylation, respectively, coincided with DHSs (see Supplementary Information section 4.5). Conversely, a proportion of DHSs seemed not to be associated with significant peaks in H3K4 mono-, di- or tri-methylation (37%, 29% and 47%, respectively), nor with peaks in H3 or H4 acetylation (both 57%). Because only a limited number of histone modification marks were assayed, the possibility remains that some DHSs harbour other histone modifications. The absence of a more complete concordance between DHSs and peaks in histone acetylation is surprising given the widely accepted notion that histone acetylation has a central role in mediating chromatin accessibility by disrupting higher-order chromatin folding.

DNA structure at DHSs. The observation that distinctive hydroxyl radical cleavage patterns are associated with specific DNA structures[78] prompted us to investigate whether DHS subclasses differed with respect to their local DNA structure. Conversely, because different DNA sequences can give rise to similar hydroxyl radical cleavage patterns[79], genomic regions that adopt a particular local structure do not necessarily have the same nucleotide sequence. Using a Gibbs sampling algorithm on hydroxyl radical cleavage patterns of 3,150 DHSs[80], we discovered an 8-base segment with a conserved cleavage signature (CORCS; see Supplementary Information section 4.6). The underlying DNA sequences that give rise to this pattern have little primary sequence similarity despite this similar structural pattern. Furthermore, this structural element is strongly enriched in promoter-proximal DHSs (11.3-fold enrichment compared to the rest of the ENCODE regions) relative to promoter-distal DHSs (1.5-fold enrichment); this element is enriched 10.9-fold in CpG islands, but is higher still (26.4-fold) in CpG islands that overlap a DHS.

Large-scale domains in the ENCODE regions. The presence of extensive correlations seen between histone modifications, DNaseI sensitivity, replication, transcript density and protein factor binding led us to investigate whether all these features are organized systematically across the genome. To test this, we performed an unsupervised training of a two-state HMM with inputs from these different features (see Supplementary Information

为小波关联的代表。y 轴从大到小表示小波分析分解的不同尺度（单位 kb）；热图中每一个点的颜色代表在给定尺度下以给定位置为中心的 20 kb 窗口内测得的相关水平。c，在 16 kb 尺度下与相应组蛋白标志物的相关性值的分布。y 轴是这些相关性值在 ENCODE 区域的密度；所有的修饰在正相关值范围内都有一个峰。

染色质开放性和组蛋白修饰的局部相关性 将尺度缩小到约 2 kb 发现了一个更加复杂的情况，H3K4m2 是跟 DNaseI 敏感性最相关的组蛋白修饰。但是，没有明显的标志物组合与在更大尺度下看到的类似的方式与 DNaseI 敏感性相关（详见补充信息 4.3）。对于在更小尺度下增加的复杂性，一种解释是它混合了不同类型的开放染色质区域，每一类型含有不同的组蛋白修饰模式。为了研究这种可能性，我们计算了组蛋白甲基化或乙酰化中有多少局部峰是在 DHS 上（详见补充信息 4.5.1）。我们发现 H3K4 单、二和三甲基化中分别有 84%，91% 和 93% 显著的峰，以及 H3ac 和 H4ac 中分别有 93% 和 81% 的显著峰，与 DHS 是一致的（详见补充信息 4.5）。相反，一部分 DHS 则好像并不与 H3K4 单、二或三甲基化中显著的峰关联（分别为 37%、29% 和 47%），也不与 H3 或 H4 乙酰化中的峰关联（均为 57%）。因为只有有限数量的组蛋白修饰标志被检测，所以一些含有其他的组蛋白修饰的 DHS 也可能存在。鉴于广泛接受的观点是组蛋白乙酰化通过破坏更高级的染色质折叠而在介导染色质开放性方面发挥核心作用，DHS 和组蛋白乙酰化峰之间缺少更完全的一致性是出乎意料的。

DHS 处的 DNA 结构 观察到不同的羟自由基切割模式与特定的 DNA 结构[78]相关，促使我们研究 DHS 亚类是否会因它们局部的 DNA 结构而不同。反过来，因为不同的 DNA 序列能产生相似的羟自由基切割模式[79]，采用特定一种局部结构的基因组区域也未必含有相同的核苷酸序列。对 3,150 个 DHS 的羟自由基切割模式应用吉布斯采样算法，我们发现了一个 8 碱基片段含有保守的切割特征（CORCS；详见补充信息 4.6）。尽管有着这种类似的结构模式，产生这种模式的潜在 DNA 序列在一级序列上却几乎没有相似性。另外，这种结构元件在靠近启动子的 DHS（跟其他的 ENCODE 区域相比有 11.3 倍富集）比在远离启动子的 DHS 明显富集（1.5 倍富集），这种元件在 CpG 有 10.9 倍的富集，而在与 DHS 重合 CpG 到则富集倍数更高（26.4 倍）。

在 ENCODE 区域内的大尺度结构域 组蛋白修饰、DNaseI 敏感性、复制、转录本密度和蛋白质因子结合间广泛的相关性的存在，促使我们研究是否所有这些特征是在基因组范围内系统地组织起来的。为了检验这种可能性，我们输入这些不同的特征数据，无监督训练一个两态的 HMM（详见补充信息 4.7 和参考文献 81）。在

section 4.7 and ref. 81). No other information except for the experimental variables was used for the HMM training routines. We consistently found that one state ("active") generally corresponded to domains with high levels of H3ac and RNA transcription, low levels of H3K27me3 marks, and early replication timing, whereas the other state ("repressed") reflected domains with low H3ac and RNA, high H3K27me3, and late replication (see Fig. 9). In total, we identified 70 active regions spanning 11.4 Mb and 82 inactive regions spanning 17.8 Mb (median size 136 kb versus 104 kb respectively). The active domains are markedly enriched for GENCODE TSSs, CpG islands and Alu repetitive elements ($P < 0.0001$ for each), whereas repressed regions are significantly enriched for LINE1 and LTR transposons ($P < 0.001$). Taken together, these results demonstrate remarkable concordance between ENCODE functional data types and provide a view of higher-order functional domains defined by a broader range of factors at a markedly higher resolution than was previously available[82].

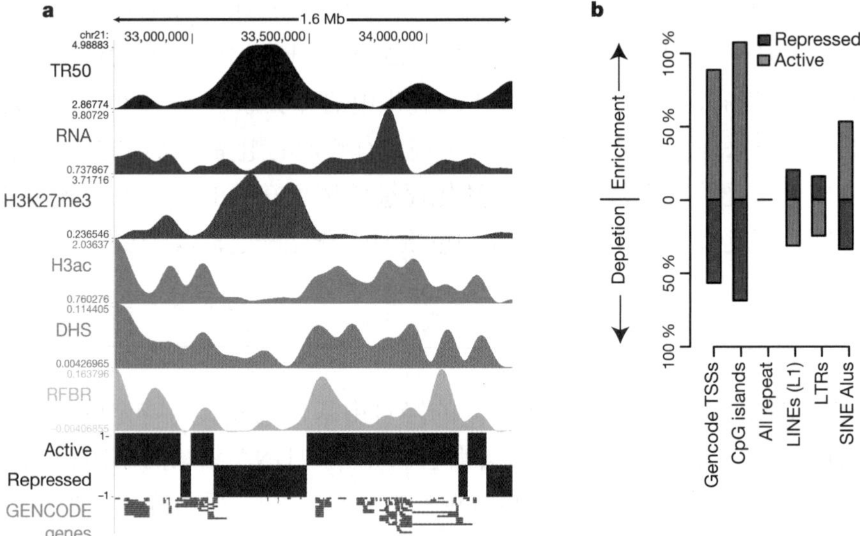

Fig. 9. Higher-order functional domains in the genome. The general concordance of multiple data types is shown for an illustrative ENCODE region (ENm005). **a**, Domains were determined by simultaneous HMM segmentation of replication time (TR50; black), bulk RNA transcription (blue), H3K27me3 (purple), H3ac (orange), DHS density (green), and RFBR density (light blue) measured continuously across the 1.6-Mb ENm005. All data were generated using HeLa cells. The histone, RNA, DHS and RFBR signals are wavelet-smoothed to an approximately 60-kb scale (see Supplementary Information section 4.7). The HMM segmentation is shown as the blocks labelled "active" and "repressed" and the structure of GENCODE genes (not used in the training) is shown at the end. **b**, Enrichment or depletion of annotated sequence features (GENCODE TSSs, CpG islands, LINE1 repeats, Alu repeats, and non-exonic constrained sequences (CSs)) in active versus repressed domains. Note the marked enrichment of TSSs, CpG islands and Alus in active domains, and the enrichment of LINE and LTRs in repressed domains.

Evolutionary Constraint and Population Variability

Overview. Functional genomic sequences can also be identified by examining evolutionary changes across multiple extant species and within the human population. Indeed, such

HMM 训练过程中，除了实验变量，没有使用其他的信息。我们不断发现一个状态（“活跃的”）通常对应于含有高水平 H3ac 和 RNA 转录、低水平 H3K27me3 标志物以及早期复制时间的结构域，而另一个状态（“抑制的”）反映的是含有低水平 H3ac 和 RNA、高水平 H3K27me3 以及晚期复制的结构域（见图 9）。总共，我们鉴定到横跨 11.4 Mb 的 70 个活跃区域和横跨 17.8 Mb 的 82 个不活跃的区域（中位数大小分别为 136 kb 和 104 kb）。活跃的结构域明显富集 GENCODE TSS、CpG 岛和 Alu 重复元件（每个都 $P < 0.0001$），而抑制的区域则明显富集 LINE1 和 LTR 转座子（$P < 0.001$）。总而言之，这些结果显示了 ENCODE 功能数据类型间显著的一致性，并且提供一个比之前已有的更高级的功能结构域图谱，该功能结构域被更广范围的因子所定义以及有着明显更高的分辨率。

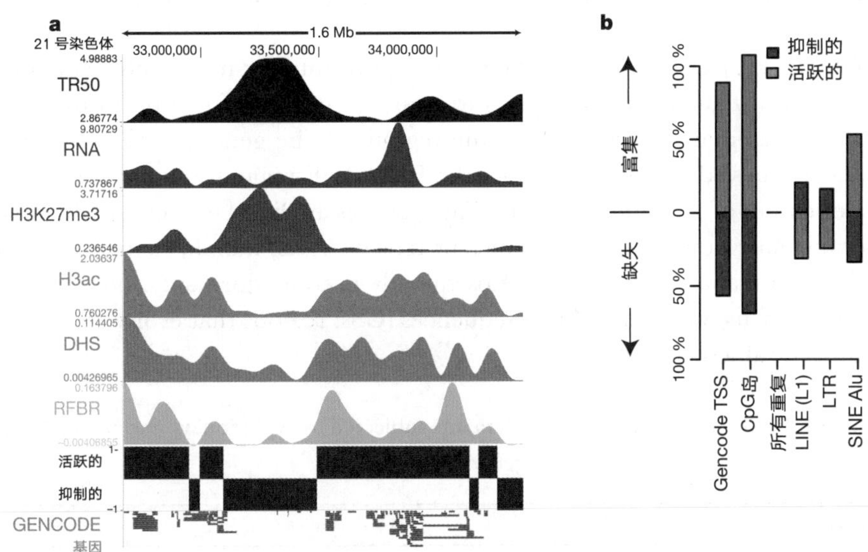

图 9. 基因组中更高级的功能结构域。本图展示的是一个用来说明 ENCODE 区域（ENm005）内多种数据类型的普遍一致性。**a**，结构域是通过对 1.6 Mb 的 ENm005 范围内对复制时间（TR50；黑色）、大量 RNA 转录（蓝色）、H3K27me3（紫色）、H3ac（橘色）、DHS 密度（绿色）和 RFBR 密度（淡蓝色）连续测量，同时进行 HMM 分段而确定的。所有的数据都是利用 HeLa 细胞产生的。组蛋白、RNA、DHS 和 RFBR 信号被小波平滑化成大约 60 kb 的尺度（详见补充信息 4.7）。HMM 分段以标为“活跃的”和“抑制的”的区块表示，最下面是 GENCODE 基因的结构（训练中没有使用）。**b**，在活跃和抑制的结构域中被注释的序列特征（GENCODE TSS、CpG 岛、LINE1 重复、Alu 重复和非外显子的限制序列（CS））的富集或缺失的比较。注意在活跃的结构域中 TSS、CpG 岛和 Alu 明显富集，以及在抑制的结构域中 LINE 和 LTR 富集。

进化的限制和群体的变异性

概况 功能性的基因组序列也能通过分析多个现存物种间和人群内部进化的变化而被鉴定。确实，这样的研究补充了鉴定特定功能元件的实验方法[83-85]。进化的

studies complement experimental assays that identify specific functional elements[83-85]. Evolutionary constraint (that is, the rejection of mutations at a particular location) can be measured by either (i) comparing observed substitutions to neutral rates calculated from multi-sequence alignments[86-88], or (ii) determining the presence and frequency of intra-species polymorphisms. Importantly, both approaches are indifferent to any specific function that the constrained sequence might confer.

Previous studies comparing the human, mouse, rat and dog genomes examined bulk evolutionary properties of all nucleotides in the genome, and provided little insight about the precise positions of constrained bases. Interestingly, these studies indicated that the majority of constrained bases reside within the non-coding portion of the human genome. Meanwhile, increasingly rich data sets of polymorphisms across the human genome have been used extensively to establish connections between genetic variants and disease, but far fewer analyses have sought to use such data for assessing functional constraint[85].

The ENCODE Project provides an excellent opportunity for more fully exploiting inter- and intra-species sequence comparisons to examine genome function in the context of extensive experimental studies on the same regions of the genome. We consolidated the experimentally derived information about the ENCODE regions and focused our analyses on 11 major classes of genomic elements. These classes are listed in Table 4 and include two non-experimentally derived data sets: ancient repeats (ARs; mobile elements that inserted early in the mammalian lineage, have subsequently become dormant, and are assumed to be neutrally evolving) and constrained sequences (CSs; regions that evolve detectably more slowly than neutral sequences).

Table 4. Eleven classes of genomic elements subjected to evolutionary and population-genetics analyses

Abbreviation	Description
CDS	Coding exons, as annotated by GENCODE
5′UTR	5′ untranslated region, as annotated by GENCODE
3′UTR	3′ untranslated region, as annotated by GENCODE
Un.TxFrag	Unannotated region detected by RNA hybridization to tiling array (that is, unannotated TxFrag)
RxFrag	Region detected by RACE and analysis on tiling array
Pseudogene	Pseudogene identified by consensus pseudogene analysis
RFBR	Regulatory factor binding region identified by ChIP-chip assays
RFBR-SeqSp	Regulatory factor binding region identified only by ChIP-chip assays for factors with known sequence-specificity
DHS	DNaseI hypersensitive sites found in multiple tissues
FAIRE	Region of open chromatin identified by the FAIRE assay
TSS	Transcription start site
AR	Ancient repeat inserted early in the mammalian lineage and presumed to be neutrally evolving
CS	Constrained sequence identified by analysing multi-sequence alignments

限制（即在特定位置排斥变异）能通过（i）比较观察到的替换与多重序列比对计算得到的中性率[86-88]，或者（ii）确定物种内多态性的存在和频率而被衡量。重要的是，两种方法无关乎限制序列可能发挥的任何特定功能。

之前的研究比较了人类、小鼠、大鼠和狗的基因组，分析了基因组中所有核苷酸大的进化特性，而对受限制的碱基的精确位置几乎没有了解。有意思的是，这些研究表明大部分受限制的碱基位于人类基因组的非编码部分。同时，人类基因组范围的多态性的数据集越来越丰富，并且已经被广泛地用于建立遗传变异与疾病的联系，但是目前利用这样的数据来评价功能限制的分析还很少[85]。

ENCODE 计划提供了一个很好的机会来更加全面地探索物种间和物种内的序列比较，进而结合在基因组相同区域广泛进行的实验研究来研究基因组的功能。我们统一了实验得到关于 ENCODE 区域的信息，将我们的分析聚焦于 11 个主要类别的基因组元件。这些类别在表 4 中被列出，其中包含两个非实验得到数据集：古老的重复（AR；一种可移动的元件，在早期就插入到哺乳动物谱系但自此休眠，被认为是中性进化）和受限制的序列（CS；可检测到比中性序列进化得更缓慢的区域）。

表 4. 十一类被用于进化和群体遗传学分析的基因组元件

缩写	描述
CDS	编码外显子，根据 GENCODE 注释
5′UTR	5′非翻译区，根据 GENCODE 注释
3′UTR	3′非翻译区，根据 GENCODE 注释
Un.TxFrag	通过 RNA 杂交到叠瓦式阵列检测到的非翻译区（即非翻译的 TxFrag）
RxFrag	通过 RACE 和叠瓦式阵列分析检测到的区域
假基因	通过共有假基因分析鉴定到的假基因
RFBR	通过 ChIP-chip 实验鉴定到的调控因子结合区域
RFBR-SeqSp	对于已知含有序列特异性的因子，仅通过 ChIP-chip 实验鉴定到的调控因子结合区域
DHS	在多种组织中发现的 DNaseI 超敏感位点
FAIRE	通过 FAIRE 实验鉴定到的开放染色质区域
TSS	转录起始位点
AR	在早期就插入到哺乳动物谱系中的古老的重复，被认为是中性进化
CS	通过多重序列比对分析鉴定到的限制序列

Comparative sequence data sets and analysis. We generated 206 Mb of genomic sequence orthologous to the ENCODE regions from 14 mammalian species using a targeted strategy that involved isolating[89] and sequencing[90] individual bacterial artificial chromosome clones. For an additional 14 vertebrate species, we used 340 Mb of orthologous genomic sequence derived from genome-wide sequencing efforts[3-8,91-93]. The orthologous sequences were aligned using three alignment programs: TBA[94], MAVID[95] and MLAGAN[96]. Four independent methods that generated highly concordant results[97] were then used to identify sequences under constraint (PhastCons[88], GERP[87], SCONE[98] and BinCons[86]). From these analyses, we developed a high-confidence set of "constrained sequences" that correspond to 4.9% of the nucleotides in the ENCODE regions. The threshold for determining constraint was set using a FDR rate of 5% (see ref. 97); this level is similar to previous estimates of the fraction of the human genome under mammalian constraint[4,86-88] but the FDR rate was not chosen to fit this result. The median length of these constrained sequences is 19 bases, with the minimum being 8 bases—roughly the size of a typical transcription factor binding site. These analyses, therefore, provide a resolution of constrained sequences that is substantially better than that currently available using only whole-genome vertebrate sequences[99-102].

Intra-species variation studies mainly used SNP data from Phases I and II, and the 10 re-sequenced regions in ENCODE regions with 48 individuals of the HapMap Project[103]; nucleotide insertion or deletion (indel) data were from the SNP Consortium and HapMap. We also examined the ENCODE regions for the presence of overlaps with known segmental duplications[104] and CNVs.

Experimentally identified functional elements and constrained sequences. We first compared the detected constrained sequences with the positions of experimentally identified functional elements. A total of 40% of the constrained bases reside within protein-coding exons and their associated untranslated regions (Fig. 10) and, in agreement with previous genome-wide estimates, the remaining constrained bases do not overlap the mature transcripts of protein-coding genes[4,5,88,105,106]. When we included the other experimental annotations, we found that an additional 20% of the constrained bases overlap experimentally identified non-coding functional regions, although far fewer of these regions overlap constrained sequences compared to coding exons (see below). Most experimental annotations are significantly different from a random expectation for both base-pair or element-level overlaps (using the GSC statistic, see Supplementary Information section 1.3), with a more striking deviation when considering elements (Fig. 11). The exceptions to this are pseudogenes, Un.TxFrags and RxFrags. The increase in significance moving from base-pair measures to the element level suggests that discrete islands of constrained sequence exist within experimentally identified functional elements, with the surrounding bases apparently not showing evolutionary constraint. This notion is discussed in greater detail in ref. 97.

可比较的序列数据集和分析。 利用包括分离 [89] 和测序 [90] 单个细菌人工染色体克隆的靶向策略，我们从 14 个哺乳动物物种中得到了 206 Mb 跟 ENCODE 区域直系同源的基因组序列。对于另外的 14 种脊椎动物，我们使用的是从全基因组测序得到的 340 Mb 直系同源的基因组序列 [3-8,91-93]。我们使用三种比对程序进行直系同源序列的比对：TBA[94]、MAVID[95] 和 MLAGAN[96]。四个独立的方法（PhastCons[88]、GERP[87]、SCONE[98] 和 BinCons[86]）得到了高度一致的结果 [97]，然后被用于鉴定受限制的序列。从这些分析中，我们形成了一组高度可信的“受限制的序列”，对应于 ENCODE 区域内 4.9% 的核苷酸。使用 5% 的 FDR 率作为受限制的阈值（见参考文献 97）；该水平跟之前估计的人类基因组中受哺乳动物限制的比例类似 [4,86-88]，但是 FDR 率并不是为了符合这个结果。这些受限制的序列长度的中位数是 19 个碱基，最短是 8 个碱基——差不多一个典型的转录因子结合位点的长度。因而，这些分析提供的受限制的序列的分辨率明显好于目前仅利用全基因组脊椎动物序列获得的 [99-102]。

物种内的变异研究主要使用来自 HapMap 计划 I 和 II 期的 SNP 数据，以及其中 48 个个体的 10 个位于 ENCODE 区域的区域的重测序数据 [103]；核苷酸的插入或者缺失（插入缺失位）数据来自 SNP 协作组和 HapMap。我们也分析了 ENCODE 区域与已知的片段复制 [104] 和 CNV 重合情况。

实验鉴定的功能元件和受限制的序列 我们首先将检测到的受限制的序列跟与实验鉴定到的功能元件进行了比较。总共 40% 受限制的碱基位于蛋白质编码的外显子和它们相关的非翻译区域（图 10），并且，与之前的全基因组估计一致，剩下的受限制的碱基不与蛋白质编码的基因的成熟转录本重合 [4,5,88,105,106]。当包括了其他实验注释时，我们发现了 20% 额外受限制的碱基与实验鉴定到的非编码功能区域重合，尽管与受限制序列重合的这些区域要远远少于编码的外显子（见下文）。无论是碱基对还是元件水平的重合，大多数的实验注释显著不同于随机预期（利用 GSC 统计，详见补充信息 1.3），当考虑元件时差异更加明显（图 11）。其中的例外是假基因、Un.TxFrag 和 RxFrag。从碱基对测量到元件水平时显著水平升高，说明受限制序列的分离的岛存在于实验鉴定到的功能元件内，而周围的碱基则明显不具有进化上的限制。这个观点在参考文献 97 中有更详细的讨论。

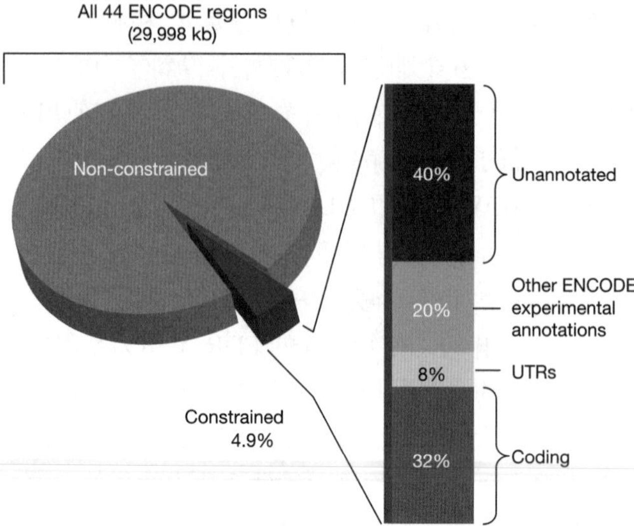

Fig. 10. Relative proportion of different annotations among constrained sequences. The 4.9% of bases in the ENCODE regions identified as constrained is subdivided into the portions that reflect known coding regions, UTRs, other experimentally annotated regions, and unannotated sequence.

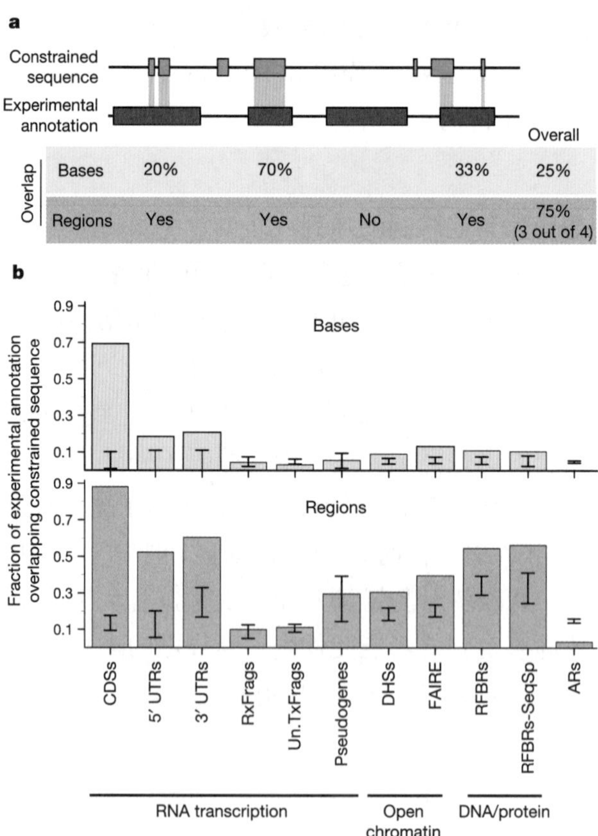

Fig. 11. Overlap of constrained sequences and various experimental annotations. **a**, A schematic depiction shows the different tests used for assessing overlap between experimental annotations and constrained

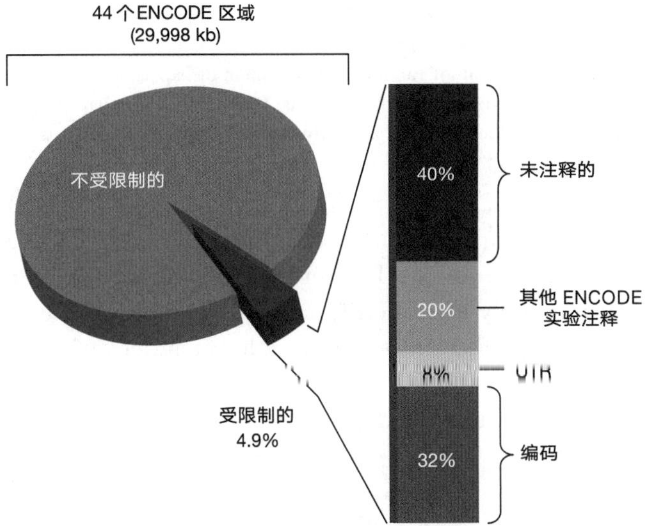

图 10. 不同注释在受限制序列中的相对比例。ENCODE 区域中有 4.9% 的碱基被鉴定为受限制的，其又被进一步分为已知的编码区域、UTR、其他实验注释的区域和未注释的序列。

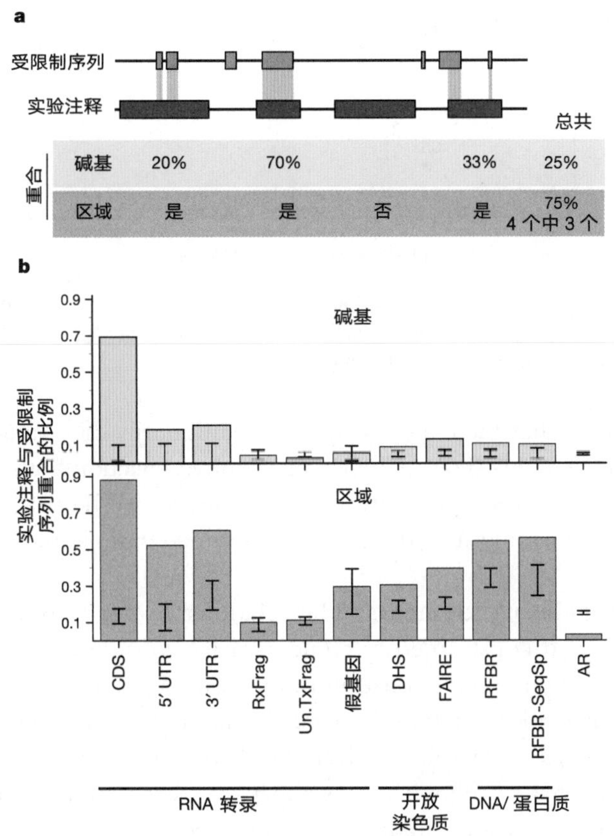

图 11. 受限制序列与各种实验注释的重合情况。**a**，示意图展示用于评价实验注释和受限制序列重合情况的不同检验方法，不只是对于单个碱基，而且针对整个区域。**b**，观察到的重合的比例，对于碱基和

sequences, both for individual bases and for entire regions. **b**, Observed fraction of overlap, depicted separately for bases and regions. The results are shown for selected experimental annotations. The internal bars indicate 95% confidence intervals of randomized placement of experimental elements using the GSC methodology to account for heterogeneity in the data sets. When the bar overlaps the observed value one cannot reject the hypothesis that these overlaps are consistent with random placements.

We also examined measures of human variation (heterozygosity, derived allele-frequency spectra and indel rates) within the sequences of the experimentally identified functional elements (Fig. 12). For these studies, ARs were used as a marker for neutrally evolving sequence. Most experimentally identified functional elements are associated with lower heterozygosity compared to ARs, and a few have lower indel rates compared with ARs. Striking outliers are 3′ UTRs, which have dramatically increased indel rates without an obvious cause. This is discussed in more depth in ref. 107.

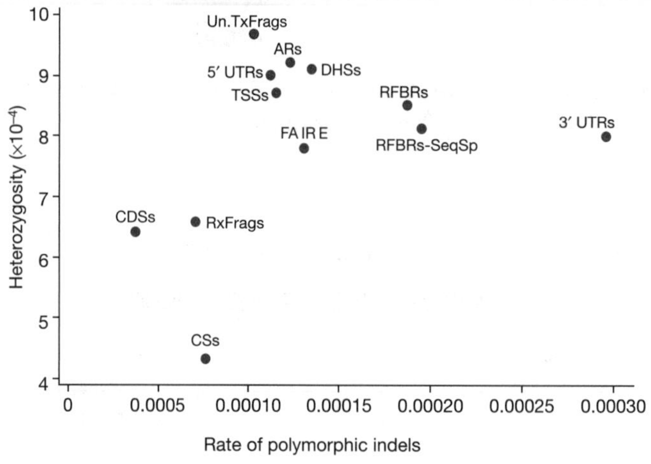

Fig. 12. Relationship between heterozygosity and polymorphic indel rate for a variety of experimental annotations. 3′ UTRs are an expected outlier for the indel measures owing to the presence of low-complexity sequence (leading to a higher indel rate).

These findings indicate that the majority of the evolutionarily constrained, experimentally identified functional elements show evidence of negative selection both across mammalian species and within the human population. Furthermore, we have assigned at least one molecular function to the majority (60%) of all constrained bases in the ENCODE regions.

Conservation of regulatory elements. The relationship between individual classes of regulatory elements and constrained sequences varies considerably, ranging from cases where there is strong evolutionary constraint (for example, pan-vertebrate ultraconserved regions[108,109]) to examples of regulatory elements that are not conserved between orthologous human and mouse genes[110]. Within the ENCODE regions, 55% of RFBRs overlap the high-confidence constrained sequences. As expected, RFBRs have many unconstrained bases, presumably owing to the small size of the specific binding site. We investigated whether the binding sites in RFBRs could be further delimited using information about evolutionary constraint. For 7 out of 17 factors with either known TRANSFAC or Jaspar

区域分开展示。展示的结果是选出来的实验注释。中间的横杠表示实验元件随机排布的 95% 的置信区间，这是利用 GSC 方法得到的，目的是为了消除数据集中的异质性。当这个横杠与观察值重合时，则不能拒绝这些重合与随机排布一致的假说。

我们也分析了实验鉴定到的功能元件中人类的变异量(杂合度、衍生的等位基因频率谱和插入缺失率)(图 12)。对于这些研究，AR 被用作中性进化的标志物。与 AR 相比，大多数实验鉴定到的功能元件与更低的杂合度相关，一小部分含有比 AR 更低的插入缺失率。显著的异常情况是 3′ UTR，它含有明显更高的插入缺失率，却没有明显的原因。在参考文献 107 中有更深入的讨论。

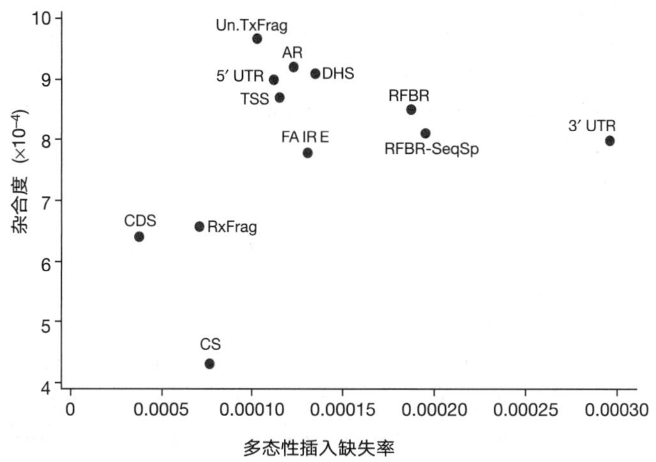

图 12. 对于各种实验注释，杂合度与多态性的插入缺失率的关系。对于插入缺失的测量，3′ UTR 是一个符合预期的离群值，原因是低复杂序列的存在(导致更高的插入缺失率)。

这些发现表明大部分进化上受限制的实验鉴定到的功能元件呈现负选择的迹象，这不仅是在哺乳动物间，在人类群体内部也是如此。再者，我们已经将 ENCODE 区域内所有受限制碱基中的大部分(60%)至少赋予了一种分子功能。

调控元件的保守性 各种调控元件与受限制序列之间的关系差别很大，有的受到强烈的进化限制(如多个脊椎动物超保守区域 [108,109])，有的调控元件则在直系同源的人和小鼠基因间不保守 [110]。在 ENCODE 区域内，55% 的 RFBR 与高可信度的受限制序列重合。正如预期的，RFBR 含有很多不受限制的碱基，可能是由于特定结合位点的长度较短。利用关于进化限制的信息，我们研究了 RFBR 中的结合位点能否被进一步划分。对于含有已知的 TRANSFAC 或 Jaspar 模序的 17 个因子中的 7 个，我们的 ChIP-chip 数据发现，相比不受限制的 RFBR，在受限制的 RFBR 中明显

motifs, our ChIP-chip data revealed a marked enrichment of the appropriate motif within the constrained versus the unconstrained portions of the RFBRs (see Supplementary Information section 5.1). This enrichment was seen for levels of stringency used for defining ChIP-chip-positive sites (1% and 5% FDR level), indicating that combining sequence constraint and ChIP-chip data may provide a highly sensitive means for detecting factor binding sites in the human genome.

Experimentally identified functional elements and genetic variation. The above studies focus on purifying (negative) selection. We used nucleotide variation to detect potential signals of adaptive (positive) selection. We modified the standard McDonald–Kreitman test (MK-test[111,112]) and the Hudson–Kreitman–Aguade (HKA)[113] test (see Supplementary Information section 5.2.1), to examine whether an entire set of sequence elements shows an excess of polymorphisms or an excess of inter-species divergence. We found that constrained sequences and coding exons have an excess of polymorphisms (consistent with purifying selection), whereas 5′ UTRs show evidence of an excess of divergence (with a portion probably reflecting positive selection). In general, non-coding genomic regions show more variation, with both a large number of segments that undergo purifying selection and regions that are fast evolving.

We also examined structural variation (that is, CNVs, inversions and translocations[114]; see Supplementary Information section 5.2.2). Within these polymorphic regions, we encountered significant over-representation of CDSs, TxFrags, and intra-species constrained sequences ($P < 10^{-3}$, Fig. 13), and also detected a statistically significant under-representation of ARs ($P = 10^{-3}$). A similar overrepresentation of CDSs and intra-species constrained sequences was found within non-polymorphic segmental duplications.

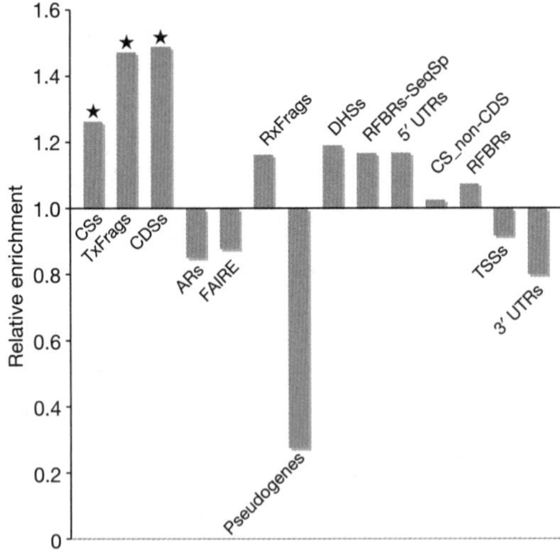

Fig. 13. CNV enrichment. The relative enrichment of different experimental annotations in the ENCODE regions associated with CNVs. CS_non-CDS are constrained sequences outside of coding regions. A value of 1 or less indicates no enrichment, and values greater than 1 show enrichment. Starred columns are cases that are significant on the basis of this enrichment being found in less than 5% of randomizations that matched each element class for length and density of features.

富集合适的模序（详见补充信息 5.1）。这种富集是在用于确定 ChIP-chip 阳性位点的阈值水平下看到的（1% 和 5%FDR 水平），说明结合序列的受限制和 ChIP-chip 数据可能提供一种高度敏感的方法来确定因子在人类基因组中的结合位点。

实验鉴定的功能元件和遗传变异　上述研究聚焦于纯化（负）选择。我们使用核苷酸的变异来检测潜在的适应性（正）选择信号。我们修改标准的麦克唐纳–克莱特曼检验（MK 检验[111, 112]）和赫德森–克莱特曼–阿瓜德（HKA）[113] 检验（详见补充信息 5.2.1），来检查一整套序列元件是否呈现过度的多态性或者过度的物种间差异。我们发现受限制序列和编码的外显子呈现过度的多态性（与纯化选择一致），而 5′UTR 呈现过度的差异迹象（其中一部分可能反映了正选择）。总的来说，非编码的基因组区域呈现更多的变异，不仅大量的片段在进行纯化选择，而且有区域正在快速地进化。

我们还检查了结构变异（即 CNV，倒位和易位[114]；详见补充信息 5.2.2）。在这些多态性的区域，我们发现 CDS、TxFrag 和物种内受限制序列显著富集（$P < 10^{-3}$，图 13），还检测到有统计学差异的 AR 的缺失（$P = 10^{-3}$）。CDS 和物种内受限制序列类似的富集在非多态性的片段重复中也被发现。

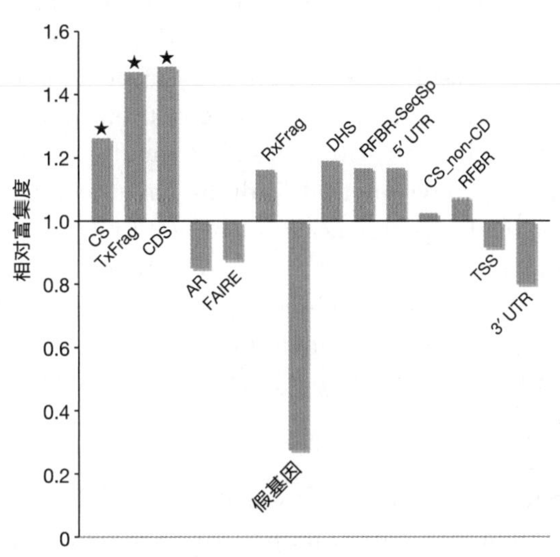

图 13. CNV 富集。在 ENCODE 区域内，与 CNV 相关的不同实验注释的相对富集程度。CS_non-CDS 是编码区域以外的受限制序列。数值为 1 或更小表示没有富集，数值大于 1 表示富集。标记星号的柱是显著的组，这种显著是基于在与相应类别元件长度和特征密度相匹配的随机组中仅有不到 5% 的富集。

Unexplained constrained sequences. Despite the wealth of complementary data, 40% of the ENCODE-region sequences identified as constrained are not associated with any experimental evidence of function. There is no evidence indicating that mutational cold spots account for this constraint; they have similar measures of constraint to experimentally identified elements and harbour equal proportions of SNPs. To characterize further the unexplained constrained sequences, we examined their clustering and phylogenetic distribution. These sequences are not uniformly distributed across most ENCODE regions, and even in most ENCODE regions the distribution is different from constrained sequences within experimentally identified functional elements (see Supplementary Information section 5.3). The large fraction of constrained sequence that does not match any experimentally identified elements is not surprising considering that only a limited set of transcription factors, cell lines and biological conditions have thus far been examined.

Unconstrained experimentally identified functional elements. In contrast, an unexpectedly large fraction of experimentally identified functional elements show no evidence of evolutionary constraint ranging from 93% for Un.TxFrags to 12% for CDS. For most types of non-coding functional elements, roughly 50% of the individual elements seemed to be unconstrained across all mammals.

There are two methodological reasons that might explain the apparent excess of unconstrained experimentally identified functional elements: the underestimation of sequence constraint or overestimation of experimentally identified functional elements. We do not believe that either of these explanations fully accounts for the large and varied levels of unconstrained experimentally functional sequences. The set of constrained bases analysed here is highly accurate and complete due to the depth of the multiple alignment. Both by bulk fitting procedures and by comparison of SNP frequencies to constraint there is clearly a proportion of constrained bases not captured in the defined 4.9% of constrained sequences, but it is small (see Supplementary Information section 5.4 and S5.5). More aggressive schemes to detect constraint only marginally increase the overlap with experimentally identified functional elements, and do so with considerably less specificity. Similarly, all experimental findings have been independently validated and, for the least constrained experimentally identified functional elements (Un.TxFrags and binding sites of sequence-specific factors), there is both internal validation and cross-validation from different experimental techniques. This suggests that there is probably not a significant overestimation of experimentally identified functional elements. Thus, these two explanations may contribute to the general observation about unconstrained functional elements, but cannot fully explain it.

Instead, we hypothesize five biological reasons to account for the presence of large amounts of unconstrained functional elements. The first two are particular to certain biological assays in which the elements being measured are connected to but do not coincide with the analysed region. An example of this is the parent transcript of an miRNA, where the current assays detect the exons (some of which are not under evolutionary selection), whereas the intronic miRNA actually harbours the constrained bases. Nevertheless, the

未解释的受限制序列 尽管有丰富的互补数据，被鉴定为受限制的 ENCODE 区域序列中仍有 40% 不与任何一项功能的实验证据有关联。没有证据表明突变冷点导致这种限制，它们与实验鉴定到的元件有类似的程度的受限制，并且含有相同比例的 SNP。为了进一步描述未解释的受限制序列的特征，我们检查了它们的聚集和系统进化分布。这些序列并非均匀分布在大多数的 ENCODE 区域，甚至在大多数的 ENCODE 区域，它们的分布跟位于实验鉴定的功能元件内的受限制序列也是不同的(详见补充信息 5.3)。考虑到目前为止仅检测了有限数量的转录因子、细胞系和生物学条件，很大一部分受限制序列不与任何实验鉴定到的元件相匹配其实并不奇怪。

不受限制的实验鉴定到的功能元件 相反，出乎意料的是很大比例实验鉴定到的功能元件中没有表现出进化受限制的证据，从 Un.TxFrag 的 93% 到 CDS 的 12%。对于大多数类型的非编码功能元件，每种元件都有大约 50% 好像在所有哺乳动物中不受限制。

有两个方法学的原因可能解释这种明显过量的不受限制的实验鉴定的功能元件：低估了受限制的序列或者高估了实验鉴定的功能元件。我们不相信这些解释中的某一个能完全解释这大量且不同水平的非受限制的实验鉴定的功能序列。由于多重比对的深度，这里分析的这组受限制的碱基是高度准确而完整的。不管是通过批量拟合方法还是通过比较 SNP 频率跟受限制性，明显有一部分受限制的碱基在 4.9% 的受限制序列中没有被捕获到，但是这个数量是很小的(详见补充信息 5.4 和 S5.5)。更激进的检测受限制的策略仅仅略微增加了与实验鉴定到的功能元件的重合度，而特异性明显降低。类似地，所有实验发现已经被独立验证，对于最不受限制的实验鉴定的功能元件(UnTxFrag 和序列特异因子的结合位点)，既有内部验证也有不同实验技术的交叉验证。这说明可能并没有显著高估实验鉴定的功能元件。因而，这两种解释可能有助于解释普遍观察到的非受限制功能元件，但是还不能完全解释它。

作为替代，我们猜测了五个生物学原因来解释大量的非受限制功能元件的存在。前两个特别针对特定的生物学实验，实验中测到的元件与分析的区域相关但是并不一致。一个这样的例子是 miRNA 的亲本转录本，现有的实验检测的是外显子(其中一些并没有受到进化的筛选)，而位于内含子的 miRNA 实际上含有受限制的碱基。尽管如此，转录本序列还是提供了被调控的启动子与 miRNA 之间关键的耦合。转

transcript sequence provides the critical coupling between the regulated promoter and the miRNA. The sliding of transcription factors (which might bind a specific sequence but then migrate along the DNA) or the processivity of histone modifications across chromatin are more exotic examples of this. A related, second hypothesis is that delocalized behaviours of the genome, such as general chromatin accessibility, may be maintained by some biochemical processes (such as transcription of intergenic regions or specific factor binding) without the requirement for specific sequence elements. These two explanations of both connected components and diffuse components related to, but not coincident with, constrained sequences are particularly relevant for the considerable amount of unannotated and unconstrained transcripts.

The other three hypotheses may be more general—the presence of neutral (or near neutral) biochemical elements, of lineage-specific functional elements, and of functionally conserved but non-orthologous elements. We believe there is a considerable proportion of neutral biochemically active elements that do not confer a selective advantage or disadvantage to the organism. This neutral pool of sequence elements may turn over during evolutionary time emerging via certain mutations and disappearing by others. The size of the neutral pool would largely be determined by the rate of emergence and extinction through chance events; low information-content elements, such as transcription factor-binding sites[110] will have larger neutral pools. Second, from this neutral pool, some elements might occasionally acquire a biological role and so come under evolutionary selection. The acquisition of a new biological role would then create a lineage-specific element. Finally, a neutral element from the general pool could also become a peer of an existing selected functional element and either of the two elements could then be removed by chance. If the older element is removed, the newer element has, in essence, been conserved without using orthologous bases, providing a conserved function in the absence of constrained sequences. For example, a common HNF4A binding site in the human and mouse genomes may not reflect orthologous human and mouse bases, though the presence of an HNF4A site in that region was evolutionarily selected for in both lineages. Note that both the neutral turnover of elements and the "functional peering" of elements has been suggested for *cis*-acting regulatory elements in *Drosophila*[115,116] and mammals[110]. Our data support these hypotheses, and we have generalized this idea over many different functional elements. The presence of conserved function encoded by conserved orthologous bases is a commonplace assumption in comparative genomics; our findings indicate that there could be a sizable set of functionally conserved but non-orthologous elements in the human genome, and that these seem unconstrained across mammals. Functional data akin to the ENCODE Project on other related species, such as mouse, would be critical to understanding the rate of such functionally conserved but non-orthologous elements.

Conclusion

The generation and analyses of over 200 experimental data sets from studies examining the 44 ENCODE regions provide a rich source of functional information for 30 Mb of

录因子的滑动（转录因子可能结合一个特定的序列但是随后沿着 DNA 移动）或者组蛋白修饰在染色质上的延伸性则是更加异乎寻常的例子。与之相关的第二个假说是基因组不受位置限制的行为，一般的染色质开放性可能通过一些生化过程而被维持（如基因间区域的转录或者特定因子的结合）不需要特定的序列元件。这两种对相连组分和扩散组分都与受限制序列相关但又不完全一致现象的解释，与相当数量未注释且未受限制的转录本特别相关。

其他三种假说——中性（或者近中性）生化元件、谱系特异的功能元件和功能保守但非直系同源元件的存在——可能更常见。我们相信存在相当大的一部分中性的生化活跃元件，并不对机体产生选择性优势或者劣势。这种中性序列元件库可能在进化过程中发生转变，通过特定突变而出现，通过其他一些突变而消失。中性库的大小很大程度上取决于元件通过随机事件而产生和消失的速率；低信息量的元件例如转录因子结合位点[110]会有更大的中性库。第二，从这个中性库中，一些元件可能偶然获得一个生物学功能，因而进入进化选择。一个新的生物学功能的获得会进而产生一个谱系特异的元件。最后，普通库中的一个中性元件也能成为一个同等的已被选择的功能元件，这两个元件中的一个会随机被去除。如果更老的元件被去除，那么更新的元件即使不使用直系同源碱基，在本质上就已经是保守的，在不存在受限制序列的情况下提供了一个保守功能。例如，在人和小鼠基因组中一个共同的 HNF4A 结合位点可能并不反映直系同源的人和小鼠碱基，尽管 HNF4A 位点在那个区域的存在在两个谱系中都是被进化选择的。需要注意的是，元件的中性转换和元件的“功能等价”在果蝇[115,116]和哺乳动物[110]中都被认为是顺式调控元件。我们的数据支持这些假说，我们已经将这个观点普遍地应用于很多不同的功能元件。由保守直系同源碱基编码的保守功能的存在，在比较基因组学中是一个很平常的假设，我们的发现提示可能在人类基因组中存在相当数量的功能保守但非直系同源的元件，这些元件在哺乳动物中似乎并不受限制。类似于 ENCODE 计划的关于其他相关物种的功能数据（例如小鼠的），对于理解这种功能保守的但非直系同源元件的速率将会至关重要。

结　论

研究针对 44 个 ENCODE 区域，产生和分析了超过 200 个实验数据集，为人类

the human genome. The first conclusion of these efforts is that these data are remarkably inform-ative. Although there will be ongoing work to enhance existing assays, invent new techniques and develop new data-analysis methods, the generation of genome-wide experimental data sets akin to the ENCODE pilot phase would provide an impressive platform for future genome exploration efforts. This now seems feasible in light of throughput improvements of many of the assays and the ever-declining costs of whole-genome tiling arrays and DNA sequencing. Such genome-wide functional data should be acquired and released openly, as has been done with other large-scale genome projects, to ensure its availability as a new foundation for all biologists studying the human genome. It is these biologists who will often provide the critical link from biochemical function to biological role for the identified elements.

The scale of the pilot phase of the ENCODE Project was also sufficiently large and unbiased to reveal important principles about the organization of functional elements in the human genome. In many cases, these principles agree with current mechanistic models. For example, we observe trimethylation of H3K4 enriched near active genes, and have improved the ability to accurately predict gene activity based on this and other histone modifications. However, we also uncovered some surprises that challenge the current dogma on biological mechanisms. The generation of numerous intercalated transcripts spanning the majority of the genome has been repeatedly suggested[13,14], but this phenomenon has been met with mixed opinions about the biological importance of these transcripts. Our analyses of numerous orthogonal data sets firmly establish the presence of these transcripts, and thus the simple view of the genome as having a defined set of isolated loci transcribed independently does not seem to be accurate. Perhaps the genome encodes a network of transcripts, many of which are linked to protein-coding transcripts and to the majority of which we cannot (yet) assign a biological role. Our perspective of transcription and genes may have to evolve and also poses some interesting mechanistic questions. For example, how are splicing signals coordinated and used when there are so many overlapping primary transcripts? Similarly, to what extent does this reflect neutral turnover of reproducible transcripts with no biological role?

We gained subtler but equally important mechanistic findings relating to transcription, replication and chromatin modification. Transcription factors previously thought to primarily bind promoters bind more generally, and those which do bind to promoters are equally likely to bind downstream of a TSS as upstream. Interestingly, many elements that previously were classified as distal enhancers are, in fact, close to one of the newly identified TSSs; only about 35% of sites showing evidence of binding by multiple transcription factors are actually distal to a TSS. This need not imply that most regulatory information is confined to classic promoters, but rather it does suggest that transcription and regulation are coordinated actions beyond just the traditional promoter sequences. Meanwhile, although distal regulatory elements could be identified in the ENCODE regions, they are currently difficult to classify, in part owing to the lack of a broad set of transcription factors to use in analysing such elements. Finally, we now have a much better appreciation of how DNA replication is coordinated with histone modifications.

基因组的这 30 Mb 区域提供了一个丰富的功能信息的资源。这些工作的第一个结论是这些数据所包含的信息量是巨大的。尽管将有进一步的工作来促进现有的实验分析、发明新的技术和开发新的数据分析方法，类似于 ENCODE 先导阶段的全基因组实验数据集的产生还是会为以后的基因组研究提供一个出色的平台。现在看来这似乎是可行的，因为很多实验的通量在提升而全基因组叠瓦式阵列和 DNA 测序费用在大幅下降。就像其他大规模基因组计划一样，这种全基因组的功能数据应该开放获取，以保证它作为一个新的基础能被所有生物学家用来研究人类基因组。正是这些生物学家将会经常为鉴定到的元件提供从生化功能到生物学功能的关键联系。

ENCODE 计划先导阶段的规模也是足够大且无偏性，进而能够反映人类基因组中功能元件的重要组织原理。在很多例子中，这些原理与现有的机制模型是一致的。例如，我们观察到 H3K4 的三甲基化在活跃基因的附近富集，基于这种以及其他组蛋白修饰，我们提高了精确预测基因活性的能力。然而，我们也发现了一些意料之外的事情，对现有的生物学机制的信条提出了挑战。已经反复发现，有大量的间插性转录本的产生，这些转录本横跨大部分的基因组 [13,14]，但是对于这种现象中这些转录本的生物学意义有不同的观点。我们通过对大量的正交数据集的分析，确凿地证实了这些转录本的存在，因而简单地认为基因组含有确定的一套分离的基因座独立地被转录好像并不是准确的。也许基因组编码一个转录本的网络，很多转录本与蛋白质编码的转录本有联系，而对于大部分的转录本，我们(还)不能确定其生物学功能。我们关于转录和基因的观点可能不得不进化，另外也产生了一些有趣的关于机制的问题。例如，当有这么多重叠的原始转录本时，剪接信号是如何协调和使用的？类似地，这在多大程度上反映了没有生物学功能的重复转录本的中性转换？

我们获得了微妙但同等重要的与转录、复制和染色质修饰相关的机制方面的发现。之前认为主要结合启动子的转录因子，其实结合更广泛，那些确定结合启动子的转录因子除了结合 TSS 上游之外，同样可能结合 TSS 的下游。有趣的是，很多元件之前被划分为远端增强子，其实靠近新鉴定到的 TSS；对于有证据表明被多个转录因子结合的位点，只有大约 35% 确实远离 TSS。这不一定说明大部分的调控信息局限于经典的启动子，但是确实暗示转录和调控并不仅是传统启动子序列的协调活动。同时，尽管远端调控元件能在 ENCODE 区域中被鉴定，但是它们目前还难以分类，部分是由于缺少一套广泛的转录因子来分析这些元件。最后，目前我们对于 DNA 复制如何与组蛋白修饰协调有了更好的认识。

At the outset of the ENCODE Project, many believed that the broad collection of experimental data would nicely dovetail with the detailed evolutionary information derived from comparing multiple mammalian sequences to provide a neat "dictionary" of conserved genomic elements, each with a growing annotation about their biochemical function(s). In one sense, this was achieved; the majority of constrained bases in the ENCODE regions are now associated with at least some experimentally derived information about function. However, we have also encountered a remarkable excess of experimentally identified functional elements lacking evolutionary constraint, and these cannot be dismissed for technical reasons. This is perhaps the biggest surprise of the pilot phase of the ENCODE Project, and suggests that we take a more "neutral" view of many of the functions conferred by the genome.

Methods

The methods are described in the Supplementary Information, with more technical details for each experiment often found in the references provided in Table 1. The Supplementary Information sections are arranged in the same order as the manuscript (with similar headings to facilitate cross-referencing). The first page of Supplementary Information also has an index to aid navigation. Raw data are available in ArrayExpress, GEO or EMBL/GenBank archives as appropriate, as detailed in Supplementary Information section 1.1. Processed data are also presented in a user-friendly manner at the UCSC Genome Browser's ENCODE portal (http://genome.ucsc.edu/ENCODE/).

(**447**, 799-816; 2007)

Received 2 March; accepted 23 April 2007.

References:

1. International Human Genome Sequencing Consortium. Initial sequencing and analysis of the human genome. *Nature* **409**, 860-921 (2001).
2. Venter, J. C. *et al.* The sequence of the human genome. *Science* **291**, 1304-1351 (2001).
3. International Human Genome Sequencing Consortium. Finishing the euchromatic sequence of the human genome. *Nature* **431**, 931-945 (2004).
4. Mouse Genome Sequencing Consortium. Initial sequencing and comparative analysis of the mouse genome. *Nature* **420**, 520-562 (2002).
5. Rat Genome Sequencing Project Consortium. Genome sequence of the Brown Norway rat yields insights into mammalian evolution. *Nature* **428**, 493-521 (2004).
6. Lindblad-Toh, K. *et al.* Genome sequence, comparative analysis and haplotype structure of the domestic dog. *Nature* **438**, 803-819 (2005).
7. International Chicken Genome Sequencing Consortium. Sequence and comparative analysis of the chicken genome provide unique perspectives on vertebrate evolution. *Nature* **432**, 695-716 (2004).
8. Chimpanzee Sequencing and Analysis Consortium. Initial sequence of the chimpanzee genome and comparison with the human genome. *Nature* **437**, 69-87 (2005).
9. ENCODE Project Consortium. The ENCODE (ENCyclopedia Of DNA Elements) Project. *Science* **306**, 636-640 (2004).
10. Zhang, Z. D. *et al.* Statistical analysis of the genomic distribution and correlation of regulatory elements in the ENCODE regions. *Genome Res.* **17**, 787-797 (2007).
11. Euskirchen, G. M. *et al.* Mapping of transcription factor binding regions in mammalian cells by ChIP: comparison of array and sequencing based technologies. *Genome Res.* **17**, 898-909 (2007).
12. Willingham, A. T. & Gingeras, T. R. TUF love for "junk" DNA. *Cell* 125, 1215-1220 (2006).
13. Carninci, P. *et al.* Genome-wide analysis of mammalian promoter architecture and evolution. *Nature Genet.* **38**, 626-635 (2006).
14. Cheng, J. *et al.* Transcriptional maps of 10 human chromosomes at 5-nucleotide resolution. *Science* **308**, 1149-1154 (2005).
15. Bertone, P. *et al.* Global identification of human transcribed sequences with genome tiling arrays. *Science* **306**, 2242-2246 (2004).
16. Guigó, R. *et al.* EGASP: the human ENCODE Genome Annotation Assessment Project. *Genome Biol.* 7, (Suppl. 1; S2) 1-31 (2006).

在 ENCODE 计划的开始时，很多人认为广泛收集的实验数据将会完美吻合来自比较多种哺乳动物序列得到的详细进化信息，进而提供一个纯粹的保守基因组元件的“词典”，每一个元件含有更多的关于它们生化功能的注释。在某种意义上，这已经达到了。在 ENCODE 区域中，大部分的受限制碱基现在至少与某些实验得到的功能信息相关联。但是，我们也遇到了大量多余的实验鉴定到的功能元件，它们缺少进化的限制，又不能因为技术原因而摒弃。这可能是 ENCODE 计划先导阶段最大的意外，提示我们要更加“中性”地看待很多基因组赋予的功能。

方　法

在补充信息中对方法进行了介绍，对于各个实验，更多的技术细节大多能在表 1 中提供的参考文献中找到。补充信息部分跟正文排列顺序相同（使用类似的标题以便于交叉参考）。补充信息的第一页也有一个索引以便于查找。原始数据可在 ArrayExpress、GEO 或者 EMBL/GenBank 档案文件获得，详见补充信息 1.1。处理过的数据也以用户友好的方式在 UCSC 基因组浏览器的 ENCODE 入口呈现（http://genome.ucsc.edu/ENCODE/）。

（李平 翻译；于军 审稿）

17. Denoeud, F. *et al.* Prominent use of distal 5′ transcription start sites and discovery of a large number of additional exons in ENCODE regions. *Genome Res.* **17**, 746-759 (2007).

18. Tress, M. L. *et al.* The implications of alternative splicing in the ENCODE protein complement. *Proc. Natl Acad. Sci. USA* **104**, 5495-5500 (2007).

19. Rozowsky, J. *et al.* The DART classification of unannotated transcription within ENCODE regions: Associating transcription with known and novel loci. *Genome Res.* **17**, 732-745 (2007).

20. Kapranov, P. *et al.* Examples of the complex architecture of the human transcriptome revealed by RACE and high-density tiling arrays. *Genome Res.* **15**, 987-997 (2005).

21. Balakirev, E. S. & Ayala, F. J. Pseudogenes: are they "junk" or functional DNA? *Annu. Rev. Genet.* **37**, 123-151 (2003).

22. Mighell, A. J., Smith, N. R., Robinson, P. A. & Markham, A. F. Vertebrate pseudogenes. *FEBS Lett.* **468**, 109-114 (2000).

23. Zheng, D. *et al.* Pseudogenes in the ENCODE regions: Consensus annotation, analysis of transcription and evolution. *Genome Res.* **17**, 839-851 (2007).

24. Zheng, D. *et al.* Integrated pseudogene annotation for human chromosome 22: evidence for transcription. *J. Mol. Biol.* **349**, 27-45 (2005).

25. Harrison, P. M., Zheng, D., Zhang, Z., Carriero, N. & Gerstein, M. Transcribed processed pseudogenes in the human genome: an intermediate form of expressed retrosequence lacking protein-coding ability. *Nucleic Acids Res.* **33**, 2374-2383 (2005).

26. Washietl, S. *et al.* Structured RNAs in the ENCODE selected regions of the human genome. *Genome Res.* **17**, 852-864 (2007).

27. Carninci, P. *et al.* The transcriptional landscape of the mammalian genome. *Science* **309**, 1559-1563 (2005).

28. Runte, M. *et al.* The IC-*SNURF–SNRPN* transcript serves as a host for multiple small nucleolar RNA species and as an antisense RNA for *UBE3A*. *Hum. Mol. Genet.* **10**, 2687-2700 (2001).

29. Seidl, C. I., Stricker, S. H. & Barlow, D. P. The imprinted *Air* ncRNA is an atypical RNAPII transcript that evades splicing and escapes nuclear export. *EMBO J.* **25**, 3565-3575 (2006).

30. Parra, G. *et al.* Tandem chimerism as a means to increase protein complexity in the human genome. *Genome Res.* **16**, 37-44 (2006).

31. Maston, G. A., Evans, S. K. & Green, M. R. Transcriptional regulatory elements in the human genome. *Annu. Rev. Genomics Hum. Genet.* **7**, 29-59 (2006).

32. Trinklein, N. D., Aldred, S. J., Saldanha, A. J. & Myers, R. M. Identification and functional analysis of human transcriptional promoters. *Genome Res.* **13**, 308-312 (2003).

33. Cooper, S. J., Trinklein, N. D., Anton, E. D., Nguyen, L. & Myers, R. M. Comprehensive analysis of transcriptional promoter structure and function in 1% of the human genome. *Genome Res.* **16**, 1-10 (2006).

34. Cawley, S. *et al.* Unbiased mapping of transcription factor binding sites along human chromosomes 21 and 22 points to widespread regulation of noncoding RNAs. *Cell* **116**, 499-509 (2004).

35. Yelin, R. *et al.* Widespread occurrence of antisense transcription in the human genome. *Nature Biotechnol.* **21**, 379-386 (2003).

36. Katayama, S. *et al.* Antisense transcription in the mammalian transcriptome. *Science* **309**, 1564-1566 (2005).

37. Ren, B. *et al.* Genome-wide location and function of DNA binding proteins. *Science* **290**, 2306-2309 (2000).

38. Iyer, V. R. *et al.* Genomic binding sites of the yeast cell-cycle transcription factors SBF and MBF. *Nature* **409**, 533-538 (2001).

39. Horak, C. E. *et al.* GATA-1 binding sites mapped in the β-globin locus by using mammalian chIp-chip analysis. *Proc. Natl Acad. Sci. USA* **99**, 2924-2929 (2002).

40. Wei, C. L. *et al.* A global map of p53 transcription-factor binding sites in the human genome. *Cell* **124**, 207-219 (2006).

41. Kim, J., Bhinge, A. A., Morgan, X. C. & Iyer, V. R. Mapping DNA–protein interactions in large genomes by sequence tag analysis of genomic enrichment. *Nature Methods* **2**, 47-53 (2005).

42. Dorschner, M. O. *et al.* High-throughput localization of functional elements by quantitative chromatin profiling. *Nature Methods* **1**, 219-225 (2004).

43. Sabo, P. J. *et al.* Genome-scale mapping of DNase I sensitivity *in vivo* using tiling DNA microarrays. *Nature Methods* **3**, 511-518 (2006).

44. Crawford, G. E. *et al.* DNase-chip: a high-resolution method to identify DNase I hypersensitive sites using tiled microarrays. *Nature Methods* **3**, 503-509 (2006).

45. Hogan, G. J., Lee, C. K. & Lieb, J. D. Cell cycle-specified fluctuation of nucleosome occupancy at gene promoters. *PLoS Genet.* **2**, e158 (2006).

46. Koch, C. M. *et al.* The landscape of histone modifications across 1% of the human genome in five human cell lines. *Genome Res.* **17**, 691-707 (2007).

47. Smale, S. T. & Kadonaga, J. T. The RNA polymerase II core promoter. *Annu. Rev. Biochem.* **72**, 449-479 (2003).

48. Mito, Y., Henikoff, J. G. & Henikoff, S. Genome-scale profiling of histone H3.3 replacement patterns. *Nature Genet.* **37**, 1090-1097 (2005).

49. Heintzman, N. D. *et al.* Distinct and predictive chromatin signatures of transcriptional promoters and enhancers in the human genome. *Nature Genet.* **39**, 311-318 (2007).

50. Yusufzai, T. M., Tagami, H., Nakatani, Y. & Felsenfeld, G. CTCF tethers an insulator to subnuclear sites, suggesting shared insulator mechanisms across species. *Mol. Cell* **13**, 291-298 (2004).

51. Kim, T. H. *et al.* Direct isolation and identification of promoters in the human genome. *Genome Res.* **15**, 830-839 (2005).

52. Bieda, M., Xu, X., Singer, M. A., Green, R. & Farnham, P. J. Unbiased location analysis of E2F1-binding sites suggests a widespread role for E2F1 in the human genome. *Genome Res.* **16**, 595-605 (2006).

53. Ruppert, S., Wang, E. H. & Tjian, R. Cloning and expression of human $TAF_{II}250$: a TBP-associated factor implicated in cell-cycle regulation. *Nature* **362**, 175-179 (1993).

54. Fernandez, P. C. *et al.* Genomic targets of the human c-Myc protein. *Genes Dev.* **17**, 1115-1129 (2003).

55. Li, Z. *et al.* A global transcriptional regulatory role for c-Myc in Burkitt's lymphoma cells. *Proc. Natl Acad. Sci. USA* **100**, 8164-8169 (2003).

56. Orian, A. *et al.* Genomic binding by the *Drosophila* Myc, Max, Mad/Mnt transcription factor network. *Genes Dev.* **17**, 1101-1114 (2003).

57. de Laat, W. & Grosveld, F. Spatial organization of gene expression: the active chromatin hub. *Chromosome Res.* **11**, 447-459 (2003).

58. Trinklein, N. D. *et al.* Integrated analysis of experimental datasets reveals many novel promoters in 1% of the human genome. *Genome Res.* **17**, 720-731 (2007).

59. Jeon, Y. *et al.* Temporal profile of replication of human chromosomes. *Proc. Natl Acad. Sci. USA* **102**, 6419-6424 (2005).

60. Woodfine, K. *et al.* Replication timing of the human genome. *Hum. Mol. Genet.* **13**, 191-202 (2004).

61. White, E. J. *et al.* DNA replication-timing analysis of human chromosome 22 at high resolution and different developmental states. *Proc. Natl Acad. Sci. USA* **101**, 17771-17776 (2004).

62. Schubeler, D. *et al.* Genome-wide DNA replication profile for *Drosophila melanogaster*: a link between transcription and replication timing. *Nature Genet.* **32**, 438-442 (2002).

63. MacAlpine, D. M., Rodriguez, H. K. & Bell, S. P. Coordination of replication and transcription along a *Drosophila* chromosome. *Genes Dev.* **18**, 3094-3105 (2004).

64. Gilbert, D. M. Replication timing and transcriptional control: beyond cause and effect. *Curr. Opin. Cell Biol.* **14**, 377-383 (2002).

65. Schwaiger, M. & Schubeler, D. A question of timing: emerging links between transcription and replication. *Curr. Opin. Genet. Dev.* **16**, 177-183 (2006).

66. Hatton, K. S. *et al.* Replication program of active and inactive multigene families in mammalian cells. *Mol. Cell. Biol.* **8**, 2149-2158 (1988).

67. Gartler, S. M., Goldstein, L., Tyler-Freer, S. E. & Hansen, R. S. The timing of *XIST* replication: dominance of the domain. *Hum. Mol. Genet.* **8**, 1085-1089 (1999).

68. Azuara, V. *et al.* Heritable gene silencing in lymphocytes delays chromatid resolution without affecting the timing of DNA replication. *Nature Cell Biol.* **5**, 668-674 (2003).

69. Cohen, S. M., Furey, T. S., Doggett, N. A. & Kaufman, D. G. Genome-wide sequence and functional analysis of early replicating DNA in normal human fibroblasts. *BMC Genomics* 7, 301 (2006).

70. Cao, R. *et al.* Role of histone H3 lysine 27 methylation in Polycomb-group silencing. *Science* **298**, 1039-1043 (2002).

71. Muller, J. *et al.* Histone methyltransferase activity of a *Drosophila* Polycomb group repressor complex. *Cell* **111**, 197-208 (2002).

72. Bracken, A. P., Dietrich, N., Pasini, D., Hansen, K. H. & Helin, K. Genome-wide mapping of Polycomb target genes unravels their roles in cell fate transitions. *Genes Dev.* **20**, 1123-1136 (2006).

73. Kirmizis, A. *et al.* Silencing of human polycomb target genes is associated with methylation of histone H3 Lys 27. *Genes Dev.* **18**, 1592-1605 (2004).

74. Lee, T. I. *et al.* Control of developmental regulators by Polycomb in human embryonic stem cells. *Cell* **125**, 301-313 (2006).

75. Karnani, N., Taylor, C., Malhotra, A. & Dutta, A. Pan-S replication patterns and chromosomal domains defined by genome tiling arrays of human chromosomes. *Genome Res.* **17**, 865-876 (2007).

76. Delaval, K., Wagschal, A. & Feil, R. Epigenetic deregulation of imprinting in congenital diseases of aberrant growth. *Bioessays* **28**, 453-459 (2006).

77. Dillon, N. Gene regulation and large-scale chromatin organization in the nucleus. *Chromosome Res.* **14**, 117-126 (2006).

78. Burkhoff, A. M. & Tullius, T. D. Structural details of an adenine tract that does not cause DNA to bend. *Nature* **331**, 455-457 (1988).

79. Price, M. A. & Tullius, T. D. How the structure of an adenine tract depends on sequence context: a new model for the structure of T_nA_n DNA sequences. *Biochemistry* **32**, 127-136 (1993).

80. Greenbaum, J. A., Parker, S. C. J. & Tullius, T. D. Detection of DNA structural motifs in functional genomic elements. *Genome Res.* **17**, 940-946 (2007).

81. Thurman, R. E., Day, N., Noble, W. S. & Stamatoyannopoulos, J. A. Identification of higher-order functional domains in the human ENCODE regions. *Genome Res.* **17**, 917-927 (2007).

82. Gilbert, N. *et al.* Chromatin architecture of the human genome: gene-rich domains are enriched in open chromatin fibers. *Cell* **118**, 555-566 (2004).

83. Nobrega, M. A., Ovcharenko, I., Afzal, V. & Rubin, E. M. Scanning human gene deserts for long-range enhancers. *Science* **302**, 413 (2003).

84. Woolfe, A. *et al.* Highly conserved non-coding sequences are associated with vertebrate development. *PLoS Biol.* **3**, e7 (2005).

85. Drake, J. A. *et al.* Conserved noncoding sequences are selectively constrained and not mutation cold spots. *Nature Genet.* **38**, 223-227 (2006).

86. Margulies, E. H., Blanchette, M., NISC Comparative Sequencing Program, Haussler D. & Green, E. D. Identification and characterization of multi-species conserved sequences. *Genome Res.* **13**, 2507-2518 (2003).

87. Cooper, G. M. *et al.* Distribution and intensity of constraint in mammalian genomic sequence. *Genome Res.* **15**, 901-913 (2005).

88. Siepel, A. *et al.* Evolutionarily conserved elements in vertebrate, insect, worm, and yeast genomes. *Genome Res.* **15**, 1034-1050 (2005).

89. Thomas, J. W. *et al.* Parallel construction of orthologous sequence-ready clone contig maps in multiple species. *Genome Res.* **12**, 1277-1285 (2002).

90. Blakesley, R. W. *et al.* An intermediate grade of finished genomic sequence suitable for comparative analyses. *Genome Res.* **14**, 2235-2244 (2004).

91. Aparicio, S. *et al.* Whole-genome shotgun assembly and analysis of the genome of *Fugu rubripes*. *Science* **297**, 1301-1310 (2002).

92. Jaillon, O. *et al.* Genome duplication in the teleost fish *Tetraodon nigroviridis* reveals the early vertebrate proto-karyotype. *Nature* **431**, 946-957 (2004).

93. Margulies, E. H. *et al.* An initial strategy for the systematic identification of functional elements in the human genome by low-redundancy comparative sequencing. *Proc. Natl Acad. Sci. USA* **102**, 4795-4800 (2005).

94. Blanchette, M. *et al.* Aligning multiple genomic sequences with the threaded blockset aligner. *Genome Res.* **14**, 708-715 (2004).

95. Bray, N. & Pachter, L. MAVID: constrained ancestral alignment of multiple sequences. *Genome Res.* **14**, 693-699 (2004).

96. Brudno, M. *et al.* LAGAN and Multi-LAGAN: efficient tools for large-scale multiple alignment of genomic DNA. *Genome Res.* **13**, 721-731 (2003).

97. Margulies, E. H. *et al.* Relationship between evolutionary constraint and genome function for 1% of the human genome. *Genome Res.* **17**, 760-774 (2007).

98. Asthana, S., Roytberg, M., Stamatoyannopoulos, J. A. & Sunyaev, S. Analysis of sequence conservation at nucleotide resolution. *PLoS Comp. Biol.* (submitted).

99. Cooper, G. M., Brudno, M., Green, E. D., Batzoglou, S. & Sidow, A. Quantitative estimates of sequence divergence for comparative analyses of mammalian genomes. *Genome Res.* **13**, 813-820 (2003).

100. Eddy, S. R. A model of the statistical power of comparative genome sequence analysis. *PLoS Biol.* **3**, e10 (2005).

101. Stone, E. A., Cooper, G. M. & Sidow, A. Trade-offs in detecting evolutionarily constrained sequence by comparative genomics. *Annu. Rev. Genomics Hum. Genet.* **6**, 143-164 (2005).

102. McAuliffe, J. D., Jordan, M. I. & Pachter, L. Subtree power analysis and species selection for comparative genomics. *Proc. Natl Acad. Sci. USA* **102**, 7900-7905 (2005).

103. International HapMap Consortium. A haplotype map of the human genome. *Nature* **437**, 1299-1320 (2005).

104. Cheng, Z. *et al.* A genome-wide comparison of recent chimpanzee and human segmental duplications. *Nature* **437**, 88-93 (2005).

105. Cooper, G. M. *et al.* Characterization of evolutionary rates and constraints in three Mammalian genomes. *Genome Res.* **14**, 539-548 (2004).

106. Dermitzakis, E. T., Reymond, A. & Antonarakis, S. E. Conserved non-genic sequences - an unexpected feature of mammalian genomes. *Nature Rev. Genet.* **6**, 151-157 (2005).

107. Clark, T. G. *et al.* Small insertions/deletions and functional constraint in the ENCODE regions. *Genome Biol.* (submitted) (2007).

108. Bejerano, G. *et al.* Ultraconserved elements in the human genome. *Science* **304**, 1321-1325 (2004).

109. Woolfe, A. *et al.* Highly conserved non-coding sequences are associated with vertebrate development. *PLoS Biol.* **3**, e7 (2005).

110. Dermitzakis, E. T. & Clark, A. G. Evolution of transcription factor binding sites in Mammalian gene regulatory regions: conservation and turnover. *Mol. Biol. Evol.* **19**, 1114-1121 (2002).

111. McDonald, J. H. & Kreitman, M. Adaptive protein evolution at the *Adh* locus in *Drosophila. Nature* **351**, 652-654 (1991).

112. Andolfatto, P. Adaptive evolution of non-coding DNA in *Drosophila. Nature* **437**, 1149-1152 (2005).

113. Hudson, R. R., Kreitman, M. & Aguade, M. A test of neutral molecular evolution based on nucleotide data. *Genetics* **116**, 153-159 (1987).

114. Feuk, L., Carson, A. R. & Scherer, S. W. Structural variation in the human genome. *Nature Rev. Genet.* **7**, 85-97 (2006).

115. Ludwig, M. Z. *et al.* Functional evolution of a *cis*-regulatory module. *PLoS Biol.* **3**, e93 (2005).

116. Ludwig, M. Z. & Kreitman, M. Evolutionary dynamics of the enhancer region of even-skipped in *Drosophila. Mol. Biol. Evol.* **12**, 1002-1011 (1995).

117. Harrow, J. *et al.* GENCODE: producing a reference annotation for ENCODE. *Genome Biol.* **7**, (Suppl. 1; S4) 1-9 (2006).

118. Emanuelsson, O. *et al.* Assessing the performance of different high-density tiling microarray strategies for mapping transcribed regions of the human genome. *Genome Res.* advance online publication, doi: 10.1101/gr.5014606 (21 November 2006).

119. Kapranov, P. *et al.* Large-scale transcriptional activity in chromosomes 21 and 22. Science 296, 916-919 (2002).

120. Bhinge, A. A., Kim, J., Euskirchen, G., Snyder, M. & Iyer, V. R. Mapping the chromosomal targets of STAT1 by Sequence Tag Analysis of Genomic Enrichment (STAGE). *Genome Res.* **17**, 910-916 (2007).

121. Ng, P. *et al.* Gene identification signature (GIS) analysis for transcriptome characterization and genome annotation. *Nature Methods* **2**, 105-111 (2005).

122. Giresi, P. G., Kim, J., McDaniell, R. M., Iyer, V. R. & Lieb, J. D. FAIRE (Formaldehyde-Assisted Isolation of Regulatory Elements) isolates active regulatory elements from human chromatin. *Genome Res.* **17**, 877-885 (2006).

123. Rada-Iglesias, A. *et al.* Binding sites for metabolic disease related transcription factors inferred at base pair resolution by chromatin immunoprecipitation and genomic microarrays. *Hum. Mol. Genet.* **14**, 3435-3447 (2005).

124. Kim, T. H. *et al.* A high-resolution map of active promoters in the human genome. *Nature* **436**, 876-880 (2005).

125. Halees, A. S. & Weng, Z. PromoSer: improvements to the algorithm, visualization and accessibility. *Nucleic Acids Res.* **32**, W191-W194 (2004).

126. Bajic, V. B. *et al.* Performance assessment of promoter predictions on ENCODE regions in the EGASP experiment. *Genome Biol.* **7**, (Suppl 1; S3) 1-13 (2006).

127. Zheng, D. & Gerstein, M. B. A computational approach for identifying pseudogenes in the ENCODE regions. *Genome Biol.* **7**, S13.1-S13.10 (2006).

128. Stranger, B. E. *et al.* Genome-wide associations of gene expression variation in humans. *PLoS Genet* **1**, e78 (2005).

129. Turner, B. M. Reading signals on the nucleosome with a new nomenclature for modified histones. *Nature Struct. Mol. Biol.* **12**, 110-112 (2005).

Supplementary Information is linked to the online version of the paper at www.nature.com/nature.

Acknowledgements. We thank D. Leja for providing graphical expertise and support. Funding support is acknowledged from the following sources: National Institutes of Health, The European Union BioSapiens NoE, Affymetrix, Swiss National Science Foundation, the Spanish Ministerio de Educación y Ciencia, Spanish Ministry of Education and Science, CIBERESP, Genome Spain and Generalitat de Catalunya, Ministry of Education, Culture, Sports, Science and Technology of Japan, the NCCR Frontiers in Genetics, the Jérôme Lejeune Foundation, the Childcare Foundation, the Novartis Foundations, the Danish Research Council, the Swedish Research Council, the Knut and Alice Wallenberg Foundation, the Wellcome Trust, the Howard Hughes Medical Institute, the Bio-X Institute, the RIKEN Institute, the US Army, National Science Foundation, the Deutsche Forschungsgemeinschaft, the Austrian Gen-AU program, the BBSRC and The European Molecular Biology Laboratory. We thank the Barcelona SuperComputing Center and the NIH Biowulf cluster for computer facilities. The Consortium thanks the ENCODE Scientific Advisory Panel for their advice on the project: G. Weinstock, M. Cherry, G. Churchill, M. Eisen, S. Elgin, J. Lis, J. Rine, M. Vidal and P. Zamore.

Author Information. Reprints and permissions information is available at www.nature.com/reprints. The authors declare no competing financial interests. The list of individual authors is divided among the six main analysis groups and five organizational groups. Correspondence and requests for materials should be addressed to the co-chairs of the ENCODE analysis groups (listed in the Analysis Coordination group) E. Birney (birney@ebi.ac.uk); J. A. Stamatoyannopoulos (jstam@u.washington.edu); A. Dutta (ad8q@virginia.edu); R. Guigó (rguigo@imim.es); T. R. Gingeras (Tom_Gingeras@affymetrix.com); E. H. Margulies (elliott@nhgri.nih.gov); Z. Weng (zhiping@bu.edu); M. Snyder (michael.snyder@yale.edu); E. T. Dermitzakis (md4@sanger.ac.uk) or collectively (encode_chairs@ebi.ac.uk).

Generation of Germline-competent Induced Pluripotent Stem Cells

K. Okita *et al.*

Editor's Note

In 2006, Japanese cell biologist Shinya Yamanaka claimed to have made stem cell-like cells by inserting a handful of genes into mouse fibroblasts. The so-called induced pluripotent stem (iPS) cells displayed many of the hallmarks of true stem cells, but failed one acid test: they were unable to contribute to live animals when injected into early embryonic mice. In this paper, Yamanaka and colleagues overcome this problem, tweaking their methodology to produce iPS cells that appear completely reprogrammed and truly stem cell-like. Two other papers, published at the same time, describe a similar feat and iPS cells remain the focus of great attention, offering the potential to make therapeutically useful stem cells from a person's own cells without the need for a donated egg or embryo.

We have previously shown that pluripotent stem cells can be induced from mouse fibroblasts by retroviral introduction of Oct3/4 (also called Pou5f1), Sox2, c-Myc and Klf4, and subsequent selection for *Fbx15* (also called *Fbxo15*) expression. These induced pluripotent stem (iPS) cells (hereafter called Fbx15 iPS cells) are similar to embryonic stem (ES) cells in morphology, proliferation and teratoma formation; however, they are different with regards to gene expression and DNA methylation patterns, and fail to produce adult chimaeras. Here we show that selection for *Nanog* expression results in germline-competent iPS cells with increased ES-cell-like gene expression and DNA methylation patterns compared with Fbx15 iPS cells. The four transgenes (*Oct3/4*, *Sox2*, *c-myc* and *Klf4*) were strongly silenced in Nanog iPS cells. We obtained adult chimaeras from seven Nanog iPS cell clones, with one clone being transmitted through the germ line to the next generation. Approximately 20% of the offspring developed tumours attributable to reactivation of the c-*myc* transgene. Thus, iPS cells competent for germline chimaeras can be obtained from fibroblasts, but retroviral introduction of c-Myc should be avoided for clinical application.

ALTHOUGH ES cells are promising donor sources in cell transplantation therapies[1], they face immune rejection after transplantation and there are ethical issues regarding the usage of human embryos. These concerns may be overcome if pluripotent stem cells can be directly derived from patients' somatic cells[2]. We have previously shown that iPS cells can be generated from mouse fibroblasts by retrovirus-mediated introduction of four transcription factors (Oct3/4 (refs 3, 4), Sox2 (ref. 5), c-Myc (ref. 6) and Klf4 (ref. 7)) and by selection for *Fbx15* expression[8]. Fbx15 iPS cells, however, have different gene

具有种系功能的诱导多能干细胞的产生

冲田奎介等

编者按

在 2006 年，日本细胞生物学家山中伸弥声称通过在小鼠成纤维细胞中插入少量基因制造出了干细胞样细胞。所谓的诱导多能干细胞（iPS 细胞）显示出真正干细胞的许多特征，但尚未通过一项严格的测试：当被注射到早期胚胎小鼠时，它们无法培养成为活体动物。在本文中，山中及其同事克服了这个问题，他们调整了实验方法，产生的 iPS 细胞呈现完全重编程和真正干细胞样。同时发表的另外两篇论文描述了类似的壮举。iPS 细胞仍然是人们关注的重点，因为它可以从一个人自己的细胞中制造出可用于治疗的干细胞而不需要捐赠的卵子或胚胎。

先前我们已经发现，通过逆转录病毒导入 Oct3/4（又称 Pou5f1）、Sox2、c-Myc 和 Klf4 并随后选择 *Fbx15*（又称 *Fbxo15*）表达，可以从小鼠成纤维细胞中诱导产生多能干细胞。这些诱导多能干细胞（iPS 细胞）（以下称为 Fbx15 iPS 细胞）与胚胎干细胞（ES 细胞）在形态、增殖和形成畸胎瘤方面类似；但是，它们在基因表达和 DNA 甲基化模式上是不同的，并不能形成成熟的嵌合体。这里我们发现选择 *Nanog* 基因的表达可以产生具有种系功能的 iPS 细胞，与 Fbx15 iPS 细胞相比增加了 ES 细胞样基因的表达和 DNA 甲基化模式。这四个转基因（*Oct3/4*、*Sox2*、*c-myc* 和 *Klf4*）在 Nanog iPS 细胞中极度沉默。我们从 7 个 Nanog iPS 细胞克隆中获得了成熟的嵌合体，其中一个克隆还通过种系培养到了下一代。由于重新激活了 *c-myc* 转基因，大约 20% 的后代出现了肿瘤。总之，具有产生种系嵌合体功能的 iPS 细胞能够从成纤维细胞中诱导获得，但是通过逆转录病毒导入 c-Myc 在临床应用时应该被避免。

尽管胚胎干细胞是细胞移植治疗很有前途的材料来源[1]，它们仍存在移植后免疫排斥的问题，而且还存在与使用人类胚胎有关的伦理学问题。如果多能干细胞能够直接来源于病人的体细胞，那么这些问题就都可以被解决[2]。先前我们已经发现 iPS 细胞能够从小鼠成纤维细胞通过逆转录病毒导入四个转录因子（Oct3/4（文献 3，4）、Sox2（文献 5）、c-Myc（文献 6）和 Klf4（文献 7））并继以选择 *Fbx15* 表达而产生[8]。但是，Fbx15 iPS 细胞具有与 ES 细胞不同的基因表达和 DNA 甲基化模式，

expression and DNA methylation patterns compared with ES cells and do not contribute to adult chimaeras. We proposed that the incomplete reprogramming might be due to the selection for *Fbx15* expression, and that by using better selection markers, we might be able to generate more ES-cell-like iPS cells. We decided to use *Nanog* as a candidate of such markers.

Although both *Fbx15* and *Nanog* are targets of Oct3/4 and Sox2 (refs 9–11), Nanog is more tightly associated with pluripotency. In contrast to *Fbx15*-null mice and ES cells that barely show abnormal phenotypes[9], disruption of *Nanog* in mice results in loss of the pluripotent epiblast[12]. *Nanog*-null ES cells can be established, but they tend to differentiate spontaneously[12]. Forced expression of *Nanog* renders ES cells independent of leukaemia inhibitory factor (LIF) for self-renewal[12,13] and confers increased reprogramming efficiency after fusion with somatic cells[14]. These results prompted us to propose that if we use *Nanog* as a selection marker, we might be able to obtain iPS cells displaying a greater similarity to ES cells.

Generation of Nanog iPS Cells

To establish a selection system for *Nanog* expression, we began by isolating a bacterial artificial chromosome (BAC, ~200 kilobases) containing the mouse *Nanog* gene in its centre. By using recombineering technology[15,16], we inserted a green fluorescent protein (GFP)-internal ribosome entry site (IRES)-puromycin resistance gene (Puro^r) cassette into the 5′ untranslated region (UTR; Fig. 1a). ES cells that had stably incorporated the modified BAC were positive for GFP, but became negative when differentiation was induced (not shown). By introducing these ES cells into blastocysts, we obtained chimaeric mice and then transgenic mice containing the Nanog-GFP-IRES-Puro^r reporter construct. In transgenic mouse blastocysts, GFP was specifically observed in the inner cell mass (Fig. 1b). In 9.5 days post coitum (d.p.c.) embryos, only migrating primordial germ cells (PGCs) showed GFP signal. In 13.5 d.p.c. embryos, GFP was specifically detected in the genital ridges of both sexes. After removing the brain, visceral tissues and genital ridges, we isolated mouse embryonic fibroblasts (MEFs) from 13.5 d.p.c. male embryos. Flow cytometry analyses showed that these MEFs did not contain GFP-positive cells, whereas ~1% of cells isolated from genital ridges showed GFP signals (Fig. 1c).

在成年鼠中无嵌合。我们认为这种不完全的重编程可能是由于通过 *Fbx15* 表达来选择细胞，因此如果选用更好的选择标记物，我们可能能够产生更加像 ES 细胞的 iPS 细胞。我们计划使用 *Nanog* 作为这种标记物的候选者。

尽管 *Fbx15* 和 *Nanog* 都是 Oct3/4 和 Sox2 的靶目标（文献 9 ~ 11），但 Nanog 与多能性的关系更加密切。与几乎没有异常表型的 *Fbx15* 敲除小鼠以及 ES 细胞[9] 相反，小鼠中 *Nanog* 基因的紊乱可以导致多能上胚层的丧失[12]。可以建立敲除 *Nanog* 的 ES 细胞，但是它们倾向于发生自发分化[12]。强制表达 *Nanog* 可以使 ES 细胞的自我更新不依赖于白血病抑制因子（LIF）[12,13]，而且在体细胞融合后赋予其更高的重编程效率[14]。这些结果促使我们假设如果使用 *Nanog* 作为选择标记物，我们可能能够获得更像 ES 细胞的 iPS 细胞。

Nanog iPS 细胞的产生

为了建立 *Nanog* 表达的选择系统，我们首先分离了在中心含有小鼠 *Nanog* 基因的细菌人工染色体（BAC，约 200 个千碱基）。通过使用重组工程技术[15,16]，我们将一个绿色荧光蛋白（GFP）–内部核糖体进入位点（IRES）–嘌呤霉素抗性基因（Puro^r）盒插入 5′非翻译区（UTR，图 1a）。那些和修饰过的 BAC 稳定整合的 ES 细胞就含有 GFP，但是如果诱导分化，就变为 GFP 阴性（没有显示）。通过将这些 ES 细胞导入胚泡，我们就获得了嵌合小鼠，然后获得含有 Nanog-GFP-IRES-Puro^r 报告基团的转基因小鼠。在转基因的小鼠胚泡中，GFP 特有地出现在内细胞团中（图 1b）。在交配后第 9.5 天的胚胎中，只有迁移的原始生殖细胞（PGC）发出 GFP 信号。在交配后第 13.5 天，GFP 特定地出现在两种性别的生殖嵴上。去除大脑、内脏组织和生殖嵴后，我们从交配后 13.5 天的雄性胚胎中分离出小鼠胚胎成纤维细胞（MEF）。流式细胞分析显示这些 MEF 不含有 GFP 阳性的细胞，然而约 1% 从生殖嵴分离出的细胞显示 GFP 信号（图 1c）。

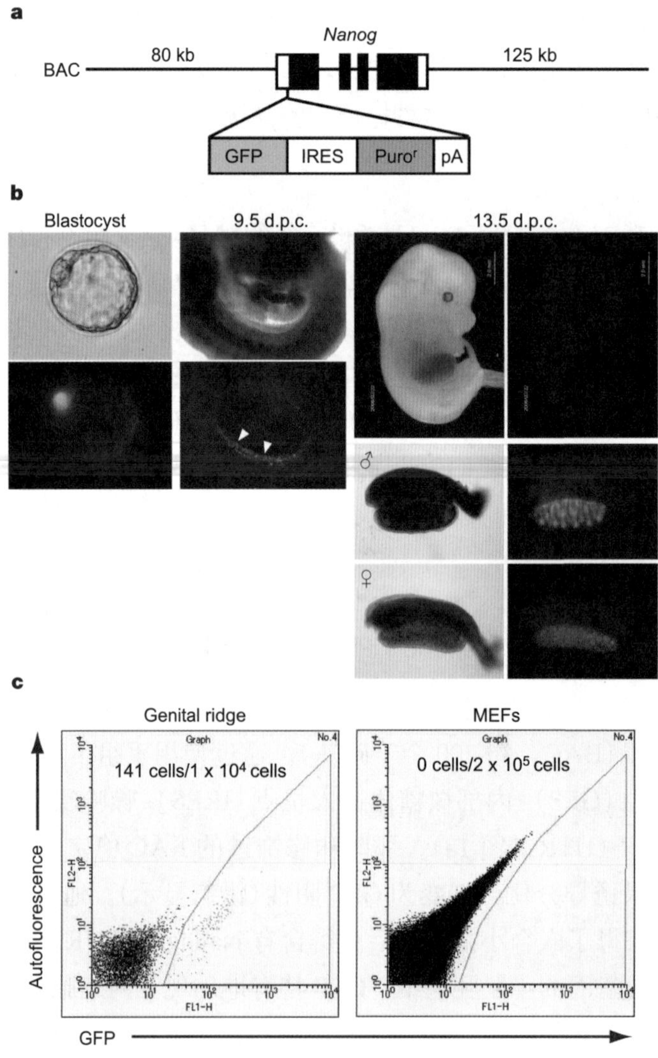

Fig. 1. Nanog-GFP-IRES-Puro^r transgenic mice. **a**, Modified BAC construct. White boxes indicate the 5′ and 3′ UTRs of the mouse *Nanog* gene. Black boxes indicate the open reading frame. **b**, GFP expression in Nanog-GFP transgenic mouse embryos. Whole embryos (top panels) and isolated genital ridges (bottom panels) from 13.5 d.p.c. mice are shown. **c**, Histogram showing GFP fluorescence in cells isolated from genital ridges of a 13.5 d.p.c. Nanog-GFP transgenic mouse embryo (left) or in MEFs isolated from the same embryo (right).

Next, we introduced the four previously described factors (Oct3/4, Sox2, Klf4 and the c-Myc mutant c-Myc(T58A)) into Nanog-GFP-IRES-Puro^r MEFs cultured on SNL feeder cells with the use of retroviral vectors. Three, five, or seven days after retroviral infection, we started puromycin selection in ES cell medium. GFP-positive cells first became apparent ~7 days after infection. Twelve days after infection, a few hundred colonies appeared, regardless of the timing of puromycin selection (Fig. 2a). By contrast, no colonies emerged from MEFs transfected with mock DNA. Among puromycin-resistant colonies, ~5% were positive for GFP (Fig. 2b). When the puromycin selection was started at 7 days after infection, we

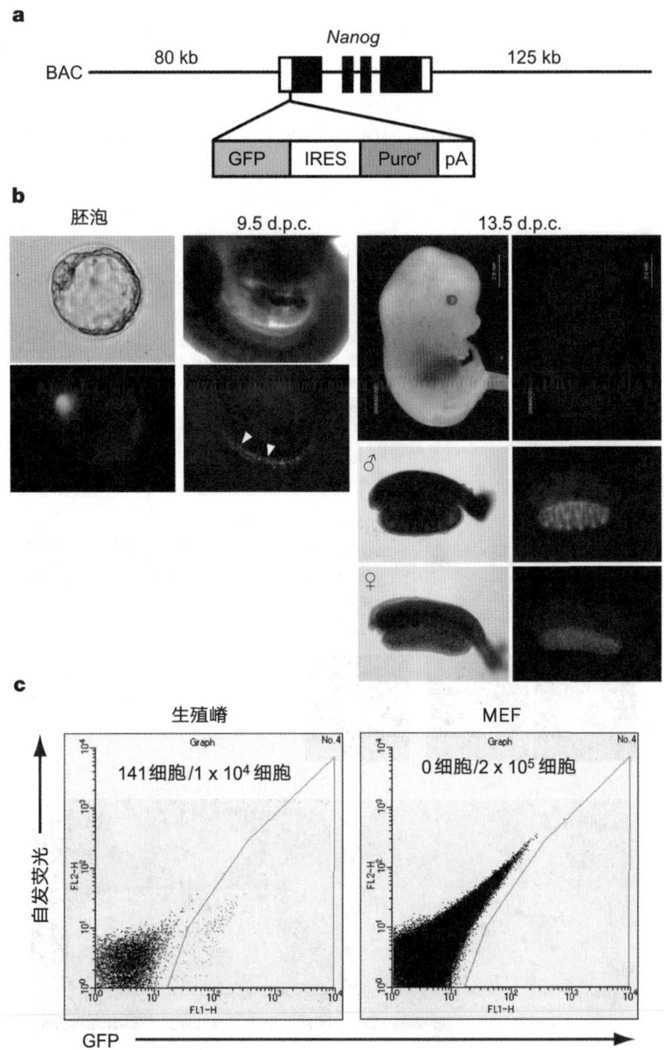

图 1. Nanog-GFP-IRES-Puro^r 转基因小鼠。**a**，修饰的 BAC 结构。白框表示小鼠 *Nanog* 基因的 5′和 3′UTR。黑框代表可读框。**b**，Nanog-GFP 转基因小鼠胚胎中的 GFP 表达。图中显示了交配后 13.5 天小鼠的整个胚胎（上部）和分离的生殖嵴（下部）。**c**，图中显示从交配后 13.5 天的 Nanog-GFP 转基因小鼠胚胎的生殖嵴中分离的细胞的 GFP 荧光（左）和从相同胚胎中分离出来的 MEF 的 GFP 荧光（右）。

然后，我们将先前所述的四种因子（Oct3/4、Sox2、Klf4 和 c-Myc 突变体 c-Myc（T58A））利用逆转录病毒载体导入到培养在 SNL 饲养细胞中的 Nanog-GFP-IRES-Puro^r MEF 内。逆转录病毒感染后的 3 天、5 天和 7 天，我们分别在 ES 细胞基质中进行嘌呤霉素选择。感染后约 7 天 GFP 阳性的细胞开始出现。感染后 12 天出现了数百个集落，并且不受嘌呤霉素选择时间的影响（图 2a）。相反地，用对照 DNA 进行转染的 MEF 没有出现集落。在嘌呤霉素抗性集落中，大约 5% 是 GFP 阳性的（图 2b）。感染 7 天后当嘌呤霉素选择开始时，我们获得了 GFP 阳性最高的集

obtained the most GFP-positive colonies. Because we used the GFP-IRES-Puro^r cassette, it is unclear why we obtained GFP-negative colonies. With increased concentrations of puromycin, we obtained fewer GFP-negative colonies (Fig. 2c). With any combination of three of the four factors, we did not obtain any GFP-positive colonies (Supplementary Fig. 1).

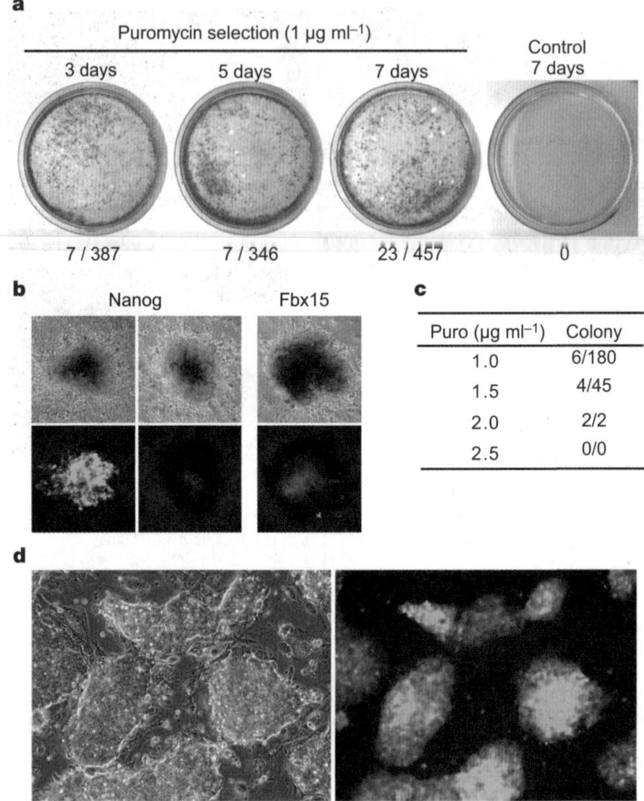

Fig. 2. Generation of iPS cells from MEFs of Nanog-GFP-IRES-Puro^r transgenic mice. **a**, Puromycin-resistant colonies. Puromycin selection was initiated at 3, 5, or 7 days after retroviral transduction. Numbers indicate GFP-positive colonies/total colonies. **b**, GFP fluorescence in resulting colonies. Phase contrast (top row) and fluorescence (bottom row) micrographs are shown. iPS cells were also generated from Fbx15 β-geo knockin MEFs. **c**, Effect of increasing concentrations of puromycin. Numbers of GFP-positive colonies/total colonies are shown on the right. **d**, Morphology of established Nanog iPS cells (clone 20D17). Phase contrast (left) and fluorescence (right) micrographs are shown.

By continuing cultivation of these GFP-positive colonies, we obtained cells that were morphologically indistinguishable from ES cells (Fig. 2d). These cells also demonstrated ES-like proliferation, with slightly longer doubling times than that of ES cells (Fig. 3a). Subcutaneous transplantation of these cells into nude mice resulted in tumours that consisted of various tissues of all three germ layers, indicating that these cells are pluripotent (Fig. 3b and Supplementary Fig. 2). We therefore refer to these cells as Nanog iPS cells in the remainder of this manuscript. Induced pluripotent stem cells were established from Fbx15 β-geo MEFs in parallel and are referred to as Fbx15 iPS cells.

落。因为我们插入的是 GFP-IRES-Puro[r] 盒，所以不清楚为什么会出现 GFP 阴性的集落。随着嘌呤霉素浓度的增加，我们得到 GFP 阴性的集落越少（图 2c）。这四个因子中的任意三个组合都不能出现 GFP 阳性的集落（补充信息图 1）。

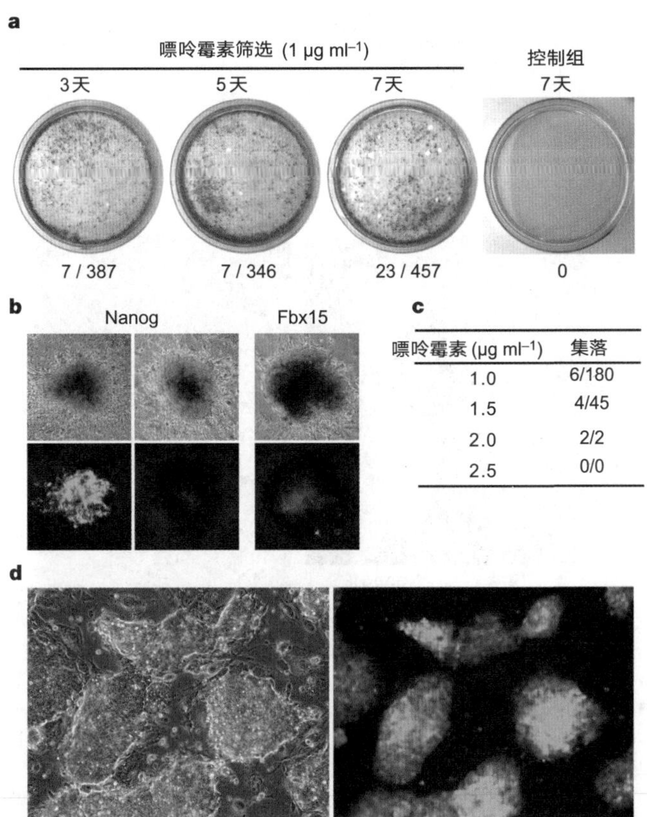

图 2. 从 Nanog-GFP-IRES-Puro[r] 转基因小鼠 MEF 中生成 iPS 细胞。**a**，嘌呤霉素抗性集落。嘌呤霉素筛选开始于逆转录病毒转导后的 3、5 和 7 天。数字代表 GFP 阳性集落/总集落。**b**，集落中的 GFP 荧光。图片为相差（顶行）和荧光（底行）显微图。iPS 细胞也从 Fbx15 β-geo 基因敲入 MEF 中产生。**c**，嘌呤霉素浓度增加的影响。右边显示 GFP 阳性集落/总集落的数目。**d**，已建立的 Nanog iPS 细胞的形态（克隆 20D17）。图片是相差（左）和荧光（右）显微图。

通过继续培养这些 GFP 阳性的集落，我们获得了形态上与 ES 细胞无法区分的细胞（图 2d）。这些细胞可以出现 ES 样的增殖，倍增时间略长于 ES 细胞（图 3a）。将这些细胞皮下移植到裸鼠中可以产生含有全部三个胚层各种组织的肿瘤，提示这些细胞是多能的（图 3b 和补充信息图 2）。因此在本文的剩余部分我们称这些细胞为 Nanog iPS 细胞。我们平行建立了来源于 Fbx15 β-geo MEF 的诱导多能干细胞，称为 Fbx15 iPS 细胞。

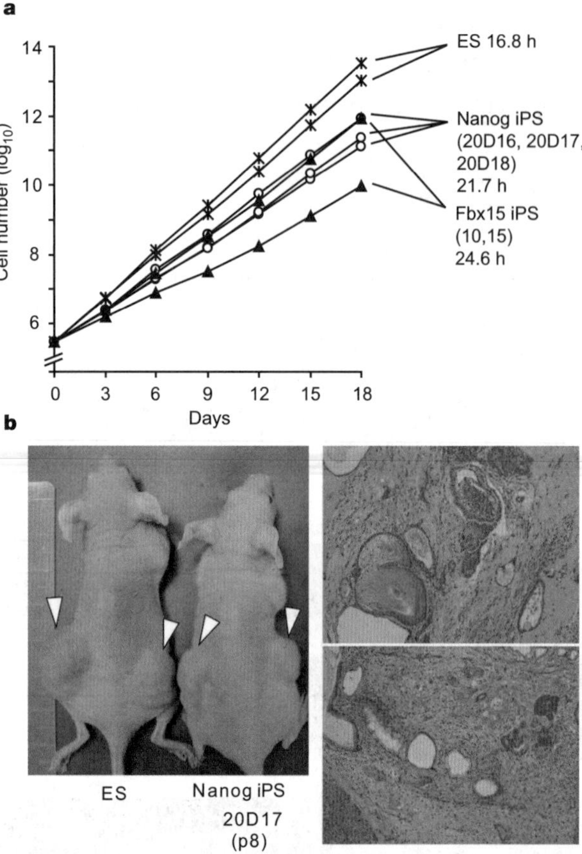

Fig. 3. Characterization of Nanog iPS cells. **a**, Proliferation. ES cells, Nanog iPS cells (clones 20D16, 20D17 and 20D18) and Fbx15 iPS cells (clones 10 and 15) were passaged every 3 days (3×10^5 cells per each well of a 6-well plate). Calculated doubling times are indicated. **b**, Teratomas. ES cells or Nanog iPS cells (clone 20D17, 1×10^6 cells) were subcutaneously transplanted into nude mice. After 8 weeks, teratomas were photographed (left) and analysed histologically with haematoxylin and eosin staining (right).

Similarity between Nanog iPS Cells and ES Cells

Polymerase chain reaction with reverse transcription (RT–PCR) showed that Nanog iPS cells expressed most ES cell marker genes, including *Nanog*, at higher and more consistent levels compared with Fbx15 iPS cells (Fig. 4a). DNA microarray analyses confirmed that Nanog iPS cells had greater ES-cell-like gene expression compared with Fbx15 iPS cells (Fig. 4b). The expression level of *Rex1* (also called *Zfp42*) in Nanog iPS cells was higher compared with Fbx15 iPS cells, but still lower than in ES cells. Thus, Nanog iPS cells show greater gene expression similarity to ES cells (without being identical) than do Fbx15 iPS cells.

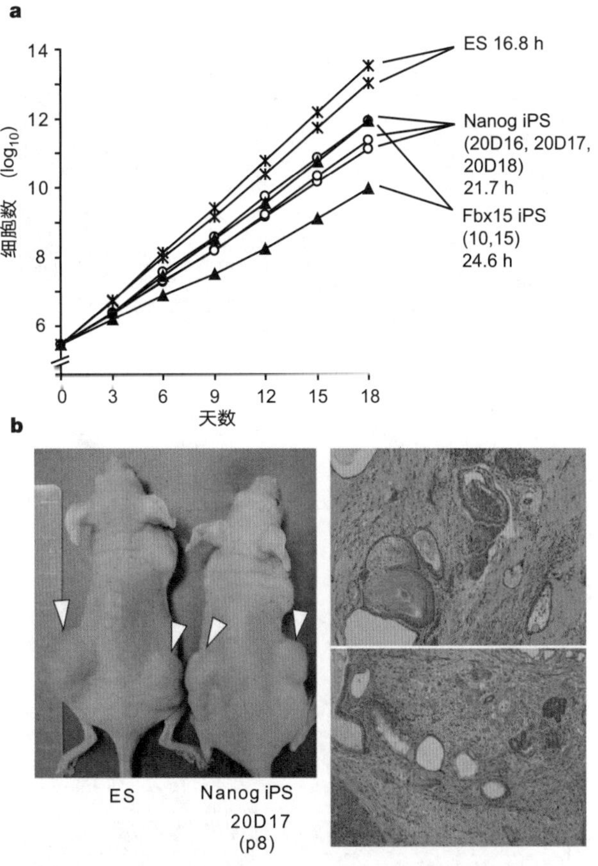

图 3. Nanog iPS 细胞的特征。**a**，增殖。ES 细胞、Nanog iPS 细胞（克隆 20D16、20D17 和 20D18）以及 Fbx15 iPS 细胞（克隆 10 和 15）每 3 天都进行传代（6 孔板上每孔 3×10^5 个细胞）。图中显示了计算出的倍增时间。**b**，畸胎瘤。ES 细胞或者 Nanog iPS 细胞（克隆 20D17，1×10^6 个细胞）皮下移植到裸鼠上。8 周以后，进行畸胎瘤照相（左图）并用苏木精–伊红染色观察组织学结构（右图）。

Nanog iPS 细胞和 ES 细胞间的相似性

反转录聚合酶链反应（RT–PCR）显示 Nanog iPS 细胞表达包括 *Nanog* 在内的大部分 ES 细胞标志物基因，而且与 Fbx15 iPS 细胞相比，Nanog iPS 细胞表达水平更高、更符合 ES 细胞（图 4a）。DNA 微阵列分析确证了 Nanog iPS 细胞与 Fbx15 iPS 细胞相比表达更多的 ES 细胞样基因（图 4b）。Nanog iPS 细胞中 *Rex1*（又称 *Zfp42*）的表达水平高于 Fbx15 iPS 细胞，但是仍然低于 ES 细胞。因此，Nanog iPS 细胞与 ES 细胞的基因表达相似程度（不是完全一致）要超过 Fbx15 iPS 细胞。

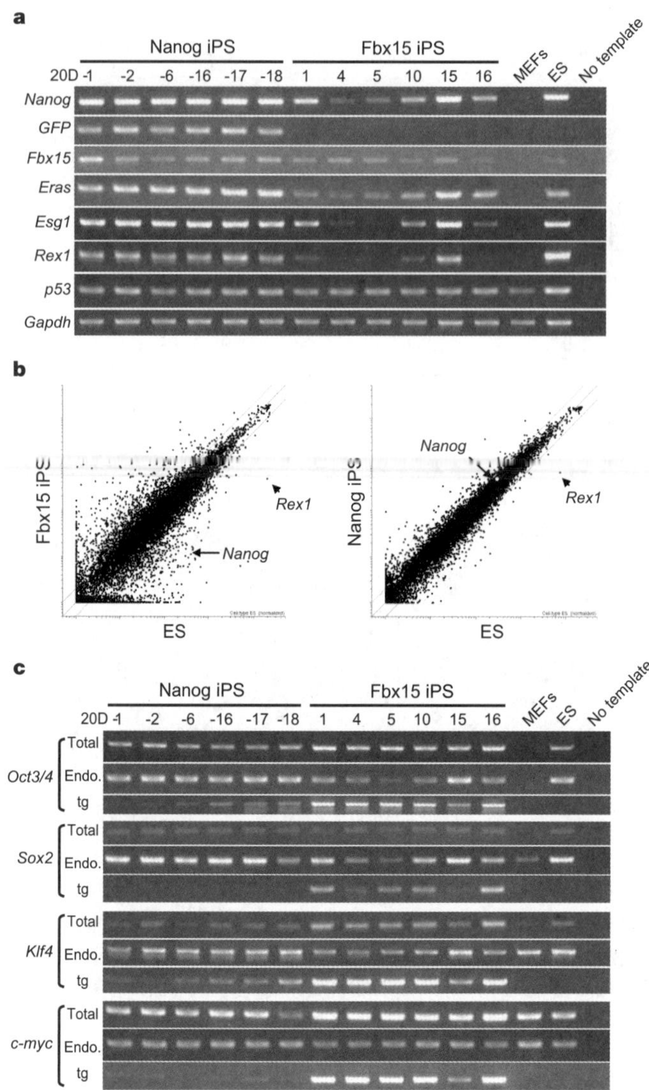

Fig. 4. Gene expression in Nanog iPS cells. **a**, RT–PCR. Total RNA was isolated from six clones of Nanog iPS cells (clones 20D1, 20D2, 20D6, 20D16, 20D17 and 20D18), six clones of Fbx15 iPS cells (clones 1, 4, 5, 10, 15 and 16), MEFs and ES cells. **b**, Scatter plots showing comparison of global gene expression between ES cells and Nanog iPS cells (right), and between ES cells and Fbx15 iPS cells (left), as determined by DNA microarrays. **c**, Expression levels of the four transcription factors. Total RNA was isolated from six clones of Nanog iPS cells (clones 20D1, 20D2, 20D6, 20D16, 20D17 and 20D18), six clones of Fbx15 iPS cells (clones 1, 4, 5, 10, 15 and 16), MEFs and ES cells. RT–PCR analyses were performed with primers that amplified the coding regions of the four factors (Total), endogenous transcripts only (Endo.), and transgene transcripts only (tg).

RT–PCR showed that Nanog iPS cells have significantly lower expression levels of the four transgenes than Fbx15 iPS cells (Fig. 4c). Real-time PCR confirmed that transgene expression was very low in Nanog iPS cells (Supplementary Fig. 4a–d). In contrast, Southern blot analyses showed similar copy numbers of retroviral integration in Nanog

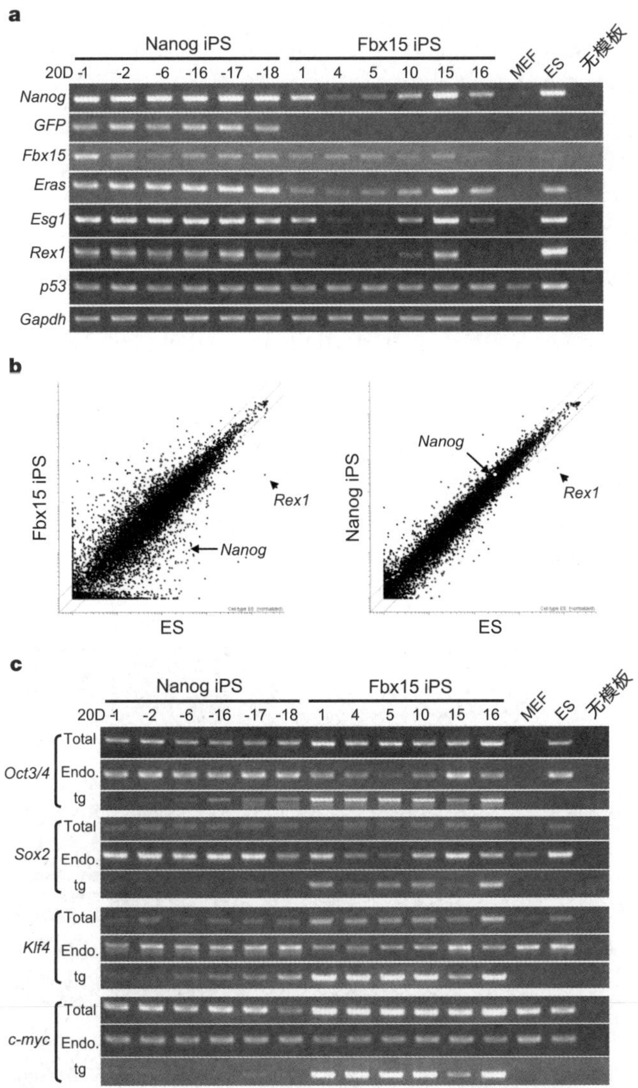

图 4. Nanog iPS 细胞的基因表达。**a**，RT–PCR。从 6 个 Nanog iPS 细胞克隆（克隆 20D1、20D2、20D6、20D16、20D17 和 20D18）、6 个 Fbx15 iPS 细胞克隆（克隆 1、4、5、10、15 和 16）、MEF 和 ES 细胞中分离出总 RNA。**b**，散点图显示了使用 DNA 微阵列得出的 ES 细胞和 Nanog iPS 细胞（右图）以及 ES 细胞和 Fbx15 iPS 细胞（左图）之间的整体基因表达的对比。**c**，四个转录因子的表达水平。从 6 个 Nanog iPS 细胞克隆（克隆 20D1、20D2、20D6、20D16、20D17 和 20D18）、6 个 Fbx15 iPS 细胞克隆（克隆 1、4、5、10、15 和 16）、MEF 和 ES 细胞中分离出总 RNA。然后进行 RT–PCR 分析，使用的引物分别扩增四个因子的编码区（Total）、仅内源转录物（Endo.）和仅转基因转录物（tg）。

RT–PCR 显示 Nanog iPS 细胞中四个转基因的表达水平远远低于 Fbx15 iPS 细胞（图 4c）。实时 PCR 确证了 Nanog iPS 细胞中转基因的表达非常低（补充信息图 4a ~ 4d）。相反，Southern 印迹分析显示 Nanog iPS 细胞和 Fbx15 iPS 细胞中逆转录

iPS cells and Fbx15 iPS cells (Supplementary Fig. 5). These data indicate that retroviral transgene expression is largely silenced in Nanog iPS cells, as has been shown in ES cells[17]. The expression levels of the transgenes are reversely correlated with *Dnmt3a2* expression, suggesting that *de novo* methyltransferase[18] may be involved in the retroviral silencing observed in iPS cells (Supplementary Fig. 6).

Bisulphite genomic sequencing analyses also revealed similarities between Nanog iPS cells and ES cells (Fig. 5). The promoter regions of *Nanog*, *Oct3/4* and *Fbx15* were largely unmethylated in Nanog iPS cells. This is in marked contrast to Fbx15 iPS cells in which the promoters of *Nanog* and *Oct3/4* were only partially unmethylated[8]. Differentially methylated regions of imprinting genes *H19* and *Igf2r* were partially methylated in Nanog iPS cells. During PGC development, imprinting is erased by 12.5 d.p.c.[19-21]. The loss of imprinting is maintained in embryonic germ cells derived from 12.5 d.p.c. PGCs[22] and cloned embryos derived from 12.5–16.5 d.p.c. PGCs[23,24]. ES cells, by contrast, showed normal imprinting patterns[25]. Thus, Nanog iPS cells show greater similarity in the methylation patterns of imprinting genes to ES cells than to embryonic germ cells.

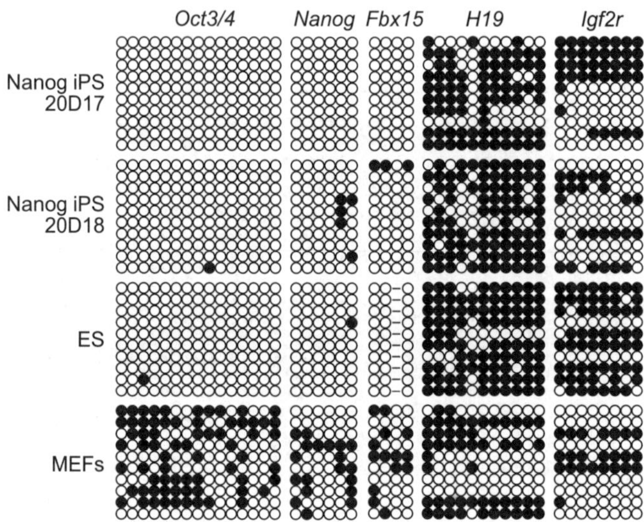

Fig. 5. DNA methylation of ES-cell-specific genes and imprinting genes. White circles indicate unmethylated CpG dinucleotides, whereas black circles indicate methylated CpG dinucleotides.

Simple sequence length polymorphism (SSLP) analyses showed that Nanog iPS cells are largely of the DBA background but also have some contribution from the C57BL/6 and 129S4 backgrounds (Supplementary Fig. 3). This result is consistent with the genetic background of the MEFs, which was 75% DBA, 12.5% C57BL/6 and 12.5% 129S4. This result also confirms that Nanog iPS cells are not a contamination of ES cells that exists in our laboratory, which are either pure 129S4 or C57BL/6.

We next compared the stability of Nanog iPS cells and Fbx15 iPS cells (Supplementary Fig. 7). Cells were cultivated in the presence of the selection drug for up to 22–26 passages.

病毒整合的拷贝数目相似(补充信息图 5)。这些数据提示 Nanog iPS 细胞中逆转录病毒转基因的表达大量地被抑制，这和 ES 细胞中出现的情况一样[17]。转基因的表达水平与 *Dnmt3a2* 的表达呈负相关，说明新生的甲基转移酶[18]可能参与了 iPS 细胞内逆转录病毒被抑制的过程(补充信息图 6)。

亚硫酸氢盐基因组测序分析也揭示了 Nanog iPS 细胞与 ES 细胞间的相似性(图 5)。Nanog iPS 细胞内的 *Nanog*、*Oct3/4* 和 *Fbx15* 的启动子区域大部分都是未甲基化的。这与 Fbx15 iPS 细胞明显相反，后者的 *Nanog* 和 *Oct3/4* 启动子仅仅是部分未甲基化[8]。印记基因 *H19* 和 *Igf2r* 的差异甲基化区域在 Nanog iPS 细胞中部分被甲基化。在 PGC 的发育过程中，印记在交配后 12.5 天消除[19-21]。这种印记消除的现象在交配后 12.5 天的 PGC 来源的胚胎生殖细胞以及交配后 12.5 ~ 16.5 天的 PGC 来源的克隆胚胎中保持[23,24]。相反，ES 细胞具有正常的印记分布[25]。因此，Nanog iPS 细胞在印记基因的甲基化分布方面与 ES 细胞的相似程度比胚胎生殖细胞更高。

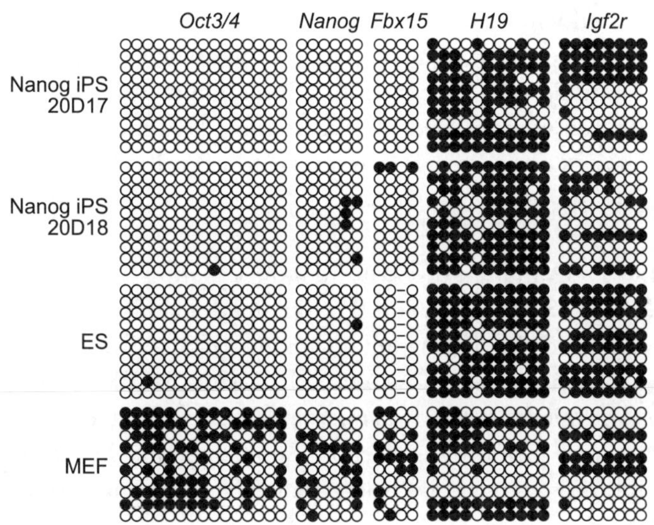

图 5. ES 细胞特异性基因和印记基因的 DNA 甲基化。白圈代表未甲基化的 CpG 二核苷酸，而黑圈代表甲基化的 CpG 二核苷酸。

简单序列长度多态性(SSLP)分析显示 Nanog iPS 细胞大部分是来源于 DBA 背景的，但是也有一部分来源于 C57BL/6 和 129S4 背景(补充信息图 3)。这个结果与 MEF 的基因背景相一致，即 75% DBA、12.5% C57BL/6 和 12.5% 129S4。该结果也证实了 Nanog iPS 细胞不是存在于我们实验室中的 ES 细胞的污染，因为那些 ES 细胞都是纯的 129S4 或者 C57BL/6 来源的。

我们接下来比较了 Nanog iPS 细胞和 Fbx15 iPS 细胞的稳定性(补充信息图 7)。

Morphologically, we did not observe significant changes over the long-term culture course. However, RT–PCR showed that Fbx15 iPS cells lost the expression of ES cell marker genes after prolonged culture. By contrast, Nanog iPS cells maintained relatively high expression levels of the ES cell marker genes. These data demonstrate that Nanog iPS cells are more stable than Fbx15 iPS cells.

We also compared the induction efficiency of Nanog iPS cells and Fbx15 iPS cells. In independent experiments, we obtained 4–125 GFP-positive colonies from 8×10^5 Nanog-reporter MEFs transfected with the four transcription factors. Because ~50% of transfected MEFs are supposed to express all four factors[8], the induction efficiency is approximately 0.001–0.03%. In contrast, from the same number of Fbx15-reporter MEFs, we obtained 47–1,800 G418-resistant colonies, with the induction efficiency of approximately 0.01–0.5%. Thus, the efficiency of Nanog iPS cell induction is approximately one-tenth that of Fbx15 iPS cells.

We then compared the responses of Nanog iPS cells and Fbx15 iPS cells to LIF or retinoic acid (Supplementary Fig. 8). As we have shown previously[8], Fbx15 iPS cells do not remain undifferentiated when cultured without feeder cells, even in the presence of LIF. Furthermore, Fbx15 iPS cells formed compact colonies when cultured without feeder cells in the presence of retinoic acid. In contrast, LIF maintained the undifferentiated state of Nanog iPS cells cultured without feeder cells. Retinoic acid induced the differentiation of Nanog iPS cells. Thus, Nanog iPS cells are similar to ES cells in their response to LIF and retinoic acid.

Initially we used the T58A mutant of c-Myc to induce Nanog iPS cells. We also tested wild-type c-Myc for Nanog iPS cell induction. We obtained a similar number of colonies with both wild-type c-Myc and the T58A mutant. Nanog iPS cells established with wild-type c-Myc were indistinguishable from those established with the T58A mutant with regards to morphology, gene expression (analysed via microarrays), teratoma formation (Supplementary Fig. 2) and stability under puromycin selection (Supplementary Fig. 9). Without puromycin selection, Nanog iPS cells induced by wild-type c-Myc were more stable (Supplementary Fig. 9).

Germline Chimaeras from Nanog iPS Cells

We next examined the ability of Nanog iPS cells to produce adult chimaeras. We injected 15–20 male Nanog iPS cells (five clones with the T58A mutant and three with wild-type c-Myc) into C57BL/6-derived blastocysts, which we then transplanted into the uteri of pseudo-pregnant mice. We obtained adult chimaeras from seven clones (four clones with the T58A mutant and three with wild-type c-Myc) as determined by coat colour (Fig. 6a and Supplementary Table 1). SSLP analyses showed that Nanog iPS cells contributed to various organs, with the level of chimaerism ranging from 10% to 90%. Chimeras from clone 20D17 showed highest iPS cell contribution in the testes. From clone 20D18, we

细胞在存在选择药物的环境中培养 22 ~ 26 代。在这段长时间的培养过程中，我们没有发现形态学上明显的改变。但是，RT–PCR 提示经过延长培养后 Fbx15 iPS 细胞丢失了 ES 细胞标志基因的表达。相反，Nanog iPS 细胞保持着相对高水平的 ES 细胞标志基因表达。这些数据说明 Nanog iPS 细胞比 Fbx15 iPS 细胞更加稳定。

我们还比较了 Nanog iPS 细胞和 Fbx15 iPS 细胞的诱导效率。在独立的实验中，我们从 8×10^5 个转染了 4 个转录因子的 Nanog 报告基因 MEF 中获得了 4 ~ 125 个 GFP 阳性的集落。因为大约 50% 的转染 MEF 被认为会表达所有的四个因子[8]，诱导效率约为 0.001% ~ 0.03%。相反，从同样数量的 Fbx15 报告基因 MEF 中，我们获得了 47 ~ 1,800 个 G418 抗性集落，诱导效率大约是 0.01% ~ 0.5%。因此，Nanog iPS 细胞的诱导效率大约是 Fbx15 iPS 细胞的十分之一。

然后我们比较了 Nanog iPS 细胞和 Fbx15 iPS 细胞对 LIF 或者视黄酸的响应（补充信息图 8）。正如我们之前所示[8]，即便是存在 LIF，Fbx15 iPS 细胞在没有饲养细胞的情况下也不会保持未分化状态。此外，Fbx15 iPS 细胞在存在视黄酸和没有饲养细胞的情况下也形成紧致的集落。相反，没有饲养细胞存在时，LIF 能保持 Nanog iPS 细胞的未分化状态。视黄酸诱导 Nanog iPS 细胞的分化。因此，Nanog iPS 细胞在对 LIF 和视黄酸的反应方面与 ES 细胞相似。

起初我们使用 c-Myc 的 T58A 突变体来诱导 Nanog iPS 细胞。我们也实验了野生型 c-Myc 诱导 Nanog iPS 细胞。两种情况下我们获得的集落数相似。在形态、基因表达（通过微阵列分析）、畸胎瘤形成（补充信息图 2）和嘌呤霉素选择下的稳定性（补充信息图 9）等方面，无法区分用野生型 c-Myc 诱导的与 T58A 突变体诱导的 Nanog iPS 细胞。没有嘌呤霉素选择的话，野生型 c-Myc 诱导的 Nanog iPS 细胞更加稳定（补充信息图 9）。

Nanog iPS 细胞来源的种系嵌合体

我们接下来检验了 Nanog iPS 细胞产生成熟嵌合体的能力。我们将 15 ~ 20 个雄性 Nanog iPS 细胞（5 个 T58A 突变体诱导的克隆和 3 个野生型 c-Myc 诱导的克隆）注射到 C57BL/6 来源的胚泡中，然后将胚泡移植到假孕小鼠的子宫内。根据毛色，我们得到了 7 个克隆的成熟嵌合体（4 个 T58A 突变体诱导的克隆和 3 个野生型 c-Myc 诱导的克隆）（图 6a 和补充信息表 1）。SSLP 分析显示 Nanog iPS 细胞参与生成多个器官，嵌合水平从 10% 到 90% 不等。克隆 20D17 来源的嵌合体在睾丸内出现最高的 iPS 细胞嵌合。对于克隆 20D18，我们仅从感染后胚泡中获得少量非嵌合

obtained only a few nonchimaeric pups from infected blastocysts; thus, whether this clone has competency for producing adult chimaeras remained to be determined. These data demonstrate that most Nanog iPS clones are competent for adult chimaeric mice.

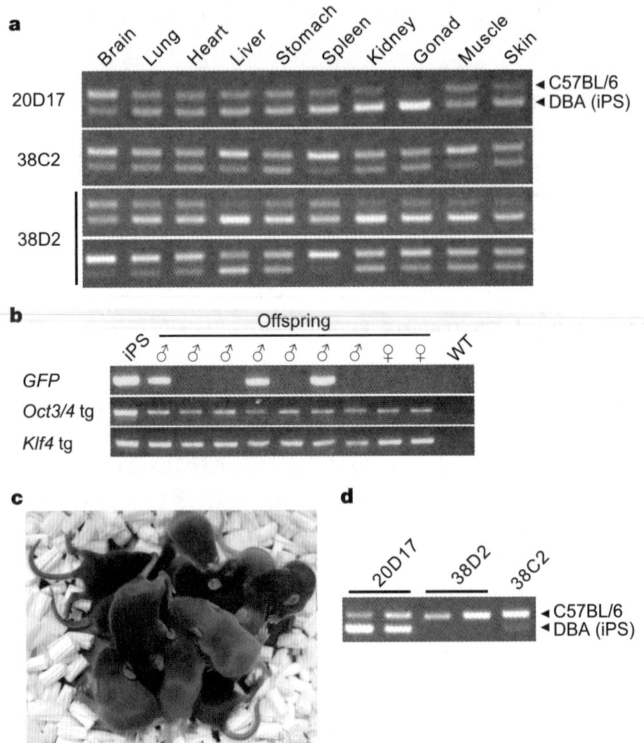

Fig. 6. Germline chimaeras from Nanog iPS cells. **a**, Tissue distribution of iPS cells in chimaeras. Genomic DNA was isolated from the indicated organs of chimaeras derived from three Nanog iPS cell clones (20D17, 38C2 and 38D2). SSLP analyses were performed for D6Mit15. **b**, PCR analyses showing the presence of the GFP cassette and retroviral transgenes in F_1 mice obtained from the intercross between a chimaeric male and a C57BL/6 female. **c**, Coat colours of F_2 mice obtained from F_1 intercrosses. **d**, Sperm contribution of iPS cells in chimaeric mice. Spermatozoa were isolated from the epididymides of chimaeric mice derived from three Nanog iPS cell clones (20D17, 38D2 and 38C2). iPS cell contribution was determined by SSLP of D6Mit15.

We then crossed three of the chimaeras from clone 20D17—for which the highest iPS cell contribution was in the testes—with C57BL/6 females. Whereas all F_1 mice showed black coat colour, all contained retroviral integration of the four transcription factors and approximately half contained the GFP-IRES-Puro^r cassette (Fig. 6b), indicating germline transmission. Furthermore, approximately half of the F_2 mice born from F_1 intercrosses showed agouti coat colour, confirming germline transmission of Nanog-iPS-20D17 (Fig. 6c).

We also examined germline competency for two other clones that produced adult chimaeras. In one chimaeric mouse from Nanog-iPS-38C2 cell line, PCR analysis detected iPS cell contribution in isolated spermatozoa (Fig. 6d), suggesting that germline competency

型幼仔。因此，该克隆是否具有产生成熟嵌合体的能力还有待确定。这些数据提示大部分 Nanog iPS 细胞具有产生成熟嵌合体小鼠的能力。

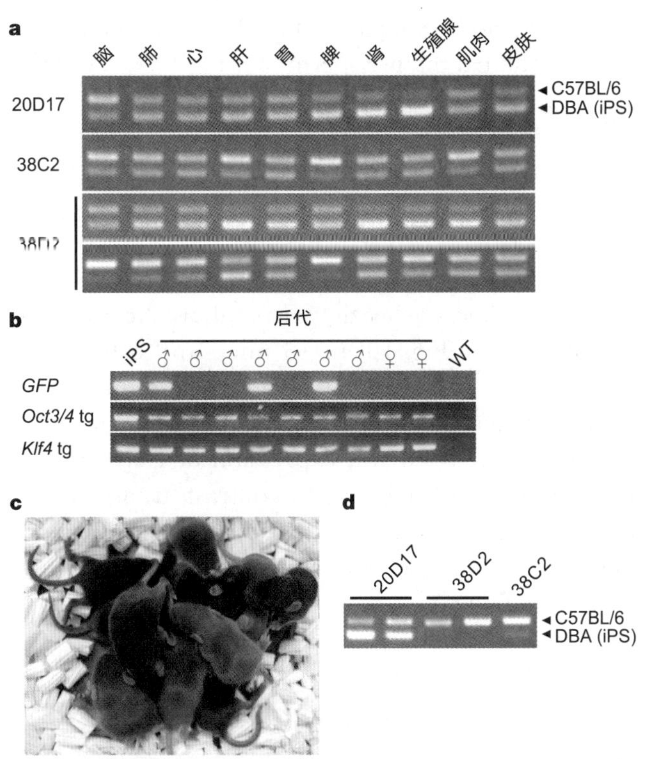

图 6. Nanog iPS 细胞来源的种系嵌合体。**a**，嵌合体中 iPS 细胞的组织分布。基因组 DNA 从三个 Nanog iPS 细胞克隆（20D17、38C2 和 38D2）来源的嵌合体的所示器官中分离出来。对 D6Mit15 进行 SSLP 分析。**b**，PCR 分析显示嵌合体雄鼠和 C57BL/6 雌鼠交配生成的 F_1 代小鼠中存在 GFP 盒和逆转录病毒转基因。**c**，F_1 代互交产生的 F_2 代小鼠的毛色。**d**，嵌合体小鼠中 iPS 细胞对精子生成的贡献。从三个 Nanog iPS 细胞克隆（20D17、38D2 和 38C2）来源的嵌合体小鼠的附睾中分离出精子。通过 D6Mit15 的 SSLP 分析确定 iPS 细胞的嵌合。

我们然后用克隆 20D17 来源的 3 只嵌合体（在这些嵌合体中，iPS 细胞在睾丸中贡献最大）与 C57BL/6 雌性小鼠杂交。所有的 F_1 代小鼠显示黑色的毛色，且都含有逆转录病毒整合的这四个转录因子，几乎一半含有 GFP-IRES-Puro^r 盒（图 6b），这标志着发生了种系的传递。此外，F_1 代互交产生的几乎一半的 F_2 代小鼠显示刺豚鼠样的毛色，证实了发生 Nanog-iPS-20D17 的种系传递（图 6c）。

我们还检测了产生成熟嵌合体的另两个克隆的种系产生能力。在一个 Nanog-iPS-38C2 细胞系来源的嵌合型小鼠中，PCR 分析检测到分离的精子中有 iPS 细胞来源（图 6d），表明其种系能力并不局限于克隆 20D17。但是，克隆 38C2 中 iPS 细

is not confined to clone 20D17. However, the iPS cell contribution to sperm of clone 38C2 is much smaller than that of clone 20D17, and no iPS-cell-derived offspring were found for 119 mice born from the cross between the 38C2 chimaera and C57BL/6 female mice. Most male mice with a high degree of chimaerism from the Nanog-iPS-38D2 cell line showed small testes and aspermatogenesis (Supplementary Fig. 13). The testes of some chimaeras from Nanog-iPS-38D2 contained mature sperm, but no iPS cell contribution was detected by PCR (Fig. 6d).

Tumour Formation by *c-myc* Reactivation

Out of 121 F_1 mice (aged 8–41 weeks) derived from the Nanog-iPS-20D17 cell line, 24 died or were killed because of weakness, wheezing or paralysis. Necropsy of 17 mice identified neck tumours (Supplementary Fig. 10) in 13 mice and other tumours in five mice, including two mice with neck tumours. Histological examination of one neck tumour showed that it was a ganglioneuroblastoma with follicular carcinoma of the thyroid gland (not shown). In these tumours, retroviral expression of *c-myc*, but not *Oct3/4*, *Sox2*, or *Klf4*, is reactivated (Supplementary Fig. 11). In contrast, transgene expression of all four transcription factors remained low in normal tissues, except for *c-myc* in muscle in one mouse (Supplementary Fig. 12). These data indicate that reactivation of *c-myc* retrovirus is attributable to tumour formation.

Discussion

Our results demonstrate that Nanog selection allows the generation of high-quality iPS cells that are comparable to ES cells in morphology, proliferation, teratoma formation, gene expression and competency for adult chimaeras. Nearly all Nanog iPS clones showed these properties, indicating that Nanog is a major determinant of quality in cellular pluripotency. However, germline competence was variable among Nanog iPS clones, indicating the existence of other important determinants of germline competency in addition to Nanog. The high quality of Nanog iPS cells underscores the possibility of using this technology to generate patient-specific pluripotent stem cells. In a separate study, we found that germline-competent iPS cells can also be obtained from adult mouse somatic cells (T. Aoi and S.Y., unpublished data). The current study, however, also reveals that reactivation of *c-myc* retrovirus may result in tumour formation. There may be ways to overcome this problem. Strong silencing of the four retroviruses in Nanog iPS cells indicates that they are only required for the induction, but not the maintenance, of pluripotency. Therefore, the retrovirus-mediated system might be eventually replaced by transient expression, such as the adenovirus-mediated system. Alternatively, high-throughput screening of chemical libraries might identify small molecules that can replace the four genes. These are crucial research areas in order to apply iPS cells to regenerative medicine.

We found that the efficiency of Nanog iPS cell induction is less than 0.1%. The low

胞产生精子的比例要远远少于克隆 20D17，而且 38C2 嵌合体与 C57BL/6 雌性小鼠间交配生出的 119 只小鼠中没有发现 iPS 细胞来源的后代。大部分具有高嵌合度 Nanog-iPS-38D2 细胞系来源的雄性小鼠都表现为小睾丸和无精子发生（补充信息图 13）。Nanog-iPS-38D2 细胞系来源的另一些嵌合型小鼠的睾丸内含有成熟的精子，但是 PCR 没有检测到其中有来源于 iPS 细胞的（图 6d）。

c-myc 再激活后的肿瘤形成

Nanog-iPS-20D17 细胞系来源的 121 只 F_1 小鼠中（年龄 8～41 周），24 只因为虚弱、喘鸣或者瘫痪而死亡或者被杀死。对 17 只小鼠进行尸检发现其中 13 只小鼠有颈部肿瘤（补充信息图 10），五只小鼠有其他肿瘤（其中两只有颈部肿瘤）。对一个颈部肿瘤进行组织学检查显示，它是滤泡性甲状腺癌的成神经节细胞瘤（未显示）。在这些肿瘤中，只有逆转录病毒的基因 *c-myc* 被重新激活，而 *Oct3/4*、*Sox2* 或者 *Klf4* 并没有（补充信息图 11）。相反，除了一只小鼠肌肉中的 *c-myc* 外，正常组织中所有四个转录因子的转基因表达都非常低（补充信息图 12）。这些数据提示逆转录病毒 *c-myc* 基因的再激活导致了肿瘤的生成。

讨　论

我们的结果显示 Nanog 选择可以产生高质量的 iPS 细胞，其在形态、增殖、畸胎瘤形成、基因表达和成熟嵌合体形成能力上与 ES 细胞相似。几乎所有的 Nanog iPS 克隆都有这些特性，表明 Nanog 是细胞多能性质量的主要决定因素。但是，Nanog iPS 克隆间的种系形成能力有差别，这提示除了 Nanog 外还存在其他重要的决定因素。Nanog iPS 细胞的这种高性能使得应用该技术产生病人特异性的多能干细胞成为可能。在另一个研究中，我们发现具有种系形成功能的 iPS 细胞也能从成年小鼠的体细胞中产生（青井贵之和山中伸弥，未发表的数据）。但是，目前的研究也揭示了 *c-myc* 基因的再激活可能导致肿瘤的形成。可能有解决这个问题的方法。Nanog iPS 细胞中这四个逆转录病毒基因的高度沉默说明，它们仅是诱导而非维持多能性所必需的。因此，逆转录病毒介导的系统可能最终被瞬时表达取代，比如腺病毒介导的系统。或者，高通量筛选化合物库可能会发现能够替代这四个基因的小分子。这是将 iPS 细胞应用于再生医学的关键研究领域。

我们发现 Nanog iPS 细胞诱导的效率低于 0.1%。这种低效率表明 iPS 细胞的起

efficiency suggests that the origin of iPS cells might be rare stem cells co-existing in MEF culture. Alternatively, activation of additional genes by retroviral integration might be required for iPS cell generation in addition to the four transcription factors. This is relevant to the fact that we have been able to obtain iPS cells only with retroviral transduction. Identification of such factor(s) may lead to the generation of iPS cells with higher efficiency, and without the need for retroviruses.

Methods Summary

To generate Nanog-reporter mice, we isolated a BAC clone containing the mouse *Nanog* gene in its centre. By using the RED/ET recombination technique (Gene Bridges), we inserted a GFP-IRES-$Puro^r$ cassette into the 5′ UTR of the mouse *Nanog* gene. We introduced the modified BAC into RF8 ES cells by electroporation[26]. We then microinjected transgenic ES cells into C57BL/6 blastocysts to generate Nanog-reporter mice containing the modified BAC. MEFs were isolated from 13.5 d.p.c. male embryos after removing genital ridges. Generation of Nanog iPS cells was performed as described[8], except that puromycin was used instead of G418 as a selection antibiotic. Retroviruses (pMXs) were generated with Plat-E packaging cells[27]. RF8 ES cells[26] and iPS cells were cultured on SNL feeder cells[28]. Analyses of iPS cells, such as RT–PCR, real-time PCR, bisulphite genomic sequencing, SSLP analyses, DNA microarrays, teratoma formation, and microinjection into C57BL/6 blastocysts, were performed as described[8]. Contribution of iPS cells in chimaeric mice was determined by PCR for the SSLP marker D6Mit15.

Full Methods and any associated references are available in the online version of the paper at www.nature.com/nature.

(**448**, 313-317; 2007)

Keisuke Okita[1], Tomoko Ichisaka[1,2] & Shinya Yamanaka[1,2]

[1] Department of Stem Cell Biology, Institute for Frontier Medical Sciences, Kyoto University, Kyoto 606-8507, Japan

[2] CREST, Japan Science and Technology Agency, Kawaguchi 332-0012, Japan

Received 6 February; accepted 22 May 2007. Published online 6 June 2007.

References:

1. Thomson, J. A. *et al.* Embryonic stem cell lines derived from human blastocysts. *Science* **282**, 1145-1147 (1998).
2. Hochedlinger, K. & Jaenisch, R. Nuclear reprogramming and pluripotency. *Nature* **441**, 1061-1067 (2006).
3. Niwa, H., Miyazaki, J. & Smith, A. G. Quantitative expression of Oct-3/4 defines differentiation, dedifferentiation or self-renewal of ES cells. *Nature Genet.* **24**, 372-376 (2000).
4. Nichols, J. *et al.* Formation of pluripotent stem cells in the mammalian embryo depends on the POU transcription factor Oct4. *Cell* **95**, 379-391 (1998).
5. Avilion, A. A. *et al.* Multipotent cell lineages in early mouse development depend on SOX2 function. *Genes Dev.* **17**, 126-140 (2003).
6. Cartwright, P. *et al.* LIF/STAT3 controls ES cell self-renewal and pluripotency by a Myc-dependent mechanism. *Development* **132**, 885-896 (2005).
7. Li, Y. *et al.* Murine embryonic stem cell differentiation is promoted by SOCS-3 and inhibited by the zinc finger transcription factor Klf4. *Blood* **105**, 635-637 (2005).
8. Takahashi, K. & Yamanaka, S. Induction of pluripotent stem cells from mouse embryonic and adult fibroblast cultures by defined factors. *Cell* **126**, 663-676 (2006).
9. Tokuzawa, Y. *et al.* Fbx15 is a novel target of Oct3/4 but is dispensable for embryonic stem cell self-renewal and mouse development. *Mol. Cell. Biol.* **23**, 2699-2708 (2003).

源可能是 MEF 培养物中共存的少量干细胞。或者，在 iPS 细胞的生成过程中除了这四个转录因子外，可能还需要通过逆转录病毒整合激活额外的基因。这与我们只能够通过逆转录病毒转导才能产生 iPS 细胞的事实相关。鉴定这些因子可能有助于更高效率地产生 iPS 细胞，而且可能不再需要借助逆转录病毒。

方法概述

为了产生 Nanog 报告小鼠，我们分离了一个在中心含有小鼠 *Nanog* 基因的 BAC 克隆。通过使用 RED/ET 重组技术（基因桥），我们将一个 GFP-IRES-Puro^r 盒插入小鼠 *Nanog* 基因的 5′UTR 区。我们通过电穿孔技术将修饰的 BAC 转到 RF8 ES 细胞内[26]。然后我们将转基因的 ES 细胞微注射到 C57BL/6 胚泡中以产生含有修饰 BAC 的 Nanog 报告小鼠。在交配后 13.5 天的雄性胚胎中移除生殖嵴后分离出 MEF。Nanog iPS 细胞的产生如文献所述[8]，但是我们用嘌呤霉素代替 G418 作为选择抗生素。逆转录病毒（pMX）用 Plat-E 包装细胞产生[27]。RF8 ES 细胞[26]和 iPS 细胞都在 SNL 饲养细胞上培养[28]。对 iPS 细胞的分析，比如 RT–PCR、实时 PCR、亚硫酸氢盐基因组测序、SSLP 分析、DNA 微阵列、畸胎瘤形成和微注射到 C57BL/6 胚泡中等，都如文献所述进行[8]。通过 SSLP 标记物 D6Mit15 的 PCR 确定嵌合体小鼠中 iPS 细胞的嵌合。

完整的方法和相关的文献可在本文的网上版本中获得。

（毛晨晖 翻译；王宇 审稿）

10. Kuroda, T. *et al.* Octamer and Sox elements are required for transcriptional cis regulation of Nanog gene expression. *Mol. Cell. Biol.* **25**, 2475-2485 (2005).

11. Rodda, D. J. *et al.* Transcriptional regulation of Nanog by OCT4 and SOX2. *J. Biol. Chem.* **280**, 24731-24737 (2005).

12. Mitsui, K. *et al.* The homeoprotein Nanog is required for maintenance of pluripotency in mouse epiblast and ES cells. *Cell* **113**, 631-642 (2003).

13. Chambers, I. *et al.* Functional expression cloning of nanog, a pluripotency sustaining factor in embryonic stem cells. *Cell* **113**, 643-655 (2003).

14. Silva, J., Chambers, I., Pollard, S. & Smith, A. Nanog promotes transfer of pluripotency after cell fusion. *Nature* **441**, 997-1001 (2006).

15. Copeland, N. G., Jenkins, N. A. & Court, D. L. Recombineering: a powerful new tool for mouse functional genomics. *Nature Rev. Genet.* **2**, 769-779 (2001).

16. Testa, G. *et al.* Engineering the mouse genome with bacterial artificial chromosomes to create multipurpose alleles. *Nature Biotechnol.* **21**, 443-447 (2003).

17. Cherry, S. R., Biniszkiewicz, D., van Parijs, L., Baltimore, D. & Jaenisch, R. Retroviral expression in embryonic stem cells and hematopoietic stem cells. *Mol. Cell. Biol.* **20**, 7419-7426 (2000).

18. Chen, T., Ueda, Y., Xie, S. & Li, E. A novel Dnmt3a isoform produced from an alternative promoter localizes to euchromatin and its expression correlates with active de novo methylation. *J. Biol. Chem.* **277**, 38746-38754 (2002).

19. Davis, T. L., Yang, G. J., McCarrey, J. R. & Bartolomei, M. S. The H19 methylation imprint is erased and re-established differentially on the parental alleles during male germ cell development. *Hum. Mol. Genet.* **9**, 2885-2894 (2000).

20. Sato, S., Yoshimizu, T., Sato, E. & Matsui, Y. Erasure of methylation imprinting of Igf2r during mouse primordial germ-cell development. *Mol. Reprod. Dev.* **65**, 41-50 (2003).

21. Brandeis, M. *et al.* The ontogeny of allele-specific methylation associated with imprinted genes in the mouse. *EMBO J.* **12**, 3669-3677 (1993).

22. Labosky, P. A., Barlow, D. P. & Hogan, B. L. Mouse embryonic germ (EG) cell lines: transmission through the germline and differences in the methylation imprint of insulin-like growth factor 2 receptor (Igf2r) gene compared with embryonic stem (ES) cell lines. *Development* **120**, 3197-3204 (1994).

23. Kato, Y. *et al.* Developmental potential of mouse primordial germ cells. *Development* **126**, 1823-1832 (1999).

24. Lee, J. *et al.* Erasing genomic imprinting memory in mouse clone embryos produced from day 11.5 primordial germ cells. *Development* **129**, 1807-1817 (2002).

25. Geijsen, N. *et al.* Derivation of embryonic germ cells and male gametes from embryonic stem cells. *Nature* **427**, 148-154 (2004).

26. Meiner, V. L. *et al.* Disruption of the acyl-CoA:cholesterol acyltransferase gene in mice: evidence suggesting multiple cholesterol esterification enzymes in mammals. *Proc. Natl Acad. Sci. USA* **93**, 14041-14046 (1996).

27. Morita, S., Kojima, T. & Kitamura, T. Plat-E: an efficient and stable system for transient packaging of retroviruses. *Gene Ther.* **7**, 1063-1066 (2000).

28. McMahon, A. P. & Bradley, A. The Wnt-1 (int-1) proto-oncogene is required for development of a large region of the mouse brain. *Cell* **62**, 1073-1085 (1990).

Supplementary Information is linked to the online version of the paper at www.nature.com/nature.

Acknowledgements. We thank K. Takahashi, M. Nakagawa and T. Aoi for scientific discussion; M. Maeda for histological analyses; M. Narita, J. Iida, H. Miyachi and S. Kitano for technical assistance; and R. Kato, R. Iyama and Y. Ohuchi for administrative assistance. We also thank T. Kitamura for Plat-E cells and pMXs retroviral vectors, and R. Farese for RF8 ES cells. This study was supported in part by a grant from the Uehara Memorial Foundation, the Program for Promotion of Fundamental Studies in Health Sciences of NIBIO, a grant from the Leading Project of MEXT, and Grants-in-Aid for Scientific Research of JSPS and MEXT (to S.Y.). K.O. is a JSPS research fellow.

Author Contributions. K.O. conducted most of the experiments in this study. T.I. performed manipulation of mouse embryos to generate Nanog-GFP transgenic mice. T.I. also maintained the mouse lines. S.Y. designed and supervised the study, and prepared the manuscript. S.Y. also performed computer analyses of DNA microarray data.

Author Information. The microarray data are deposited in GEO under accession number GSE7841. Reprints and permissions information is available at www.nature.com/reprints. The authors declare no competing financial interests. Correspondence and requests for materials should be addressed to S.Y. (yamanaka@frontier.kyoto-u.ac.jp).

In Vitro Reprogramming of Fibroblasts into a Pluripotent ES-cell-like State

M. Wernig *et al.*

Editor's Note

Here, American cell biologist Rudolf Jaenisch and colleagues manage to convert mature mouse fibroblasts into stem-like cells called induced pluripotent stem (iPS) cells. The study builds on the work of Japanese cell biologist Shinya Yamanaka who showed, the previous year, that turning on the expression of just four genes in adult mouse cells could partially reprogram them to become stem cell-like. Jaenisch tweaked Yamanaka's egg- and embryo-free "recipe" to produce cells that could contribute to a whole organism—a key hallmark of a true stem cell. Yamanaka simultaneously published an essentially identical study, as did a third group of scientists based in America. Jaenisch's group has since made dopamine-producing neurons from iPS cells derived from patients with Parkinson's disease, highlighting the potential of these cells for reproductive medicine.

Nuclear transplantation can reprogramme a somatic genome back into an embryonic epigenetic state, and the reprogrammed nucleus can create a cloned animal or produce pluripotent embryonic stem cells. One potential use of the nuclear cloning approach is the derivation of "customized" embryonic stem (ES) cells for patient-specific cell treatment, but technical and ethical considerations impede the therapeutic application of this technology. Reprogramming of fibroblasts to a pluripotent state can be induced *in vitro* through ectopic expression of the four transcription factors Oct4 (also called Oct3/4 or Pou5f1), Sox2, c-Myc and Klf4. Here we show that DNA methylation, gene expression and chromatin state of such induced reprogrammed stem cells are similar to those of ES cells. Notably, the cells—derived from mouse fibroblasts—can form viable chimaeras, can contribute to the germ line and can generate live late-term embryos when injected into tetraploid blastocysts. Our results show that the biological potency and epigenetic state of *in-vitro*-reprogrammed induced pluripotent stem cells are indistinguishable from those of ES cells.

EPIGENETIC reprogramming of somatic cells into ES cells has attracted much attention because of the potential for customized transplantation therapy, as cellular derivatives of reprogrammed cells will not be rejected by the donor[1,2]. Thus far, somatic cell nuclear transfer and fusion of fibroblasts with ES cells have been shown to promote the epigenetic reprogramming of the donor genome to an embryonic state[3-5]. However, the therapeutic application of either approach has been hindered by technical complications as well as ethical objections[6]. Recently, a major breakthrough was reported whereby

体外重编程技术将成纤维细胞诱导到 ES 样细胞的多能性状态

沃尼格等

编者按

在本文中，美国细胞生物学家鲁道夫·耶尼施和他的同事们成功地将成熟的小鼠成纤维细胞转化为干细胞样细胞，称为“诱导多能性干细胞”(iPS)。这项研究建立在日本细胞生物学家山中伸弥工作的基础上。山中伸弥在前一年的研究中发现，在胎鼠或者成年小鼠的成纤维中，只要激活四个转录因子的表达，就能将它们部分重编程，使其变成干细胞样的细胞。耶尼施调整了山中伸弥的不直接操作卵子和胚胎的“配方”，从而产生了可以形成整个个体的细胞——这是真正干细胞的一个关键特征。山中伸弥和美国的第三组科学家同时发表了一项基本相同的研究。自那以后，耶尼施的团队从帕金森病患者的 iPS 细胞分化得到能够形成多巴胺的神经元，凸显了这些细胞在再生医学上的潜力。

核移植能够将体细胞的基因组重新编程回到胚胎表观遗传状态，而重新编程的细胞核能够形成克隆动物或者产生多能性胚胎干细胞。这种核克隆方法的一个潜在应用价值就是衍生出“定制的”胚胎干(ES)细胞，用于患者的特异性的细胞治疗，但是技术和伦理学方面的考虑限制了该技术的治疗应用。成纤维细胞能够在体外通过四个转录因子 Oct4(又称 Oct3/4 或 Pou5f1)、Sox2、c-Myc 和 Klf4 的异位表达诱导至多能性状态。在此，我们证明了这些诱导多能性干细胞在 DNA 甲基化、基因表达和染色质状态上都与 ES 细胞相似。尤为重要的是，这些来自小鼠成纤维细胞的细胞可以形成可存活的嵌合体，促进生殖系的发育，并可以在注射到四倍体囊胚时产生活的晚期胚胎。我们的结果显示了体外重编程得到的诱导多能性干细胞在生物学功能和表观遗传状态上与 ES 细胞没有明显区别。

由于重编程得到的细胞衍生物不会被供体排斥，体细胞经过表观遗传重编程形成的 ES 细胞具有定制移植治疗的潜力，因此引起了广泛关注 [1,2]。到目前为止，体细胞核转移以及成纤维细胞与 ES 细胞的融合被证实能够促进供体基因组的表观遗传重编程，使其处于胚胎状态 [3-5]。但是，这两种方法的治疗应用都受到技术并发症以及伦理问题的阻碍 [6]。最近，有一个重大的突破，即转录因子 Oct4、Sox2、

expression of the transcription factors Oct4, Sox2, c-Myc and Klf4 was shown to induce fibroblasts to become pluripotent stem cells (designated as induced pluripotent stem (iPS) cells), although with a low efficiency[7]. The iPS cells were isolated by selection for activation of *Fbx15* (also called *Fbxo15*), which is a downstream gene of *Oct4*. This important study left a number of questions unresolved: (1) although iPS cells were pluripotent they were not identical to ES cells (for example, iPS cells injected into blastocysts generated abnormal chimaeric embryos that did not survive to term); (2) gene expression profiling revealed major differences between iPS cells and ES cells; (3) because the four transcription factors were transduced by constitutively expressed retroviral vectors it was unclear why the cells could be induced to differentiate and whether continuous vector expression was required for the maintenance of the pluripotent state; and (4) the epigenetic state of the endogenous pluripotency genes *Oct4* and *Nanog* was incompletely reprogrammed, raising questions about the stability of the pluripotent state.

Here we used activation of the endogenous *Oct4* or *Nanog* genes as a more stringent selection strategy for the isolation of reprogrammed cells. We infected fibroblasts with retroviral vectors transducing the four factors, and selected for the activation of the endogenous *Oct4* or *Nanog* genes. Positive colonies resembled ES cells and assumed an epigenetic state characteristic of ES cells. When injected into blastocysts the reprogrammed cells generated viable chimaeras and contributed to the germ line. Our results establish that somatic cells can be reprogrammed to a pluripotent state that is similar, if not identical, to that of normal ES cells.

Selection of Fibroblasts for *Oct4* or *Nanog* Activation

Using homologous recombination in ES cells we generated mouse embryonic fibroblasts (MEFs) and tail-tip fibroblasts (TTFs) that carried a neomycin-resistance marker inserted into either the endogenous *Oct4* (Oct4-neo) or *Nanog* locus (Nanog-neo) (Fig. 1a). These cultures were sensitive to G418, indicating that the *Oct4* and *Nanog* loci were, as expected, silenced in somatic cells. These MEFs or TTFs were infected with Oct4-, Sox2-, c-Myc- and Klf4-expressing retroviral vectors and G418 was added to the cultures 3, 6 or 9 days later. The number of drug-resistant colonies increased substantially when analysed at day 20 (Fig. 1i). Most colonies had a flat morphology (Fig. 1h, right) and between 11% and 25% of the colonies were "ES-like" (Fig. 1h, left) when selection was applied early (Fig. 1k), a percentage that increased at later time points. At day 20, ES-like colonies were picked, dissociated and propagated in G418-containing media. They gave rise to ES-like cell lines (designated as Oct4 iPS or Nanog iPS cells, respectively) that could be propagated without drug selection, displayed homogenous Nanog, SSEA1 and alkaline phosphatase expression (Fig. 1b–g and Supplementary Figs 1 and 5), and formed undifferentiated colonies when seeded at clonal density on gelatincoated dishes (see inset in Fig. 1b). Four out of five analysed lines had a normal karyotype (Supplementary Table 1).

c-Myc和Klf4的表达可以诱导成纤维细胞成为诱导多能性干细胞，尽管其效率很低[7]。这些iPS细胞是通过选择激活*Oct4*的下游基因*Fbx15*(又称*Fbxo15*)而分离出来的。这个重要的研究遗留了一系列没有解决的问题：(1)尽管iPS细胞具有多能性，但是它们与ES细胞并不相同(比如，注射到囊胚中的iPS细胞产生异常的嵌合体胚胎，无法长久存活)；(2)基因表达分析显示iPS细胞和ES细胞之间存在较大差异；(3)由于这四个转录因子是通过组成性表达的逆转录病毒载体转导的，目前尚不清楚为什么可以诱导细胞分化，以及是否需要持续的载体表达才能维持这种多能状态；(4)内源性多能基因*Oct4*和*Nanog*的表观遗传状态并不是被完全重编程的，研究者们对多能状态的稳定性提出了质疑。

在这里，我们使用内源性基因*Oct4*或*Nanog*的激活作为更加严格的选择策略来分离重编程的细胞。我们用转导这四个因子的逆转录病毒感染成纤维细胞，并选择激活内源性基因*Oct4*或*Nanog*的细胞。阳性克隆酷似ES细胞，呈现出ES细胞的表观遗传特性。当被注射到囊胚中后，这些重编程的细胞能够形成存活的嵌合体并对生殖系的形成具有贡献。我们的研究结果证明体细胞能够被重编程成与ES细胞相似(尽管不是完全一致)的多能性状态。

选择用于*Oct4*或*Nanog*激活的成纤维细胞

通过在ES细胞中进行同源重组，我们得到了小鼠胚胎成纤维细胞(MEF)和尾尖纤维母细胞(TTF)，这些成纤维细胞携带新霉素耐药标记物，并将其插入到内源性*Oct4*(Oct4-neo)或者*Nanog*(Nanog-neo)位点中(图1a)。这些细胞对G418敏感，表明*Oct4*和*Nanog*位点在体细胞中如预期的那样沉默。分别用表达Oct4、Sox2、c-Myc和Klf4的逆转录病毒载体包装得到的病毒感染MEF或TTF，然后分别在第3、6、9天后加入G418进行培养。第20天进行分析时，耐药克隆的数量显著增加(图1i)。大部分克隆的形态都是扁平的(图1h，右侧)，当早期进行选择时(图1k)，大约11%～25%的克隆形态像ES细胞(图1h，左侧)，这个比例随着药物处理时间延长而增加。在第20天，挑出ES样的克隆，将其消化后用含有G418的培养基进行培养。它们能够形成ES样的细胞系(分别称为Oct4 iPS细胞或者Nanog iPS细胞)，而且这些细胞系能够在没有药物选择的情况下繁殖，展示出均质的Nanog、SSEA1和碱性磷酸酶的表达(图1b～g和补充信息图1和5)，当在明胶包被的培养皿上克隆密接种时，这些细胞能够形成未分化的克隆(见图1b插图)。进行分析的细胞系中有五分之四具有正常的核型(补充信息表1)。

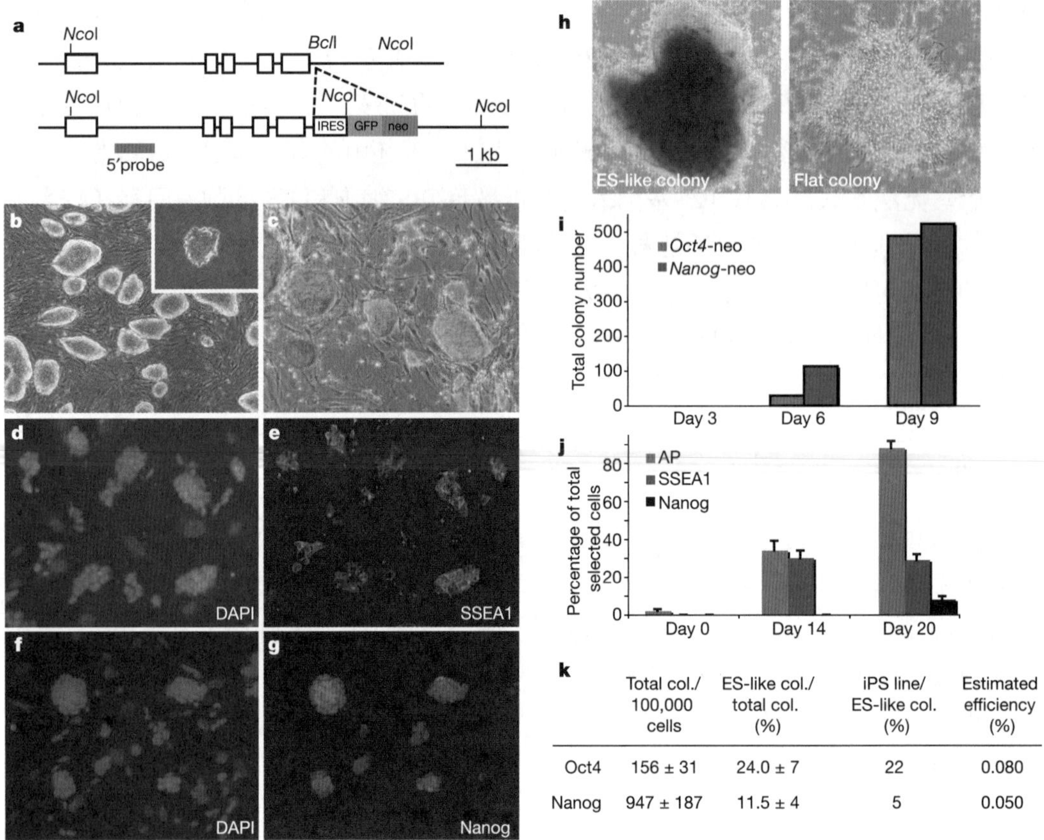

	Total col./ 100,000 cells	ES-like col./ total col. (%)	iPS line/ ES-like col. (%)	Estimated efficiency (%)
Oct4	156 ± 31	24.0 ± 7	22	0.080
Nanog	947 ± 187	11.5 ± 4	5	0.050

Fig. 1. Generation of Oct4- and Nanog-selected iPS cells. **a**, Targeting strategy to generate an Oct4-IRES-GFPneo allele. The resulting GFPneo fusion protein has sufficient neomycin-resistance activity in ES cells; GFP fluorescence, however, is not visible. **b**, Phase-contrast micrograph of Oct4 iPS cells (clone 18) grown on irradiated MEFs. Inset: an ES-cell-like colony 5 days after seeding in clonal density without feeder cells. iPS clone 18 cells exhibited strong alkaline phosphatase activity (**c**) and were homogenously labelled with antibodies against SSEA1 (**d**, **e**) and Nanog (**f**, **g**). **h**, One example of an ES-like colony 16 days after infection (left). Most G418-resistant colonies, however, consisted of flat non-ES-like cells (right): **b**, 10 ×; **c**–**g**, 20 ×; **h**, 4 ×. **i**, Gradual activation of the Nanog and Oct4-neo alleles. Shown are the total colony numbers of one experiment at day 20 after infection starting neomycin selection at day 3, 6 and 9. **j**, Fraction of total selected cells expressing alkaline phosphatase, SSEA1 and Nanog 0, 14, and 20 days after infection (counted were more than ten visual fields containing $n > 1{,}000$ total cells for every time point; error bars indicate s.d.). **k**, Estimated reprogramming efficiency of Oct4 selection and Nanog selection ($n = 3$ different experiments; s.e.m. is shown). Indicated are the total number of drug-resistant colonies per 100,000 plated MEFs 20 days after infection; the fraction of ES-like colonies per total number of colonies; the fraction of iPS cell lines that could be established from picked ES-like colonies as defined by homogenous alkaline phosphatase, SSEA1 and Nanog expression. After determining the fraction of *Sox2*- (83.4%), *Oct4*- (53.2%) and *c-myc*- (46.3%) infected MEFs 2 days after infection by immunofluorescence and assuming 50% were infected by *Klf4* viruses, we estimated the overall reprogramming efficiency as the ratio of quadruple-infected cells and the extrapolated total number of iPS cell lines that could be established with G418 selection starting at day 6 after infection.

Although the timing and appearance of colonies were similar between the Oct4 and Nanog selection, we noticed pronounced quantitative differences between the two

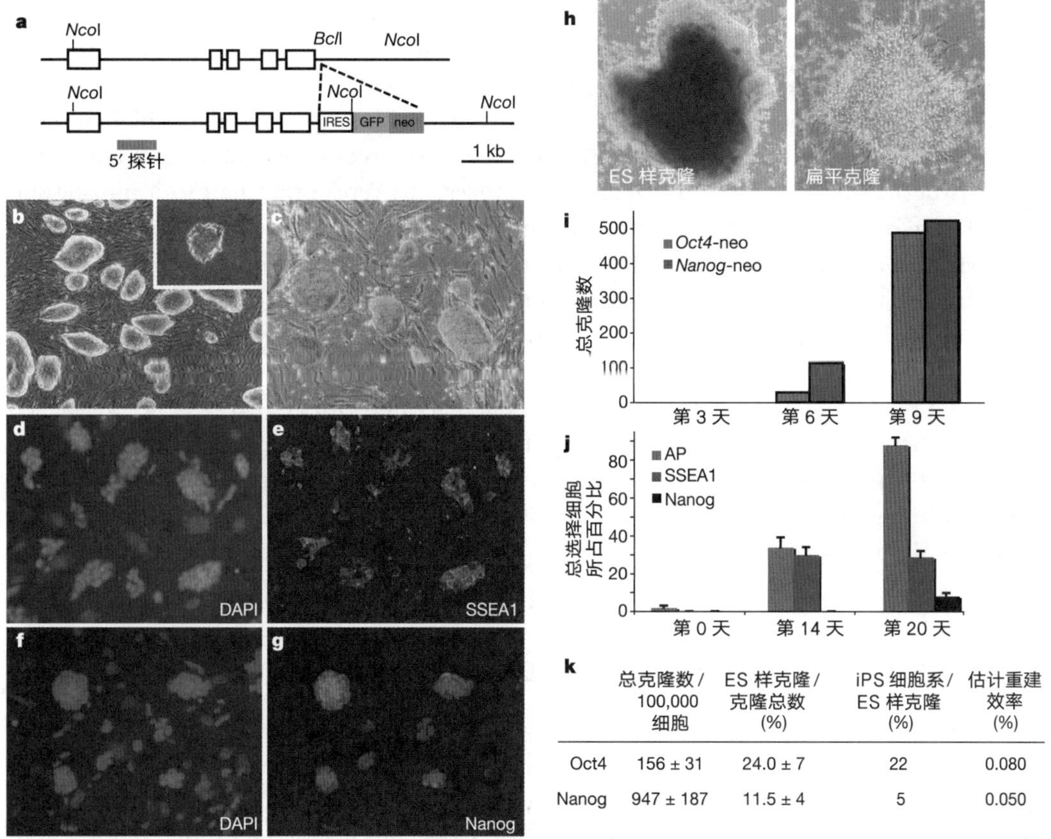

	总克隆数 / 100,000 细胞	ES 样克隆 / 克隆总数 (%)	iPS 细胞系 / ES 样克隆 (%)	估计重建效率 (%)
Oct4	156 ± 31	24.0 ± 7	22	0.080
Nanog	947 ± 187	11.5 ± 4	5	0.050

图 1. 选择性表达 *Oct4* 和 *Nanog* 的 iPS 细胞的产生。**a**，产生 Oct4-IRES-GFPneo 等位基因的靶向策略。得到的 GFPneo 融合蛋白在 ES 细胞内具有足够的新霉素抗性活性；但是看不到 GFP 荧光。**b**，生长在辐射过的 MEF 上的 Oct4 iPS 细胞(克隆 18)的相差显微镜图像。插图：在没有饲养细胞的情况下，将细胞以克隆的密度接种 5 天后形成的 ES 细胞样克隆。iPS 克隆 18 的细胞表现出强烈的碱性磷酸酶活性(**c**)，并用抗 SSEA1(**d,e**) 和 Nanog(**f,g**) 的抗体均匀标记。**h**，感染 16 天后的 ES 样克隆的一个例子(左图)。然而，大多数 G418 耐药细胞集落都由扁平的非 ES 样细胞组成(右图)：**b**, 10 × ；**c** ~ **g**, 20 × ；**h**, 4 × 。**i**, Nanog 和 Oct4-neo 等位基因的逐渐激活。图为在第 3、6 和 9 天开始用新霉素进行筛选，病毒感染后第 20 天，一个实验的克隆总数。**j**，感染后第 0 天、第 14 天和第 20 天表达碱性磷酸酶、SSEA1 和 Nanog 的细胞的比例(每个时间点只有超过 10 个视野含有 $n > 1{,}000$ 个总细胞的才计数，误差条指代 s.d.)。**k**，估算 *Oct4* 和 *Nanog* 选择的重建效率($n = 3$ 个不同的实验，s.e.m. 如图所示)。图中表示的是感染 20 天后每 100,000 个 MEF 重编程得到的总数；ES 样克隆占克隆总数的百分比；从挑取的 ES 样克隆中能够均质表达碱性磷酸酶，SSEA1 和 Nanog 的 iPS 克隆比例。通过免疫荧光确定了感染 2 天后 *Sox2*(83.4%)、*Oct4*(53.2%) 和 *c-myc*(46.3%) 感染 MEF 的比例后，并且假设 50% 受到 *Klf4* 病毒的感染，我们估计总体的重建效率为四倍感染的细胞的比例，并推算出在感染后第 6 天用 G418 选择获得的 iPS 细胞系的总数。

虽然 *Oct4* 和 *Nanog* 选择性表达的克隆出现的时间和外观相似，但我们注意到这两种选择策略之间存在明显的数量差异：*Oct4* 选择性表达的 MEF 能够形成的克

selection strategies: whereas Oct4-selected MEF cultures had 3- to 10-fold fewer colonies, the fraction of ES-like colonies was 2- to 3-fold higher than in Nanog-selected cultures. Accordingly, approximately four times more Oct4-selected ES-like colonies gave rise to stable and homogenous iPS cell lines compared with Nanog-selected ES-like colonies (Fig. 1k). This suggests that although the *Nanog* locus was easier to activate, a higher fraction of the drug-resistant colonies in Oct4-neo cultures was reprogrammed to a pluripotent state. Therefore, the overall estimated efficiency of 0.05–0.1% to establish iPS cell lines from MEFs was similar between Oct4 selection and Nanog selection, despite the larger number of total Nanog-neo resistant colonies (Fig. 1k). Next we investigated the time course of reprogramming by studying the fraction of alkaline-phosphatase-, SSEA1- and Nanog-positive cells in Oct4-selected MEF cultures. Fourteen days after infection some cells had already initiated alkaline phosphatase activity and SSEA1 expression, but lacked detectable amounts of Nanog protein (Fig. 1j), whereas by day 20, alkaline phosphatase and SSEA1 expression had increased and ~8% of the cells were Nanog-positive. Thus, the reprogramming induced by the four transcription factors (Oct4, Sox2, c-Myc and Klf4) is a gradual and slow process.

Expression and DNA Methylation

To characterize the reprogrammed cells on a molecular level we used quantitative polymerase chain reaction with reverse transcription (qRT–PCR) to measure the expression of ES-cell- and fibroblast- specific genes. Figure 2a shows that in Oct4 iPS cells the total level of Nanog and Oct4 was similar to that in ES cells but decreased on differentiation to embryoid bodies. MEFs did not express either gene. Using specific primers for endogenous or total *Sox2* transcripts showed that most *Sox2* transcripts originated from the endogenous locus rather than the viral vector (Fig. 2b). In contrast, *Hoxa9* and *Zfpm2* were highly expressed in MEFs but were expressed at very low levels in iPS or ES cells (Fig. 2c). Western blot analysis showed that multiple iPS clones expressed Nanog and Oct4 proteins at similar levels compared to ES cells (Fig. 2d). Finally, we used microarray technology to compare gene expression patterns on a global level. Figure 2f shows that the iPS cells clustered with ES cells in contrast to wild-type or donor MEFs.

To investigate the DNA methylation level of the *Oct4* and *Nanog* promoters we performed bisulphite sequencing and combined bisulphite restriction analysis (COBRA) with DNA isolated from ES cells, iPS cells and MEFs. As shown in Fig. 2g, both loci were demethylated in ES and iPS cells and fully methylated in MEFs. To assess whether the maintenance of genomic imprinting was compromised we assessed the methylation status of the four imprinted genes *H19*, *Peg1* (also called *Mest*), *Peg3* and *Snrpn*. As shown in Fig. 2e, bands corresponding to an unmethylated and methylated allele were detected for each gene in MEFs, iPS cells and TTFs. In contrast, embryonic germ cells, which have erased all imprints[8], were unmethylated. Our results indicate that the epigenetic state of the *Oct4* and *Nanog* genes was reprogrammed from a transcriptionally repressed (somatic) to an active (embryonic) state and that the pattern of somatic imprinting was maintained in iPS cells.

隆是*Nanog*选择性表达的1/3～1/10，但是其能够形成ES样克隆的比例却是*Nanog*选择性表达的MEF的2～3倍。因此，与*Nanog*选择性表达的ES样克隆相比，*Oct4*选择性表达的ES样克隆能够产生的稳定且均质的iPS细胞系的数量约是前者的4倍（图1k）。这表明，尽管*Nanog*的基因位点更加容易被激活，但是Oct4-neo的细胞能够被重编程到多能性状态的比例更高。因此，尽管Nanog-neo的细胞总数更多（图1k），但从*Oct4*选择性表达和*Nanog*选择性表达的MEF重编程获得iPS克隆的比例几乎都是0.05～0.1（图1k）。接下来，我们通过研究*Oct4*选择性表达的MEF重编程后获得的碱性磷酸酶、SSEA1和Nanog阳性克隆的比例，研究重编程的时间进程。感染后14天，部分细胞已经出现了碱性磷酸酶活性和SSEA1表达，但检测不到Nanog的蛋白表达（图1j），而到了第20天左右，碱性磷酸酶和SSEA1的表达增加了，大约8%的细胞是Nanog阳性的。因此，这四种转录因子（Oct4、Sox2、c-Myc和Klf4）诱导的重编程是一个渐进而缓慢的过程。

基因表达和DNA甲基化

为了在分子水平描述重编程后细胞的特征，我们使用定量反转录聚合酶链式反应（qRT–PCR）测量ES细胞和成纤维细胞特异性表达的基因。图2a显示了在Oct4 iPS细胞中，*Nanog*和*Oct4*的总水平与ES细胞中相似，但分化成EB球后表达量就减少了。MEF不表达这两个基因。使用内源或者总*Sox2*转录的特异性引物进行检测基因的表达，结果显示大部分*Sox2*转录起源于内源性位点而不是病毒载体（图2b）。相反，*Hoxa9*和*Zfpm2*在MEF中高表达，但是在iPS或者ES细胞中表达水平很低（图2c）。Western bolt分析显示多个iPS克隆表达的Nanog和Oct4蛋白水平与ES细胞相近（图2d）。最后，我们使用微阵列技术在整体水平上比较了基因表达模式。图2f显示了iPS细胞与ES细胞聚集在一起，而不是与野生型或者供者MEF聚集在一起。

为了研究*Oct4*和*Nanog*启动子的DNA甲基化水平，我们对从ES细胞、iPS细胞和MEF中分离出的DNA进行了亚硫酸氢盐测序和联合亚硫酸氢盐限制分析（COBRA）。如图2g所示，两个位点在ES和iPS细胞中都去甲基化，而在MEF中完全甲基化。为了评估基因组印迹的维持是否受到损害，我们评价了4个印记基因*H19*、*Peg1*（又称为*Mest*）、*Peg3*和*Snrpn*的甲基化状态。如图2e所示，在MEF、iPS细胞和TTF中，每个基因都检测到与未甲基化和甲基化等位基因对应的条带。相反，已经消除所有印记的胚胎生殖细胞[8]没有甲基化。我们的结果表明，*Oct4*和*Nanog*基因的表观遗传状态从转录抑制（体细胞）状态被重编辑为活跃（胚胎）状态，

Furthermore, the presence of imprints suggests a non-embryonic-germ-cell origin of iPS cells.

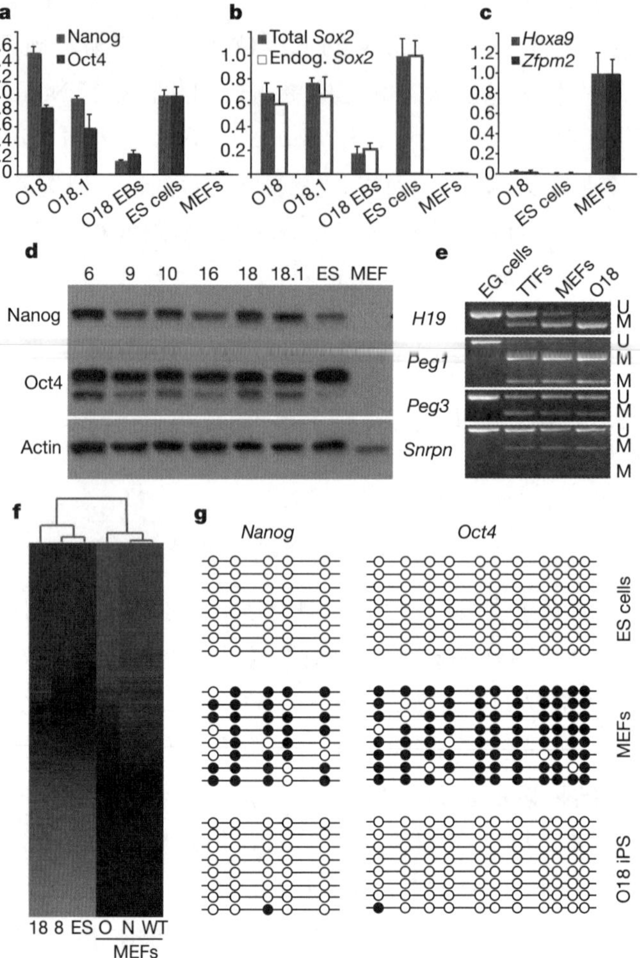

Fig. 2. Expression and promoter methylation analysis of iPS cells. **a–c**, qRT–PCR analysis ($n = 3$ independent PCR reactions; error bars indicate s.d.) of Oct4 iPS clone 18, subclone 18.1, 2-week-old embryoid bodies (EBs) derived from clone 18, V6.5 ES cells and Oct4-neo MEFs shows similar Nanog (red bars) and total Oct4 (blue bars) levels as in ES cells (**a**); slightly lower total Sox2 levels (filled red bars), mostly due to expression of endogenous Sox2 transcripts (open red bars, **b**); and strong downregulation of Hoxa9 (red) and Zfpm2 (blue) transcripts in iPS cells (**c**). Transcript levels were normalized to Gapdh expression, with expression levels in ES cells (**a**, **b**) and MEFs (**c**) set as 1. **d**, Western blot analysis for Oct4 and Nanog expression of different Oct4 iPS clones (6, 9, 10, 16, 18) and a GFP-labelled subclone of clone 18 (18.1). **e**, COBRA methylation analysis[32] of imprinted genes *H19* (maternally expressed), *Peg1* (paternally expressed), *Peg3* (paternally expressed) and *Snrpn* (paternally expressed). Upper band, unmethylated (U); lower band, methylated (M). **f**, Unsupervised hierarchical clustering of averaged global transcriptional profiles obtained from Oct4-neo iPS clone 18, Nanog-neo iPS clone 8, genetically matched ES cells (V6.5;129SvJae/C57Bl/6), Oct4-neo MEFs (O), Nanog-neo MEFs (N) and wild-type 129/B6 F1 MEFs (WT). **g**, Analysis of the methylation state of the *Oct4* and *Nanog* promoters using bisulphite sequencing. Open circles indicate unmethylated and filled circles methylated CpG dinucleotides. Shown are eight representative sequenced clones from ES cells (V6.5), Oct4-neo MEFs and Oct4-neo iPS clone 18.

并在 iPS 细胞中保持了体细胞印记的模式。此外，印记基因的存在进一步表明 iPS 细胞的非胚胎生殖细胞起源。

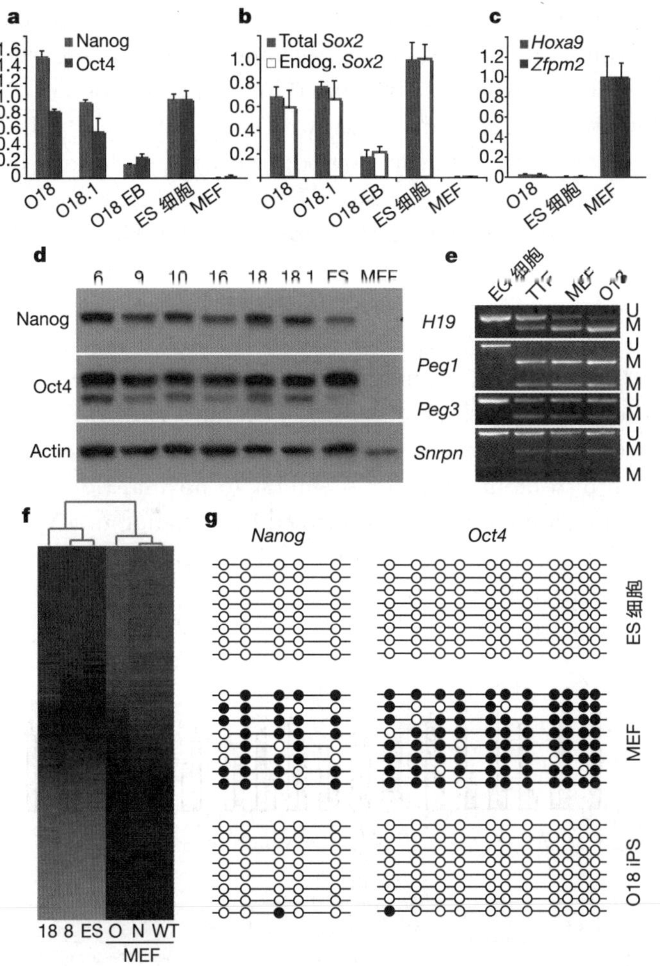

图 2. iPS 细胞的基因表达和启动子甲基化分析。**a**～**c**，对 Oct4 iPS 克隆 18、亚克隆 18.1、来源于克隆 18 的 2 周龄胚胎体、V6.5 ES 细胞和 Oct4-neo MEF 进行 qRT–PCR 分析（$n = 3$ 个独立 PCR 反应，误差条表示 s.d.），显示 *Nanog*（红条）和总 *Oct4*（蓝条）水平与 ES 细胞（**a**）相似；*Sox2* 总水平略低（实心红条），主要是由于内源性 *Sox2* 转录本的表达（空心红条，**b**）；以及 iPS 细胞中 *Hoxa9*（红）和 *Zfpm2*（蓝）转录本的明显下调（**c**）。转录本水平根据 Gapdh 的表达进行了标化，即将 ES 细胞（**a**,**b**）和 MEF 细胞（**c**）内的表达水平设置为 1。**d**，不同 Oct4 iPS 克隆（6，9，10，16，18）和 GFP 标记的克隆 18 的亚克隆（18.1）中 *Oct4* 和 *Nanog* 表达的 Western 印迹分析。**e**，印记基因 *H19*（母方表达）、*Peg1*（父方表达）、*Peg3*（父方表达）和 *Snrpn*（父方表达）的 COBRA 甲基化分析 [32]。上面的条带，未甲基化（U）；下面的条带，甲基化（M）。**f**，Oct4-neo iPS 克隆 18、Nanog-neo iPS 克隆 8、与之遗传信息匹配的 ES 细胞（V6.5；129SvJae/C57BL/6）、Oct4-neo MEF（O）、Nanog-neo MEF（N）和野生型 129/B6 F1 MEF（WT）中的平均整体转录情况的自发分级集合。**g**，使用亚硫酸氢盐测序分析 *Oct4* 和 *Nanog* 启动子的甲基化状态。空圈表示未甲基化，实圈表示甲基化的 CpG 二核苷酸。图中所示的是 ES 细胞（V6.5）、Oct4-neo MEF 和 Oct4-neo iPS 克隆 18 中的 8 个代表性的测序方法。

Chromatin Modifications

Recently, downstream target genes of *Oct4*, *Nanog* and *Sox2* have been defined in ES cells by genome-wide location analyses[9,10]. These targets include many important developmental regulators, a proportion of which is also bound and repressed by PcG (Polycomb-Group) complexes[11,12]. Notably, the chromatin at many of these non-expressed target genes adopts a bivalent conformation in ES cells, carrying both the "active" histone H3 lysine 4 (H3K4) methylation mark and the "repressive" histone H3 lysine 27 (H3K27) methylation mark[13,14]. In differentiated cells, those genes tend instead to carry either H3K4 or H3K27 methylation depending on their expression state. We used chromatin immunoprecipitation (ChIP) and real-time PCR to quantify H3K4 and H3K27 methylation for a set of genes reported to be bivalent in pluripotent ES cells[13]. Figure 3a shows that the fibroblast-specific genes *Zfpm2* and *Hoxa9* carried stronger H3K4 methylation than H3K27 methylation in the donor MEFs, whereas the silent genes *Nkx2.2*, *Sox1*, *Lbx1* and *Pax5* primarily carried H3K27 methylation. In contrast, in the Oct4 iPS cells, all of these genes showed comparable enrichment for both histone modifications, similar to normal ES cells (Fig. 3a). Identical results were obtained in Nanog iPS clones selected from Nanog-neo MEFs (Supplementary Fig. 2). These data suggest that the chromatin configuration of somatic cells is re-set to one that is characteristic of ES cells.

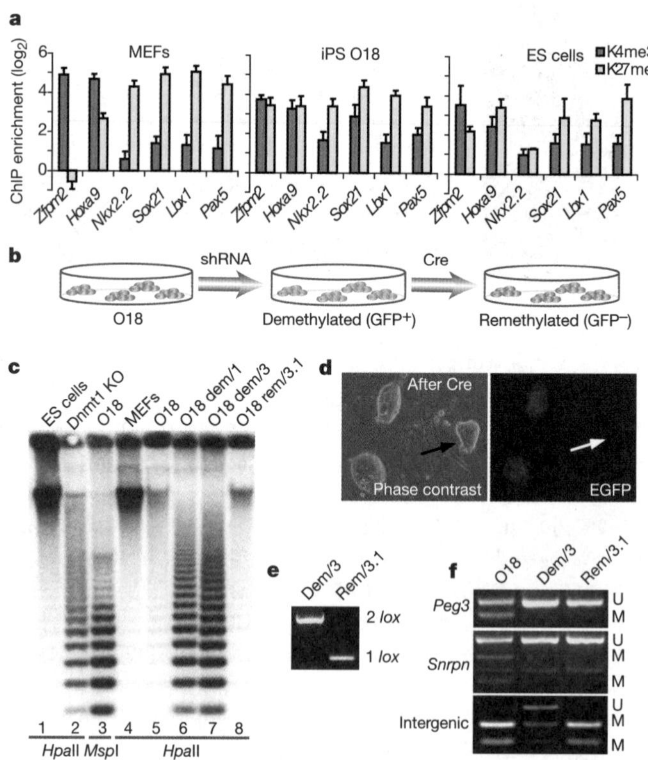

Fig. 3. Reprogrammed MEFs acquire an ES-cell-like epigenetic state. **a**, Real-time PCR after chromatin immunoprecipitation using antibodies against tri-methylated histone H3K4 and H3K27. Shown are the

染色质修饰

最近，通过全基因组的定位分析[9,10]，在 ES 细胞中发现了 *Oct4*、*Nanog* 和 *Sox2* 的下游目的基因。这些目的基因包括了许多重要的发育调节因子，其中一部分还受到 PcG（多聚硫）复合物的限制和抑制[11,12]。值得注意的是，这些未表达的目的基因中的染色质状态在 ES 细胞中形成双标构型，同时携带激活基因表达的组蛋白 3 赖氨酸 4（H3K4）甲基化标志和抑制基因表达的组蛋白 3 赖氨酸 27（H3K27）甲基化标志[13,14]。在分化的细胞中，这些基因根据倾向于携带 H3K4 或 H3K27 甲基化，这取决于它们的表达状态。我们用染色质免疫沉淀法（ChIP）和实时 PCR 对多能 ES 细胞内一系列被认为是双标基因的 H3K4 和 H3K27 甲基化进行了定量分析[13]。图 3a 显示了在供体 MEF 中成纤维细胞中特异性表达的基因 *Zfpm2* 和 *Hoxa9* 携带的 H3K4 甲基化水平强于 H3K27，而沉默基因 *Nkx2.2*、*Sox1*、*Lbx1* 和 *Pax5* 主要带有 H3K27 甲基化标志。相反，在 Oct4 iPS 细胞内，所有这些基因在两种组蛋白修饰上都表现出类似于正常 ES 细胞的相似富集（图 3a）。从 Nanog-neo MEF 中获得的 Nanog iPS 克隆得到了相同的结果（补充信息图 2）。这些数据表明，体细胞的染色质结构被重置成了具有 ES 细胞特征的模式。

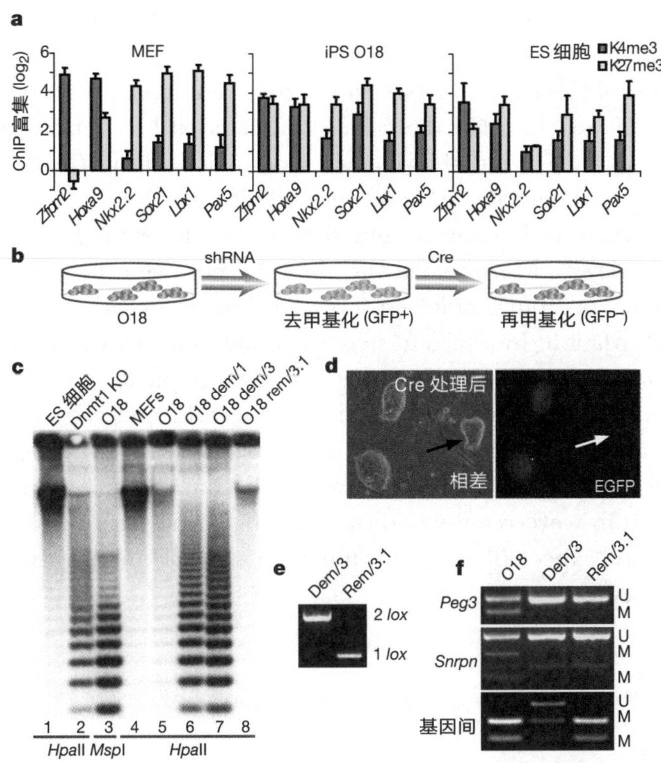

图 3. 重编程的 MEF 获得了 ES 细胞样的表观遗传状态。**a**，用三甲基化组蛋白 H3K4 和 H3K27 抗体进行染色质免疫沉淀后的实时 PCR。图中所示的是先前报道的 ES 细胞中数个“双标”位点的 $\log_2$ 富集

log_2 enrichments for several previously reported "bivalent" loci in ES cells ($n = 3$ experiments; error bars indicate s.d.). *Zfpm2* and *Hoxa9* show enrichment for the active (H3K4) mark in MEFs and are expressed (Fig. 1c and microarray data), whereas the other tested genes remain silent (microarray data). All loci tested in iPS clone O18 show enrichment for both H3K4 and H3K27 tri-methylation ("bivalent"), as seen in ES cells (V6.5). (See Supplementary Fig. 2 for H3K4 and H3K27 tri-methylation analysis of a subclone (clone O18.1) and Nanog-neo iPS clone N8.) **b**, Experimental design to de- and remethylate genomic DNA. Clone O18 was infected with the *Dnmt1*-hairpin-containing lentiviral vector pSicoR-GFP. The shRNA and GFP marker in the pSicoR vector are flanked by *loxP* sites[18]. Green colonies were expanded and passaged four times. Tat-Cre protein transduction was used to remove the shRNA[33]. **c**, Southern blot analysis of the minor satellite repeats using a methylation-sensitive restriction enzyme (*Hpa*II) and its methyl-insensitive isoschizomer (*Msp*I) as a control. Loss of methylation in two different clones (lanes 6 and 7) is comparable to Dnmt1 knockout ES cells (lane 2). After Cre-mediated recombination, complete remethylation (lane 8) of the repeats is observed within four passages. **d**, **e**, Successful loop out after Tat-Cre treatment was identified by disappearance of EGFP fluorescence (arrow) and verified by PCR analysis (**e**). **f**, COBRA assay of the imprinted genes *Peg3* and *Snrpn* and a random intergenic region close to the *Otx2* locus (Intergenic), demonstrating the expected resistance to *de novo* methylation of imprinted genes in contrast to non-imprinted intergenic sequences. U, unmethylated band; M, methylated band.

iPS Cells Tolerate Genomic Demethylation

Tolerance of genomic demethylation is a unique property of ES cells in contrast to somatic cells, which undergo rapid apoptosis on loss of the DNA methyltransferase Dnmt1 (refs 15–17). We investigated whether iPS cells would be resistant to global demethylation after Dnmt1 inhibition and would be able to re-establish global methylation patterns after restoration of Dnmt1 activity. To this end, we used a conditional lentiviral vector harbouring a *Dnmt1*-targeting short hairpin (sh)RNA and a green fluorescence protein (GFP) reporter gene (Fig. 3b and ref. 18). Infected iPS cells were plated at low density and GFP-positive colonies were picked and expanded. Southern blot analysis using *Hpa*II-digested genomic DNA showed that global demethylation of infected iPS cells (Fig. 3c, lanes 6, 7) was similar to $Dnmt1^{-/-}$ ES cells (lane 2). In contrast, uninfected iPS cells or MEFs (lanes 4, 5) displayed normal methylation levels. Morphologically, the GFP-positive cells were indistinguishable from the parental line or from uninfected sister subclones, indicating that iPS cells tolerate global DNA demethylation. In a second step, the *Dnmt1* shRNA was excised through Cre-mediated recombination and GFP-negative clones were picked (Fig. 3d). The cells had excised the shRNA vector (Fig. 3e) and normal DNA methylation levels were restored (Fig. 3c, lane 8) and were able to generate chimaeras (see below, Table 1), as has been reported previously for ES cells[19]. These observations imply that the *de novo* methyltransferases Dnmt3a and Dnmt3b were reactivated in iPS cells[20], leading to restoration of global methylation levels. As expected[19], the imprinted genes *Snrpn* and *Peg3* were unmethylated and resistant to remethylation (Fig. 3f).

($n = 3$ 个实验，误差条表示 s.d.)。*Zfpm2* 和 *Hoxa9* 显示了 MEF 中活性(H3K4)标记的富集和表达(图 1c 和微阵列数据)，而其他被检测的基因都保持沉默(微阵列数据)。iPS 克隆 O18 中检测的所有位点都显示 H3K4 和 H3K27 三甲基化的富集(双标性)，与 ES 细胞(V6.5)一样。(亚克隆(克隆 18.1)和 Nanog-neo iPS 克隆 N8 的 H3K4 和 H3K27 三甲基化分析见补充信息图 2)。**b**，基因组 DNA 去甲基化和再甲基化的实验。克隆 O18 感染了含有 *Dnmt1* 发夹结构的慢病毒载体 pSicoR-GFP。pSicoR 载体中 shRNA 和 GFP 标记物的两侧是 *loxP* 位点[18]。绿色的克隆进行扩增并且传代四次。Tat-Cre 蛋白转导被用于去除 shRNA[33]。**c**，用甲基化敏感的限制性内切酶(*Hpa*II)进行小卫星重复序列的 Southern 印迹分析，以甲基化不敏感型异构酶(*Msp*I)作为对照。两个不同的克隆(泳道 6 和 7)中甲基化的缺失与 Dnmt1 敲除的 ES 细胞(泳道 2)相似。Cre 介导的重组之后，在四代内观察到了重复序列的完全再甲基化(泳道 8)。**d, e**，通过 EGFP 荧光的消失(箭头)验证了 Tat-Cre 处理后成功的消除作用，并用 PCR 进行验证(**e**)。**f**，印记基因 *Peg3* 和 *Snrpn* 以及邻近 *Otx2* 位点的随机基因间区域(基因间的)的 COBRA 实验，说明印记基因相对于非印记的基因间序列而言具有预期的对从头合成甲基化的耐受作用。U，未甲基化条带，M，甲基化条带。

iPS 细胞耐受基因组去甲基化

与体细胞不同，基因组去甲基化的耐受是 ES 细胞的特有性质，而体细胞在 DNA 甲基转移酶 Dnmt1 缺失后迅速凋亡(参考文献 15 ~ 17)。我们研究了在 Dnmt1 抑制后，iPS 细胞是否会对全面去甲基化产生耐药性，以及在 Dnmt1 活性恢复后能否重建甲基化模式。为此，我们使用了条件慢病毒载体，它含有一个 *Dnmt1* 靶向短发夹(sh)RNA 和一个绿色荧光蛋白(GFP)报告基因(图 3b 和参考文献 18)。将感染后的 iPS 细胞以较低的密度种下，并挑出 GFP 阳性克隆进行扩增。使用 *Hpa*II 消化的基因组 DNA 进行的 Southern 印迹分析显示，感染后 iPS 细胞的整体去甲基化状态(图 3c，6，7 道)与 $Dnmt1^{-/-}$ 的 ES 细胞相似(2 道)。相反，未感染的 iPS 细胞或者 MEF(4，5 道)显示正常的甲基化水平。形态学上，GFP 阳性的细胞与其父代或者未受感染的姐妹亚克隆无明显区别，表明 iPS 细胞能够耐受整体的 DNA 去甲基化。第二步，通过 Cre 介导的重组将 *Dnmt1* shRNA 去除，并选择 GFP 阴性克隆(图 3d)。这些细胞已经去除了 shRNA 载体(图 3e)，并恢复了正常的 DNA 甲基化水平(图 3c，8 道)，能够产生嵌合体(见下，表 1)，正和之前报道的 ES 细胞性质相同[19]。这些观察结果表明，在 iPS 细胞中从头合成甲基转移酶 Dnmt3a 和 Dnmt3b 被重新激活[20]，导致了整体甲基化水平的恢复。如预期所示[19]，印记基因 *Snrpn* 和 *Peg3* 未发生甲基化，并且能够抵御再甲基化(图 3f)。

Table 1. Summary of blastocyst infections

Cell line	2N injections				4N injections		
	Injected blastocysts	Live chimaeras	Chimaerism (%)	Germ line	Injected blastocysts	Dead embryos (arrested)	Live embryos (analysed)
O6	ND	ND	ND	ND	13	0	2 (E12.5)
O9	30	5	30–70	Yes	90	3 (E11–13.5)	12 (E10–12.5)
O16	15	3	10–30	Yes	ND	ND	ND
O18	95	8	5–50	No	134	7 (E9–11.5)	4* (E10–12.5)
O3-2	ND	ND	ND	ND	25	2 (E8,11.5)	0
O4-16	ND	ND	ND	ND	35	4 (E11–13.5)	3 (E14.5)
N7	30	1	30	ND	ND	ND	ND
N8	90	14	5–50	No	118	9 (E9–11.5)	1* (E12.5)
N14	30	5	5–20	ND	46	2 (E8,11.5)	1 (E12.5)
TT-O25	50	2	30†	ND	39	3 (E9.5)	0
O18 rem/3.1	25	1	30	ND	ND	ND	ND

The extent of chimaerism was estimated on the basis of coat colour or EGFP expression. ND, not determined. 4N injected blastocysts were analysed between embryonic day E10.5 and E14.5.

"Analysed" indicates the day of embryonic development analysed; "arrested" indicates the estimated stage of development of dead embryos.

* Developmentally retarded or abnormal. O18 rem/3.1 is a de- and remethylated iPS clone (Fig. 3c).

† On the basis of GFP fluorescence.

Maintenance of the Pluripotent State

Southern blot analysis indicated that Oct4-neo iPS clone 18 carried four to six copies of the *Oct4*, *c-myc* and *Klf4* retroviral vectors and only one copy of the *Sox2* retroviral vector (Fig. 4a). Because these four factors were under the control of the constitutively expressed retroviral long terminal repeat, it was unclear in a previous study why iPS cells could be induced to differentiate[7]. To address this question, we designed primers specific for the four viral-encoded transcription factor transcripts and compared expression levels by qRT–PCR in MEFs 2 days after infection in iPS cells, in embryoid bodies derived from iPS cells, and in demethylated and remethylated iPS cells (Fig. 4b). Although the MEFs represented a heterogeneous population composed of uninfected and infected cells, virally encoded RNA levels of *Oct4*, *Sox2* and *Klf4* RNA were 5-fold higher and of *c-myc* more than 10-fold higher than in iPS cells. This suggests silencing of the viral long terminal repeat by *de novo* methylation during the reprogramming process. Accordingly, the total *Sox2* and *Oct4* RNA levels in iPS cells were similar to those in wild-type ES cells, and the *Sox2* transcripts in iPS cells were mostly, if not exclusively, transcribed from the endogenous gene (compare Fig. 2b). On differentiation to embryoid bodies, both viral and endogenous transcripts were downregulated. All viral *Sox2*, *Oct4* and *Klf4* transcripts were upregulated by approximately twofold in Dnmt1 knockdown iPS cells, and again downregulated on restoration of Dnmt1 activity. This is consistent with previous data that Moloney virus is efficiently *de*

表 1. 囊胚感染总结

细胞系	2N 注射				4N 注射		
	注射的囊胚	活的嵌合体	嵌合率 %	胚系	注射的囊胚	死胚胎（停止）	活胚胎（已分析）
O6	ND	ND	ND	ND	13	0	2(E12.5)
O9	30	5	30 ~ 70	Yes	90	3(E11 ~ 13.5)	12(E10 ~ 12.5)
O16	15	3	10 ~ 30	Yes	ND	ND	ND
O18	95	8	5 ~ 50	No	134	7(E9 ~ 11.5)	4*(E10 ~ 12.5)
O3-2	ND	ND	ND	ND	25	2(E8,11.5)	0
O4-16	ND	ND	ND	ND	35	4(E11 ~ 13.5)	3(E14.5)
N7	30	1	30	ND	ND	ND	ND
N8	90	14	5 ~ 50	No	118	9(E9 ~ 11.5)	1*(E12.5)
N14	30	5	5 ~ 20	ND	46	2(E8,11.5)	1(E12.5)
TT-O25	50	2	30†	ND	39	3(E9.5)	0
O18 rem/3.1	25	1	30	ND	ND	ND	ND

嵌合的程度根据毛色或者 EGFP 的表达估计。ND：不确定。4N 注射的囊胚在胚胎发育的 E10.5 到 E14.5 天之间进行了分析。

“已分析”表示已分析的胚胎发育天数；“停止”表示死亡胚胎的估计发育阶段。

* 发育迟缓或者异常。O18rem/3.1 是一个去甲基化和再甲基化的 iPS 克隆（图 3c）。

† 基于 GFP 荧光。

多能状态的保持

Southern 印迹分析显示 Oct4-neo iPS 克隆 18 带有 4 到 6 个拷贝的 *Oct4*、*c-Myc* 和 *Klf4* 逆转录病毒载体，只带有一个拷贝的 *Sox2* 逆转录病毒载体（图 4a）。因为这四个因子都是在组成性表达的逆转录病毒长末端重复的控制之下，所以在以前的研究中并不清楚 iPS 细胞能够被诱导分化的原因 [7]。为了解决这个问题，我们设计了针对这四个病毒编码转录因子特异性的引物，并用 qRT–PCR 比较 iPS 细胞感染后 2 天的 MEF、iPS 细胞衍生的 EB 球、去甲基化和再甲基化的 iPS 细胞中的表达水平（图 4b）。虽然 MEF 是由未感染和感染细胞组成的异质性群体，但 *Oct4*、*Sox2* 和 *Klf4* RNA 的病毒编码 RNA 水平是 iPS 细胞的 5 倍，*c-Myc* 则是 10 倍以上。这表明在重编程的过程中发生了新的甲基化使病毒的长末端重复沉默。因此，iPS 细胞内的 *Sox2* 和 *Oct4* RNA 的总水平与野生型 ES 细胞内相似，而且 iPS 细胞内的 *Sox2* 转录本不完全是大部分来源于内源性基因（对照图 2b）。在向 EB 球分化的过程中，病毒和内源性的转录产物都发生了下调。在 Dnmt1 敲降的 iPS 细胞中，所有的病毒 *Sox2*、*Oct4* 和 *Klf4* 转录本都上调至原来的将近 2 倍，而在 Dnmt1 功能恢复时又下调回来。这与先前的数据一致，在胚胎中莫洛尼病毒能够有效地被甲基化和沉默，

novo methylated and silenced in embryonic but not in somatic cells[21,22]. Transcript levels of *c-myc* were about 20-fold lower in iPS cells than in infected MEFs, and did not change on differentiation or demethylation.

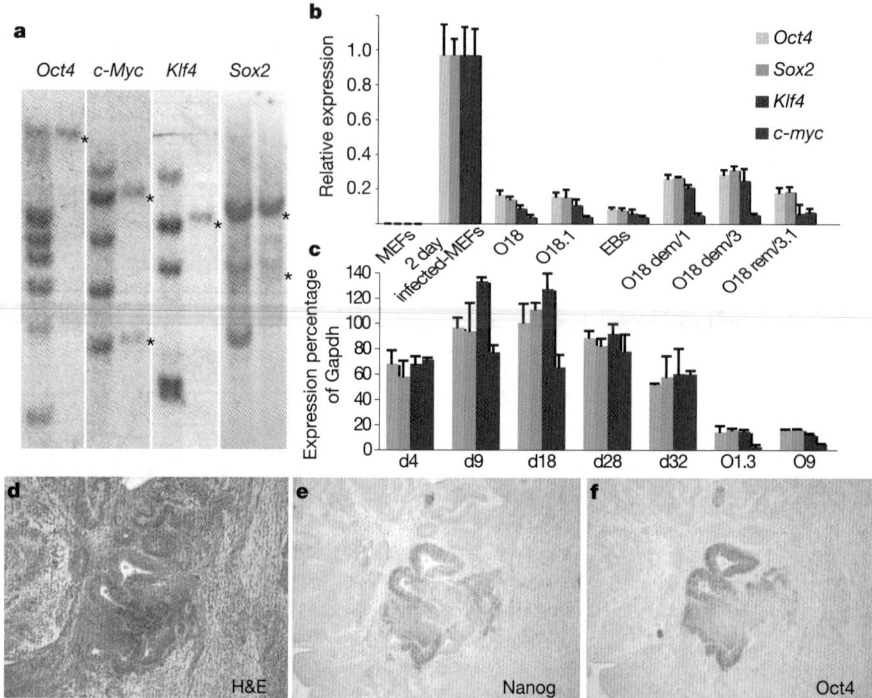

Fig. 4. Efficient silencing of retroviral transcripts in induced pluripotent cells. **a**, Southern blot analysis of proviral integrations in iPS clone O18 (left lanes) for the four retroviral vectors. Uninfected ES cells (right lanes) show only one or two bands corresponding to the endogenous gene (marked by an asterisk). **b**, Quantitative RT–PCR using primers specifically detecting the four viral transcripts. Shown are Oct4-neo iPS clone 18 and a GFP-labelled subclone, Oct4-neo MEFs, 2-week-old embryoid bodies generated from clone 18, two demethylated clones (18 dem/1 and 18 dem/3), a remethylated clone (18 rem/3.1), and Oct4-neo MEFs 2 days after infection with all four viruses but not selected with G418 ($n = 3$ independent experiments; error bars indicate s.d.). **c**, Viral transcript levels at various time points in cell populations after infection and Oct4 selection and in the two Oct4 iPS cell lines O1.3 and O9 ($n = 3$ independent experiments; error bars indicate s.d.). **d–f**, Paraffin sections of a teratoma 26 days after subcutaneous injection of Oct4 iPS clone 18 cells into SCID mice. H&E, haematoxylin and eosin. Nanog (**e**) and Oct4 (**f**) expression was confined to undifferentiated cell types as indicated an immunohistochemical analysis.

To follow the kinetics of vector inactivation during the reprogramming process, we isolated RNA from drug-resistant cell populations at different times after infection. Figure 4c shows that the viral-vector-encoded transcripts were gradually silenced during the transition from MEFs to iPS cells with a time course that corresponded to the gradual appearance of pluripotency markers (compare Fig. 1j). Finally, to visualize directly Oct4 and Nanog expression during differentiation, we injected Oct4 iPS cells into SCID mice to induce teratoma formation (Fig. 4d). Immunostaining revealed that Oct4 and Nanog were expressed in the centrally located undifferentiated cells but were silenced in the differentiated parts of the teratoma (Fig. 4e, f). Our results suggest that the retroviral vectors

而在体细胞中不行[21,22]。iPS细胞中的*c-Myc*转录本水平大约为感染的MEF的二十分之一，而且在分化或者去甲基化过程中没有变化。

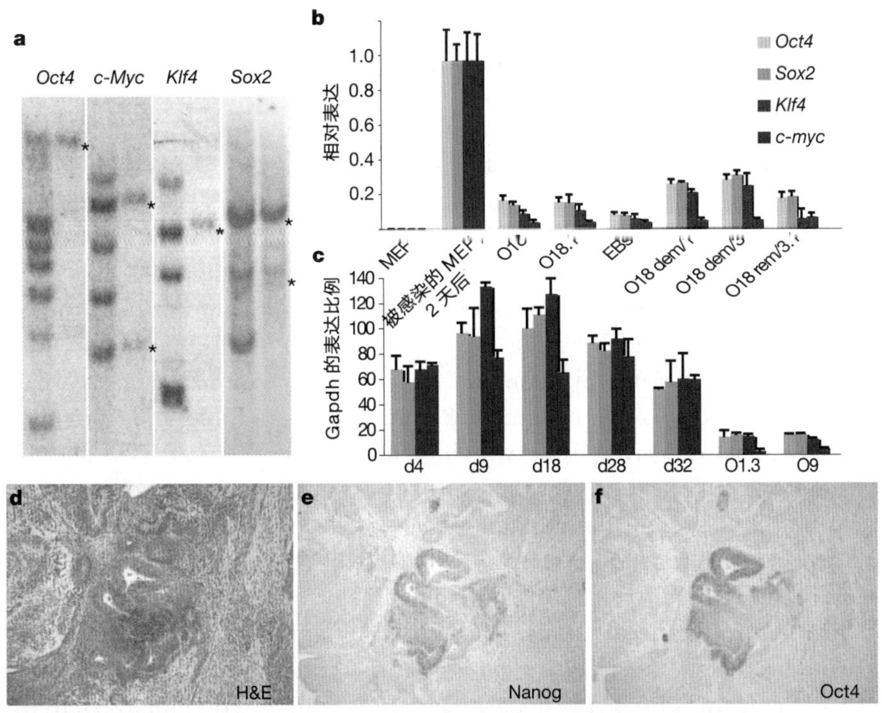

图4. 诱导的多能细胞中逆转录病毒转录本的有效沉默。**a**，iPS克隆O18(左侧泳道)中四个逆转录病毒载体的前病毒整合的Southern印迹分析。未感染的ES细胞(右侧泳道)只显示与内源性基因对应的一条或者两条带(星号标记)。**b**，使用识别四个病毒转录子的引物进行的定量RT–PCR。显示的分别是Oct4-neo iPS克隆18和GFP标记的亚克隆，OCT4-neo MEF，来源于克隆18的2周龄胚胎体，两个去甲基化的克隆(18dem/1和18dem/3)，一个再甲基化的克隆(18dem/3.1)和用所有四种病毒感染2天后但是不用G418进行选择得到的OCT4-neo MEF($n = 3$独立的实验，误差条表示s.d.)。**c**，经过感染和Oct4选择的细胞群以及两个Oct4 iPS细胞系O1.3和O9在不同时间点的病毒转录子水平($n = 3$独立的实验，误差条表示s.d.)。**d~f**，皮下注射Oct4 iPS克隆18到SCID小鼠中26天后形成的畸胎瘤的石蜡切片。H&E，苏木精和伊红。免疫组化分析显示，Nanog(**e**)和Oct4(**f**)表达仅限于未分化的细胞类型。

为了研究重编程过程中外源基因失活的动力学，我们从感染后不同时间的耐药细胞中分离出RNA。图4c显示了从MEF向iPS转化的过程中，病毒载体编码的转录本逐渐沉默，这一过程与多能性标志物的逐渐出现相对应(对照图1j)。最后，为了直接观察Oct4和Nanog在分化过程中的表达，我们将Oct4 iPS细胞注射到SCID小鼠体内诱导畸胎瘤形成(图4d)。免疫染色显示Oct4和Nanog在中心位置的未分化细胞中表达，但在畸胎瘤的已分化部分沉默(图4e、f)。我们的结果表明，在重

are subject to gradual silencing by *de novo* methylation during the reprogramming process. The maintenance of the pluripotent state and induction of differentiation strictly depends on the expression and normal regulation of the endogenous *Oct4* and *Nanog* genes.

Developmental Potency

We determined the developmental potential of iPS cells by teratoma and chimaera formation. Histological and immunohistochemical analysis of Oct4- or Nanog-iPS-cell-induced teratomas revealed that the cells had differentiated into cell types representing all three embryonic germ layers (Supplementary Figs 3 and 4). To assess more stringently their developmental potential, various iPS cell lines were injected into diploid (2N) or tetraploid (4N) blastocysts. After injection into 2N blastocysts both Nanog iPS and Oct4 iPS clones derived from MEFs (Fig. 5a) or from TTFs (Fig. 5b, c), as well as iPS cells that had been subjected to a consecutive cycle of demethylation and remethylation (compare Fig. 3b, c), efficiently generated viable high-contribution chimaeras (summarized in Table 1). To test for germline transmission, chimaeras derived from two different iPS lines (Oct4 iPS O9 and O16) were mated with normal females, and blastocysts were isolated and genotyped by three different PCR reactions for the presence of the multiple viral *Oct4* and *c-myc* genes and for the single-copy GFPneo sequences inserted into the *Oct4* locus of the donor cell (Fig. 1a). Figure 5f shows that 9 out of 16 embryos from two chimaeras were positive for the viral copies. As expected, only half of the viral-positive blastocysts contained the GFPneo sequences (5 out of 9 embryos, Fig. 5f, left panel). When embryonic day (E)10 embryos derived from an Oct4 iPS line O16 chimaera were genotyped, three out of eight tested embryos were transgenic (Fig. 5f, right panel). Finally, we injected iPS cells into 4N blastocysts as this represents the most rigorous test for developmental potency, because the resulting embryos are composed only of the injected donor cells ("all ES embryo"). Figure 5d, e shows that both Oct4 and Nanog iPS cells could generate mid- and late-gestation "all iPS embryos" (summarized in Table 1). These findings indicate that iPS cells can establish all lineages of the embryo and thus have a similar developmental potential as ES cells.

建的过程中逆转录病毒载体通过从头合成甲基化逐渐沉默。保持多能状态和诱导分化严格依赖于内源性 *Oct4* 和 *Nanog* 基因的表达和正常调控。

发育潜力

我们用畸胎瘤和嵌合体的形成来确定 iPS 细胞的发育潜力。对 Oct4 或 Nanog iPS 细胞诱导的畸胎瘤进行组织学和免疫组化分析显示，这些细胞已经分化成代表三个胚层（补充信息图 3 和 4）的细胞类型。为了更加严格地评估其发育潜力，不同的 iPS 细胞系被注射到二倍体（2N）或者四倍体（4N）囊胚中。注射到 2N 囊胚后，来源于 MEF（图 5a）或者 TTF（图 5b、c）的 Nanog iPS 和 Oct4 iPS 细胞克隆以及经过连续去甲基化和再甲基化循环的 iPS 细胞（对照图 3b、c）都有效地产生高度嵌合的嵌合体（总结见表 1）。为了检验胚系传递的能力，来源于两个不同 iPS 细胞系（Oct4 iPS O9 和 O16）的嵌合体与正常雌性小鼠进行交配，分离出囊胚并用三种不同的 PCR 反应进行基因分型，验证是否存在多种病毒 *Oct4* 和 *c-Myc* 基因以及为了将单拷贝 GFPneo 序列插入到供者细胞 *Oct4* 位点内（图 1a）。图 5f 显示两个嵌合体共 16 个胚胎中的 9 个含有病毒拷贝。正如所期望的，只有一半的病毒阳性囊胚含有 GFPneo 序列（9 个胚胎中有 5 个，图 5f，左侧栏）。当对 Oct4 iPS 细胞系 O16 嵌合体的 10 个胚胎进行基因型鉴定时，8 个被测胚胎中有 3 个是转基因的（图 5f，右侧栏）。最后，我们将 iPS 细胞注射到 4N 囊胚中，这代表了发育潜力最严格的检测，因为得到的胚胎仅仅由注射的供者细胞组成（全 ES 胚胎）。图 5d、e 显示了 Oct4 和 Nanog iPS 细胞都能够产生孕中期和孕晚期的全 iPS 胚胎（总结见表 1）。这些结果表明 iPS 细胞能够建立胚胎的所有细胞系，因此具有与 ES 细胞类似的发育潜力。

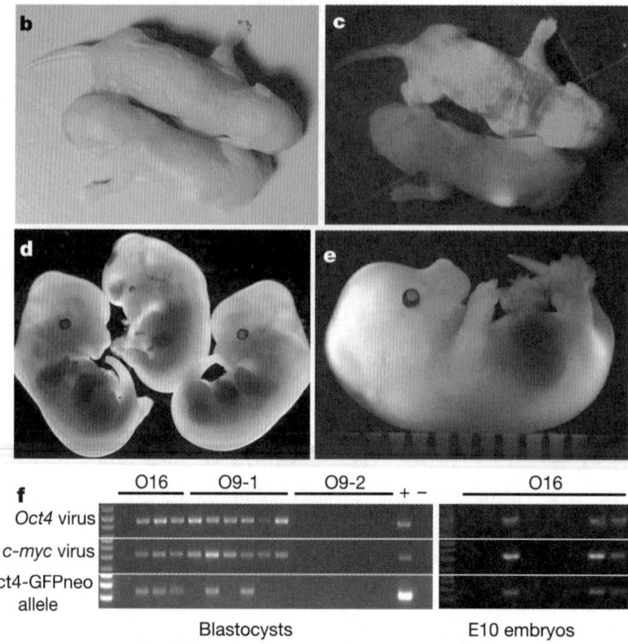

Fig. 5. Developmental pluripotency of reprogrammed fibroblasts. **a**, A 6-week-old chimaeric mouse. Agouti-coloured hairs originated from Oct4 iPS cell line O18.1. **b**, **c**, Two live pups after 2N blastocyst injection, one of which shows a high contribution (**c**) of the TTF-derived Oct4 iPS cell line TT-O25, which had been GFP-labelled with a lentiviral ubiquitin-EGFP vector. **d**, "All iPS cell embryos" were generated by injection of iPS cells into 4N blastocysts[34]. Live E12.5 embryos generated from Oct4 iPS line O6 (left), from Nanog iPS line N14 (middle) and from V.6.5 ES cells (right) are shown. **e**, A normally developed E14.5 embryo was derived from Oct4 iPS cell line O4-16 after tetraploid complementation and was isolated by screening MEFs for activation of GFP inserted into the *Oct4* locus. **f**, Germline contribution of Oct4 iPS clones O9 and O16. Genotyping of blastocysts from females mated with three chimaeric males demonstrated the presence of *Oct4* and *c-myc* virus integrations and the Oct4-IRES-GFPneo allele (left panel). Because of the multiple integrations (Fig. 4a) all embryos with iPS cell contribution are expected to be positive for proviral sequences in this assay. In contrast, the single-copy Oct4-IRES-GFPneo allele segregated into only 5 of the 9 virus-positive embryos. All six blastocysts from O9 chimaera 1 were iPS-cell-derived, suggesting that this chimaera was a pseudo-male. Additional genotyping identified 13 out of 72 tested blastocysts derived from iPS line O9 and 4 out of 13 blastocysts derived from iPS line O16 chimaeras carrying the viral transgenes. The right panel shows that 3 out of 8 tested E.10 mid-gestation embryos were sired by a chimaera derived from the donor iPS line O16. +, positive control; −, negative control.

Discussion

The results presented here demonstrate that the four transcription factors Oct4, Sox2, c-myc and Klf4 can induce epigenetic reprogramming of a somatic genome to an embryonic pluripotent state. In contrast to selection for Fbx15 activation[7], fibroblasts that had reactivated the endogenous *Oct4* (Oct4-neo) or *Nanog* (Nanog-neo) loci grew independently of feeder cells, expressed normal Oct4, Nanog and Sox2 RNA and protein levels, were epigenetically identical to ES cells by a number of criteria, and were able to generate viable chimaeras, contribute to the germ line and generate viable late-gestation embryos after

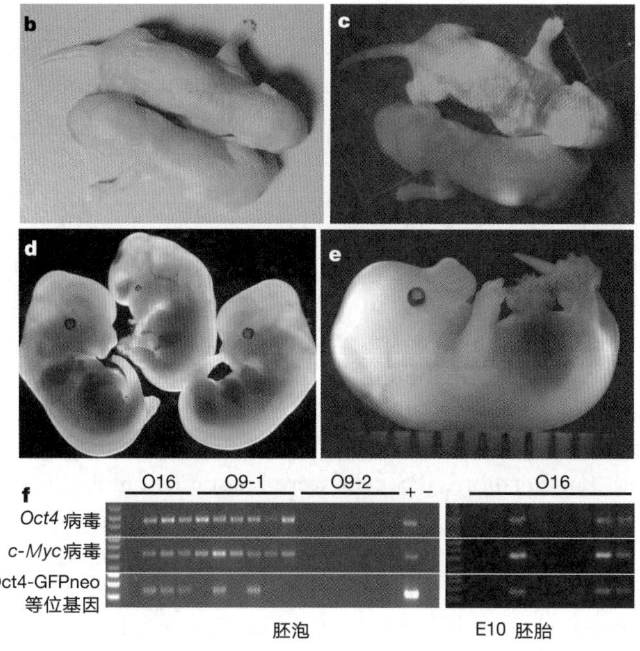

图 5. 重编程后成纤维细胞的发育多能性。**a**，一只 6 周龄的嵌合体小鼠。刺豚鼠样的毛色来源于 Oct4 iPS 细胞系 O18.1。**b, c**，2N 囊胚注射后的两只活幼鼠，其中一只显示了高比例(**c**)的 TTF 来源的 Oct4 iPS 细胞系 TT-O25，该细胞系已被慢病毒泛素化 -EGFP 进行 GFP 标记。**d**，所有的 iPS 细胞胚胎都是通过将 iPS 细胞注入 4N 囊胚生成的 [34]。图中显示了来源于 Oct4 iPS 细胞系 O6(左侧)、Nanog iPS 细胞系 N14(中间)和 V.6.5 ES 细胞(右侧)的活的 E12.5 胚胎。**e**，经过四倍体互补后，从 Oct4 iPS 细胞系 O4-16 中获得一个正常发育的 E14.5 胚胎，通过筛选插入到 Oct4 位点的 MEF，激活 GFP，将其分离出来。**f**，Oct4 iPS 克隆 O9 和 O16 的生殖系嵌合贡献。对雌鼠和三只嵌合体雄鼠交配产生的囊胚进行基因型鉴定，显示存在 *Oct4* 和 *c-Myc* 病毒整合以及 Oct4-TRES-GFP 等位基因(左侧泳道)。由于多重整合(图 4a)，在本实验中，所有含有 iPS 细胞成分的胚胎的前病毒序列均呈阳性。相反，9 只病毒阳性的胚胎中只有 5 只具有单拷贝 Oct4-TRES-GFPneo 等位基因。O9 嵌合体 1 的 6 个囊胚都是 iPS 细胞来源，表明该嵌合体是假雄性。另外的基因分型在 72 个测试的囊胚中鉴定出 13 个来自 iPS 细胞系 O9，在 13 个来自携带病毒转基因的 iPS 系 O16 嵌合体的囊胚中鉴定出 4 个。右侧泳道显示了 8 个检测的 E10 孕中期胚胎中有 3 个来源于供体 iPS 细胞系 O16 的嵌合体。+，阳性对照；−，阴性对照。

讨　论

本研究表明，Oct4、Sox2、c-Myc 和 Klf4 这四个转录因子能够诱导体细胞基因组的表观遗传重编程成胚胎多能状态。相对于 *Fbx15* 活化选择 [7]，激活内源性 *Oct4* (Oct4-neo) 或者 *Nanog*(Nanog-neo) 位点的成纤维细胞可以不依赖饲养细胞而独立生长，能够表达正常的 *Oct4*、*Nanog* 和 *Sox2* RNA 和蛋白水平。在许多标准下，成纤维细胞与 ES 细胞在表观遗传学上是相同的，它能够产生可以存活的嵌合体，在注

injection into tetraploid blastocysts. Transduction of the four factors generated significantly more drug-resistant cells from Nanog-neo than from Oct4-neo fibroblasts but a higher fraction of Oct4-selected cells had all the characteristics of pluripotent ES cells, suggesting that *Nanog* activation is a less stringent criterion for pluripotency than *Oct4* activation.

Our data suggest that the pluripotent state of Oct4 iPS and Nanog iPS cells is induced by the virally transduced factors but is largely maintained by the activity of the endogenous pluripotency factors including Oct4, Nanog and Sox2, because the viral-controlled transcripts, although expressed highly in MEFs, become mostly silenced in iPS cells. The total levels of Oct4, Nanog and Sox2 were similar in iPS and wild-type ES cells. Consistent with the conclusion that the pluripotent state is maintained by the endogenous pluripotency genes is the finding that the *Oct4* and the *Nanog* genes became hypomethylated in iPS cells as in ES cells, and that the bivalent histone modifications of developmental regulators were re-established. Furthermore, iPS cells were resistant to global demethylation induced by inactivation of Dnmt1, similar to ES cells but in contrast to somatic cells. Re-expression of Dnmt1 in the hypomethylated ES cells resulted in global remethylation, indicating that the iPS cells had also reactivated the *de novo* methyltransferases Dnmt3a and Dnmt3b. All these observations are consistent with the conclusion that the iPS cells have gained an epigenetic state that is similar to that of normal ES cells. This conclusion is further supported by the recent observation that female iPS cells, similar to ES cells, reactivate the somatically silenced X chromosome[23].

Expression of the four transcription factors proved to be a robust method to induce reprogramming of somatic cells to a pluripotent state. However, the use of retrovirus-transduced oncogenes represents a serious barrier to the eventual use of reprogrammed cells for therapeutic application. Much work is needed to understand the molecular pathways of reprogramming and to eventually find small molecules that could achieve reprogramming without gene transfer of potentially harmful genes.

Methods Summary

Cell culture, gene targeting and viral infections. ES and iPS cells were cultivated on irradiated MEFs. Using homologous recombination we generated ES cells carrying an IRES-GFPneo fusion cassette downstream of *Oct4* exon 5 (Fig. 1a). The *Nanog* gene was targeted as described[24]. Transgenic MEFs were isolated and selected from E13.5 chimaeric embryos after blastocyst injection of Oct4-IRES-GFPneo- or Nanog-neo-targeted ES cells. MEFs were infected overnight with the Moloney-based retroviral vector pLIB (Clontech) containing the murine complementary DNAs of *Oct4*, *Sox2*, *Klf4* and *c-myc*.

Southern blot, methylation and chromatin analyses. To assess the levels of DNA methylation, genomic DNA was digested with *Hpa*II and hybridized to a probe for the minor satellite repeats[25] or with an IAP probe[26]. Bisulphite treatment was performed with the Qiagen EpiTect Kit. For the methylation status of *Oct4* and *Nanog* promoters, bisulphite sequencing analysis was performed

射到四倍体囊胚中后，对生殖系的形成有促进作用，能够产生孕晚期的胚胎。这四个因子的转导使 Nanog-neo 成纤维细胞中的耐药细胞明显比 Oct4-neo 成纤维细胞中的多，但 *Oct4* 选择性表达的细胞中具有多能 ES 细胞所有的特征的比例较高，这表明 *Nanog* 活化是比 *Oct4* 活化更加不严格的多能性标准。

我们的数据表明 Oct4 iPS 和 Nanog iPS 细胞的多能状态是通过病毒转导的因子诱导的，但大部分是由内源多能性因子包括 *Oct4*、*Nanog* 和 *Sox2* 的活性来维持的，因为病毒控制的转录子尽管在 MEF 中高度表达，却在 iPS 细胞内大部分都沉默了。iPS 细胞和野生型 ES 细胞中的 *Oct4*、*Nanog* 和 *Sox2* 总水平相似。与内源性多能基因保持多能状态的结论相一致的是发现了 *Oct4* 和 *Nanog* 基因在 iPS 细胞内与在 ES 细胞一样低甲基化，而且发育调节因子的双标组蛋白修饰发生了重建。此外，iPS 细胞对 Dnmt1 失活诱导的整体去甲基化具有抗性，这与 ES 细胞相似但与体细胞相反。Dnmt1 在低甲基化 ES 细胞中的重新表达导致了整体的再甲基化，这表明 iPS 细胞也重新激活了从头合成甲基转移酶 Dnmt3a 和 Dnmt3b。所有的这些发现与 iPS 细胞已经获得了类似于正常 ES 细胞的表观遗传状态的结论相一致。最近的发现进一步支持了这一结论，即雌性 iPS 细胞，类似于 ES 细胞，能够重新激活体细胞中沉默的 X 染色体 [23]。

四个转录因子的表达被证实是一种能将体细胞重编程成多能状态的可靠的方法。但是，使用逆转录病毒转导的致癌基因意味着最终将重编程的细胞用于临床治疗存在严重的障碍。我们需要做更多的工作来了解重编程的分子机制，并最终找到小分子物质，可以在没有潜在有害基因的基因转移情况下实现重编程。

方法概述

细胞培养、基因靶向和病毒感染　在辐射过的 MEF 上进行 ES 和 iPS 细胞培养。通过同源重组，我们获得了在 *Oct4* 的 5 号外显子下游产生携带 IRES-GFPneo 融合盒的 ES 细胞系（图 1a）。*Nanog* 基因如所描述的是靶向的 [24]。将 Oct4-IRES-GFPneo 或者 Nanog-neo 靶向 ES 细胞注入囊胚后，从 E13.5 嵌合体胚胎中选择和分离出转基因 MEF。MEF 用含有 *Oct4*、*Sox2*、*c-Myc* 和 *Klf4* 的小鼠互补 DNA 的莫洛尼逆转录病毒载体 pLIB（Clontech）产生的病毒感染过夜。

Southern 印迹、甲基化和染色质分析　为了评估 DNA 甲基化水平，基因组 DNA 用 *Hpa*II 进行消化，并与小卫星重复序列探针 [25] 或者 IAP 探针 [26] 进行杂交。用 Qiagen EpiTect 试剂盒进行亚硫酸氢盐处理。如先前所述的用亚硫酸氢盐序列分析法研究 *Oct4* 和 *Nanog* 启

as described previously[27]. For imprinted genes, a COBRA assay was performed. PCR primers and conditions were as described previously[28]. The status of bivalent domains was determined by chromatin immunoprecipitation followed by quantitative PCR analysis, as described previously[12].

Expression analysis. Total RNA was reverse-transcribed and quantified using the QuantTtect SYBR green RT–PCR Kit (Qiagen) on a 7000 ABI detection system. Western blot and immunofluorescence analysis was performed as described[29,30]. Microarray targets from 2 μg total RNA were synthesized and labelled using the Low RNA Input Linear Amp Kit (Agilent), hybridized to Agilent whole-mouse genome oligonucleotide arrays (G4122F) and analysed as previously described[31].

Full Methods and any associated references are available in the online version of the paper at www.nature.com/nature.

(**448**, 318-324; 2007)

Marius Wernig[1]*, Alexander Meissner[1]*, Ruth Foreman[1,2]*, Tobias Brambrink[1]*, Manching Ku[3]*, Konrad Hochedlinger[1]†, Bradley E. Bernstein[3,4,5] & Rudolf Jaenisch[1,2]

[1] Whitehead Institute for Biomedical Research and [2] Department of Biology, Massachusetts Institute of Technology, Cambridge, Massachusetts 02142, USA

[3] Molecular Pathology Unit and Center for Cancer Research, Massachusetts General Hospital, Charlestown, Massachusetts 02129, USA

[4] Broad Institute of Harvard and MIT, Cambridge, Massachusetts 02142, USA

[5] Department of Pathology, Harvard Medical School, Boston, Massachusetts 02115, USA

†Present address: Center for Regenerative Medicine and Cancer Center, Massachusetts General Hospital, Harvard Medical School and Harvard Stem Cell Institute, Boston, Massachusetts 02414, USA

*These authors contributed equally to this work

Received 27 February; accepted 22 May 2007. Published online 6 June 2007.

References:

1. Hochedlinger, K. & Jaenisch, R. Nuclear transplantation, embryonic stem cells, and the potential for cell therapy. *N. Engl. J. Med.* **349**, 275-286 (2003).
2. Yang, X. *et al.* Nuclear reprogramming of cloned embryos and its implications for therapeutic cloning. *Nature Genet.* **39**, 295-302 (2007).
3. Hochedlinger, K. & Jaenisch, R. Nuclear reprogramming and pluripotency. *Nature* **441**, 1061-1067 (2006).
4. Tada, M., Takahama, Y., Abe, K., Nakatsuji, N. & Tada, T. Nuclear reprogramming of somatic cells by *in vitro* hybridization with ES cells. *Curr. Biol.* **11**, 1553-1558 (2001).
5. Cowan, C. A., Atienza, J., Melton, D. A. & Eggan, K. Nuclear reprogramming of somatic cells after fusion with human embryonic stem cells. *Science* **309**, 1369-1373 (2005).
6. Jaenisch, R. Human cloning—the science and ethics of nuclear transplantation. *N.Engl. J. Med.* **351**, 2787-2791 (2004).
7. Takahashi, K. & Yamanaka, S. Induction of pluripotent stem cells from mouse embryonic and adult fibroblast cultures by defined factors. *Cell* **126**, 663-676 (2006).
8. Labosky, P. A., Barlow, D. P. & Hogan, B. L. Mouse embryonic germ (EG) cell lines: transmission through the germline and differences in the methylation imprint of insulin-like growth factor 2 receptor (*Igf2r*) gene compared with embryonic stem (ES) cell lines. *Development* **120**, 3197-3204 (1994).
9. Boyer, L. A. *et al.* Core transcriptional regulatory circuitry in human embryonic stem cells. *Cell* **122**, 947-956 (2005).
10. Loh, Y. H. *et al.* The Oct4 and Nanog transcription network regulates pluripotency in mouse embryonic stem cells. *Nature Genet.* **38**, 431-440 (2006).
11. Lee, T. I. *et al.* Control of developmental regulators by Polycomb in human embryonic stem cells. *Cell* **125**, 301-313 (2006).
12. Boyer, L. A. *et al.* Polycomb complexes repress developmental regulators in murine embryonic stem cells. *Nature* **441**, 349-353 (2006).
13. Bernstein, B. E. *et al.* A bivalent chromatin structure marks key developmental genes in embryonic stem cells. *Cell* **125**, 315-326 (2006).
14. Azuara, V. *et al.* Chromatin signatures of pluripotent cell lines. *Nature Cell Biol.* **8**, 532-538 (2006).
15. Jackson-Grusby, L. *et al.* Loss of genomic methylation causes p53-dependent apoptosis and epigenetic deregulation. *Nature Genet.* **27**, 31-39 (2001).
16. Li, E., Bestor, T. H. & Jaenisch, R. Targeted mutation of the DNA methyltransferase gene results in embryonic lethality. *Cell* **69**, 915-926 (1992).

动子的甲基化水平[27]。对于印记基因，进行COBRA实验。PCR引物和条件如前所述[28]。通过染色质免疫沉淀然后通过PCR分析确定双标结构域的状态[12]。

基因表达分析 用QuantTtect SYBR green RT−PCR试剂盒（Qiagen）将总RNA进行反转录，之后再通过7000ABI检测系统完成定量检测。按照上述方法进行Western印迹和免疫荧光分析[29,30]。基因微阵列的靶标用Law RNA Input Linear Amp试剂盒，由2μg总的RNA合成，并进行标记，然后与Agilent全鼠基因组寡核苷酸阵列（G4122F）进行杂交和分析[31]。

完整的方法和相关的文献可在本文的网上版本中获得。

（毛晨晖 翻译；裴端卿 审稿）

17. Meissner, A. *et al.* Reduced representation bisulfite sequencing for comparative high-resolution DNA methylation analysis. *Nucleic Acids Res.* **33**, 5868-5877 (2005).
18. Ventura, A. *et al.* Cre-lox-regulated conditional RNA interference from transgenes. *Proc. Natl Acad. Sci. USA* **101**, 10380-10385 (2004).
19. Holm, T. M. *et al.* Global loss of imprinting leads to widespread tumorigenesis in adult mice. *Cancer Cell* **8**, 275-285 (2005).
20. Okano, M., Bell, D. W., Haber, D. A. & Li, E. DNA methyltransferases Dnmt3a and Dnmt3b are essential for *de novo* methylation and mammalian development. *Cell* **99**, 247-257 (1999).
21. Stewart, C. L., Stuhlmann, H., Jähner, D. & Jaenisch, R. *De novo* methylation, expression, and infectivity of retroviral genomes introduced into embryonal carcinoma cells. *Proc. Natl Acad. Sci. USA* **79**, 4098-4102 (1982).
22. Jähner, D. *et al.* *De novo* methylation and expression of retroviral genomes during mouse embryogenesis. *Nature* **298**, 623-628 (1982).
23. Maherali, N. *et al.* Global epigenetic remodeling in directly reprogrammed fibroblasts. *Cell Stem Cells* (in the press).
24. Mitsui, K. *et al.* The homeoprotein Nanog is required for maintenance of pluripotency in mouse epiblast and ES cells. *Cell* **113**, 631-642 (2003).
25. Chapman, V., Forrester, L., Sanford, J., Hastie, N. & Rossant, J. Cell lineage specific undermethylation of mouse repetitive DNA. *Nature* **307**, 284-286 (1984).
26. Walsh, C. P., Chaillet, J. R. & Bestor, T. H. Transcription of IAP endogenous retroviruses is constrained by cytosine methylation. *Nature Genet.* **20**, 116-117 (1998).
27. Blelloch, R. *et al.* Reprogramming efficiency following somatic cell nuclear transfer is influenced by the differentiation and methylation state of the donor nucleus. *Stem Cells* **24**, 2007-2013 (2006).
28. Lucifero, D., Mertineit, C., Clarke, H. J., Bestor, T. H. & Trasler, J. M. Methylation dynamics of imprinted genes in mouse germ cells. *Genomics* **79**, 530-538 (2002).
29. Hochedlinger, K., Yamada, Y., Beard, C. & Jaenisch, R. Ectopic expression of Oct-4 blocks progenitor-cell differentiation and causes dysplasia in epithelial tissues. *Cell* **121**, 465-477 (2005).
30. Wernig, M. *et al.* Functional integration of embryonic stem cell-derived neurons *in vivo*. *J. Neurosci.* **24**, 5258-5268 (2004).
31. Brambrink, T., Hochedlinger, K., Bell, G. & Jaenisch, R. ES cells derived from cloned and fertilized blastocysts are transcriptionally and functionally indistinguishable. *Proc. Natl Acad. Sci. USA* **103**, 933-938 (2006).
32. Eads, C. A. & Laird, P. W. Combined bisulfite restriction analysis (COBRA). *Methods Mol. Biol.* **200**, 71-85 (2002).
33. Peitz, M., Pfannkuche, K., Rajewsky, K. & Edenhofer, F. Ability of the hydrophobic FGF and basic TAT peptides to promote cellular uptake of recombinant Cre recombinase: a tool for efficient genetic engineering of mammalian genomes. *Proc. Natl Acad. Sci. USA* **99**, 4489-4494 (2002).
34. Eggan, K. *et al.* Hybrid vigor, fetal overgrowth, and viability of mice derived by nuclear cloning and tetraploid embryo complementation. *Proc. Natl Acad. Sci. USA* **98**, 6209-6214 (2001).
35. Naviaux, R. K., Costanzi, E., Haas, M. & Verma, I. M. The pCL vector system: rapid production of helper-free, high-titer, recombinant retroviruses. *J. Virol.* **70**, 5701-5705 (1996).

Supplementary Information is linked to the online version of the paper at www.nature.com/nature.

Acknowledgements. We thank H. Suh, D. Fu and J. Dausman for technical assistance; J. Love for help with the microarray analysis; S. Markoulaki for help with blastocyst injections; F. Edenhofer for a gift of Tat-Cre; and S. Yamanaka for the Nanog-neo construct. We acknowledge L. Zagachin in the MGH Nucleic Acid Quantitation core for assistance with real-time PCR. We also thank C. Lengner, C. Beard and M. Creyghton for constructive criticism. M.W. was supported in part by fellowships from the Human Frontiers Science Organization Program and the Ellison Foundation; B.B. by grants from the Burroughs Wellcome Fund, the Harvard Stem Cell Institute and the NIH; and R.J. by grants from the NIH.

Author Contributions. M.W., A.M. and R.J. conceived and designed the experiments and wrote the manuscript; M.W. derived all iPS lines; M.W. and A.M. performed the *in vitro* and *in vivo* characterization of the iPS lines (teratoma, 2N and 4N injections and IHC) and the conditional Dnmt1 experiment; A.M. investigated the promoter and imprinting methylation; M.K. and B.B. performed and analysed the real-time PCRs and ChIP experiments; R.F. and K.H. generated the selectable MEFs and TTFs; R.F. performed western blot and PCR analyses; and T.B. performed the microarray analysis and the proviral integration Southern blots.

Author Information. All microarray data from this study are available from Array Express at the EBI (http://www.ebi.ac.uk/arrayexpress) under the accession number E-MEXP-1037. Reprints and permissions information is available at www.nature.com/reprints. The authors declare no competing financial interests. Correspondence and requests for materials should be addressed to R.J. (jaenisch@wi.mit.edu).

Appendix: Index by Subject
附录：学科分类目录

Physics
物理学

Chemistry
化学

Biology
生物学

Astronomy
天文学

Geoscience
地球科学